Franz Rothlauf et al. (Eds.)

Applications of Evolutionary Computing

EvoWorkshops 2005: EvoBIO, EvoCOMNET, EvoHOT, EvoIASP, EvoMUSART, and EvoSTOC
Lausanne, Switzerland, March 30 – April 1, 2005
Proceedings

Springer

Volume Editor
see next page

Cover illustration: *Triangular Urchin*, by Chaps (www.cetoine.com).
Chaps has obtained an MSc in Physics at the Swiss Federal Institute of Technology. He is the developer of the ArtiE-Fract software that was used to create *Triangular Urchin*. *Triangular Urchin* (an Iterated Functions System of 2 polar functions) emerged from an urchin structure after a few generations using ArtiE-Fract. The evolutionary process was only based on soft mutations, some of them directly induced by the author.

Library of Congress Control Number: 2005922824

CR Subject Classification (1998): F.1, D.1, B, C.2, J.3, I.4, J.5

ISSN 0302-9743
ISBN-10 3-540-25396-3 Springer Berlin Heidelberg New York
ISBN-13 978-3-540-25396-9 Springer Berlin Heidelberg New York

Springer is a part of Springer Science+Business Media

springeronline.com

Printed in Germany

Typesetting: Camera-ready by author, data conversion by Scientific Publishing Services, Chennai, India
Printed on acid-free paper SPIN: 11402169 06/3142 5 4 3 2 1 0

Volume Editors

Franz Rothlauf
Dept. of Business Administration and Information Systems
University of Mannheim
Schloss, 68131 Mannheim, Germany
rothlauf@uni-mannheim.de

Jürgen Branke
Institute AIFB
University of Karlsruhe
76128 Karlsruhe, Germany
branke@aifb.uni-karlsruhe.de

Stefano Cagnoni
Dept. of Computer Engineering
University of Parma
Parco Area delle Scienze 181/a
43100 Parma, Italy
cagnoni@ce.unipr.it

David W. Corne
Department of Computer Science
University of Exeter
North Park Road
Exeter EX4 4QF, UK
d.w.corne@ex.ac.uk

Rolf Drechsler
Institute of Computer Science
University of Bremen
28359 Bremen, Germany
drechsle@informatik.uni-bremen.de

Yaochu Jin
Honda Research Institute Europe
Carl-Legien-Str.30
63073 Offenbach/Main, Germany
yaochu.jin@honda-ri.de

Penousal Machado
Dep. de Engenharia Informática
University of Coimbra
Polo II, 3030 Coimbra, Portugal
machado@dei.uc.pt

Elena Marchiori
Dept. of Mathematics and Computer Science
Free University of Amsterdam
de Boelelaan 1081a
1081 HV, Amsterdam, The Netherlands
elena@cs.vu.nl

Juan Romero
Facultad de Informatica
University of A Coruña
A Coruña, CP 15071, Spain
jj@udc.es

George D. Smith
School of Computing Sciences
University of East Anglia
UEA Norwich
Norwich NR4 7TJ, UK
gds@sys.uea.ac.uk

Giovanni Squillero
Dip. di Automatica e Informatica
Politecnico di Torino
Corso Duca degli Abruzzi 24
10129 Torino, Italy
squillero@polito.it

Preface

Evolutionary computation (EC) techniques are efficient nature-inspired planning and optimization methods based on the principles of natural evolution and genetics. Due to their efficiency and the simple underlying principles, these methods can be used for a large number of problems in the context of problem solving, optimization, and machine learning. A large and continuously increasing number of researchers and practitioners make use of EC techniques in many application domains. The book at hand presents a careful selection of relevant EC applications combined with thorough examinations of techniques for a successful application of EC. The presented papers illustrate the current state of the art in the application of EC and should help and inspire researchers and practitioners to develop efficient EC methods for design and problem solving.

All papers in this book were presented during EvoWorkshops 2005, which was a varying collection of workshops on application-oriented aspects of EC. Since 1999, the format of the EvoWorkshops has proved to be very successful and well representative of the advances in the application of EC. Consequently, over the last few years, EvoWorkshops has become one of the major events addressing the application of EC. In contrast to other large conferences in the EC field, the EvoWorkshops focus solely on application aspects of EC and are an important link between EC research and the application of EC in a large variety of different domains. The EvoWorkshops are combined with EuroGP, the main European event dedicated to genetic programming, and EvoCOP, which has become the main European conference on EC in combinatorial optimization. The proceedings for both events, EuroGP and EvoCOP, are also available in the LNCS series (numbers 3447 and 3448).

EvoWorkshops 2005, of which this volume contains the proceedings, was held in beautiful Lausanne, Switzerland, on March 30–April 1, 2005, jointly with EuroGP 2005 and EvoCOP 2005. EvoWorkshops 2005 consisted of the following individual workshops:

- *EvoBIO*, the Third European Workshop on Evolutionary Bioinformatics,
- *EvoCOMNET*, the Second European Workshop on Evolutionary Computation in Communications, Networks, and Connected Systems,
- *EvoHOT*, the Second European Workshop on Hardware Optimization,
- *EvoIASP*, the Eighth European Workshop on Evolutionary Computation in Image Analysis and Signal Processing,
- *EvoMUSART*, the Third European Workshop on Evolutionary Music and Art, and
- *EvoSTOC*, the Second European Workshop on Evolutionary Algorithms in Stochastic and Dynamic Environments.

EvoBIO was concerned with the exploitation of EC and related techniques in bioinformatics and computational biology. For analyzing and understanding biological data, EC plays an increasingly important role in pharmaceuticals, biotechnology, and associated industries, as well as in scientific discovery.

EvoCOMNET addressed the application of EC techniques to problems in communications, networks, and connected systems. New communication technologies, the creation of interconnected communication and information networks such as the Internet, new types of interpersonal and interorganizational communication, and the integration and interconnection of production centers and industries are the driving forces on the road towards a connected, networked society. EC techniques are important tools for facing these challenges.

EvoHOT highlighted the latest developments in the field of EC applications to hardware and design optimization. This includes various aspects like the design of electrical and digital circuits, the solving of classical hardware optimization problems like VLSI floorplanning, the application of EC to antenna array synthesis, or the use of ant colony optimization as a hardware-oriented metaheuristic.

EvoIASP, which was the first international event solely dedicated to the applications of EC to image analysis and signal processing, has been a traditional meeting since 1999. This year it addressed topics ranging from solutions for problems in the context of image and signal processing to the adaptive learning of human vocalization in robotics, and the design of multidimensional filters.

EvoMUSART focused on the use of EC techniques for the development of creative systems. There is a growing interest in the application of these techniques in fields such as art, music, architecture and design. The goal of EvoMUSART was to bring together researchers who use EC in this context, providing an opportunity to promote, present and discuss the latest work in the area, fostering its further developments and collaboration among researchers.

EvoSTOC addressed the application of EC in stochastic environments. This includes optimization problems with noisy and approximated fitness functions that are changing over time, the treatment of noise, and the search for robust solutions. These topics recently gained increasing attention in the EC community, and EvoSTOC was the first workshop that provided a platform to present and discuss the latest research in this field.

EvoWorkshops 2005 continued the tradition of providing researchers in these fields, as well as people from industry, students, and interested newcomers, with an opportunity to present new results, discuss current developments and applications, or just become acquainted with the world of EC, besides fostering closer future interaction between members of all scientific communities that may benefit from EC techniques.

This year, EvoWorkshops had the highest number of submissions ever. The number of submissions increased from 123 in 2004 to 143 in 2005. Therefore, EvoWorkshops introduced a new presentation format and accepted a limited number of posters with a reduced number of pages (six pages). In contrast to regular papers, which were presented orally, the posters were presented and discussed in a special poster session during the workshops. The acceptance rate of 39.1% for EvoWorkshops is an indicator of the high quality of the papers presented at the workshops and included in these proceedings. The following table gives some numbers for the different workshops (accepted posters are in parentheses). Of further importance for the statistics is the acceptance rate of the EvoWorkshops 2004, which was 44.7%.

Workshop	submitted	accepted	acceptance ratio
EvoBIO 2005	32	13	40.6%
EvoCOMNET 2005	22	5	22.7%
EvoHOT 2005	11	7	63.6%
EvoIASP 2005	37	17	45.9%
EvoMUSART 2005	29	10(6)	34.5%
EvoSTOC 2005	12	4(4)	33.3%
Total	143	56(10)	39.1%

We would like to thank all members of the program committees for their quick and thorough work. Furthermore, we would like to acknowledge the support from the University of Lausanne, which provided a great place to run a conference, and from EvoNet, the European Network of Excellence in Evolutionary Computing. The success of EvoWorkshops 2005 shows that the EvoWorkshops, as well as EuroGP and EvoCOP, have reached a degree of maturity and scientific prestige that will allow them to continue their success even without the active support from EvoNet. Over the years, the EvoWorkshops have become major EC events that have been important not only for Europeans but have also attracted large numbers of international EC researchers.

Finally, we want to say a special thanks to everybody who was involved in the preparation of the event. Special thanks are due to Jennifer Willies, whose work and support is a great and invaluable help for scientists who are planning to organize an international conference, and to the local organizers Marco Tomassini, Mario Giacobini, Leonardo Vanneschi, Leslie Luth and Denis Rochat. Without their hard work and continuous support, it would not have been possible to be in such a nice place and to have such a great conference.

April 2005

Franz Rothlauf, Jürgen Branke, Stefano Cagnoni
David W. Corne, Rolf Drechsler, Yaochu Jin
Penousal Machado, Elena Marchior, Juan Romero
George D. Smith, Giovanni Squillero

Organization

EvoWorkshops 2005 was jointly organized with EuroGP 2005 and EvoCOP 2005.

Organizing Committee

EvoWorkshops chair	Franz Rothlauf, University of Mannheim, Germany
Local chair	Marco Tomassini, University of Lausanne, Switzerland
Publicity chair	Jano van Hemert, Napier University, Edinburgh, UK
EvoBIO co-chairs	David W. Corne, University of Exeter, UK
	Elena Marchiori, Free University Amsterdam, The Netherlands
EvoCOMNET co-chairs	Franz Rothlauf, University of Mannheim, Germany
	George D. Smith, University of East Anglia, UK
EvoHOT co-chairs	Giovanni Squillero, Politecnico di Torino, Italy
	Rolf Drechsler, University of Bremen, Germany
EvoIASP chair	Stefano Cagnoni, University of Parma, Italy
EvoMUSART co-chairs	Juan Romero, University of A Coruña, Spain
	Penousal Machado, University of Coimbra, Portugal
EvoSTOC co-chairs	Jürgen Branke, University of Karlsruhe, Germany
	Yaochu Jin, Honda Research Institute Europe, Germany

Program Committees

EvoBIO Program Committee

Francisco J. Azuaje, Ireland
Jesus S. Aguilar-Ruiz, University of Seville, Spain
Wolfgang Banzhaf, Memorial University of Newfoundland, Canada
Jacek Blazewicz, Institute of Computing Science, Poznan, Poland
David W. Corne, University of Exeter, UK
Carlos Cotta-Porras, University of Malaga, Spain
Alfredo Ferro, University of Catania, Italy
Bogdan Filipic, Jozef Stefan Institute, Ljubljana, Slovenia
David Fogel, Natural Selection, Inc., USA
Gary B. Fogel, Natural Selection, Inc., USA

James Foster, University of Idaho, USA
Alex Freitas, University of Kent, UK
Rosalba Giugno, University of Catania, Italy
Jin-Kao Hao, LERIA, Université d'Angers, France
Jaap Heringa, Free University Amsterdam, The Netherlands
Visakan Kadirkamanathan, University of Sheffield, UK
Antoine van Kampen, AMC University of Amsterdam, The Netherlands
W.B. Langdon, UCL, UK
Bob MacCallum, Stockholm University, Sweden
Elena Marchiori, Vrije Universiteit Amsterdam, The Netherlands
Brian Mayoh, Aarhus University, Denmark
Andrew C.R. Martin, University of Reading, UK
Jason H. Moore, Vanderbilt University Medical Center, USA
Pablo Moscato, University of Newcastle, Australia
Ajit Narayanan, University of Exeter, UK
Jagath C. Rajapakse, Technology University, Singapore
Jon Rowe, University of Birmingham, UK
Jem Rowland, University of Wales, Aberystwyth, UK
Vic J. Rayward-Smith, University of East Anglia, UK
El-ghazali Talbi, Laboratoire d'Informatique Fondamentale de Lille, France
Eckart Zitzler, Swiss Federal Institute of Technology, Switzerland

EvoCOMNET Program Committee

Stuart Allen, Cardiff University, UK
Alexandre Caminada, France Télécom R&D, France and
Université de Technologie Belfort-Montbéliard, France
Jin-Kao Hao, University of Angers, France
Bryant Julstrom, St. Cloud State University, USA
Geoff McKeown, UEA Norwich, UK
Günther R. Raidl, Vienna University of Technology, Austria
Franz Rothlauf, University of Mannheim, Germany
Giovanni Squillero, Politecnico di Torino, Italy
George D. Smith, University of East Anglia, UK
Andrew Tuson, City University, London, UK

EvoHOT Program Committee

Gabriella Kokai, Friedrich-Alexander Universität Erlangen, Germany
Ernesto Sanchez, Politecnico di Torino, Italy
Lukas Sekanina, Brno University of Technology, Czech Republic
George Smith, School of Information Systems, UK
Tan Kay Chen, National University of Singapore, Singapore
Luca Sterpone, Politecnico di Torino, Italy
Massimo Violante, Politecnico di Torino, Italy

EvoIASP Program Committee

Giovanni Adorni, University of Genoa, Italy
Lucia Ballerini, Sweden
Bir Bhanu, University of California, Riverside, USA
Dario Bianchi, University of Parma, Italy
Leonardo Bocchi, University of Florence, Italy
Alberto Broggi, University of Parma, Italy
Stefano Cagnoni, University of Parma, Italy
Ela Claridge, University of Birmingham, UK
Ernesto Costa, University of Coimbra, Portugal
Laura Dipietro, Massachusetts Institute of Technology, USA
Marc Ebner, University of Würzburg, Germany
Terry Fogarty, South Bank University, UK
Daniel Howard, QinetiQ, UK
Mario Koeppen, FhG IPK Berlin, Germany
Evelyne Lutton, INRIA, France
Gustavo Olague, CICESE, Mexico
Riccardo Poli, University of Essex, UK
Conor Ryan, University of Limerick, Ireland
Stephen Smith, University of York, UK
Giovanni Squillero, Politecnico di Torino, Italy
Wolfgang Stolzmann, DaimlerChrysler, Germany
Kiyoshi Tanaka, Shinshu University, Japan
Ankur M. Teredesai, Rochester Institute of Technology, USA
Andy Tyrrell, University of York, UK
Robert Vanyi, Siemens PSE, Hungary
Hans-Michael Voigt, Center for Applied Computer Science (GFaI), Germany
Stewart Wilson, Prediction Dynamics, USA
Mengjie Zhang, Victoria University of Wellington, New Zealand

EvoMUSART Program Committee

Mauro Annunziato, Plancton Art Studio, Italy
Paul Brown, University of London, UK
Amlcar Cardoso, University of Coimbra, Portugal
Pierre Collet, Université du Littoral Côte d'Opale, France
John Gero, University of Sydney, Australia
Andrew Gartland-Jones, University of Sussex, UK
Carlos Grilo, School of Technology and Management of Leiria, Portugal
Matthew Lewis, Ohio State University, USA
Bill Manaris, College of Charleston, USA
Eduardo R. Miranda, University of Plymouth, UK
Ken Musgrave, Pandromeda Inc., USA
Francisco C. Pereira, University of Coimbra, Portugal

Luigi Pagliarini, Academy of Fine Arts of Rome, Italy and
University of Southern Denmark, Denmark
Juan Romero, Universidade da Coruña, Spain
Celestino Soddu, Politecnico de Milano, Italy
Tim Taylor, University of Edinburgh, UK
Jorge Tavares, CISUC, Centre for Informatics and Systems, Portugal
Stephen Todd, IBM, UK
Tatsuo Unemi, University of Zurich, Switzerland
Geraint Wiggins, City University, London, UK

EvoSTOC Program Committee

Dirk Arnold, Dalhousie University, Canada
Hans-Georg Beyer, Vorarlberg University of Applied Sciences, Austria
Tim Blackwell, Goldsmiths College, UK
Ernesto Costa, University of Coimbra, Portugal
Kalyan Deb, IIT Kanpur, India
Martin Middendorf, University of Leipzig, Germany
Ferrante Neri, Bari Polytechnic, Italy
Markus Olhofer, Honda Research Institute, Germany
Yew Soon Ong, Nanyang Technical University, Singapore
Khaled Rasheed, University of Georgia, USA
Christian Schmidt, University of Karlsruhe, Germany
Holger Ulmer, Robert Bosch GmbH, Germany
Sima Uyar, Istanbul Technical University, Turkey
Karsten Weicker, Leipzig University of Applied Sciences, Germany
Lars Willmes, NuTech GmbH, Germany
Shengxiang Yang, University of Leicester, UK

Sponsoring Institutions

EvoNet, the European Network of Excellence in Evolutionary Computing
University of Lausanne, Lausanne, Switzerland

Table of Contents

EvoBIO Contributions

EvoCOMNET Contributions

EvoHOT Contributions

EvoIASP Contributions

EvoMUSART Contributions

EvoSTOC Contributions

Evolutionary Biclustering of Microarray Data

Jesus S. Aguilar–Ruiz and Federico Divina

University of Seville, Seville, Spain
{aguilar, federico}@lsi.us.es

Abstract. In this work, we address the biclustering of gene expression data with evolutionary computation, which has been proven to have excellent performance on complex problems. In expression data analysis, the most important goal may not be finding the maximum bicluster, as it might be more interesting to find a set of genes showing similar behavior under a set of conditions. Our approach is based on evolutionary algorithms and searches for biclusters following a sequential covering strategy. In addition, we pay special attention to the fact of looking for high quality biclusters with large variation. The quality of biclusters found by our approach is discussed by means of the analysis of yeast and colon cancer datasets.

1 Introduction

Microarray data are widely used in genomic research due to the enormous potential in gene expression profiling, facilitating the prognosis and the discovering of subtypes of diseases. The gene expression data are organized in matrices, where rows represent genes and columns represent experimental conditions. Each element in the matrix refers to the expression level of a particular gen under specific conditions. A basic approach to the study of expression data consists of applying traditional statistical techniques. In many problems these methods have shown to be unable to extract relevant knowledge from data.

Clustering has been applied to gene expression data [1], which usually refers to conditions or patients, although genes can also be grouped in order to search for functional similarities. However, relevant genes are not necessarily related to every condition, or in other words, there are genes that can be relevant for a subset of conditions [11]. On the contrary, it is also possible to discriminate groups of conditions by using different groups of genes. From this point of view, clustering can not only be addressed horizontally (conditions) or vertically (genes), but also in the two dimensions simultaneously. This approach, named *biclustering*, identify groups of genes that show "similar" level expression under a specific subset of experimental conditions.

Biclustering [4] was first introduced by [9], as a way to cluster simultaneously rows and columns of a matrix, and it was named "direct clustering". The goal was to find biclusters with minimum variance, what ideally provided biclusters of size 1, since they looked for constant biclusters (constant values within the submatrix). Hartigan tried to avoid this problem by searching for k biclusters

F. Rothlauf et al. (Eds.): EvoWorkshops 2005, LNCS 3449, pp. 1–10, 2005.

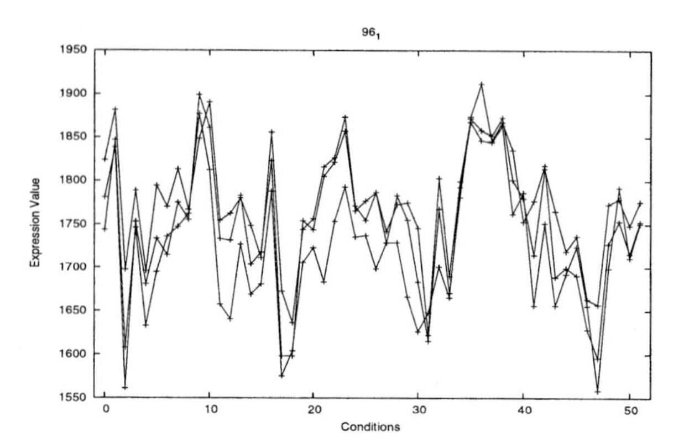

Fig. 1. Example of bicluster obtained by our approach from the colon cancer dataset

at a time. Later in 2000, Cheng and Church [6] proposed the biclustering of gene expression, introducing the *residue* of an element in the bicluster and the *mean squared residue* of a submatrix. In addition, they adjusted that measure to reject trivial biclusters by means of the *row variance.* Getz et al. [8] presented the coupled two–way clustering. It uses hierarchical clustering applied separately to each dimension and then they defined the process to combine both results. Obviously, the quality of biclusters depends on the clusters generated at each dimension, which in turn, allow us to experiment with different types of clustering algorithms. Lazzeroni and Owen [10] used "plaid models" in the same context, where the concept of "layers" (bicluster) is used to compute the values in the data matrix, which is described as a linear function of layers. Basically, each element is seen as a superposition of layers. Yang et al. [12] presented δ–clusters, and a year later, the same authors improved the Cheng and Church's approach in FLOC [13], paying attention to missing values. FLOC follows the same technique as Cheng and Church's algorithm, by adding/removing each row/column to a set of initial biclusters, improving its quality iteratively. Also in 2002, Tanay et al. [2] identified biclusters by means of a bipartite graph–based model and using a greedy approach to add/remove vertices in order to find maximum weight subgraphs, which is related to its statistical significance. Recently, Bleuler et al. [5] proposed a EA framework for biclustering gene expression data. An example of bicluster is shown in Figure 1, where gene expression values (y–axis) are plotted for three genes through seventy conditions (x–axis).

In this work, we address the biclustering problem with evolutionary computation (EC), which has been proven to have an excellent performance on highly complex optimization problems. Our approach, named `SEBI` is based on evolutionary algorithms (EA) and search for biclusters following a sequential covering strategy. As the algorithm partially uses the *squared mean residue*, the results have been compared to those of Cheng and Church. In expression data analysis, the most important goal may not be finding the maximum bicluster or even finding a bicluster cover for the data matrix. It is more interesting to find a set of genes showing strikingly similar up–regulation and down–regulation under a set of conditions. A low mean squared residue score plus a large variation from the constant may be a good criterion for identifying these genes and conditions. Therefore, our goal is to find biclusters of maximum size, with mean squared residue lower than a given δ, with a relatively high row variance, and with a low level of overlapping among biclusters.

The paper is organized as follows: in Section 2 the definitions related to biclustering are presented; the description of the algorithm is illustrated in Section 3, together with all the evolutionary features and the evaluation of the quality of a bicluster; experimental results are discussed in Section 4, comparing the quality to those generated by Cheng and Church's algorithm; finally, the most interesting conclusions are summarized in Section 5.

2 The Model of Biclusters

We follow the biclustering model proposed in [6]. A bicluster is defined on a gene–expression matrix. Let $G = \{g_1, \ldots, g_N\}$ be a set of genes and $C = \{c_1, \ldots, c_M\}$ a set of conditions. The data can be viewed as an $N \times M$ expression matrix EM. EM is a matrix of real numbers, with possible null values, where each entry el_{ij} corresponds to the logarithm of the relative abundance of the mRNA of a gene g_i under a specific condition c_j.

A bicluster essentially corresponds to a sub–matrix that exhibits some coherent tendency. Each bicluster can be identified by a unique set of genes and conditions, that determine the sub–matrix. Thus a bicluster is a matrix $I \times J$, denoted as (I, J), where I and J are set of genes (rows) and conditions (columns), respectively, and $|I| \leq |N|$ and $|J| \leq |M|$. We define the volume of a bicluster (I, J) as the number of elements el_{ij} such that $i \in I$ and $j \in J$.

In the following, we give some definitions that are used in a measure for assessing the quality of a bicluster, most of which are taken from [13].

Definition 1. Let (I, J) be a bicluster, then we define the base of a gene g_i as $e_{iJ} = \frac{\sum_{j \in J} e_{ij}}{|J|}$. In the same way we define the base of a condition c_j as $e_{Ij} = \frac{\sum_{i \in I} e_{ij}}{|I|}$. The base of a bicluster is the mean of all the entries contained in (I, J), $e_{IJ} = \frac{\sum_{i \in I, j \in J} e_{ij}}{|I| \cdot |J|}$.

Definition 2. The residue of an entry e_{ij} of a bicluster (I, J) is

$$r_{ij} = e_{ij} - e_{iJ} - e_{Ij} + e_{IJ}$$

The residue is an indicator of the degree of coherence of an element with respect to the remaining ones in the bicluster, given the tendency of the relevant gene and the relevant condition. The lower the residue, the stronger the coherence.

Definition 3. The mean squared residue of a bicluster (I, J) is $r_{IJ} = \frac{\sum_{i \in I, j \in J} r_{ij}^2}{|I| \cdot |J|}$.

The mean squared residue is the variance of the set of all elements in the bicluster, plus the mean row variance and the mean column variance. The lower the mean squared residue, the stronger the coherence exhibited by the bicluster, and the better the quality of the bicluster. If a bicluster has a mean squared residue lower than a given value δ, then we call the bicluster a δ-bicluster. The problem of finding the largest square δ–bicluster is NP–hard [6]. In addition to the mean squared residue, we may prefer the row variance to be relatively large to reject trivial bicluster.

Our goal is to find biclusters of maximum size, with mean squared residue lower than a given δ, with a relatively high row variance, and with a low level of overlapping among biclusters.

3 Description of the Algorithm

The algorithm adopts a sequential covering strategy: a EA, called `EBI` (for Evolutionary BIclustering), is called several times, until an *end condition* is met. `EBI` takes as input the expression matrix and the δ value and returns either a bicluster with mean squared residue lower than δ or nothing. In the former case, the returned bicluster is stored in a list called *Results*, and `EBI` is called again. The end condition is also met when `EBI` is called a maximum number of times. When the end condition is met, the list *Results* is returned.

After a bicluster is returned, weights associated with the expression matrix are adjusted. This operation is performed in order to to bias the search towards biclusters that do not overlap with already found biclusters. In order to do so, we associate a weight to each element of the expression matrix. The weight of an element depends on the number of biclusters in *Results* containing the element. The more biclusters cover an element, the higher the weight of the element will be.

The aim of `EBI` is to find δ–biclusters with maximum volume, with a relatively high row variance, and minimizing the effect of overlapping among biclusters.

The initial population consists of biclusters containing only one element of the expression matrix. These biclusters have the property of having a mean squared residue equal to 0. Tournament selection is used for selecting parents. Selected pairs of parents are recombined with a crossover operator with a given probability p_c (default value 0.9), and the resulting offspring is mutated with a probability p_m (default value 0.1). Three crossover operators can be applied with equal probability: one-point, two-point and uniform crossover. Three mutation operators are used, a standard mutation operator, a mutation operator that adds a row and a mutation operator that adds a column to the bicluster. Elitism is applied with a probability p_e (default value 0.75). At the end of the evolutionary process, if the best individual, according to the fitness, encodes a δ–bicluster, then it is returned, otherwise `EBI` does not return anything.

Each individual of the population encodes one bicluster. Biclusters are encoded by means of binary strings of length $N+M$, where N and M are the number of rows (genes) and of columns (conditions) of the expression matrix, respectively. Each of the first N bits of the binary string is related to the rows, in the order in which the bits appear in the string. In the same way, the remaining M bits are related to the columns. If a bit is set to 1, it means that the relative row or column belongs to the encoded bicluster; otherwise it does not. It is worth to note that given the large value of M, the search space size for the evolutionary algorithm is huge, and therefore more emphasis has to be placed on the performance of genetic operators.

The fitness function rewards individuals encoding biclusters with low mean squared residue, with high volume and row variance and covering elements of the expression matrix that are not covered by biclusters found by previous executions of `EBI`. The final objective of the `EBI` is to minimize the fitness.

4 Experimental Results

In order to asses the goodness of the proposed method for finding biclusters in expression data, we conducted experiments on two well known datasets: *Saccharomyces Cerevisiae* cell cycle expression dataset and the *Colon Cancer* dataset. The first dataset, here referred as yeast dataset, originated from [7], while the colon cancer from [3]. The expression matrix contained in the yeast dataset consists of 2884 genes and 17 conditions. For this dataset the δ was set to 300. The dataset is taken from [6], where the original data is preprocessed regarding missing values. The expression matrix contained in the colon cancer dataset consists of 2000 genes and 62 conditions. This dataset was preprocessed as in [6], where each entry x of the original dataset was substituted by the value $100 \cdot log(10^5 \cdot x)$. For this dataset the value of δ was set to 500, because the expression matrix contained in the dataset has a size that is about the double of that contained in the yeast dataset. A similar reasoning was adopted in [6] for determining values of δ for some datasets.

In Figure 2, nine out of one hundred biclusters found by `EBI` on the yeast dataset are shown. Information about these biclusters is given in Table 1. These biclusters were found with the following parameters for `EBI`: generation 100, population size 200, crossover probability 0.85, mutation probability 0.20, elitism probability 0.90, weight for column 10, weights for rows 1.0, weight for volume 1.0. From a visual inspection of the biclusters proposed in Figure 2, one can notice that the genes present a similar behavior under a set of conditions. This is especially evident for biclusters labeled 37_1 and 52_1. Many biclusters found on the yeast dataset contain all the seventeen conditions, indicating that these conditions form a good cluster, with respect to the genes included in the biclusters. A similar result was also obtained in [6]. Bicluster 37_1 and 52_1 are specially interesting because of their high row variance, which shows the good performance of the evolutionary algorithm.

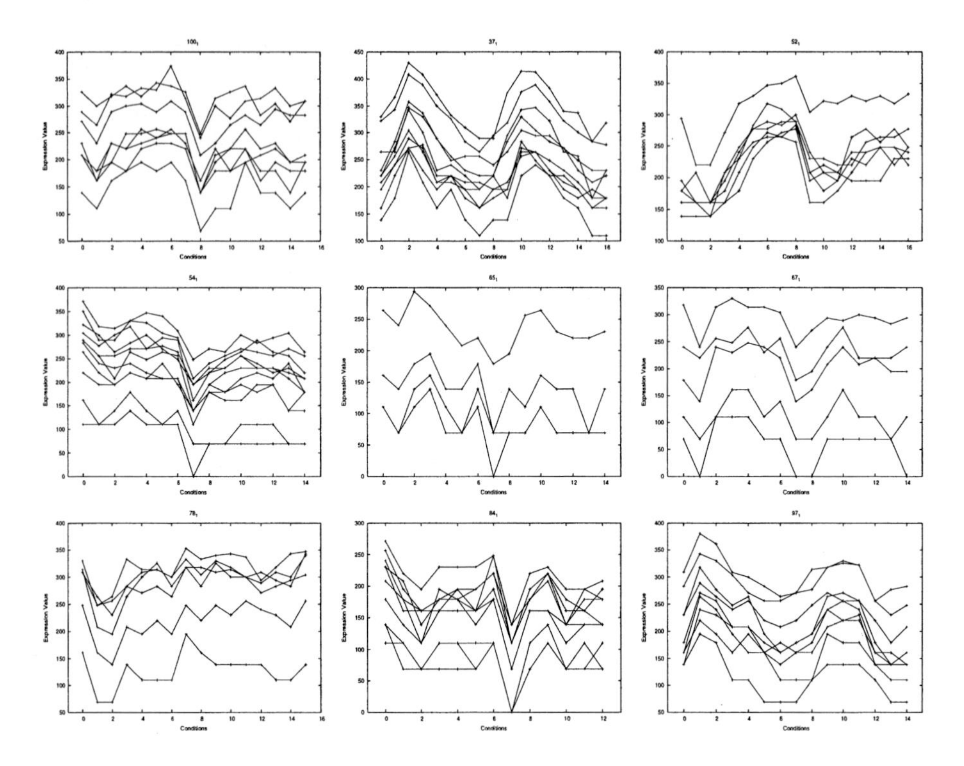

Fig. 2. Nine biclusters found for the yeast dataset. All the mean squared residues of the biclusters are lower than 220

Table 1. Information about biclusters of Figure 2. In the first column the identifier of each bicluster is reported. The second and third columns report the number of rows and of columns of the bicluster, respectively, the fourth column reports the mean squared residues, and the last column report the row variance of the biclusters

Bicluster	Genes	Conditions	Residue	Row Variance
101_1	8	16	203.98	910.09
37_1	10	17	206.80	1656.38
52_1	7	17	209.83	1730.06
54_1	11	15	205.55	1158.58
65_1	4	15	203.65	1009.16
67_1	5	15	208.56	1013.62
78_1	6	16	200.06	945.05
84_1	11	13	205.24	1099.91
97_1	9	15	199.28	1469.4

The one hundred biclusters found on the yeast dataset cover 42.16% of the elements of the expression matrix, 46.22% of the genes and 100% of the conditions. These results confirm the effectiveness of the adopted method for

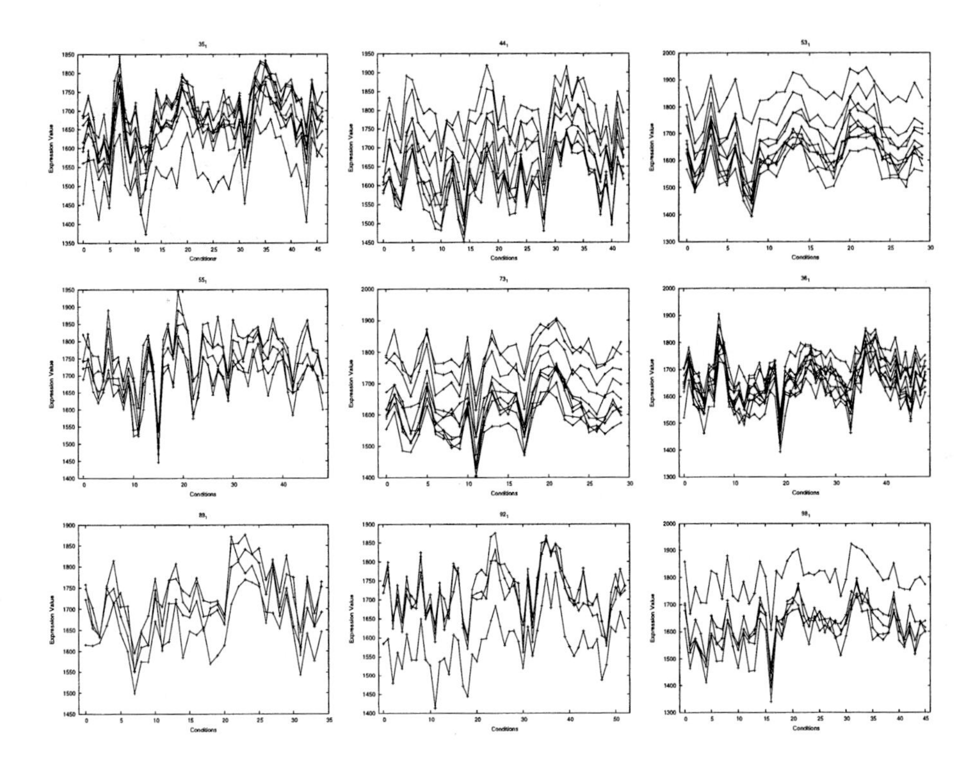

Fig. 3. Nine biclusters found for the colon cancer dataset. All the mean squared residues of the biclusters are lower than 500

avoiding overlapping. Each call to EBI on the yeast dataset requires about 72 seconds on a Pentium IV 3.06 Ghz

In Figure 3 nine out of one hundred biclusters found by EBI on the colon cancer dataset are shown. Table 2 reports information about the biclusters shown. These biclusters were found with the same parameters setting as for the yeast dataset. It can be noticed that these biclusters contain genes that present a similar behavior under some conditions. For example, the bicluster labeled 36_1 contains ten genes that show strikingly similar up–regulations and down–regulations under the same forty-nine conditions.

Figure 4 shows three graphs relative to a typical run of EBI on the colon cancer dataset. The graphs are relative to a first call of EBI. In 4(a), the average fitness and the best fitness present in the population at each generation are shown. It can be noticed that also in this case, the fitness decreases rapidly in the first generations, until about the 22nd generation, and then it keeps decreasing even if more slowly. In Figure 4(b), the average volume and the best volume of the biclusters encoded in the population are given at each generation. Also this graph confirms the successfulness of the fitness function in promoting biclusters with greater volume. Finally, Figure 4(c) reports the average mean squared residue and the lowest residue are given at each generation. A similar behavior of the

Table 2. Information about biclusters of Figure 3. In the first column the identifier of each bicluster is reported. The second and third columns report the number of rows and of columns of the bicluster, respectively, the fourth column reports the mean squared residues, and the last column report the row variance of the biclusters

Bicluster	Genes	Conditions	Residue	Row Variance
35_1	9	47	496.26	4298.55
44_1	8	43	498.10	4279.42
53_1	10	30	498.63	4691.73
55_1	5	49	494.30	5057.84
73_1	10	30	492.63	4011.11
36_1	10	49	495.02	4538.34
89_1	4	35	492.06	4524.06
92_1	4	53	491.69	4752.67
98_1	5	46	493.06	4949.33

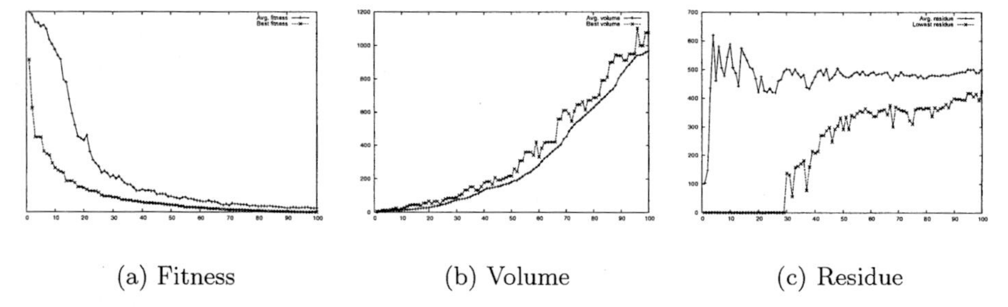

(a) Fitness (b) Volume (c) Residue

Fig. 4. Graphs relative to a typical run of EBI on the colon cancer dataset. In 4(a) the average and best fitness at each generation are shown. In 4(b) the average and best volumes are shown for each generation. In 4(c) the average residue and the lowest residue at each generation are plotted

one observed for the yeast dataset can be noticed from this graph: the residue increases in the first generations for then assessing to a almost stable value.

The one hundred biclusters found on the yeast dataset cover 41.60% of the elements of the expression matrix, 49.22% of the genes and 100% of the conditions. These results confirm again the effectiveness of the adopted method for avoiding overlapping. Each call of the procedure EBI takes about 86 seconds on a Pentium IV 3.06 Ghz.

In Table 3, we compare the performance of SEBI with that of Cheng and Church's algorithm (here named CC), for what concerns the average residue and the average dimension of the biclusters found on the yeast dataset. We also report the average information obtained by SEBI on the colon dataset. We can see that CC is capable of finding biclusters characterized by a higher volume than the ones found by SEBI. This is probably due to the overlapping policy adopted by SEBI. In fact, the first biclusters found by SEBI have volumes comparable with those of the biclusters found by CC. After some runs, when the most trivial

Table 3. Performance comparison between SEBI and CC

Algorithm	Dataset	Avg. residue	Avg. Volume	Avg. gene num.	Avg. cond. num
SEBI	Yeast	202.68 (8.81)	204.67 (172.03)	13.20 (10.52)	15.44 (1.33)
	Colon	492.46 (6.23)	403.48 (215.70)	9.86 (4.51)	40.91 (8.00)
CC	Yeast	204.29 (42.78)	1576.98 (2178.46)	166.71 (226.37)	12.09 (4.39)
	Colon	–	–	–	–

biclusters have been found, SEBI focuses on elements of the expression matrix that are not contained in already found biclusters. However, after CC has found a bicluster, the covered elements of the expression matrix are substituted by randomly generated values, in the range of the original data. This may cause the biclusters to overlap much more than what happens in SEBI where overlapping is avoided as much as possible. As far as the residue is concerned, the results obtained by the two systems are comparable on the yeast dataset.

5 Conclusions

In this paper we have introduced an algorithm based on evolutionary computation, called SEBI, for finding biclusters on expression data. The proposed algorithm adopts a sequential covering strategy, and an evolutionary algorithm in order to find biclusters. The experimental results show that CC is capable of finding bigger clusters. This could be due to the overlapping policy adopted by CC. In SEBI overlapping is avoided by means of the weights assigned to the elements of the expression matrix, while CC replaces covered entries of the expression matrix by random values. This strategy could not prevent overlapping as efficiently as the one adopted in SEBI allowing the system to find biclusters with higher volume.

We can conclude that SEBI is successful in finding set of genes that show strikingly similar up–regulations and down–regulations under a set of conditions, as shown from the presented results. Thus, evolutionary computation represents a useful framework for addressing the challenges of gene expression analysis.

Acknowledgment

The research was supported by the Spanish Research Agency CICYT under grant TIN2004–00159 and Junta de Andalucia (III Research Program).

References

1. A. Ben-Dor and R. Shamir and Z. Yakhini. Clustering gene expression patterns. *Journal of Computational Biology*, 6(3–4):281–297, 1999.
2. A. Tanay and R. Sharan and R. Shamir. Discovering statistically significant biclusters in gene expression data. *Bioinformatics*, 19 (Sup. 2):196–205, 2002.

3. U. Alon, N. Barkai, D. A. Notterman, and et al. Broad patterns of gene expression revealed by clustering analysis of tumor and normal colon tissues probed by oligonucleotide arrays. *National Academy of Sciences of the United States of America*, 96:6745–6750, 1999.
4. B. Mirkin. *Mathematical classification and Clustering.* Kluwer, 1996.
5. S. Bleuler, A. Prelić, and E. Zitzler. An EA framework for biclustering of gene expression data. In *Congress on Evolutionary Computation (CEC-2004)*, pages 166–173, Piscataway, NJ, 2004. IEEE.
6. Y. Cheng and G. M. Church. Biclustering of expression data. In *Proceedings of the 8th International Conference on Itelligent Systems for Molecular Biology (ISMB'00)*, pages 93–103, 2000.
7. R. Cho, M. Campbell, E. Winzeler, L. Steinmetz, A. Conway, L. Wodicka, T. Wolfsberg, A. Gabrielian, D. Landsman, D. Lockhart, and R. Davis. A genome-wide transcriptional analysis of the mitotic cell cycle. *Molecular Cell*, 2:65–73, 1998.
8. G. Getz and E. Levine and E. Domany. Coupled two–way clustering analysis of gene microarray data. *Proceedings of the Natural Academy of Sciences USA*, pages 12079–12084, 2000.
9. J. A. Hartigan. Direct clustering of a data matrix. *Journal of the American Statistical Association*, 67(337):123–129, 1972.
10. L. Lazzeroni and A. Owen. Plaid models for gene expression data. Technical report, Stanford University, 2000.
11. H. Wang, W. Wang, J. Yang, and P. S. Yu. Clustering by pattern similarity in large data sets. In *Proceedings of the ACM SIGMOD International Conference on Management of Data*, pages 394–405, 2002.
12. J. Yang, W. Wang, H. Wang, and P. S. Yu. δ–clusters: Capturing subspace correlation in a large data set. In *Proceedings of the 18th IEEE Conference on Data Engineering*, pages 517–528, 2002.
13. J. Yang, W. Wang, H. Wang, and P. S. Yu. Enhanced biclustering on expression data. In *Proceedings of the 3rd IEEE Conference on Bioinformatics and Bioengineering*, pages 321–327, 2003.

A Fuzzy Viterbi Algorithm for Improved Sequence Alignment and Searching of Proteins

N. P. Bidargaddi, M. Chetty, and J. Kamruzzaman

Gippsland School of Computing and Information Technology,
Monash University, Gippsland campus, Vic –3842, Australia
{niranjan.bidargaddi, madhu.chetty,
joarder.kamruzzaman}@infotech.monash.edu.au

Abstract. Profile HMMs based on classical hidden Markov models have been widely studied for identification of members belonging to protein sequence families. Classical Viterbi search algorithm which has been used traditionally to calculate log-odd scores of the alignment of a new sequence to a profile model is based on the probability theory. To overcome the limitations of the classical HMM and for achieving an improved alignment and better log-odd scores for the sequences belonging to a given family, we propose a fuzzy Viterbi search algorithm which is based on Choquet integrals and Sugeno fuzzy measures. The proposed search algorithm incorporates ascending values of the scores of the neighboring states while calculating the scores for a given state, hence providing better alignment and improved log-odd scores. The proposed fuzzy Viterbi algorithm for profiles along with classical Viterbi search algorithm has been tested on globin and kinase families. The results obtained in terms of log-odd scores, Z-scores and other statistical analysis establish the superiority of fuzzy Viterbi search algorithm.

Keywords: protein sequence alignment, profiles, HMM, fuzzy HMM.

1 Introduction

Proteins can be classified into families based on related sequences and structures wherein different residues are subjected to different selective pressures. Multiple alignment of protein sequences reveal the pattern of conservation. Some regions of multiple alignment tolerate more insertions and deletions compared to other regions [1]. Since the conservation across positions is not uniform it is desirable to use position specific information from multiple alignments when searching databases for homologous sequences. Taylor *et al.* [2] used 'profile' methods to build position-specific scoring methods for multiple alignments. Krogh [3] used hidden Markov models (HMMs) for multiple sequence alignment since they provide a coherent theory for profile methods. HMMs are a class of probabilistic models used in time series applications such as speech recognition, image processing and handwritten word recognition [4-7]. HMMs were initially introduced to computational biology by Churchill [8] and since then have been widely used in various ways. Large libraries of

F. Rothlauf et al. (Eds.): EvoWorkshops 2005, LNCS 3449, pp. 11–21, 2005.

profile based HMM and multiple alignments are available in many database servers [9]. Recently profile HMM methods have also been used in the area of protein structure prediction by fold recognition [10].

Traditionally Viterbi algorithm is used in HMM to find the most likely sequence of hidden states that result in a sequence of observed amino-acid symbols. The algorithm makes a number of assumptions. First, both the observed events and hidden events must be in a sequence. This sequence often corresponds to time. Second, these two sequences need to be aligned, and an observed event needs to correspond to exactly one hidden event. Third, computing the most likely hidden sequence up to a certain point t must only depend on the observed event at point t, and the most likely sequence at point $t-1$.

Due to these assumptions, Viterbi algorithm has inherent limitations which are observed in HMMs. An HMM assumes that the identity of an amino acid at a particular position is independent of the identity of all other positions [11]. HMM models are also constrained by the statistical independence assumptions in the formulation of the forward and backward variables which are used to compute the matching scores of an unknown sequence to a known family. Due to the statistical independence assumptions, the joint measure variables (forward and backward) are decomposed as a combination of two measures defined on amino acid emission probabilities and state probabilities. To relax the statistical independence assumptions and achieve improved performance and flexibility Magdi *et al.* [12] defined a fuzzy hidden Markov model based on fuzzy forward and backward variables. This model does not require the assumption of decomposing the measures. The fuzzy model offers more flexibility and robustness in the sense that it does not require fixing the lengths of the sequences and the need for large training data sets in order to learn the transition parameters as required by classical hidden Markov models. Fuzzy hidden Markov models have been successfully investigated on various domains such as speech and image processing [4-7]. The authors have investigated its potential in forward-backward algorithm for decoding protein sequence in [13]. This paper investigates the feasibility of fuzzifying the Viterbi algorithm based on the above ideas.

2 Profile Hidden Markov Models

Profiles introduced by Gribskov *et al.* [14] are statistical descriptions of the consensus of multiple sequence alignment. They use position-specific scores for amino acids and position specific penalties for opening and extending an insertion or deletion. Due to the use of position-specific scores for amino acids, profiles capture important information about the degree of conservation at various positions in the multiple alignments, and the varying degree to which gaps and insertions are permitted. Due to its probabilistic nature, a profile HMM can be trained from unaligned sequences, if a trusted alignment is not yet known. Another benefit of using probability theory is that HMMs have a consistent theory behind gap and insertion scores. Profile HMM architecture is characterized by Mx (Match state x, with k emission probabilities), Dx (Delete state x, non-emitter) and Ix (Insert state x, with k emission probabilities).

2.1 Classical Viterbi Algorithm for Profile Alignment

Let $V_j^M(i), V_j^I(i)$ and $V_j^D(i)$ be the log-odds score of the best paths matching subsequence $x_{1,\ldots,i}$ to state j of the submodel ending with x_i being emitted by state M_j, I_j and D_j respectively which are calculated by the following equations in profile HMM.

$$V_j^M(i) = \log\frac{e_{M_j}(x_i)}{q_{x_i}} + \max\begin{cases} V_{j-1}^M(i-1) + \log a_{M_{j-1}M_j} \\ V_{j-1}^I(i-1) + \log a_{I_{j-1}M_j} \\ V_{j-1}^D(i-1) + \log a_{D_{j-1}M_j} \end{cases} \quad (1)$$

$$V_j^I(i) = \log\frac{e_{I_j}(x_i)}{q_{x_i}} + \max\begin{cases} V_j^M(i-1) + \log a_{M_jI_j} \\ V_j^I(i-1) + \log a_{I_jI_j} \\ V_j^D(i-1) + \log a_{D_jI_j} \end{cases} \quad (2)$$

$$V_j^M(i) = \max\begin{cases} V_{j-1}^M(i) + \log a_{M_{j-1}D_j} \\ V_{j-1}^I(i) + \log a_{I_{j-1}D_j} \\ V_{j-1}^D(i) + \log a_{D_{j-1}D_j} \end{cases} \quad (3)$$

q_{x_i} is the emission probability of the ith amino acid with respect to a standard model, a represents the transition probabilities (e.g., $a_{I_iM_j}$ represents the transition probability from i^{th} I state to j^{th} M state) and e represents the amino acid emission probabilities.

3 Fuzzy HMM

Fuzzy measures introduced by Sugeno [15] provide a more flexible way to overcome the drawbacks of the additive hypothesis of the probability model which is not well suited for modeling systems that manifest a high degree of interdependencies among sources of information. Let X be an arbitrary set and Ω be power set of X. A set function f: Ω→[0,1] defined on Ω satisfying the property of boundary conditions and monotonicity is called a fuzzy measure. Magdi *et al.* fuzzified the hidden Markov model through fuzzy measures and fuzzy integrals. The fuzzy hidden Markov model $\hat{\lambda} = (\hat{A}, \hat{B}, \hat{\Pi})$ is characterized by the following parameters [12].

O	Observation sequence
T	Length of the observation sequence
N	Number of states in the model
S	$(S_1, S_2, \ldots, S_N)$ states
Ω	Observation sequence space
$X = \{x_1, x_2, \ldots, x_N\}$	States at time t

$Y = \{y_1, y_2, \ldots y_N\}$	States at time t+1
$\hat{\Pi}_s(.)$	Initial state fuzzy measure
$\hat{\Pi}_i = \hat{\Pi}_s^i = \hat{\Pi}_s(\{S_i\})$	Initial state fuzzy density
$\Pi_i = [\hat{\Pi}_i]$	Vector of initial state fuzzy densities
$\hat{b}_j(O_t)$	Fuzzy densities of symbols
$\hat{B} = [\hat{b}_j(O_t)]$	Symbol density matrix
$\hat{a}_y(. \mid X)$	Transition fuzzy measure
$\hat{a}_{ij} = \hat{a}_y(\{y_j \mid x_i\})$	Transition fuzzy density
$\overline{\hat{A}} = [\overline{\hat{a}}_{ij}]$	Matrix of transition fuzzy densities

Fuzzy HMM incorporates the joint fuzzy measures $\hat{\alpha}_{\Omega_y}(\{O_1,\ldots,O_t\} \times \{y_j\})$ which can be written as a combination of two measures defined on $O_1, O_2..., O_t$ and on the states respectively without any assumptions about the decomposition [11-12].

In classical HMM's, the joint probability measure is a product of independent probability measures, $P(O_1,O_2,...,O_t, q_{t+1}= S_j) = P(O_1,O_2,...,O_t) \bullet P(q_{t+1}= S_j)$ necessitating the assumption of statistical independence.

4 Proposed Fuzzy Viterbi Algorithm for Profile HMM

The Viterbi algorithm computes the negative logarithm of the probability of the single most likely path, $\hat{\delta}$ for the sequence s which can be written as

$$\hat{\delta} = \text{-log} \max_{\hat{\pi}} P(s, \hat{\pi} \mid \hat{\lambda}) \tag{4}$$

where $\hat{\pi}$ represents the combination of states emitting the symbols in sequence s. Magdi *et al.* proposed a fuzzy Viterbi algorithm by modifying the classical Viterbi algorithm using fuzzy measures to estimate the quantity $\hat{\delta}_t(i)$ which is given by

$$\hat{\delta}_t(i) = \max_{q_1, q_2, \ldots, q_{t-1}} P(q_1, q_2, \ldots, q_t = S_i, O_1 O_2, \ldots, O_t \mid \lambda) \tag{5}$$

where $O_1, O_2, \ldots\ Ot$ are the constituent symbols of sequence s, q_t represents the state at time t, S_i is the ith state. The classical Viterbi algorithm is modified to perform the maximization of $\hat{\delta}_t(i)$ as shown below

$$\hat{\delta}_t(i) = \max_{q_1, q_2, \ldots, q_{t-1}} \left\{ \hat{\pi}_{q_1} \hat{b}_{q_1}(O_1) \prod_{\tau=2}^{t} [\hat{a}_{q_{\tau-1}, q_\tau} \rho_\tau(q_{\tau-1}, q_\tau)] \hat{b}_{q_\tau}(O_\tau) \right\} \tag{6}$$

$\hat{\pi}_{q_i}$ is the initial state fuzzy density for state q_i, $\hat{b}_{q_\tau}(O_\tau)$ is the emission density of symbol O_τ for state q_τ and $\hat{a}_{q_{\tau-1}, q_\tau}$ represents the transition probability from state $q_{\tau-1}$

to q_T. ρ_τ is the fuzzy measure difference and is explained further later in this section, $\hat{\delta}_{t+1}(i)$ is computed recursively using the fuzzy Viterbi algorithm as shown below.

(i) Initialization (for $1 \le i \le N$)

$$\hat{\delta}_1(i) = \hat{\pi}_i \hat{b}_i(O_1) \qquad \hat{\varphi}_1(i) = 0 \tag{7}$$

(ii) Recursion (for $2 \le t \le T$ and $1 \le j \le N$)

$$\hat{\delta}_t(j) = \max_{1 \le i \le N} [\hat{\delta}_{t-1}(i)\hat{a}_{ij}\rho_t(i,j)]\hat{b}_j(O_t) \quad \hat{\varphi}_t(j) = \arg\max_{1 \le i \le N} [\hat{\delta}_{t-1}(i)\hat{a}_{ij}\rho_t(i,j)] \tag{8}$$

(iii) Termination

$$\hat{P}^* = \max_{1 \le i \le N}[\hat{\delta}_T(i)] \qquad \hat{q}_T^* = \arg\max_{1 \le i \le N}[\hat{\delta}_T(i)] \tag{9}$$

The optimal path can be obtained by backtracking for all $1 \le t \le T\text{-}1$

$$\hat{q}_t^* = \hat{\varphi}_{t+1}(\hat{q}_{t+1}^*) \tag{10}$$

The above equations for the fuzzy Viterbi algorithm are very similar to the classical ones except for the introduction of a new term ρ. The new term ρ in the above equations represents the fuzzy measure difference which is calculated using the Choquet integral. According to the definition of fuzzy measures and fuzzy integrals on a discrete set X, with a function h:X→[0,1]; g:2^X→[0,1], Choquet-integral ($e_{choquet}$) is given by Eq. (13) after satisfying the constraints in Eq. (11-12) [12].

$$h(x_1) \le h(x_2) \le \ldots \le h(x_i) \le \ldots \le h(x_N) \tag{11}$$

$$K_i = \{x_i, x_{i+1}, \ldots, x_N\} \tag{12}$$

$$e_{choquet} = \sum_{i=1}^{N} h(x_i)[g(K_i) - g(K_{i+1})] = \sum_{i=1}^{N} h(x_i) d_i \tag{13}$$

where d_i represents the difference between successive fuzzy measures and $g(K_i)$ represents the fuzzy measure. For the fuzzy profile HMMs, the matrix A containing all the transition parameters (Insert, Match and Delete) represents the function h which is sorted at j^{th} row to obtain $K_i(j)$ as

$$K_i(j) = \{S_i, S_{i+1}, \ldots, S_N\} \tag{14}$$

where S_i is the state number at i^{th} position according to constraints in Eq. (11)-(12) based on transition to j^{th} state from all other states. The two most suitable fuzzy measures for profile HMMs are λ-fuzzy measures and possibility measures. After obtaining $K_i(j)$ for the fuzzy profile HMM, fuzzy measure $g(K_i(j))$ is calculated using Eq. (15) for λ- fuzzy measures and Eq. (16) for possibility measures. The calculation of fuzzy integral with respect to either of the above fuzzy measure only requires the knowledge of the fuzzy densities, $g(\{S_i\})$ which are calculated as shown in Eq. (17).

$$g(K_i(j)) = g(\{S_i, S_{i+1}, \ldots, S_N\})$$

$$g(K_i(j)) = g(\{S_i\}) + g(K_{i+1}(j)) + \lambda g(\{S_i\})\, g(K_{i+1}(j)) \tag{15}$$

$$g(K_i(j)) = \max(g(\{S_i\}), g(K_{i+1}(j))) \tag{16}$$

$$g(\{S_i\}) = \alpha_t(i) \tag{17}$$

To avoid data overflows and to speed up the computing process, we use possibility measure with *max* operator for computing the fuzzy measures. After obtaining the fuzzy measures the difference between successive fuzzy measures $d_t(i, j)$ is calculated as

$$d_t(i, j) = g\,(K_i\,(j)) - g(K_{i+1}(j)) \tag{18}$$

$d_t(i, j)$ is normalized with respect to fuzzy densities and stored in $\rho_t(i, j)$ as shown in equation (19).

$$\rho_t(i, j) = d_t(i, j) / \alpha_t(i) \tag{19}$$

$$V_j^M(i) = \log\frac{e_{M_j}(x_i)}{q_{x_i}} + \max\begin{cases} V_{j-1}^M(i-1) + \log a_{M_{j-1}M_j}\,\rho_{j-1}(M_{j-1}, M_j) \\ V_{j-1}^I(i-1) + \log a_{I_{j-1}M_j}\,\rho_{j-1}(M_{j-1}, M_j) \\ V_{j-1}^D(i-1) + \log a_{D_{j-1}M_j}\,\rho_{j-1}(M_{j-1}, M_j) \end{cases} \tag{20}$$

$$V_j^I(i) = \log\frac{e_{I_j}(x_i)}{q_{x_i}} + \max\begin{cases} V_j^M(i-1) + \log a_{M_jI_j}\,\rho_{j-1}(M_{j-1}, M_j) \\ V_j^I(i-1) + \log a_{I_jI_j}\,\rho_{j-1}(M_{j-1}, M_j) \\ V_j^D(i-1) + \log a_{D_jI_j}\,\rho_{j-1}(M_{j-1}, M_j) \end{cases} \tag{21}$$

$$V_j^M(i) = \max\begin{cases} V_{j-1}^M(i) + \log a_{M_{j-1}D_j}\,\rho_{j-1}(M_{j-1}, M_j) \\ V_{j-1}^I(i) + \log a_{I_{j-1}D_j}\,\rho_{j-1}(M_{j-1}, M_j) \\ V_{j-1}^D(i) + \log a_{D_{j-1}D_j}\,\rho_{j-1}(M_{j-1}, M_j) \end{cases} \tag{22}$$

We incorporate the fuzzy Viterbi algorithm described above in profile HMM with an aim to achieve improved protein alignment and more accurate classification. Accordingly we reformulate the Viterbi variables used to calculate log-odd scores defined in Eq. (1)-(3) for classical Viterbi algorithm as follows to construct fuzzy Viterbi algorithm for profile HMMs. It can be observed from Eq. (20)-(22) that the fuzzy measure for each state is calculated using all the states which have greater forward variable. As a result fuzzy measure term introduced in the above equations takes into account the local interaction among neighboring states while calculating the Viterbi variables for each state. Since j^{th} M state has transitions from $j\text{-}1^{th}$ I, D and M states the fuzzy measure calculation for j^{th} state will take into account of all the three states.

5 Evaluation Methods

One commonly used technique for scoring a match of the sequence x to hidden Markov model is to calculate the log-odd scores by using Viterbi equations. Log-odd score gives the most probable alignment π^* of the sequence x for a given model λ and is given by

$$\text{Log-odd Score} = P(x\,\pi^*, |\,\lambda) \tag{23}$$

Equations (20)-(22) are used to calculate the log-odd scores for the fuzzy profile Markov model. Since the profile HMM model is concentrated on a subset of the protein family, we also calculate Z-scores for each sequence to have a clear view of match scores. The Z-score is a measure of how far an observation is from the mean, as measured in units of standard deviation. For a given data set X with mean μ and standard deviation σ, Z-score is given by

$$\text{Z-score} = (X - \mu) / \sigma \tag{24}$$

To calculate Z-score, a smooth curve is fitted for the log-odd score plot using the local window technique [11]. A standard deviation is estimated for each length and Z-score is calculated for each score by estimating its distance from the curve in terms of standard deviation.

6 Results

To evalulate the performance of fuzzy Viterbi algorithm on profile HMM, we performed the experiment on two different families namely globins and kinases. Profile HMM of the globin and kinase families from pfam built using HMMER is used as the start point. The estimation of this profile model was done several times and the model with the highest overall log-likelihood (LL) score was chosen according to the default settings of SAM package [16]. Log-odd scores and Z-scores are calculated for the sequences using both the classical and fuzzy Viterbi algorithms. All the simulations were carried out using the Bioinformatics toolbox in Matlab environment. The Viterbi algorithm is suitably modified to incorporate the fuzzy measures.

6.1 Globin

The modeling was first tested on the widely studied globins, a large family of heme containing proteins involved in the storage and transport of oxygen that have different oligomeric states and overall architecture. Globin sequences were extracted from the Pfam database by searching for the keyword "globin". The globin data set used in the experiment consists of 126 different globin sequences. The sequences in the family vary in length from 109 to 428 amino acids. We divided the data set into training set of 76 sequences which were used in building the profile model, and the remaining 50 sequences along with 1953 non-globin sequences as the test data set (1969 sequences in total). We align the sequences in the training and test set against the existing globin model from the profile library of *pfam* database [accession id: PF00042] using classical and fuzzy Viterbi algorithm. Figures 1(a) and (b) respectively show the normalized log-odd scores for the sequences using classical and fuzzy Viterbi algorithms. The Z-scores for classical and fuzzy Viterbi are shown in Fig. 2(a) and Fig. 2(b) respectively. As seen from the plots fuzzy Viterbi performs better than classical Viterbi algorithm in distinguishing the proteins. Choosing a Z-score cut off = 2.0 distinguishes all the globins from non globins with fuzzy Viterbi, whereas the classical Viterbi produces 10 false positive globins and 2 globins (protozoan/cyanobacterial globin and Hypothetical protein R13A1.8) are missed out. Table 1 shows the classification counts of globins (positives) and non-globins (negatives) for classical and fuzzy Viterbi for different values of Z-score cut offs.

It can be observed from Table 1 that fuzzy Viterbi has produced no false positives and false negatives for Z-score cut off ranging from 0.5 to 2.0. It can also be observed that there are no false positives for the fuzzy Viterbi in entire range of Z-score cut off from 0.0 to 3.0. The 'close to best' performance for classical Viterbi is observed with the Z-score cut off = 2.5 producing 2 false-positives and 5 false-negatives. Choosing a Z-score cut off anywhere in the range of 0.5 to 2.0 distinguishes all the globins from non globins with fuzzy Viterbi, whereas the classical Viterbi yields no clear classification in this range. With a Z-score cut-off = 3.0, classical Viterbi search yields 17 false-negatives and hence correctly identifies 109 globin sequences. With the same cut-off, fuzzy Viterbi yields 12 false negatives which is 5 less than the classical Viterbi search. With 0 chosen as the Z-score cut off, classical Viterbi performance drops down drastically since it wrongly identifies 1004 non-globin sequences as globin sequences, whereas fuzzy Viterbi produces only 4 false negatives. Figures 3(a)-(b) show a section of the first 8 of the 625 globin sequences aligned using classical and fuzzy Viterbi algorithms respectively.

Table 1. Performance of classical and fuzzy Viterbi with Z-score cut/off

Z-Score cut-off value	Classical-Viterbi search				Fuzzy-Viterbi search			
	FP	FN	TP	TN	FP	FN	TP	TN
0.0	1004	0	126	949	0	4	126	1949
0.5	950	0	126	1003	0	0	126	1953
1.0	473	0	126	1480	0	0	126	1953
1.5	62	0	126	1891	0	0	126	1953
2.0	10	2	124	1943	0	0	126	1953
2.5	2	5	119	1951	0	3	123	1953
3.0	0	17	109	1953	0	12	114	1953

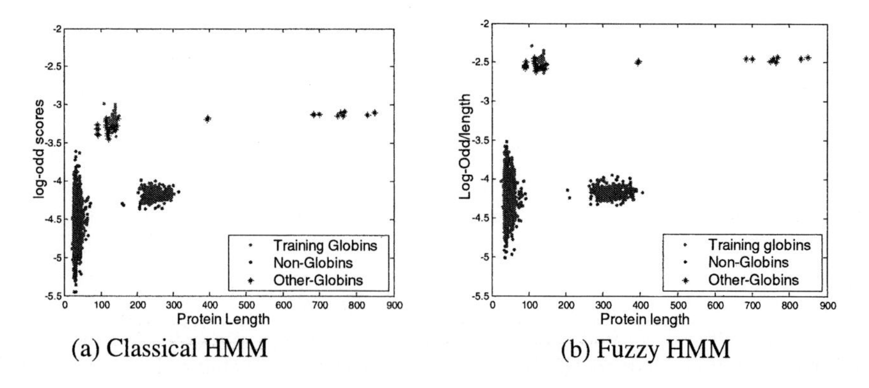

(a) Classical HMM (b) Fuzzy HMM

Fig. 1. Normalized Log-odd scores

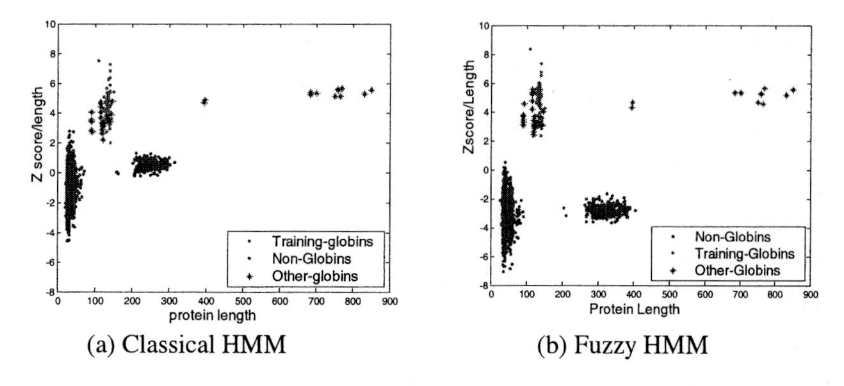

(a) Classical HMM (b) Fuzzy HMM

Fig. 2. Z-Score calculated from Log-odd scores

```
HBB1_VAREX   Vh........WTAEEKQLICSLW....GKIDV..GLIG.........GET
HBB2_TRICR   Vh........LTAEDRKEIAAIL....GKVNV..DSLG.........GQC
HBB2_XENTR   Vh........WTAEEKATIASVW....GKVDI..EQDG.........HDA
HBBL_RANCA   Vh........WTAEEKAVINSVW....QKVDV..EQDG.........HEA
HBB_CALAR    Vh........LTGEEKSAVTALW....GKVNV..DEVG.........GEA
HBB_COLLI    Vh........WSAEEKQLITSIW....GKVNV..ADCG.........AEA
HBB_EQUHE    Vq........LSGEEKAAVLALW....DKVNE..EEVG.........GEA
HBB_LARRI    Vh........WSAEEKQLITGLW....GKVNV..ADCG.........AEA
```

(a) Classical Viterbi

```
HBB1_VAREX   V.HWtAEEKQLICSLWGKI..DVGLIGGETLAGLLVIYPWTQRQFSHF..
HBB2_TRICR   V.HLtAEDRKEIAAILGKV..NVDSLGGQCLARLIVVNPWSRRYFHDF..
HBB2_XENTR   V.HWtAEEKATIASVWGKV..DIEQDGHDALSRLLVVYPWTQRYFSSF..
HBBL_RANCA   V.HWtAEEKAVINSVWQKV..DVEQDGHEALTRLFIVYPWTQRYFSTF..
HBB_CALAR    V.HLtGEEKSAVTALWGKV..NVDEVGGEALGRLLVVYPWTQRFFESF..
HBB_COLLI    V.HWsAEEKQLITSIWGKV..NVADCGAEALARLLIVYPWTQRFFSSF..
HBB_EQUHE    VqLS.GEEKAAVLALWDKV..NEEEVGGEALGRLLVVYPWTQRFFDSF..
HBB_LARRI    V.HWsAEEKQLITGLWGKV..NVADCGAEALARLLIVYPWTQRFFASF..
```

(b) Fuzzy Viterbi

Fig. 3. Alignment of the first 8 of the 650 globin sequences

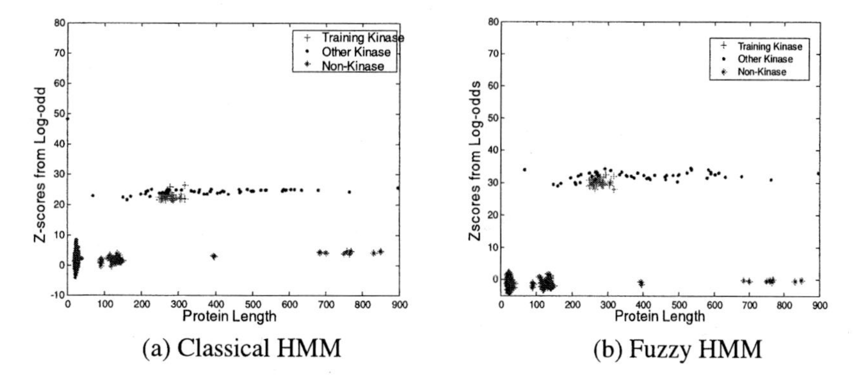

(a) Classical HMM (b) Fuzzy HMM

Fig. 4. Z-Score calculated from Log-odd scores

6.2 Protein Kinase

Protein kinases have been extensively studied in [1-3]. Complete protein kinase catalytic domains range from 250 to 300 residues. The training set is made up of 54 kinase representative sequences and the test set consists of 72 different kinase sequences obtained by searching for the keyword 'kinase' in NCBI database along with 1141 random non-kinase sequences. The kinase sequences in the test set range in length from 100 to 800 amino acids. Due to space limitations we have shown only the Z- score plots for kinase family using classical and fuzzy Viterbi in Fig. 4(a) and Fig. 4(b) respectively.

7 Conclusion

Classical Viterbi search algorithm which has been used to calculate log-odd scores of the alignment of a new sequence to a profile model is based on the probability theory. To achieve an improved alignment and better log-odd scores for the sequences belonging to a given family we propose a fuzzy Viterbi search algorithm which is based on Choquet integrals and Sugeno fuzzy measures. The proposed search algorithm takes into account ascending values of the scores of the neighboring states while calculating the scores for a given state, hence providing better alignment, improved log-odd scores and Z-scores. The proposed fuzzy Viterbi algorithm for profiles along with classical Viterbi search algorithm has been tested both on globin and kinase families. The results obtained in terms of log-odd scores and other statistical analyses establish the superiority of fuzzy Viterbi search algorithm in both cases. Future work will concentrate on further improvement of the model by using fuzzy expectation-maximisation (EM) algorithm to estimate the parameters of the profiles instead of classical Baum-Welch algorithm and an intensive analysis of the effect of different fuzzy measures.

References

[1] S. R. Eddy, "Profile hidden Markov models", *Bioinformatics*, vol. 14, pp. 755–763, 1998.

[2] W. R. Taylor, "Identification of protein sequence homology by consensus template alignment", *J. Mol. Biol*, 188, pp. 233-258, 1986.

[3] A. Krogh, "An introduction to hidden Markov models for biological sequences", *Computational Methods in Molecular Biology, Elsevier*, pp. 45–63, 1998.

[4] Z. Valsan, I. Gavat and B. Sabac, "Statistical and hybrid methods for speech recognition in Romanian", *International Journal of Speech Technology 5*, vol. 5, pp. 259-268, 2002.

[5] H. Shi and P. D. Gader, "Lexicon-driven handwritten word recognition using Choquet Fuzzy Integral", *IEEE Conf.*, pp. 412-417, 1996.

[6] A. D. Cheok *et al.*, "Use of a novel generalized Fuzzy Hidden Markov Model for speech recognition", *IEEE Conf. Fuzzy System*, pp. 1207-1210, 2001.

[7] D. Tran and M. Wagner, "Fuzzy Hidden Markov Models for Speech and Speaker Recognition", *IEEE Conf. Speech Processing*, pp. 426-430, 1999.

[8] G. A. Churchill, "Stochastic models for heterogeneous DNA sequence", *Bull. Math. Biol.* vol. 51, pp. 79-94, 1989.

[9] E. L. Sonnhammer, S. R. Eddy, and R. Durbin, "Pfam: A comprehensive database of protein families based on seed alignments", *Proteins*, vol. 28, pp. 405–420, 1997.

[10] M. Levitt, "Competitive assessment of protein recognition and alignment accuracy", *Proteins*, 1 (Suppl.) pp. 92-104, 1997.

[11] R. Durbin, S.Eddy, A. Krogh, G. Mitchison, Biological sequence analysis- Probabilistic models of proteins and nucleic acids, *Cambridge University Press*, Cambridge, 2003.

[12] M. A. Magdi and P. Gader, "Generalized hidden Markov models- part I: theoretical frameworks", *IEEE Trans. Fuzzy Systems*, vol. 8, no.1, pp. 67-80, 2000.

[13] N.P. Bidargaddi, M. Chetty and J. Kamruzzaman, "Fuzzy decoding in profile hidden Markov models for protein family identification", to appear in *proc. Intl. Conf. Bioinformatics and its Applications*, 2004.

[14] M. Gribskov *et al.*, "Profile analysis: Detection of distantly related proteins", *Proc. Natl. Acad. Sci. USA*, vol. 84, pp. 4355–4358, 1987.

[15] M. Sugeno, "Fuzzy measures and fuzzy integrals- A survey", *Fuzzy Automata and Decision Processes*, M.M. Gupta, G. N. Saridis, and B. R. Gaines, Eds, New York: North-Holland, pp. 89-102, 1977.

[16] J. Gough *et al.*, "Assignment of homology to genome sequences using a library of hidden Markov models that represent all proteins of known structure", *J. Mol. Biol.*, vol. 313(4), pp. 903-919, 2001.

[17] A. Bateman *et al.*, "The Pfam protein families database", *Nucleic Acids Research*, vol. 30, pp. 276-280, 2002.

[18] N. J. Mulder *et al.*, "The InterPro database", *Nucleic Acids Res.*, vol.31, pp. 315-318, 2003.

Tabu Search Method for Determining Sequences of Amino Acids in Long Polypeptides

Jacek Błażewicz[1,2], Marcin Borowski[1],
Piotr Formanowicz[1,2], and Maciej Stobiecki[2]

[1] Institute of Computing Science, Poznań University of Technology,
Piotrowo 3A, 60-965 Poznań, Poland
[2] Institute of Bioorganic Chemistry, Polish Academy of Sciences,
Noskowskiego 12/14, 61-704 Poznań, Poland
piotr@cs.put.poznan.pl

Abstract. The amino acid sequences of proteins determine their structure and functionality, hence methods for reading such sequences are crucial for many areas of biological sciences. Since direct methods for reading amino acid sequences allow for determining only very short fragments, some methods for assembly of these fragments are required. In this paper, tabu search algorithm solving this problem is proposed. Computational tests show its usefulness in the process of determining sequences of amino acids in long polypeptides.

1 Introduction

Problems of reading nucleotide sequences are crucial for molecular and computational biology since the whole genetic information is encoded in nucleotide sequences. Hence, reading of DNA or RNA sequences is the first stage in possible broader research projects.

The nucleotide sequences may play three main roles in living organisms: they encode proteins, they encode RNAs which are not translated into proteins like tRNA or rRNA and they also perform some regulatory roles, respectively. Most of the DNA sequences whose roles have been identified are genes encoding proteins. Since the genetic code is known, it is easy to determine amino acid sequences of proteins whose genes are known. However, it is also the case that not always the information about the nucleotide sequence is sufficient. Indeed, proteins are often subjected to some post-translational modifications which are not explicitly encoded in the nucleotide sequence of the protein's gene. It means that these modifications are caused by some other molecules encoded by their own genes. The relationships among products of groups of genes are usually very hard to determine, so the prediction of the exact amino acid sequence of the protein being post-translational modified is also extremely difficult. So, in this case some direct method for amino acid sequence determining would be useful.

Fortunately, such methods exists. One approach requires an application of mass spectrometry. This approach usually allows for a determination of some

F. Rothlauf et al. (Eds.): EvoWorkshops 2005, LNCS 3449, pp. 22–32, 2005.

short fragments of peptides which are compared with databases of known proteins in the next step. Of course, such approaches do not allow for a discovery of unknown proteins but in spite of that in many cases they are useful. Recently, there are also developed some methods based on mass spectrometry which allow for sequencing unknown polypeptides.

The other group of methods is based on Edman's degradation. This degradation is a biochemical procedure which allows for a direct determination of short peptide sequences.

Both types of peptide sequencing methods allow for determining rather short amino acid sequences. So, if the whole protein is to be sequenced, then it must be cleaved into smaller pieces at the beginning of the reading process. Such a cleavage can be done by enzymes called proteases which cut the amino acid chain in some specific places. Next, each of the pieces is sequenced by one of the methods mentioned above. Obviously, the task which remains to be done is to assemble short amino acid chains of known sequence into one polypeptide. This can be done by some algorithmic methods. In the paper, we propose tabu search algorithm for one of the variants of the peptide sequencing problem belonging to the class of strongly NP-hard search problems [3]. Computational tests show its usefulness in the process of determining sequences of amino acids in long polypeptides.

The organization of the paper is as follows. In the next Section a biochemical stage of the peptide sequencing procedure will be briefly described. In Section 3 the combinatorial problem which must be solved in the second stage of the procedure is formulated while in Section 4 the tabu search algorithm solving this problem is described. Section 5 contains results of the computational experiment. The paper ends with conclusions in Section 6.

2 The Biochemical Stage of the Peptide Sequencing Process

The biochemical stage of the method may be based on *Edman's degradation* [10] which allows for detaching of consecutive amino acid residues from an amino end of the analyzed polypeptide. The most important property of the method is the fact that only the residue which is at the amino end of the peptide is detached while the rest of the peptide remains unchanged. The amino acid composition (but not the sequence, of course) of the shortened peptide can be compared with the composition of the analyzed peptide. As a result of this comparison one gets information what residue has been detached.

The same Edman's procedure can be applied again to the shortened peptide which will lead to identification of the second amino acid residue at the amino end of the target polypeptide. Repeating this procedure many times allows for determining the amino acid sequence of the whole polypeptide.

Nevertheless, the described above procedure has a serious limitation, i.e. it can be applied to polypeptides composed of not more than 50 amino acid residues [10]. The cause of this limitation is the efficiency of the reaction of detaching single residues. In practice, not from every peptide in the solution the residue is detached at each stage of the method.

This shortage of the method can be overcome by dividing the analysed polypeptide into shorter peptides of lengths not exceeding 50 amino acids. Specific cleavage of a polypeptide chain can be obtained by enzymatic or chemical methods [10]. Some proteases cut polypeptide chains after some specific amino acid residues. As a result one obtains a collection of short polypeptides which can be fully analyzed by Edman's degradation. However, the sequence information discovered in this way do not suffice for determining the amino acid sequence of the whole polypeptide since the order of the short peptides remains unknown. To determine this order it is necessary to use more than one protease cutting the polypeptide in different places. If two such proteases are used then the obtained short peptides will overlap which allows for determining their order in the analysed polypeptide.

As mentioned in the previous Section sequences of short peptides can be also determined using a mass spectrometer [9, 2]. Roughly speaking in a mass spectrometer short peptide is broken into shorter fragments and masses of these fragments are measured. In an ideal case the examined peptide is broken between any two consecutive amino acid residues. As a result masses of these fragments are obtained. The problem of determining the sequence is relatively simple in the case of an ideal fragmentation, but in practice there may occur many errors and the output from the spectrometer may be far from being an ideal one, what makes the sequencing problem difficult. However, the problem of determining peptide sequences using mass spectrometers is recently intensively investigated which may lead to a development of some effective algorithms for the real world problems.

It must be also noted that due to the current technology constraints, the amino acid sequences which can be determined using mass spectrometers are rather short and their lengths usually falls into the range between 10 and 20 amino acid residues. It follows that when a whole protein is to be sequenced, it must be first cleaved into short pieces, analogously like in the case where Edman's degradation is used. Obviously, it leads to the same assembling problem.

3 Formulation of the Peptide Assembly Problem

Let us assume that the analysed polypeptide is cut by two proteases each of them cleaving the peptide after a given amino acid which will be denoted, respectively, by α and β. So, when the solution of many copies of the examined polypeptide is subjected to the two proteases, a collection of short peptides is obtained. In the last position of every such a peptide, i.e. in the amino end, there is amino acid α or β, which follows from the way the target polypeptide is cleaved by the proteases. It may also happen that some of the peptides will be substrings of some other peptides. These short peptides completely included in some other ones can be excluded from the collection and further consideration. As a result two sets of amino acid sequences can be created – S_α containing peptides being cut out from the target sequence by the protease recognizing amino acid α, and S_β which contains all the other peptides, i.e. those being cut out by the protease recognizing amino acid β. Determining the sequence of the target polypeptide is equivalent to determining some permutation of the elements of the sets S_α and S_β which should satisfy the following conditions:

1) in the permutation any two consecutive elements should not belong to the same set,
2) every two consequtive elements (i.e. sequences) of the permutation should maximally overlap, so the polypeptide sequence corresponding to this permutation should be as short as possible.

It is possible to formulate the problem as a graph theoretical one in the following way [1]. Let $G = (V, A)$ be a directed graph, where the set of vertices $V = V_1 \cup V_2$, $|V| = n$, $V_1 \cap V_2 = \emptyset$ and the set of directed arcs is defined as $(v_i, v_j) \in A \iff (v_i \in V_1 \wedge v_j \in V_2) \vee (v_i \in V_2 \wedge v_j \in V_1)$. With every vertex $v_i \in V$ there is associated label s_i over alphabet Σ_a, where $|\Sigma_a| = 20$. Moreover, function w called the weight of the arc is defined as $w : A \to \mathbb{N}$, $w(v_i, v_j) = \max_{k\in\{1,2,\ldots,|s_i|\}}\{p : \forall_{q\in\{1,2,\ldots,p\}} s_i(k-1+q) = s_j(q)\}$ if there exist k and p such that $\forall_{q\in\{1,2,\ldots,p\}} s_i(k-1+q) = s_j(q)$. Otherwise, $w(v_i, v_j) = 0$.

One can notice that sets V_1 and V_2 correspond to sets S_α and S_β containing sequences of peptides obtained using proteases cleaving the target sequence after amino acids α and β, respectively. We are looking for a permutation of elements of set $S = S_\alpha \cup S_\beta$ such that elements from S_α and S_β alternate in this permutation. It is easy to see that it is equivalent to looking for a path in graph G since it is a bipartite graph which does not contain arcs between any pair of vertices corresponding to elements of only one of the sets S_α and S_β. Moreover, function w is defined in such a way that its value determines the biggest possible overlapping of any pair of vertices' labels (corresponding to peptides). So, looking for a Hamiltonian path in graph G minimizing the total sum of w values is equivalent to looking for the peptide permutation satisfying the conditions previously formulated.

Let us also observe that the graph model can be extended such that it will cover situations where more than two proteases are used [1]. Let $G_p = (V_p, A_p)$ be a directed graph, where $V_p = V_1 \cup V_2 \cup \cdots \cup V_N$, $|V| = n$, $\forall_{x,y\in\{1,2,\ldots,N\},x\neq y} V_x \cap V_y = \emptyset$ and $(v_i, v_j) \in A \iff [(v_i \in V_x \wedge v_j \in V_y) \vee (v_i \in V_y \wedge v_j \in V_x)] \wedge x \neq y$. Function w is defined as in the case of graph G (obviously G is a special case of G_p, where $N = 2$). In both cases, i.e. in graphs G and G_p we are looking for a directed path $P = v_{i_1} v_{i_2} v_{i_3} \cdots v_{i_{n-1}} v_{i_n}$ such that $\sum_{r=1}^{n-1} f(v_{i_r}, v_{i_{r+1}})$ is maximal, and $\forall_{r\in\{1,2,\ldots,n-1\}}(v_{i_r} \in V_x \wedge v_{i_{r+1}} \in V_y)$.

Note, that the described peptide assembly problem differs considerably from the DNA assembly one [7]. The most important difference follows from the method used for obtaining the fragments to be sequenced and then assembled. As described above, in the case of peptides the long polypeptide is cut by proteases, which makes the obtained short peptides to be ended by some specific amino acid. This property of the short molecules is crucial for the mathematical models of the assembly process and also for the possible algorithms. On the other hand, in the case of DNA the short fragments are usually obtained by random breaking the long molecule, so there is no specific form of the obtained short fragments.

Let us also observe, that in the problem described here, there is assumed, that the input data contain no errors. This assumption is justified by the fact, that the results presented here establish a starting point for a possible further research, where more realistic data would be assumed. In general, the input data may contain errors following from imperfection of the sequencing process, similarly like in the case of DNA assembly.

The variant of the peptide assembly problem for which the tabu search algorithm will be proposed, can be formulated in the following way [3].

Let C be a subset of alphabet Σ, containing symbols called cutters. Let us consider a particular cutter $c \in C$ and string s. Then a fragment obtainable from the c-cutter is a substring x of s having two properties. First, c is the last symbol of x. Second, if x is found at position i in s, for $i > 1$, then at position $i-1$ in s there is symbol c. If c occurs exactly once in s then it is said that x results from a full digest of s, i.e. the one in which all cuts after c are made; otherwise, s results from a partial digest. Further, it is said that a string is *obtainable from set* C if there is some element of C from which it is obtainable.

Now, the peptide assembly problem can be formulated in the following way.
PEPTIDE ASSEMBLY PROBLEM – SEARCH VERSION:

Instance: Multiset S of strings over alphabet Σ, a distribution D of letters from alphabet Σ, i.e. a set of pairs (x, i) for all symbols $x \in \Sigma$, where i is a positive integer.
Answer: Superstring for S satisfying D such that all elements of S are obtainable from $C \subseteq \Sigma$.

We see that the above formulated problem differs from the one previously described by the requirement concerning the amino acid composition of the reconstructed sequence. This requirement is justified by the information which can be obtained in the biochemical stage of the process. Since the peptide assembly problem is strongly NP-hard [3], in order to solve it in a resonable time, one needs an efficient on average exact algorithm (like branch and bound) or a good polynomial time heuristic. In the next section, a method belonging to the latter class of algorithms will be proposed.

4 Tabu Search Algorithm

The algorithm proposed to solve instances of this problem is based on tabu search heuristic method being one of the most frequently used in combinatorial optimization. The tabu search method is a kind of a local search procedure [4, 5]. The general step of an iterative procedure consists in constructing next solution j from the current solution i and checking whether one should stop there or perform another step. In the tabu search method neighbourhood $N(i)$ is defined for each feasible solution i, and the next solution j is searched among the solutions in $N(i)$. In order to improve the efficiency of the search process, one needs to keep track not only of local information (like the current value of the objective function) but also of some information related to the search process. Like other methods, the tabu search method also keeps in memory value $f(i^*)$ of the best solution visited. However, there are some mechanisms to prevent the method being stuck in a local optimum. One of them is using tabu list which can contain moves already performed by the algorithm. None of the moves can be performed if it is on the tabu list unless it leads to the solution better than the best already found. The next mechanism preventing the method being stuck in a local optimum is a mechanism of random moves, which results in moving the search process to another area of the search space.

The method described above is a general framework of the tabu search approach. In order to solve a particular problem, it must be adopted by using problem-specific definition of moves and possibly by adding to the method some other components not mentioned above.

Let us observe that in the case of the peptide assembly problem, we are looking for a kind of a Hamiltonian path in a directed graph, so we deal with a permutation problem. However, not all permutations of the vertices are feasible solutions. From this follows that any solution can be viewed as a sequence of elements composed of N permutations of elements of sets V_x, $x = 1, 2, \ldots, N$, where the permutations are interlaced. More formally, the elements of sets V_x, $x = 1, 2, \ldots, N$ can be permuted, so N permutations $\pi_1, \pi_2, \ldots, \pi_N$ can be created corresponding to the sets. On the basis of these permutations, permutation Π consisting of all elements of $V = \bigcup_{x=1}^{N} V_x$ can be created in such a way that for the i-th element of permutation π_x position $\alpha_{x,i}$ in permutation Π is assigned. If we denote the set of positions in Π assigned to elements of permutation π_x by Z_x, then the assignment has to satisfy two conditions:

$$\forall_{x,y \in \{1,2,\ldots,N\}, x \neq y} Z_x \cap Z_y = \emptyset \tag{1}$$

and

$$\forall_{x \in \{1,2,\ldots,N\}} \forall_{i,j \in \{1,2,,\ldots,|V_x|\}, i \neq j} \alpha_{x,i} \neq \alpha_{x,j} \tag{2}$$

The algorithm starts with some randomly generated initial solution. The solution is created in such a way that for each element of $V = \bigcup_{x=1}^{N} V_x$ a position in permutation Π is randomly generated. Obviously, the complete set of these positions has to satisfy conditions (1) and (2).

The algorithm tends to maximize objective function w - a sum of overlaps of peptides from multiset S according to a permutation of integers representing these peptides. On the other hand, a distribution of amino acids in the constructed sequence tends to take value of the assumed distribution D.

The neighbourhood solutions are all permutations which can be attained from the permutation of the current solution by exchanging only one pair of integers representing strings from multiset S and satisfying conditions (1) and (2).

While selecting the best candidate solution from the generated neighbourhood solutions, the algorithm checks if a move leading to it is not on $TabuList$.

In case the *TabuList* contains this move, the algorithm can select the solution if its criterion value is better (greater) than the best solution found so far (A).

After performing a number of moves not leading to an improvement of the solution quality a series of random moves is executed. Moreover, after a given number of such series the method is restarted, i.e. some randomly generated feasible solution becomes an initial solution. If a specified number of restarts is performed the algorithm stops.

Below the Tabu Search algorithm is presented in a pseudo-code form.

Tabu Search

```
var : A, // best solution
      a, // current solution
      tmp : object; // temporary solution
      restarts, // number of restarts of algorithm
      moves1, // number of moves without improvement of solutions
      moves2, // number of random moves at one series
      series, // number of series of random moves
      i, j, k, l : integer;
begin
  w(A) := 0;
  d(A) := 1000;//integer much greater than amino acid sequence length
  for k := 1 to restarts
  begin
    initialize TabuList; // clearing Tabu List
    a :=create start solution; // a permutation of integers representing
                               // strings from multiset S
    j := 0;
    while j is not greater than or equal to series do
    begin
       w(a) :=value of solution a; // w(a):= sum of max overlaps of
                                   // peptides from multiset S
       d(a) :=value of a distribution of amino acids in solution a;
       if w(A) < w(a) then
       begin
         A := a; w(A) := w(a); d(A) := d(a);
       end
       if w(A) = w(a) and d(a) < d(A) then
       begin
         A := a; w(A) := w(a); d(A) := d(a);
       end
       i := 0;
       do
       begin
           generate neighbourhood solutions of solution a;
           evaluate solutions from the neighbourhood;
           tmp =select the best solution from the neighbourhood;
           // m(a, tmp) - move from solution a to tmp
           // (a pair of exchanged integers in the permutation)
           while m(a, tmp) is on TabuList or w(tmp) <= w(a) do
```

```
            begin
               i := i + 1;
               tmp =next best solution from the neigbourhood;
            end
            add m(a, tmp) on TabuList;
            a := tmp; w(a) := w(tmp); d(a) := d(tmp);
         end
         while  i is not greater than or equal to moves_1
         // random moves
         for l := 1 to  moves_2
         begin
            tmp =a random feasible solution;
            if w(a) < w(tmp) then
            begin
               a = tmp; w(a) := w(tmp); d(a) := d(tmp);
            end
            if  w(a) = w(tmp) and d(tmp) < d(a) then
            begin
               a = tmp; w(a) := w(tmp); d(a) := d(tmp);
            end
         end
         j := j + 1;
      end
   end
end.
```

The algorithm was implemented in C++ language with a standard library. It has been tested on a number of test instances. The computational experiment is described in the next section.

5 Computational Experiment

The computational experiment has been performed on a PC computer with Pentium 4, 2.4 GHz processor and 512 MB of RAM. The experiment has been divided into three stages. The results are shown in Tables 1 - 3. Each entry in the tables corresponds to computations performed on three amino acid sequences. The time and similarity values are the averages of these three sequences.

In the first stage of the experiment 150 amino acid sequences have been randomly generated. Among these sequences 30 have length of 100 amino acids, 30 of 150, 30 of 200, 30 of 250, and 30 of 300 amino acids. Based on these groups of sequences, three sets of test instances have been created (15 sets have been obtained) by substituting the previously generated amino acid by the one recognized by one of the proteases in randomly chosen 10, 15 and 20 positions, respectively. The proteases have been randomly chosen as well from the set of the two proteases used. In this way the size of the instance, i.e. the number of short peptides to be assembled, has been controlled.

The instances for the second stage have been generated in a similar way, except for there were no substitutions made in the generated sequences. It means that the number of short sequences resulting from the cleavage has been determined by the random process of sequence generation. So, here this number has not been controlled. In addition, it has been assumed that 2, 3 and 4 proteases, respectively, have been used.

In the last stage of the experiment 10 sequences of length of 120 – 220 amino acids have been taken from GenBank. The test instances have been obtained by cutting them by 2, 3 and 4 proteases, respectively.

A similarity to the original sequence is calculated using Needleman-Wunsch algorithm [8] (see also [6]). The algorithm compares two sequences: the one generated by the tabu search algorithm s_t and the original sequence s_o. The similarity of the sequences is determined according to the following formula: $\sigma = 100\frac{\delta-\psi}{\chi-\psi}$, where δ is a scoring for the two sequences, being a sum of scores for all columns in an optimal alignment (1 point for a match, -1 for mismatch or gap), and

$$\psi = \begin{cases} l(s_o)d + (l(s_t) - l(s_o))g \text{ if } l(s_t) > l(s_o) \\ l(s_t)d + (l(s_o) - l(s_t))g \text{ otherwise} \end{cases}$$

$$\chi = \begin{cases} l(s_t)m \text{ if } l(s_t) > l(s_o) \\ l(s_o)m \text{ otherwise} \end{cases}$$

where $l(s_t)$ and $l(s_o)$ are lengths of sequence s_t and s_o, respectively, and $m = 1$, $d = -1$, $g = -1$.

In the experiment the following values of the algorithm parameters have been used: tabu list length – 10, number of moves without improvement of solution quality – 10, number of random moves – 5, number of random move series – 3, number of restarts – 10.

Analysing the results of the computational experiment we see that the proposed algorithm assembled amino acid sequences with a high accuracy for a

Table 1. Results of stage I

Sequence length	Number of cuts	Similarity [%]	Time [s]
100	10	95.0	3.5
100	15	93.0	7.5
100	20	94.0	10.8
150	10	93.0	4.2
150	15	92.0	9.8
150	20	95.0	12.9
200	10	93.0	6.3
200	15	94.0	11.3
200	20	96.0	16.7
250	10	94.0	8.2
250	15	91.0	13.9
250	20	92.0	19.1
300	10	91.0	10.1
300	15	93.0	15.4
300	20	92.0	21.7

Table 2. Results of stage II

Sequence length	Number of proteases	Similarity [%]	Time [s]
100	2	92.0	2.2
100	3	94.0	4.1
100	4	95.0	6.5
150	2	91.0	3.5
150	3	93.0	7.2
150	4	94.0	9.1
200	2	96.0	5.6
200	3	92.0	8.7
200	4	93.0	11.9
250	2	94.0	6.2
250	3	91.0	9.4
250	4	93.0	12.1
300	2	91.0	8.2
300	3	93.0	11.3
300	4	92.0	13.5

Table 3. Results of stage III

Number of proteases	Similarity [%]	Time [s]
2	95.0	3.4
3	94.0	7.4
4	92.0	11.4

broad range of parameters assumed. Computational times were small, thus, allowing for a practical use of the algorithm.

6 Conclusions

In this paper, an application of the tabu search method to the problem of assembling long polypeptide sequences, has been described. This new approach can be used to reconstruct long amino acid chains regardless of the method used for identifying short peptide fragments obtained after digestion by proteases. The computational tests confirmed high efficiency of the algorithm proposed and its usefulness in the protein identification process.

References

1. J. Błażewicz, M. Borowski, P. Formanowicz, T. Głowacki, On graph theoretical models for peptide sequence assembly, submitted for publication
2. V. Dančík, T.A. Addona, K. R. Clauser, J.E. Vath, P. A. Pevzner, De novo peptide sequencing via tandem mass spectometry, *Journal of Computational Biology* 6 (1999) 327–342.

3. J. K. Gallant, The complexity of the overlap method for sequencing biopolymers, *Journal of Theoretical Biology* 101 (1983) 1–17.
4. F. Glover, Tabu Search, Part I, ORSA Journal on Computing 1 (1989) pp. 190–206.
5. F. Glover, Tabu Search, Part II, ORSA Journal on Computing 1 (1990) pp. 4–32.
6. O. Gotoh, An improved algorithm for matching biological sequences, *Journal of Molecular Biology* 162 (1982) 705–708.
7. J. Meidanis, J. Setubal, *Introduction to Computational Biology*, Boston, PWS Publishing Company, 1997.
8. S.B Needleman, C.D. Wunsch, A general method applicable to the search for similarities in the amino acid sequence of two proteins, *Journal of Molecular Biology* 48 (1970) 443–453.
9. P. A. Pevzner, *Computational molecular biology. An algorithmic approach*, Cambridge, Massachusetts, The MIT Press, 2000.
10. L. Stryer, *Biochemistry*, 4th edition, W.H. Freeman and Company, New York 1995.

Order Preserving Clustering over Multiple Time Course Experiments

Stefan Bleuler * and Eckart Zitzler

Swiss Federal Institute of Technology Zurich,
Computer Engineering and Networks Laboratory (TIK),
Gloriastrasse 35, CH–8092 Zürich, Switzerland
{bleuler, zitzler}@tik.ee.ethz.ch

Abstract. Clustering still represents the most commonly used technique to analyze gene expression data—be it classical clustering approaches that aim at finding biologically relevant gene groups or biclustering methods that focus on identifying subset of genes that behave similarly over a subset of conditions. Usually, the measurements of different experiments are mixed together in a single gene expression matrix, where the information about which experiments belong together, e.g., in the context of a time course, is lost. This paper investigates the question of how to exploit the information about related experiments and to effectively use it in the clustering process. To this end, the idea of order preserving clusters that has been presented in [2] is extended and integrated in an evolutionary algorithm framework that allows simultaneous clustering over multiple time course experiments while keeping the distinct time series data separate.

1 Motivation

A central goal in the analysis of genome wide gene expression measurements is to identify groups of genes with shared functions or shared regulatory mechanisms. To this end different clustering concepts have been developed. Standard clustering methods such as k-means, hierarchical clustering [6], self organizing maps [11], partition the set of genes into disjoint groups according to the similarity of their expression patterns over *all* conditions. Thereby, they may fail to uncover processes that are active only over some but not all conditions. In contrast, biclustering aims at finding subsets of genes which are similarly expressed over a *subset* of conditions, which often better reflects biological reality. The usefulness of this concept in the context of microarray measurements has been demonstrated in different studies [5, 12, 10].

A promising biclustering approach, which is especially useful in the context of time course experiments, is the order preserving submatrix (OPSM) method

* Stefan Bleuler has been supported by the SEP program at ETH Zürich under the poly project TH-8/02-2.

F. Rothlauf et al. (Eds.): EvoWorkshops 2005, LNCS 3449, pp. 33–43, 2005.

by Ben-Dor et al. [2]. This method concerns the discovery of one or several submatrices in a gene expression matrix in the which the expression levels of the selected genes induce the same linear ordering of the selected experiments. However, there are several potential drawbacks of this approach: the algorithm (i) does not allow to relax the clustering criterion, i.e., deviations from the perfect linear ordering will be not considered as clusters, (ii) needs excessive computing resources if applied to large gene expression matrices, and (iii) does not provide means to keep different types of experiments separate from each other. The first issue has been addressed in [7] by assigning similar expression levels equal ranks but still searching for perfect linear orderings. The third issue, which applies to most of the existing clustering approaches, is important insofar as the mixture of different types of experiments on the one hand side may blur the clustering outcome and on the other hand does not allow to study similarities and differences between the distinct experiment groups. Therefore in this paper, we

- propose a cluster scoring scheme that represents a relaxation of the strict order preserving criterion introduced in [2],
- present an evolutionary algorithm for clustering that uses the above scoring scheme, allows to treat strongly related experiments such as time series separately, and can be applied to arbitrarily large data sets, and
- demonstrate the usefulness of the suggested approach on various data sets for *Arabidopsis thaliana*.

The proposed method can generally be applied to combine data sets from different experiment groups which cannot be compared directly. It even is possible to use separate scores for measuring similarity in the different groups. In this paper, though, we focus on the analysis of time courses as an example for such data sets. This approach is conceptually different from other methods for the analysis of time course data as it focuses on the relation between different time courses while most other methods focus on the relation between the different time points within a single experiment, cf. [1].

In the following, we will first discuss the underlying clustering concept, before a corresponding implementation of an evolutionary algorithm is presented. Later, the proposed method is compared with the OPSM approach on various time series data, and especially the issue of mixed and separated time courses is investigated. The last section summarizes the main results of the study.

2 Order Preserving Clustering

In the remainder of this paper, we will assume that the measurements of several experiments are given in terms of an $m \times n$-matrix, E, where m is the number of considered genes and n the number of experiments. A cell e of E contains a real value that reflects the abundance of mRNA for gene i under condition j relative to a control experiment (absolute mRNA concentrations are seldom used). A time course stands for a sequence of measurements that have been performed at different points in time under the same condition for the same organism. Such

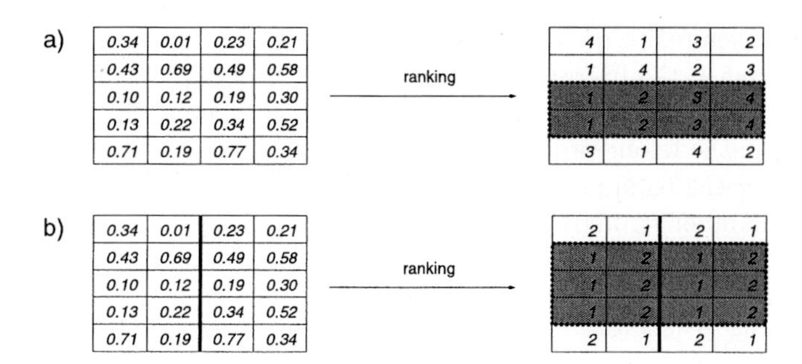

Fig. 1. (a) On the left hand side, a gene expression matrix is shown, on the right hand side, the corresponding expression levels are replaced by their ranks within each row; the shaded area marks the largest order preserving submatrix. (b) The same gene expression matrix as in a) is now divided into two time courses, each consisting of two experiments; the resulting ranking induces a larger order preserving submatrix compared to a), as each time course is treated separately

a time course measurement is represented by a gene expression matrix E where each column stands for one point in time and where the order of the columns reflects the order of the experiments in time.

2.1 The Order Preserving Submatrix Problem

Given the above notation, the order preserving submatrix problem, which has been introduced by Ben-Dor et al. [2], can be described as follows: find a subset G of genes ($|G| \leq m$) and a subset C of experiments ($|C| \leq n$) such that the submatrix D of E defined by G and C maximizes a given score $f(G, C)$ and is an order preserving submatrix (OPSM), see below. The score f reflects the probability of observing an OPSM of size $|G| \cdot |C|$ in a randomly chosen matrix of the same dimensions as E. A submatrix D is order preserving if there is a permutation of its columns (experiments) such that the sequence of (gene expression) values for each row (gene) is strictly increasing; this concept is illustrated in Fig. 1a.

A potential problem with the approach presented in [2] is that the running time of the proposed algorithm can increase considerably with the size of E. Another drawback with the OPSM concept is that the homogeneity criterion (perfect order) is a hard constraint. However, if slight deviations from a perfect ordering would be allowed, this often may better reflect biological reality.

2.2 A Score for the Degree of Order Preservation

The first question we address is how to relax the condition of order preservation such that slight disagreements between the genes are still acceptable. If the allowed error is adjustable by the user, it is possible to account for errors in the measurements and to adapt the cluster criterion to the current biological data set.

Several measures to quantify the unsortedness of a sequence of integers have been suggested in the literature, see, e.g., [9]. One potential measure is to compare all possible pairs of the sequence elements and count the number of pairs that appear in the wrong order. When we extend this concept to a submatrix, one could consider the total number of mismatches over all genes. However, the corresponding number strongly depends on the actual order of the selected columns and finding the order that minimizes this scores is itself an NP-hard problem [8]. Therefore, we propose a scoring scheme that is independent of the actual order of the columns; it is based on another biclustering approach proposed by Cheng and Church [5] that uses the mean squared residue score.

The optimization task in [5] is to find the largest bicluster that does not exceed a certain homogeneity constraint. The size $f(G,C)$ is simply defined as the number of cells in E covered by a bicluster (G,C), while the homogeneity $g(G,C)$ is given by the *mean squared residue score*. Formally, the problem is to maximize $f(G,C) = |G| \cdot |C|$ subject to $g(G,C) \leq \delta$ with $(G,C) \in X$, and where

$$g(G,C) = \frac{1}{|G||C|} \sum_{\in \quad \in} (e \quad - e \quad - e \quad \cdot + e \quad)^2$$

is called the mean squared residue score and

$$e \quad = \frac{1}{|C|} \sum_{\in} e \quad , \; e \quad = \frac{1}{|G|} \sum_{\in} e \quad , \; e \quad = \frac{1}{|G||C|} \sum_{\in \quad \in} e$$

denote the mean column and row expression values for (G,C) and the mean over all cells, respectively. The threshold δ needs to be set by the user and defines the maximum allowable dissimilarity within the cells of a bicluster. Roughly speaking, the residue expresses how well the value of an element in the bicluster is determined by the column and row it lies in. A set of genes whose expression levels change in accordance to each other over a set of conditions can thus form a perfect bicluster even if the actual values lie far apart.

In our scenario, we use a scoring function h which combines the mean squared residue score with the OPSM concept. Given a submatrix D, we first rank the values in D per row and then replace the expression values with their ranks; afterwards, we apply the mean squared residue score to the transformed submatrix D' in order to assign a score to D. Since the ranks in each row of D' sum up to the same value, the row mean e and the total mean e cancel each other out and the scoring scheme is reduced to measure the unsortedness of the column means of the ranks. It can be easily shown that a score of 0 using this modified scheme h is equivalent to D being an OPSM. Additional tests with small random matrices showed that h correlates well with the abovementioned count of unordered pairs.

2.3 Clustering Scores for Time Course Data

This paper focuses on time series data, although the presented concepts can be used for other types of experiments as well. As time series often consist of a few experiments only (usually 6 to 10), in the following we will consider the optimization task to find the largest subset G of genes that has a score $h(G) \leq \delta$, where δ is a constraint set by the user and all experiments in E are considered, i.e., the submatrix D extends over the rows of E specified by G and all columns of E. As we will show later in Section 4.2, this restriction is reasonable if n is small.

The situation changes, though, if multiple time series data are taken into account. The common way is to combine several time courses into a single matrix; however, thereby information is lost and the resulting OPSM can be small. Here, we propose to treat each of the time courses separately as depicted in Fig. 1b. We still aim at maximizing the number of genes in the cluster, but the constraint on h is now computed for each time series separately. That is, given G, for each time course the resulting submatrix D is computed and it is checked whether the corresponding h score is lower than or equal to the constraint δ; only if the score constraint is fulfilled for all time series, the cluster G is considered a feasible solution. In the next section, we present an EA implementation for this problem.

3 Evolutionary Algorithm

The main idea is to use an evolutionary algorithm to explore the search space of possible gene sets systematically. As we will see, the representation and most of the operators are generic while the local search procedure and the environmental selection are more specific to the proposed optimization problem.

Each individual encodes one cluster. For reasons of simplicity we have chosen to use a binary representation with a bit string of length m where a bit is set to 1 if the corresponding gene is included in the cluster. We apply uniform crossover and independent bit mutation. As to environmental selection, a special diversity maintenance mechanism was used which is described later in this section. For mating selection, a tournament selection is used.

Since the objective is to find large clusters the fitness of an individual is calculated as the inverse of the number of genes included in the cluster which leads to a minimization problem. The threshold on h is used as constraint and a local search is performed before the fitness assignment which produces only feasible solutions.

Local Search. In order to increase the efficiency of the EA a local search procedure is applied to each individual before evaluation. This procedure is based on a greedy heuristic which tries to reduce h while keeping a maximum number genes in the cluster. The algorithm is similar to the one proposed in [5] and consists of two main steps: First genes are removed from the cluster until the homogeneity constraints $h(G) < \delta$ for all time courses are met and in a second step all genes which can be added without increasing $h(G)$ are included in the cluster. This procedure guarantees to produce a feasible solution because in the extreme case a cluster is reduced to a single gene which always has perfect homogeneity ($h = 0$).

while number of genes threshold **do**
calculate $_{Gj}$ for all columns for all time courses
calculate for all time courses
for all genes in the cluster **do**
calculate for all time courses separately:
$= \frac{1}{m} \sum_{j=1..m}(\ _{ij} - \ _{Gj})^2$
if for any time course **then**
remove the gene
end if
end for
if nothing was removed **then**
switch to Single Gene Deletion
end if
end while

Fig. 2. Algorithm for Multiple Gene Deletion

while constraint violated for any time course **do**
calculate $_{Gj}$ for all columns for all time courses
for all genes in the cluster **do**
calculate for all time courses separately:
$= \frac{1}{m} \sum_{j=1..m}(\ _{ij} - \ _{Gj})^2$
sum up the for all time courses into
end for
remove gene with maximal
end while
continue with Gene Addition

Fig. 3. Algorithm for Single Gene Deletion

The removal and addition of the genes is generally done one by one in a greedy fashion which requires to update the homogeneity score h after the addition or removal of each gene. If many genes need to be removed this recalculation can increase the running time heavily. To prevent this multiple genes are removed in each iteration while the number of genes in the cluster is above a certain threshold (default = 100). The complete local search thus consists of the three steps described in Figures 2, 3 and 5.

Additionally, it is necessary to specify what happens with the result of the local search. In this study, we use Lamarckian evolution which means that the solution found by the local search is kept and replaces the individual the local search started from as opposed to Baldwinian evolution where the locally optimized solution is just used to calculate the fitness of the individual.

Diversity Maintenance. As a whole population of clusters is evolved simultaneously it is possible not only to optimize one cluster but also to find a set of clusters which fulfill a desired property like total coverage or minimum overlap. To this end a special diversity maintenance mechanism is used. The general idea is to select the biggest cluster first and in each following step the cluster which adds most to the coverage of the set of all genes. The algorithm is described in detail in [3].

4 Results

In the simulation runs mainly two questions were investigated: (i) How does the EA compare to the OPSM algorithm proposed in [2] when applied to find perfectly ordered clusters, and (ii) what is the effect of separating the different time series?

4.1 Data Preparation and Experimental Setup

All simulations were performed on gene expression data generated with Affymetrix GeneChips from *Arabidosis thaliana*, a small plant. The data set

Table 1. Default parameter settings for this study

	1.2
probability of 1 in initalization	0.001
mutation rate	0.001
crossover rate	0.1
tournament size	3
population size	100
number of generations	100

used in this study was provided by the ATGenExpress consortium [1] and used to investigate the response of Arabidopsis to different kinds of stresses. It consists of 8 time series with 6 time points each. The total expression matrix thus contains 22746 genes and 48 chips. All expression values were preprocessed using RMA [4] and the logratios with the measurement from an untreated control plant were calculated.

The EA parameter settings used in the following simulations are described in Table 1. The crossover rate refers to the percentage of parents involved in crossover. The mutation rate is the probability for bit flips in the independent bit mutation.

The EA was programmed in C++ while the OPSM algorithm was implemented in Java. The implementation closely follows the description in [2]. The OPSM algorithm takes a parameter l describing how many candidate solutions should be further investigated during the greedy search for OPSMs, see [2] for the details. Consistent with the value used in [2] we set l to 100.

All simulation runs were performed on a 3 GHz Intel Xeon machine. For each run, 30 replicates with different random number generator seeds were performed.

4.2 Single Time Series

As mentioned above, searching for perfect OPSMs which extend over all chips in the data set corresponds to setting the constraint on the inhomogeneity h to zero. It is thus interesting to compare the clusters found by the EA to the ones found by the OPSM algorithm. To this end we ran both algorithms on six time course data sets. The largest cluster found by the EA equaled the one found by the OPSM algorithm in all cases with sizes ranging from 290 to 662 genes. Often this cluster was found after only a few generations.

While the OPSM algorithm is tailored to find one maximal OPSM for each number of chips the EA can find several clusters in one run. Without the diversity mechanism described in Section 3 the population quickly converges to a set of clusters with large overlap with the best cluster. The diversity mechanism prevents this: for one data set, as an example, all 100 clusters in the final population were non overlapping and consisted of an average of 90 genes. These groups have different orderings of the expression levels. It makes sense to investigate these clusters as well and not just concentrate on finding the biggest OPSM.

[1] See http://web.uni-frankfurt.de/fb15/botanik/mcb/AFGN/atgenex.htm

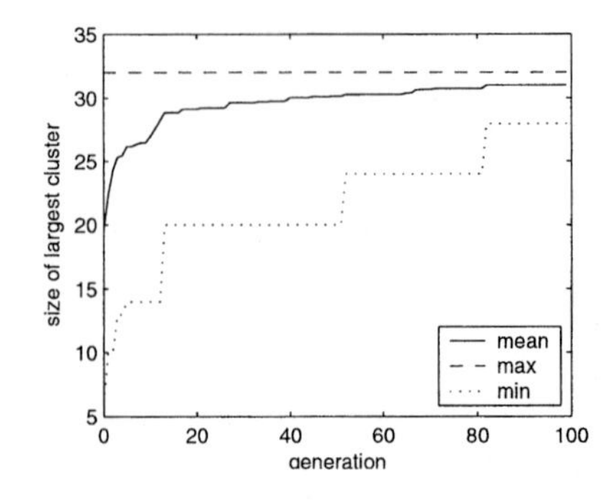

Fig. 4. Size of the largest cluster found by the EA on the two "cold" data sets. Mean, max and min over 30 runs

```
while a gene was added in the last iteration
do
  calculate  Gj for all columns   for all time
  courses
  calculate   for all time courses
  for all genes not in the cluster do
    calculate   = 1/m Σ_{j=1..m}( ij − Gj)^2
    if        for all time courses then
      add gene
      recalculate all  Gj and
    end if
  end for
end while
```

Fig. 5. Algorithm for Gene Addition

The high number of large clusters found by the EA makes it unnecessary to relax the constraint in the case of such small data sets. The effect of relaxing the constraint will be discussed in the following section where several time courses are combined into one data set.

4.3 Multiple Time Series

In a second set of experiments we investigated the effect of keeping the time courses separated during the search for order preserving clusters. To this end we first combined two data sets and then tested the algorithms on the combination of all eight data sets.

Two Time Series. When comparing the performance of the OPSM algorithm to the EA on the combination of both data sets in same sense as in the previous section, both algorithms found OPSMs consisting of two genes and all twelve chips. When relaxing the constraint for the EA, larger clusters can be found; for instance, if the constraint on h is set to 0.5 which means that the average difference between the actual rank and the column mean of the ranks must be smaller than 0.5 the largest cluster found by the EA contained 6 genes (mean over 30 runs).

We then used the EA to search for groups of genes which fulfill the homogeneity constraint of $h = 0$ on both data sets separately. As shown in Figure 4 the largest cluster found by the EA contained 31 genes. Additionally the final population contained 100 clusters with an average size of 7 genes and no overlap between them. It is obvious that many significant clusters are missed if the data sets are mixed and the proposed EA is able to retrieve many of them.

Eight Time Series. When applying the algorithms to the total of 48 chips another drawback of the OPSM algorithm becomes apparent: the running time increases rapidly with the number of chips in the data set. While the OPSM

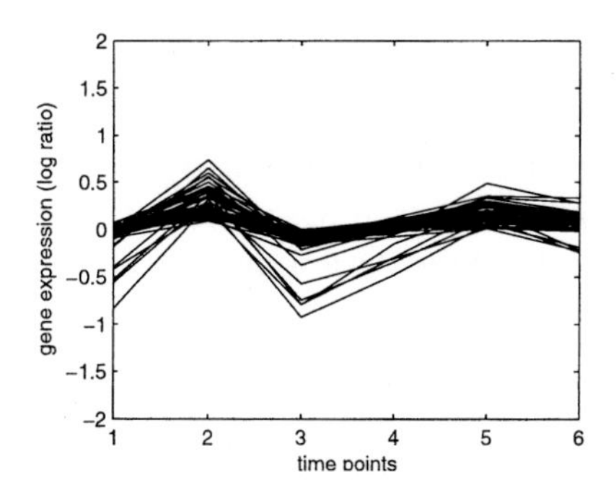

Fig. 6. Identifying differences: Expression values for the well ordered time course

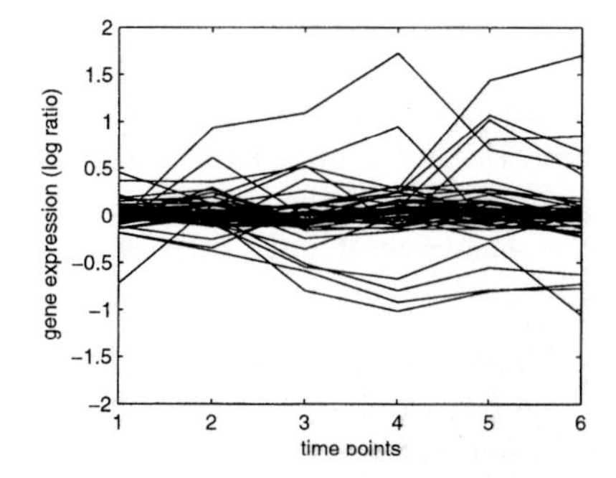

Fig. 7. Identifying differences: Expression values for one of the unordered time courses

algorithm takes 30 seconds to run on a data set of six chips is takes more than two hours to finish on the full data set. The EA run time lies between 10 and 20 mins. For the EA, however, the running time can be adjusted by changing the number of generations and the population size. The EA is thus still applicable for large data sets which cannot be analyzed using the OPSM algorithm.

As expected neither OPSM nor EA found perfect cluster over all 48 chips. Also for the separated data sets no cluster was found which had a perfect ordering for all time series. Relaxing the constraint to 1 for each time course allowed the EA to find clusters of a maximal size of 9 genes. Higher constraints could be used to find larger but less coherent clusters. Also for the combined data set large clusters can be found but the constraint must be relaxed to about 25 which does not provide a good ordering anymore.

4.4 Identifying Differences in Multiple Time Series

Another problem which is of high biological relevance is to identify groups of genes that behave similarly in some of the time courses and exhibit inhomogeneous expression patterns in other time courses. There are different possibilities to include these objectives into the EA depending on the biological scenario. We chose the following: the expression values for at least one time course must be ordered, i. e., h must be zero. Under this constraint the maximum h over all time series is maximized. The local search procedure was changed accordingly so that only the constraint of the best ordered time course must be fulfilled.

We tested this method on the combination of all eight data sets. In the resulting clusters in general only the expression levels of one time course are well ordered. Figures 6 and 7 show one example where time course on the left hand side is well ordered and the one on the right hand side stands as example for all the others which are not well ordered.

This shows that the problem of finding differences in multiple time series can be address using the proposed approach, however, further investigations and biological analysis of the results are necessary.

5 Conclusions

The order preserving submatrix problem as defined in [2] consists of finding large submatrices in a gene expression matrix such that for each submatrix there is a permutation of the columns under which the sequence of expression values is for each row strictly increasing. We have extended this approach in three aspects:

- A more flexible scoring scheme was proposed that allows to arbitrarily scale the degree of orderedness required for a cluster.
- Based on this scoring scheme, a methodology to handle different set of experiments such as distinct time series in a systematic way was introduced.
- An evolutionary algorithm for this order preserving problem was presented that is capable of finding multiple non-overlapping clusters in a single optimization run.

The effectiveness of the suggested approach was demonstrated on eight time series experiments covering overall 48 measurements for about 20000 genes of *Arabidopsis thaliana*. Especially, the separation of different time series experiments has proven to be a valuable concept: the cluster sizes could be improved substantially in comparison to the common approach where all time series experiments are combined in a single gene expression matrix.

References

1. Z. Bar-Joseph. Analizing time series gene expression data. *Bioinformatics*, 20(16):2493–2503, 2004.
2. A. Ben-Dor, B. Chor, R. Karp, and Z. Yakhini. Discovering local structure in gene expression data: The order-preserving submatrix problem. In *Conference on Computational Biology (RECOMB 02)*, pages 49–57. ACM Press, 2002.
3. S. Bleuler, A. Prelić, and E. Zitzler. An EA framework for biclustering of gene expression data. In *Congress on Evolutionary Computation (CEC 2004)*, pages 166–173, Piscataway, NJ, 2004. IEEE.
4. B. M. Bolstad, R. A. Irizarry, M. Astrand, and T. P. Speed. A comparison of normalization methods for high density oligonucleotide array based on variance and bias. *Bioinformatics*, 19(2):185–193, Jan. 2003.
5. Y. Cheng and G. M. Church. Biclustering of gene expression data. In *ISMB 2000*, pages 93–103, 2000. http://cheng.ecescs.uc.edu/biclustering.
6. M. B. Eisen, P. T. Spellman, P. O. Brown, and D. Botstein. Cluster analysis and display of genome-wide expression patterns. *PNAS*, 95:14863–14868, Dec. 1998.
7. J. Liu, J. Yang, and W. Wang. Biclustering in gene expression data by tendency. In *Computational Systems Bioinformatics Conference (CSB 2004)*. IEEE, 2004.
8. G. Reinelt. *The Linear Ordering Problem*. Heldermann, Berlin, 1985.
9. J. Scharnow, K. Tinnefeld, and I. Wegener. Fitness landscapes based on sorting and shortest path problems. In *Parallel Problem Solving from Nature (PPSN VII)*, number 2439 in LNCS, pages 54–63. Springer-Verlag, 2002.

10. E. Segal, A. Battle, and D. Koller. Decomposing gene expresision into cellular processes. In *Pacific Symposium on Biocomputing*, pages 8:89–100, 2003.
11. P. Tamayo et al. Interpreting patterns of gene expression with self-organizing maps: Methods and applicataion to hematopoietic differentiation. *PNAS*, 96:2907 – 2912, March 1999.
12. A. Tanay, R. Sharan, and R. Shamir. Discovering statistically significant biclusters in gene expression data. *Bioinformatics*, 18(Suppl. 1):S136–S144, 2002.

Can Neural Network Constraints in GP Provide Power to Detect Genes Associated with Human Disease?

William S. Bush, Alison A. Motsinger, Scott M. Dudek, and Marylyn D. Ritchie

Center for Human Genetics Research, Department of Molecular Physiology & Biophysics, Vanderbilt University, Nashville, TN, USA 37232
{wbush, motsinger, dudek, ritchie}@chgr.mc.vanderbilt.edu
http://chgr.mc.vanderbilt.edu/ritchielab

Abstract. A major goal of human genetics is the identification of susceptibility genes associated with common, complex diseases. Identifying gene-gene and gene-environment interactions which comprise the genetic architecture for a majority of common diseases is a difficult challenge. To this end, novel computational approaches have been applied to studies of human disease. Previously, a GP neural network (GPNN) approach was employed. Although the GPNN method has been quite successful, a clear comparison of GPNN and GP alone to detect genetic effects has not been made. In this paper, we demonstrate that using NN evolved by GP can be more powerful than GP alone. This is most likely due to the confined search space of the GPNN approach, in comparison to a free form GP. This study demonstrates the utility of using GP to evolve NN in studies of the genetics of common, complex human disease.

1 Introduction

In the search for disease susceptibility genes in common diseases, human geneticists are faced with numerous challenges. For rare, Mendelian diseases like cystic fibrosis or Huntington disease, a single mutation in a single gene results in disease. Unfortunately, the genetic architecture of common disease is not that simple. Common diseases, such as diabetes, cancer, and hypertension are complex with a variety of genetic and environmental factors leading to disease risk [1, 2]. The potential for many genes with independent effects in addition to interaction effects makes the detection of disease susceptibility genes far more difficult. Traditional analysis approaches were designed for the situation of single gene, single mutation disorders. Therefore, they are not sufficient for the detection of multiple genetic and environmental factors or gene-gene and gene-environment interactions associated with human disease. Thus, novel statistical and computational approaches are needed for the study of common, complex human disease.

A variety of methods are being explored for the detection of gene-gene and gene-environment interactions associated with complex disease These include logic regression [3], penalized logistic regression [4], automated detection of informative combined effects (DICE) [5], combinatorial partitioning method (CPM) [6] which has recently been extended by Culverhouse et al. in a method called RPM [7] and

F. Rothlauf et al. (Eds.): EvoWorkshops 2005, LNCS 3449, pp. 44–53, 2005.

multifactor dimensionality reduction (MDR) [8-10] to name a few. Neural networks are an additional class of computational approaches that have been utilized in human genetics. Several researchers have applied NN to studies of human disease [11-12]. While the applications of NN have met with some success, limitations were present in most of the previous studies. One of the most prominent challenges with NN analysis is the design of the NN architecture. NN architecture must be pre-specified before NN analysis can commence. Machine learning methods such as genetic programming [13] and genetic algorithms [14] have been explored to optimize the selection of NN architecture. To circumvent this trial and error process, Ritchie et al. developed a genetic programming optimized NN (GPNN) [15]. GPNN was developed in an attempt to improve upon the trial-and-error process of choosing an optimal architecture for a pure feed-forward back propagation neural network. The GPNN optimizes the inputs from a larger pool of variables, the weights, and the connectivity of the network including the number of hidden layers and the number of nodes in the hidden layer. Thus, the algorithm attempts to generate optimal neural network architecture for a given data set and has an advantage over the traditional back propagation NN in which the inputs and architecture are pre-specified and only the weights are optimized.

Although previous empirical studies suggest GPNN has excellent power for identifying gene-gene interactions, one must question whether constraining GP to build NN is an improvement over using GP alone. GP can be used to build discriminant functions, as a combination of symbolic regression and linear discriminant analysis. Using GP to build discriminant functions has been performed and applied to studies of human genetics [16, 17]. This method, called SDA (symbolic discriminant analysis), has been used in microarray studies [16, 17]. Since GP can be used to build discriminant functions, is there an advantage to defining a set of rules to constrain GP trees to conform to the structure of a NN? While prior studies demonstrate the power of GPNN, a comparison of GPNN with unconstrained GP has not yet been performed. The goal of the present study was to compare the power of GPNN and an unconstrained GP for identifying gene-gene interactions using data simulated from a variety of gene-gene interaction models. This study is motivated by the previous applications of GPNN and the question of the benefits of constraining the tree structure. We find that GPNN has higher power and lower false positive rates to detect gene-gene interactions than an unconstrained GP alone in the data simulated. These results demonstrate that GPNN may be an important pattern recognition tool for studies in genetic epidemiology.

2 Methods

2.1 Genetic Programming Neural Network (GPNN)

GPNN was developed to improve upon the trial-and-error process of choosing an optimal architecture for a pure feed-forward back propagation neural network (NN) [15]. Optimization of NN architecture using genetic programming (GP) was first proposed by Koza and Rice [13]. The goal of this approach is to use the evolutionary features of GP to evolve the architecture of a NN. The use of binary expression trees allows the GP to evolve a tree-like structure that adheres to the components of a NN. Figure 1 shows an example of a binary expression tree representation of a NN generated by GPNN. The GP is constrained such that it uses standard GP operators but retains the typical structure of a feed-forward NN. While GP could be implemented without constraints, the goal was to evolve NN since they were being explored as a tool for genetic epidemiology. Therefore, we wanted to improve a method already being used. Rules are defined prior to network evolution to ensure that the GP tree is constrained in such a way that it functionally represents a NN. These rules are consistent with those described by Koza and Rice [13]. The flexibility of the GPNN allows optimal network architectures to be generated that consist of the appropriate inputs, connections, and weights for a given data set.

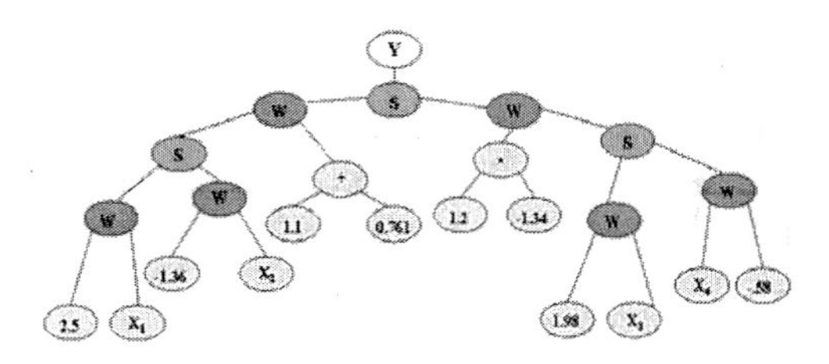

Fig. 1. An example of a NN evolved by GPNN. The Y is the output node, S indicates the activation function, W indicates a weight, and X_1-X_4 are the NN inputs

The GPNN method has been described in detail [15, 18]. The steps of the GPNN method are described in brief as follows. First, GPNN has a set of parameters that must be initialized before beginning the evolution of NN models. These include an independent variable input set, a list of mathematical functions, a fitness function, and finally the operating parameters of the GP. These operating parameters include number of demes (or populations), population size, number of generations, reproduction rate, crossover rate, mutation rate, and migration [15]. Second, the data are divided into 10 equal parts for 10-fold cross-validation. Here, we will train the GPNN on 9/10 of the data to develop a NN model. Later, we will test this model on the 1/10 of the data left out to evaluate the predictive ability of the model.

Third, training of the GPNN begins by generating an initial population of random solutions. Each solution is a binary expression tree representation of a NN, similar to that shown in Figure 1. Fourth, each GPNN is evaluated on the training set and its fitness recorded. Fifth, the best solutions are selected for crossover and reproduction using a fitness-proportionate selection technique, called roulette wheel selection, based on the classification error of the training data [19]. Classification error is defined as the proportion of individuals where the disease status was incorrectly specified. A predefined proportion of the best solutions will be directly copied (reproduced) into the new generation. Another proportion of the solutions will be used for crossover with other best solutions. The new generation, which is equal in size to the original population, begins the cycle again. This continues until some criterion is met at which point the GPNN stops. This criterion is either a classification error of zero or a limit on the number of generations. An optimal solution is identified after each generation. At the end of the GPNN evolution, the overall best solution is selected as the optimal NN. Sixth, this best GPNN model is tested on the 1/10 of the data left out to estimate the prediction error of the model. Prediction error is a measure of the ability to predict disease status in the 1/10 of the data. Steps two through six are performed ten times with the same parameters settings, each time using a different 9/10 of the data for training and 1/10 of the data for testing.

The results of a GPNN analysis include 10 GPNN models, one for each split of the data. In addition, a classification error and prediction error are recorded for each of the models. A cross-validation consistency can be measured to determine those variables which have a strong signal in the gene-gene interaction model [20, 21]. Cross-validation consistency is the number of times a particular combination of variables are present in the GPNN model out of the ten cross-validation data splits. Thus a high cross-validation consistency, ~10, would indicate a strong signal, whereas a low cross-validation consistency, ~1, would indicate a weak signal and a potential false positive result. An open source version of the GPNN software will be available at http://chgr.mc.vanderbilt.edu/ritchielab/gpnn.

2.2 Unconstrained Genetic Programming (GP)

Using GP to build binary classifiers is not a new technique. It has been used for many years and suggested as an extension to symbolic regression. Here, binary expression trees will be evolved that will classify samples into two groups. This implementation is much like the work on symbolic discriminant analysis [16, 17] that has been applied to microarray data. Similarly, we will use binary expression trees to build classifiers. Koza [22, 23] and Koza et al. [24] give a detailed description of GP. Evolutionary computation strategies such as GP have been shown to be effective for both variable and feature selection with methods such as symbolic discriminant analysis for microarray studies [16, 17].

The implementation of GP in this study is simply the GPNN methodology with the rules governing tree construction removed. Thus, the algorithm is identical in every other way. This was done to ensure that the only comparison here was whether constraining GP trees to build NN was more powerful than GP alone.

2.3 Data Simulation

The utility of data simulation is the ability to know the solution to the problem, and determine if the method can identify the solution. This allows one to investigate the comparison of methods and decide which method is preferred. This process is not possible when real data is analyzed where the correct solution is unknown. Here, if the two methods differ in their solutions, one cannot determine which method, if any, is correct. The goal of the simulation in this study was to generate data sets that exhibit gene-gene interactions for the purpose of evaluating the power of GPNN in comparison to the power of GP. We simulated a collection of models varying several conditions including number of interacting genes, allele frequency, and heritability.

Heritability is defined in the broad sense as the proportion of phenotypic variation that is attributed to genetic factors. Loosely, this means the strength of the genetic effect. Thus a higher heritability will be a larger effect and easier to detect. Heritability is calculated using equations described in [25]. Additionally, we used a constant sample size for all simulations. We selected the sample size of 400 cases (individuals with disease) and 400 controls (individuals without disease) because this is a typical sample that is used in many genetic epidemiology studies.

To evaluate the power of GPNN and GP for detecting gene-gene interactions, we simulated case-control data using a variety of epistasis models in which the functional genes are single-nucleotide polymorphisms (SNPs). We selected models that exhibit interaction effects in the absence of any main effects. Interactions without main effects are desirable because they provide a high degree of complexity to challenge the ability of a method to identify gene-gene interactions. If main effects are present, it could be difficult to evaluate whether particular genes were detected due to the main effects or the interactions or both. In addition, it is likely that a method that can detect interacting genes in the absence of main effects will be able to detect main effect genes as well.

To generate a variety of epistasis models for this study, we selected three criteria for variation. First, we selected epistasis models with a varying number of interacting genes: either two or three. We speculate that common diseases will be comprised of complex interactions among many genes. The number of interacting genes simulated here may still be too few to be biologically relevant. Next, we selected two different allele frequencies to represent both a common and a rare minor allele frequency which are both possible for disease susceptibility genes. Finally, we selected a range of heritability values including 3%, 2%, 1.5%, 1%, and 0.5%. These heritability values fall into the realm of very small genetic effects. We chose to simulate data using epistasis models with such small heritability values to test the lower limits of GPNN. Based on previous studies, GPNN has over 80% power when the heritability is between 2%-5% [15]. For this particular study, we wanted to explore even smaller genetic effects to identify the point at which GPNN loses power.

We generated models using software described by Moore et al. [26]. We selected models from all possible combinations of number of interacting genes, allele frequency, and heritability, resulting in twenty total models. These twenty models have been reported previously in a study comparing GPNN to stepwise logistic regression [18], thus we will not show the models in this paper. The models are available at http://chgr.mc.vanderbilt.edu/ritchielab/gpnn.

Each data set consisted of 400 cases and 400 controls. We simulated 100 data sets of each model consisting of the functional SNPs and either seven or eight non-functional SNPs for a total of ten SNPs. This resulted in 2000 total datasets. We used a dummy variable encoding for the genotypes where *n-1* dummy variables are used for *n* levels (or genotypes) [27]. Based on the dummy coding, these datasets had 20 input variables. All datasets are available for download at http://chgr.mc.vanderbilt.edu/ritchielab/gpnn.

2.4 Data Analysis

Next, we used GPNN and GP to analyze 100 data sets for each of the epistasis models. The GP parameter settings for GPNN and GP included 10 demes, population size of 200 per deme, 50 generations, reproduction rate of 0.10, crossover rate of 0.90, mutation rate of 0.0, and migration every 25 generations. Neither GP nor GPNN are required to use all the variables as inputs. Here, random variable selection is performed in the initial population of solutions. Through evolution, the methods select the most relevant variables. We calculated a cross-validation consistency for each model. This measure is defined as the number of times each set of genes is in the best model produced across the ten cross validation intervals. Thus, one would expect a strong signal to be consistent across all ten or most of the data splits, where a false positive signal may be present in only one or a few of the cross validation intervals. We selected the best model for each dataset based on the combination of genes with the highest cross-validation consistency. We then estimated the power of GPNN or GP as the number of times the correct functional genes were selected as the best model in the datasets, divided by the total number of datasets for each epistasis model. Either one or both of the dummy variables could be selected to consider a gene present in the model.

Table 1. Power and False Positive Rates of GPNN and Unconstrained GP

Model			Power		False Positives	
Number genes	Allele freq	Heritability	GPNN	Unconstrained GP	GPNN	Unconstrained GP
2	.2/.8	3.0%	100	84	0	16
2	.2/.8	2.0%	99	64	1	36
2	.2/.8	1.5%	100	80	0	20
2	.2/.8	1.0%	94	61	6	39
2	.2/.8	0.5%	66	30	34	70
2	.4/.6	3.0%	100	76	0	24
2	.4/.6	2.0%	99	96	1	4
2	.4/.6	1.5%	100	96	0	4
2	.4/.6	1.0%	97	78	3	22
2	.4/.6	0.5%	66	42	34	58
3	.2/.8	3.0%	89	63	11	37
3	.2/.8	2.0%	89	68	11	32
3	.2/.8	1.5%	1	1	99	99
3	.2/.8	1.0%	0	2	100	98
3	.2/.8	0.5%	9	3	91	97
3	.4/.6	3.0%	65	54	35	46
3	.4/.6	2.0%	14	11	86	89
3	.4/.6	1.5%	8	4	92	96
3	.4/.6	1.0%	2	5	98	95
3	.4/.6	0.5%	0	3	100	97

3 Results

In this study, we compared GPNN to an unconstrained GP using simulated data. Table 1 and Figure 2 show the power comparison of each method. Here, power refers to the method correctly identifying the functional genes. For the two-gene models, the power of GPNN was consistently higher than the unconstrained GP. For the three-gene models, GPNN had higher power than GP in seven out of ten models, although many of those models had very low power with both methods. Table 1 and Figure 3 show the false positive results using each method. Here false positives include all models identified as the best model that included genes other than the functional genes. For the two-gene models, GPNN consistently had fewer false positive genes identified than the unconstrained GP. For the three-gene models, GPNN had fewer or equal numbers of false positive genes identified in seven out of ten models. In contrast to the power results, the false positive results for the three-gene models are quite high. This is the reciprocal of power so when the method fails to identify the correct genes, it must have identified false positive genes.

4 Discussion

In this study, we have compared a GP approach constrained to build NN to an unconstrained GP approach in an application to data in human genetics. Studies of common, complex disease are complicated by the likelihood of gene-gene and gene-environment interactions associated with disease risk. Identifying these complex interactions poses a difficult challenge and creates the need for novel computational techniques. This led to the development of GPNN. While GPNN has shown to be powerful in the detection of gene-gene interactions [15, 18], the question remained as to whether GPNN was more powerful than an unconstrained GP alone. This was the goal of the present study.

Here we tested the power of GPNN and GP on simulated datasets generated under twenty different epistasis models. Based on these results, it is evident that under the set of parameters chosen, GPNN is more powerful than GP alone, and results in fewer false positive findings. Unfortunately, both methods suffered lower power and high false positive rates in the three-gene epistasis models. This can be anticipated, however due to the small genetic effect and only moderate sample size. In addition, the size of the GP population and number of generations in both the GP and GPNN analyses could have been expanded to maximize the search of the space.

While these results are not striking, it is important to consider the source of the differential power results achieved by the two methods. Both methods are using a stochastic GP and identical genetic operators. The only difference in the techniques is the set of rules governing the constraints on GPNN to ensure that each binary expression tree represents a NN. Thus, these constraints are limiting the search space for GPNN and this could lead to the improved power of the GPNN approach. Given the same number of model evaluations, GP has a much larger number of possible solutions to test. As such, GPNN may be more powerful only because of the limited

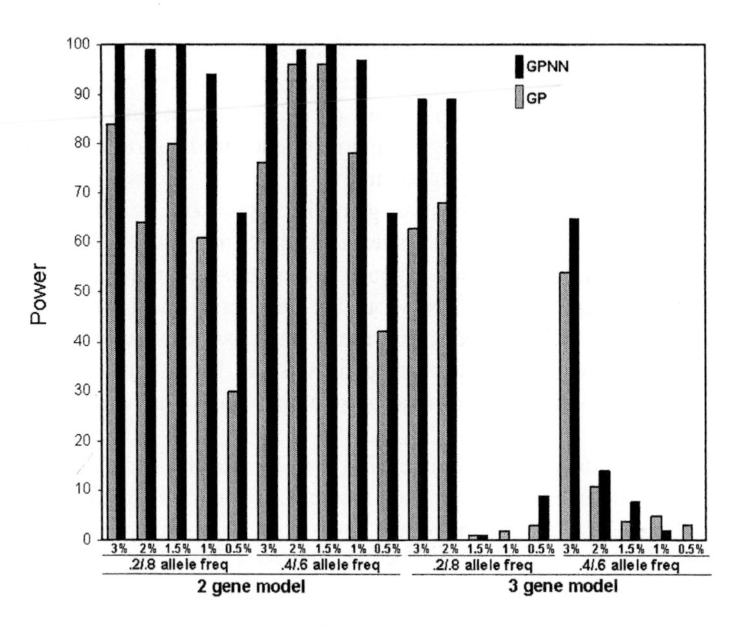

Fig. 2. Power of GPNN (black) and GP (grey) for each epistasis model

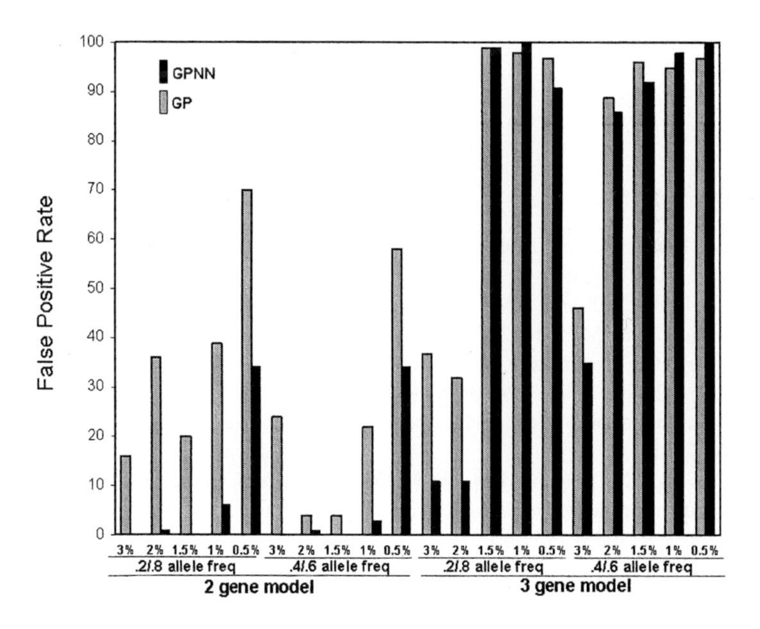

Fig. 3. False positive rates for GPNN (black) and GP (grey) for each epistasis model

search space. If both methods were able to run through subsequent generations, they may ultimately conclude with the same optimal model. This was not performed in these studies, however, because the goal was to compare the approaches with the same parameter set. In addition, these results were produced in one set of experiments. It would be optimal to generate standard errors on the power and false positive estimates based on multiple runs of these experiments so that one can determine if there is a statistically significant difference in the power or false positive rate. This is an area of future investigation.

The results of this study demonstrate that GP approaches for studies of human genetics are worthwhile. GPNN is a powerful method for the detection of gene-gene and gene-environment interactions. The beauty of this technique is the ability to explain the solutions in terms of NN which are familiar to many scientists. GP, in methods such as SDA, is not inferior however. The possible solution space is much greater for these methods, and therefore care must be taken to set the appropriate population size and number of generations so that the optimal solutions can be identified.

Novel techniques such as GPNN and SDA will be vital for the detection of gene-gene and gene environment interactions in studies of common, complex disease. The power of GP to analyze both genetic and microarray data will provide the field of human genetics with the ability to explore datasets that have previously been incomprehensible.

Acknowledgements

This work was supported by National Institutes of Health grants HL65962 and AG20135.

References

1. Kardia S.L.R.: Context-dependent genetic effects in hypertension. Curr. Hypertens. Reports. 2 (2000) 32-38
2. Moore J.H. and Williams S.M.: New strategies for identifying gene-gene interactions in hypertension. Ann. Med. 34 (2002) 88-95
3. Kooperberg C, Ruczinski I, LeBlanc ML, Hsu L. Sequence Analysis using Logic Regression, Genetic Epidemiology, S1 (2001) 626-631.
4. Zhu, J. Classification of gene microarrays by penalized logistic regression. Biostatistics 5 (2003) 427-443.
5. Tahri-Daizadeh N, Tregouet DA, Nicaud V, Manuel N, Cambien F, Tiret L. Automated detection of informative combined effects in genetic association studies of complex traits. Genome Res. 8 (2003) 1952-60.
6. Nelson M, Kardia SLR, Ferrell RE, Sing CF. A combinatorial partitioning method to identify multilocus genotypic partitions that predict quantitative trait variation. Genome Res 11 (2001) 458-470.
7. Culverhouse R, Klein T, Shannon W. Detecting epistatic interactions contributing to quantitative traits. Genet Epidemiol. 27 (2004) 141-52.
8. Ritchie MD, Hahn LW, Roodi N, Bailey LR, Dupont WD, Parl FF, Moore JH. Multifactor dimensionality reduction reveals high-order interactions among estrogen metabolism genes in sporadic breast cancer. Am J Hum Genet 69 (2001) 138-147.

9. Hahn LW, Ritchie MD, Moore JH. Multifactor dimensionality reduction software for detecting gene-gene and gene-environment interactions.Bioinformatics 19 (2003) 376-382.
10. Ritchie MD, Hahn LW, Moore JH. Power of multifactor dimensionality reduction for detecting gene-gene interactions in the presence of genotyping error, missing data, phenocopy, and genetic heterogeneity. Gen Epi 24 (2003) 150-157.
11. Marinov M, Weeks D: The complexity of linkage analysis with neural networks. Hum Hered 51 (2001) 169-176
12. Lucek P., Hanke J., Reich J., Solla S.A., Ott, J. Multi-locus nonparametric linkage analysis of complex trait loci with neural networks. Hum Hered 48 (1998), 275-84
13. Koza JR, Rice JP: Genetic generation of both the weights and architecture for a neural network. IEEE Press (1991) Vol II.
14. Gruau FC: Cellular encoding of genetic neural networks. Master's thesis Ecole Normale Superieure de Lyon (1992) 1-42
15. Ritchie M.D., White B.C., Parker J.S., Hahn L.W., Moore J.H.: Optimization of neural network architecture using genetic programming improves detection of gene-gene interactions in studies of human diseases. BMC Bioinformatics, 4 (2003) 28
16. Moore J.H., Parker J.S., Olsen N.J., Aune T.S.: Symbolic discriminant analysis of microarray data in autoimmune disease. Genet Epidemiol 23 (2002) 57-69
17. Moore J.H, Parker, J.S., Hahn L.W.: Symbolic Discriminant Analysis for Mining Gene Expression Patterns. In: Lecture Notes in Artificial Intelligence Vol. 2167, (2001) 372-381
18. Ritchie M.D, Coffey C.S, Moore J.H.: Genetic Programming Neural Networks: A Bioinformatics Tool for Human Genetics. Lecture Notes in Computer Science, Vol. 3102 (2004) 438-448
19. Mitchell M.: An Introduction to Genetic Algorithms. MIT Press, Cambridge (1996)
20. Moore J.H.: Cross validation consistency for the assessment of genetic programming results in microarray studies. In: Lecture Notes in Computer Science Vol 2611 ed. by: Raidl, G, et al. Springer-Verlag, Berlin (2003) 99-106
21. Ritchie M.D., Hahn, L.W., Roodi N., Bailey L.R., Dupont W.D., Parl F.F., Moore J.H.: Multifactor dimensionality reduction reveals high-order interactions among estrogen metabolism genes in sporadic breast cancer. Am. J. Hum. Genet. 69 (2001) 138-147
22. Koza JR: Genetic Programming: On the programming of computers by means of natural selection. MIT Press, Cambridge (1993)
23. Koza JR: Genetic Programming II: Automatic discovery of reusable programs. MIT Press, Cambridge (1998)
24. Koza JR, Bennett FH, Andre D, Keane, MA: Genetic Programming III: Automatic programming and automatic circuit synthesis. MIT Press, Cambridge (1999)
25. Culverhouse R., Suarez B.K., Lin J., Reich T.: A Perspective on Epistasis: Limits of Models Displaying No Main Effect. Am J Hum Genet 70 (2002) 461-471
26. Moore, J.H., Hahn L.W., Ritchie M.D., Thornton T.A., White B.C.: Application of genetic algorithms to the discovery of complex genetic models for simulations studies in human genetics. In: Proceedings of the Genetic and Evolutionary Algorithm Conference ed. by W.B. Langdon, et al. Morgan Kaufman Publishers San Francisco (2002) 1150-1155
27. Ott J.: Neural networks and disease association. Am. J. Med. Genet. 105 (2001) 60-61

A Class of Pareto Archived Evolution Strategy Algorithms Using Immune Inspired Operators for Ab-Initio Protein Structure Prediction

Vincenzo Cutello[1], Giuseppe Narzisi[1], and Giuseppe Nicosia[1,2]

[1] Department of Mathematics and Computer Science,
University of Catania, V.le A. Doria 6, 95125 Catania, Italy
{vctl, narzisi, nicosia}@dmi.unict.it
[2] Computing Laboratory, University of Kent Canterbury, Kent CT2 7NF

Abstract. In this work we investigate the applicability of a multiobjective formulation of the Ab-Initio Protein Structure Prediction (PSP) to medium size protein sequences (46-70 residues). In particular, we introduce a modified version of Pareto Archived Evolution Strategy (PAES) which makes use of immune inspired computing principles and which we will denote by "I-PAES". Experimental results on the test bed of five proteins from PDB show that PAES, (1+1)-PAES and its modified version I-PAES, are optimal multiobjective optimization algorithms and the introduced mutation operators, mut_1 and mut_2, are effective for the PSP problem. The proposed I-PAES is comparable with other evolutionary algorithms proposed in literature, both in terms of best solution found and computational cost.

Keywords: Protein structure, tertiary fold prediction, multi-objective evolutionary algorithms, genetic algorithms, clonal selection principle, hypermutation operators.

1 Introduction

Proteins are long sequences of 20 different amino acids. The amino acids composition of a protein will usually uniquely determine its 3D structure [1], to which the protein's functionality is directly related.

The Protein Structure Prediction problem (PSP) is simply defined as the problem of predicting the native conformation of a protein given the amino acid sequence. Common methods for finding protein 3D structures (such as x-ray crystallographic and NMR - Nuclear Magnetic Resonance) are slow and costly, and may take up to several months of lab work. As a consequence, there has been a continuously growing interest in the design of ad hoc algorithms for the PSP problem. The main computational strategies employed today are of two type: knowledge-based and Ab-Initio. The hypothesis of knowledge-based methods (homology modelling, threading) is that similar sequences will fold similarly. Ab-initio strategies are required when no homology is available so that one is forced to fold the proteins from scratch.

F. Rothlauf et al. (Eds.): EvoWorkshops 2005, LNCS 3449, pp. 54–63, 2005.

Successful structure prediction requires a free energy function sufficiently close to the true potential for the native state, as well as a method for exploring conformational space. Protein structure prediction is a challenging problem because current potential functions have limited accuracy and the conformational space is vast. Several algorithmic approaches have been applied to the PSP problem in the last 50 years [2, 3, 4, 5, 6]. In spite of all these efforts, PSP remains a challenging and computationally open problem.

2 PSP as a Multi-objective Optimization Problem

Historically, the Protein Folding problem (PF) and the Protein Structure Prediction problem (PSP), central problems in molecular biology, have been faced as a large single-objective optimization problem: given the primary sequence find the 3D native conformation with minimum energy (PSP), and the pathway(s) to reach the native conformation (PF), using a single-objective potential Energy Function. Molecular dynamics, monte carlo methods and evolutionary algorithms are today's state of the art methodologies to tackle the protein folding problem as single-objective optimization problem.

We conjecture and partially verify by computational experiments that, instead, it could be more suitable and less time consuming to model the PSP problem as a Multi-Objective Optimization problem (MOOP).

When an optimization problem involves more than one objective function, the task of finding one (or more) optimum solution, is known as multi-objective optimization. The Protein folding problem naturally involves multiple objectives. Different solutions (the 3D conformations) may involve a trade-off (the conflicting scenario in the funnel landscape) among different objectives. An optimum solution with respect to one objective may not be optimum with respect to another objective. As a consequence, one cannot choose a solution which is optimal with respect to only one objective. In general, in problems with more than one conflicting objective, there is no single optimum solution. There exist, instead, a number of solutions which are all optimal (the Optimal Pareto front). This is the fundamental difference between a single-objective and multi-objective optimization task. Hence, for a multi-objective optimization problem we can define the following procedure:

1. find the Optimal Pareto Front with a wide range of values for objectives;
2. choose one of the obtained solutions using some "higher-level information".

Lamont *et al.* [5] reformulated PSP as a MOOP and used a multi-objective evolutionary algorithm (MO fmGA) for the structure prediction of two small protein sequences: [Met]-Enkephelin (5 residues), Polyalanine (14 residues). In this work we investigate for the first time the applicability of such a multi-objective approach to medium size protein sequences (46-70 residues).

The most difficult task when using evolutionary algorithms, or any other type of stochastic search, for the PSP problem, is to come up with "good"

– *representation* of the conformations,
– *operators* for creating new conformations, and
– a *cost function* for evaluating conformations.

Representation of the Polypeptide Chain. Few conformation-representations are commonly used: all-atom 3D coordinates, all-heavy-atom coordinates, backbone atom coordinates + sidechain centroids, C_α coordinates, backbone and sidechain torsion angles. Some algorithms use multiple representations and move between them for different purpose.

In this work, we use an internal coordinates representation (torsion angles): each residue type requires a fixed number of torsion angles to fix the 3D coordinates of all atoms. Bond lengths and angles are fixed at their ideal values. All the ω torsion angles are fixed at their ideal value 180°. The degrees of freedom in this representation are the backbone and sidechain torsion angles (ϕ, ψ and χ_i). The number of χ angles depends on the residue type[1]. Angles are represented by real numbers approximated to the third decimal digit.

Potential Energy Function. The literature on cost functions (often called energy functions) is enormous.

In this work, in order to evaluate the conformation of a protein, we use the CHARMM (version 27) energy function. CHARMM (Chemistry at HARvard Macromolecular Mechanics) is a popular all-atom force field used mainly for the study of macromolecules. It is a composite sum of several molecular mechanics equations grouped into two major types: *bonded* (stretching, bending, torsion, Urey-Bradley, impropers) and *non-bonded* (van-der-Walls, electrostatics). The various tools (CHARMM, TINKER, ECEPP, etc) for the evaluation of the conformations can produce different energy values. The energy of the predicted and native structures are calculated using TINKER[2] Molecular Modeling Package [7]. For the mathematical definition of the CHARMM energy function see reference [8].

Constraints. In order to reduce the size of the conformational space, backbone torsion angles are bounded in regions derived from secondary and supersecondary structure prediction (table 1).

Sidechain torsion angles are constrained in regions derived from the backbone-independent rotamer library of Roland L. Dunbrack[3][9]. Supersecondary structure is defined as the combination of two secondary structural elements with a short connecting peptide between one to five residues in length. A short connecting peptide can have a large number of conformations. They play an important role in defining protein structures. The conformations of the residues in the short connecting peptides are classified into five major types, namely, *a*, *b*, *e*, *l*, or *t* [10] each represented by a region on the ϕ-ψ map. Sun *et al.* [11] developed

[1] Introduction to mathematical biophysics, J. R. Quine (2004).
[2] http://dasher.wustl.edu/tinker
[3] www.fccc.edu/research/labs/dunbrack

Table 1. Corresponding regions of the Secondary and Supersecondary Structure Constraints

Supersecondary Structures	ϕ	ψ
H (α-helix)	$[-75°, -55°]$	$[-50°, -30°]$
E (β-strand)	$[-130°, -110°]$	$[110°, 130°]$
a	$[-150°, -30°]$	$[-100°, 50°]$
b	$[-230°, -30°]$	$[100°, 200°]$
e	$[30°, 130°]$	$[130°, 260°]$
l	$[30°, 150°]$	$[-60°, 90°]$
t	$[-160°, -50°]$	$[50°, 100°]$
undefined	$[-180°, 0°]$	$[-180°, 180°]$

an artificial neural network method to predict the 11 frequently occurring supersecondary structure: H-*b*-H, H-*t*-H, H-*bb*-H, H-*ll*-E, E-*aa*-E, E-*ea*-E, H-*lbb*-H, H-*lba*-E, E-*aal*-E, E-*aaal*-E, and H-*l*-E, where H and E represent α-helix and β-strand, respectively. Sidechain constraint regions are of the form: $[m-\sigma, m+\sigma]$; where m and σ are the mean and the standard deviation for each sidechain torsion angle computed from the rotamer library. Under these constraints the conformation is still highly flexible and the structure can take on various shapes that are vastly different from the native shape.

Multi-objective Formulation. The energy function CHARMM is decomposed in two partial sums: bonded and non-bonded atom energies: $f_1 = E_{bond}, f_2 = E_{non-bond}$. These two functions represent our minimization *objectives*, the torsion angles of the protein are the *decision variables* of the multi-objective problem, and the constraint regions are the *variable bounds.*

The Metrics: DME and RMSD. To evaluate how similar is the predicted conformation to the native one, we employ two frequently used metrics: Root Mean Square Deviation (RMSD) and Distance Matrix Error (DME).

For a particular pair of structures, the RMSD, which measures the similarity of atomic positions, is usually larger than DME, which measures the similarity of interatomic distances.

3 The Pareto Archived Evolutionary Strategy Algorithms

The algorithm PAES (Pareto Archived Evolutionary Strategy) was proposed for the first time by Knowles and Corne in 1999 [12]. PAES is a multi-objective optimizer which uses a simple (1+1) local search evolution strategy. Nonetheless, it is capable of finding diverse solutions in the Pareto optimal set because it maintains an archive of nondominated solutions which it exploits to accurately estimate the quality of new candidate solutions. At any iteration t, there are a candidate solution c_t and a mutated solution m_t which must be compared for dominance. Acceptance is simple if one solution dominates the other. In case neither solution dominates the other, the new candidate solution is compared with the reference population of previously archived nondominated solutions. If comparison to the population in the archive fails to favour one solution over the other, the tie is split to favour the solution which resides in the least crowded region of the space. A maximum size of the archive is always maintained. The crowding procedure is based on recursively dividing up the M-dimensional objective space in 2^d equal-sized hypercube, where d is a user defined depth parameter. The algorithm proceeds until an input given, fixed number of *iterations* is reached.

3.1 I-PAES

I-PAES is a modified version of (1+1)-PAES[12, 13] with a different solution representation (polypeptide chain) and immune inspired operators: *cloning* and *hypermutation* [14]. Hypermutation can be seen as a search procedure that leads to a fast maturation during the affinity maturation phase. The clonal expansion phase triggers the growth of a new population of high-value solutions centered on a higher affinity value. The algorithm starts by initializing a random conformation. The torsion angles (ϕ, ψ, χ_i) are generated randomly from the constraint regions. After that, the energy of the conformation (a point in the landscape) is evaluated using routines from TINKER. First, the protein structure in internal coordinates (torsion angles) is transformed in cartesian coordinates using the PROTEIN routine. Then the conformation is evaluated using the ANALYZE routine, that gives back the CHARMM energy potential of the structure.

At this point, we have the main loop of the algorithm. From the current solution, a number δ of clones will be generated, producing the population (Pop^{clo}) which will be mutated into (Pop^{hyp}) and then evaluated reselecting the best one. From this moment on, the algorithm proceeds following the standard structure of $(1+1)$-PAES. Figure 1 shows the pseudo-code of the algorithm.

```
I-PAES(δ, depth, archive_size, objectives)
1. t := 0;
2. Initialize(c); /*Generate initial random solution*/
3. Evaluate(c); /*Evaluation of initial solution*/
4. AddToArchive(c); /*Add c to archive*/
5. while(not(Termination()))
          /*Start Immune phase*/
6.        Pop^clo := Cloning(c, δ); /*Clonal expansion phase*/
7.        Pop^hyp := Hypermutation(Pop^clo); /*Affinity maturation phase*/
8.        Evaluate(Pop^hyp); /*Evaluation phase*/
9.        m := SelectBest(Pop^hyp); /*Selection phase*/
          /*End Immune phase*/
          /*Start (1+1)-PAES*/
10.        if(c dominates m) discard m;
11.       else if(m dominates c)
12.             AddToArchive(m);
13.             c := m;
14.       else if(m is dominated by any member of the archive) discard m;
15.       else test(c, m, archive_size, depth);
16. t := t + 1;
17. endwhile
```

Fig. 1. Pseudo-code of I-PAES

Hypermutation Operators. Two kinds of mutation operators were used together in the affinity maturation phase (line 6 of I-PAES). The first mutation operator, mut_1, may change the conformation dramatically. When this operator acts on a peptide chain, all the values of the backbone and sidechain torsion angles of a randomly chosen residue are reselected from their corresponding constrained regions. The second mutation operator, mut_2, performs a local search of the conformational space. It will perturb all torsion angles (ϕ, ψ, χ_i) of a randomly chosen residue with the law: $\theta' = \theta + N(0,3)$, where θ is the generic torsion angle, and $N(0,3)$ is a real number generated by a gaussian distribution of mean $\mu = 0$ and standard deviation $\sigma = 3$. The first half of Pop^{clo} is mutated using mut_1 and the second half using mut_2.

Two mutation rates are studied. The first one is a static scheme where each clone is mutated only once using one of the two possible mutation operators. We call I-PAES(1-mut) the algorithm version that uses this mutation rate. The second mutation rate instead is similar to the scheme presented in [6]. The number of mutations decreases as the search method proceeds following the law: $M = 1 + (L/k) \times e^{\left(\frac{-t}{\gamma}\right)}$ where L is the number of residues, γ is a constant set to 3×10^4, and k is set to 6 for mut_1 and 4 for mut_2. We call I-PAES(M) the algorithm version that uses this second mutation rate.

4 Results and Comparisons with Other Approaches

Tables 2 and 3 show results applying I-PAES(M), I-PAES(1-mut) and (1+1)-PAES to the five PDB proteins on ten independent runs. We set the maximum number of fitness function evaluations ($Tmax$) to 2.255×10^5 (so to compare it to [6]), a minimal duplication value (δ=2), archive size of 300 and $depth = 4$. Two possible versions of (1+1)-PAES were implemented. The first one perturbs the angles following the mut_1 operator scheme, the second one perturbs the angles by applying both mut_1 and mut_2 schemes. In both cases the residue for mutation is chosen using the standard PAES operator with probability of $1/\ell$ (ℓ protein length). From table 2 is clear that both versions of I-PAES perform better than (1+1)-PAES versions. Minima energy values obtained by I-PAES are closer to those of the native structure. Moreover, the high value of the standard deviation for (1+1)-PAES shows worse convergence than I-PAES. The best energy values found for 1CTF are below the native conformation value. Probably for the limited accuracy of CHARMM energy function: near native conformations can have smaller energy values than that of the native one. In table 3 the best DME and RMSD values are always obtained with I-PAES(M), although I-PAES(1-mut) reaches better energy values. The examples in figures 2 and 3 show the predicted structures and the Pareto fronts calculated using I-PAES algorithms versus the native structure for two of the five proteins examined.

We compared our Immune Algorithms I-PAES and their results to other works in literature and others MOEA's, in particular NSGA2 [15], that we implemented and tested on PSP. Table 3 shows the best DME values obtained by the genetic algorithm proposed in [6] (no RMS values were presented by the au-

Table 2. Results obtained by I-PAES (both versions) and (1+1)-PAES (both versions) applied to the test bed of five proteins used in [6]. For each algorithm, the minimum, the mean and the standard deviation on ten independent runs of best energy values are reported. For each protein were reported, the Protein Data Bank (PDB) identifier, the length, the approximate class (α-helix, β-sheet) and the energy value calculated with TINKER

Protein	Algorithm	min(kcal/mol)	mean(kcal/mol)	σ(kcal/mol)
1ROP(56 aa)	I-PAES(M)	-526.9542	-417.4685	98.2774
class: α	I-PAES(1-mut)	**-661.4819**	**-554.9819**	**82.9940**
energy: *-667.0515 kcal/mol*	(1+1)-PAES	2640.7719	833976.1875	1497156.7511
	(1+1)-PAES($mut_1 + mut_2$)	-409.9522	534.6435	192.4232
1UTG(70 aa)	I-PAES(M)	357.9829	619.8551	174.8500
class: α	I-PAES(1-mut)	**282.2497**	**511.4623**	**142.1591**
energy: *202.7321 kcal/mol*	(1+1)-PAES	7563.0714	53937.0271	55304.4139
	(1+1)-PAES($mut_1 + mut_2$)	397.1290	1231.2125	432.7464
1CRN(46 aa)	I-PAES(M)	410.0382	464.2972	**42.4524**
class: $\alpha + \beta$	I-PAES(1-mut)	**232.2967**	**357.2083**	75.9134
energy: *-142.4612 kcal/mol*	(1+1)-PAES	1653.9359	27995.0374	43275.1845
	(1+1)-PAES($mut_1 + mut_2$)	509.5245	1221.9564	687.9383
1R69(63 aa)	I-PAES(M)	264.5602	397.6853	74.9013
class: α	I-PAES(1-mut)	**211.2640**	**290.0966**	**46.3440**
energy: *-676.5322 kcal/mol*	(1+1)-PAES	9037.8915	2636441.5872	4462510.0991
	(1+1)-PAES($mut_1 + mut_2$)	659.4954	8733.4471	782.6253
1CTF(68 aa)	I-PAES(M)	**218.9968**	281.27994	64.3010
class: $\alpha + \beta$	I-PAES(1-mut)	71.5572	**161.4119**	**48.8140**
energy: *230.0890 kcal/mol*	(1+1)-PAES	1424.3397	52109.3556	44669.0231
	(1+1)-PAES($mut_1 + mut_2$)	617.6945	5632.4211	923.2351

Table 3. Results obtained by I-PAES (both versions) and (1+1)-PAES (both versions) applied to the test bed of five proteins used in [6]. For each algorithm, the minimum, the mean and the standard deviation on ten independent runs for DME and RMSD values are reported. For each protein we report, the Protein Data Bank (PDB) identifier, the length and the approximate class (α-helix, β-sheet). Also shown is the best DME value obtained for each protein respect the GA proposed in [6] (no RMSD values were presented by the authors)

Protein	Algorithm	min		mean		σ	
		DME(Å)	RMSD(Å)	DME(Å)	RMSD(Å)	DME(Å)	RMSD(Å)
1ROP(56 aa)	I-PAES(M)	**1.684**	**3.740**	4.444	6.462	2.639	2.661
class: α	I-PAES(1-mut)	2.016	4.110	**3.405**	**5.592**	**1.036**	**1.128**
	(1+1)-PAES	4.919	6.312	9.465	10.111	3.866	3.468
	(1+1)-PAES($mut_1 + mut_2$)	5.997	8.665	6.954	9.422	4.871	3.620
	GA[6]	**1.48**	-	-	-	-	-
1UTG(70 aa)	I-PAES(M)	**3.474**	**4.272**	5.417	7.404	1.484	2.330
class: α	I-PAES(1-mut)	4.498	5.117	**5.221**	**6.351**	**0.817**	**1.066**
	(1+1)-PAES	4.708	6.047	6.637	8.936	1.242	1.647
	(1+1)-PAES($mut_1 + mut_2$)	4.826	5.566	7.848	8.788	1.331	1.432
	GA[6]	**3.47**	-	-	-	-	-
1CRN(46 aa)	I-PAES(M)	**3.436**	**4.316**	**5.057**	5.874	1.278	0.960
class: $\alpha + \beta$	I-PAES(1-mut)	4.137	4.731	5.156	**5.817**	**0.758**	**0.726**
	(1+1)-PAES	4.676	6.181	6.700	7.778	2.164	1.404
	(1+1)-PAES($mut_1 + mut_2$)	6.055	7.895	7.837	8.547	3.744	2.564
	GA[6]	**2.73**	-	-	-	-	-
1R69(63 aa)	I-PAES(M)	**4.091**	**5.057**	7.867	9.630	0.815	0.911
class: α	I-PAES(1-mut)	5.932	8.425	**7.218**	**9.557**	**0.669**	**0.551**
	(1+1)-PAES	5.167	7.599	7.589	9.607	2.544	1.809
	(1+1)-PAES($mut_1 + mut_2$)	6.886	8.521	7.681	9.912	1.232	1.998
	GA[6]	**4.48**	-	-	-	-	-
1CTF(68 aa)	I-PAES(M)	**6.822**	**10.121**	10.773	13.559	1.351	0.727
class: $\alpha + \beta$	I-PAES(1-mut)	8.081	10.691	**9.192**	**11.303**	0.988	**0.468**
	(1+1)-PAES	9.609	12.092	10.534	12.957	**0.936**	0.832
	(1+1)-PAES($mut_1 + mut_2$)	8.845	10.214	10.662	11.948	1.131	1.227
	GA[6]	**4.00**	-	-	-	-	-

thors) on the same protein test bed. Results obtained by GA [6] are comparable to those obtained I-PAES. I-PAES did not perform well for 1CTF where GA[6]

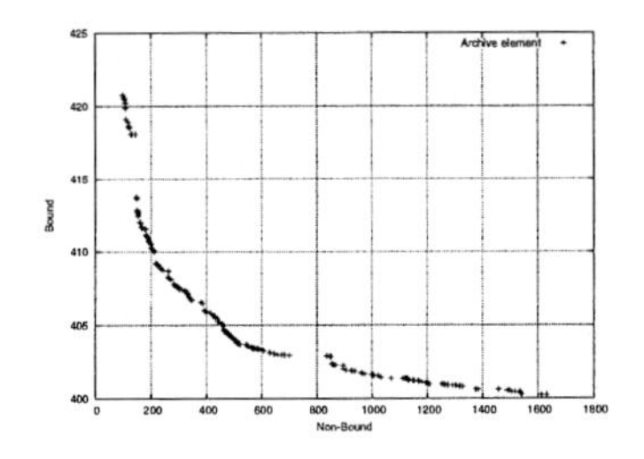

Fig. 2. 1CRN: predicted structure (left image) with DME = 3.436Å and RMSD = 4.375Å, pareto front (center plot), native structure (right image)

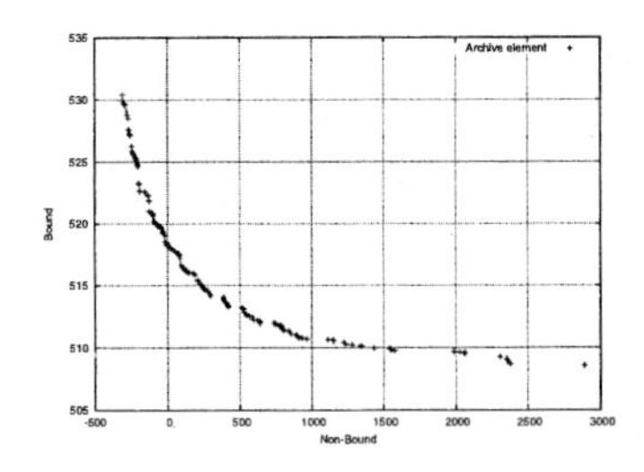

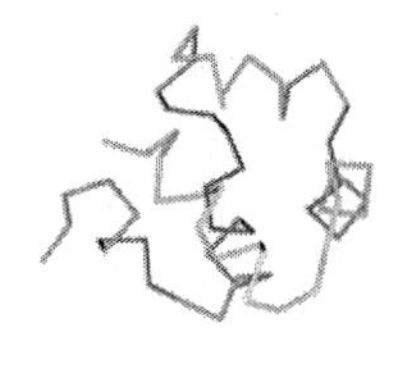

Fig. 3. 1R69: predicted structure (left image) with DME = 4.091Å and RMSD = 5.057Å, pareto front (center plot), native structure (right image)

Table 4. Best results between (1+1)-PAES, I-PAES, NSGA2, Hill-climbing GA [16] and Dandekar and Argos' GA [17] on 1CRN

Algorithm	RMSD	AES
I-PAES(M)	**4.316Å**	2.255×10^5
I-PAES(1-mut)	4.731Å	2.255×10^5
Dandekar and Argos' GA[17]	5.4Å	4×10^5
Hill-climbing GA [16] (with hydrophobic term)	5.6Å	5×10^6
(1+1)-PAES($mut_1 + mut_2$)	6.055Å	2.255×10^5
(1+1)-PAES	6.181Å	2.255×10^5
NSGA2 (with high-level operators)	6.447Å	2.5×10^5
Hill-climbing GA [16] (without hydrophobic term)	6.8Å	10^6
NSGA2 (with low-level operators)	10.34Å	2.5×10^5

reached a better DME value. However, for the other proteins, I-PAES obtained results of the same quality, and sometimes better, as in the case of 1R69 (DME of 4.091Å). For 1CRN instead the quality of the result is a bit inferior than GA[6], probably for the presence of both α and β classes, similarly to 1CTF. In [16] the best RMSD found for 1CRN, using a Hill-climbing genetic algorithm, is 5.6 and with average number of evaluation to solution (AES) equal to 5×10^6 fitness function evaluations. In this case, our method performed better both in terms of best solution found and time efficiency. Inspecting the results reported in table 4, both versions of I-PAES outperform the good RMSD value obtained by GA designed by Dandekar and Argos [17] which uses 4×10^5 fitness function evaluations. Two possible versions of NSGA2 were implemented. The first one uses standard low-level operators (SBX crossover and polynomial mutation), and the protein is considered as a long sequence of torsion angles (real numbers). The second one uses high-level operators (naive crossover and the scheme of mutation used by I-PAES). In this case, the protein is manipulated at the amino acid level. The dimension of the population is 300. The better performance of the high-level version is very clear. Table 4 shows the comparison between I-PAES, (1+1)-PAES, NSGA2, Hill-climbing GA [16] and Dandekar *et al.*'s GA [17] on 1CRN.

5 Conclusions and Future Works

We proposed a modified version of the algorithm PAES that uses immune inspired principles (Clonal Expansion and Hypermutation operators) as a new search method for PSP. The obtained class of Pareto Archived Evolution Strategy Algorithms, denoted by I-PAES, has better performance than the standard (1+1)-PAES for PSP both in terms of best energy and metrics (DME, RMSD) solutions. Moreover, I-PAES has better convergence than PAES as shown by the smaller values of the standard deviation in ten independent runs. For the first time, a multi-objective approach was used to fold medium size proteins (46-70 residues), and the results are comparable to other approaches in literature. Lamont *et al.* were the first to study PSP as multi-objective problem, but their work [5] was related only to two short protein sequences (5 and 14 residues). Experimental results on 1CRN protein show also a better performance of the PAES algorithms (I-PAES and (1+1)-PAES) with respect to NSGA2.

As reported in [18] "the folded state is a small ensemble of conformational structures compared to the conformational entropy present in the unfolded ensemble". The multiobjective approach used in this work, allows us to obtain good Pareto fronts of non-dominated compact solutions near the folded state.

We mention in conclusion some possible future lines of investigation: 1)adding a third objective that includes hydrophobic interactions, one of the most important driving force in the protein folding; 2)studying the relationship between high and low level representations and mutation operators for PSP; 3)using dynamic representation/operators during the folding process of the protein.

Acknowledgements. we are grateful to the anonymous referees for their valuable comments.

References

1. Anfinsien C.B.: Principles that govern the folding of protein chains. Science, 181, pp. 223-230, (1973).
2. Simons K.T., Kooperberg C., Huang E., Baker D.: Assembly of of protein tertiary structures from fragments with similar local sequences using simulated annealing and Bayesian scoring function. J Mol Biol, 306, pp. 1191-9, (1997).

3. Hansmann U.H., Okamoto Y.: Numerical comparisons of three recently proposed algorithms in the protein folding problem. J Comput Chem, 18, pp. 920-33, (1997).
4. Pendersen J.T., Moult J.: Protein folding simulations with genetic algorithms and a detailed molecular description. J Mol Biol, 169, pp.240-59, (1997).
5. Day R. O., Zydallis J. B., Lamont G. B.: Solving the protein structure prediction problem through a multiobjective genetic algorithm. ICNN. 2, pp. 32-35, (2001).
6. Cui Y., Chen R.S., Wong W.H.: Protein Folding Simulation using Genetic Algorithm and Supersecondary Structure Constraints. Proteins: Structure, Function and Genetics, 31(3), pp. 247-257, (1998).
7. Huang Enoch S., Samudrala R., Ponder Jay W.: Ab Initio Folding Prediction of Small Helical Proteins using Distance Geometry and Knowledge-based Scoring Functions. J. Mol. Biol., 290, pp. 267-281, (1999).
8. Foloppe N, MacKerell A. D., Jr: All-Atom Empirical Force Field for Nucleic Acids: I. Parameter Optimization Based on Small Molecule and Condensed Phase Macromolecular Target Data.
J. Comput. Chem., 21, pp. 86-104, (2000).
9. R. L. Dunbrack, Jr. and F. E. Cohen.: Bayesian statistical analysis of protein sidechain rotamer preferences. Protein Science, 6, pp. 1661-1681 (1997).
10. Sun, Z., Jang, B.J.: Patterns and conformations commonly occurring supersecondary structures (basic motifs) in Protein Data Bank. J Protein Chem. 15, pp. 675-690, (1996).
11. Sun Z., Rao X., Peng L., Xu D.: Prediction of protein supersecondary structures based on the artificial neural network method. Protein Eng. 10, pp. 763-769, (1997).
12. Knowles, J.D., Corne, D.W.: The Pareto Archived Evolution Strategy : A New Baseline Algorithm for Pareto Multiobjective Optimisation. In Proceedings of the 1999 Congress on Evolutionary Computation (CEC'99), pp. 98-105, (1999).
13. Knowles, J.D., Corne, D.W.: Approximating the Nondominated Front Using the Pareto Archived Evolution Strategy. Evolutionary Computing., 8(2), pp. 149-172, (2000).
14. Cutello V., Nicosia G.: The Clonal Selection Principle for in silico and in vitro Computing. In L. N. de Castro and F. J. Von Zuben editors, Recent Developments in Biologically Inspired Computing, (2004).
15. Deb K., Pratap A., Agarwal S., Meyarivan T.: A fast and elitist multiobjective genetic algorithm: NSGA-II IEEE Trans Evol Computat., 6, pp. 182-197, (2002).
16. Cooper L. R., Corne D. W., Crabbe M. J.: Use of a novel Hill-climbing genetic algorithm in protein folding simulations. Comp. Bio. Chem., 27, pp. 575-580, (2003).
17. Dandekar,T., Argos,P.: Identifying the tertiary fold of small proteins with different topologies from sequence and secondary structure using the genetic algorithm and extended criteria specific for strand regions. J. Mol. Biol., 256, pp. 645-660, (1996).
18. S. S. Plotkin, J. N. Onuchic, Understanding protein folding with energy landscape theory, Quarterly Reviews of Biophysics, 35(2), pp. 111-167, 2002.

Neural Networks and Temporal Gene Expression Data

A. Krishna[1], A. Narayanan[2], and E.C. Keedwell[2]

[1] Bioinformatics Laboratory, Washington Singer Laboratories,
School of Biological and Chemical Sciences
[2] School of Engineering, Computer Science and Mathematics,
University of Exeter, EX4 3PT, UK
{A.Krishna, A.Narayanan, E.C.Keedwell}@exeter.ac.uk

Abstract. Temporal gene expression data is of particular interest to systems biology researchers. Such data can be used to create gene networks, where such networks represent the regulatory interactions between genes over time. Reverse engineering gene networks from temporal gene expression data is one of the most important steps in the study of complex biological systems. This paper introduces sensitivity analysis of systematically-perturbed trained neural networks to both select a smaller and more influential subset of genes from a temporal gene expression dataset as well as reverse engineer a gene network from the reduced temporal gene expression data. The methodology was applied to the rat cervical spinal cord development time-course data, and it is demonstrated that the method not only identifies important genes involved in regulatory relationships but also generates candidate gene networks for further experimental study.

1 Introduction

Temporal gene expression analysis is an active area of research and is undergoing a transition in response to a shift from traditional biology to systems biology due to advances in high throughput technologies. Current approaches to the problem of temporal gene expression analysis include cluster analysis (e.g. [1]), principal component analysis (PCA) and singular value decomposition (SVD) [2]. A general intro-duction to the reverse engineering problem can be found in [3]. Current methods struggle to extract regulatory relationships between genes because of the large num-ber of genes typically measured as well as low number of samples. Much prior knowledge concerning known regulatory relationships has to be included in many current reverse engineering algorithms to ensure that the final 'gene circuit diagrams' make sense biologically. On the other hand, artificial neural networks (ANNs) have been shown to be effective in extracting biologically-plausible classificatory knowledge from non-temporal gene expression data [4], and an ANN-Genetic Algorithm hybrid [5] has also been applied to temporal data. One of the problems in the application of artificial neural networks to temporal gene analysis concerns the difficulty in creating complex regulatory networks from an analysis of high-dimensional weight matrices that represent the individual connections between pairs of genes over time. Also, a problem

F. Rothlauf et al. (Eds.): EvoWorkshops 2005, LNCS 3449, pp. 64–73, 2005.

with genetic algorithms in this domain is that they can find spurious relationships between genes due to the relatively low numbers of samples (time-points) in comparison to the number of genes measured (minimally hundreds, typically thousands).

Narayanan *et al.* [4] demonstrated the applicability of single-layer ANNs (perceptrons) for non-temporal gene expression data for classification purposes, where weight values were used to identify the biological importance of genes for classification. According to that approach, there are as many input nodes in the neural network as genes, and only one output node if binary classification (e.g. cancer/normal) is required. This paper extends this approach in two ways. First, in addition to the input layer containing as many input nodes as genes, the output layer also contains as many output nodes as genes, with full connection between every input node and every output node. That is, if 100 genes are measured over time, the neural network will contain 100 input nodes and 100 output nodes, where input node 1 and output node 1 stand for gene 1, input node 2 and output node 2 stands for gene 2, etc. The input layer represents one time-point and the output layer the immediately following time point. This architecture requires the temporal data to be presented for training and testing in an appropriate way, to be described below. Secondly, once the neural networks has been successfully trained, the novel technique of sensitivity analysis is introduced to systematically perturb the expression values of genes to check for the effect of 'gene silencing' or 'gene activation' on genes at the output layer, with the weights between input and output genes clamped. Sensitivity analysis consists of experimenting with the expression values and letting the weights connecting every input gene to every output gene determine how varying the expression of an input gene affects the expression value of an output gene. It has been previously shown [4] that large weight values (both positive and negative) of individual input genes represent greater contribution to classification when only one output (class) node is used at the output layer. The hypothesis adopted in this paper is that, with full connection between n input nodes and n output nodes, it is not the weight values that need further analysis but the expression values of genes. If a gene passes the sensitivity analysis as either an input (affecting gene) or as an output (affected gene), it is kept for subsequent participation in a gene network.

2 The Methodology

The methodology, represented as a flowchart in Figure 1, provides an iterative method for reverse engineering the regulatory relationships from temporal gene expression data. The data must first be transformed into a format appropriate for the architecture (Figure 2). At the heart of the methodology lies sensitivity analysis of the trained neural network. Sensitivity analysis measures the effect of small changes (perturbations) in the input channel (gene expression at time t) on the output (gene expression at time t+1). Sensitivity analysis is performed to determine the effect that each of the network inputs (gene expression values at one time step) has on the network output (gene expression at subsequent time step). This provides feedback as to which input and output channels are the most significant. If perturbing a gene's

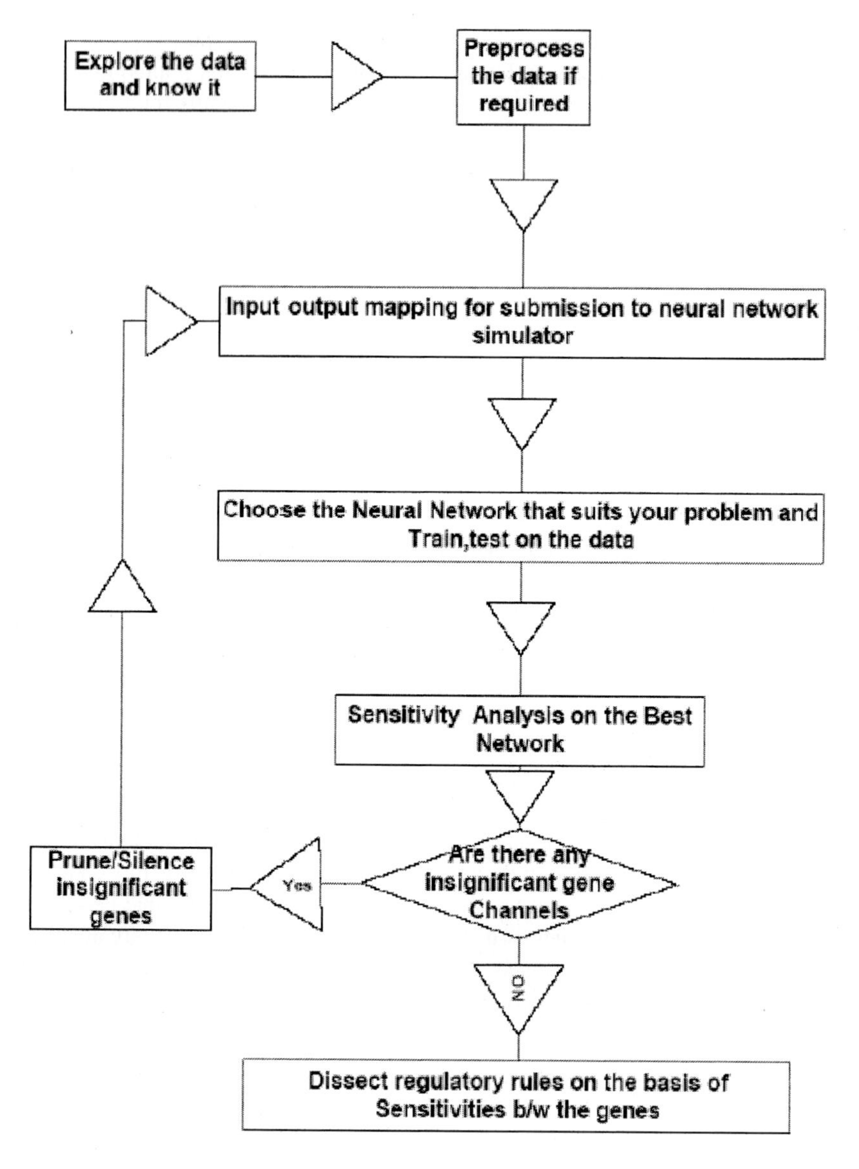

Fig. 1. The iterative methodology for reverse engineering using neural networks. After exploring and preprocessing the data, a single layer neural network (one input node per gene, one output node per gene) is constructed, trained and tested. A sensitivity analysis is then carried out on how every input gene affects every output gene to identify insignificant genes

expression value has little effect on genes at the output layer, this is considered an 'insignificant' channel that can be removed (pruned) from the input space. The gene's expression value in these experiments is varied between its mean and ± 1 standard deviation while all other inputs are fixed at their respective means. This range of sensitivity can be varied if desired. The network output is computed for a number of steps above and below the mean. This process is repeated for each input gene to determine which genes have most effect on which output genes.

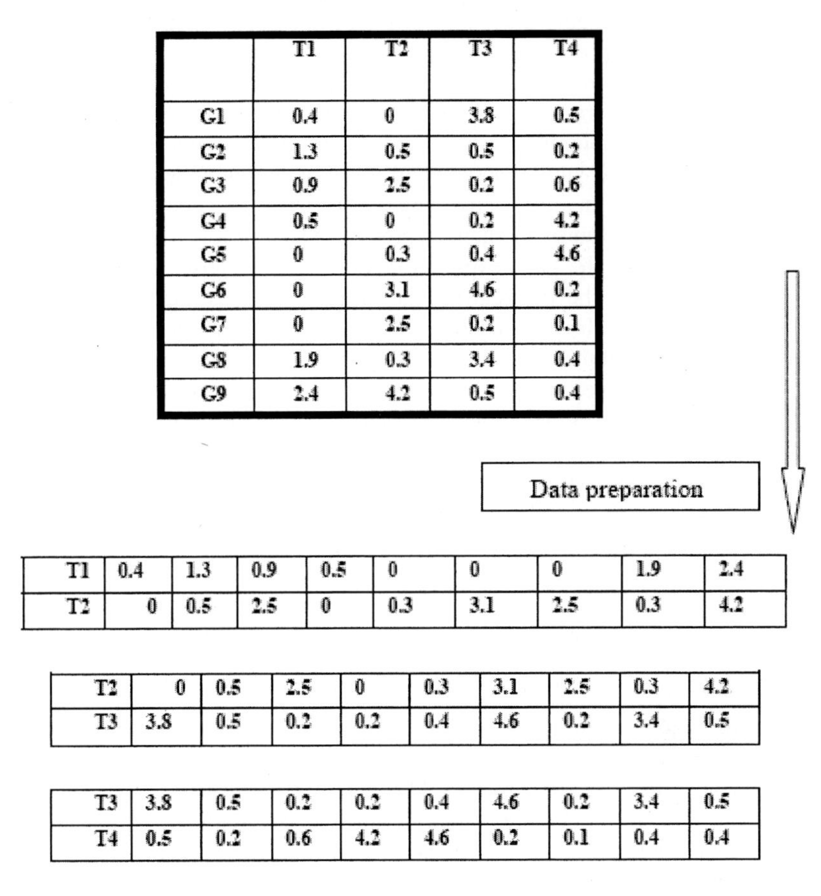

	T1	T2	T3	T4
G1	0.4	0	3.8	0.5
G2	1.3	0.5	0.5	0.2
G3	0.9	2.5	0.2	0.6
G4	0.5	0	0.2	4.2
G5	0	0.3	0.4	4.6
G6	0	3.1	4.6	0.2
G7	0	2.5	0.2	0.1
G8	1.9	0.3	3.4	0.4
G9	2.4	4.2	0.5	0.4

T1	0.4	1.3	0.9	0.5	0	0	0	1.9	2.4
T2	0	0.5	2.5	0	0.3	3.1	2.5	0.3	4.2

T2	0	0.5	2.5	0	0.3	3.1	2.5	0.3	4.2
T3	3.8	0.5	0.2	0.2	0.4	4.6	0.2	3.4	0.5

T3	3.8	0.5	0.2	0.2	0.4	4.6	0.2	3.4	0.5
T4	0.5	0.2	0.6	4.2	4.6	0.2	0.1	0.4	0.4

Fig. 2. Data transformation. Imagine that we measure 9 genes (G1-G9) over four time steps (T1-T4, top table). The data is transformed into three data-pairs for ANN training, where the first row of each pair represents the values of the 9 genes at one time step and the second row of each pair represents the values of these 9 genes at the next time step. The first row of each pair is used for input to the input layer of a single layer ANN, and the second row is used for supervised training at the output layer

The variation of each output with respect to the variation in each input is analyzed and insignificant gene inputs can be 'silenced', or pruned, resulting in regulatory relationships being reverse engineered. Genes that are not affected at the output layer by any input gene can also be removed.

Sensitivity analysis has an analogy with the systems biology approach described by Ideker *et al.* [6], in which they observe the effect of genetic and environmental perturbations in a system and then analyse the effect on other system components, thereby producing predictions and hypotheses about the system. In our case, a particular temporal gene expression experiment is regarded as a system, and the components affected are genes. A single layer neural network is first trained to produce the appropriate output values for each set of input values, with the delta rule identifying differences between desired target values for each gene and actual output values. An error is calculated with each presentation and a small difference to the weights made connecting each input node to each output node so that, the next time the same data is input, the output values are closer to the target. Network training consists of repeated presentation of all input-output pairs until no further improvement in the error is identified (typically, when the sum of squared error on the output nodes is below a certain threshold, such as 0.01 ideally). Sensitivity analysis is then executed on the trained ANN. The algorithm perturbs the gene expression value of a gene at a previous time step to see the effect on the gene expression value of other genes at subsequent time steps, thereby revealing the inherent regulatory rules in the temporal data. We perform sensitivity analysis by training the network as we normally do and then fixing the weights. The next step is to randomly perturb, one at a time, each channel of the input vector around its mean value, while keeping the other inputs at their mean values, and then measure the change in the output. The change in the input is normally done by adding a random value of known variance to each sample and computing the output. The sensitivity for a particular input gene can be expressed as:

$$S_{Input-gene} = \frac{\sum_{p=1}^{P}\sum_{i=1}^{n}(y_{ip} - \overline{Y}_{ip})^2}{\sigma_{Input-gene}^2}$$

, where y_{ip} is the i_{th} output obtained with the fixed weights for the p_{th} pattern, $\overline{Y}_{ip}$ is the mean value for the ith output on pattern p, P is the total number of patterns, n is the number of network outputs, and $\sigma_{Input-gene}^2$ is the variance of input perturbation. In the following experiments, the methodology was applied on RT-PCR temporal data of rat spinal cord development [7]. The Rat (Sprague-Dawley albino rat) data is an RT-PCR[1] study of 112 genes each measured on cervical spinal cord tissue at nine different time points during the development of the rat central nervous system. This gene expression data is accepted to be non-noisy, small and accurate (each gene was measured in triplicate), and is ideal for testing a new strategy because of previous work in literature.

3 Experimentation

There are nine different time points in the spinal cord study namely E11, E13, E15, E18, E21, P0, P7, P14, A. The only pre-processing required for the data is preparing the data for submitting to the neural network simulator. The input-output mappings

[1] Reverse transcriptase - polymerase chain reaction, A technique commonly employed in molecular genetics through which it is possible to produce copies of DNA sequences rapidly. It is an extraordinarily sensitive method for analysis of gene expression in cells available in only limited quantities. This technique is often employed to validate microarray experiments. See http://www.bio.davidson.edu/courses/Immunology/Flash/RT_PCR.html

(data pairs) are E11 input-E13 output, E13 Input-E15 output, E15 input-E18 output, E18 input-E21 output, E21 input-P0 output, P0 input-P7output, P7 input-P14 output, P14 input-A output. There are therefore eight temporal transition 'exemplars' for the network to learn. Each exemplar has 112 input gene expression values of the prior time step and 112 desired gene expression output values of the subsequent time step. The data was formatted according to the exemplar format in Figure 2. Since the data consists of temporal measurements on one 'sample' rat, there is no need for complex data-splitting schemes in this domain, except to check on whether order of presentation of the temporal data matters.

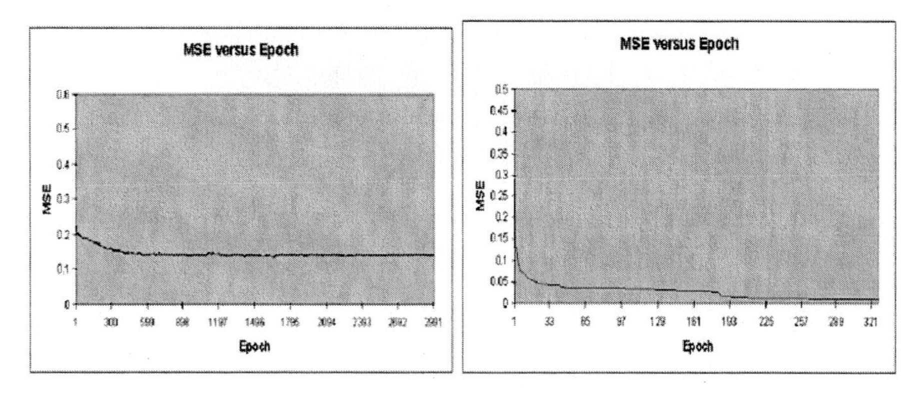

Fig. 3. Mean squared error behaviour. The left chart shows the MSE (calculated from the error of each output node) for the full 112-gene ANN, with reduction after the first 500 or so epochs tailing off and remaining around 0.15 on average for each output node/gene. The 112 genes were reduced to 24 after sensitivity analysis and a new ANN initialized for just these 24 genes. The right chart shows major improvement in the MSE for the second network, with output nodes now producing values much closer to the target values (on average 0.01 error for each output gene)

The first stage was to identify the optimal neural network in terms of learning parameters. A number of single layer perceptrons (that is, full connectivity between the 112 input and 112 output nodes, with no hidden layers) were tried, adopting different learning rule parameters and transfer functions. Sequential and random order of data-pair presentation were tried for 1000 epochs and no difference found (that is, the neural network was not concerned that the data-pairs were presented in shuffled order). Training was carried out on 6 exemplars (E11 to P7), cross-validation on one exemplar (P7-P14), and testing on one exemplar (P14-A). This method was adopted to identify the best learning parameters. Once these were identified, the *second stage* was to re-initialise the single-layer neural network with the chosen learning parameters and train it on all 8 exemplars for 3000 epochs so that the neural network was given all the temporal information. The best network at the first stage with minimum MSE for training as well as cross validation set was a perceptron with the activation/transfer function as the hyperbolic Tanh function. This will squash the range of each neuron in the layer to between -1 and 1.The activation function was $f(x_i, w_i) =$

$\tanh[x_i^{\text{lin}}]$, where $x_i^{\text{lin}} = \beta\, x_i$ is the scaled and offset activity inherited from the Linear Axon, x_i is gene i's expression value and w_i is the weight for gene i.

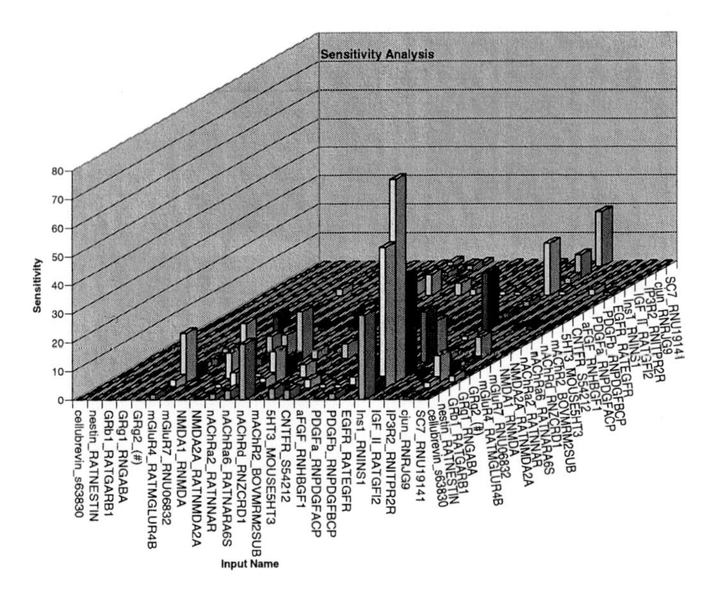

Fig. 4. Sensitivity analysis for 24 genes in Rat spinal cord development. Sensitivity analysis consists of each input being varied between its mean and ±1 standard deviation. The network output was computed for 50 steps below and above the mean and the sensitivity was calculated. The x- axis denotes gene expression at time t, y-axis depicts the gene expression at time t+1 and z-axis the sensitivities between genes in time: the higher the sensitivity value, the more important the gene. These genes survived the pruning

Sensitivity analysis was then performed on the 112x112 node perceptron successfully trained on all eight exemplar pairs (9 time steps, 3000 epochs) after the second stage. The MSE for this network during training is in Figure 3 (left chart). The non-influential input gene channels were pruned from the input space as were all the genes at the output layer that were not also affected. The 112 gene input space was thereby reduced to the 24 most influential genes (either input or output), with a gene channel pruned if the sensitivity is less than an arbitrarily chosen value of 5. This threshold can be varied if desired. The 24 genes and their input and output values were extracted from the full dataset and entered into a reconfigured ANN (all 8 exemplars, 24 input nodes, and 24 output nodes, with full connection and the same learning parameters and transfer function). The MSE showed great improvement for the reduced gene set, providing some evidence that many non-influential genes had been pruned (Figure 3, right chart). Sensitivity analysis was also applied to this reduced ANN. The results of the sensitivity analysis for just these 24 genes are provided in Figure 4. A gene network was then extracted from the top 50 connections that accounted for 80% of the total sensitivity of the network, as calculated for this second round of sensitivity analysis, in the 24x24 gene network (576 connections altogether). The specific up-

and down-regulations can be identified by referring back to the original dataset and comparing the original gene values. Figure 5 shows only those genes that, according to the sensitivity analysis, have a strong regulatory relationship (either up or down) between them.

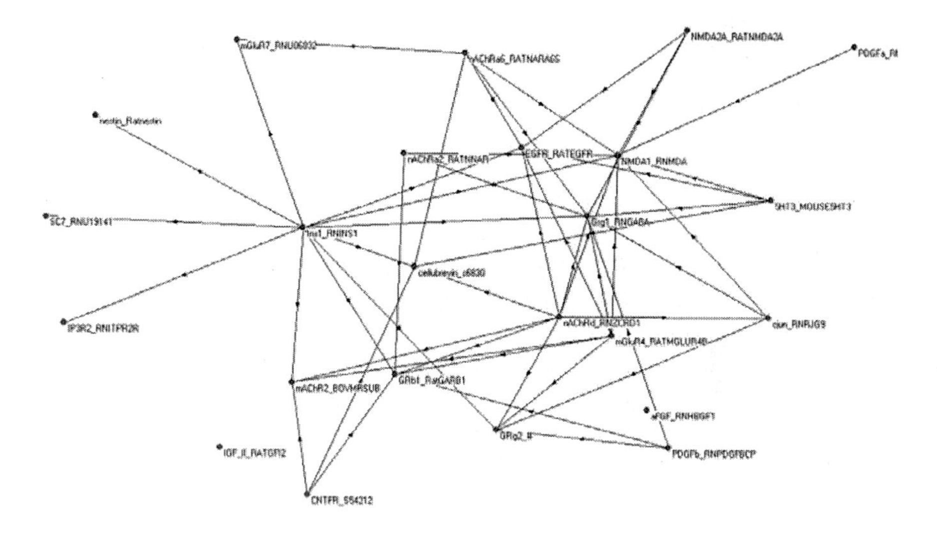

Fig. 5. Gene relationships for the top 50 connections (out of 576) for the reduced set of 24 genes, as ranked by the sensitivity analysis. As is typical of gene networks in the literature, only an affecting relationship is shown here (that is, up-regulation and down-regulation are not specifically distinguished) and specific up- and down-regulatory links can be extracted from the original dataset. The top 14 connections account for just over 50% of the total sensitivity, whereas the top 150 connections account for about 95% of the total sensitivity. The top 50 connections here account for 80% of total sensitivity. Different networks will result depending on how much sensitivity the user wishes to account for

4 Results

Analysis of the results is within the context of the 112 genes measured by Wen *et al.* [7] already being considered important by biologists for rat spinal cord development. Hence, a second round of gene selection was not considered important in this domain. Nevertheless, some specific and novel regulatory relationships between small subsets of genes within the reduced 24-gene set can be identified. The most strongly affecting gene according to the sensitivity analysis is *Ins1_RN1NS1* (insulin 1 gene), which affects: *Cellubrevin* (a vesicle-associated protein present in nerve terminals); *Grg1_RNGABA* (Groucho-related gene 1 that is closely associated with epidermal growth factor receptors and neurogenesis), the strongest signal in Figure 5; and *Grg2_#* (another Groucho-related gene). Thus, *Ins1_RN1NS1* appears to be involved in several pathways. The gene expression of this gene is prominent only in the initial time points of E11-E13. It may be the case that this gene being subsequently switched off (perhaps reflecting nutrition) starts the process of cervical spinal cord development. The other prominent relationships as evident from the sensitivity analysis and represented in Figure 5 corroborate previous suggestions that cholinergic transmission forms the leading edge in neurotransmitter signalling in cervical spinal cord [7]. The most strongly affected gene in the gene expression system is *Grg1_RNGABA*, which seems to be biologically plausible owing to the importance of Gamma amino butyric acid (GABA) in neuro-signalling. Overall the sensitivity analysis extracts the regulatory importance of many of the genes described as activators of the development of the rat spinal cord [8].

5 Conclusions

The reverse engineering of gene networks and extraction of regulatory relationships between genes in temporal gene expression data is a major obstacle in systems biology. Our experiments have demonstrated that a neural network approach, combined with sensitivity analysis, can reverse engineer biologically plausible relationships from real-world data. While more work is required to determine the effect of altering the threshold, strength and scale of perturbations on sensitivity analysis, the results presented here provide evidence of a novel, alternative approach to reverse engineering and pruning (gene silencing) that can lead to the automatic extraction of regulatory knowledge from temporal gene expression data.

More importantly, a single-layer neural network with full connectivity between input and output genes can be considered a 'universal' model of all possible gene interactions in the data. After successful training, gene expression values can then be perturbed through sensitivity analysis to remove redundant links (i.e. genes that have little or no effect on other genes and genes that are barely affected by other genes) to produce a 'leaner' and 'fitter' ANN for further, more detailed analysis with regard to specific potential regulatory relationships between small numbers of genes. Also, sensitivity thresholds can be varied to produce networks with high or low connectivity. As with all gene reduction methods, great care must be taken to ensure that important genes are not discarded because they fail to reach arbitrarily defined thresholds. Our method allows a number of parameters to be adjusted (selecting genes with sensitivity greater than 5; choosing the number of connections depending on the amount of sensitivity to be accounted for) and further work is required to identify appropriate and optimal parameter values, especially when dealing with much larger gene sets where it can be expected that a significant proportion of genes are irrelevant to the phenotype measured. Finally, all *in silico* predictions must be tested *in vitro* with genuine gene models on isolated cells or tissues to provide empirical support for the accuracy and reliability of proposed gene networks.

References

[1] Eisen, M.B., Spellman, P.T., Brown, P.O. and Botstein, D. (1998) Cluster analysis and display of genome-wide expression patterns. *Proc. Natl. Acad. Sci.* 95, pp. 14863-14868.

[2] Wall, M.E., Rechtsteiner, A., Rocha, L.M. (2003) Singular value decomposition and principal component analysis. In *A Practical Approach to Microarray Data Analysis*. D.P. Berrar, W. Dubitzky, M. Granzow, eds. pp. 91-109, Kluwer: Norwell, MA.

[3] Liang, S., Fuhrman, S. and Somogyi, R. (1998) REVEAL, a general reverse engineering algorithm for inference of Genetic network architectures. *Pacific Symposium on Biocomputing,* Vol. 3, Hawaii, USA, 1998, pp. 18–29.

[4] Narayanan, A., Keedwell,E.C., Tatineni, S.S. and Gamalielsson, J. (2004) Single-layer artificial neural networks for gene expression analysis. *Neurocomputing* 61, pp 217-240.

[5] Keedwell, E.C. and Narayanan, A. (2003) Genetic algorithms for gene expression analysis. *Applications of Evolutionary Computation: Proceedings of the 1st European Workshop on Evolutionary Bioinformatics*, G. Raidl et al., Springer Verlag LNCS 2611, pp 76-86.

[6] Ideker, T., Thorsson, V., Ranish, J.A., Christmas, R., Buhler, J., Eng, J.K., Bumgarner, R., Goodlett, D.R., Aebersold, R. and Hood, L. (2001) Integrated genomic and proteomic analysis of a systematically perturbed metabolic networks. *Science* 292:929-934.

[7] Wen, X., Fuhrman, S., Michaels, G.D., Carr, D.B., Smith, S., Barker, J.L. and Somogyi, R. (1998) Large–scale temporal gene expression mapping of central nervous system development, *PNAS* 95, pp.334-339.

[8] D'haeseleer, P., Wen, X., Fuhrman, S. and Somogyi, R. (1998) Mining the gene expression matrix: Inferring gene relationships from large scale gene expression data. In *Information Processing in Cells and Tissues,* R.C. Paton and M. Holcombe (Eds), Plenum, pp 203-212.

Bayesian Learning with Local Support Vector Machines for Cancer Classification with Gene Expression Data

Elena Marchiori[1] and Michèle Sebag[2]

[1] Department of Computer Science,
Vrije Universiteit Amsterdam, The Netherlands
elena@cs.vu.nl
[2] Laboratoire de Recherche en Informatique,
CNRS-INRIA, Université Paris-Sud Orsay, France
sebag@lri.fr

Abstract. This paper describes a novel method for improving classification of support vector machines (SVM) with recursive feature selection (SVM-RFE) when applied to cancer classification with gene expression data. The method employs pairs of support vectors of a linear SVM-RFE classifier for generating a sequence of new SVM classifiers, called local support classifiers. This sequence is used in two Bayesian learning techniques: as ensemble of classifiers in Optimal Bayes, and as attributes in Naive Bayes. The resulting classifiers are applied to four publically available gene expression datasets from leukemia, ovarian, lymphoma, and colon cancer data, respectively. The results indicate that the proposed approach improves significantly the predictive performance of the baseline SVM classifier, its stability and robustness, with satisfactory results on all datasets. In particular, perfect classification is achieved on the leukemia and ovarian cancer datasets.

1 Introduction

This paper deals with tumor classification with gene expression data. Microarray technology provides a tool for estimating expression of thousands of genes simultaneously. To this end, DNA arrays are used, consisting of a large number of DNA molecules spotted in a systematic order on a solid substrate. Depending on the size of each DNA spot on the array, DNA arrays are called microarrays when the diameter of DNA spot is less than 250 microns, and macroarrays when the diameter is bigger than 300 microns. DNA microarrays contain thousands of individual DNA sequences printed in a high density array on a glass microscope slide using a robotic instrument. The relative abundance of these spotted DNA sequences in the two DNA and RNA samples may be assessed by monitoring the differential hybridization of the two samples to the sequences on the array. For mRNA samples, the two samples are reverse-transcribed into cDNA, labeled using different fluorescent dyes mixed (red-fluorescent dye Cy5 and green-fluorescent dye Cy3).

F. Rothlauf et al. (Eds.): EvoWorkshops 2005, LNCS 3449, pp. 74–83, 2005.

After these samples are hybridized with the arrayed DNA probes, the slides are imaged using scanner that makes fluorescence measurements for each dye. The log ratio between the two intensities of each dye is used as the gene expression data (cf. [11]) $expression(gene) = log_2(int(Cy5)/int(Cy3))$, were $int(Cy5)$ and $int(Cy3)$ are the intensities of the two fluorescent dyes.

Four main machine learning tasks are used to analyze DNA microarray data: clustering, e.g. for identifying tumor subtypes, classification, e.g. for tumor diagnostic, feature selection for potential tumor biomarker identification, and gene regulatory network modeling. This paper deals with classification.

Many machine learning techniques have been applied to classify gene expression data, including Fisher linear discriminat analysis [10], k-nearest neighbour [18], decision tree, multi-layer perceptron [17, 25], support vector machine (SVM) [6, 13, 15, 21], boosting and ensemble methods [14, 7, 23, 3, 9]. A recent comparison of classification and feature selection algorithms applied to tumor classification can be found in [7, 23].

This paper introduces a method that improves the predictive performance of a linear SVM with Recursive Feature Elimination (SVM-RFE) [15] on four gene expression datasets. The method is motivated by previous work on aggregration of classifiers [4, 5], where it is shown that gains in accuracy can be obtained by aggregrating classifiers built from perturbed versions of the train set, for instance using bootstrapping. Application of aggregration of classifiers to microarray data is described e.g. in [10, 7, 3, 9].

In this paper a novel approach is proposed, for generating a sequence of classifiers from the support vectors of a baseline linear SVM-RFE classifier. Each pair of support vectors of the same class are used to generate an element of the sequence, called local support classifier (*lsc*). Such classifier is obtained by training SVM-RFE on data consisting of the two selected support vectors and all the support vectors of the other class.

The sequence of *lsc*'s provides an approximate description of the data distribution by means of a set of linear decision functions, one for each region of the input space in a small neighbourhoods of two support vectors having equal class label.

We propose to use this sequence of classifiers in Bayesian learning (cf. [20]). The first technique applies Naive Bayes to the transformed data, where an example is mapped into the binary vector of its classification values. The resulting classifier is called Naive Bayes Local Support Classifier (NB-LSC). The second technique applies Optimal Bayes to the sequence of *lsc*'s classifiers. The resulting classifier is called Optimal Bayes Local Support Classifier (OB-LSC).

The two classifiers are applied to four publically available datasets for cancer classification with gene expression. The results show a significant improvement in predictive performance of OB-LSC over the baseline linear SVM-RFE classifier, and a gain in stability. In particular, on the leukemia and ovarian cancer datasets perfect classification is obtained, and on the other datasets performance comparable to the best published results we are aware of.

The rest of the paper is organized as follows. The next two sections describe the baseline and new methods. Sections 4 contains a short description of the data. Section 5 reports results of experiments and discuss them. Finally, the paper ends with conclusive considerations on research issues to be tackled in future work.

2 Support Vector Machines

This section describes in brief SVM-RFE, the local support classifier construction procedure, and the integration of the resulting classifier sequence in Naive Bayes and Optimal Bayes classification.

2.1 SVM

In linear SVM binary classification [24, 8] patterns of two classes are linearly separated by means of a maximum margin hyperplane, that is, the hyperplane that maximizes the sum of the distances between the hyperplane and its closest points of each of the two classes (the margin). When the classes are not linearly separable, a variant of SVM, called soft-margin SVM, is used. This SVM variant penalizes misclassification errors and employs a parameter (the soft-margin constant C) to control the cost of misclassification.

Training a linear SVM classifier amounts to solving the following constrained optimization problem:

$$min_{\mathbf{w},b,\xi_k} \frac{1}{2}||\mathbf{w}||^2 + C\sum_{i=1}^{m} \xi_i \quad \text{s.t.} \quad \mathbf{w} \cdot \mathbf{x}_i + b \geq 1 - \xi_i$$

with one constraint for each training example $\mathbf{x}_i$. Usually the dual form of the optimization problem is solved:

$$min_{\alpha_i} \frac{1}{2} \sum_{i=1}^{m} \sum_{j=1}^{m} \alpha_i \alpha_j y_i y_j \mathbf{x}_i \cdot \mathbf{x}_j - \sum_{i=1}^{m} \alpha_i$$

such that $0 \leq \alpha_i \leq C$, $\sum_{i=1}^{m} \alpha_i y_i = 0$. SVM requires $O(m^2)$ storage and $O(m^3)$ to solve.

The resulting decision function $f(\mathbf{x}) = \mathbf{w} \cdot \mathbf{x} + b$ has weight vector $\mathbf{w} = \sum_{k=1}^{m} \alpha_k y_k \mathbf{x_k}$. Examples $\mathbf{x}_i$ for which $\alpha_i > 0$ are called *support vectors*, since they define uniquely the maximum margin hyperplane.

Maximizing the margin allows one to minimize bounds on generalization error. Because the size of the margin does not depend on the data dimension, SVM are robust with respect to data with high input dimension. However, SVM are sensitive to the presence of (potential) outliers, (cf. [15] for an illustrative example) due to the regularization term for penalizing misclassification (which depends on the choice of C).

2.2 SVM-RFE

The weights w_i provide information about feature relevance, where bigger weight size implies higher feature relevance. In this paper feature x_i is scored by means of the absolute value of w_i. Other scoring functions based on weight features are possible, like, e.g., w_i^2, which is used in the original SVM-RFE algorithm [15].

SVM-RFE is an iterative algorithm. Each iteration consists of the following two steps. First feature weights, obtained by training a linear SVM on the training set, are used in a scoring function for ranking features as described above. Next, the feature with minimum rank is removed from the data. In this way, a chain of feature subsets of decreasing size is obtained.

In the original SVM-RFE algorithm one feature is discarded at each iteration. Other choices are suggested in [15], where at each iteration features with rank lower than a user-given theshold are removed. In general, the threshold influences the results of SVM-RFE [15]. In this paper we use a simple instance of SVM-RFE where the user specifies the number of features to be selected, 70% of the actual number of features are initially removed, and then 50% at each further iteration. These values are chosen after cross-validation applied to the training set.

3 Local Support Classifiers

We propose to describe the distribution of the two classes by means of a sequence of classifiers, generated from pairs of support vectors ofSVM-RFE. Each of these classifiers, called local support classifier (*lsc*), is obtained using data generated from two support vectors of the same class, and all support vectors of the other class. In this way, each classifier uses only a local region near the two selected support vectors when separating the two classes. Each classifier generated from two (distinct) support vectors of the same class provides an approximate description of the distribution of the other class *given* the two selected support vectors.

Before describing the procedure for constructing *lsc*'s, some notation used throughout the paper is introduced.

- D denotes the training set,
- c denotes the classifier obtained by training a linear SVM on D,
- S_p and S_n denote the set of positive and negative support vectors of c, respectively,
- $Pair_p$ and $Pair_n$ denote the set of pairs of distinct elements of S_p and S_n, respectively.

The following procedure, called `LSC`, takes as input one $(\mathbf{s}, \mathbf{s}')$ in $Pair_p$ and outputs a linear SVM classifier $C_{s,s'}$ by means of the following two steps.

1. Let $Xp = \{\mathbf{s}, \mathbf{s}'\}$. Assign positive class label to these examples.
2. Let $C_{s,s'}$ be the classifier obtained by training a linear SVM on data $Xp \cup Sn$.

An analogous procedure is applied to generate $C_{s,s'}$ from pairs $(\mathbf{s}, \mathbf{s}')$ in $Pair_n$.

When applied to all pairs of support vectors in $Pair_p, Pair_n$, LSC produces a sequence of *lsc*'s. Such sequence of classifiers induces a data transformation, called seq_D, which maps example $\mathbf{x}$ in the sequence $seq_D(\mathbf{x})$ of class values $C_{s,s'}(\mathbf{x})$, with $(\mathbf{s}, \mathbf{s}')$ in $Pair_p \cup Pair_n$.

The construction of the sequence of*lsc*'s requires computation that grows quadratically with the number of support vectors. However, this is not a severe problem, since the number of examples, hence of support vectors, is small for this type of data. Furthermore, LSC is applied to each pair of support vectors independently, hence can be executed in parallel.

3.1 Naive Bayes and Optimal Bayes Classification

Naive Bayes (NB) is based on the principle of assigning to a new example the most probable target value, given the attribute values of the example. In order to apply directly NB to the original gene expression data, gene values need to be discretized, since NB assumes discrete-valued attributes. Examples transformed using seq_D contain binary attributes, hence discretization is not necessary.

Let $\mathbf{x}$ be a new example. Suppose $seq_D(\mathbf{x}) = (x_1, \ldots, x_N)$.

First, the prior probabilities p_y of the two target values are estimated by means of the frequency of positive and negative examples occurring in the train set D, respectively. Next, for each attribute value x_i, the probability $P(x_i \mid y)$ of x_i given target value y is estimated as the frequency with which x_i occurs as value of i-th attribute among the examples of D with class value y. Finally, the classification of $\mathbf{x}$ is computed as the y that maximizes the product

$$p_y \prod_{i=1}^{N} P(x_i \mid y).$$

The resulting classifier is denoted by NB-LSC.

Optimal Bayes (OB) classifier is based on the principle of maximizing the probability that a new example is classified correctly, given the available data, classifiers, and prior probabilities over the classifiers.

OB maps example $\mathbf{x}$ to the class that maximizes the weighted sum

$$\sum_{C_{s,s'}} w_{s,s'} I(C_{s,s'}(\mathbf{x}) = y),$$

where $w_{s,s'}$ is the accuracy of $C_{s,s'}$ over D, and I is the indicator function, which returns 1 if the test contained in its argument is satisfied and 0 otherwise. The resulting classifier is denoted by OB-LSC.

4 Datasets

There are several microarray datasets from published cancer gene expression studies, including leukemia cancer dataset, colon cancer dataset, lymphoma dataset, breast cancer dataset, NCI60 dataset, ovarian cancer, and prostate

Table 1. Datasets description

Name	Tot	Positive	Negative	Genes
Colon	62	22	40	2000
Leukemia	72	25	47	7129
Lymphoma	58	26	32	7129
Ovarian	54	30	24	1536

dataset. Among them four datasets are used in this paper, available e.g. at `http://sdmc.lit.org.sg/GEDatasets/Datasets.html`. The first and third dataset contain samples from two variants of the same disease, the second and last dataset consist of tumor and normal samples of the same tissue. Table 1 shows input dimension and class sizes of the datasets. The following short description of the datasets is partly based on [7].

4.1 Leukemia

The Leukemia dataset consists of 72 samples: 25 samples of acute myeloid leukemia (AML) and 47 samples of acute lymphoblastic leukemia (ALL). The source of the gene expression measurements is taken from 63 bone marrow samples and 9 peripheral blood samples. Gene expression levels in these 72 samples are measured using high density oligonucleotide microarrays [2]. Each sample contains 7129 gene expression levels.

4.2 Colon

The Colon dataset consists of 62 samples of colon epithelial cells taken from colon-cancer patients. Each sample contains 2000 gene expression levels. Although the original data consists of 6000 gene expression levels, 4000 out of 6000 were removed based on the confidence in the measured expression levels. 40 of 62 samples are colon cancer samples and the remaining are normal samples. Each sample is taken from tumors and normal healthy parts of the colons of the same patients and measured using high density oligonucleotide arrays [1].

4.3 Lymphoma

B cell diffuse large cell lymphoma (B-DLCL) is a heterogeneous group of tumors, based on significant variations in morphology, clinical presentation, and response to treatment. Gene expression profiling has revealed two distinct tumor subtypes of B-DLCL: germinal center B cell-like DLCL and activated B cell-like DLCL [19]. Lymphoma dataset consists of 24 samples of germinal center B-like and 23 samples of activated B-like.

4.4 Ovarian

Ovarian tissue from 30 patients with cancer and 23 without cancer were analyzed for mRNA expression using glass arrays spotted for 1536 gene clones. Attribute

i of patient j is the measure of the mRNA expression of the i-th gene in that tissue sample, relative to control tissue, with a common control employed for all experiments [22].

5 Numerical Experiments

The two classifiers NB-LSC and OB-LSC, described in Section 3.1, are applied to the four gene expression datasets the baseline SVM-RFE algorithm. In all experiments the same value of the SVM parameter $C = 10$ is used, while the number of selected genes was set to 30 for the lymphoma dataset and 50 for all other datasets. These values are chosen by means of cross-validation applied to the training set.

Because of the small size of the datasets, Leave One Out Cross Validation (LOOCV) is used to estimate the predictive performance of the algorithms [12].

Table 2 reports results of LOOCV. They indicate a statistically significant improvement of OB-LSC over the baseline SVM-RFE classifier, and a gain in stability, indicated by lower standard deviation values. In particular, on the ovarian and leukemia datasets both NB-LSC and OB-LSC achieve perfect classification.

Moreover, while the performance of SVM on the Lymphoma dataset is rather scare (possibly due to the fact that we did not scale the data), OB-LSC obtains results competitive to the best results known (see Table 3).

Table 3 reports results of OB-LSC and the best result among those contained nine papers on tumor classification and feature selection using different machine learning methods [7]. Note that results reported in this table have been obtained using different cross-validation methods, mainly by repeated random partitioning the data into train and test set using 70 and 30 % of the data, respectively.

Table 2. Results of LOOCV: average sensitivity, specificity and accuracy (with standard deviation between brackets)

Method Dataset	Sensitivity	Specificity	Accuracy
SVM-RFE Colon	0.90 (0.3038)	1.00 (0.00)	0.9355 (0.2477)
NB-LSC	0.75 (0.4385)	1.00 (0.00)	0.8387 (0.3708)
OB-LSC	0.90 (0.3038)	1.00 (0.00)	0.9355 (0.2477)
SVM-RFE Leukemia	0.96 (0.20)	1.00 (0.00)	0.9861 (0.1179)
NB-LSC	1.00 (0.00)	1.00 (0.00)	1.00 (0.00)
OB-LSC	1.00 (0.00)	1.00 (0.00)	1.00 (0.00)
SVM-RFE Ovarian	0.7000 (0.4661)	0.9583 (0.2041)	0.8148 (0.3921)
NB-LSC	1.00 (0.00)	1.00 (0.00)	1.00 (0.00)
OB-LSC	1.00 (0.00)	1.00 (0.00)	1.00 (0.00)
SVM-RFE Lymphoma	0.6923 (0.4707)	0.6562 (0.4826)	0.6724 (0.4734)
NB-LSC	1.00 (0.00)	0.6562 (0.4826)	0.8103 (0.3955)
OB-LSC	1.00 (0.00)	0.8750 (0.3360)	0.9310 (0.2556)

Table 3. Comparison of results with best average accuracy reported in previous papers on tumor classification. The type of classifiers considered in the paper are given between brackets. An entry '-' means that the corresponding dataset has not been considered

	Colon	Leukemia	Lymphoma	Ovarian
Furey et al (SVM)	0.90	0.94	-	-
Li et al 00 (Logistic regression)	-	0. 94	-	-
Li et al 01 (KNN)	0.94	-	0.94	-
Ben-Dor et al (Quadratic SVM, 1NN, AdaBoost)	0.81	0.96	-	-
Dudoit et al (1NN, LDA, BoostCART)	-	0.95	0.95	-
Nguyen et al (Logistic discriminant, QDA)	0.94	0.96	0.98	-
Cho et al (Ensemble SVM, KNN)	0.94	0.97	0.96	-
Liu et al 04 (Ensemble NN)	0.91	-	-	-
Dettling et al 03 (Boosting)	0.85	-	-	-
OB-LSC	0.94	1.00	0.93	1.00

Because the resulting estimate of predictive performance may be more biased than the one of LOOCV [12], those results give only an indication for comparing the methods. Only the results on the colon dataset from Liu et al 04 and Dettling et al 03 [9, 3] are obtained using LOOCV. The methods proposed in these latter papers use boosting and bagging, respectively. The results they obtain seem comparable to OB-LSC.

The results indicate that OB-LSC is competitive with most recent classification techniques for this task, including non-linear methods.

6 Conclusion

This paper introduced an approach that improves predictive performance and stability of linear SVM for tumor classification with gene expression data on four gene expression datasets.

We conclude with two considerations on research issues still to be addressed. Our approach is at this stage still an heuristic, and needs further experimental and theoretical analysis. In particular, we intend to analyze how performance is related to the number of support vectors chosen to generate *lsc*'s. Moreover, we intend to investigate the use of this approach for feature selection, for instance whether the generated *lsc*'s can be used for ensemble feature ranking [16].

References

1. U. Alon, N. Barkai, and D.A. Notterman et al. Broad patterns of gene expression revealed by clustering analysis of tumor and normal colon tissues probed by oligonucleotide arrays. *PNAS*, 96:6745–50, 1999.
2. A. Ben-Dor, L. Bruhn, and N. Friedman et al. Tissue classification with gene expression profiles. *Journal of Computational Biology*, 7:559–584, 2000.

3. B.Liu, Q. Cui, and T.Jiang et al. A combinational feature selection and ensemble neural network method for classification of gene expression data. *BMC Bioinformatics*, 5(136), 2004.
4. L. Breiman. Bagging predictors. *Machine Learning*, 24(2):123–140, 1996.
5. L. Breiman. Arcing classifiers. *The Annals of Statistics*, 26(3):801–849, 1998.
6. M. Brown, W. Grundy, D. Lin, N. Cristianini, C. Sugnet, T. Furey, M. Jr, and D. Haussler. Knowledge-based analysis of microarray gene expression data by using support vector machines. In *Proc. Natl. Acad. Sci.*, volume 97, pages 262–267, 2000.
7. S. Cho and H. Won. Machine learning in DNA microarray analysis for cancer classification. In *Proceedings of the First Asia-Pacific bioinformatics conference on Bioinformatics 2003*, pages 189 – 198. Australian Computer Society, 2003.
8. N. Cristianini and J. Shawe-Taylor. *Support Vector machines.* Cambridge Press, 2000.
9. M. Dettling and P. Buhlmann. Boosting for tumor classification with gene expression data. *BMC Bioinformatics*, 19:1061–1069, 2003.
10. S. Dudoit, J. Fridlyand, and T.P. Speed. Comparison of discrimination methods for the classification of tumors using gene expression data. *Journal of the American Statistical Association*, 97(457), 2002.
11. M.B. Eisen and P.O. Brown. Dna arrays for analysis of gene expression. *Methods Enzymbol*, (303):179–205, 1999.
12. T. Evgeniou, M. Pontil, and A. Elisseeff. Leave one out error, stability, and generalization of voting combinations of classifiers. *Mach. Learn.*, 55(1):71–97, 2004.
13. T. S. Furey, N. Christianini, N. Duffy, D. W. Bednarski, M. Schummer, and D. Hauessler. Support vector machine classification and validation of cancer tissue samples using microarray expression data. *Bioinformatics*, 16(10):906–914, 2000.
14. T. R. Golub, D. K. Slonim, and P. Tamayo et al. Molecular classification of cancer: class discovery and class prediction by gene expression monitoring. *Science*, 286:531–7, 1999.
15. I. Guyon, J. Weston, S. Barnhill, and V. Vapnik. Gene selection for cancer classification using support vector machines. *Mach. Learn.*, 46(1-3):389–422, 2002.
16. K. Jong, J. Mary, A. Cornuejols, E. Marchiori, and M. Sebag. Ensemble feature ranking. In *Proceedings Eur. Conference on Machine Learning and Principles and Practice of Knowledge Discovery in Databases, ECML/PKDD 2004*, 2004.
17. J. Khan, J. S. Wei, M. Ringner, and et al. Classification and diagnostic prediction of cancers using gene expression profiling and artificial neural networks. *Nature Medicine*, 7:673–679, 2001.
18. Li, Darden, Weinberg, Levine, and Pedersen. Gene assessment and sample classification for gene expression data using a genetic algorithm/k-nearest neighbor method. *Combinatorial Chemistry and High Throughput Screening*, 4(8):727–739, 2001.
19. I. Lossos, A. Alizadeh, and M. Eisen et al. Ongoing immunoglobulin somatic mutation in germinal center b cell-like but not in activated b cell-like diffuse large cell lymphomas. In *Proc Natl Acad Sci U S A*, volume 97, pages 10209–10213, 2000.
20. Tom Mitchell. *Machine Learning.* McGraw Hill, 1997.
21. W.S. Noble. Support vector machine applications in computational biology. In B. Schoelkopf, K. Tsuda, and J.-P. Vert, editors, *Kernel Methods in Computational Biology*, pages 71–92. MIT Press, 2004.

22. M. Schummer, W. V. Ng, and R. E. Bumgarnerd et al. Comparative hybridization of an array of 21,500 ovarian cdnas for the discovery of genes overexpressed in ovarian carcinomas. *Gene*, 238(2):375–85, 1999.
23. A.C. Tan and D. Gilbert. Ensemble machine learning on gene expression data for cancer classification. *Applied Bioinformatics*, 2((3 Suppl)):75–83, 2003.
24. V.N. Vapnik. *Statistical Learning Theory.* John Wiley & Sons, 1998.
25. Y. Xu, F. M. Selaru, and J. Yin et al. Artificial neural networks and gene filtering distinguish between global gene expression profiles of barrett's esophagus and esophageal cancer. *Cancer Research*, 62(12):3493 – 3497, 2002.

Genes Related with Alzheimer's Disease: A Comparison of Evolutionary Search, Statistical and Integer Programming Approaches

Pablo Moscato[1], Regina Berretta[1], Mou'ath Hourani[1], Alexandre Mendes[1], and Carlos Cotta[2]

[1] Newcastle Bioinformatics Initiative,
School of Electrical Engineering and Computer Science,
The University of Newcastle, Callaghan, NSW, 2308, Australia
[2] Dept. Lenguajes y Ciencias de la Computación, University of Málaga
ETSI Informática, Campus de Teatinos, 29071, Málaga, Spain
moscato@cs.newcastle.edu.au

Abstract. Three different methodologies have been applied to microarray data from brains of Alzheimer diagnosed patients and healthy patients taken as control. A clear pattern of differential gene expression results which can be regarded as a molecular signature of the disease. The results show the complementarity of the different methodologies, suggesting that a unified approach may help to uncover complex genetic risk factors not currently discovered with a single method. We also compare the set of genes in these differential patterns with those already reported in the literature.

1 Introduction

Alzheimer's disease affects ten percent of the population aged over 65 and nearly half of all individuals aged above 85 will experience its effects. Since the first clear onset of the symptoms, a sufferer of Alzheimer's would have an average life span of eight years and occasionally up to two decades. Since this disease is a progressive brain disorder that affects the patient's ability to learn new things, to make judgements, and to accomplish simple daily activities, Alzheimer also has an emotional impact on families. This is aggravated by the fact that the sufferers may have an increasingly complex pattern of changes in personality and behavior as well as increasing anxiety, suspiciousness, agitation and forms of delusions or hallucinations. Memory loss is generally the first and worst symptom, but others include difficulty performing familiar tasks, communication, temporal/spatial disorientation, poor or decreased judgment, difficulty with tasks that require abstract thinking, misplacing objects, and passivity and loss of initiative[1].

After a certain point, people with Alzheimer's generally require 24-hour care. Estimations of total cost abound but differ very little, with the average lifetime

[1] http://www.alz.org/Resources/FactSheets.asp

F. Rothlauf et al. (Eds.): EvoWorkshops 2005, LNCS 3449, pp. 84–94, 2005.

cost of care per patient around US$ 200,000. Alzheimer's is the third most expensive disease in the United States, only second to heart disease and cancer, with the latter being a generic name for many different disease forms. According to figures provided by the Alzheimer's Association, the cost of 24-hour care plus diagnosis, treatment, and paid care costs, is estimated to be US$100 billion annually. In the US, only a small percentage is covered, in almost equal parts by the federal government and the states (US$8.5 billion total), the rest is paid by both patients and their families, putting a severe stress, which apart from the emotional cost, may also compromise their finances. With an increasingly aging population in the developed world, there is a need for more research on the causes of Alzheimer's disease. While a skilled physician would be able to diagnose it with 90 percent accuracy, an early genetic diagnosis of risks would help enormously. Quoting Zaven Khachaturian, *"If we can push back the onset of Alzheimer's for just five years, we can reduce by 50 percent the number of people who get the disease, add years of independent functioning to people's lives, reduce the amount of care they need, and save this country billions of dollars in health-care costs."* [2] Accordingly, more research on the genetic basis of this disease is needed as demographically the picture is not good for developed countries, with a large number of their aging individuals getting the disease thus impacting on society as a whole.

We have conducted an extensive search in the scientific literature to try to identify which genes have already been reported as linked to Alzheimer's disease, resulting in a set of 95 genes. These genes have been identified using public available web search engines as well as PubMed, *Web of Science* and other bibliographic databases. This process did not employ a single software tool, but a thorough investigation and critical reading of articles on the genetics of Alzheimer. For 29 of the 95 genes identified, we have found that there is a microarray data study in the public domain that contains their gene expression in control and Alzheimer's affected brains [3]. A visual inspection of the relative gene expression of this dataset (containing approximately 2,000 genes), clustered with our memetic algorithm [1], clearly shows a pattern of differentiated gene expression in healthy and Alzheimer's affected brains (see Fig. 1). From the set of 29 genes (out of those 95 identified as somewhat related), eight have already been reported in [2] (COX7B, IDI1, MAPK10, PRKCB1, RARS, SMS, WASF1 and YWHAH). The others are: ABCB1 [3], ADAM10 [4], ATOX1 [5], BCL2L2 [6], BRD2 [7], CRH [8], CTCF [9], GSK3B [10], HFE and TF [11], HTR2A [12], LAMC1 [13], NCSTN [14], NRG1 [15], NUMB [16], PRDX2 [17], **PRDX5** [18], PRKR [19], PSEN1 [20], **MAPK14** [21], and VSNL1 [22]. The two in boldface have been found by the methods we will present in this paper and were not reported in [2]. It is clear that a differential pattern of expression exists for these genes between Normal and AD brains, as shown in Fig. 1(b). In

[2] `http://www.fda.gov/fdac/features/1998/398_alz.html`

[3] `http://labs.pharmacology.ucla.edu/smithlab/genome_research_data/index.html`

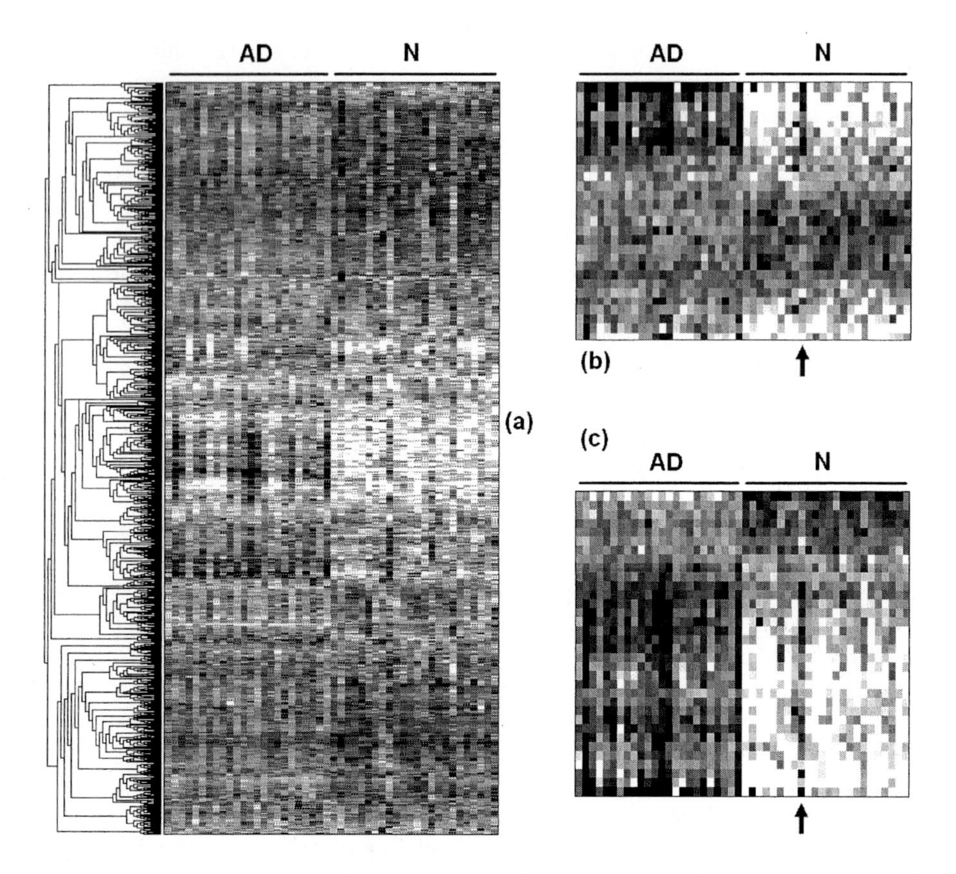

Fig. 1. **(a)** Gene expression of 2,100 genes in both Alzheimer (AD) and normal (N) brains (dataset from [2]). The columns correspond to different voxels (as described in [2]). It is clear that there exist a relatively large number of genes which are differentially expressed within different regions in AD and N brains. **(b)** Gene expression of the 29 genes found in our literature search (from a total of 95 we identified as possibly related) which are also present in the dataset. **(c)** The 34 genes highlighted in Table 3 of Ref [2]. These 34 genes are those of the union of the four subsets (I, II, III, and V) from Fig. 2 and have been obtained using a singular-value decomposition approach [2]. In all figures, we provide a high-quality clustering of the gene expression patterns using the memetic algorithm described in [1]

this paper we present a set of 70 genes which also show correlated patterns that may be useful to understand the genetic risk factors of the disease.

2 Modeling the Gene Subset Selection Problem

In order to model the problem of finding gene subsets of interest, we will proceed in two steps. Firstly, we will introduce the $(\alpha, \beta) - k-$FEATURE SET Problem, so as to provide a combinatorial setting for the target problem. Then we will address how to discretize microarray measurements to obtain a problem instance of the $(\alpha, \beta) - k-$FEATURE SET Problem.

2.1 The $(\alpha, \beta) - k-$Feature Set Problem

The $(\alpha, \beta) - k-$FEATURE SET Problem is a generalization of $k-$FEATURE SET and it has been introduced with the aim of obtaining subsets of features of robust discriminatory power [23]. Its use coupled with standard data mining algorithms has led us to successfully predict the outcome of the 2004 US Presidential election, two months in advance of the actual voting, only based on historical information from previous elections [24]. The problem can be formally defined as follows:

- **Instance**: A set of m examples $X = \{x^{(1)}, \ldots, x^{(m)}\}$, such that for all i, $x^{(i)} = \{x_1^{(i)}, x_2^{(i)}, \ldots, x_n^{(i)}, t^{(i)}\} \in \{0,1\}^{n+1}$, and three integers $k > 0$, and $\alpha, \beta \geqslant 0$.
- **Question:** Does there exist an $(\alpha, \beta) - k$-feature set S, $S \subseteq \{1, \cdots, n\}$, with $|S| \leqslant k$ and such that:
  - for all pairs of examples $i \neq j$, if $t^{(i)} \neq t^{(j)}$ there exists $S' \subseteq S$ such that $|S'| \geqslant \alpha$ and for all $l \in S'$ $x_l^{(i)} \neq x_l^{(j)}$?
  - for all pairs of examples $i \neq j$, if $t^{(i)} = t^{(j)}$ there exists $S' \subseteq S$ such that $|S'| \geqslant \beta$ and for all $l \in S'$ $x_l^{(i)} = x_l^{(j)}$?

We remark that the set S' is not fixed for all pairs of examples, but it is a function of the pair of examples chosen, so in the definition we mean $S' = S'(i, j)$. Obviously, the problem is $NP-$hard as it contains the $k-$Feature Set Problem as special case [25]. Furthermore, the $(\alpha, \beta) - k-$FEATURE SET problem is not likely to be fixed-parameter tractable for parameter k as Cotta and Moscato have recently proved that the $k-$FEATURE SET problem is $W[2]$-complete [26].

We mentioned before that robustness is the goal. Indeed, robust feature identification methods are essential since microarray data measurements are notoriously prone to errors. This robustness comes at a price though. When this problem is used as a modelling tool for pattern recognition, robustness comes with redundancy in the number of features required for discrimination of a pair of examples. This may appear as counter-intuitive at first sight. A large number of approaches in data mining, and particularly in Bioinformatics, are concerned with finding "minimal" cardinality solutions. In the area of microarray data analysis, however, the true requirement is different. A small number of examples, as compared with the number of features, means that by just random chance a certain feature could dichotomize a set of examples. This said, the problem is how to preserve in our solutions a potentially useful set of features that could explain the examples, since they could be left aside due to the requirements of finding a minimal cardinality solution. Given a set of measurements obtained by means of a microarray experiment on m samples/conditions, the 0-1 values for each one of the features would correspond to under- or over-expressed genes respectively after a threshold value is determined.

2.2 Threshold Selection Issues

An instance of the $(\alpha, \beta) - k-$FEATURE SET problem can be obtained once thresholds for discretizing gene-expression values have been set. The associated problem (finding appropriate thresholds given the particular values of α, β and k in the instance sought) can be formalized as follows:

$(\alpha, \beta)-$THRESHOLD SELECTION

- **Input**: A $m \times n$ $\mathbb{R}-$matrix $\tilde{X}$, class identifiers $t^{(i)} \in \mathbb{N}$ for every row i, $1 \leqslant i \leqslant m$, and two integers $\alpha, \beta \geqslant 0$.
- **Question**: Does there exist an array of m thresholds $\theta_1, \cdots, \theta_m$ (i.e., one for each of the rows in $\tilde{X}$) such that each entry in the ith row of $\tilde{X}$ greater than the θ_i is given the value 1, and 0 otherwise, and such that
  1. $\forall i, j$, $t^{(i)} \neq t^{(j)}$, the number of columns where $\tilde{X}_l^{(i)} \neq \tilde{X}_l^{(j)}$ (disagree) is at least α, and
  2. $\forall i, j$, $t^{(i)} = t^{(j)}$, the number of columns where $\tilde{X}_l^{(i)} = \tilde{X}_l^{(j)}$ (agree) is at least β ?

We note that this is a necessary but not a sufficient condition to create a yes-instance of the $(\alpha, \beta) - k-$FEATURE SELECTION problem. Unfortunately, it is unlikely that an efficient algorithm for $(\alpha, \beta)-$THRESHOLD SELECTION would be found as it is $NP-$complete [23]. However, Cotta, Sloper and Moscato have shown that evolutionary search strategies may help in practice to find thresholds allowing $(\alpha, \beta) - k-$feature sets to be found in microarray data in lymphomas, opening the possibility of using the methodology in other domains as well.

3 Complementary Methodologies for Gene Subset Selection

3.1 The Statistical Approach and the Microarray Dataset

The gene expression dataset is obtained from samples of normal and Alzheimer-affected diseased humans (for the complete description see [2]). Samples are obtained from spatially registered voxels (cubes) which produce multiple volumetric maps of gene expression. The technique is analogous to the reconstructed images obtained in biomedical imaging systems. A total of 24 voxel images of coronal hemisections at the level of the hippocampus of both the normal human brain and Alzheimer's disease affected brain were acquired for 2,100 genes. The statistical methods involve the use of a standard singular value decomposition (SVD) analysis. They show the most strongly differentially expressed genes between Alzheimer's affected and normal brains (having p-values $\leqslant 0.05$). Notably, the SVD results allow to produce images which correlate well with the neuroanatomy, including cortex, caudate, and hippocampus. This suggests that this technique will be a useful approach for understanding how the genome, and gene expression, constructs and regulates the brain.

3.2 The Evolutionary Search Approach

The evolutionary method used is similar to the one described in [23]. Therein, the authors present results of an evolution strategy that allowed to find, on a microarray dataset of two different types of diffuse large B-cell lymphoma (each one containing 4 samples, and gene-expression profiles for 2,984 genes), an $\alpha = \beta = k = 100$ feature set. For $\alpha = \beta = 200$ and $\alpha = \beta = 300$, the ES found gene subsets of 227.3 and 360.5 genes on average respectively, 25% smaller gene subsets that those provided by a greedy heuristic.

In this case, we have utilized a (1,10)-ES with binary tournament selection, gaussian mutation with independent self-adaptive stepsizes for each variable, and no recombination. The algorithm evolves adequate thresholds for each gene; for each of these candidate solutions, the ES algorithm generates an $(\alpha, \beta) - k-$FEATURE SELECTION problem instance, and uses a combination of kernelization techniques and greedy heuristics to solve it. The particular dataset we have considered seems to be difficult in practice for this algorithm, due to the fact that the size of this underlying $(\alpha, \beta) - k-$FEATURE SELECTION problem that is being continuously generated and solved scales quadratically with the number of columns. Nevertheless, we have been able to identify several $(10, 10) - k-$feature sets with $17 \leqslant k \leqslant 19$ (see Fig. 3(c)).

3.3 The Integer Programming Approach

For the initial exploratory tests, we have used an standard integer programming (IP) formulation of the $(\alpha, \beta) - k-$FEATURE SET problem as described in [27]. For some values of α and β we have been able to solve the instances to optimality using the CPLEX 9.0 mathematical software package. Treating all voxels' samples in the Alzheimer and control brains as 48 examples of two different classes (24 from each), allows us to find groups of genes differentially expressed in all regions of the brain (and expressed consistently within a class due to the large values of $\beta > 0$ obtained). However, we expect a degree of gene expression variation within different parts of a brain (both AD and control) to be present (due to normal functional differentiation). As a consequence, we adapted the IP formulation to look at the problem from a different perspective.

The aim of our new IP approach is now to find genes that are diffentially expressed in the same voxel in both the AD and control brains. This said, the number of pairs of examples corresponding to different classes drops from 24^2 to just 24 and the number of pairs of examples that belong to the same class drops to zero. This said, the IP model reflects our aim to find $(\alpha) - k-$feature sets with large values of α and small values of k (the parameter β makes no sense here as we are treating any individual example as a member of a different class). This allows us as to find minimal sets of genes that are differentially expressed in both the AD and control brains in the same voxel.

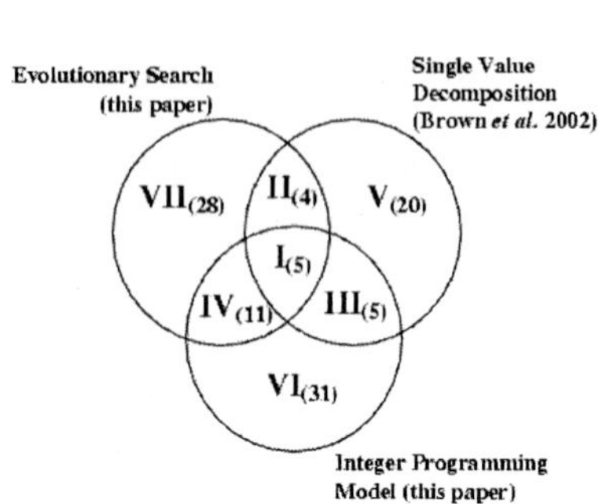

I – DNCI1, KIAA0069, LOC51628, NR1I3, TAF2F
II – IDI1, MAPK10, WASF1, RAP2A
III – ICAP-1A, FOXJ3, KIAA0992, LOC51235, **YWHAH**
IV – HAX1, LOC54460, **PRDX5** + 8 ESTs (Clone IDs 377827, 395436, 669471, 858450, 884653, 1032362, 1161775, 1500241)
V – BICD1, CCS, **COX7B**, DRAP1, DSCR1L1, IDH3A, LIMS1, NFATC3, **PRKCB1**, PSCA, PSCD2, PTPRN2, RAB2, **RARS**, SALL2, SEPW1, **SMS**, TIF1, XPO1, ZNF142
VI – ADD1, ATP6F, CANPX, CYBA, EIF2C1, EIF4B, FLJ11132, FLJ11200, GLG1, GNG10, INSL4, KIAA0154, KIAA0608, **MAPK14**, MCF2, NFIX, THBS1, TPD52, USP16 + 12 ESTs (Clone IDs 246116, 308788, 768324, 824479, 867751, 868188, 1034472, 1291971, 1292501, 1292893, 1493181, 1505783)
VII – APOC4, ATP5G3, FLJ11220, FLJ12895, FLJ20323, GAPDH, IL11RA, KIAA0308, LOC153561, NFKBIB, PPP2R1A, S100A11, SIAH1, SLC2A5, SLC9A6, SMAP, SRI, SRP46, Z39IG + 9 ESTs (Clone IDs 48906, 147192, 462944, 469379, 796548, 813813, 126858, 1493137, 1505240)

Fig. 2. A Venn diagram helps us to present the results of our comparison. We have uncovered a total of 70 genes not reported in [2] obtained from the solution of two different methodolgies, one is a variant of the evolutionary search methodology presented in [23] and the other is a truncated complete anytime algorithm based on the integer programming model discussed in Sect. 3.3 (with $\alpha = 40$ and 52 genes) based on the method presented in [27] (union of the subsets IV, VI, and VII in the diagram). Our solutions also contained 14 genes already reported in [2] (subsets I, II and III). Ten genes from these sets, marked in boldface, have been also linked to Alzheimer's and neurodegenerative diseases (references are provided in Sect. 1)

In addition, in [27] good results were obtained by using an IP model in which we fix the number of features required to be in the solution with the objective of finding those features that maximize the "coverage". The coverage represents the sum of the cardinalities of all the sets $S'(i,j)$ as defined in Sect. 2.1. Due to our good experience with this model, we fixed $k = 52$ and we have been able to find a feature set with $\alpha = 40$ and maximum coverage (998). The thresholds were fixed in this case, unlike the ES methodology, at the median value of the expression of each individual gene on the 48 samples.

4 Results

The main results are described with the aid of Fig. 2 and Fig. 3 and their accompanying captions. Another result worth mentioning is that the combined use of these three methodologies has uncovered that, in the union of all the genes (see Fig. 3) that provide a clear pattern of differential gene expression, there exists a peculiarity in the gene expression of area D2 of the normal brain (following the grid labeling used of [2]) [4]. The gene expression pattern (for this subset of the genes) for region D2 in normals seems to be highly similar to the pattern of activities for other regions of the Alzheimer's brain. Puzzled by this

[4] http://labs.pharmacology.ucla.edu/smithlab/genome_research_data/voxelgrid2.htm

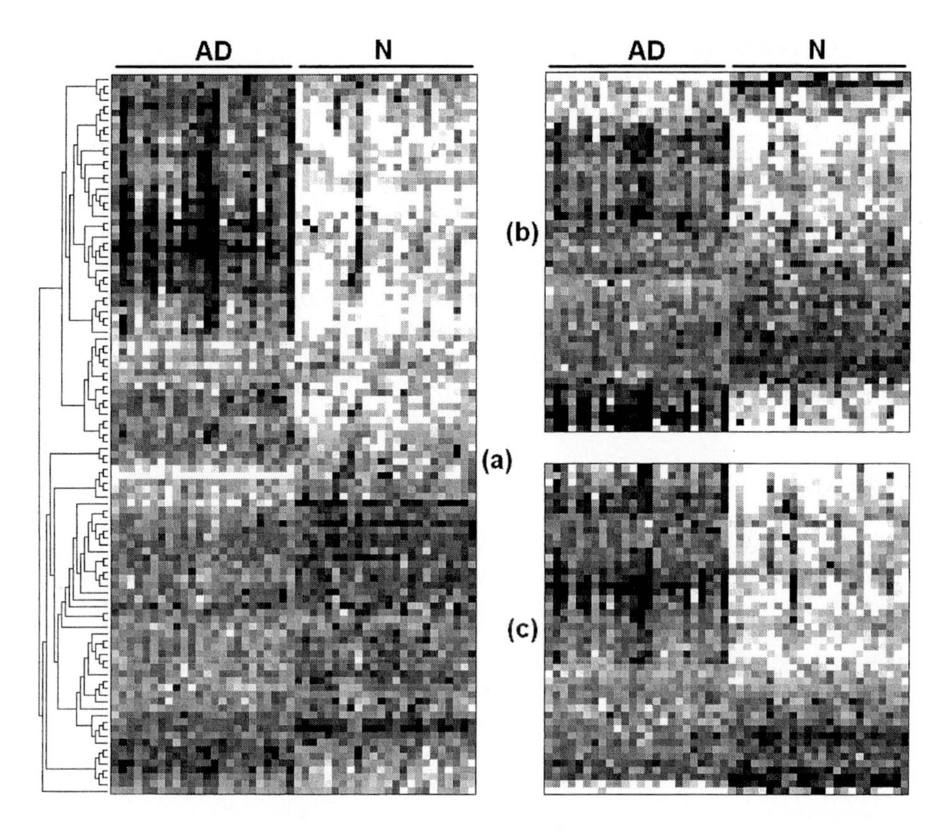

Fig. 3. **(a)** A clustering using the memetic algorithm of [1] of the union of all the genes reported in Fig. 2. **(b)** Clustering of the 52 genes found with the truncated exact search method based on the integer programming model (subsets I, III, IV, and VI of Fig. 2). **(c)** Clustering of the 48 genes found with the evolutionary search strategy proposed in this paper (subsets I, II, IV, and VII of Fig. 2). The ES finds appropriate thresholds that will allow an $\alpha = \beta = 10$-feature set to exist

fact, we conducted an experiment where we performed a hierarchical clustering of the columns as well as genes and indeed the pattern of activities for region D2 in normals was clustered together to those of Alzheimer's, though it appeared as an outgroup. This is intriguing, as this differential pattern of activity for this voxel is also clear in Fig. 1(b) and Fig. 1(c) (an arrow indicates the voxel D2 in these figures) where a distinctive dark column clearly stands out within a pattern of under expressed genes in the normal brain. All the methods revealed a similar characteristic and it could be visually appreciated even with our clustering of the entire set of 2,100 genes (Fig. 1(a)). While more research is needed, we believe that new analysis on this area, as differential to other areas in the normal and AD brains, may help to provide a bridge between genomics, functional differentiation and disease.

5 Conclusions

The combination of results obtained in separately from evolutionary search, statistical and integer programming methods, allows the identification of a large number of genes differentially expressed in normal and Alzheimer's affected brain. Our analysis show that there are at least one hundred candidates for further exploration which have strong correlations with those found with our methods. This issue should be further tested in additional, specifically designed, microarray experiments. Also, the combined methodology would encompass the three methods so far applied to this dataset, plus a user-defined bias based on annotated information from biologists (and the biomedical literature) will be the subject of further studies.

References

1. Cotta, C., Mendes, A., Garcia, V., França, P., Moscato, P.: Applying memetic algorithms to the analysis of microarray data. In Raidl, G., *et al.*, eds.: Applications of Evolutionary Computing. Volume 2611 of Lecture Notes in Computer Science. Springer-Verlag, Berlin (2003) 22–32
2. Brown, V., Ossadtchi, A., Khan, A., Cherry, S., Leahy, R., Smith, D.: High-throughput imaging of brain gene expression. Genome Research **12** (2002) 244–254
3. Vogelgesang, S., Cascorbi, I., Schroeder, E., Pahnke, J., Kroemer, H., Siegmund, W., Kunert-Keil, C., Walker, L., Warzok, R.: Deposition of Alzheimer's beta-amyloid is inversely correlated with P-glycoprotein expression in the brains of elderly non-demented humans. Pharmacogenetics **12** (2002) 535–541
4. Colciaghi, F., Borroni, B., Pastorino, L., Marcello, E., Zimmermann, M., Cattabeni, F., Padovani, A., Di Luca, M.: Alpha-secretase ADAM10 as well as alpha-APPs is reduced in platelets and CSF of Alzheimer disease patients. Journal of Molecular Medicine **8** (2002) 67–74
5. Bellingham, S., Lahiri, D., Maloney, B., La Fontaine, S., Multhaup, G., Camakaris, J.: Copper depletion down-regulates expression of the Alzheimer's disease amyloid-beta precursor protein gene. Journal of Biological Chemistry **279** (2004) 20378–20386
6. Zhu, X., Wang, Y., Ogawa, O., Lee, H., Raina, A., Siedlak, S., Harris, P., Fujioka, H., Shimohama, S., Tabaton, M., Atwood, C., Petersen, R., Perry, G., Smith, M.: Neuroprotective properties of BCL-W in Alzheimer disease. Journal of Neurochemistry **89** (2004) 1233–1240
7. Johnson, N., Bell, P., Jonovska, V., Budge, M., Sim, E.: NAT gene polymorphisms and susceptibility to Alzheimer's disease: identification of a novel NAT1 allelic variant. BMC Medical Genetics **5** (2004) 6
8. Rehman, H.: Role of CRH in the pathogenesis of dementia of Alzheimer's type and other dementias. Current Opinion in Investigational Drugs **3** (2002) 1637–1642
9. Burton, T., Liang, B., Dibrov, A., Amara, F.: Transforming growth factor-beta-induced transcription of the Alzheimer beta-amyloid precursor protein gene involves interaction between the CTCF-complex and Smads. Biochemical and Biophysical Research Communications **295** (2002) 713–723
10. Bhat, R., Budd, S.: GSK3beta signalling: casting a wide net in Alzheimer's disease. Neurosignals **11** (2002) 251–261

11. Robson, K., Lehmann, D., Wimhurst, V., Livesey, K., Combrinck, M., Merryweather-Clarke, A., Warden, D., Smith, A.: Synergy between the C2 allele of transferrin and the C282Y allele of the haemochromatosis gene (HFE) as risk factors for developing Alzheimer's disease. Journal of Medical Genetics **41** (2004) 261–265
12. Assal, F., Alarcon, M., Solomon, E., Masterman, D., Geschwind, D., Cummings, J.: Association of the serotonin transporter and receptor gene polymorphisms in neuropsychiatric symptoms in Alzheimer disease. Archives of Neurology **61** (2004) 1249–1253
13. Palu, E., Liesi, P.: Differential distribution of laminins in Alzheimer disease and normal human brain tissue. Journal of Neuroscience Research **69** (2002) 243–256
14. Dermaut, B., Theuns, J., Sleegers, K., Hasegawa, H., Van den Broeck, M., Vennekens, K., Corsmit, E., George-Hyslop, P., Cruts, M., Van Duijn, C., Van Broeckhoven, C.: The gene encoding nicastrin, a major gamma-secretase component, modifies risk for familial early-onset Alzheimer disease in a Dutch population-based sample. The American Journal of Human Genetics **70** (2002) 1568–1574
15. Chaudhury, A., Gerecke, K., Wyss, J., Morgan, D., Gordon, M., Carroll, S.: Neuregulin-1 and erbB4 immunoreactivity is associated with neuritic plaques in Alzheimer disease brain and in a transgenic model of Alzheimer disease. Journal of Neuropathology and Experimental Neurology **62** (2003) 42–54
16. Chan, S., Pedersen, W., Zhu, H., Mattson, M.: NUMB modifies neuronal vulnerability to amyloid beta-peptide in an isoform-specific manner by a mechanism involving altered calcium homeostasis: implications for neuronal death in Alzheimer's disease. Neuromolecular Medicine **1** (2002) 55–67
17. Kim, S., Fountoulakis, M., Cairns, N., Lubec, G.: Protein levels of human peroxiredoxin subtypes in brains of patients with Alzheimer's disease and Down syndrome. Journal of Neural Transmission. Supplementum **61** (2001) 223–235
18. Krapfenbauer, K., Engidawork, E., Cairns, N., Fountoulakis, M., Lubec, G.: Aberrant expression of peroxiredoxin subtypes in neurodegenerative disorders. Brain Research **967** (2003) 152–160
19. Peel, A., Bredesen, D.: Activation of the cell stress kinase PKR in Alzheimer's disease and human amyloid precursor protein transgenic mice. Neurobiology of Disease **14** (2003) 52–62
20. Goldman, J., Reed, B., Gearhart, R., Kramer, J., Miller, B.: Very early-onset familial Alzheimer's disease: a novel presenilin 1 mutation. International Journal of Geriatric Psychiatry **17** (2002) 649–651
21. Johnson, G., Bailey, C.: The p38 MAP kinase signaling pathway in Alzheimer's disease. Experimental Neurology **183** (2003) 263–268
22. Schnurra, I., Bernstein, H., Riederer, P., Braunewell, K.: The neuronal calcium sensor protein VILIP-1 is associated with amyloid plaques and extracellular tangles in Alzheimer's disease and promotes cell death and tau phosphorylation in vitro: a link between calcium sensors and Alzheimer's disease? Neurobiology of Disease **8** (2001) 900–909
23. Cotta, C., Sloper, C., Moscato, P.: Evolutionary search of thresholds for robust feature selection: application to microarray data. In Raidl, G., *et al.*, eds.: Proc. of the 2nd European Workshop in Evolutionary Computation and Bioinformatics (EvoBIO-2004), Universidade de Coimbra, Portugal. Volume 3005 of Lecture Notes in Computer Science., Springer (2004) 31–40
24. Moscato, P., Mathieson, L., Mendes, A., Berretta, R.: The Electronic Primaries: Predicting the U.S. presidency using feature selection with safe data reduction. In: (ACSC 2005), Newcastle, Australia, Australian Computer Society (2005)

25. Davies, S., Russell, S.: NP-completeness of searches for smallest possible feature sets. In Greiner, R., Subramanian, D., eds.: AAAI Symposium on Intelligent Relevance, New Orleans, AAAI Press (1994) 41–43
26. Cotta, C., Moscato, P.: The k-FEATURE SET problem is $W[2]$-complete. Journal of Computer and Systems Science **67** (2003) 686–690
27. Berretta, R., Mendes, A., Moscato, P.: Integer programming models and algorithms for molecular classification of cancer from microarray data. In: Proceedings of the 28th Australasian Computer Science Conference (ACSC 2005), Newcastle, Australia. (2005)

Syntactic Approach to Predict Membrane Spanning Regions of Transmembrane Proteins

Koliya Pulasinghe[1] and Jagath C. Rajapakse[2]

[1] Sri Lanka Institute of Information Technology, Sri Lanka
[2] BioInformatics Research Centre, Nanyang Technological University, Singapore
koliya@sliit.lk, asjagath@ntu.edu.sg

Abstract. This paper exploits "biological grammar" of transmembrane proteins to predict their membrane spanning regions using hidden Markov models and elaborates a set of syntactic rules to model the distinct features of transmembrane proteins. This paves the way to identify the characteristics of membrane proteins analogous to the way that identifies language contents of speech utterances by using hidden Markov models. The proposed method correctly predicts 95.24% of the membrane spanning regions of the known transmembrane proteins and correctly predicts 79.87% of the membrane spanning regions of the unknown transmembrane proteins on a benchmark dataset.

1 Prediction of Membrane Spanning Regions of the Transmembrane Proteins

Transmembrane Proteins (TMPs), which traverse the phospholipid bi-layer of the membrane one to many times, as illustrated in Fig. 1 and Fig. 2, are integral membrane proteins, i.e., proteins which are attached to the cell membrane to keep their hydrophobic regions intact with aqueous cytosol. Thus, they make a channel between cytosome and extracellular environment, which transports various ions and proteins to and from cytosol. In addition, TMPs take part in vital cell functions such as cleavage of substances for metabolic functions, functioning as receptors, recognition and mediation in specific cell signaling, and participation in intercellular communication. Therefore, they are good therapeutic targets and the knowledge of the topography of the TMPs is of paramount importance to the design new drugs.

TMPs with experimentally verified structures are limited to about 1% of the total entries in most of the protein databanks though they amount to 20-30% of all open reading frames of the genomic sequences of several organisms [1][2]. Verifying TMP structures using experimental methods, such as X-ray crystallography and nuclear magnetic resonance spectroscopy, is not only expensive but also requires a lot of efforts due to the difficulties in protein expression, purification, and crystallization. Especially, TMPs have hydrophobic regions, which are buried inside the membrane, i.e., membrane spanning regions (MSRs), to

F. Rothlauf et al. (Eds.): EvoWorkshops 2005, LNCS 3449, pp. 95–104, 2005.

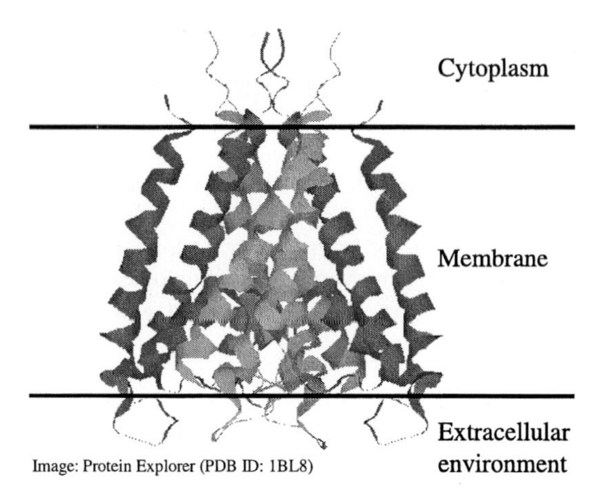

Fig. 1. Typical structure of a transmembrane protein: Potassium channel protein (KCSA) from *Streptomyces lividans*

keep the hydrophobic residues intact with aqueous cytosol and extracellular environment, and do not dissolve properly in aqueous solvents in the process of purification. Consequently, the prediction of MSRs of the TMPs became a classical problem in bioinformatics. Experimentally verified TMPs have two different motifs: membrane spanning α-helix bundles and β-barrels. Usually the α-helix bundles (Fig. 1) are predominant [3]. This paper focuses its attention on predicting the α-helix bundles of the TMPs.

Early MSR prediction methods of the TMPs were based on the hydrophobicity analysis of the constituent amino acids [4][5][6]. Because, hydrophobicity values of the amino acids in MSRs are relatively high compared to the other regions. As illustrated in Fig. 2, high presence of Isoleucine, Valine, and Leucine,

```
MLYGF SGVIL QGAIV TLELA LSSVV LAVLI GLVGA GAKLS
ooooo ooooo ooMMM MMMMM MMMMM MMMMM MMMii iiiii

QNRVT GLIFE GYTTL IRGVP DLVLM LLIFY GLQIA LNVVT
iiiii iiiii iiiii iiiMM MMMMM MMMMM MMMMM MMMMo

DSLGI DQIDI DPMVA GIITL GFIYG AYFTE TFRGA FMAVP
ooooo ooMMM MMMMM MMMMM MMMMM MMMii iiiii iiiii

KGHIE AATAF GFTHG QTFRR IMFPA MMRYA LPGIG NNWQV
iiiii iiiii iiiii iiiii iiiii iiiii iiMMM MMMMM

ILKAT ALVSL LGLED VVKAT QLAGK STWEP FYFAV VCGLI
MMMMM MMMMM MMMoo ooooo ooooo ooooo ooooM MMMMM

YLVFT TVSNG VLLLL ERRYS VGVKR ADL
MMMMM MMMMM MMMMM iiiii iiiii iii
```

Fig. 2. Amino acid sequence and topography information of Histidine transport system permease protein (hisQ) of *Salmonella typhimurium*: `o`, `M`, `i` indicate outer (extracellular), membrane, and inner (cytoplasmic) residues respectively

which are relatively high hydrophobic amino acids according to Kyte-Doolittle hydrophobicity indices [4], can be observed in MSRs (denoted by M's). Accordingly, frequent occurrences of highly hydrophobic amino acids is a good guess for detecting MSRs. This technique is employed in hydropathy plots, an early technique that is still popular in recognizing the MSRs in TMPs [4][7]. Among more recent methods, which predict the topography of transmembrane proteins, hidden Markov model (HMM) based methods claims the highest accuracy [8]. Among them, TMHMM [2][9], HMMTOP [10], and MEMSAT [11] can predict the membrane bounded region of the transmembrane proteins upto 65% to 80% accuracy [8]. The above methods are different due to the structure of the HMMs, i.e., the domains and segments that the HMM represents, and the training method used.

Among the non-HMM methods, PHDhtm predicts MSRs of TMPs by using an artificial neural network (ANN) [12]. A special feature of the PHDhtm is that the ANN learns the patterns of the evolutionary information (homology). In Toppred, the approach combines hydrophobicity analysis and positive inside rule to predict the putative transmembrane helices [6]. A general dynamic programming-like algorithm, MaxSubSeq (stands for Maximal Sub-Sequence), optimizes the MSRs predicted by other methods [13]. An evaluation of methods for the prediction of MSRs can be found in [8]. Protein sequences of the TMPs verified by the imperial methods can be found in several databases such as MPtopo database [14], TMPDB [15], and TMHMM site [2]. In TMHMM, a state was designed to absorb the properties of one residue except in self-looping globular state. All other states are designed without self-transition probabilities. Contrary to that, in HMMTOP, each characteristic region is represented by a self-looping single state. Approach taken in the proposed method used moderate number of states to represent various characteristic regions. This approach is motivated by the fact that each turn of the helix in MSR consist of 3-4 residues. Accordingly, a state is designed to represent one turn of a α-helix rather than a one residue, as in TMHMM, or one characteristic segment, as in HMMTOP, of the TMPs. Length of an MSR is ranging from 15 residues to 30 residues.

The our approach to MSR predication of TMP is also based on HMMs. Unlike previous approaches, in our HMM model (see Fig. 4), self transitions and transition between every other states can align different length MSRs in the training process as well as in the recognition process. The proposed method correctly predicts 95.24% of the MSRs of the known TMPs and correctly predicts 79.87% of the MSRs of the unknown TMPs on a bench mark dataset.

The organization of this paper is as follows. In Section 2, the syntactic rules derived by observing the various segments of the TMPs are described as a syntactic network where each HMM model is aimed at recognizing an allowable segment combinations of the TMPs. In Section 3, we described the HMMs and training algorithm based on Viterbi segmentation. Section 4 describes the data used in training and testing the proposed method along with results obtained. Finally, a brief discussion about the proposed method and the future directions are given in Section 5.

2 Syntactic Rules of the "Biological Grammar" of TMPs

The presence of alternate sequences of *i*nner (i.e., inside or cytoplasmic), *m*embrane spanning, and *o*uter (outside or extra-cytoplasmic) regions of the transmembrane proteins follows a simple rule of grammar [16]. These regions have unique features inherent to them and do not occur randomly. The syntactical rules of these occurrences of different regions are derived and given below in Fig. 3. The establishment of these rules has great importance to our study, which follows a similar approach that used to identify the language contents of the unknown speech utterances. TMPs with unknown regional boundaries are analogous to unknown speech utterances.

The following symbols lay down the syntactic rules in the biological grammar:

| denotes alternatives
[ ] encloses options
{ } denotes zero or more repetitions
⟨ ⟩ denotes one or more repetitions
`$var` denotes a variable word.

The two different orientations, outer-membrane-inner (i.e., *omi*) and inner-membrane-outer (i.e., *imo*), with respect to the cell membrane can be observed in the helix core of the MSRs and are defined them as separate literals. Inner and outer residue sequences can be observed in different lengths. They are categorized into three groups, each according to their length. As an example, in inner loops, the literal "*i*" represents a protein sequence with 1-6 residues. The literal "*ii*" represents protein sequences with 7-20 residues, while the literal "*iii*" represent the very long protein sequences with more than 20 residues. Same procedure is applied in defining literals "*o*", "*oo*", and "*ooo*". Accordingly syntactical rules governing on possible TMP configuration can be symbollically described as follows:

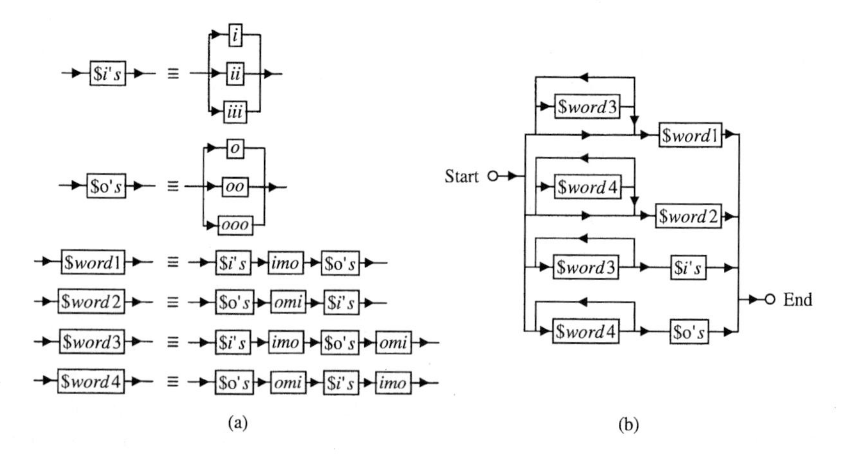

Fig. 3. A graphical illustration of characteristic regions of transmembrane proteins

$\texttt{\$i's} = i \mid ii \mid iii$;
$\texttt{\$o's} = o \mid oo \mid ooo$;
$\texttt{\$word1} = \texttt{\$i's}\ imo\ \texttt{\$o's}$;
$\texttt{\$word2} = \texttt{\$o's}\ omi\ \texttt{\$i's}$;
$\texttt{\$word3} = \texttt{\$i's}\ imo\ \texttt{\$o's}\ omi$;
$\texttt{\$word4} = \texttt{\$o's}\ omi\ \texttt{\$i's}\ imo$;

A pictorial view of these syntactic rules is shown in Fig. 3(a). As illustrated in the Fig. 3(b), any continuous path from start node to end node, along the direction of arrows, generate a possible segment sequence of any TMP. According to our symbolic notations, syntax of any TMP can be given by:

$$(\langle \texttt{\$word3} \rangle\ \texttt{\$i's} \mid \langle \texttt{\$word4} \rangle\ \texttt{\$o's} \mid \{\texttt{\$word3}\}\ \texttt{\$word1} \mid \{\texttt{\$word4}\}\ \texttt{\$word2})$$

These grammatical rules can generate numerous syntactically correct TMPS, as illustrated below. To derive the following sequences, the word network shown in Fig. 3(b) can be used. According to our interpretation, these are possible topological structures of the TMPs, where each literal represents a characteristic feature of a TMP segment:

- *i imo o omi i imo oo omi ii*
- *ooo omi iii*
- *ii imo ooo omi i imo oo omi iii imo oo omi i*
- *oo omi i imo o omi ii imo ooo*
- *i imo o omi i imo o*

A set of HMMs to represent these literals are described in the Section 3.

3 Methodology

Several HMMs are defined, in which each HMM represent a literal, e.g., *imo*, described in the previous section. A special kind of HMM called left-to-right HMM is defined as shown in the Fig. 4 with the intention that all HMMs can be tied parallelly by using first state and last state, to make a single large HMM. The motive behind is that the combination of a giant HMM and syntactical networks described above can be used to recognize unknown segments of a TMP by training and using testing algorithm as described in [18].

3.1 Definition of HMMs

In our design, each literal is designed by a separate HMM; all HMMs share the same configuration as illustrated in Fig. 4. In this type of HMMs, no transitions are allowed to the states whose indices are lower than the current state. In what follows, we give a definition for the HMM to be used, by using the same notation used in Rabinar's seminal paper [19].

1) N: the number of states in the model. We denote the set of individual states as $S = \{S_1, S_2, \ldots, S_N\}$, and the state at site t (or tth observation or tth residue) as q_t.

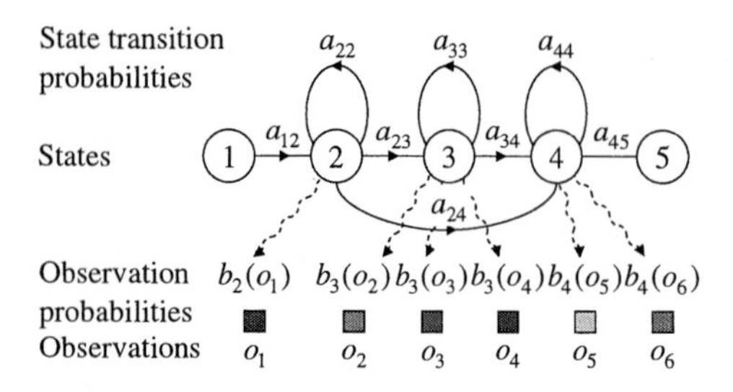

Fig. 4. A left-to-right hidden Markov model

2) M: the number of distinct observation symbols per state, i.e., the number of amino acids (#20) in our study. Though we are not interested about the physical output of the system, to model the system, this must be taken into consideration. We denote the set of individual residues as $R = \{r_1, r_2, \ldots, r_M\}$.

3) State transition probability distribution $A = \{a_{ij}\}_{N \times N}$ where,

$$a_{ij} = P\left[q_{t+1} = S_j \mid q_t = S_i\right], \qquad 1 \leq i, j \leq N. \tag{1}$$

In left-to-right HMM model, the state transition coefficients have the property

$$a_{ij} = 0, \qquad j < i \tag{2}$$

That is, no transitions are allowed to states whose indices are lower than the current state. It should be also noted that, for the last state in a left-to-right model, the state transition coefficients are specified as

$$a_{NN} = 1 \tag{3}$$

$$a_{Ni} = 0, \quad i < N. \tag{4}$$

4) The observation symbol probability distribution in state j, $B = \{b_j(k)\}$, where

$$b_j(k) = P\left[r_k \mid q_t = S_j\right], \qquad \begin{array}{l} 1 < j < N \\ 1 \leq k \leq M \end{array} \tag{5}$$

5) The initial state distribution $\pi = \{\pi_i\}$ where

$$\pi_i = \begin{cases} 0, & i \neq 1 \\ 1, & i = 1 \end{cases} \tag{6}$$

Once parameters are estimated using a proper algorithm, this HMM can generate observation sequence $O = (O_1 O_2 \ldots O_T)$, where each observation O_t is the residue at site t, and T is the number of observations in the sequence.

3.2 HMM Parameter Estimation

Parameter estimation of the HMMs is done by Viterbi alignments [20].To initialize the model parameters Viterbi training is replaced by a uniform segmentation, i.e., each training observation is divided into N equal segments. In Viterbi training, each training sequence is segmented using a state alignment procedure which results from maximizing

$$\phi_N(T) = \max_i \{\phi_i(T) a_{iN}\} \tag{7}$$

for $1 < i < N$ where

$$\phi_j(t) = \max_i \{\phi_i(t-1) a_{ij}\} b_j(r_t) \tag{8}$$

with initial conditions given by

$$\phi_1(1) = 1 \tag{9}$$

$$\phi_j(1) = a_{1j} b_j(r_1). \tag{10}$$

for $1 < j < N$.

If A_{ij} represents the total number of transitions from state i to state j and $b_i(k)$ represents the observation probabilities of emitting symbol k in state i, by performing the above maximization, the transition probabilities can be estimated from the relative frequencies:

$$\hat{a}_{ij} = \frac{A_{ij}}{\sum_{k=2}^{N} A_{ik}} \tag{11}$$

$$\hat{b}_i(k) = \frac{\sum^{N}_{\substack{k=2 \\ s.t. O_t = r_k}} A_{ik}}{\sum_{k=2}^{N} A_{ik}} \tag{12}$$

As a by-product of above calculation the maximum likelihood $\hat{P}(O|M)$ is given by Eq. (7). The above process can be iteratively carried out until the change of the maximum likelihood between two consecutive iteration reached to an acceptable level.

4 Experiments

In this section, we demonstrate the accuracy and efficacy of the proposed approach, using the dataset that used in training TMHMM [2]. And for the testing, 73 TMPs unknown to the system is extracted from dataset C, which contribute maximum number of unknown proteins to the comparison of different methods including TMHMM 2.0, TMHMM 1.0, HMMTOP, and MEMSAT 1.5 [21]. The labeled data was used to estimate the parameters of each HMM separately. The number of states in each literal, which denotes an HMM, is given in the Table 1.

After training separately, all HMMs are tied parallelly by using first state and the last state to make a single large HMM. The combination of this giant HMM and a syntactical network described in Section 2 above is used to recognize unknown segments of TMPs by using a token passing algorithm described in [18].

Table 1. The number of states in each literal in the training dataset

Literal	Number of States
imo	7
omi	7
i and *o*	2
ii and *oo*	6
iii and *ooo*	9

Table 2. The performance of the present method in the prediction of topology of both training and testing dataset

	Training Set		Testing Set	
	Number	Percentage	Number	Percentage
Number of Proteins	159		73	
MSRs found	132	83%	46	63%
Additionally correct sidedness	110	85%	34	74%
Total number of helices	694		328	
Predicted helices	661	95.24%	262	79.87%
Over-predicted helices	20	2.88%	21	6.40%
Under-predicted helices	20	2.88%	53	16.15%
Shifted helix prediction	13	1.87%	14	4.27%
Falsely merged helices	24	3.46%	21	6.40%

Tools provided with HTK toolkit, a toolkit primarily designed for modeling and manipulating HMMs in speech processing, was used in training process as well as in testing process [20].

Results of the prediction can be found in the Table 2, which shows the performance of the proposed method for the training dataset as well as for the test dataset. Performance of the method is evaluated on two different bases, firstly as a complete topography predictor and secondly as an MSR predictor. The present method predicts all the MSRs of 46 TMPs out of 73 unknown TMPs. In addition, it predicts correct positioning of start region in 34 TMPs out of 46 TMPs. As an MSR predictor, it predicts the 95.24% of MSRs (true positive predictions) from the total number of 694 MSRs in training data set, 79.87% of MSRs from the total number of 328 MSRs in test data set. It reported about 3% of over-predicted helices (false positives) and under-predicted helices (false negatives) in training data set, while those values were 6.4% and 16.15% in the test data set respectively. Shift helix prediction represents the regions, which share less than 9 residues with the reference annotation's MSRs. Falsely merged helices shows the regions, where adjacent helices are predicted as a single helix. Here, an MSR to be evaluated as predicted, it must share at least nine residues with the reference annotation's MSR. The other methods compared in Table 3 was evaluated on this basis in [8]. A test data set consists of 73 TMPs retrieved from the same data set that is used to evaluate the other methods. Table 3

Table 3. Comparison of performance of the present method compared to the previous approaches

Method	No. of proteins	All MSRs found	Additionally correct sidedness
Prposed Method	73	46 (63%)	34 (74% of 44)
TMHMM 2.0	108	64 (59%)	40 (63% of 64)
TMHMM 1.0	108	57 (53%)	21 (37% of 57)
HMMTOP	106	54 (51%)	42 (78% of 54)
MEMSAT 1.5	159	80 (50%)	58 (73% of 80)

compares the proposed method with previous approaches to TMP topology prediction. The performance figures of the TMHMM 2.0, TMHMM 1.0, HMMTOP, and MEMSAT 1.5 were obtained from [8]. The present method showed the best performance on the tested dataset.

5 Discussion and Future Directions

We have trained and have tested a new algorithm to predict the membrane spanning regions (α-helices) of the transmembrane proteins by looking at the protein in a syntactic point of view. The proposed model is a dynamic one which adjusts to the protein structure according to the characteristics of its segments. The hidden Markov models of the proposed method contain states which represent properties of small segments rather than a single residue and automatically adjust to the segment lengths.

On the tested dataset, the present method showed better performance over the reported accuracy measures of previous methods in both identification of MSR and description of their sidedness. The methods predicting protein topology with high accuracy has high pharmaceutical applications as membrane proteins are good therapeutic targets.

The syntactic rule set is flexible to absorb new characteristics such as sequences belong to the signal peptides which hamper the prediction accuracy, when they are inserted in the transmembrane proteins. The performance of the present method can be improved either by removing the signal peptides before the prediction process or by introducing new HMM model trained with signal peptide data.

References

1. Wallin, E., von Heijne, G.: Genome-wide analysis of integral membrane proteins from eubacterial archaean, and eukaryotic organisms. Protein Sci. **7** (1998) 1029–1038
2. Krogh, A., et al.: Predicting transmembrane protein topology with a hidden Markov model: Application to complete genomes. J. Mol. Biol. **305** (2001) 567–580

3. White, H. W., Wimley, C. W.: Membrane protein folding and stability: Physical principles. Annu. Rev. Biophys. Biomol. Struct. **28** (1999) 319–365
4. Kyte, J., Doolittle, R. F.: A simple method for displaying the hydropathic character of a protein. J. Mol. Biol. **157** (1982) 105–132
5. von Heijne, G.: The distribution of positively charged residues in bacterial inner membrane proteins correlates with the transmembrane topology. EMBO J. **5** (1986) 3021–3027
6. von Heijne, G.: Membrane protein structure prediction: Hydrophobicity analysis and the positive inside rule. J. of Mol. Bio. **225**(2) (1992) 487–494
7. Claros, M. G., von Heijne, G.: TopPred II: An improved software for membrane protein structure predictions. Computer Applications in the Biosciences. **10**(6) (1994) 685–686
8. Moller, S., et al.: Evaluation of methods for the prediction of membrane spanning regions. Bioinformatics **17**(7) (2001) 646–653
9. Sonnhammer, E. L. L., et al.: A Hidden Markov Model for Predicting Transmembrane Helices in Protein Sequences. Proc. on Intelligent Systems in Mol. Biology. **6** (1998) 175–182
10. Tusnady, G. E., Simon, I.: Principles governing amino acid composition of integral membrane proteins: Applications to topology prediction. J. of Mol. Bio. **283** (1998) 489–506
11. Jones, D. T., et al.: A model recognition approach to the prediction of all helical membrane protein structure and topology. Biochemistry **33** (1994) 3038–3049
12. Rost, B., et al.: Topology prediction for helical transmembrane proteins at 86% accuracy. Protein Science. **5** (1996) 1704–1718
13. Fariselli, P., et al.: MaxSubSeq: an algorithm for segment-length optimization. The case study of the transmembrane spanning segments. Bioinformatics. **19**(4) (2003) 500–505
14. Jayasinghe S. et al.: MPtopo: A database of membrane protein topology. Protein Science. **10** (2001) 455–458
15. Ikeda, M., et al.: TMPDB: a database of experimentally-characterized transmembrane topologies. Nucleic Acids Research, **31**(1) (2003) 406–409
16. Melen, K., et al.: Reliability measures for membrane protein topology prediction algorithms. J. Mol. Biol. **327** (2003) 735–744
17. Grundy, W. N.: Modelling Biological Sequences Using HTK. Technical Report, prepared for Entropic Research Laboratory, Inc. (1997)
18. Young, S. J., et al.: Token Passing: a conceptual model for connected speech recognition systems. CUED Technical Report F_INFENG/TR38, Cambridge University (1989)
19. Rabinar, L. R.: A tutorial on hidden Markov models and selected applications in speech recognition. Proceedings of the IEEE **77**(2) (1989) 257–286
20. Young, S., et al.: The HTK book (for HTK version 3.2.1). Cambridge university Engineering Department. (2002)
21. Moller, S., et al.: A collection of well characterised integral membrane proteins. Bioinformatics **16**(12) (2000) 646–653

An Evolutionary Approach for Motif Discovery and Transmembrane Protein Classification

Denise F. Tsunoda[1], Heitor S. Lopes[1], and Alex A. Freitas[2]

[1] Lab. Bioinformática, Centro Federal Educ. Tecnol. do Paraná, Curitiba, Brazil
dtsunoda@brturbo.com, hslopes@cpgei.cefetpr.br
[2] Computing Laboratory, University of Kent, Canterbury, UK
A.A.Freitas@ukc.ac.uk

Abstract. Proteins can be grouped into families according to their biological functions. This paper presents a system, named GAMBIT, which discovers motifs (particular sequences of amino acids) that occur very often in proteins of a given family but rarely occur in proteins of other families. These motifs are used to classify unknown proteins, that is, to predict their function by analyzing the primary structure. To search for motifs in proteins, we developed a GA with specially tailored operators for the problem. GAMBIT was compared with MEME, a web tool for finding motifs in the TransMembrane Protein DataBase. Motifs found by both methods were used to build a decision tree and classification rules, using, respectively, C4.5 and Prism algorithms. Motifs found by GAMBIT led to significantly better results, when compared with those found by MEME, using both classification algorithms.

1 Introduction

After unveiling the DNA sequence of an organism, researchers turn to the laborious task of annotation. Afterwards, the proteome of the organism is seen as one of the main products of genome sequencing projects. In recent years researchers have witnessed an exponential growth of biological databases, thanks to the many genome sequencing projects in the world.

Proteins are essential for life since they are responsible for most functions in an organism, such as: transport of small molecules (e.g., hemoglobin), regulation (e.g., insulin), sustentation (e.g., collagen), increase of reaction speed (e.g., enzymes) and others. Biological organisms have thousands of different types of proteins, which are constituted basically of amino acids linked in linear chains through peptide connections. Active intra-molecular forces cause the proteins to assume specific three-dimensional shapes that are directly related to their biological functions [8]. Proteins are grouped into super families, families and subfamilies according to their biological function. The classification of proteins is an important task for the molecular biologist, and, ultimately, it is aimed to identify the function of the protein.

There are several protein databases available, for instance, Swiss-Prot and Protein Data Bank (PDB) [1]. In this work we used the TMPDB (TransMem-

F. Rothlauf et al. (Eds.): EvoWorkshops 2005, LNCS 3449, pp. 105–114, 2005.

brane Protein DataBase) [13],[7],[6], a transmembrane subset extracted from some public databases that contains information about the primary structure of 302 transmembrane proteins. The choice for this subset was due to the extremely important functions that these proteins plays in life as pumps, channels, receptors, catalyzers, energy transducers, etc., and have been reported recently to share approximately 20-30% of genes in a whole genome. The transmembrane protein molecules are difficult to crystallize due to their amphiphilic characteristics – they present hydrophobic transmembrane segments (TMSs) but also hydrophilic loops.

The protein-classification problem (PCP) is a very important research area in bioinformatics. As mentioned before, the many genome sequencing projects have been unveiling a growing number of gene products whose function is unknown or barely estimated by homology techniques. The prediction of protein function has been done basically in two ways: prediction of the protein structure and then prediction of function from the structure, or else, classifying proteins into functional families and supposing that similar sequences will have similar functions. Notwithstanding, most proteins share similar structures (in particular, considering the primary structure), since many of them have a common evolutionary origin [11]. Common structures may be characteristic of a given family of proteins but, on the other hand, unrelated families can also share common structures. This two-fold characteristic makes the PCP a challenging problem, for which many methods have been suggested; see, for instance [5],[9],[10],[14],[15].

This paper reports the development and application of a computational tool, named GAMBIT (Genetic Algorithm-based Motif Browsing and Identification Tool), specially devised for the automatic discovery of motifs (short sequences of amino acids). This tool is based on genetic algorithms and uses as input only the information about the primary structure of proteins. The system finds variable-length motifs that occur very often in proteins of a given class (family) but rarely occur in proteins of other classes. Those discovered motifs can be further used to discriminate families of (known) proteins and for the automatic classification of unknown proteins. That is, using the motifs discovered by the proposed system, one can estimate function of an unknown protein by analysing only its primary structure.

2 Methodology

2.1 Data Preprocessing

The version of the TMPDB used in this work was #6.3, from November 2003. A TMPDB file uses the same format as Swiss-Prot and it has information about the primary sequence of a protein. For the purposes of this work we used only the following fields: ID (identification code in other databases), ME (membrane in which the protein exists) and SQ (Sequence header and its length, followed by the amino acids sequence).

The TMPDB contributors [6] have collected 1,074 articles reporting TM topology, by using MEDLINE search using keywords "transmembrane" and "topology" and they found 895 articles. By searching the web directly without using MEDLINE they found 46 articles, and by searching for Swiss-Prot and TrEMBL entries whose RP line contains the annotations of "X-ray crystallography", "structure by neutron diffraction", "structure by electron cryomicroscopy"', "structure by NMR" or "topology", they found 133 articles. After the validation of each article, they extracted the 302 experimentally-characterized transmembrane proteins. To obtain the complete sequence, they made a cross-reference to public databases (using the protein name or the partial sequences). Finally, by combining the information contained in the articles and other information of the public databases, they constructed TMPDB.

The transmembrane proteins are distributed across 25 classes. In this work, aiming to have statistically meaningful results, we used only 6 classes, those with 10 or more proteins. The number of proteins in each class was: 144 proteins in class Inner Membrane (IM), 64 in class Plasma Membrane (PM), 22 in class Mitochondrial Inner Membrane (MM), 10 in class Chloroplast Thylakoid Membrane (CM), 25 in class Endoplasmic Reticulum Membrane (EM) and 16 in class Outer Membrane (OM). Therefore, we used 281 out of 302 proteins of TMPDB, and this data set is available at *http://bioinfo.cpgei.cefetpr.br/en/softwares.htm.*

2.2 Encoding and Fitness Function

Genetic Algorithms (GA) were used in this work due to its ability to perform adaptive, powerful and robust searches. Besides, their intrinsic parallelism allows the simultaneous exploration of different regions of the search space. The use of GA for real-world problems encompasses two important definitions: the encoding scheme of an individual and the fitness function. In the implemented system, individuals represent a single motif, that is, a variable-length string of characters, over the alphabet used for encoding the 20 standard amino acids [8].

Recall that our goal is to find a sequence of amino acids (motif) with a high discriminatory power – i.e., a pattern that occurs in most proteins of a given class and occurs in few or no proteins of all other classes. Therefore, this pattern can be characteristic of a given family, allowing it to be discriminated from all others – the essence of classification.

In order to discriminate an individual, we developed a special fitness function that is computed as follows. Given a motif found by the GA, for each class i, i=1,...,6 (for the transmembrane dataset used in this work), the relative frequency of occurrence of the motif in that class is computed. This is simply the number of proteins of the i-th class where the motif occurs anywhere in the protein´s sequence divided by the number of proteins in the i-th class. Next, for each class i, a measure of the ability of the motif to discriminate between class i and the other classes is given by the equation (1):

$$Disc_i = F_i. \left(1 - \frac{\sum_{j=1,j\neq i}^{n} F_j}{k-1}\right) \quad (1)$$

where F_i is the relative frequency of the motif in the i-th class, n is the number of classes ($n = 6$ in this work), and k is the number of classes that contain at least one protein whose primary sequence contains the given motif. The rightmost term of the formula simply computes the average relative frequency of the motif in all classes $j \neq i$ containing at least one occurence of the motif. This term is subtracted from 1, so that the term between brackets is to be maximized. Similarly, the value of F_i is also to be maximized. Therefore, a high value of $Disc_i$ means that the motif occurs very often in class i but rarely in the remaining classes. If $k = 1$, in order to avoid division by zero in equation (1), the fraction in the formula is considered to collapse to zero, so that the term between brackets collapses to 1 and $Disc_i$ collapses to F_i. This reflects the desirable case where the motif occours only in class i (and in no class j, $j \neq i$), so that the motif quality depends only on F_i.

Once the value of $Disc_i$ has been computed for all n classes ($i = 1, ..., n$), the individual is associated with the class having the largest value of $Disc_i$. In other words, the motif represented by the individual is considered as a characteristic pattern for proteins of the class with the largest value of $Disc_i$. The proposed fitness function is normalized in the range [0..1], making the interpretation of results somewhat easier, since 1 is the best possible value, meaning maximum discrimination.

2.3 Selection Method and Genetic Operators

In this work, the selection method used was the well-known stochastic tournament (with tournament size $k \geq 2$). The usual one-point crossover operator is stochastically applied with a predefined probability, using two individuals of the selected pool. Since the length of the chromosome is variable, the traditional concept of crossover point was slightly modified and adapted to our individual representation. The crossover point is a percentage (of the length of the individual) that defines the starting point from where the crossover breaks the string. The same percentage is used for both parents. For instance, if the percentage is 80%, the rightmost 20% of the amino acids contained in the parents are crossed-over.

As usual, the mutation operator is used to further explore the search space and to avoid unrecoverable loss of genetic material that leads to premature convergence to some local minima. Due to the specific purpose of our system, we devised four different types of mutation (herein, sub-operations), as follows:

1. Left-adding: one randomly generated character (corresponding to an amino acid) is added to the left of the motif.
2. Right-adding: one randomly generated character (corresponding to an amino acid) is added to the right of the motif.
3. Random-changing: all the amino acids from a randomly selected starting point up to the end of the motif are changed, except the first and the last position.
4. Cutting-out: it removes a single character from the amino acid sequence. The removal position is randomly generated.

The mutation probability is a user-defined parameter, as usual in GA. Once the system has decided to do a mutation, all sub-operations have the same probability of being chosen, in a random fashion.

Both crossover and mutation operators are also "hill-climbing-based operators" because they are implemented in such a way that a new individual is immediately evaluated after it has been created and, if its fitness is lower than the parent's fitness, the parent (rather than the child) is copied to the next generation. This procedure does not increase significantly the computational cost and makes the evolutionary process faster in terms of number of generations necessary for convergence, since the generated offspring will be always better than their parents (or will not be generated otherwise). Hence, after a genetic operator is selected according to a given probability, it can be applied in the usual way (inserting the children in the new population regardless of their fitness) or as a hill-climbing-based operator. This choice is done probabilistically according to another user-defined parameter – hill climbing-based operator rate.

The expansion operator is a new operator specifically designed for the motif discovery and protein classification problem. This operator starts by accessing the first protein of the class associated with the individual (this class was determined during the computation of the fitness) and locating the position, in that protein, where the individual's amino acid sequence occurs. Then, it tentatively adds the immediately preceding amino acid (in the protein) to the individual's amino acid sequence (candidate motif). The relative frequency of the individual's amino acid sequence in that class is recomputed. If the new relative frequency is equal to or higher than the previous relative frequency, the just-added amino acid is effectively added to the motif. This operation corresponds to expansion of the individual's genotype. This process is iteratively repeated until the relative frequency becomes lower than the previous one. At this point the above-described expansion process is applied to the amino acid immediately subsequent in the protein. Finally, the whole process (expansion to the left and expansion to the right) is repeated for all the other proteins of the class associated with the individual. Note that this is a computationally expensive operator, but our preliminary experiments have shown that it effectively leads to motifs with a higher predictive power for protein classification.

2.4 Running Parameters

The implemented GA has several parameters and many preliminary runs were done to adjust these parameters. This task was done using an enzyme dataset with 6 classes and 100 enzimes per class. These results will be published in [14]. In these runs the expansion operator was always turned on, and those tests produced the following optimal values of parameters: number of generations = 300, population size = 200, hill-climbing-based operator rate = 10%, tournament size = 1%; crossover probability = 20%; mutation probability = 70%. The hill climbing-based operator rate is low to avoid losing population diversity and to prevent a premature convergence.

A conventional GA returns, as its result, the best individual (the one with highest fitness) found during the run. However, in our system the desired result is not a single individual, but rather, a set of individuals. That is, each individual represents a single amino acid sequence (motif), associated with a single class, and this kind of pattern will be used further to classify proteins. Therefore, it is necessary to discover many patterns, associated with as many different classes as possible during the GA search. In each generation, after the fitnesses of all individuals have been computed, some high-quality motifs for each class are saved in a separated file, called the set of discovered patterns - SDP. In fact, the individuals representing those patterns still remain in the population; only a copy of them is saved into SDP. The criterion to select these individuals is their fitness – only those with fitness greater than a user-defined minimum quality threshold will be saved. This procedure results in the discovery of many motifs, associated with different classes, as desired. However, special care is taken to prevent adding motifs that are substrings of other motifs already in the SDP.

3 Computational Experiments and Results

Using the data described in Section 2.1, motifs were discovered using two different tools: GAMBIT and MEME (Multiple EM for Motif Elicitation) [2]. MEME is a well-known freely-available web tool supported by the San Diego Supercomputer Center (*http://meme.sdsc.edu/meme/website/intro.html*). MEME essentially uses statistical modeling techniques to automatically choose the best width and description for each motif. In our experiments, we used all default parameters of MEME, except the number of motifs, set to 15.

After running GAMBIT and MEME, the top fifteen motifs discovered by each of those tools were set aside as designated results for each of those tools. The goodness of a motif was measured by its class-discrimination ability, as defined in equation 1. Recall that both GAMBIT and MEME are intended to discover motifs in sets of sequences and are not designed as classification tools. Hence, in order to evaluate the effectiveness of the discovered motifs in predicting the functional class of proteins, we have used the discovered motifs as predictor attributes in two classification algorithms available in WEKA (Waikato Environment for Knowledge Analysis) [16] , version 3.4.3. WEKA is a well-known Java-based data mining toolkit freely-available on the internet (*http://www.cs.waikato.ac.nz/ml/weka*).

The two classification algorithms used in the experiments were J4.8 (the WEKA implementation of the very well-known C4.5 decision tree induction algorithm [12]) and Prism [4], a rule induction algorithm that discovers classification rules directly from the data, without producing a decision tree. In our experiments, we used the default parameters of both J4.8 and Prism.

The predictive accuracies obtained by J4.8 and Prism were measured using a well-known 3-fold cross validation procedure [16], as follows. The data set was partitioned into 3 mutually-exclusive and exhaustive partitions. In the i-th iteration of the cross-validation procedure, i=1,2,3, the i-th partition was used as the test set and the other two partitions were grouped and used as the training set. In each of the 3 iterations, first GAMBIT and MEME were used to discover motifs from the training set. Then, as mentioned earlier, those motifs were used as predictor attributes in J4.8 and Prism, which were also run on the training set. Each motif was used as a binary attribute, indicating whether or not the motif occurred in a given protein (training example).

Note that each of the two classification algorithms, J4.8 and Prism, was run twice: one run used motifs discovered by GAMBIT, and the other run used motifs discovered by MEME. This produced four classification models – two decision trees produced by J4.8 and two rule sets produced by Prism. Finally, the four classification models were evaluated on the test set – which was never accessed during training – in order to measure the predictive accuracy (generalization ability) of the discovered classification models. This procedure was carried out 3 times (corresponding to the 3 iterations of the cross-validation procedure), and the reported results are the average of the accuracy rate on the test set across the 3 iterations.

Figure 1 shows the decision tree generated by J4.8 and Table 1 shows the rules generated by Prism. Due to space limitations, both Figure 1 and Table 1 show only the classification models built from the motifs discovered by GAMBIT. In Figure 1, each internal node tests for the presence (1) or absence (0) of an attribute (a motif). Similarly, in Table 1 the conditions in the rule antecedents refer to the presence or absence (indicated by a "not" operator) of motifs. The predicted classes – represented in the leaf nodes of the decision tree and in the consequents of the rules – are the membrane classes defined in Section 2.1. For instance, the top-right part of the decision tree in Figure 1 corresponds to the rule: IF motif GHL is absent (0) AND motif AQS is present (1) THEN class = PM (Plasma Membrane).

Although there are many ways to measure classification accuracy (see, for instance, [3],[16]), in this work, the final performance was measured using the accuracy rate. The average accuracy rates (on the test set) computed by the cross-validation procedure were: 73.4% using J4.8 with motifs found by GAMBIT, 58.0% using J4.8 with motifs found by MEME, 99% using Prism with motifs found by GAMBIT, and 65.4% using Prism with motifs found by MEME.

Therefore, the motifs found by GAMBIT were clearly much more effective in predicting protein class than the motifs found by MEME, in both the classification algorithms used in the experiment (J4.8 and Prism).

Note that Prism obtained considerably better results than J4.8. A likely explanation for this result is that Prism is more flexible, in the sense that it can select only one relevant value of an attribute (motif) – either its presence or its absence. By contrast, J4.8 has to select both values of an attribute (motif) – both "1" (presence) and "0" (absence) – to be included in the tree (in different branches coming out from the same parent). In this kind of data set, intuitively the presence of a motif is a more relevant attribute value than its absence, which gives an advantage to the more flexible rule representation of Prism. Indeed, out of the 20 rule conditions in Table 1, only 4 refer to the absence of a motif (using

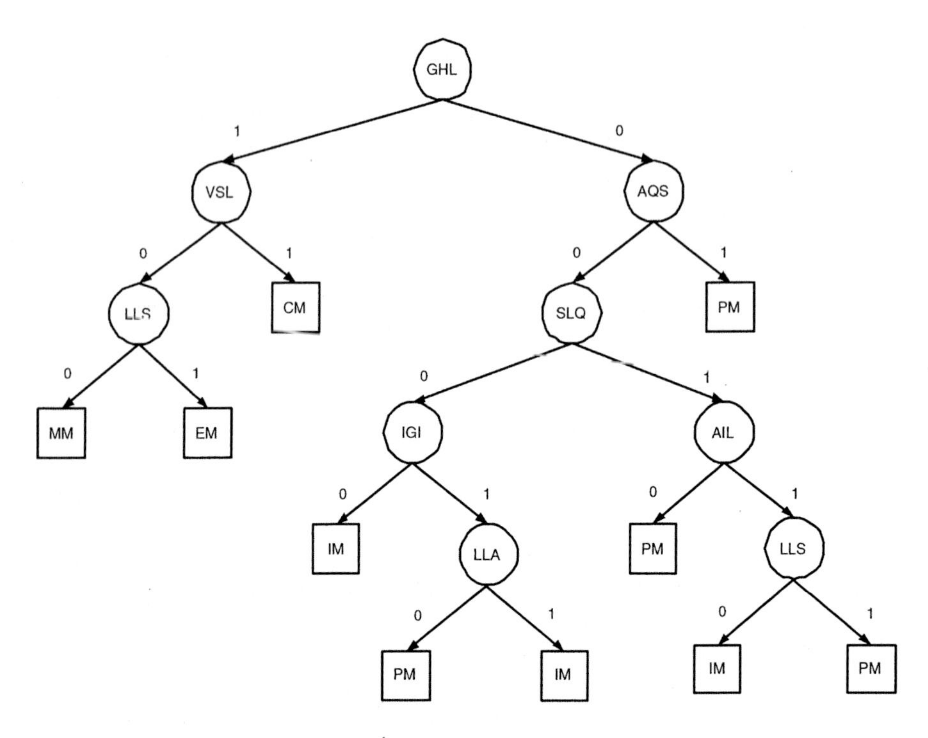

Fig. 1. Decision tree constructed by J48 using motifs found by GAMBIT

Table 1. Subset of the best rules found by PRISM using motifs found by GAMBIT

If (SRR) then IM
If (SNN) then IM
If (APML) then IM
If (MNNM) then IM
If (EWR) then PM
If (LIG and VLG and SLK) then PM
If (LWK and not(MKK)) then MM
If (RGYWQE) then CM
If (VTV and GFV and not(TN)and not(LWA)) then EM
If (VDY and DGD) then OM
If (DPT and LID and not(GDI)) then OM

the operator "not"). The other 16 conditions refer to the presence of a motif. In addition, note that the class OM does not appear in the decision tree of Figure 1, which is a clear disadvantage of that classification model. Finally, note also that the decision tree of Figure 1 uses only short motifs (with 3 amino acids), whereas the rules in Table 1 have a somewhat wider diversity of motif size: two rules use motifs with four amino acids, and one rule uses a motif with six amino acids (a motif produced by the expansion operator).

4 Conclusions and Future Work

We described a system based on a Genetic Algorithm specifically designed for motif discovery, aiming to classify unknown-class proteins. The system was evaluated using a transmembrane protein dataset.

The genetic operators of GAMBIT, specifically designed for the PCP, have played an important role in the positive results achieved, since they allowed the GA to obtain motifs with high discriminatory power.

Comparing results obtained by GAMBIT with MEME, it can be seen that the latter did not find good motifs to discriminate one class from the others. On the other hand, this is a remarkable characteristic of GAMBIT, an innate ability accomplished by its fitness function. It is a matter of fact that MEME was not projected for the same purpose as GAMBIT but, to the best of our knowledge, it is the tool that most closely can be compared with our system. In short, MEME discovers motifs in a group of proteins, while GAMBIT discovers motifs that discriminate a group of proteins from another.

Using the discovered motifs found by both systems, the J48 and Prism algorithms generated comprehensive classifiers, useful to biologists. It is possible that those discovered motifs are related to known specific secondary or tertiary structures (this investigation was left to future work).

Finding groups of amino acids that uniquely characterize protein families is a very important issue in molecular biology. Results for the transmembrane dataset using GAMBIT and WEKA strongly suggest the efficiency of the method to find motifs capable of discriminating between groups or proteins, offering a feasible solution to the PCP.

Future work includes more exhaustive tests of the GA control parameters for fine-tuning and development of biologically-inspired genetic operators. We intend to improve GAMBIT so as to find motifs based on regular expressions. Also, it is intended to apply this system to find motifs for classification of other protein families of biological interest.

Acknowledgments

This work was partially supported by a research grant from the Brazilian National Research Council – CNPQ (350053/02-0 and 475049/03-9).

References

1. Abola EE, Sussman JL, Prilusky J and Manning NO (1997) Protein data bank archives of three-dimensional macromolecular structures. *Meth Enzymol* 277:556-571
2. Bailey TL and Elkan C (1994) Fitting a mixture model by expectation maximization to discover motifs in biopolymers. *Proc. 2nd Int.Conf. on Intelligent Systems for Molecular Biology*, pp. 28-36

3. Bojarczuk CC, Lopes HS, Freitas AA (2004) A constrained-syntax genetic programming system for discovering classification rules: application to medical data sets. *Artif Intell Med* 30(1):27-48
4. Cendrowska J (1987) Prism: an algorithm for inducing modular rules. *Int J Man-Mach Stud* 27:349-370
5. Hanke J, Beckmann G, Bork P and Reich JG (1996) Self-organizing hierarchic networks for pattern recognition in protein sequence. *Protein Sci* 5(1):72-82
6. Ikeda M, Arai M, Lao DM and Shimizu T (2002) Transmembrane topology prediction methods: a re-assessment and improvement by a consensus method using a dataset of experimentally-characterized transmembrane topologies. *In Silico Biol* 2:19-33
7. Kihara D, Shimizu T and Kanehisa M (1998) Prediction of membrane proteins based on classification of transmembrane segments. *Protein Eng* 11:961-970
8. Lehninger AL, Nelson DL and Cox MM (1998) *Principles of Biochemistry.* 2nd ed. Worth Publishers, New York, pp. 134-137
9. Manning AM, Brass A, Goble CA, and Keane JA (1997) Clustering techniques in biological sequence analysis. In: *Proc.*1st *Europ. Symp. on Principles of Data Mining and Knowledge Discovery*, pp. 315-322
10. Mathura VS, Schein CH and Braun W (2003) Identifying property based sequence motifs in protein families and superfamilies: application to DNase-1 related endonucleases. *Bioinformatics* 19(11):1381-1390
11. Murzin AG, Brenner SE, Hubbard T and Chothia C (1995) A structural classification of proteins database for the investigation of sequences and structures. *J Mol Biol* 247:536-40
12. Quinlan JR (1993) *C4.5: Programs for Machine Learning.* Morgan Kaufmann Publishers, San Francisco
13. Shimizu T and Nakai K (1994) Construction of a membrane protein database and an evaluation of several prediction methods of transmembrane segments. In *Proc. Genome Informatics Workshop*, pp. 148-149
14. Tsunoda DF and Lopes HS (2005) Automatic motif discovery in an enzyme database using a genetic algorithm-based approach. To appear.
15. Weinert WR and Lopes HS (2004) Neural networks for protein classification. *Appl. Bioinformatics* 3:41-48
16. Witten IH and Frank E (2000) *Data Mining: Practical Machine Learning Tools and Techniques with Java Implementations.* Morgan Kaufmann, San Francisco

Differential Evolution and Its Application to Metabolic Flux Analysis

Jing Yang[1], Sarawan Wongsa[1], Visakan Kadirkamanathan[1],
Stephen A. Billings[1], and Phillip C. Wright[2]

[1] Department of Automatic Control and Systems Engineering,
University of Sheffield, Sheffield, United Kingdom
[2] Department of Chemical and Process Engineering,
University of Sheffield, Sheffield, United Kingdom
visakan@shef.ac.uk

Abstract. Metabolic flux analysis with measurement data from ^{13}C tracer experiments has been an important approach for exploring metabolic networks. Though various methods were developed for ^{13}C positional enrichment or isotopomer modelling, few researchers have investigated flux estimation problem in detail. In this paper, flux estimation is formulated as a global optimization problem by carbon enrichment balances. Differential evolution, which is a simple and robust evolutionary algorithm, is applied to flux estimation. The algorithm performances are illustrated and compared with ordinary least squares estimation through simulation of the *cyclic pentose phosphate metabolic network* in a noisy environment. It is shown that differential evolution is an efficient approach for flux quantification.

1 Introduction

The action and regulation of metabolic networks are complicated due to large number of enzymes and various control mechanisms involved in the networks. Significant contributions have been made in this area in recent years[1][2][3][4]. Metabolic flux analysis, i.e., the quantification of all intracellular metabolic fluxes in a given model of the cellular metabolism, has long been noted as an important computational tool in metabolic engineering [5]. Despite their importance, the intracellular metabolic fluxes are *per se* nonmeasurable quantities [6], which require additional computation effort for their quantification.

A systematic method for determining metabolic fluxes is carried out by a stoichiometric model. By combining data on substrate uptake rates from the medium, secretion rates of products from the cells and quasi-steady-state mass balances on metabolic intermediates, the intracellular fluxes can be calculated [7]. A limited number of measurable extracellular fluxes and a number of linear constraints from stoichiometric matrix often lead to an underdetermined system. Additional assumptions about enzyme activities *in vivo* or certain objective functions are often introduced in order to derive quantities for intracellular

F. Rothlauf et al. (Eds.): EvoWorkshops 2005, LNCS 3449, pp. 115–124, 2005.

metabolic fluxes, which is called metabolic flux balancing [8]. The correctness of the computed fluxes largely depend on the validity of these assumptions and the appropriate choices of objective functions [6].

The potential drawbacks of metabolic flux balancing method based on stoichiometry can be overcome by using ^{13}C labelling experiments, whose labelling states are able to be measured by nuclear magnetic resonance (NMR) and/or gas chromatography mass spectrometry (GC-MS). The measurement quantities from such experiments provide isotopomer or positional enrichment data of intracellular metabolites when carbon isotopic balance is reached, which offer constraints for intracellular fluxes. More measurement data than the required constraints are often available in such experiments, which lead to an overdetermined system. The accurate calculation of fluxes from such a system, therefore, will greatly benefit metabolic system analysis.

Despite the importance of deriving fluxes quantities from the overdetermined system acquired through ^{13}C labelling experiments, most authors simply apply least squares estimation to solve the problem [5][9], whose results may be quite unsatisfactory. In [6], the problem was solved using evolutionary algorithms, however, no details about the algorithm was given. In this paper, *differential evolution* (DE), is applied for fluxes quantification from ^{13}C labelling experiment and its performance is shown to be superior to the least squares estimation method.

2 Metabolic System Description

2.1 Metabolic System Modelling

Here the metabolic system and principles of formulating labeling balance equations are described through a simple metabolic network shown in Fig. 1, in which the left part of the figure shows metabolic reactions for metabolites A, B, C, D and the right part shows the carbon atom transitions for the above reactions. Metabolites A and D are substrate and product of the example system respectively. B and C are intracellular metabolites. A and B each has three carbon atoms. C and D both have two carbon atoms. $\mathbf{v_1} - \mathbf{v_4}$ are fluxes associated with the reactions. Fluxes $\mathbf{v_1}$ and $\mathbf{v_4}$ represent substrate consumption and product formation rates, which are always unidirectional. The intracellular flux $\mathbf{v_3}$ is assumed to be unidirectional. The flux $\mathbf{v_2}$ is set to be bidirectional. In practice, extracellular fluxes are often measurable. Consequently, $\mathbf{v_1}$ and $\mathbf{v_4}$ are assumed to be known quantities, whereas $\mathbf{v_2}$ and $\mathbf{v_3}$ are unknown intracellular fluxes.

In the paper we assume that the carbon enrichment data available are the positional enrichments for metabolites $P_M=(P_M(1),P_M(2),\cdots, P_M(N))$, giving information about the concentration of labelled carbon in each position of the atom carbon backbone, where $P_M(i)$ is the positional enrichment data for the i_{th} carbon atom of metabolite M with N carbon atoms.

When the steady-state is reached, the rate of consumption must equal the rate of production for each metabolite. Hence, the stoichiometric matrix can be formulated as follows:

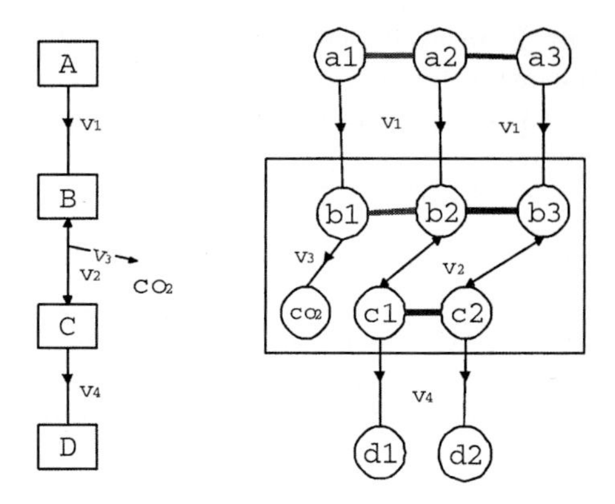

Fig. 1. A metabolic network (left) and the carbon atoms fate in the network(right)

$$\begin{aligned} \overrightarrow{v}_1 + \overleftarrow{v}_2 &= \overrightarrow{v}_2 + \overrightarrow{v}_3 \\ \overrightarrow{v}_2 &= \overleftarrow{v}_2 + \overrightarrow{v}_4 \end{aligned} \tag{1}$$

where $\overrightarrow{v}$ and $\overleftarrow{v}$ represent the forward and backward fluxes respectively. When considering the carbon isotopic balances for B and C, the balances equations are:

$$\begin{aligned} \text{b1}: \; & \overrightarrow{v}_1 P_{\text{A}}(1) &&= \overrightarrow{v}_3 P_{\text{B}}(1) \\ \text{b2}: \; & \overrightarrow{v}_1 P_{\text{A}}(2) + \overleftarrow{v}_2 P_{\text{C}}(1) &&= \overrightarrow{v}_2 P_{\text{B}}(2) \\ \text{b3}: \; & \overrightarrow{v}_1 P_{\text{A}}(3) + \overleftarrow{v}_2 P_{\text{C}}(2) &&= \overrightarrow{v}_2 P_{\text{B}}(3) \\ \text{c1}: \; & \overrightarrow{v}_2 P_{\text{B}}(2) &&= (\overleftarrow{v}_2 + \overrightarrow{v}_4) P_{\text{C}}(1) \\ \text{c2}: \; & \overrightarrow{v}_2 P_{\text{B}}(3) &&= (\overleftarrow{v}_2 + \overrightarrow{v}_4) P_{\text{C}}(2) \end{aligned} \tag{2}$$

Assume that $\mathbf{x}$ are the intracellular positional enrichments, $\tilde{\mathbf{x}}$ are the extracellular positional enrichments, $\mathbf{v}$ are intracellular fluxes and $\tilde{\mathbf{v}}$ are extracellular fluxes respectively:

$$\mathbf{x} = \begin{pmatrix} P_{\text{B}} \\ P_{\text{C}} \end{pmatrix} \qquad \tilde{\mathbf{x}} = \begin{pmatrix} P_{\text{A}} \\ P_{\text{D}} \end{pmatrix} \qquad \mathbf{v} = \begin{pmatrix} \overrightarrow{v}_2 \\ \overleftarrow{v}_2 \\ \overrightarrow{v}_3 \end{pmatrix} \qquad \tilde{\mathbf{v}} = \begin{pmatrix} \overrightarrow{v}_1 \\ \overrightarrow{v}_4 \end{pmatrix}$$

Equation (2) can be rewritten concisely as:

$$\mathbf{A}(\mathbf{x})\mathbf{v} = \mathbf{b}(\tilde{\mathbf{x}}, \tilde{\mathbf{v}}) \tag{3}$$

or

$$\bar{\mathbf{A}}(\mathbf{v})\mathbf{x} = \bar{\mathbf{b}}(\tilde{\mathbf{x}}, \tilde{\mathbf{v}}) \tag{4}$$

and (1) as:

$$\mathbf{B}\mathbf{v} = \mathbf{d}(\tilde{\mathbf{v}}) \tag{5}$$

2.2 Nonlinear System Model and Cost Function

In subsequent flux estimation problem, extracellular quantities are treated as inputs to metabolic system, therefore, they are not considered as part of the measurement equations. In terms of the noise involved in ^{13}C enrichment experiments, as not all of the intracellular carbon enrichments, i.e. $\mathbf{x}$, can be measured, we can formulate following equation of the measured intracellular carbon enrichments, i.e. $\mathbf{y}$, as follows:

$$\mathbf{y} = \mathbf{C}\mathbf{x} + \mathbf{n} \tag{6}$$

where $\mathbf{n}$ denotes the noise associated with measurements. In the case where there is no influence between measurements of different intracellular metabolites, the matrix $\mathbf{C}$ will be such that only one element in each row will be non-zero, with a value of 1 associated with the measured metabolites.

Considering the linear matrix equation with respect to intracellular fluxes $\mathbf{A}(\mathbf{x})\mathbf{v} = \mathbf{b}(\tilde{\mathbf{x}}, \tilde{\mathbf{v}})$, its counterpart $\bar{\mathbf{A}}(\mathbf{v})\mathbf{x} = \bar{\mathbf{b}}(\tilde{\mathbf{x}}, \tilde{\mathbf{v}})$ and system stoichiometric equation $\mathbf{B}\mathbf{v} = \mathbf{d}(\tilde{\mathbf{v}})$, above estimation problem can be posed as a constrained minimization problem given below:

$$(\hat{\mathbf{x}}, \hat{\mathbf{v}}) = \arg\min_{\mathbf{x},\mathbf{v}}(\|\mathbf{A}(\mathbf{x})\mathbf{v} - \mathbf{b}(\tilde{\mathbf{x}}, \tilde{\mathbf{v}})\|_2^2 + \lambda\|\mathbf{y} - \mathbf{C}\mathbf{x}\|_2^2)$$
$$\text{subject to} \quad \mathbf{B}\mathbf{v} = \mathbf{d}(\tilde{\mathbf{v}}) \tag{7}$$

where $\hat{\mathbf{x}}$ and $\hat{\mathbf{v}}$ are estimates to positional enrichment data and fluxes respectively and λ is a positive weight parameter. This can be transformed to an unconstrained optimization problem via the penalty function approach:

$$(\hat{\mathbf{x}}, \hat{\mathbf{v}}) = \arg\min_{\mathbf{x},\mathbf{v}} \mathrm{J}(\mathbf{x}, \mathbf{v})$$

in which

$$\mathrm{J}(\mathbf{x}, \mathbf{v}) = \|\mathbf{A}(\mathbf{x})\mathbf{v} - \mathbf{b}(\tilde{\mathbf{x}}, \tilde{\mathbf{v}})\|_2^2 + \|\mathbf{B}\mathbf{v} - \mathbf{d}(\tilde{\mathbf{v}})\|_2^2 + \lambda\|\mathbf{y} - \mathbf{C}\mathbf{x}\|_2^2 \tag{8}$$

where we give the same weight priority to both equations on $\mathbf{v}$, as is typically done in least squares estimation approach [9].

3 Differential Evolution

3.1 Introduction

Evolutionary algorithms (EAs) have long been recognized for their ability to solve global optimization problem within various situations. For a long time, the application of EAs in high dimensional spaces have been restricted due to a lack of efficient strategies to mutate evolutionary populations smoothly. In addition to self-adaptive evolutionary algorithm which we have adopted in [10], differential evolution is another efficient EA method to handle optimization problems in high-dimensional spaces [11], which is notable for its simplicity and good convergence properties.

3.2 Basic Procedures of Differential Evolution

Contrary to most EAs that deal with selection, combination and mutation individually, DE combines the three steps together in a compact way.

Assume that $\mathbf{X}_1^K, \mathbf{X}_2^K, \ldots, \mathbf{X}_N^K$ are N populations in the K_{th} generation and each $\mathbf{X}_i^K$ is a vector with M dimensions. The new i_{th} population in the $(K+1)_{th}$ generation $\mathbf{X}_i^{K+1}$ is created according to following strategies:

1. Three different parents r_1, r_2 and r_3 are randomly chosen from all the N populations, who are also different from the running index i;
2. A perturbed vector $\mathbf{Y}$ is created according to following equation;

$$\mathbf{Y} = \mathbf{X}_{r_1}^K + \mathbf{F} \cdot (\mathbf{X}_{r_2}^K - \mathbf{X}_{r_3}^K)$$

   where $F \in [0, 2]$ is a constant weight parameter [11].
3. Combination is introduced into DE by comparison between an uniform random number indicator R and the predefined combination probability P_c for each dimension of the perturbed population. To make sure that at least one parameter is from perturbed vector $\mathbf{Y}$, an integer index number **index** lying between [1,M] is randomly selected and the combination works according to following strategy:

$$\mathbf{Z}(j) = \begin{cases} \mathbf{Y}(j) & \text{if } (R \le P_c) \quad \text{or} \quad j = \text{index} \\ \mathbf{X}_i^K(j) & \text{otherwise} \end{cases}$$

4. To decide if the generated population $\mathbf{Z}$ is accepted, its fitness value is compared with that of $\mathbf{X}_i^K$. If $\mathbf{Z}$ yields a better fitness value than $\mathbf{X}_i^k$, then $\mathbf{X}_i^{K+1} = \mathbf{Z}$; otherwise, $\mathbf{X}_i^{K+1} = \mathbf{X}_i^K$.

4 Different Methods for Metabolic Flux Quantification

In what follows, we describe a least squares method (LS), a DE-based method and two hybrid-DE methods referred to as DE-LS. We assume that the positional enrichment data for all intracellular metabolites, in the metabolic system we are investigating, are measurable.

1. Method I: LS estimation method. Equations (3) and (5) can be written in the form:

$$\begin{pmatrix} \mathbf{A}(\mathbf{x}) \\ \mathbf{B} \end{pmatrix} \mathbf{v} = \begin{pmatrix} \mathbf{b}(\tilde{\mathbf{x}}, \tilde{\mathbf{v}}) \\ \mathbf{d}(\tilde{\mathbf{v}}) \end{pmatrix} \tag{9}$$

   In the case where there are no additional constraints on system measurement, so that $\mathbf{C}$ in (6) is in the format of identity matrix, we can substitute $\mathbf{x}$ with its measurement $\mathbf{y}$. Therefore, with $\mathbf{G} = \begin{pmatrix} \mathbf{A}(\mathbf{y}) \\ \mathbf{B} \end{pmatrix}$ and $\mathbf{f} = \begin{pmatrix} \mathbf{b}(\tilde{\mathbf{x}}, \tilde{\mathbf{v}}) \\ \mathbf{d}(\tilde{\mathbf{v}}) \end{pmatrix}$ defined, the least squares estimate of fluxes $\mathbf{v}$ is given by

$$\hat{\mathbf{v}} = (\mathbf{G}^T\mathbf{G})^{-1}\mathbf{G}^T\mathbf{f} \ . \tag{10}$$

2. Method II: DE on fluxes and LS on metabolites method. In case that the estimation to fluxes $\hat{\mathbf{v}}$ is available, the estimation of positional enrichment data $\hat{\mathbf{x}}$ can be calculated from (4) by least squares estimation:

$$\hat{\mathbf{x}} = (\bar{\mathbf{A}}(\hat{\mathbf{v}})^T \bar{\mathbf{A}}(\hat{\mathbf{v}}))^{-1} \bar{\mathbf{A}}(\hat{\mathbf{v}})^T \bar{\mathbf{b}}(\tilde{\mathbf{x}}, \tilde{\mathbf{v}}) \tag{11}$$

If we revise the cost function into fitness function:

$$\text{fitness}(\mathbf{x}, \mathbf{v}) = \exp(-\eta \mathrm{J}(\mathbf{x}, \mathbf{v})) \tag{12}$$

where η is a weight parameter, $\hat{\mathbf{v}}$ and $\hat{\mathbf{x}}$ can be substituted into the fitness function for fitness evaluation. Therefore, we can apply DE for fluxes estimation and LS for positional enrichment estimation. Populations of fluxes are iteratively generated by DE and their adaptability to the system are evaluated by the fitness calculated from fluxes themselves and positional enrichment data estimated by the least squares method.

3. Method III: DE on metabolites and LS on fluxes method. Alternatively, if considering (3) and (5), we can apply DE on positional enrichments, while utilizing least squares estimation on fluxes:

$$\hat{\mathbf{v}} = \left(\begin{pmatrix} \mathbf{A}(\hat{\mathbf{x}}) \\ \mathbf{B} \end{pmatrix}^T \begin{pmatrix} \mathbf{A}(\hat{\mathbf{x}}) \\ \mathbf{B} \end{pmatrix} \right)^{-1} \begin{pmatrix} \mathbf{A}(\hat{\mathbf{x}}) \\ \mathbf{B} \end{pmatrix}^T \begin{pmatrix} \mathbf{b}(\tilde{\mathbf{x}}, \tilde{\mathbf{v}}) \\ \mathbf{d}(\tilde{\mathbf{v}}) \end{pmatrix} \tag{13}$$

4. Method IV: DE on both fluxes and metabolites: However, when introducing least squares estimation into fitness evaluation, fitness calculation will inevitably be influenced or even biased by the misleading least squares estimation. A solution to the problem is to apply DE on both quantities, that is fluxes and positional enrichment data are both evolved by DE and their outcomes are evaluated by the combined fitness function.

5 Simulation Results

The example metabolic system *cyclic pentose phosphate pathway* shown in Fig. 2, which was utilized in [12], was used to test the proposed four algorithms: LS-based method (method I) , DE on fluxes and LS on positional enrichments method (method II), DE on positional enrichments and LS on fluxes method (method III) and DE on both positional enrichments and fluxes method (method IV). Please refer to [12] for chemical reactions of the example network. The example system is composed of seven intracellular metabolites, two extracellular metabolites and sixteen fluxes. The fluxes $\mathbf{v_1}, \mathbf{v_3}, \mathbf{v_4}$ and $\mathbf{v_5}$ are assumed to be unidirectional.

The system intracellular and extracellular positional enrichments and fluxes are give by,

$$\mathbf{x} = (P_{\text{g6p}}, P_{\text{f6p}}, P_{\text{gap}}, P_{\text{p5p}}, P_{\text{s7p}}, P_{\text{e4p}}, P_{\text{co2}})^T$$
$$\mathbf{v} = (\overrightarrow{v}_2, \overrightarrow{v}_3, \overrightarrow{v}_4, \overrightarrow{v}_5, \overrightarrow{v}_6, \overrightarrow{v}_7, \overrightarrow{v}_8, \overleftarrow{v}_2, \overleftarrow{v}_3, \overleftarrow{v}_4, \overleftarrow{v}_5, \overleftarrow{v}_6, \overleftarrow{v}_7, \overleftarrow{v}_8)^T$$
$$\tilde{\mathbf{x}} = (P_{\text{xyl}})^T \qquad \tilde{\mathbf{v}} = (\overrightarrow{v}_1, \overleftarrow{v}_1)^T$$

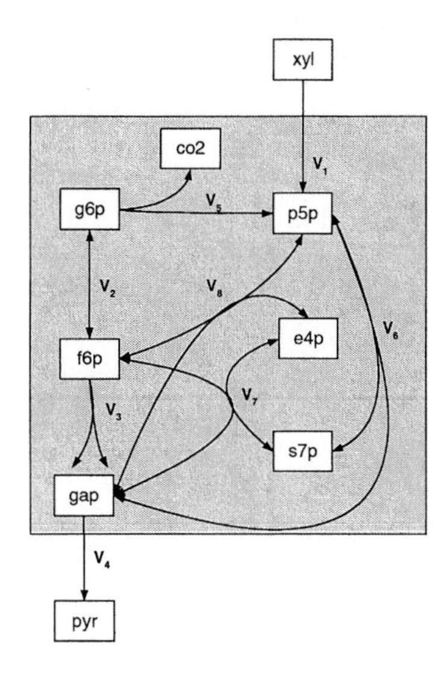

Fig. 2. Cyclic pentose phosphate pathway. Abbreviations: xyl, xylose; g6p, glucose-6-phosphate; f6p, fructose-6-phosphate; gap, glyceraldehyde phosphate; p5p; pentose phosphate; s7p, sedoheptulose-7-phosphate; e4p, erythrose-4-phosphate; co2, CO_2; pyr, pyruvate

In the simulation, we assumed that the data available were positional ^{13}C enrichments synthesized from a metabolic model with input ^{13}C enrichment $\tilde{\mathbf{x}} = (1\ 0\ 0\ 0\ 0)^T$, the fluxes $\tilde{\mathbf{v}}=(1, 0)^T$ and $\mathbf{v}=(1.6667, 0.3333, 1.3333, 1.0000, 1.3333, 1.3333, 1.3333, 0.6667, 0, 0, 0, 0.6667, 0.6667, 0.6667)^T$.

The system identifiability can be determined by the auxiliary matrix $\begin{pmatrix} \mathbf{A}(\mathbf{x}) \\ \mathbf{B} \end{pmatrix}$. In case that the auxiliary matrix is full rank, unknown variables involved in the system can be uniquely identified. In the example system, we found that rank of the system auxiliary matrix was 15, implying that one flux was unidentifiable in the system. It was shown in [12] that the flux $\mathbf{v_2}$ was not identifiable from labeling data and extracellular fluxes. In the following analysis, we apply the four methods to both identifiable and unidentifiable systems.

5.1 Differential Evolution Parameters

The population number was chosen to be ten times more than dimensions of individual population vector [13]. The dimensions for fluxes and metabolites in the simulation were 16 and 32 respectively, therefore, by trial and error, the population numbers for DE on fluxes and metabolites were set to 150 and 300. In method IV, we set population number to 200 in order to compromise between both fluxes and metabolites parts of DE. The weight parameters F for DE on fluxes and metabolites were set to 0.1 and 0.2. The maximum iteration number was 500. The combination probability P_c was 0.3. The weight parameters for λ and η were set to 0.4 and 5.0.

5.2 Simulation Results for Identifiable and Unidentifiable Systems

In order to identify the adaptability of the methods proposed, we revised the system to be identifiable by fixing $\overrightarrow{v_2}$ at its original value, i.e. 1.6667. We ran 50 simulations for each noise level from 2%–10%. The mean square error (MSE) and variance of the MSE are displayed in Table 1 and Table 2. Due to small variances involved in the results, we illustrate the simulation results in two separate figures shown in Fig. 3 instead of using error bars. It is illustrated that all the three DE-based methods produce lower MSE compared to that of LS when the noise level becomes higher. Method II and IV provide nearly the same MSE in all noise

Table 1. MSE performances in identifiable system

Noise Level	2%	4%	6%	8%	10%
Method I	0.0010	0.0088	0.0309	0.0599	0.0997
Method II	0.0080	0.0046	0.0047	0.0057	0.0056
Method III	0.0002	0.0010	0.0066	0.0065	0.0133
Method IV	0.0032	0.0043	0.0040	0.0056	0.0077

Table 2. Variance of MSE performances in identifiable system

Noise Level	2%	4%	6%	8%	10%
Method I	0.0001	0.0006	0.0020	0.0040	0.0027
Method II	0.0004	0.0004	0.0005	0.0008	0.0009
Method III	0.0001	0.0003	0.0021	0.0087	0.0076
Method IV	0.0006	0.0007	0.0013	0.0021	0.0027

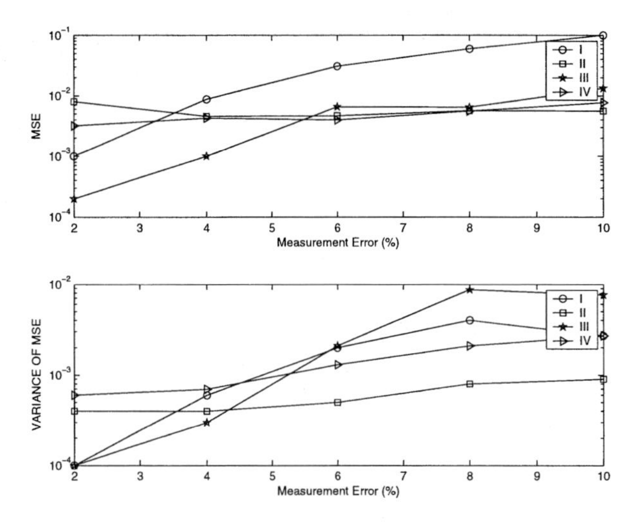

Fig. 3. Simulation performances of LS and DEs in identifiable system

Table 3. MSE performances in unidentifiable system

Noise Level	2%	4%	6%	8%	10%
Method I	0.3180	0.3606	0.4106	0.3917	0.3972
Method II	0.0055	0.0095	0.0078	0.0065	0.0070
Method III	0.3030	0.2874	0.4362	0.3332	0.3250
Method IV	0.0079	0.0091	0.0144	0.0105	0.0190

Table 4. Variance of MSE performances in unidentifiable system

Noise Level	2%	4%	6%	8%	10%
Method I	0.0248	0.0078	0.0142	0.0040	0.0163
Method II	0.0008	0.0009	0.0013	0.0014	0.0015
Method III	0.0034	0.0447	0.0139	0.0576	0.0481
Method IV	0.0018	0.0016	0.0017	0.0026	0.0019

levels above 2%, nevertheless, method IV involves higher variance compared to that of method II. On the other hand, the performance of method III is greatly affected by increased noise levels.

When no additional constraints are applied, the example system is unidentifiable. We again ran 50 simulations for every noise level of 2%–10%. The simulation data are displayed in Table 3 and Table 4. This time, method III and I give nearly the same MSE. Method II and IV show good performances in the unidentifiable system, where method II is slightly better in terms of MSE and variance of the MSE. If we consider the computation efforts involved with the four methods, method II will be the best choice for flux quantification. It provides good accuracy and manageable computational complexity. The reason for the poorer performance of method III is due to a lack of efficient constraint handling techniques in the algorithm. It is found that random positional enrichment data are prone to producing negative fluxes in the unidentifiable system, so that constraints are needed in method III in order to restrict DE search space.

6 Conclusions

Metabolic flux analysis is carried out by formulating the problem as a nonlinear estimation problem. Four methods combining least squares estimation with differential evolution are developed and applied to flux quantification of a simulated metabolic network in both identifiable and unidentifiable systems. It is found that the method that applies differential evolution on fluxes and least squares estimation on positional enrichment data gives the best results among the four methods when noise level becomes high. In practice, it is often the case that not all positional enrichment measurements can be obtained. Further work will be focused on this issue.

References

1. Mendes, P., Kell, D.B.: Non-linear optimization of biochemical pathways: applications to metabolic engineering and parameter estimation. Bioinformatics **14** (1998) 869–883
2. Koza, J.R., Mydlowec, W., Lanza, G., Jessen, Y., Keane, M.A.: Automated reverse engineering of metabolic pathways from observed data using genetic programming. In: Foundations of Systems Biology. (2001) 95–117
3. Iba, H., Mimura, A.: Inference of a gene regulatory network by means of interactive evolutionary computing. Inf. Sci. **145** (2002) 225–236
4. Kitagawa, J., Iba, H.: Identifying metabolic pathways and gene regulation networks with evolutionary algorithms. In Fogel, G.B., Corne, D.W., eds.: Evolutionary Computation in bioinformatics, San Francisco, CA (2003) 95–117
5. Stephanopoulos, G.N., Aristidou, A.A., Nielsen, J.: Metabolic Engineering – Principles and Methodologies. Academic Press, San Diego, USA (1998)
6. Schmidt, K., Nørregaard, L.C., Pedersen, B., Meissner, A., Duus, J.O., Nielsen, J.O., Villadsen, J.: Quantification of intracellular metabolic fluxes from fractional enrichment and ^{13}C-^{13}C coupling constraints on the isotopomer distribution in labeled biomass components. Metab. Eng. **1** (1999) 166–179
7. Wiechert, W.: ^{13}C metabolic flux analysis. Metab. Eng. **3** (2001) 195–206
8. Varma, A., Palsson, B.O.: Metabolic flux balancing: Basic concepts, scientific and practical use. Bio/Technology **12** (1994) 994–998
9. Marx, A., de Graaf, A.A., Wiechert, W., Eggeling, L., Sahm, H.: Determination of the fluxes in the central metabolism of *corynebacterium glutamicum* by nuclear magnetic resonance spectroscopy combined with metabolite balancing. Biotechnol. Bioeng. **49** (1996) 111–129
10. Yang, J., Wongsa, S., Kadirkamanathan, V., Billings, S.A., Wright, P.C.: Self-adaptive evolutionary algorithm based methods for quantification in metabolic systems. In: 2004 IEEE Symp. on Comp. Int. in Bio. (2004) 260–267
11. Storn, R., Price, K.: Differential evolution – a simple and efficient heuristic for global optimization over continuous spaces. J. Glob. Optim. **11** (1997) 341–359
12. Wiechert, W., de Graaf, A.A.: Bidirectional reactiona steps in metabolic networks: I. modeling and simulation of carbon isotope labeling experiments. Biotechnol. Bioeng. **55** (1997) 101–117
13. Storn, R.: On the usage of differential evolution for function optimization. In: 1996 Biennial Conference of the North American Fuzzy Information Processing Society (NAFIPS 1996). (1996) 519–523

GEMPLS: A New QSAR Method Combining Generic Evolutionary Method and Partial Least Squares

Yen-Chih Chen[1], Jinn-Moon Yang[2], Chi-Hung Tsai[1], and Cheng-Yan Kao[1]

[1] Department of Computer Science and Information Engineering,
National Taiwan University, Taipei, Taiwan
{r91051, d90008, cykao}@csie.ntu.edu.tw
[2] Department of Biological Science and Technology & Institute of Bioinformatics,
National Chiao Tung University, Hsinchu, Taiwan
moon@cc.nctu.edu.tw

Abstract. We have proposed a new method for quantitative structure-activity relationship (QSAR) analysis. This tool, termed GEMPLS, combines a genetic evolutionary method with partial least squares (PLS). We designed a new genetic operator and used Mahalanobis distance to improve predicted accuracy and speed up a solution for QSAR. The number of latent variables (*lv*) was encoded into the chromosome of GA, instead of scanning the best *lv* for PLS. We applied GEMPLS on a comparative binding energy (COMBINE) analysis system of 48 inhibitors of the HIV-1 protease. Using GEMPLS, the cross-validated correlation coefficient (q^2) is 0.9053 and external *SDEP* ($SDEP_{ex}$) is 0.61. The results indicate that GEMPLS is very comparative to GAPLS and GEMPLS is faster than GAPLS for this data set. GEMPLS yielded the QSAR models, in which selected residues are consistent with some experimental evidences.

1 Introduction

QSAR techniques are commonly regarded as a key role to computational molecular design. The major goal of QSAR is to formulate mathematical relationships between physicochemical properties of compounds and their experimentally determined *in vitro* biological activities. Thus the derived QSAR model can be subsequently used to predict the biological activities of new derivatives. A good QSAR model both enhances our understanding of the specifics of drug action and provides a theoretical foundation for lead optimization [1].

Many QSAR methodologies have been studied, such as comparative molecular field analysis (CoMFA) [2], the partial least square (PLS) [3], comparative molecular binding energy analysis (COMBINE) [4,5]. Among those methodologies, the PLS analysis is able to deal with strongly collinear input data and make no restriction on the number of variables used. Unfortunately, the predictive performance of PLS model drops and the PLS model becomes complicated when the number of features increases. Several feature selection methods for PLS have been proposed, in which genetic algorithm (GA) combined with PLS approach (GAPLS) has demonstrated the improvement on the prediction and interpretation of model [6]. The essence of GA is to

F. Rothlauf et al. (Eds.): EvoWorkshops 2005, LNCS 3449, pp. 125–135, 2005.

mimics the metaphor of natural biological evolution. GA operates on a population of potential solutions applying the principle of survival of the fittest to produce successively better approximations to optimum solution. Hasegawa *et al.* [7] examined a set of 48 human immunodeficiency virus type I (HIV-1) protease inhibitors by applying GAPLS to the variables derived from COMBINE. Several improved GAPLS models with significantly better predictability than the original study were formulated [5,7,8].

However, the accuracy of GAPLS was still blemished by many features especially deriving from CoMFA or COMBINE. The numerous noise features, which GAPLS did not eliminated completely, were interfered with the significant features strongly correlating with biological activity. Besides, with regard to each possibly select feature set GAPLS needed to spend additional time to decide the optimum number of latent variables (*lv*) through PLS.

Here, we have developed an efficient method for evolving QSAR models by introducing a number of successive refinements which can be summarized as follows: 1) An extra bit *lv*, representing the number of latent variables, was appended to the chromosome of GA and expected to efficiently solve the problem of the optimum number of latent variables though evolutionary process; 2) Mahalanobis distance was adopted to select significant features from numerous features from COMBINE; 3) A new genetic operator, called biased mutation, was designed to lead the evolution of GA toward significant feature set and to reduce the interference of noise features. In this paper, we proposed a new QSAR method by integrating a generic evolutionary method, modified and enhanced from our previous works [9,10] and above issues, and PLS (GEMPLS). GEMPLS is general able to evolve the relationship between biological activities and compound features generated by other methods, such as CoMFA and COMBINE. Here we applied GEMPLS to evolve the QSAR models according to the interaction energy features generated by the COMBINE method on 48 HIV-1 protease inhibitors. Experiments show that GEMPLS is able to improve the predictability and efficiency, at the same time, the selected residues in the yielded QSAR model are consistent with some experimental evidences.

2 Material and Methods

Fig. 1 shows the main steps of applying GEMPLS in the COMBINE analysis: 1) prepare the inhibitor set and model protein-inhibitor complexes; 2) refine protein-inhibitor complexes and calculate features (i.e., energy interactions); 3) select important features by Mahalanobis distance; 4) select features and evolve QSAR models. Each step is described in the following subsections.

2.1 COMBINE: Feature Extraction

The COMBINE analysis is the use of structural information about ligand-receptor complexes [4,5]. When the three-dimensional structure of macromolecule is available, ligand-receptor interaction energies could be calculated as features, which are subjected to statistical analysis in COMBINE. A subset of these features will be account for the ligand affinity. The critical interaction patterns between ligands and the receptor could be identified and be used to derive the correlation of binding affinities.

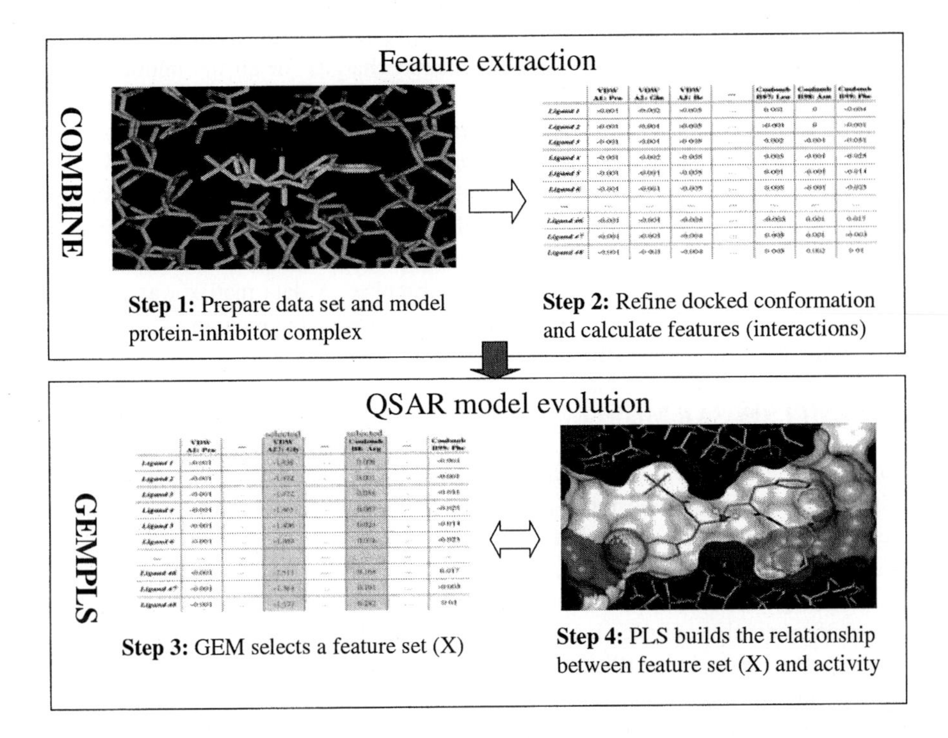

Fig. 1. The framework and steps of GEMPLS applied in the COMBINE analysis

2.1.1 Prepare Data Sets and Model Protein-Inhibitor Complexes

Here, we have chosen 48 inhibitors of human immunodeficiency virus type I (HIV-1) protease studied in previous works [5,7,8]. The chemical structures of HIV-1 protease inhibitors and the 48 complexes were modeled on the crystallographic structure of the complex of HIV-1 protease with L-689,502 solved at 2.25A resolution [11] using the interactive graphics program Insight II. All crystallographic waters were removed with the exception, which is involved in hydrogen bonding with the NH's of the flap residues IleA50 and IleB50 and the oxygen of the inhibitors [12]. All inhibitors were built using L-689, 502 as a template except for the more differential inhibitors 39-45, which employed the inhibitor saquinavir from the HIV-1 protease complex (protein data bank is 1c6z) as a template.

2.1.2 Refine Protein-Inhibitor Complexes and Calculate Features (Interactions)

Each complex model was performed a mild and progressive refinement. The flexibility of each inhibitor was manually explored as necessary to obtain a satisfactory conformation in the enzyme active site, which also corresponded to a low energy conformer. The docked conformation of an inhibitor was energy minimized in the three stages using the consistent valence force field, CVFF [13]. In the first stage, the hydrogen atoms of each complex were allowed reorient. Then the geometry of the inhibitor was optimized while the atoms of both the protein and the water were held fixed. Finally, the whole complex was energy minimized but the protein atoms were restrained to their crystallographic positions. Atom-centered charges for all the inhibitors were derived by fitting the molecular electrostatic potential calculated with the AM1 Hamiltonian [14] to a monopole-monopole expression [15].

The calculated ligand-receptor interaction energies in the refined complexes were partitioned on a per residue basis. Since HIV-1 protease has two protein subunits (subunit A and subunit B) of 99 amino acids, and two energy contributions (van der Waals and electrostatic) are considered for each residue. There are 396 features were yielded to characterize each protein-inhibitor complex. A data matrix was built with 396 columns representing each of the interaction energy features and with 48 rows representing each inhibitor in the data set.

2.2 GEMPLS: QSAR Model Evolution

PLS has played a critical role in the derivation of QSAR in CoMFA or COMBINE studies. Recently, more and more people recognize the benefits of feature selection before PLS regression. GAPLS has been shown as a practical solution. But when the number of features becomes large, GAPLS still has difficulty in driving out noises. And scanning for best *lv* is too inefficient and time consuming. Here, we introduce a number of successive enhancements, which are described in the following paragraphs, to construct our model GEMPLS to overcome the drawbacks of GAPLS.

The general idea of PLS is to try to extract these latent variables, accounting for as much of the manifest feature variation as possible while modeling the inhibitory activities well. To decide both the optimum number of latent variables and prediction error of a QSAR model, we defined the weighted standard deviation error of the predictions (WSDEP) as the scoring function of our GEMPLS:

$$WSDEP = \sqrt{\frac{\sum\left(y_i - y_{pred,i}\right)^2}{N - lv - 1}\left(\frac{100}{95}\right)^{lv}}, \tag{1}$$

where y_i and $y_{pred,i}$ are the observed and predicted inhibitory activities belong to inhibitor *i*, *N* is the total number of samples, and *lv* is the number of latent variables in the current model. In order to improve on the efficiency, we append an extra bit *lv*, representing the number of latent variables, to the original chromosome and expect GEMPLS model to efficiently solve the problem of the optimum number of latent variables though evolutionary process.

2.2.1 Select Features by Mahalanobis Distance

Mahalanobis distance is able be used to measure the deviation of a sample from the mean of the distribution in multivariable calculus. Therefore, the Mahalanobis distance is adopted to identify significant features from all of those.

$$M^2 = (v - \mu)' \Sigma^{-1} (v - \mu) . \tag{2}$$

M is the Mahalanobis distance from the feature vector v (column vector of data matrix here) to the mean vector μ, where Σ is the covariance matrix of the features.

2.2.2 Feature Selections and QSAR Models Evolution

The inhibitory activity usually correlates with few important interaction energy features, that is, most of interaction energy features are meaningless or not apparently distinct from each other. GEM was applied to find out the significant interaction energy features and PLS was used to build the QSAR models based on these selected features. *WSDEP* was used as the objective function to provide a measure of how the internal predictability with respect to the selected features. The fittest individual will have the lowest *WSDEP*.

GEM, modified and enhanced from our previous works [9,10], consists of five steps briefly described in the following:

(1) **Initiation and evaluation of the initial population** ($G_{t=0}$). Each chromosome is composed by an array of feature set and an *lv* value. For example, a chromosome has n+3 bits if the number of candidate feature is n and three bits for *lv* value. The initial population ($G_{t=1}$) of population size (N_p) is created by setting feature bits (0 denote the absence of corresponding feature, and 1 denote its presence) and an *lv* value (denote the number of latent variables and range in [1~5]) of each chromosome to random values and one, respectively. Then PLS is used to build a quasi-QSAR model, and evaluated by the scoring function (*WSDEP*), for each chromosome.

(2) **Selection of the reproductive population**. The chromosomes of reproductive population (P_sG_t) are selected from the population (G_t) with a fixed proportion (P_s) according to the stochastic universal sampling [16].

(3) **Crossover and mutate the reproductive population** (P_sG_t). The offspring population (G_{off}) is generated by uniform crossover with a probability (crossover rate: P_c) and mutation operators, including uniform and biased mutation operators, with a probability (mutation rate: P_m).

(4) **Evaluation of the offspring population** (G_{off}). PLS is then used to build a quasi-QSAR model, evaluated by *WSDEP*, for each chromosome in the offspring population.

(5) **Reinsertion of the child population.** To form the population of the next generation (G_{next}), the chromosomes of the current population (G_t) with lower objectives in the preceding (1- P_s) proportion are protected to the next generation, while the others are replaced with better ones from the offspring population (G_{off}). Let $t = t+1$ and $G_t = G_{next}$.

(6) The cycle of above four steps (from step 2 to 5) is repeated until the number of generation reaches to the maximum number of generations (N_{max}). The values of empirical parameters are defined as follows: $N_p = 100$, $N_{max} = 200$, $P_s = 0.9$, $P_c = 0.6$, and $P_m = 0.05$.

Biased Mutation. The uniform mutation may incur a risk of local convergence and slow evolution because plenty of features will raise the combinatorial complexity of feature space. To reduce the phenomena, the uniform mutation was cooperated with biased mutation to lead the evolution of GA toward significant feature set and to reduce the interference of noise features.

$$F(x_i) = MIN + (MAX - MIN) \times \left(\frac{N_f - x_i}{N_f - 1} \right), \quad (3)$$

where $F(x_i)$ is the probability of selection of feature i; x_i is the rank of feature i in the descending order of Mahalanobis distance of all features, *MIN* and *MAX* are the lower and upper bounds, respectively, of probability of biased mutation; N_f is the number of significant features. The value of $F(x_i)$ is derived from x_i only when x_i is ahead of N_f, otherwise $F(x_i)$ is set to *MIN*. The meaning of $F(x_i)$ is that the more significant feature, the more higher probability of selection. In this study, MAX=0.8, MIN=0.2 and N_f=39.

2.3 Performance Evaluation

The predictability of QSAR model was assessed by the conventional correlation coefficient (r^2), the cross-validated correlation coefficient (q^2), the cross-validated *SDEP* ($SDEP_{cv}$), and external *SDEP* ($SDEP_{ex}$):

$$q^2 = 1 - \frac{\sum (y_i - y_{pred,i})^2}{\sum (y_i - \overline{y})^2}, \quad (4)$$

$$SDEP = \sqrt{\frac{\sum (y_i - y_{pred,i})^2}{N}}, \quad (5)$$

where y_i and $y_{pred,i}$ are the observed and predicted activity of inhibitor i, $y_{pred,i}$, respectively, $\overline{y}$ is the average activity value of the inhibitor set, and N is the total number of inhibitors. The model with more remarkable predictability can provide the higher correlation coefficient (r^2, q^2) and the lower *SDEP* between the observed and predicted inhibitory activities.

3 Results and Discussion

To evaluate the performance of PLS, GAPLS, and GEMPLS, 48 compounds shown in Table A (see appendix) were randomly divided into 6 subsets, and a six-fold cross validation was performed. For each round, one subset (8 compounds) was used as evaluation set, and other subsets (40 compounds) were used to train a QSAR model by Leave-One-Out method to optimize *WSDEP*. Table 1 shows the results, which were the average values of the six-fold cross validation. Five filter conditions (M>0, M>1, M>5, M>10, and M>15) of Mahalanobis distance of features were used to prescreen candidate features before GA feature selection steps. That is, there were five kinds of data matrices (48-by-396 (M>0), 48-by-188 (M>1), 48-by-85 (M>5), 48-by-59 (M>10), and 48-by-39 (M>15)) to be examined on those QSAR models according to these five conditions.

Table 1 shows the execution times, the numbers of selected features, and the values of *lv* of PLS, GAPLS and GEMPLS with five different Mahalanobis distance criteria. Several widely used performance measures, correlation coefficient (r^2), cross-validated correlation coefficient (q^2), cross-validated *SDEP* ($SDEP_{cv}$), external correlation coefficient (r^2_{ex}), and external *SDEP* ($SDEP_{ex}$), were also summarized in Table 1. With the increasing of the degree of filter criteria, the better results were obtained for GEMPLS when the Mahalanobis distance threshold is less than 10. The highest $SDEP_{ex}$ (0.6080) was obtained by GEMPLS with Mahalanobis distance threshold 10. These results reveal that the usage of Mahalanobis distance could successfully discriminate significant features and reduce the ill effect of numerous features generated by the COMBINE method. But when the Mahalanobis distance threshold get higher, the performance degraded due to some important features were filtered out. The adjustment of proper amount of significant features would further improve the predictability and interpretation of QSAR models.

Table 1. The average predictive accuracies of PLS as well as GAPLS and GEMPLS with five different Mahalanobis distances for the HIV-1 protease by six-fold cross validation

Model[a]	Time(s)[b]	M-Features[c]	Features[d]	*lv*	r^2	q^2	$SDEP_{cv}$	r^2_{ex}	$SDEP_{ex}$
PLS	0.047	396	396	3	0.9177	0.8576	0.6031	0.7433	0.7454
GAPLS	11233.5	396	117.8	1	0.9107	0.9045	0.4718	0.6958	0.6582
GEMPLS-M0	1471.1	396	98.8	1	0.9091	0.9029	0.4754	0.7944	0.6464
GEMPLS-M1	766.7	188	42.3	1	0.9101	0.9040	0.4722	0.8030	0.6231
GEMPLS-M5	649.1	85	21.5	1	0.9110	0.9053	0.4691	0.8107	0.6115
GEMPLS-M10	**485.3**	**59**	**19**	**1**	**0.9109**	**0.9046**	**0.4708**	**0.8126**	**0.6080**
GEMPLS-M15	427.7	39	16.7	1	0.9084	0.9027	0.4757	0.7994	0.6284

[a] GEMPLS-M0, M1, M5, M10, M15 mean that GEMPLS analysis performed with feature sets filtered by different Mahalanobis distance thresholds (i.e., 0, 1, 5, 10, and 15).
[b] The executing time is measured on a single-processor of 1.4GHz/PentiumIV PC in seconds.
[c] The number of candidate features is selected by the Mahalanobis distance.
[d] The number of selected features is finally selected by GEMPLS and Mahalanobis distance.

The COMBINE method essentially generates numerous interaction energy features and the usage of Mahalanobis distance is able to reduce the number of these features. One of the evolutionary forces of GEMPLS is come from Mahalanobis distance between a wide distribution of features. At the same time, GEMPLS could decide the optimum number of latent variables for each chromosome though evolutionary process since the *lv* bit was encoded in the chromosome. Table 1 shows that GEMPLS is much faster than GAPLS and slightly better than GAPLS on this data set. Both GEMPLS and GAPLS outperform PLS.

Figure 2 shows a typical QSAR model yielded by GEMPLS. This model reveals some experimental evidences. Figure 2(b) shows the features selected by GEMPLS and Figure 2(a) indicates the pseudo coefficients of the QSAR model evolved by PLS according to these selected features. This evolved QSAR model reflects some important residues of HIV-1 protease shown in Figure 2(c). Residues Asp25, Thr26, and Gly27 are highly conserved catalytic triad and Asp25 is essential to both catalytically and structurally. Residues Ala28 and Asp30 located at subsite S2. The mobile flap, residues 46-54, contains three characteristic regions: side chains that extend outward (Met46, Phe53), hydrophobic chains extending inward (Ile47, Ile54), and a glycine rich region. Residues Pro81, Val82, and Ile84 form the binding pocket. Residues Arg8 and Asp29 at the subsite S3, potentially bind polar residues. These results show that our QSAR model is able to yield many biological meanings.

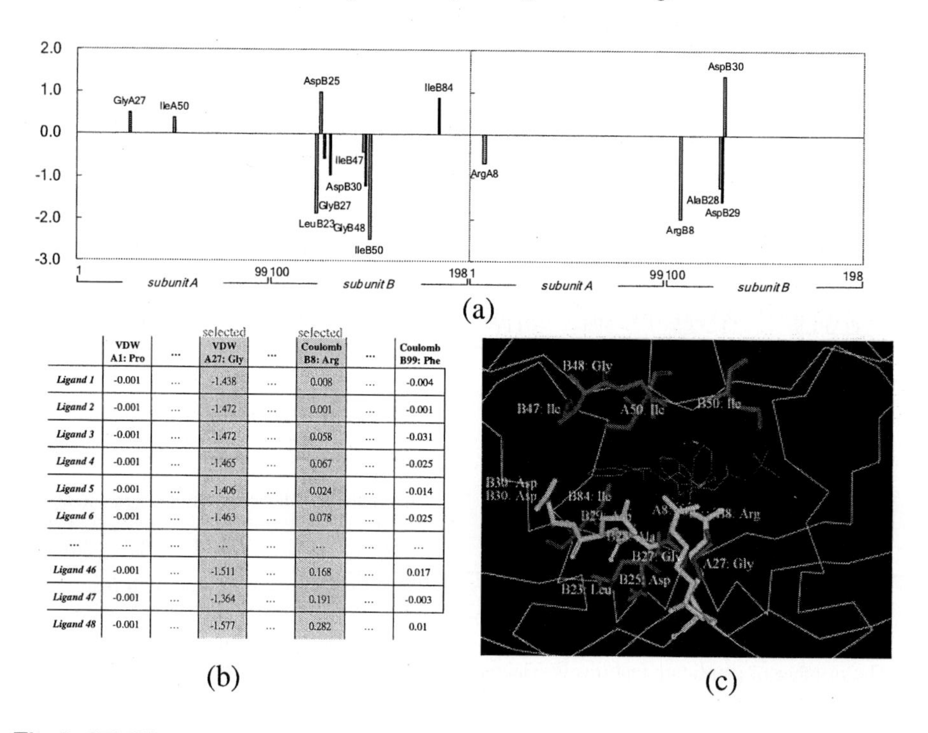

	VDW A1: Pro	...	VDW A27: Gly	...	Coulomb B8: Arg	...	Coulomb B99: Phe
Ligand 1	-0.001	...	-1.438	...	0.008	...	-0.004
Ligand 2	-0.001	...	-1.472	...	0.001	...	-0.001
Ligand 3	-0.001	...	-1.472	...	0.058	...	-0.031
Ligand 4	-0.001	...	-1.465	...	0.067	...	-0.025
Ligand 5	-0.001	...	-1.406	...	0.024	...	-0.014
Ligand 6	-0.001	...	-1.463	...	0.078	...	-0.025
...	...	...	...	...	...	...	...
Ligand 46	-0.001	...	-1.511	...	0.168	...	0.017
Ligand 47	-0.001	...	-1.364	...	0.191	...	-0.003
Ligand 48	-0.001	...	-1.577	...	0.282	...	0.01

Fig. 2. GEMPLS evolves a typical QSAR model of the HIV-1 protease. a) The pseudo coefficients of QSAR model, b) features selected by GEMPLS, and c) the important residues of the QSAR model are consistent with some experimental evidences

4 Conclusions

In summary, we have developed an evolutionary method with a novel scoring function for evolving QSAR models. By integrating a number of genetic operators, each having a unique search mechanism, GEMPLS blends the local and global searches so that they work cooperatively. Our scoring function is indeed able to enhance the prediction accuracy. GEMPLS not only increases the predictability and interpretation of a QSAR model, but also improves the performance and efficiency for feature selection. Our results demonstrate the applicability and adaptability of GEMPLS for QSAR models.

References

(1) Kubinyi, H. QSAR and 3-D QSAR in drug design. 2. Applications and problems. *Drug Discovery Today* **1997**, *2*, 538-546.

(2) Cramer, R. D.; Patterson, D. E.; Bunce, J. D. Comparative molecular field analysis (CoMFA). 1. Effect of shape on binding of steroids to carrier proteins. *Journal of the American Chemical Society* **1988**, *110*, 5959-5967.

(3) Wold, S.; Johansson, E.; Cocchi, M. PLS - Partial least-squares projections to latent structures. In H.Kubinyi Ed., 3D QSAR in Drug Design; Theory, Methods and Applications. *ESCOM Science Publishers, Leiden Holland* **1993**.

(4) Ortiz, A. R.; Pisabarro, M. T.; Gago, F.; Wade, R. C. Prediction of drug binding affinities by comparative binding energy analysis. *Journal of Medicinal Chemistry* **1995**, *38*, 2681-2691.

(5) Perez, C.; Pastor, M.; Ortiz, A. R.; Gago, F. Comparative binding energy analysis of HIV-1 protease inhibitors: incorporation of solvent effects and validation as a powerful tool in receptor-based drug design. *Journal of Medicinal Chemistry* **1998**, *41*, 836-852.

(6) Wanchana, S.; Yamashita, F.; Hashida, M. QSAR analysis of the inhibition of recombinant CYP 3A4 activity by structurally diverse compounds using a genetic algorithm-combined partial least squares method. *Pharmaceutical Research* **2003**, *20*, 1401-1408.

(7) Hasegawa, K.; Kimura, T.; Funatsu, K. GA Strategy for Variable Selection in QSAR Studies: Enhancement of Comparative Molecular Binding Energy Analysis by GA-Based PLS Method. *Quantitative Structure-Activity Relationships* **1999**, *18*, 262-272.

(8) Holloway, B. A prior prediction of activity for HIV-1 protease inhibitors employing energy minimization in the active site. *Journal of Medicinal Chemistry* **1995**, *38*, 305-317.

(9) Yang, J.-M.; Chen, C.-C. GEMDOCK: a generic evolutionary method for molecular docking. *Proteins: Structure, Function, and Bioinformatics* **2004**, *55*, 288-304.

(10) Yang, J. M.; Kao, C. Y. An evolutionary algorithm for the synthesis of multilayer coatings at oblique light incidence. *IEEE/OSA Journal of Lightwave Technology* **2001**, *19*, 559-570.

(11) Thompson, W. J.; Fitzgerald, P. M.; Holloway, M. K.; Emini, E. A.; Darke, P. L. et al. Synthesis and antiviral activity of a series of HIV-1 protease inhibitors with functionality tethered to the P1 or P1' phenyl substituents: X-ray crystal structure assisted design. *Journal of Medicinal Chemistry* **1992**, *35*, 1685-1701.

(12) Wang, Y. X.; Freedberg, D. I.; Yamazaki, T.; Wingfield, P. T.; Stahl, S. J. et al. Solution NMR Evidence That the HIV-1 Protease Catalytic Aspartyl Groups Have Different Ionization States in the Complex Formed with the Asymmetric Drug KNI-272. *Biochemistry* **1996**, *35*, 9945-9950.

(13) Lifson, S.; Hagler, A. T.; Dauber, P. Consistent Force Field Studies of Intermolecular Forces in Hydrogen Bonded Crystals I: Carboxylic Acids, Amides, and the C=O...H Hydrogen Bonds. *Journal of the American Chemical Society* **1979**, *101*, 5111-5120.

(14) Dewar, M. J. S.; Zoebisch, E. G.; Healy, E. F.; Stewart, J. J. P. AM1: A New General Purpose Quantum Mechanical Molecular Model. *Journal of the American Chemical Society* **1985**, *107*, 3902-3909.

(15) Besler, B. H.; Merz, K. M.; Kollman, P. A. Atomic Charges Derived from Semiempirical Methods. *Journal of Computational Chemistry* **1990,** *11,* 431-439.
(16) Baker, J. E. Reducing Bias and Inefficiency in the Selection Algorithm. *The Second International Conference on Genetic Algorithms and their Application* **1987**, 14-21.

Appendix:

Table A. HIV-1 protease inhibitors used in training set (1-32) and test set (33-48), and their corresponding observed inhibitory activities (pIC_{50})

No.	Chemical Structure	pIC_{50}	No.	Chemical Structure	pIC_{50}	No.	Chemical Structure	pIC_{50}
1	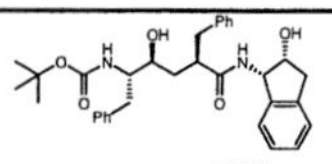 Ph, OH, OH	9.6	2	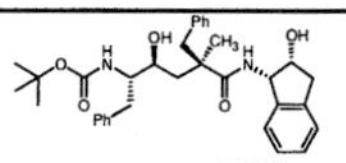 Ph, CH_3, OH, OH	8.11	3	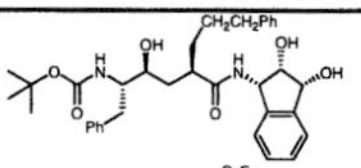 CH_2CH_2Ph, OH, OH, OH	9.72
4	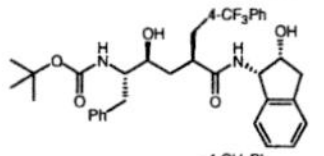 4-CF_3Ph, OH, OH	9.59	5	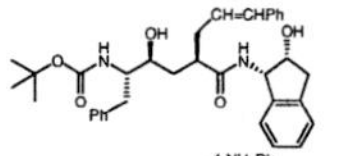 CH=CHPh, OH, OH	9.64	6	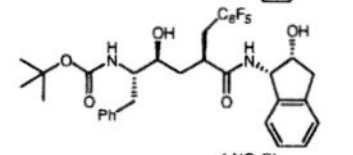 C_6F_5, OH, OH	9.22
7	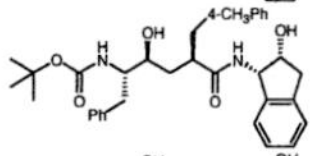 4-CH_3Ph, OH, OH	9.54	8	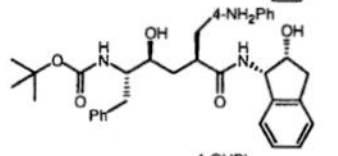 4-NH_2Ph, OH, OH	9.51	9	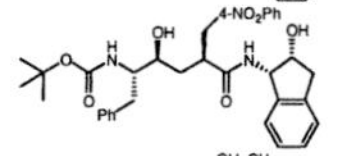 4-NO_2Ph, OH, OH	9.57
10	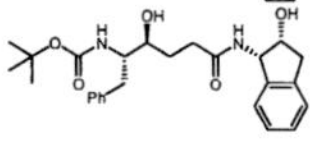 OH, OH	5.53	11	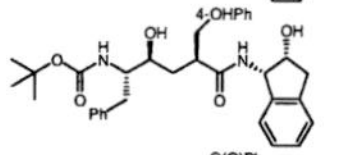 4-OHPh, OH, OH	9.8	12	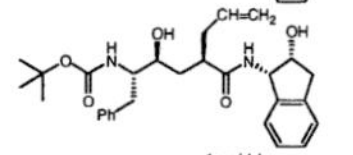 $CH=CH_2$, OH, OH	7.56
13	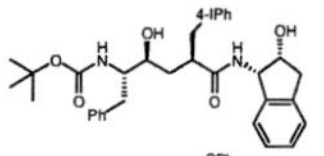 4-IPh, OH, OH	9.14	14	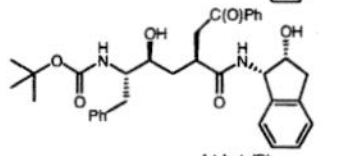 C(O)Ph, OH, OH	8.27	15	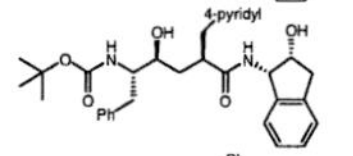 4-pyridyl, OH, OH	9.28
16	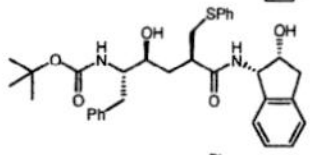 SPh, OH, OH	9.6	17	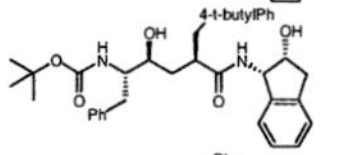 4-t-butylPh, OH, OH	9.77	18	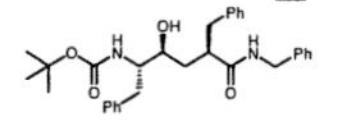 Ph, OH, Ph	6.94
19	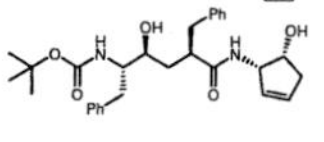 Ph, OH, OH	8.02	20	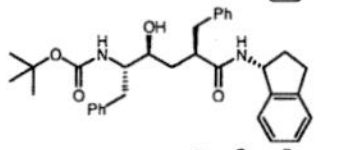 Ph, OH	7.47	21	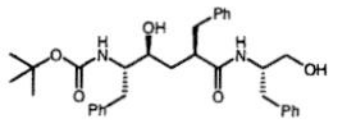 Ph, OH, OH, Ph	6.16
22	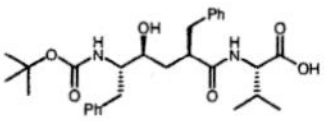 Ph, OH, OH	6.79	23	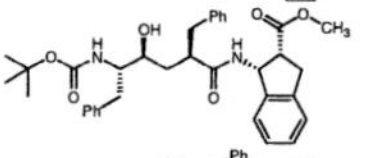 Ph, OH, CH_3	7.18	24	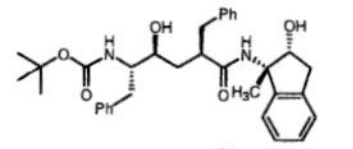 Ph, OH, OH, H_3C	6.67
25	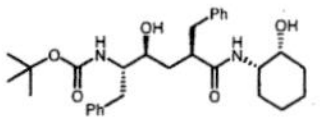 Ph, OH, OH	6.91	26	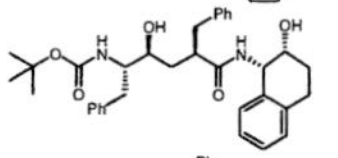 Ph, OH, OH	9.16	27	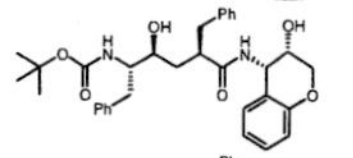 Ph, OH, OH	9.75
28	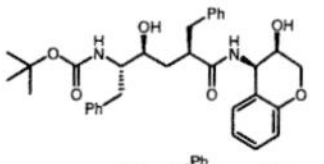 Ph, OH, OH	7.39	29	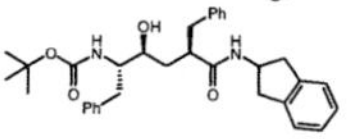 Ph, OH	6.89	30	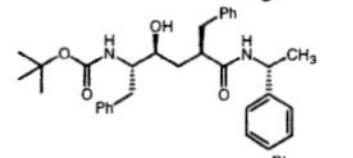 Ph, OH, CH_3	6.84
31	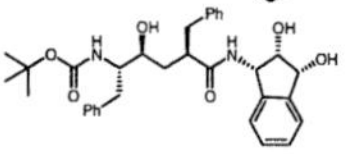 Ph, OH, OH, OH	10	32	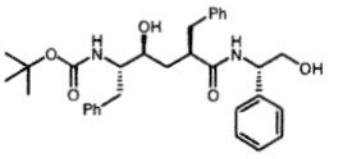 Ph, OH, OH	7.41	33	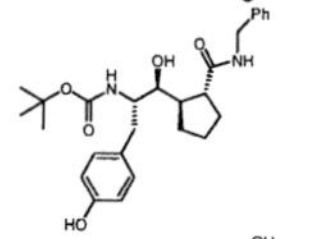 Ph, OH, NH, HO	6.23
34	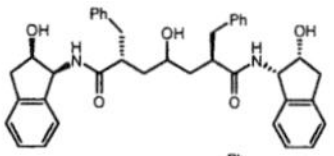 Ph, OH, OH, Ph, OH	9.16	35	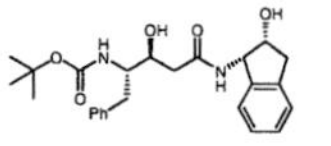 OH, OH	6.25	36	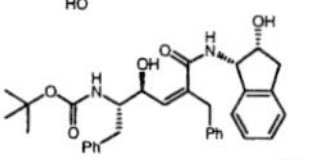 OH, OH, Ph	8.89
37	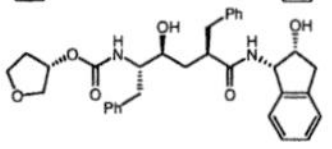 Ph, OH, OH	10.22	38	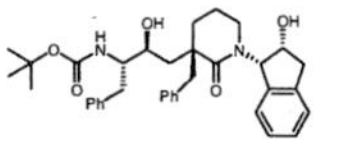 OH, OH, Ph	5.9	39	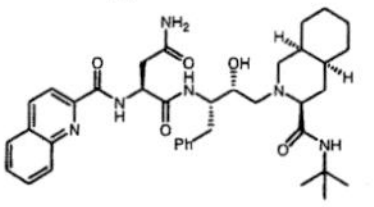 NH_2, OH, NH	9.64

No.	Chemical Structure	pIC_{50}	No.	Chemical Structure	pIC_{50}	No.	Chemical Structure	pIC_{50}
40	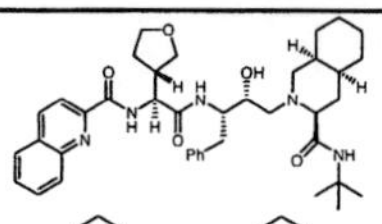	8.27	41	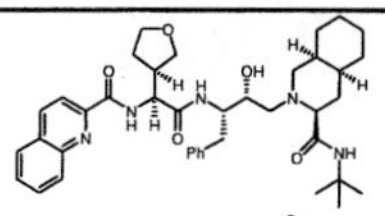	10.27	42	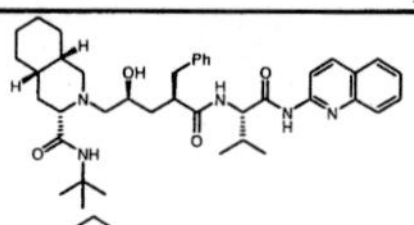	7.28
43	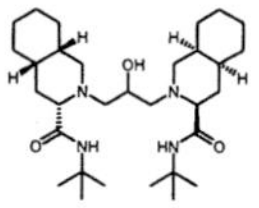	5.17	44	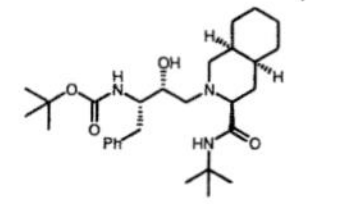	5.52	45	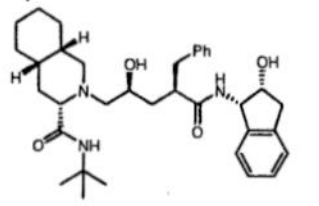	8.12
46	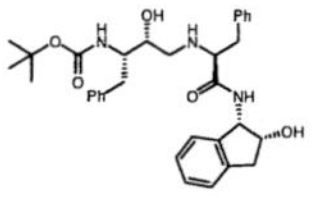	6.64	47	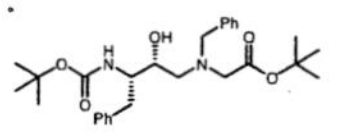	5.33	48	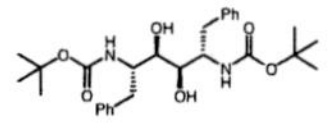	5.86

A Performance Evaluation Framework for Nature Inspired Routing Algorithms

Horst F. Wedde and Muddassar Farooq

Informatik III, University of Dortmund, 44221, Dortmund, Germany
{wedde, farooq}@ls3.cs.uni-dortmund.de

Abstract. Performance evaluation of routing protocols is an important area of research that deals with the analysis and investigation of such protocols. A performance evaluation framework unveils different facets of a protocol and explores its behavior under diversified network operations. The nature inspired routing community, at the moment, lacks such a framework. Therefore, in this paper we propose a comprehensive performance evaluation framework that will empower the routing protocol designers to design state-of-the-art algorithms and extensively evaluate their performance. Using our framework, we exhaustively evaluated three state-of-the-art nature inspired routing algorithms. The results show some undiscovered aspects of the algorithms and provide valuable understanding about their merits and demerits. We believe that this will be the first major step in designing, standardizing and developing a performance evaluation library that will facilitate an extensive and unbiased evaluation of nature inspired routing algorithms.

1 Introduction

Performance evaluation of communication networks is an important area of research that provides a framework for analysis and investigation of the performance of the network by changing different parameters of the network [1]. One such important parameter is the routing protocol used for transporting packets from their sources to destinations. The network engineer who studies the subject in depth gets two benefits, one he is able to make reasoned and educated decisions during the design phase of a routing protocol, and two he is able to design a performance evaluation framework that unveils different facets of the protocol and explore its behavior under diversified network operations. The latter benefit provides a valuable feedback that helps in the re-engineering of the protocol; as a result, the algorithm becomes more robust, dynamic and adaptive.

Over the past decade, researchers in the field of natural computing have developed an interest in designing nature inspired routing algorithms. However, to our knowledge, little effort has been made to come up with a standard performance evaluation framework that provides an unbiased platform for an extensive evaluation of the performance parameters of the algorithms. In the absence of such a framework, researchers focus on optimizing the basic performance parameters, as a result, the readers of the papers never get a complete picture about

F. Rothlauf et al. (Eds.): EvoWorkshops 2005, LNCS 3449, pp. 136–146, 2005.

the behavior of the algorithms. This observation provided us with the motivation to design a comprehensive performance evaluation framework which assists the researchers in extensive and unbiased evaluations of the routing algorithms. The main contributions of the work are implementation of three state-of-the-art nature inspired routing algorithms and then extensively evaluating them with the help of the performance framework. Our results from the experiments show some undiscovered aspects of the algorithms and provide valuable understanding about their merits and demerits. We hope that the framework will act as a guideline for the nature inspired routing algorithm developers and they will report all the parameters suggested in the paper. This will help other researchers in understanding the complete behavior of the algorithms. We believe that this will be the first major step in designing, standardizing and developing a performance evaluation library that will facilitate an extensive and unbiased evaluation of nature inspired routing algorithms.

Organization of the Paper. In the next section we provide a brief overview of three state-of-the-art nature inspired routing algorithms. In Section 3 we will introduce our performance evaluation framework and in doing so we will emphasize the motivation for different parameters in the framework. We will discuss the experimental and simulation set up in Section 4 and then discuss the results obtained from the experiments. Finally we conclude the paper and provide an outlook for our future research.

2 A Review of Nature Inspired Routing Algorithms

In this section we will provide a brief overview of three state-of-the-art nature inspired routing algorithms, namely *AntNet*, DGA and *BeeHive.*

AntNet. *AntNet* was proposed by Di Caro and Dorigo in [2]. In *AntNet* the network state is monitored through two ant agents: *Forward_Ant and Backward_Ant.* A Forward_Ant agent is launched at regular intervals from a source to a certain destination. It uses the same queues as data packets to monitor the real traffic situation. Forward_Ant agent is equipped with a stack memory on which the address and entrance time of each node on its path are pushed. Once the Forward_Ant agent reaches its destination it creates a Backward_Ant agent and transfers all information to it. Backward_Ant visits the same nodes as Forward_Ant in reverse order and modifies the entries in the routing tables based on the trip time from the nodes to the destination. At each node the average trip time, the best trip time and the variance of the trip times for each destination are maintained. The trip time values are calculated by taking the difference of entrance times of two subsequent nodes pushed onto the stack. Backward_Ant agent uses the system priority queues so that it disseminates the information to the nodes as soon as possible. The interested reader may find more details in [2]. Later on the authors of [3] made significant improvements in the routing table initialization algorithm of *AntNet*, bounded the number of Forward_Ant agents during congestion, and proposed a mechanism to handle routing table entries at the neighbors of crashed routers. We made yet another improvement that significantly reduced the number of data packets following cyclic paths: we do not send a data packet to a neighbor from where it has been received.

Distributed Genetic Algorithm-DGA. The authors of [4] showed that the information needed by *AntNet* for each destination is difficult to obtain in real networks. Their idea of *global information* is that there is an entry in the routing table for each destination.This shortcoming motivated the authors to propose in [5] a Distributed Genetic Algorithm (*DGA*) that eliminates the need for having an entry for each destination node in the routing table. In this algorithm ants are asked to traverse a set of n nodes in a particular order, known as a *chromosome*. Once an agent visits the nth node then it is converted into a backward agent that returns to its source node. The authors believe that a value of 6 is good enough for n (chromosome length). In contrast to *AntNet* the backward agents only modify the routing tables at the source node. The source node also measures the fitness of this agent based on the trip time value, and then it generates a new population using single point crossover. New agents enter the network and evaluate the assigned paths. The routing table stores the agents' IDs, their fitness values and trip times to the visited nodes. Routing of a data packet is done through the path that has the shortest trip time to the destination. If no entry for a particular destination is found then a data packet is routed with the help of an agent that has the maximum fitness value. *DGA* was designed assuming that the routers could crash during network operations. We have made a small change in the algorithm: at initialization we launch only four agents rather than half of the population as suggested by the authors. Through this improvement, we have been able to reduce the number of agents on the network without a significant degradation of the performance. Please refer to [5] for details.

BeeHive. This algorithm has been proposed by Wedde, Farooq and Zhang in [6]. The algorithm has been inspired by the communication language of honey bees. Each node periodically sends a *bee agent* by broadcasting the replicas of it to each neighbor site. The replicas explore the network using priority queues and they use an estimation model to estimate the propagation and queuing delay from a node, where they are received, to their launching node. Once the replicas of the same agent arrive at a node via different neighbor sites of the node, they exchange routing information to model the network state at this node. Through this exchange of information by the replicas at a node, the node is able to maintain a quality metric for reaching destinations via its neighbor sites. The algorithm utilizes just forward moving agents and, as opposed to *AntNet*, no statistical parameters are stored in the routing tables. In *BeeHive* a network is divided into *Foraging Regions* and *Foraging Zones*. Each node belongs to only one *Foraging Region*. Each *Foraging Region* has a representative node. A *Foraging Zone* of a node consists of all the nodes from whom a replica of an agent could reach this node in 6 hops. This approach significantly reduces the size of the routing table as compared to *AntNet* because each node maintains detailed routing information only about reaching the nodes within its *Foraging*

Zone and for reaching the representative nodes of the *Foraging Regions*. In this way, a data packet, whose destination is beyond the *Foraging Zone* of a node, is forwarded in the direction of the representative node of the *Foraging Region* containing the destination node. The next hop for a data packet at a node is selected in a probabilistic fashion depending upon the goodness of each neighbor for reaching the destination. *BeeHive* is also fault-tolerant to crashing of routers. The interested reader will find more details in [6].

3 A Performance Evaluation Framework for Nature Inspired Routing Algorithms

We now define our performance evaluation framework that we used for an unbiased evaluation of the algorithms presented in Section 2. We used the guidelines suggested by Higginbottom in [1] and our discussions with the Cisco network engineers in our system management group for defining the important performance parameters of our framework. The parameters and their symbolic representation are shown in Table 1. The first two parameters (MSIA and MPIA) are given as an input to the framework while others are calculated by the framework.

Offered Load. We present two types of traffic to the algorithms, one is session-oriented and another is session-less. In session-oriented traffic, all packets of a session have the same destination. This type of traffic is realistic and tests the congestion control behavior of a routing algorithm. In session-less traffic, the destination of each packet is selected from a uniform distribution. This traffic pattern simulates static network conditions. Generally the researchers use one of the two, though we believe that a good routing algorithm should be able to do congestion control and be competitive under static network loads as well.

Table 1. Symbols used in the paper

Symbol	Description
MSIA	Mean of sessions inter-arrival times (sec)
MPIA	Mean of packets inter-arrival times (sec)
T_{av}	Average throughput (Mbits/sec)
P_d	Percentage of packets delivered
P_{drop}	Percentage of packets dropped
P_{loop}	Percentage of packets that followed a cyclic path
S_c	Percentage of sessions completed
T_d	Average packet delay (msec)
T_{90d}	90th percentile of packet delays (msec)
S_d	Average session delay (msec)
S_{90d}	90th percentile of session delays (msec)
R_o	Routing overhead
S_o	Suboptimal overhead
h_i^{sd}	hops packet i took to reach from node s to node d
h_o^{sd}	minimum hops needed to reach from node s to node d

Average Throughput. Throughput is a measure of how much traffic is successfully received at the intended destination in a unit interval of time [1]. A routing protocol should try to maximize this value.

Packet Delay. We report for all the algorithms the average packet delay and 90th percentile of the packet delays. A good algorithm should be able to deliver packets with minimum delay and with minimum standard deviation of delays.

Session Delay. Our Cisco engineers suggested that in case of session-oriented traffic, the most important parameter is time needed to complete a session. An application layer at the destination node only gets the packets after all the packets are received in the correct order. Packet delay factors out this waiting time and hence favors multi-path algorithms which deliver packets in an out of order manner but with smaller delays.

Sessions Completed. The percentage of sessions that are able to complete without any support from transport layer protocols. For example if only one packet in a session is dropped due to congestion or TTL expiration, we report the session as an incomplete one. We believe that this parameter reports the way packets were deleted on the face of congestion. Our results substantiate that it is more difficult to maximize this parameter than throughput.

Packet Delivery Ratio. This measure tells us how much of the data packets are successfully delivered at their destinations. Under saturated loads a 1% improvement in packet delivery ratio at times means about few 100,000 more data packets delivered at their destinations, however, one can not observe this improvement via throughput values only (see Table 2).

Packet Drop Ratio. The percentage of data packets that are dropped because their time to live timer (TTL) value expired or the queue buffers were full.

Packet Loop Ratio. The percentage of data packets that followed a cyclic path. A cyclic path is an error in an algorithm and should be reported but we do not kill these packets the way the authors of [5] did.

Routing Overhead. The ratio of the bandwidth occupied by the routing/control packets and the total available bandwidth in the network [2]. Generally, the authors of the papers report this parameter to show the control overhead of their routing algorithm.

Suboptimal Overhead. This metric was introduced by [7] in the context of MANETS but we believe that it is equally relevant in fixed networks as well. It is defined as *"The difference between the bandwidth consumed when transmitting data packets from all the sources to destinations and the bandwidth that would have been consumed should the data packets have followed the shortest hop count path"*. Formally we could define the parameter as

$$S_o = \frac{\sum_{d=1}^{n} \sum_{s=1}^{n} \sum_{i=1}^{k} (h_i^{sd} - h_o^{sd}) \times L_i^{sd}}{B_t} \quad , s \neq d \tag{1}$$

where n is total number of nodes in the network, k is total number of packets generated, L_i^{sd} is length of packet i from source s to destination d, and B_t is the total bandwidth of the network. We report this parameter because it implicitly includes the overhead of loops.

4 Simulation Environment and Experimental Findings

In order to evaluate the algorithms *AntNet*, *DGA*, *BeeHive*, and *OSPF*, we implemented all of them in the OMNeT++ simulator [8]. The object oriented design of the simulator allows to prototype algorithms quite easily. For *OSFP* we implemented a static link state routing that implements the deterministic Dijkstra Algorithm [9] which selects the next hop according to the shortest path from a source to a destination. For *AntNet*, *DGA* and *BeeHive* we used the same parameters that were reported by the authors in [2] ,[5] and [6] respectively. The network instance that we used in our simulation framework is the Japanese Internet Backbone (NTTNet). It is a 57 node, 162 bidirectional links network. The link bandwidth is 6 Mbits/sec and propagation delay is from 2 to 5 milliseconds.

As suggested in the Section 3 we have used two types of traffic generators: session-oriented and session-less. Session-oriented traffic is defined in terms of open sessions between two different nodes. Each session is characterized completely by sessionSize (2 Mbits), MSIA , source, destination, and MPIA. The size of data packet is 512 bytes. To inject dynamically changing traffic patterns, we have defined two states: uniform and weighted. Each state lasts 10 seconds and then a state transition to another state occurs. In *Uniform* state (U) a destination is selected from a uniform distribution. While in *Weighted* state (W), a destination selected in *Uniform* state is favored over other destinations. In session-less traffic, the destination of each packet is chosen from a uniform distribution. Such a traffic pattern with low MPIA models static network conditions. Please recall that OSPF is a state-of-the-art algorithm for such a scenario and we compare it with three other algorithms for this scenario.

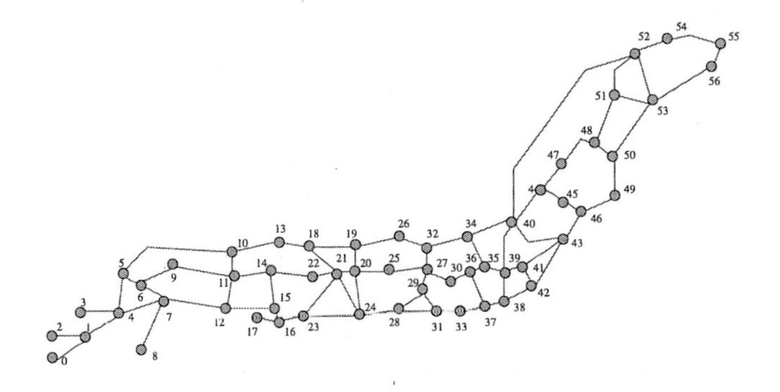

Fig. 1. Japanese Backbone NTTNet

Table 2. Performance Parameters under Saturated Loads

MSIA	Algorithm	T_{av}	P_d	S_c	T_d	T_{90d}	S_d	S_{90d}
6.8	DGA	13.6	86.0	38.0	577	2147	2842	3748
	OSPF	15.5	99.9	98.1	44	188	2643	2845
	AntNet	15.7	99.8	99.7	30	82	2629	2785
	BeeHive	15.8	99.9	99.6	25	68	2620	2771
4.8	DGA	20.8	81.2	43.0	693	2404	2834	3558
	OSPF	24.8	97.9	82.5	262	838	2748	3135
	AntNet	24.9	99.8	97.1	75	327	2674	2938
	BeeHive	25.3	99.9	99.4	28	86	2627	2783
2.8	DGA	32.0	73.6	30.0	1449	3259	2891	3623
	OSPF	36.1	83.2	49.9	706	1930	2680	2987
	AntNet	43.5	99.2	90.1	210	699	2797	3256
	BeeHive	43.3	99.8	95.4	110	391	2746	3110
1.8	DGA	40.7	60.1	21.7	1268	3324	2857	3419
	OSPF	47.6	70.6	41.6	849	2473	2692	2984
	AntNet	63.2	93.8	49.0	721	1762	2992	3653
	BeeHive	62.8	92.9	45.0	602	1595	2887	3479

Saturated Loads. The purpose of the experiments was to study the behavior of the algorithms by gradually increasing the traffic load, through decreasing MSIA from 4.8 to 1.8 seconds. We summarize the results in Table 2 due to page limitations. One could easily conclude that *BeeHive* and *AntNet* are able to maintain higher throughput because they deliver more packets to their destinations. However, the packet delay and session delay, both average and 90th percentile, of *BeeHive* are the best. The difference in session delays, however, is less significant as compared to other parameters. The most striking difference is in the average packet delay, 90th percentile of packet delays and percentage of sessions completed, especially at MSIA=2.8. One could conclude from Table 2 that *BeeHive* and *AntNet* scale better to the increased network load.

Control/Suboptimal Overhead. Figure 2 shows the control overhead and suboptimal overhead of the algorithms. It is quite interesting to note that the suboptimal overhead is much higher than the control overhead (please note that axis have different scales on the two figures). OSPF has the smallest suboptimal overhead though it is able to deliver less packets and complete less sessions (see Table 2). The routing overhead of DGA decreases with an increase in the load and vice versa. This happens due to the genetic algorithm. Recall that the next generation of agents are launched once four agents are received. Under low load, the return times for the agents are smaller, as a result, the agents are launched at a higher rate and vice versa. Since in *AntNet*, Forward_Ant agents use the same queues that data packets also use, therefore more ants were dropped under increased network load and this explains the decrease in routing overhead behavior of the algorithm. Figure 2 justifies the motivation to find a route with

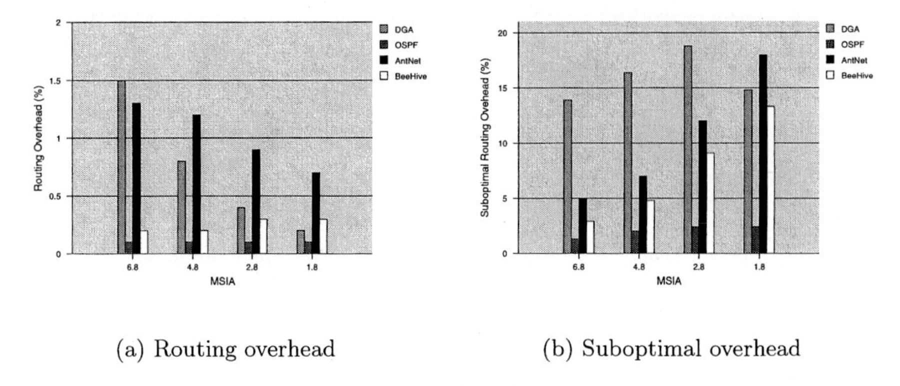

(a) Routing overhead (b) Suboptimal overhead

Fig. 2. Saturated Loads

less hops from source to destination, a parameter that generally received little attention by the nature inspired routing community.

Queue Control Behavior. The purpose of the experiments is to investigate the behavior of performance parameters by varying the size of the queue buffers. Researchers typically show the results with large queue buffers only. Figure 3 shows the effect of varying the size of the queue buffer from 50 packets to 4000 packets. Please remember that MSIA=2.8 and MPIA=0.005 were kept constant during all the experiments. *BeeHive* maintains its superiority over other algorithms when the queue sizes are small, however, *AntNet* achieves a similar performance at queue size of 1000. The figures clearly indicate the shortcoming of OSFP under saturated loads. The increase in the size of queue buffer only results in rise of the packet delay without any significant improvement in the packet delivery ratio. DGA is able to improve its performance with an increase in the size of the queue buffer. These figures indirectly also provide an insight into the queue control behavior of the algorithms. Please remember that at this load 1% increase in packet delivery ratio at times might result in an increase of about 15% completed sessions (see Figure 3).

Size of Routing Table. *AntNet* on the average has 162 entries in the routing table of the nodes in NTTNet as compared to 78 and 57 for *BeeHive* and OSPF respectively. However, such a parameter for DGA is not available because it routes data packets with agents and the memory needed to store them depends on a number of different parameters [5].

Session-less Low Traffic. The purpose of these experiments were twofold, first to test the algorithms in an operation domain where the performance of OSPF is the best and two to be confident that our implementation of DGA is functionally correct. The authors of DGA only published their results under this traffic load. Table 3 shows that our changes in the DGA algorithm rather improved the packet delay and still delivered the same number of packets as that of original DGA. Please closely monitor the performance parameters of OSPF.

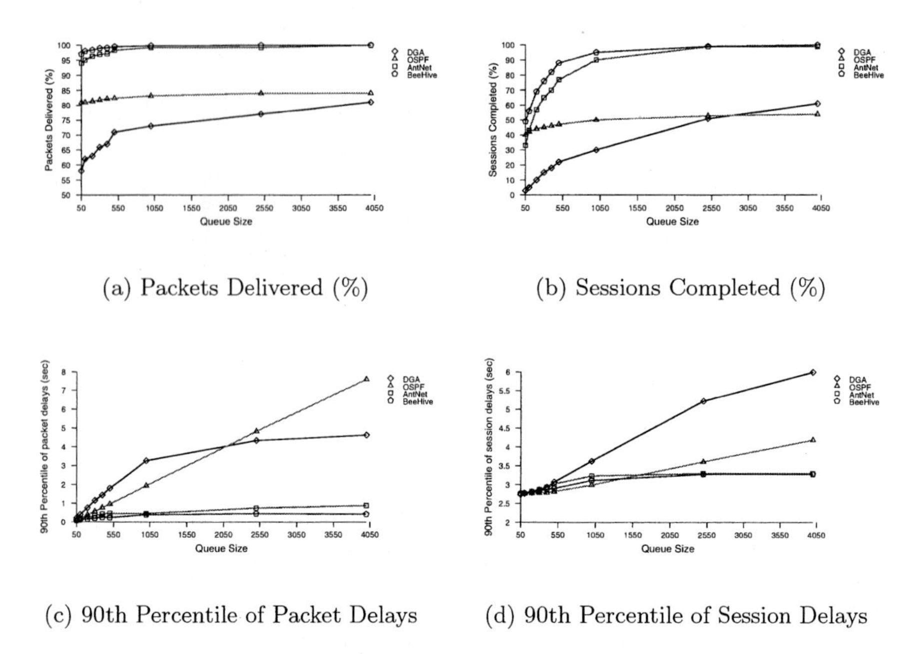

(a) Packets Delivered (%)

(b) Sessions Completed (%)

(c) 90th Percentile of Packet Delays

(d) 90th Percentile of Session Delays

Fig. 3. Queue Control Behavior

Table 3. Performance Parameters under session-less low traffic

MPIA	Algorithm	P_d	P_{drop}	P_{loop}	T_d	T_{90d}	R_o	S_o
0.035	DGA [a]	99.99	0	20	742	-	-	-
	AntNet(Local) [b]	99.99	0	51	398	-	-	-
	DGA(Our)	99.92	0.08	35	137	735	8	27
	AntNet(Global)	99.7	0.3	9	30	65	2.8	5.2
	BeeHive	99.99	0	2.7	23	42	0.37	2.7
	OSPF	99.99	0	0	20	36	0.1	1.02

[a] These results were published in [5]
[b] These results were published in [5]

Only *BeeHive* comes somewhat nearer to it. *AntNet* even drops about 0.3% packets. We investigated the problem and found out that it happened because of loops. We drop a packet once it has taken about 100 hops and is still not at the destination. If we put the same restriction on DGA then it also drops about 6% packets. The results in table 3 clearly demonstrate the superiority of OSPF over all other algorithms, a fact seldom reported by the nature inspired routing community. This also shows that the stochastic distribution of packets under low loads is not a promising approach. *BeeHive* finds a good compromise between static algorithms like OSFP and dynamic algorithms like *AntNet*.

5 Conclusion and Future Research

We have proposed a comprehensive performance evaluation framework that would help the nature inspired routing community to characterize the behavior of a routing algorithm. The power of the framework is that it provides an unbiased testbed over a divergent operations landscape where one could see the benefits/shortcoming of a routing protocol. We have been able to report, through this framework, that all nature inspired routing algorithms perform inferior to OSPF under static network conditions. Another important conclusion is that suboptimal overhead of nature inspired routing protocols is much higher than the routing overhead. But this parameter, to our knowledge, has received no/little attention by the algorithm developers. On the basis of our experience, we believe that the real challenges of designing any new routing protocol are

- to match the performance of OSPF under low loads
- to achieve the performance of a daemon algorithm (see [2] for the algorithm) under saturated loads
- to have suboptimal routing overhead as close to OSPF as possible.
- to design a routing table whose size is of the order of OSPF's routing table.
- to have minimum overhead of processing agents and data packets

In the near future we would like to extend our performance framework for quality of service (QoS) parameters and then evaluate the algorithms with this perspective. Moreover, we would also like to design a profiler to find out the complexity of processing agents and data packets. The processing complexity, measured in number of processor cycles, of an algorithm helps in defining the hardware requirements for the router that will run the algorithm. We also plan to evaluate the algorithms on large topologies up to 1000 nodes to investigate the scalability issues.

References

1. G.N. Higginbottom. *Performance evaluation of communication networks.* Artech house Inc, Norwood, MA, 1989.
2. G. Di Caro and M. Dorigo. AntNet: Distributed stigmergetic control for communication networks. *Journal of Artificial Intelligence*, 9:317–365, December 1998.
3. B. Barán and R. Sosa. A new approach for antnet routing. In *Proceedings of the Ninth International Conference on Computer, Communications and Networks*, 2000.
4. S. Liang, A.N. Zincir-Heywood, and M.I. Heywood. The effect of routing under local information using a social insect metaphor. In *Proceedings of IEEE Congress on Evolutionary Computing*, May 2002.
5. S. Liang, A.N. Zincir-Heywood, and M.I. Heywood. Intelligent packets for dynamic network routing using distributed genetic algorithm. In *Proceedings of Genetic and Evolutionary Computation Conference.* GECCO, July 2002.
6. Horst F. Wedde, Muddassar Farooq, and Yue Zhang. Beehive: An efficient fault-tolerant routing algorithm inspired by honey bee behavior. In *Proceedings of ANTS Workshop, LNCS 3172.* Springer, Sept 2004.

7. C. Santivanez, B. McDonald, I. Stavrakakis, and R. Ramanathan. On the scalability of ad hoc routing protocols. In *Proceedings of IEEE INFOCOM 2002*. IEEE, June 2002.
8. A. Varga. OMNeT++: Discrete event simulation system: User manual. http://www.omnetpp.org.
9. E.W. Dijkstra. A note on two problems in connection with graphs. *Numerical Mathematics*, 1:269–271, 1959.

Empirical Models Based on Hybrid Intelligent Systems for Assessing the Reliability of Complex Networks

Douglas E. Torres D. and Claudio M. Rocco S.

Facultad de Ingeniería, Universidad Central de Venezuela, Venezuela
douglastd@cantv.net, crocco@reacciun.ve

Abstract. This paper describes the application of Hybrid Intelligent Systems (HIS) in a new domain: the reliability of complex networks. The reliability of a network is assessed by employing two algorithms, TREPAN and Adaptive Neuro-Fuzzy Inference Systems ANFIS belonging to the HIS paradigm. TREPAN is a technique to extract linguistic rules from a trained Neural Network, and ANFIS is a method that combines fuzzy inference systems and neural networks. A numerical example, related to a complex network, illustrates the application of the approach and shows that HIS is a promising approach for reliability assessment. The structure function of the complex network analyzed is properly emulated by training both algorithms on a subset of possible system configurations, generated by a Monte Carlo simulation and an appropriate Evaluation Function. Both algorithms successfully describe the network status through a set of rules, which allows the reliability assessment.

1 Introduction

In communication networks, in addition to satisfying constraints that specify links and performance, an important consideration is reliability.

A convenient way of modeling any system is to adopt an undirected or a directed connected graph, called Reliability Block Diagram (RBD) [1], in which every block or link is associated with a system component. A typical RBD can be viewed as a network of components that interact to comply with the system purpose. Each block in an RBD assumes one of two possible states, operating or failed; therefore, a binary variable can be associated with each connection in the system, assuming value 1 if the corresponding connection behaves correctly or value 0 otherwise. In this way, the whole system can be described by a Boolean vector $\boldsymbol{x}$, having as many components as the number of edges in the RBD. Typical RBDs include series and parallel components. If a network has components neither in series nor in parallel, it is considered as a complex network [2]. Many communication networks are modeled as an RBD with a source and a terminal node (*s-t* network).

The state of the whole system, uniquely determined by the Boolean vector $\boldsymbol{x}$, can be operating or failed, and is therefore described by a binary variable y [2]. The Boolean mapping that associates every input vector $\boldsymbol{x}$ to its corresponding output y is called the Structure Function (*SF*) [3]. The procedure employed to retrieve the value of y that

F. Rothlauf et al. (Eds.): EvoWorkshops 2005, LNCS 3449, pp. 147–155, 2005.

corresponds to a given $\boldsymbol{x}$ is usually referred to as an Evaluation Function (*EF*) [4]. The most widely studied reliability measure (*s-t* reliability) assumes that the system is operating if there exists at least one working path from the source node *s* to the terminal node *t*. In this case a depth-first procedure [5,6] can be employed as an *EF*.

There are also situations where the reliability assessment is more complicated (e.g. electric power system), since the success of a given state requires the evaluation of a more complex *EF* (e.g. load-flow evaluations) [7].

An important issue is that the *SF* determination and therefore the reliability assessment, requires the solution of an NP-hard problem [8]. A possible way to reduce the computational burden is to employ Monte Carlo techniques, which attempt to produce an estimate of the network reliability by analyzing a subset of possible system states $\boldsymbol{x}$.

Generally, Monte Carlo techniques require a large number of *EF* evaluations to establish the reliability of a system; therefore, it seems to be convenient to employ a machine learning method for approximating the reliability expression through a reduced collection of *EF* values. To this aim several different approaches have been considered in the literature: Neural Networks [9], Decision Trees [10], Support Vector Machines [9,11] and Hamming Clustering [12]. However, in search of possible more comprehensible models, novel investigation areas are developed, by integrating several intelligent systems. This operative synergy, called Hybrid Intelligent Systems (HIS) [13], seeks to improve the efficiency, reasoning power and comprehensibility of the integrand systems.

The combination of different intelligent techniques such as neural networks, genetic algorithms, decision trees, systems based on fuzzy rules, reasoning based on cases, among others, represents an important area of investigation that has been used in diverse application domains [14]. This paper presents, under the integrative perspective of HIS, an approach for the reliability assessment of complex networks. To this aim empirical models induced by two techniques (TREPAN [15] and ANFIS [16]) are compared, when applied to the samples generated by a Monte Carlo simulation for a given *EF*. To our best knowledge, this approach, based on HIS paradigm, has not hitherto been used to assess the reliability of complex systems. The paper is organized as follows: Section 2 presents some definitions. Section 3 introduces the machine learning methods considered for approximating the reliability of a network, while Section 4 presents the proposed approach to assess the system reliability. Section 5 compares the results obtained through TREPAN and ANFIS, on an example related to a complex network with 21 links.

2 Definitions

It is assumed that the system components have two states (operating and failed) and that component failures are independent events. The state x_i of the *i*th component is defined by Billinton and Allan [2] as:

$$x_i = \begin{cases} 1 \text{ (operating state)} & \text{with probability } P_i \\ 0 \text{ (failed state)} & \text{with probability } Q_i = 1 - P_i \end{cases} \tag{1}$$

where P_i is the probability of success of component *i*.

The state of a system containing d components is expressed by a vector $\boldsymbol{x}=(x_1,x_2,\ldots, x_d)$. To establish if $\boldsymbol{x}$ is an operating or a failed state for the network, a proper Evaluation Function (*EF*) is defined:

$$y = EF(\boldsymbol{x}) = \begin{cases} 1 & \text{if the system is operating in this state} \\ 0 & \text{if the system is failed in this state} \end{cases} \tag{2}$$

A depth-first procedure [5,6] can be employed as an *EF*, if the criterion to be used for establishing reliability is simple connectivity. In the case of capacity requirements, the *EF* could be given by the max-flow-min-cut algorithm [5,6]. For other metrics, special *EFs* may be used.

3 Hybrid Intelligent Systems Models

Hybrid intelligent systems are computational systems, which are based mainly on the integration of soft-computing techniques (especially artificial neural networks and fuzzy systems). This integration allows exploring their advantages in order to increase the overall system performance for a given task [17]. In this paper we study two of such integration examples.

3.1 Extraction of Knowledge from Trained Neural Networks

The Extraction of Knowledge from Neural Networks consists of the development of techniques that allow the comprehensible representation of the knowledge acquired by a trained network. This can be expressed in diverse ways, through symbolic rules, fuzzy rules or decision trees.

The Extraction of Knowledge allows the validation and refinement of the neural networks, as well as the integration of connectionist and symbolic systems. TREPAN [15] is a technique to extract decision trees from a trained neural network.

TREPAN differs from other algorithms that extract information from neural networks in several ways [15]:

1. **The Oracle.** It is used to determine the class of each instance that is presented as a query. The Oracle is used for three different purposes: to determine the class labels for the network's training examples; to determine the class labels for the tree's leaves; and to select the splits that create each of the tree's internal nodes.
2. **Split Types.** That is, the way the input space is partitioned. TREPAN forms trees that use M-of-N expressions for its splits, that is a Boolean expression specified by an integer threshold, m, and a set of n Boolean conditions. An M-of-N expression is satisfied when at least m of its n conditions are satisfied.
3. **Split Selection.** Split selection involves deciding how to partition the input space at a given internal node in the tree. TREPAN uses a special heuristic search process to build its splitting test.
4. **Tree expansion.** TREPAN grows trees using a best-first expansion that chooses the node where there is the greatest potential to increase the fidelity of the extracted tree to the network.

5. **Stopping Criteria.** TREPAN uses local and global stopping criteria. A local criterion considers the state of only a single node to decide whether or not it should be made a leaf, and a global criterion considers the state of the entire tree to decide if the tree-growing process should stop.

TREPAN requires as input the weights and biases of the trained neural network (NN) and a training data set. As output it produces a decision tree that provides an approximation to the function represented by the network.

Figure 1b presents a decision tree extracted by TREPAN, from the system network in Figure 1a. Note that from the extracted tree, it is easy to obtain rules. It is interesting to note that in general the first node in the tree refers to the most important component of the network.

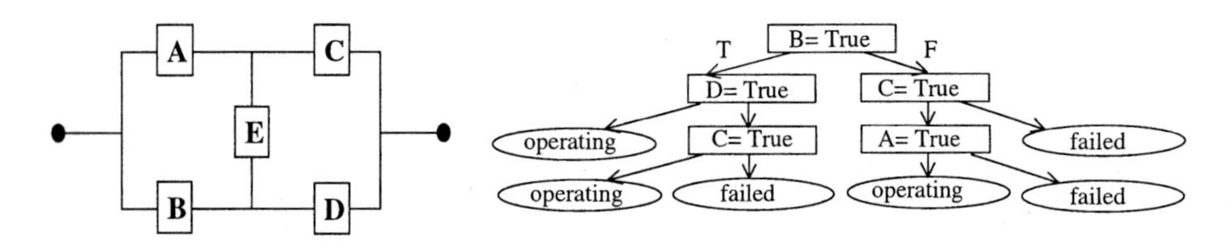

Fig. 1.a: A Complex Reliability Network; **1.**b: Tree extracted by TREPAN

3.2 Fuzzy Model Identification

Fuzzy system identification is the process of identifying the structure and the parameters of a fuzzy model. The construction of the fuzzy model requires two phases.

The first phase is model structure identification, that is the identification of the input variables and rules. The rules identification sub phase consists of the identification of the rule and structure type to be used by the fuzzy model to represent a given input-output data relation. As a result, the first phase produces a set of IF/THEN rules [18]. The second phase corresponds to the adaptation of the parameters (membership functions and coefficients).

In the example presented in section 5, the first phase was performed through the construction of knowledge-based neural networks (KBNN) [19], using the procedure Neural Fuzzy Networks (FuNN) [20]. This procedure combines elements of fuzzy modeling and neural network computations into single connectionist architecture.

The FuNN procedure [21,22] consists of five layers: input variable layer, condition elements (input fuzzy membership function) layer, rule layer, action elements (output fuzzy member function) layer, and output layer.

The rules obtained by FuNN are of the linguistic type. For example, referred to the system shown in Figure 1, the rules developed by FuNN are:

- IF (B is Failed) and (C is Failed) and (E is Failed) THEN (System is Failed)
- IF (B is Operating) and (D is Operating) and (E is Operating) THEN (System is Operating)

- IF (A is Operating) and (C is Operating) THEN (System is Operating)
- IF (A is Failed) and (B is Failed) and (C is Failed) and (D is Failed) THEN (System is Failed)

3.3 Parameter Optimization

The fuzzy inference system described in Section 3.2 presents knowledge in the form of IF/THEN rules. These rules represent the structure and a first approach to the *SF* to be estimated. An additional phase is required for tuning the parameters of a preliminary fuzzy system and then to carry out the evaluation of the final system.

The Adaptive Neuro-Fuzzy Inference Systems (ANFIS) [16,18] uses a hybrid learning algorithm to identify parameters of Sugeno-type fuzzy inference systems (A type of fuzzy inference in which the consequence of each rule is a linear combination of the inputs)[23]. The ANFIS procedure can construct an input-output mapping based on both human knowledge (in the form of fuzzy if-then rules) and input-output data pairs. The parameters that define membership functions are adjusted through the learning process by a back-propagation algorithm.

4 The Proposed Approach

To evaluate the performance of the methods presented in the previous section, the network shown in Figure 2 has been considered [24]. It is assumed that each link has reliability P_i and the goal is to obtain models that approximate the *s-t* reliability metrics.

In order to apply the HIS paradigm, such as TREPAN or ANFIS, it is first necessary to collect a set of examples $(\boldsymbol{x},y)$, where $y = EF(\boldsymbol{x})$, to be used in the training phase and in the subsequent performance evaluation of the resulting models. To this aim, N_T system states have been randomly selected without replacement and for each of them the evaluation of the corresponding value of the *EF* has been performed. In the case to be analyzed, only connectivity is checked to assess if a selected state $\boldsymbol{x}$ corresponds to an operating or to a failed state; thus, the *EF* is given by a depth-first procedure [5-6].

To select the appropriate models a 10-fold cross-validation (CV) was performed. The N_T system states have been divided into 10 subsets of equal size. Every method was trained 10 times, each time leaving out one of the subsets from training and using only this omitted subset to evaluate the obtained model. The performance of each method is measured using sensitivity, specificity and accuracy indexes [25]:

$$\text{sensitivity} = \frac{\text{TP}}{\text{TP}+\text{FN}}; \qquad \text{specificity} = \frac{\text{TN}}{\text{TN}+\text{FP}} \qquad \text{accuracy} = \frac{\text{TP}+\text{TN}}{\text{TP}+\text{TN}+\text{FP}+\text{FN}} \tag{3}$$

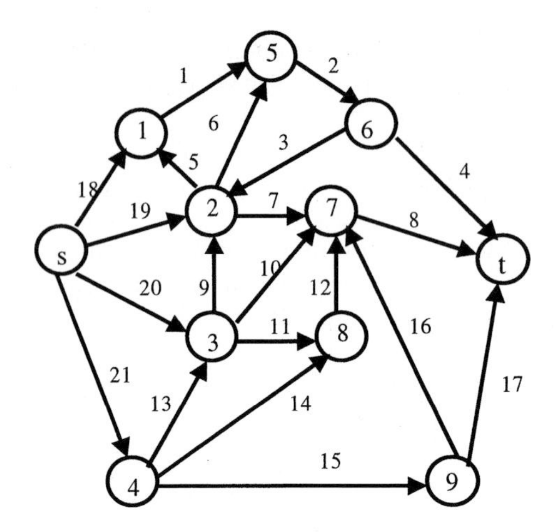

Fig. 2. Complex Network to be evaluated [24]

where:

TP = Number of True Positive classified cases (the method correctly classifies)
TN = Number of True Negative classified cases (the method correctly classifies)
FP = Number of False Positive classified cases (the method labels a case as positive while it is a negative)
FN = Number of False Negative classified cases (the method labels a case as negative while it is a positive)

For system reliability assessment, sensitivity index gives the percentage of correctly classified operating states and the specificity index the percentage of correctly classified failed states.

The average accuracy across the CV is computed, whereas the model with the highest value of accuracy and with the lower complexity is selected for evaluating the system reliability. To this aim, a system state $\boldsymbol{x}^*_i$ is generated at random and it is decided if it is a failed state or not using the induced models. The process is repeated by analyzing N_M system states, which yields the following estimate for the system:

$$\text{Reliability} = \frac{\text{Total number of Operating States}}{N_M} \tag{4}$$

5 Models Determination

The state space associated to the system shown in Figure 2 (2^{21} possible states) is randomly sampled and a data set with 2000 different $(\boldsymbol{x},y)$ pairs is generated.

The neural network (NN) used was an MLP network composed of one input layer, one hidden layer and one output layer. The architecture of the network is denoted by '*i:h:o*' indicating *i* neurons in input layer, *h* neurons in hidden layer and *o* neurons in output layer. The activation function used was a sigmoid function. Different numbers of neurons in the hidden layer were evaluated, using a constructive process, adding neurons to the hidden layer one at time until there is no further improvement in network performance. The network was trained using the Levenberg-Marquardt optimization method combined with Bayesian optimization of the regularization parameters. The aim of regularization is to avoid over-fitting of the model by minimizing the sum of squares of errors and the sum of squares of model parameters [26]. The average accuracy during testing (92.25 %) was obtained with 14 hidden units. The average accuracy during training was 99.99 %. The Matlab Neural Network Toolbox was used to train the network [27].

The trained neural network (21:14:1) is then integrated to the TREPAN model, previously described. The fidelity, that is the percentage of predictions made by the extracted tree that agrees with the predictions made by the network, was 96.39% during the training and 92.75% during the testing. In order to obtain the ANFIS model, a preliminary fuzzy system was induced, using the FuNN model [20]. Some rules extracted by FuNN are:

- **If** x_{16} is Operating and x_{17} is Failed and x_{21} is Operating **then** System is Operating
- **If** x_8 is Failed and x_{17} is Failed and x_{18} is Failed **then** System is Failed
- **If** x_{15} is Operating and x_{17} is Operating and x_{21} is Operating **then** System is Operating
- **If** x_5 is Operating and x_7 is Failed and x_8 is Failed and x_{10} is Operating and x_{12} is Failed and x_{16} is Failed and x_{18} is Failed and x_{19} is Failed and x_{20} is Operating and x_{21} is Failed **then** System is Failed

Some of the rules generated have a physical meaning, related to the minimal path sets (e.g. rule c) and cut sets (e.g. rule b) of the network [2].

Finally, the optimization of the previous fuzzy model was performed by ANFIS. The Matlab Fuzzy Logic Toolbox [27] was used for training.

Table 1 shows the average number of rules along with the average performance results obtained using the 10-fold cross-validation for the algorithms compared, for the training and testing phases. It is interesting to note that although the NN model presents the highest performance indexes during the training, the performance indexes of the TREPAN model during the testing, are superior to the NN and ANFIS models, even if the complexity of the induced ANFIS model (rules generated) is lower. Table 1 also shows that, during the testing phase, TREPAN performs better that the ANFIS model.

Table 1. Average performance results for NN, TREPAN and ANFIS models

Model	Rules	Sensitivity %		Specificity %		Accuracy %	
		Train	Test	Train	Test	Train	Test
NN	14[1]	100	91.67	99.99	92.78	99.99	92.25
TREPAN	108.4	95.61	93.31	97.12	96.21	96.40	94.80
ANFIS	58.3	93.52	91.93	94.36	93.50	93.96	92.75

[1] number of neurons

Once TREPAN and ANFIS are trained, their models are used to estimate the network reliability. A random data set with N_M = 10000 data pairs ($\boldsymbol{x}_i, y_i$) was generated using P_i=0.90. Each system state is evaluated using the *EF* previously selected (that is, a depth-first procedure) and both trained models. The average system reliability based on *EF* was 0.9940. TREPAN and ANFIS models produced the same average system reliability.

Rocco and Moreno [11] and Rocco and Muselli [12] analyze the same network using two machine learning techniques: Support Vector Machines (SVM) and Hamming Clustering (HC). The SVM approach does not produce intelligible rules, so a fair comparison with TREPAN or ANFIS is not possible. HC is a logical synthesis method able to obtain excellent approximations of a Boolean function through the use of IF/THEN rules. Average performance results obtained by HC are better that those obtained in these preliminary results. Future researches, such as parameterizations, different NN training methods, different numbers of layers and nodes among others, are required to obtain more concrete conclusions about the performance of TREPAN and ANFIS.

6 Conclusions

This paper has presented the reliability assessment of a complex system based on two methodological approaches (TREPAN and ANFIS) that uses the hybridization of different soft computing techniques. For the case analyzed, both models, built from a small sample of the state space, produce approximations with a satisfactory accuracy but, in average, the TREPAN model outperforms the best ANFIS model. Nevertheless, from a complexity point of view, the ANFIS model is better since it produces a smaller number of rules.

The use of hybrid intelligent systems seems to be a promising approach for assessing the reliability of complex networks. It not only improves the efficiency of the integrand systems, but also increases the capacity of understanding, since it produces useful topological information about the network, such as minimal path and/or cut sets.

Acknowledgements

The authors are grateful to the anonymous referees for their comments and valuable suggestions on the earlier version of the paper.

References

1. Lynn, N., Singpurwalla, N. (1998): Bayesian assessment of network reliability. SIAM Review, 40: 202-227.
2. Billinton R., Allan R. (1992): Reliability Evaluation of Engineering Systems, Concepts and Techniques, 2nd ed. Plenum Press, New York
3. Dubi A. (2001): Modeling of realistic system with the Monte Carlo method: A unified system engineering approach. Proc. Annual Reliability and Maintainability Symposium – Tutorial Notes.

4. Grimaldi R.P., Shier D.R.: Redundancy and reliability of communication networks. www.math.clemson.edu/ ~shierd/ Shier/abstracts/randr.html.
5. Reingold E., Nievergelt J., Deo N. (1977): Combinatorial Algorithms: Theory and Practice. Prentice Hall, New Jersey.
6. Papadimitriou C. H., Steiglitz K. (1982): Combinatorial Optimization: Algorithms and Complexity, Prentice Hall, New Jersey.
7. Aggarwal K.K., Chopra Y.C., Bajwa J.S. (1982): Capacity consideration in reliability analysis of communication systems. IEEE Trans on Reliability, 31
8. Billinton R. Allan R.N. (1992): Reliability Evaluation of Engineering Systems, Concepts and Techniques, 2nd Ed. Plenum Press, New York
9. Rocco C.M., Moreno J.A. (2002): Machine Learning Models for On-Line Dynamic Security Assessment of Electric Power Systems. In: IBERAMIA 2002 Lecture Notes in Artificial Intelligence, Vol. 2527, Springer-Verlag, Berlin.
10. Rocco C.M. (2003): A rule induction approach to improve Monte Carlo system reliability assessment. Reliability Engineering and System Safety, 82: 87-94.
11. Rocco C.M., Moreno J.M. (2002): Reliability evaluation using Monte Carlo simulation and support vector machine. Lecture Notes in Computer Science, Vol. 2329, Springer-Verlag, Berlin.
12. Rocco C., Muselli M (2004): Empirical models based on machine learning techniques for determining approximate reliability expressions. Reliability Engineering and System Safety, 83:301-309.
13. Jacobsen H.A. (1998): A Generic Architecture for Hybrid Intelligent Systems. IEEE Fuzzy Systems IEEE Fuzzy Systems. Anchorage, Alaska.
14. Jain L.C, Martin N. M.(1998): Fusion of Neural Networks, Fuzzy Sets, and Genetic Algorithms, Industrial Applications. Ed. CRC Press.
15. Craven M.W. (1996): Extracting Comprehensible Models from Trained Neural Networks. Ph.D. Thesis. University of Wisconsin- Madison.
16. Jang J.S, Sun C.T., Mizutani E. (1997): Neuro-Fuzzy and Soft Computing. A Computational Approach to Learning and Machine Intelligence. Ed. Prentice Hall, NJ
17. http://www.comp.nus.edu.sg/~pris/HybridSystems/DescriptionDetailed1.html
18. Jang J.S. (1993): ANFIS Adaptive-Network-based Fuzzy Inference Systems. IEEE Transactions on Systems, Man and Cybernetics, 23:665-685.
19. Lin C.T, Lee C.S.G. (1996): Neural Fuzzy Systems: A Neuro-Fuzzy Synergism to Intelligent Systems. Prentice Hall, NJ
20. http://divcom.otago.ac.nz/infoscie/kel/cbiis.htm
21. Kasabov N. (1996): Investigating the Adaptation and Forgetting in Fuzzy Neural Networks Through a Method of Training and Zeroing. *Proc.of ICNN'96.* Vol 1996: 118-123
22. Kasabov N. (2001): On-line learning, reasoning, rule extraction and aggregation in locally optimized evolving fuzzy neural networks. Neurocomputing Issues 1-4: 25-45
23. Takagi T., Sugeno M.(1985): Fuzzy identification of systems and its application to modelling and control, *IEEE Trans. Systems Man Cybernet.* **15** (1985) 116-132.
24. Yoo Y.B, Deo N. (1988): A Comparison of Algorithm for Terminal-Pair Reliability, IEEE Transaction on Reliability, 37(2): 210-215
25. Veropoulos K, Camppbell, Cristianini N. (1999): Controlling the Sensitivity of Support Vector Machines. Proceedings of the IJCAI99
26. Foresee F.D., Hagan M.T. (1997): Gauss-Newton Approximation to Bayesian Regularization Proc. of IJCNN´97
27. MathWorks (2002) *MatLab 6.5* R13

A Study of an Iterated Local Search on the Reliable Communication Networks Design Problem

Dirk Reichelt, Peter Gmilkowsky, and Sebastian Linser

Institute of Information Systems, Ilmenau Technical University,
Helmholtzplatz 3, P.O. Box 100565, 98684 Ilmenau, Germany
{Dirk.Reichelt, Peter.Gmilkowsky}@tu-ilmenau.de,
Sebastian.Linser@stud.tu-ilmenau.de

Abstract. The reliability of network topologies is an important key issue for business success. This paper investigates the reliable communication network design problem using an iterated local search (ILS) method. This paper demonstrates how the concepts of local search (LS) and iterated local search can be applied to this design problem. A new neighborhood move that finds cheaper networks without violating the reliability constraint is proposed. Empirical results show that the ILS method is more efficient than a genetic algorithm.

1 Introduction

For many network and internet based IT-applications the error-free operation of the underlying network topology is a key issue for business success. Also, the ongoing integration of IT-systems along the value chain requires high-speed communication networks with low failure probability. Therefore, the availability of communication network topologies is an important factor of design. During the designing process, the designer tries to balance the investments made in the network with the services and benefits provided to its users. One important service measurement of network topology is its all-terminal reliability. This is defined as the probability that all nodes in the network will remain connected, given the probability of success/failure for each node and link in the network [1]. The network design problem dealt with in this paper focuses on choosing those links from a given set of communication links, which minimize the network costs under a given network reliability constraint. The design problem itself, and the calculation of network reliability have been proven as a NP-hard problem[2, 3]. In the past, metaheuristics were successfully applied to the network design process [4, 5, 6, 7, 8].

It is known that the calculation of the all-terminal reliability is the most time-sensitive part of the evaluation of the problems solution. Most of the existing heuristics for this problem require a considerable computational effort in order to evaluate several solutions. For example, population based metaheuristics such

F. Rothlauf et al. (Eds.): EvoWorkshops 2005, LNCS 3449, pp. 156–165, 2005.

as genetic algorithms (GA) proposed in [4, 9, 10, 8] perform a high number of reliability evaluations in each generation. In this paper, we investigate a random restart local search (RRLS) and an iterated local search (ILS). Both methods need significantly fewer fitness/reliability evaluations during the optimization process. For the neighborhood search we define a new 1-by-2-move which minimizes the total network costs with respect to the reliability constraint in each step. With the application of this local search strategy in an ILS, we are able to overcome local optima found by a local search and converge into global optimal solutions. In the empirical results presented, the RRLS and the ILS are compared to a GA using a repair heuristic. We show that an ILS finds optimal solutions with less computational effort when compared to the GA approach.

2 Problem Definition

The work presented here investigates the reliable communication network design (RCND) problem. The challenge is to generate network topologies that satisfy a given reliability measurement while minimizing network costs. The design problem has been proven as NP-hard [2]. Several papers have already been published about this problem and others like it. Dengiz et al. [9] propose a GA using a penalty function to incorporate the reliability constraint into the fitness function. Baran and Laufer [11] build upon the work of Dengiz et al. in order to treat bigger problem situations by using a parallel GA. In [6], Dengiz and Alabap introduce a simulated annealing algorithm. In [8], Reichelt et al. propose a genetic algorithm using a repair heuristic. Baran et al. [12] investigate topology design by a GA with multiple objectives.

In this paper the communication network N is modeled as an undirected simple graph $G(E, V)$, where E is the set of edges and V the set of vertices. Each element of the graph (edge or vertex) represents a link or node in the network. It is assumed that the location of each node is given, setup costs for network nodes are not considered and that, for each possible network link l_{ij} between node i and j, cost c_{ij} and reliability $r(l)$ are known. We do not consider repair of failed edges. It is proposed that nodes are perfectly reliable, and edges are either in an operational or failed state. The failures of the edges are statistically independent with known failure probabilities. The reliability of edge e_{ij} in G is $r(e_{ij})$. A network N as a solution for the problem is represented by subgraph $G_N(E_N, V)$ with $E_N \subset E$. The objective function may be stated as:

$$C(N) = \sum_{i=1}^{|V|} \sum_{j=i+1}^{|V|-1} c_{ij} x_{ij} \rightarrow min \qquad (1)$$
$$\text{subject to: } R_{All}(G) \geq R_0$$

where $C(N)$ is the total cost for the network topology and c_{ij} is the cost for a network link between node i and node j. The variable $x_{ij} \in \{0, 1\}$ indicates whether edge e_{ij} from G representing the network link l_{ij} exists in G_N.

For the reliability measurement we use the all-terminal reliability R_{All}. To determine the all-terminal reliability $R_{All}(G_N)$ we consider a set of states $St = (St_1, \ldots, St_n)$ of the graph G_N. Each state St_i represents a subgraph G_{N_i} of G_N when z edges in G_N fail. A state $St_i \in St$ is operational if G_{N_i} is connected. We define $\Phi(St_i)$=1 if St_i is operational, otherwise $\Phi(St_i)$=0. $Pr(St_i)$ is the probability for state St_i. The all-terminal reliability is:

$$R_{All}(G) = \sum_{St_i \in St} \Phi(St_i) \cdot Pr(St_i) \qquad (2)$$

The constraint for $R_{All}(G)$ determines the minimum reliability requirement R_0 for G_N. The calculation of the all-terminal reliability has been proven as NP-hard [3]. For calculation of network reliability the literature proposes exact algorithms [1, 13] for networks with few edges, and Monte Carlo based estimation [14] for large network topologies. In this paper we use a decomposition approach from [1], and an upper bound method from [15] to calculate and estimate the all-terminal reliability.

3 Applying an Iterated Local Search for the Communication Network Design Problem

The concept of iterated local search is a well-known metaheuristic for combinatorial optimization problems. A complete introduction to iterated local search is given in [16]. This section first introduces a new neighborhood move for the RCND problem. Afterwards an iterated local search procedure with the new move is presented.

3.1 A 1-by-2-Neighborhood Operator

To apply a local search to a problem, one has to define a move that generates a solution in the neighborhood $N(s)$. In this paper we propose the 1-by-2-move for the RCND problem. This move decreases the total cost of the network with respect to a given reliability constraint R_0. In order to generate the neighborhood $N(s)$ the move tries to delete the most costly links from the network. The complete 1-by-2-move procedure for a given configuration s represented by G_N and an edge $e_{ij} \in G_N$ is shown in Figure 1.

The procedure first tries to delete the edge e_{ij} from G_N. If the resulting graph is a valid solution, the move is accepted. If the 1-by-2-move cannot construct a valid neighbor by deleting the edge e_{ij} from G_N the procedure searches for a pair of edges that connects the vertices i and j over a vertex l using an overall cheaper pair of edges. A reliability check ensures that the replacement of the most cost-intensive edge by two overall cheaper edges represents a valid solution under the given constraint. To generate the neighborhood for a solution the 1-by-2-move is applied to all cost-intensive edges until an edge cannot be deleted or replaced. Figure 2(a) shows a sample graph. Each edge is ranked by the edge costs c_{ij}. The solid lines denote a subgraph G_N representing a solution (network)

procedure 1-by-2-move
input: e_{ij}, G_N, G , R_0
if $(R_{All}(G_N) \geq R_0$ with : $E_N \setminus \{e_{ij}\})$
    $E_N \leftarrow E_N \setminus \{e_{ij}\}$
    return G_N
candEdgesPair $= \emptyset$
for all $(\{(e_{ik}, e_{jl}) | (e_{ik}, e_{jl} \in E) \wedge (e_{ik}, e_{jl} \notin E_N)\})$
    if $(k = l) \wedge (c_{ij} > (c_{ik} + c_{jl}))$
        add $\{e_{ik}, e_{jl}\}$ to candEdgesPair
sort candEdgesPair by $(c_{ik} + c_{jl})$ ascending
for $\{e_1, e_2\} \in$ candEdgesPair do
    if $(R_{All}(G_N) \geq R_0$ with: $E_N \setminus \{e_{ij}\} \wedge E_N \cup \{e_1, e_2\})$
        $E_N \leftarrow E_N \setminus \{e_{ij}\} \wedge E_N \cup \{e_1, e_2\}$
        return G_N

Fig. 1. 1-by-2-neighbor move

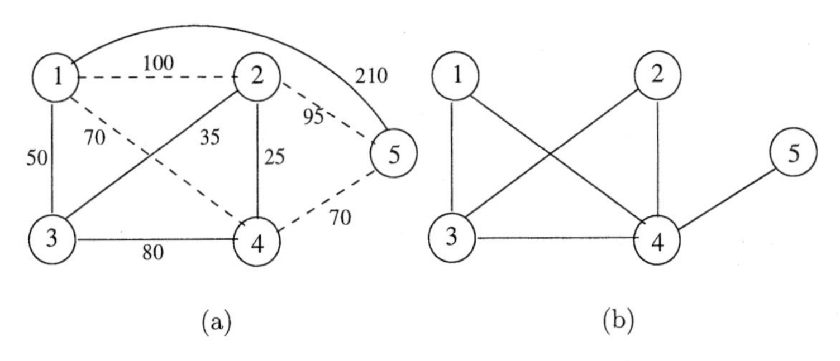

Fig. 2. Example 1-by-2-neighborhood

s_{sample} for the problem. The dashed lines in G indicate those edges currently not used in the network. For this example we use a 1-by-2-neighborhood solution for the edge e_{15}. We assume that $R_{All}(G_N) < R_0$ after the removal of e_{15}. The 1-by-2-heuristic searches two edges with total costs less than c_{15}. Candidate edges for the replacement are the edges $\{\{e_{12}, e_{25}\}, \{e_{14}, e_{45}\}\}$ (assuming that $R_{All}(G_N) > R_0$ for the candidate edges). Since the total costs of $\{e_{12}, e_{25}\}$ is 195 and the total cost of $\{e_{14}, e_{45}\}$ is 140 the 1-by-2-heuristic replaces the edge between vertex 1 and 5 with the edges $\{e_{14}, e_{45}\}$. The resulting neighbor for s_{sample} is shown in figure 2(b).

3.2 An Iterated Local Search with the 1-by-2-Move for the RCND Problem

The concept of ILS is simple and easy to implement. In order to apply ILS to a problem, one has to define an initialization method, a local search operator, a mutation operator, an acceptance criteria and a termination condition. ILS could also be used very easily for problems with a previously defined local search

```
procedure Iterated Local Search
input: G_N, G, R_0
s_0 = RandInitNetwork()
s* = 1-by-2-LocalSearch(s_0)
do
    s' = mutate(s*)
    s''= 1-by-2-LocalSearch(s')
    if (C(N_s'') < C(N_s*))
            set s* = s''
while (s* improved in last 10 iterations)
```

Fig. 3. ILS procedure

operator. The steps performed using the ILS for the RNCD problem are shown in Figure 3. The ILS starts from an arbitrarily randomly-generated solution s_0. For each randomly-generated solution a reliability check to ensure that the solution is valid under the given reliability constraint. Using the 1-by-2-move the local search procedure tries to find a solution s in the neighborhood of s_0 with $C(N_s) < C(N_{s_0})$ and $R(N_s) > R_0$. The solution with the smallest $C(N_s)$ is saved as $s*$. Clearly, a solution $s*$ is a local optimal for a RCND problem.

Afterwards, the ILS procedure enters an inner loop, which iteratively starts a mutation followed by a local search using the best found solution $s*$. In order to bypass the local optima and to arrive at the global optimal solution, a mutation operator generates s' by perturbing the current best solution $s*$. The ILS mutation operator used here randomly adds currently unused edges from G to G_N. The new solution s' generated by the mutation operator is used as a starting solution for the local search procedure. With the 1-by-2-move the local search performs a local search in the neighborhood of s'. The best result found by the local search is saved in s''. A new solution s'' is accepted as a starting solution for the next ILS inner loop iteration when the total cost of the new solution $C(N_{s''})$ is less than the current best network cost $C(N_{s*})$ and the solution does not violate the reliability constraint. Otherwise the inner loop iterates with the best solution previously found. An ILS-run stops if there is no improvement for $C(N_{s*})$ in the ten previous iterations.

4 Experiments

4.1 Experimental Design

For our experiments we used a random restart local search (RRLS), an ILS and a Steady State GA (with overlapping populations). All experiments are done on a PIV- 2Ghz Linux PC. For each heuristic an initial solution is randomly generated. We use the decomposition approach from [1] and an upper bound method from [15] to calculate the all-terminal reliability. To accelerate the algorithms, the reliability calculation procedure first estimates the reliability upper bound by [15]. Only for networks with a reliability upper bound greater than R_0 the exact reliability is calculated using the method from [1]. We perform 1000 independent runs for each problem with the RRLS, and 10 independent runs for each problem with the ILS and GA. The RRLS and LS used the 1-by-2-move from Section 3.1 to generate the neighborhood for a solution. The RRLS and ILS are implemented in C++. The GA uses the repair heuristic from [8] and is implemented in C++ using the GALib[17]. For the GA, we use a population size of 100, 50% replacement, a uniform crossover with a crossover probability of p_{cross}=0.9 and a mutation probability of p_{mut}=0.01.

4.2 Results

Table 1 summarizes the results obtained by the RRLS, the ILS and the GA. The table shows the number of nodes $|V|$ and the number of edges $|E|$, the edge reliabilities $r(l)$ and the reliability constraint R_0. The test problems are taken from [9]. Networks with the same number of nodes and same number of edges differ in the node positions and edge costs. C_{best} for the 11-nodes-problem with $r(l) = 0.9, R_0 = 0.95$ and $r(l) = 0.95, R_0 = 0.95$ are the best costs ever found. The optimal solutions C_{best} for all other test problems are published in [9]. We call the best fitness at the end of a run $C_{bestrun}$. We define D_{AVG} as the average difference (in %) for all runs between the best fitness at the end of one run and C_{best} :

$$D_{AVG} = \frac{\sum_{i=1}^{|runs|} \left(\frac{C_{bestrun_i} * 100}{C_{best}} - 100\right)}{|runs|} \tag{3}$$

where $|runs|$ is the number of runs. A high value for D_{AVG} means that there are many runs with a high difference between $C_{bestrun}$ and C_{Best}. A small value for D_{AVG} shows that an algorithm finds solutions close to C_{Best} in all runs .

D_{Best} (in %) is the difference between the best fitness of all runs and C_{best} for a test problem. Defined as:

$$D_{Best} = \frac{\min\{C_{bestrun_1} \ldots C_{bestrun_{|runs|}}\} * 100}{C_{best}} - 100 \tag{4}$$

D_{best} shows the ability of an algorithm to find C_{best} in at least one run. $D_{Best} = 0$ means that the algorithm finds a solution with $C(N) = C_{best}$ in at least one run. The average number of fitness evaluations over all runs is shown by $\#Eval$.

For each problem the size of the search space is $2^{|E|}$. While a 6-nodes-test problem has only 32768 solutions the search space for a 10 nodes problem with 45 edges is already $2^{45} \approx 3.5 \cdot 10^{13}$. The results show that the RRLS is able to find optimal solutions for small problem instances (up to 7 nodes). An increase of D_{Best} for larger problems indicates that the RRLS remained in a local optimum. This can be explained by the fact that a local search method is unable to leave a local optima during a search. The higher value D_{Best} for larger problems shows that the best solutions found by the RRLS have a higher fitness difference $C_{bestrun} - C_{Best}$. This means that the RRLS often converges in a local optima. For the RRLS, the high value of D_{AVG} for all test problems shows that the RRLS only finds global optima in few of the runs, while most runs end up with a local optimal solution.

Table 1. Comparison of RRLS, ILS and GA

					RRLS			ILS			GA		
$\|V\|$	$\|E\|$	$r(l)$	R_0	C_{best}	D_{AVG}	D_{Best}	#Evals	D_{AVG}	D_{Best}	#Evals	D_{AVG}	D_{Best}	#Evals
6	15	0,9	0,9	231	23,33%	0,00%	10	9,52%	0,00%	49	0,00%	0,00%	288
6	15	0,9	0,9	239	41,63%	6,28%	10	6,02%	0,00%	62	0,00%	0,00%	184
6	15	0,9	0,9	227	41,35%	0,00%	10	6,17%	0,00%	55	0,00%	0,00%	171
6	15	0,9	0,9	212	50,35%	0,00%	10	2,40%	0,00%	48	0,00%	0,00%	278
6	15	0,9	0,9	184	42,13%	0,00%	10	5,00%	0,00%	44	0,00%	0,00%	193
6	15	0,95	0,95	227	20,40%	0,44%	10	13,00%	0,00%	36	1,10%	0,00%	758
6	15	0,95	0,95	213	46,63%	0,00%	10	1,31%	0,00%	46	0,00%	0,00%	284
6	15	0,95	0,95	190	60,53%	0,00%	10	13,58%	0,00%	64	0,00%	0,00%	201
6	15	0,95	0,95	200	52,29%	0,00%	10	9,50%	0,00%	33	0,50%	0,00%	436
6	15	0,95	0,95	179	41,03%	0,00%	10	6,70%	0,00%	77	0,00%	0,00%	762
7	21	0,9	0,9	189	31,72%	0,00%	14	7,67%	0,00%	124	0,00%	0,00%	598
7	21	0,9	0,9	184	57,85%	0,00%	14	1,25%	0,00%	92	0,00%	0,00%	403
7	21	0,9	0,9	243	38,03%	3,29%	14	6,42%	0,00%	94	2,18%	0,00%	418
7	21	0,9	0,9	129	53,36%	2,33%	15	4,14%	0,00%	99	0,85%	0,00%	657
7	21	0,9	0,9	124	112,22%	11,29%	15	17,10%	0,00%	90	0,00%	0,00%	319
7	21	0,95	0,95	185	30,60%	0,00%	14	7,73%	0,00%	67	0,00%	0,00%	977
7	21	0,95	0,95	182	54,57%	0,00%	14	4,51%	0,00%	75	0,00%	0,00%	782
7	21	0,95	0,95	230	40,67%	2,17%	14	4,65%	0,00%	72	1,04%	0,00%	416
7	21	0,95	0,95	122	52,44%	5,74%	14	4,84%	0,00%	79	0,57%	0,00%	846
7	21	0,95	0,95	124	104,97%	5,65%	14	13,55%	0,00%	70	0,00%	0,00%	291
8	28	0,9	0,9	208	48,49%	4,81%	17	4,71%	0,00%	149	0,00%	0,00%	401
8	28	0,9	0,9	203	55,27%	4,93%	18	0,00%	0,00%	137	0,00%	0,00%	507
8	28	0,9	0,9	211	58,97%	18,96%	14	6,93%	0,00%	84	0,00%	0,00%	685
8	28	0,9	0,9	291	42,97%	0,00%	16	2,75%	0,00%	159	0,10%	0,00%	710
8	28	0,9	0,9	178	54,02%	0,00%	19	1,01%	0,00%	143	0,84%	0,00%	887
8	28	0,95	0,95	179	59,26%	0,00%	17	3,02%	0,00%	149	0,28%	0,00%	522
8	28	0,95	0,95	194	52,40%	4,12%	18	4,28%	0,00%	128	0,31%	0,00%	836
8	28	0,95	0,95	197	46,25%	0,00%	19	5,69%	0,00%	88	0,46%	0,00%	1070
8	28	0,95	0,95	276	42,97%	0,36%	16	4,78%	0,00%	108	2,17%	0,00%	805
8	28	0,95	0,95	173	51,65%	2,31%	19	8,79%	0,00%	89	1,62%	0,00%	1133
9	36	0,9	0,9	239	57,93%	2,93%	21	6,78%	0,00%	136	0,00%	0,00%	790
9	36	0,9	0,9	191	64,26%	1,57%	23	5,85%	0,00%	134	1,57%	0,00%	979
9	36	0,9	0,9	257	42,59%	8,95%	23	6,85%	0,00%	129	2,18%	0,00%	1051
9	36	0,9	0,9	171	73,73%	4,09%	21	5,15%	0,00%	153	0,00%	0,00%	714
9	36	0,9	0,9	198	58,57%	0,51%	22	0,05%	0,00%	148	0,00%	0,00%	809
9	36	0,95	0,95	209	65,01%	0,00%	21	1,96%	0,00%	151	0,00%	0,00%	683
9	36	0,95	0,95	171	70,97%	0,00%	23	11,29%	0,00%	165	0,70%	0,00%	1261
9	36	0,95	0,95	233	48,61%	6,87%	22	8,28%	0,00%	137	0,69%	0,00%	1103
9	36	0,95	0,95	151	80,00%	0,00%	21	12,38%	0,00%	137	1,79%	0,00%	757
9	36	0,95	0,95	185	55,62%	0,00%	22	6,92%	0,00%	111	0,16%	0,00%	1062
10	45	0,9	0,9	131	42,97%	1,53%	30	4,05%	0,00%	222	0,53%	0,00%	1282
10	45	0,9	0,9	154	84,15%	10,39%	27	11,30%	0,00%	224	0,00%	0,00%	940
10	45	0,9	0,9	267	52,02%	1,87%	28	0,75%	0,00%	215	0,22%	0,00%	1207
10	45	0,9	0,9	263	45,37%	0,00%	29	2,55%	0,00%	158	0,00%	0,00%	791
10	45	0,9	0,9	293	48,08%	14,33%	26	8,58%	0,00%	194	2,87%	0,00%	1208

Table 1. (*Continued*)

					RRLS			ILS			GA		
$\|V\|$	$\|E\|$	$r(l)$	R_0	C_{best}	D_{AVG}	D_{Best}	#Evals	D_{AVG}	D_{Best}	#Evals	D_{AVG}	D_{Best}	#Evals
10	45	0,95	0,95	121	42,42%	3,31%	28	5,04%	0,00%	184	2,64%	0,00%	1614
10	45	0,95	0,95	136	88,20%	5,88%	26	15,59%	0,00%	245	0,00%	0,00%	891
10	45	0,95	0,95	236	57,28%	5,93%	27	5,44%	0,00%	198	2,29%	0,00%	1096
10	45	0,95	0,95	245	45,58%	0,00%	29	0,57%	0,00%	228	0,12%	0,00%	1037
10	45	0,95	0,95	268	49,03%	12,60%	30	10,90%	0,00%	164	0,90%	0,00%	1352
11	55	0,9	0,9	246	48,08%	9,35%	31	3,46%	0,00%	217	0,00%	0,00%	1200
11	55	0,9	0,95	277	61,35%	12,64%	34	8,45%	0,00%	230	0,00%	0,00%	1049
11	55	0,95	0,95	210	52,91%	0,48%	31	10,14%	0,00%	195	1,33%	0,00%	1543

For all problem instances the average number of fitness/reliability evaluations done by RRLS is less than that of the results obtained from the ILS and the GA. The average number of fitness evaluations for all test problems done by the RRLS is less than 35 runs. This shows the ability of the 1-by-2-move to rapidly guide the search to a local optimal solution. The results show that the extension of the LS by an additional mutation operator in an ILS method bypasses the local optima, and finally ends up with global optimal solutions. An analysis of D_{Best} for the ILS points out that the heuristic found optimal solutions for all test problems. When comparing the number of fitness/reliability evaluations (*#eval*), one finds that the ILS requires more evaluations than the RRLS, but significantly less computational effort than the GA. An analysis of the results obtained by the ILS and the GA shows that D_{Best} for the GA is equal to the D_{Best} for the ILS. But the GA performed more fitness evaluations than the ILS. The D_{AVG} measure shows that the GA, compared to the ILS, has a low diversification of the best solution in all ten runs. Compared to the D_{AVG} results for the ILS, the GA converges more often than the ILS does with the global optimal solution in all ten runs. This result can be explained by the nature of the GA heuristic. Over a GA run, solutions in the population with a low fitness quality are replaced by better solutions. At the end of a GA run only the best of the 100 solutions in the population is used for the D_{AVG} measure. In the ILS, the search process is driven by only one solution, which is not always the optimal solution of the problem in all runs.

Figure 4 shows a comparison of the running time for the ILS and GA heuristics. Each plot shows the average running time (in seconds) for all runs for the same problem class (same number of nodes) and the same configuration (link reliability and R_0). The plots in Figure 4(a) point out that the ILS is faster than the GA for all test problems. One can see that the difference between the ILS and GA running times increase as the problem size (number of nodes) grows. Figure 4(b) shows a similar result with only one exception (10 nodes). Although the GA performed more fitness evaluations than the ILS for the 10-nodes-test problems (with $r(l) = 0.95$ and $R_0 = 0.95$) the GA is faster than the ILS. This can be explained by the implemented all-terminal reliability calculation procedure which is based on a decomposition approach (see [1]). For highly reliable networks the procedure stops after a few decomposition steps and requires low

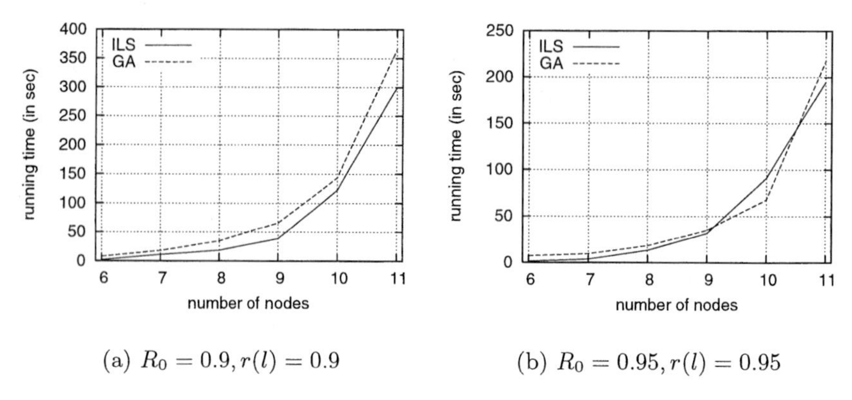

(a) $R_0 = 0.9, r(l) = 0.9$ (b) $R_0 = 0.95, r(l) = 0.95$

Fig. 4. Comparison of running time for ILS and GA

computational effort. For networks with $R_{All} \approx R_0$ the procedure must perform more decomposition steps which require a higher computational effort. This fact speeds up the GA, when compared to the ILS, as it generates many high reliable networks. The all-terminal reliability evaluation for these highly reliable networks can be done very quickly. The all-terminal reliability calculation procedure used here also explains an other interesting fact. A comparison of the running times for different configurations (link reliability and R_0) and the same problem class (same number of nodes) shows, that both heuristics run faster for $r(l) = 0.95$ and $R_0 = 0.95$ than for $r(l) = 0.9$ and $R_0 = 0.9$ although the heuristics performed more or approximately the same number of fitness evaluations. As mentioned before, this is caused by the reliability calculation procedure. If the all-terminal reliability evaluation procedure is replaced by a Monte Carlo simulation, always drawing the same number of samples for each reliability evaluation, the procedure has a constant computational effort. In this case the running time of a heuristic is proportional to the number of fitness evaluations.

5 Conclusions

This paper investigated a local search and iterated local search approach for the reliable communication network design problem. Existing approaches are capable of finding good solutions, but call for high computational effort levels. Due to the application of local search methods, the number of fitness evaluations was decreased while maintaining the same quality of solutions. We presented a new 1-by-2-move to generate the neighborhood for a solution. By dropping and replacing the most cost-intensive network links by two overall cheaper links under the given reliability constraint, the move found fitter neighbor solutions. The move connected two nodes that were previously connected directly via a third node because the indirect connection was cheaper and did not violate the reliability constraint. The empirical results showed that a local search with the 1-by-2-move often converges in a local optimum. We proposed an iterated local search that is more efficient than existing approaches. This iterated local search with the 1-by-2-move finds global optimal solutions and outperforms a GA using a repair heuristic for a set of test problems.

References

1. Yunbin Chen, Jiandong Li, and Jiamo Chen. A new algorithm for network probabilistic connectivity. In *Military Communications Conference Proceedings*, volume 2, pages 920–923, 1999.
2. M. R. Garey and D. S. Johnson. *Computers and Intractibility: A Guide to the Theory of NP-Completeness.* W. H. Freeman and Company, San Fransisco, 1979.
3. Li Ying. Analysis method of survivability of probabilistic networks. *Military Communication Technology Magazine*, 48, 1993.
4. L.T.M. Berry, B.A. Murtagh, G. McMahon, S. Sudgen, and L. Welling. An integrated GA-LP approach to communication network design. *Telecommunication Systems*, 12:265–280, 1999.
5. Baoding Liu and K. Iwamura. Topological optimization model for communication network with multiple reliability goals. *Computer and Mathematics with Applications*, 39:59–69, 2000.
6. B. Dengiz and C. Alabap. A simulated annealing algorithm for design of computer communication networks. In *Proceedings of World Multiconference on Systemics, Cybernetics and Informatics, SCI 2001*, volume 5, 2001.
7. B. Fortz and M. Labbe. A tabu search heuristic for the design of two-connected networks with bounded rings. Working Paper 98/3, Universit Catholique de Louvain, 2002.
8. Dirk Reichelt, Franz Rothlauf, and Peter Gmilkowsky. Designing reliable communication networks with a genetic algorithm using a repair heuristic. In *Proceedings 4th European Conference, EvoCOP 2004*, pages 177–187. Springer, 2004.
9. B. Dengiz, F. Altiparmak, and A. E. Smith. Local search genetic algorithm for optimal design of reliable networks. *IEEE Trans. on Evolutionary Computation*, 1(3):179–188, 1997.
10. Darren Deeter and Alice E. Smith. Economic design of reliable networks. *IIE Transactions*, 30:1161–1174, 1998.
11. Benjamin Baran and Fabian Laufer. Topological optimization of reliable networks using a-teams. In *Proceedings of World Multiconference on Systemics, Cybernetics and Informatics - SCI '99 and ISAS '99*, volume 5, 1999.
12. S. Duarte and B. Barn. Multiobjective network design optimisation using parallel evolutionary algorithms. In *Proccedings of XXVII Conferencia Latinoamericana de Informtica CLEI'2001*, Merida, Venezuela, 2001.
13. K.K. Aggarwal and Suresh Rai. Reliability evaluation in computer-communication networks. *IEEE Transactions on Reliability*, 30(1):32–35, 1981.
14. E. Manzi, M. Labbe, G. Latouche, and F. Maffioli. Fishman's sampling plan for computing network reliability. *IEEE Transactions on Reliability*, 50(1):41–46, 2001.
15. A. Konak and A. Smith. An improved general upperbound for all-terminal network reliability, 1998.
16. Helena R. Loureno, Olivier C. Martin, and Thomas Stuetzle. Iterated local search. Economics Working Papers 513, Department of Economics and Business, Universitat Pompeu Fabra, November 2000.
17. M. Wall. Galib: A c++ library of genetic algorithm components, 1998.

Unsupervised Anomaly Detection Based on an Evolutionary Artificial Immune Network[1]

Liu Fang and Lin Le-Ping

School of Computer Science and Engineering,
Xidian University, Xi'an 710071, China
f63liu@163.com

Abstract. To solve the problem of unsupervised anomaly detection, an unsupervised anomaly-detecting algorithm based on an evolutionary artificial immune network is proposed in this paper. An evolutionary artificial immune network is "evolved" by using unlabeled training sample data to represent the distribution of the original input data set. Then a traditional hierarchical agglomerative clustering method is employed to perform clustering analysis within the algorithm. It is shown that the algorithm is feasible and effective with simulations over the 1999 KDD CUP dataset.

1 Introduction

Anomaly detection has been an active field of intrusion detection research since it was originally proposed by Denning [1]. Anomaly detection is one of the two major paradigms for training data mining-based intrusion detection systems. Anomaly detection approaches build models of normal data and attempt to detect deviations from the normal models in data, while the other paradigm, misuse detection, detects activities that match attack signatures. Anomaly detection has the advantage over misuse detection in that it can detect new types of intrusions.

Most anomaly detection algorithms available require a training data set in which data points are either purely normal or attacks with correct labels. If some data in the training data set is not labelled correctly, the algorithm may not work effectively. However, we do not normally have either correctly labelled or purely normal data readily available. And it is impractical to manually classify and label the enormous amount of audit data available. One could obtain labelled data by simulating intrusions, but then could only detect attacks that were able to be simulated.

Researchers have addressed this in a variety of ways. In [2], it was proposed to detect intrusions over noisy data. In [3], *unsupervised anomaly detection* was presented to address this problem. An unsupervised anomaly detection algorithm takes a set of unlabelled data as inputs and attempts to find intrusions buried within the data. It can

[1] Supported by the National Natural Science Foundation of China under Grant Nos. 60372045, 60133010; National High Technology Development 863 Program of China under Grant No. 2002AA135080.

F. Rothlauf et al. (Eds.): EvoWorkshops 2005, LNCS 3449, pp. 166–174, 2005.

either provide labelled data for those algorithms that require labelled data or train an intrusion detection classifier.

Clustering is a typical unsupervised learning method. Traditional clustering algorithms have been applied to unsupervised anomaly detection [4, 5, 6]. In [4] an algorithm with low time complexity is presented .The algorithm scans the training data set only once. Initially, a cluster is created and the first training datum is its representative sample. And then for each datum, the distances between it and the representative samples of the existing clusters are calculated respectively. If all these distances are larger than a pre-defined cluster's width threshold, a new cluster is created and the datum is the cluster's representative sample, otherwise we find an existing cluster whose representative sample is closer to this datum than representatives of any other existing clusters. In [5], the steps are similar to those in [4], except that the representative sample of each cluster is the average of the data existing in the cluster.

De Castro et al. [7, 8] proposed an evolutionary artificial immune network (aiNet) with the main goals of clustering and filtering an unlabelled numerical data set. The algorithm is based on a combination of the graph clustering method, immune network theory and the idea of clonal selection. It has the advantages of graph clustering methods of being able to handle data independent of distribution, automatically compressing large-scale data, and not requiring a predetermined number of clusters. Furthermore, it possesses the learning and memorizing ability by using the concepts and mechanisms of vertebrate immune systems. An evolutionary artificial immune network is "evolved" by using the training sample data. The construction of the evolutionary artificial immune network represents the distribution of the input data set, and then the clusters are generated by the minimal spanning tree (MST).

Motivated by aiNet, an unsupervised anomaly detection algorithm based on an evolutionary artificial immune network is proposed in this paper. We evaluate our method over the 1999 KDD CUP IDS data [9]. The main idea of our method is to compress the training data with an evolutionary artificial immune network, and then to employ a traditional hierarchical agglomerative clustering algorithm to perform clustering analysis.

2 Evolutionary Artificial Immune System

2.1 Clonal Selection Principle and Immune Network Theory [10]

It was Burnet [11] who originally proposed the famous clonal selection principle in 1959. The principle describes the basic features of an immune response to an antigenic stimulus. The main properties of clonal selection are:

1. Elimination of self antigens;
2. Proliferation and differentiation on stimulation of cells with antigen;
3. Restriction of one pattern to one differentiated cell and retention of that pattern by clonal descendants;
4. Generation of new random genetic changes subsequently expressed as diverse antibody patterns by a form of accelerated somatic mutation.

The immune network theory was originally proposed by Jerne [12] in 1974, which suggests that the immune system maintains a network of interconnected B-cells for antigen recognition. These cells both stimulate and suppress each other in certain ways that lead to the dynamic stabilization of the network. Two B cells are connected if the affinities they share exceed a certain threshold, and the strength of connection is directly proportional to the affinity they share.

Clonal selection is a dynamic process in which the immune system adaptively responds to an antigenic stimulus. The diversity of antibody, the abilities of learning and memorizing, network tolerance and suppression in this process are used for reference in an artificial immune system.

2.2 Evolutionary Artificial Immune Network

De Castro proposed an evolutionary artificial immune network algorithm aiNet [7], which simulates the process of the immune network response to an antigenic stimulus. The stimulated mechanisms include antibody-antigen recognition, clonal proliferation, affinity maturation, and network suppression. An immune network is an edge-weighted graph. It is formally defined as:

Definition 1. An immune network is defined as an edge-weighted graph, not necessarily fully connected, composed of a set of nodes called cells, and one node is connected with another node by an edge. Each connected edge has a number assigned, called weight or connection strength.

The network serves as an internal image of the original training data set. The number of network cells is much smaller than that of training data, so the network performs data compression. The KDD CUP 1999 data is a typical kind of intrusion audit data with large scale, high dimensions, and heterogeneous attributes. The aiNet compresses data and requires little expert knowledge, which motivates us to propose an unsupervised anomaly detection algorithm based on an evolutionary immune network.

3 Unsupervised Anomaly Detection Algorithm Based on an Evolutionary Immune Network

Definition 2. Matrix A ($N_A \times m$) is used to represent a network. Each row of A is a network cell. The affinity between two cells A_i and A_j is the distance between them, i.e. $s_{ij} = d(A_i, A_j)$. The smaller s_{ij} is, the less difference between A_i and A_j.

Definition 3. $f(x_i, A_j)$ is the affinity of a sample x_i in the training data set with a network cell A_j:

$$f(x_i, A_j) = 1/(1 + d(x_i, A_j)) \qquad i = 1,2,\cdots,N; j = 1,2,\cdots,N_A. \tag{1}$$

Algorithm 1. Learning algorithm based on an evolutionary immune network.

1. Initialise the parameters. Randomly generate the network cells. $i=1$.
2. For antigen x_i determine its affinities to all the network cells, and do:
   2.1 Select n network cells of higher affinity to clone. The higher the affinity, the more cloned cells are generated;
   2.2 Mutate the cloned cells to improve the affinity of x_i and network cells.
   2.3 Re-select several of the highest affinity network cells and create a memory cell matrix M_p
   2.4 Eliminate cells in M_p whose affinities are inferior to threshold σ_d. Eliminate some cells to make the affinities among cells in M_p larger than σ_s.
   2.5 Concatenate A and M_p, $A \leftarrow [A; M_p]$.
   2.6 If i is divided exactly by 2000 or $i=N$, go to Step 3; otherwise, $i=i+1$, go to Step2.
3. Calculate the affinities among the network cells and eliminate some cells to make all the affinities among cells larger than σ_s.
4. If $i=N$, stop and output A; otherwise $i=i+1$, go to Step2.

end{Algorithm1}

In Step 2.2, let the selected cells be $\{A_{r_1}, A_{r_2}, \cdots, A_{r_n}\}$, the clone size q_j for A_{r_j} is given by

$$q_j = \text{int}\left(N_c \times \frac{f(x_i, A_{r_i})}{\sum_{k=1}^{n} f(x_i, A_{r_k})}\right) \quad j=1,2,\cdots,n, \tag{2}$$

where N_c is a predetermined clone size, and $\text{int}(a)$ is a function that returns the minimum integer larger than a. The real total clone size is $N_c' = \sum_{j=1}^{n} q_j$. Let the cloned cell be a matrix CA ($N_c' \times m$).

In Step 2.3, apply a mutation operation to CA as follows,

$$CA_i \leftarrow CA_i - \alpha_i * (CA_i - x_i) \quad i=1,2,\cdots,N_c'. \tag{3}$$

Round the discrete attributes of CA_i: $CA_{ij} = round(CA_{ij})$, $j \in C$, $round(a)$ is a function that rounds a into an integer, '*' is an operator that computes the product of the corresponding attributes of two vectors and α_i is the mutation rate of CA_i. The value of α_i is set according to the antigen-antibody affinity, the higher the affinity the smaller the α_i. This step improves the affinities of x_i and network cells.

Step 3 is a suppression step to reduce the number of cells. In the experiments the network is suppressed once after 2000 samples is learned. In practice, it can be performed once the system is idle or the network is too large.

After Algorithm 1, we get the internal image of the data set. Now we analyse the obtained network. The main goals are to determine the number of clusters and the network cells belonging to each of the identified clusters. A traditional hierarchical agglomerative clustering method is employed to analyse the network $A(N_A \times m)$. A cluster is represented by a set ω_i, which contains the cells belonging to it.

Definition 4. $d(\omega_i, \omega_j)$ is the distance between clusters ω_i and ω_j, and it is equal to the minimum distance of the cells belonging to them respectively:

$$d(\omega_i, \omega_j) = \min d(x_p, x_q) \qquad x_p \in \omega_i, x_q \in \omega_j . \tag{4}$$

Definition 5. $d(y, \omega_i)$ is the distance between vector y and cluster ω_i, and it is decided by the minimum distance of y and cells of ω_j,

$$d(y, \omega_i) = \min d(y, x_j) \qquad x_j \in \omega_i . \tag{5}$$

Algorithm 2. Label clusters.

1. Treat each cell A_i as a cluster. The number of clusters $L = N_A$.
2. Calculate all the distances among clusters.
3. If all the distances are larger than the threshold σ_w, go to Step5;
4. otherwise
   4.1 merge ω_i and ω_j, which satisfy $d(\omega_i, \omega_j) = \min d(\omega_p, \omega_q)$ for $p, q = 1, 2, \cdots, L$,
   4.2 $L \leftarrow L - 1$,
   4.3 go to Step2.
5. Label the clusters. The k-th cluster ($k = 1, 2, \cdots, L$) is labelled k.

end{Algorithm2}

When the network is used to classify the data set $Y = \{y_1, y_2, \cdots\}$, y_i is labelled j, if $d(y_i, \omega_j) = \min d(y_i, \omega_l), l = 1, 2, \cdots, L$, and $d(y_i, \omega_j) < \sigma_n$. If $d(y_i, \omega_j) \geq \sigma_n$, y_i is considered an attack of unknown type. σ_n is a pre-defined threshold.

Classify the training data set with the network. If the number of data in some clusters is larger than 10 percent of the number of the whole training data, this cluster is considered a normal cluster. The normal network cells are considered the representatives of normal data, and they are the models of normal behaviours. The models can perform anomaly detection. Supposed the data set of the normal network cells is $S = \{s_1, s_2, \cdots\}$. Anomaly detection is performed on data set $Y = \{y_1, y_2, \cdots\}$. If there is some $s_k \in S$, which makes $d(y_i, s_k) \leq \sigma_n$, y_i is considered a normal data point, otherwise an anomalous one.

4 Simulations and Results

4.1 Data Description and Pre-processing

The dataset used is the KDDCUP1999 data, which consists of about 4,900,000 data instances. Each instance is a 42-dimensional vector (label included). Each dimension represents an extracted feature value from a connection record obtained from the raw network data. Some features are nominal-valued, some are continuous-valued, and some are discrete-valued.

Process a data set of size N. The nominal attributes are converted into linear discrete values (integers). For example, 'http' protocol is represented by 1 and 'ftp' by 2. Then, the attributes fall into two types: discrete-valued and continuous-valued. After eliminating labels, the data set is described as a matrix X, which has N rows (data samples) and m =41 columns (attributes). There are m_d =8 discrete-value attributes and m_c =33 continuous-value attributes. Supposed $X = \{x_1, x_2, \cdots, x_N\}'$, $x_i = (x_{i1}, x_{i2}, \cdots, x_{im})$, $i = 1,2,\cdots,N$. Define the set C= {k | the k-th attribute of X is continuous-valued attribute}, and set D= {k | the k-th attribute of X is discrete-valued attribute}.

Now normalization is performed, since, if one of the input attributes has a relatively large range, it can overpower the other attributes. The method is distance-based. The data set is converted, but for simplicity the data set is still represented by X. The vectors below are normalized vectors.

The distance between two vectors is given by

$$d(x_i, x_j) = \sqrt{\sum_{k=c_1}^{c_{m_c}} (x_{ik} - x_{jk})^2 + \lambda \sum_{l=d_1}^{d_{m_d}} \delta(x_{il}, x_{jl})} \,. \tag{6}$$

where $\delta(\cdot)$ is a function as $\delta(x, y) = \begin{cases} 0, & x = y \\ 1, & x \neq y \end{cases}$.

Unsupervised anomaly detection algorithms make two assumptions about the data. The first is normal instances vastly outnumber intrusions. The second is intrusions are qualitatively different from normal instances.

4.2 The Results and Analysis

Six data sets are selected for the experiments. Each data set contains 12,000-15,000 data and 6-8 types of attacks. The number of intrusions in each set meets the first assumption requirement. The data sets are divided into two groups. Each group has one training set and two test sets. Attacks in one of the test set are the same types as in the training set, whilst the other test set contains attack types that do not appear in the training set and is used to test the capability of detecting unknown attacks.

The performance of an intrusion detection algorithm is usually measured by the detection rate (DR, the number of intrusion instances detected divided by the total number of intrusion instances) and the false positive rate (FR, the number of normal instances misclassified as intrusions divided by the total number of normal instances). The trade-off between them is inherently present in many methods, including ours.

Now the parameters need to be fixed. The key parameter in Algorithm 2 is σ_w which decides the number of output classes. As for Algorithm 1, there are quite a few parameters, and it is difficult to give a theoretical guide on how to decide them. Here we discuss two of them, σ_d and σ_s, which have a great effect on the size of the output network, i.e. the number of output cells. A network with large size represents the data set better, but needs more resources for computation and storage, while a small-size network saves resources with more loss of precision of representing a data set. When σ_d is between $[0.1, 0.5]$, the results are acceptable.

An experiment is made to decide the choice of parameters. It is made over training set 1 and one of its test sets, which has the same attack types as training set 1. The training set has 6 types of attacks. With the normal class, there are 7 classes. In Table~1 the influence of σ_s on the network size is presented. For a fixed σ_w (10), the larger the output network is, the higher the detection rate. Considering the trade-off between detection rate and resources, we find 150 a good selection for network size. Accordingly 1.25 is used for σ_s to keep the size around 150.

Table 1. Influence of network size

Network size	Class Number	Example Training set 1		Test set	
		DR (%)	FR (%)	DR (%)	FR (%)
106	10	66.3	2.3	63.3	2.8
135	11	75.5	4.6	72.1	4.9
151	12	90.5	11.8	85.6	12.2
180	12	90.8	12.2	86.1	12.5

Table 2 presents the comparison of our method with the method in [3]. Comparing the results of the two methods, we see that our method achieves a higher detection rate when the false positive rates of the two methods are similar. And our method has a lower false positive rate while both the detection rates are similar. Besides, our method performs better in the detection of unknown attacks. Now we can state that our method performs better than the method in [4] and compares well with other approaches [13].

Both methods are able to learn incrementally. In our method the obtained network cells can be used as the initial cells of Algorithm 1. Our method is more complicated in computation than the method in [4], but simpler than the direct use of Algorithm 2. The complexity of Algorithm 2 is $O(cN^2m^2)$, in which c is the number of clusters. We can always get $N_A << N$ (N_A is less than 0.5% of N). If Algorithm 2 were used directly in training data set, the complexity would be unacceptable.

Comparing the two methods, one point is saved for one cluster in the method of [4], while several points are saved for each individual cluster in our method. That is one of the reasons why our method approximates the data of irregular shape distribution better and hence performs better.

Table 2. The comparison of the method in this paper and in reference [4]

		The performance in test sets								
		Method in the paper					Method in [4]			
Training set	σ_w	Known attacks		Unknown attacks		w	Known attacks		Unknown attacks	
		DR	FR	DR	FR		DR	FR	DR	FR
Training set 1	10	90.6	13.2	86.1	13.9	15	90.5	15.1	85.8	15.4
	12	85.3	4.6	83.2	5.6	20	82.1	4.5	82.1	5.5
	15	72.1	2.2	69.5	3.1	22	75.2	3.2	72.5	5.9
	18	57.5	0.9	55.8	1.5	25	56.3	1.5	53.5	2.3
Training set 2	10	92.5	12.5	87.5	12.9	15	91.8	15.4	87.5	13
	12	86.7	4.2	83.3	5.3	20	87.6	5.5	82.8	5.5
	15	73.8	1.5	73.1	2.1	22	74.5	2.4	73.1	2.9
	18	45.5	0.8	42.5	1.2	25	45.8	1.2	42.2	1.5

Just like other distance-based unsupervised anomaly detection algorithms [4,5,6], our method cannot detect some kinds of attacks (e.g. some DOS and R2U attacks) for they are not qualitatively different from the normal instances. Another disadvantage of our method is that a lot of parameters and thresholds have to be pre-defined, and the algorithm is sensitive to the parameters and the thresholds.

5 Conclusion

This paper addresses the problem of unsupervised anomaly detection, which has the advantage of processing unlabeled raw data directly. An evolutionary artificial immune network is first "evolved" by using the unlabelled training sample data, and then a traditional cluster method is used. The algorithm performs well over the KDD CUP

1999 data sets, which are typical of the kind of large-scale, high dimensional, heterogeneous-attributes data in IDS. The method will help detect new unknown type attacks.

Our further work includes defining a better distance metric, automatically determining parameters and increasing the computing speed.

References

1. D. E. Denning: An intrusion detection model. IEEE Transactions on Software Engineering, SE-13: 222-232, 1987.
2. E. Eskin: Anomaly detection over noisy data using learned probability distribution. Proceedings of the International Conference on Machine Learning, 2000.
3. Eleazar Eskin, Salvatore Stolfo, etc.: A Geometric Framework for Unsupervised Anomaly Detection: Detecting Intrusions in Unlabeled Data. Data Mining for Security Applications, Kluwer, 2002.
4. Leonid Portnoy: Intrusion Detection with Unlabeled Data using Clustering. Undergraduate Thesis, Columbia University: December 2000.
5. LUO Min, WANG Li-na, ZHANG Huan-guo: An Unsupervised Clustering-Based Intrusion Detection Method. ACTA ELECTRONICA SINICA, 2003, vol.30 (11): 1713-1716.
6. Michael J. Prerau, Eleazar Eskin: Unsupervised Anomaly Detection Using an Optimized K-Nearest Neighbors Algorithm. Undergraduate Thesis, Columbia University: December 2000
7. Leandro Nunes de Castro, Fenando J. Von Zuben: An Evolutionary Immune Network for Data Clustering. Proc. of the IEEE SBRN, pp.84-89, November, 2000
8. Leandro N. de Castro, Jon Timmis. Hierarchy and Convergence of Immune Networks: Basic Ideas and Preliminary Results, 1stICARIS, 2002
9. KDD99.KDD99 cup dataset: http://kdd.ics.uci.edu/databases/kddcup99/kddcup.html, 1999
10. Jiao Licheng, Du Haifeng: An Artificial Immune System: Progress and Prospect, ACTA ELECTRONICA SINICA, 2003, 31(10): 1540~1549
11. F. M. Burnett: The Clonal Selection Theory of Immunity. Vanderbilt University Press, Nashville, 1959.
12. N. K. Jerne: Towards a Network Theory of the Immune System. Ann. Immunol. (Inst. Pasteur) 125C, p373-389, 1974.
13. Results of the KDD'99 Classifier Learning Contest. http://wwwcse.ucsd.edu/users/elkan/clresults.html

Evolutionary Algorithms for Location Area Management

Bahar Karaoğlu[1], Haluk Topçuoğlu[2], and Fikret Gürgen[1]

[1] Department of Computer Engineering,
Boğaziçi University, Bebek, Istanbul, Turkey
{bahar.karaoglu,gurgen}@boun.edu.tr
[2] Department of Computer Engineering,
Marmara University, Goztepe, Istanbul, Turkey
haluk@eng.marmara.edu.tr

Abstract. Location area (LA) management is a very important problem in mobile networks. In general, registration and paging costs are associated with tracking the current location of a mobile user. Considering minimizing the total of paging and registration costs as the main objective, the aim is to provide corresponding cell-to-switch and cell-to-LA assignments. This paper compares three well-known evolutionary algorithms to measure their suitability for solving location area management problems; these are genetic algorithms, multi-population genetic algorithms and memetic algorithms. To handle multiple objectives of paging and registration, a two-stage multi-population GA is developed. A memetic algorithm is introduced in order to improve the performance of a GA with the local search techniques. The effectiveness of these methods is shown for a number of test problems with different network size and characteristics.

1 Introduction

Mobile communication becomes more prominent every day as the globalization and the speed of daily life increases. One of the challenges in mobile networks is *Location Management*, which is to be able to track the current locations of the users. In a typical cellular mobile network, the area of coverage is divided into cells and in general, the shapes of these cells are modeled pictorially as hexagons. Each cell is associated with a base station, which is responsible for communicating with the users in its coverage area, i.e. the associated cell. In order to route incoming calls to appropriate base stations, the network should be aware of which cell the user is currently located in.

The two main operations of location management are *registration* and *paging*. When an incoming call arrives to a mobile station, the network searches the mobile station in all possible cells to find out the cell on which the mobile station is located, so the call can be routed to the corresponding base station. This search operation is called paging. The set of base stations in which a mobile is paged is called the location area (LA). The registration operation is performed

F. Rothlauf et al. (Eds.): EvoWorkshops 2005, LNCS 3449, pp. 175–184, 2005.

by the mobile station. When a mobile station changes its location, it updates its location information on the system. The number of all possible cells to be paged depends on the performance of the registration of mobile stations.

Location area management techniques are classified according to their use of zone, time, movement, distance or profile information [1]. In zone-based schemes, the cellular network is divided into location areas; if the mobile station changes its location area, the registration operation is applied. Whenever an incoming call arrives, the mobile user is paged in its current location area.

Optimal location area planning in zone-based schemes is not widely studied in the literature. Saraydar *et al.* [2] proposed a one-dimensional location area design. In their work, they consider a linear service area like a highway, which connects cities or railway lines. They aim to divide the highway into location areas along its length so that paging and registration cost is minimized. Demirkol *et al.* [3, 1, 4] developed a simulated annealing (SA)-based solution for the cellular network structuring problem. Subrata and Zomaya [5, 6] proposed three different techniques (genetic algorithm, tabu search and ant colony optimization) for determining the optimal reporting cell locations efficiently. Reporting cells are used to keep track of the location of mobile users in the network. Quintero and Pierre [7, 8] proposed a multi-population memetic algorithm for assigning cells to switches in a mobile network. When compared with our method, there are differences in these methods with respect to the network structure, set of constraints and/or the objective function.

In this paper, we study the zone-based scheme due to its wide usage in GSM systems. Evolutionary algorithms have been applied successfully in various domains of search and optimization. This paper presents a comparison of three evolutionary algorithms applied in location area management problem, namely genetic algorithm (GA), multi-population genetic algorithm (MPGA), and memetic algorithm (MA).

The rest of the paper is organized as follows. Problem definition, constraints and objective function are given in Section 2. Section 3 presents the details of evolutionary algorithms applied in location area management problems. Section 4 provides a set of computational experiments for comparing the performance of the proposed methods, and Section 5 concludes the paper.

2 Problem Definition

In this study, a model that is consistent with the GSM hierarchy is presented, based on the model given in [3, 1, 4]. It consists of base stations (BSs), base station controllers (BSCs) and mobile switching centers (MSCs). The base stations (BSs) are grouped in a number of location areas (LAs). The objective is to provide cell-to-switch and cell-to-LA assignments so that the total cost is minimized. Figure 1 gives a 3x3 network and a sample solution for cell-to-switch and cell-to-LA assignments.

The total cost (C_T) is formally represented as

$$C_T = P \cdot C_P + R \cdot C_R \tag{1}$$

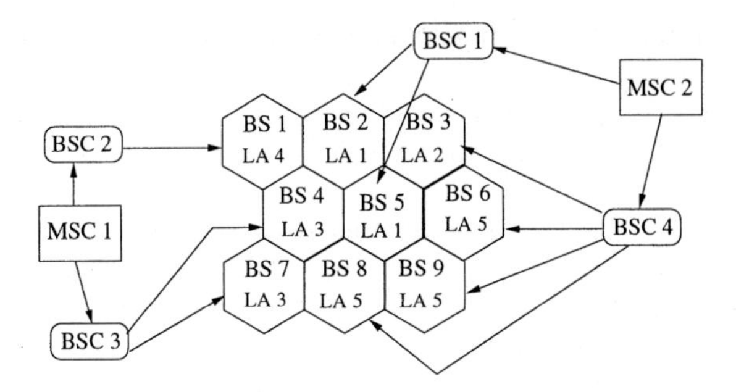

Fig. 1. A 3x3 network with cell-to-switch and cell-to-LA assignments

where C_P is the paging cost and C_R is the registration cost; These costs do not have comparable units. Although in reality the cost of paging and registration to the network varies from cell to cell, previous researchers use some assumptions for the relative values of these costs such as assuming one unit cost for each paging-event is equal to 10 units cost for each registration-event [1].

The paging cost is a result of incoming calls to mobiles. The paging cost on a cell is the total of paging rates of the cells that belong to the same location area with that cell, which is calculated by,

$$C_P = P \cdot \sum_i \lambda_i (1 - d_{is}) \tag{2}$$

where λ_i is the paging rate per unit time for each base station in the network. The term d_{is} is a function that returns 1 if the ith base station and s^{th} base station reside in the same location area; and it is equal to 0, otherwise.

Each time a user leaves a location area and enters a new one, it sends a registration message on the uplink control channel. Therefore the rate at which mobile users cross a boundary determines the cost of registration signalling. This cost is calculated by:

$$C_R = R \cdot \sum_i q_i \tag{3}$$

where q_i is the rate at which users pass by location i (i.e., handover rate). It should be noted that only the active-state handovers between cells at LA boundaries are considered, since it is difficult for a cellular network to collect idle-state cell boundary crossings. The unified objective given in equation 1 is subject to the constraints given below:

- **Paging Capacity of BS:** Maximum number of paging for one base station in one time slot.

$$\lambda_i < Pi^{BS}, \forall i \tag{4}$$

P_i^{BS}: The paging capacity of each base station.

- **Paging Capacity of BSC:** Maximum number of paging for one base station controller in one time slot.

$$\sum_i x_{ij}\lambda i < P_j^{BSC}, \forall j \tag{5}$$

x_{ij}:x_{ij} returns 1 if i^{th} cell is a member of jth BSC, otherwise 0.
P_i^{BSC}: The paging capacity of each base station controller.

- **Call Traffic Capacity of BSC:** Maximum number of calls for one base station controller in one time slot.

$$\sum_i x_{ij}c_i < C_j^{BSC}, \forall j \tag{6}$$

c_i : call traffic rate for cell i.
C_j^{BSC}: Maximum call traffic capacity of BSC j .

- **Call Traffic Capacity of MSC:** Maximum number of calls for one mobile services switching center in one time slot.

$$\sum_j \sum_i x_{ij}y_{jk}c_i < C_k^{MSC}, \forall k \tag{7}$$

y_{jk} : returns 1 if j^{th} BSC is a member of kth MSC, otherwise 0.
C_k^{MSC}: Maximum call traffic capacity of each MSC.

- **BHCA Capacity of BSC:**BHCA stands for busy hour call attempt rate. This call arrival rate includes not only established connections but also the failed attempts.

$$\sum_i x_{ij}d_i < D_j^{BSC}, \forall j \tag{8}$$

d_i: peak call attempt rate of cell i per unit time.
D_j^{BSC}: Busy hour call attempt capacity of each BSC.

- **BHCA Capacity of MSC:** In order to have a feasible cellular network, the limited call processing capability mobile services switching centers may create a limit on the peak call arrival rate.

$$\sum_j \sum_i x_{ij}y_{jk}d_i < D_k^{BSC}, \forall k \tag{9}$$

- **TRX Capacity of BSC:** TRX stands for transmitters. Each base station controller has a finite number of transmitters.

$$\sum_i x_{ij}r_i < R_j^{BSC}, \forall j \tag{10}$$

r_i: number of TRXs in each cell i. R_j^{BSC}: number of TRXs in each BSC j.

- **TRX Capacity of MSC:** Each mobile service switching center has a finite number of transmitters, which defines the number of channels used in that cell.

$$\sum_j \sum_i x_{ij}y_{jk}r_i < R_k^{MSC}, \forall k \tag{11}$$

R_k^{MSC}: Maximum number of TRXs for each MSC.

Additionally, each BS should be assigned to exactly one BSC; each BSC should be assigned to exactly one MSC; and each BS should be assigned to exactly one LA. There are also proximity constraints. All base stations in a location area must be adjacent to each other. Also the location areas in base station controllers must be adjacent to each other.

On the other hand, the maximum number of location areas that can exist in a given network is equal to the number of cells in that network. The same constraints are applied to base station controllers. Additionally, maximum number of mobile services switching centers that can exist in a network is equal to the number of base station controllers in a network. The minimum number of mobile services switching center is equal to 1.

3 Evolutionary Algorithms for Location Area Management

This paper presents a comparison of three evolutionary algorithms applied to the location area management problem, namely genetic algorithm (GA), multi-population genetic algorithm (MPGA) and memetic algorithm (MA). In this section we first present the common parts in these three methods, including string representation, initial population generation and variation operators, which is followed by extensions in MPGA and MA.

3.1 String Representation

The string representation of a given network structure is represented by pointer arrays in an hierarchical way. Each individual is represented with an array that holds all mobile switching centers within the given network. Each mobile switching center points to a base station controller array, which includes all base station controllers associated with the given mobile switching center. Similarly, each base station controller is linked to the corresponding location areas, where each location area points to all base stations within the given location area. Figure 2 gives the string representation of the solution given in Figure 1.

3.2 Initial Population Generation

First, the number of location areas in a solution is determined randomly, and base stations in the network are placed in these location areas. The location areas are placed into the base station controllers and the base station controllers are placed into mobile services switching centers, randomly. Then, the validation phase is performed in order to satisfy the constraints presented in Section 2. An important condition for validation of BS-to-LA, BS-to-BSC and BSC-to-MSC mappings is the consideration of proximity constraints. The base stations mapped to a given location area must be neighbors. After the validation phase is performed, all BSCs, MSCs and LAs in the network will have at least one mapping.

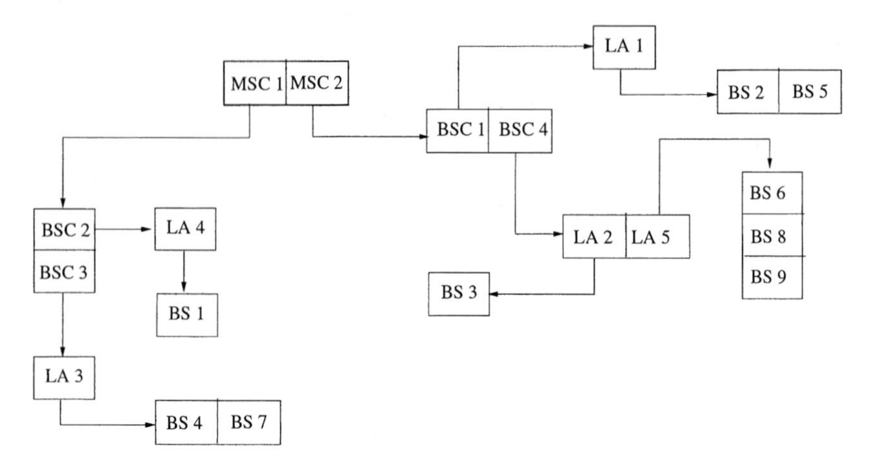

Fig. 2. String representation of the solution given in Figure 1

3.3 Fitness Function and Parameters of GA

The fitness value of any individual is equal to $(1/C_T)$ where C_T is the total cost of an individual. Roulette wheel selection is performed after the fitness values are normalized to the [0..1] range. Since an elitist strategy is considered in our solutions, the best 10% solutions of a given population are passed to the next generation. The population size and number of generations are set according to the size of the network from the set {50, 100, 250, 500}. After a number of experiments with various network sizes are performed in order to determine the best crossover and mutation rates, the minimum total costs are observed for the cases when crossover rate is set to 0.70 and mutation rate is set to 0.15.

3.4 Variation Operators

The crossover operator is performed on location area assignments of the given solutions. The crossover point is randomly selected, which should be in the range between 1 and min(LA^{P_1},LA^{P_2}) -1, where LA^{P_1} and LA^{P_2} are the number of location areas in parent solutions P_1 and P_2, respectively. Then, parent solutions change their location areas according to the crossover point. The number of location areas in each solution does not change after the crossover, instead base stations in location areas will change. A validation phase is required after the crossover operator, due to the following conditions:

- A base station may disappear in the offspring after crossover is performed.
- A base station may occur twice in the offspring.
- The location areas in base station controllers may not preserve proximity constraints.

In our mutation operator, first, a location area is selected randomly. Then, a randomly selected base station in this location area is swapped with a base station from another location area. After the mutation operation is performed, a location area may include a new base station which does not preserve the proximity constraints of the location area. A similar validation phase is performed to adjust both proximity and capacity constraints.

3.5 Extensions for Multi-population Genetic Algorithm

We consider a two-stage multi-population genetic algorithm (MPGA) in this study. In the first stage, a population is divided into two sub-populations, and each sub-population is evolved separately using a genetic algorithm by considering its own objective, which is either minimization of the paging cost or minimization of the registration cost.

After the completion of the first phase, an elitist strategy is used to generate a population from the two sub-populations, which becomes the initial population of the second stage. In the second stage, the objectives are combined by weighting to form a single objective function for optimization. We consider the total cost function given in the Equation 1, as the objective of the second phase.

According to this representation, our approach can be considered as a method of using a modified version of VEGA (vector evaluated genetic algorithm) in the first stage and a modified version of MOGA (multiple-objective genetic algorithm) in the second stage.

3.6 Extensions for Memetic Algorithm

Memetic Algorithms, also known as hybrid genetic algorithms, are population-based heuristic search approaches for combinatorial optimization problems based on cultural evolution [9, 10, 11]. Local search plays a significant role in memetic algorithms. We consider local search in our method for generating the initial population and applying the mutation operator. It is considered to improve randomly generated initial populations with a predefined number of neighborhood solutions.

Similarly, a predefined number of exchanges (i.e., switching the LA-assignments of two base stations) are performed as part of the mutation operator. The one with the minimum total cost is the output of the mutation operator. As in the GA-based version, the output of the mutation should be validated according to the given constraints.

4 Experimental Study

The tests presented in this section are performed on a Pentium IV 2GHz PC with 2.6GB RAM. A scenario generator is implemented in order to generate data sets with different characteristics. It assigns paging rate, handover rate and busy hour call attempt rates for each base station in the network, according to the inputs that specify the ranges of each parameter. The number of base stations in a given network defines the size of the problem.

In order to determine the values of GA parameters in our comparison study, a set of experiments is performed using different parameter values on different data

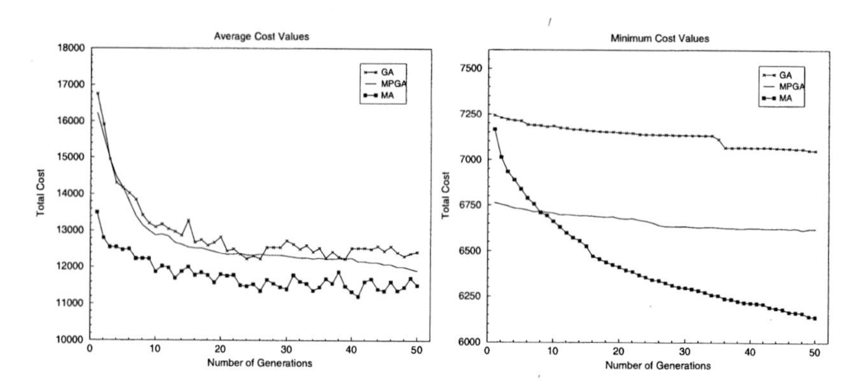

Fig. 3. Performance comparison of three algorithms for average and best costs

sets. Experiments are repeated 5 times for each data and parameter set. Based on these tests, the crossover rate is set to 0.7 and mutation rate is set to 0.15.

4.1 Varying the Number of Generations

The first group of experiments measure the performance of algorithms by varying the number of generations for a network with 256 base stations. Figure 2 and Figure 3 present the average total costs and minimum total costs of solutions generated by the three algorithms for the given network size, respectively. As can be seen from the figures, memetic algorithm outperforms the other methods.

When the running time of the algorithms is considered for the given network, the MA takes 69.99 seconds to provide the solution, MPGA takes 11.21 seconds, and the GA takes 5.72 seconds. The running time of MA is significantly higher than the others, which is due to the local search phase added for generating the initial population and applying the mutation operator.

4.2 Varying the Network Sizes

The performance of algorithms are compared with respect to different network sizes and input parameters. We consider two different network sizes: a network with 256 base stations and a network with 576 base stations. The number of location areas, base station controllers and mobile services switching numbers are determined within the program. Average total costs of solutions that are generated from the evolutionary algorithms are presented in Table 1, which includes results of eight random data sets (four with the 256 base stations and four with the 576 base stations). According to the results, a memetic algorithm outperforms the other algorithms for all given data sets (i.e. both small size and large size networks).

Additionally, the algorithms are compared with different number of base stations, base station controllers and mobile services switching centers. A total of 9 experiments are done. Six of them are for 256 BSs, 8 BSCs and 8 MSCs; and three of them are for 576 BSs, 16 BSCs and 16 MSCs. As in the previous experiments, the best results are obtained from the memetic algorithm [12].

Table 1. Average total costs of solutions for different data sets

Data	GA	MPGA	MA
Random Set 1 (256 BS)	7077	6738	6190
Random Set 2 (256 BS)	7090	6761	6164
Random Set 3 (256 BS)	7059	6730	6171
Random Set 4 (256 BS)	7129	6696	6146
Large Random Set 1 (576 BS)	13214	12552	11388
Large Random Set 2 (576 BS)	13100	12513	11350
Large Random Set 3 (576 BS)	13055	12418	11404
Large Random Set 4 (576 BS)	13068	12496	11340

Finally, a set of experiments are done for observing the effects of network characteristics. Handover rate, paging rate and call traffic rates are varied from a given range for the given data sets. For each network parameter, experiments are performed by setting the value of the parameter with a predefined low and high values. It was observed that memetic algorithm outperforms other algorithms for both high and low values of all three network parameters [12].

5 Conclusions

In this paper we present a formulation for cell-to-switch and cell-to-LA assignments for a given network structure. Considering the minimization of the total paging and registration costs is the main objective, the aim is to provide feasible BS-to-LA, BS-to-BSc and BSC-to-MSC assignments. We compare three well-known evolutionary algorithms to measure their suitability for solving location area management problem. When the algorithms are compared with respect to the unified cost value, memetic algorithm always gives the best results. When the running times of the algorithms are considered, the memetic alorithm requires more computation due to the local search phase.

Acknowledgments: The authors would like to thank to Prof. Cem Ersoy and Ilker Demirkol for their support on the model and the input data sets.

References

1. I. Demirkol, Location Area Planning and Cell to Switch Assignment in Cellular Networks, M.Sc. Thesis, Bogazici University, 2002.
2. C. Saraydar, O. Kelly and C. Rose, One-dimensional Location Area Design, IEEE Transactions on Vehicular Technology, vol. 49 (5), 1626-1632 , 2000.
3. I. Demirkol, C. Ersoy, U. Caglayan and H. Delic, Location Area Planning and Cell to Switch Assignment in Cellular Networks Using Simulated Annealing, IEEE Transactions on Wireless Communications, Vol.3, No.3, pp.880-890, May 2004.
4. I. Demirkol, C. Ersoy, U. Caglayan and H. Delic, Location Area Planning in Cellular Networks Using Simulated Annealing, INFOCOM 2001:13-2.

5. R. Subrata and A. Zomaya, A Comparison of Three Artificial Life Techniques for Reporting Cell Planning in Mobile Computing, IEEE Transactions Parallel and Distributed Systems, vol. 14(2): 142-153, 2003.
6. R. Subrata, and A. Zomaya, Evolving Cellular Automata for Location Management in Mobile Computing Networks, IEEE Transactions on Parallel and Distributed Systems, vol. 14(1): 13-26, 2003.
7. A. Quintero, and S.Pierre, Sequential and Multi-population Memetic Algorithms for Assigning Cells to Switches in Cellular Mobile Networks, Computer Networks, Vol. 43, No. 3, pp. 247-261, 2003.
8. A. Quintero, and S. Pierre, Evolutionary Approach to Optimize the Assignment of Cells to Switches in Personal Communication Networks, Computer Communications 26(9): 927-938, 2003.
9. D. Corne, M. Dorigo and F. Glover (editors), New Ideas In Optimization, McGraw Hill, 1999.
10. P. A. Moscato, On evolution, search, optimization, genetic algorithms and martial arts: Towards memetic algorithms, Tech. Rep. Caltech Concurrent Computation Program Report 826, Caltech, 1989
11. P. Moscato, Memetic Algorithms' Home Page: www.densis.fee.unicamp.br/~moscato/memetic_home.html
12. B. Karaoglu, Location Area Management for Mobile Networks with Evolutionary Algorithms, M.Sc. Thesis, Bogazici University, 2004.

Evolutionary Design of Gate-Level Polymorphic Digital Circuits

Lukáš Sekanina

Faculty of Information Technology, Brno University of Technology,
Božetěchova 2, 612 66 Brno, Czech Republic
sekanina@fit.vutbr.cz

Abstract. A method for the evolutionary design of polymorphic digital combinational circuits is proposed. These circuits are able to perform different functions (e.g. to switch between the adder and multiplier) only as a consequence of the change of a sensitive variable, which can be a power supply voltage, temperature etc. However, multiplexing of standard solutions is not utilized. The evolved circuits exhibit a unique structure composed of multifunctional polymorphic gates considered as building blocks instead. In many cases the area-efficient solutions were discovered for typical tasks of the digital design. We demonstrated that it is useful to combine polymorphic gates and conventional gates in order to obtain the required functionality.

1 Introduction

Evolutionary algorithms have been utilized to design analog as well digital circuits in the recent years [1, 2, 4]. In many cases the evolutionary approach discovered new and creative solutions to hard problems. New solutions, methods and techniques have been developed that designers did not know before. For example, the intrinsic evolution allowed engineers to exploit physical features of reconfigurable devices in order to utilize hardware effectively in a given particular situation [8]. One of these achievements is called *polymorphic electronics*.

The papers [5, 6, 7] show that it is possible to design and effectively implement multifunctional digital gates whose functionally can be controlled in a non-traditional way: by temperature, power supply voltage (Vdd), some external signals etc. As an example, we can mention a novel topology created for the multifunctional NAND/NOR gate which operates as NOR in the case that Vdd=1.8V and as NAND in the case that Vdd=3.3V. No conventional design is available with this logic function controlled by Vdd [5]. There is a great potential for various applications of polymorphic electronics in many areas because the devices composed of these gates are inherently adaptable to the changes of a particular environment and this feature is practically for free; with no reconfiguration overhead.

The available literature describes only the implementations of polymorphic gates. According to the knowledge of the author of this paper no concrete circuits composed of these gates have been reported so far. The design using these

F. Rothlauf et al. (Eds.): EvoWorkshops 2005, LNCS 3449, pp. 185–194, 2005.

unconventional gates seems to be a difficult task because no conventional design technique is available. We believe that compact circuits can be created by using the multifunctional gates as building blocks rather than by trivial multiplexing the outputs of several conventional modules using a polymorphic multiplexer. Hence the use of evolutionary algorithms could be a promising approach for the design of compact and useful adaptive circuits.

The objective of this research is to evolve multifunctional digital combinational circuits using the multifunctional gates as building blocks. It is assumed that suitable multifunctional gates exist. Only the bi-functional gates are considered in this paper; however, the proposed concept is general. The evolved circuits will perform the first required function in the first environment and the second required function in the second environment. The change of their functionality will be determined by the change of a control variable. For example, a circuit should operate as the adder for the temperature 30C and as the multiplier for 200C, i.e. the temperature is the sensitive variable in that case. The problem will be approached using cartesian genetic programming (CGP) which has already demonstrated its success for designing digital electronic circuits [4].

The rest of the paper is organized as follows. Section 2 introduces the area of polymorphic electronics. In Section 3 the polymorphic circuit design problem is formulated formally and a design approach is specified in detail. While Section 4 summarizes the obtained results, Section 5 discusses them. Conclusions are given in Section 6.

2 Polymorphic Electronics

In polymorphic electronics a function change does not require reconfiguration as in traditional approaches in which n different modules and a switch are needed to perform n different functions. Instead the change comes from modifications in the characteristics of components (e.g. in the transistor's operation point) involved in the circuit in response to controls such as temperature, power supply voltage, light, etc. [6]. Polymorphic circuits are able to work in several modes of operation. In the most straightforward approach, there are only two modes (it will also be our case). The existence of digital polymorphic circuits is based on polymorphic gates. Table 1 gives examples of the polymorphic gates reported in literature. Most of them have been designed by means of evolutionary techniques.

The NAND/NOR gate is the most famous example [5]. The evolution obtained a creative novel topology more compact than by multiplexing NAND/NOR gate which is a conventional solution using a standard digital library with external voltage control. No conventional design is available with the logic function controlled by Vdd for this task. The design of a 6-transistor NAND/NOR gate controlled by Vdd is a complex task for a human designer. The circuit was fabricated in a 0.5-micron CMOS technology and silicon tests showed a good correspondence with the simulations. The circuit is stable for $\pm 10\%$ variations of Vdd and for temperatures in the range 20C – 200C.

Table 1. Examples of existing polymorphic gates

Gate	control values	control method	ref.
AND/OR	27/125C	temperature	[7]
AND/OR/XOR	3.3/0.0/1.5V	external voltage	[7]
AND/OR	1.2/3.3V	Vdd	[6]
NAND/NOR	3.3/1.8V	Vdd	[5]

There are also conventionally designed multifunctional gates (e.g. a four transistor XOR/OR/AND gate described in the US patent 042335245); however these are not usually considered as polymorphic (see discussion in [6]).

Potential applications involve special circuits that are able to decrease resolution of digital/analog converters or speed/resolution of a data transmission when a battery voltage decreases, circuits with a hidden/secret function that can be used to ensure security, intelligent sensors, novel solutions for reconfigurable cells and function generators in reconfigurable devices (such as FPGA and CPLD), circuits of random number generators changing distributions of the probability according to the external environment and some others circuits as discussed in [6].

The design of complex polymorphic circuits is a difficult task for engineers since these circuits typically utilize normally unused characteristics of electronic devices and working environment. A. Thompson has shown that unconstrained evolutionary design is able to produce innovative designs that effectively utilize these characteristics [8]. However, only relatively small circuits have been evolved successfully in the field of evolvable hardware so far. Hence the evolution of useful polymorphic circuits (i.e. the circuits larger than a simple gate) is also difficult directly at the transistor level. So we will use the polymorphic gates as building blocks to evolve larger polymorphic circuits.

3 Problem Formulation and Design Approach

In this work we assume that suitable polymorphic gates exist and can be used as standard building blocks. We propose an EA-based approach to design nontrivial combinational modules. An alternative approach could be to design a complete complex module as a polymorphic circuit with an undistinguishable internal structure at the gate level. This is a more challenging task since very compact solutions could be obtained. However, it is difficult to manually control the design process and, furthermore, the evolutionary design is not scalable.

3.1 Problem Formulation

Let P be a set of polymorphic gates. Each of them is able to implement up to K functions (K is specified beforehand) according to a control signal which holds up to K different values. A gate is in *mode* j (and so performing the j-th function) in the case that j-th value of the control signal is activated. For purposes of this paper we will denote a polymorphic gate as $X_1/X_2/\ldots/X_K$, where X_i is its i-th logic function. For example, NAND/NOR denotes the gate operating as NAND in *mode 1* and as NOR in *mode 2*. Note that some gates can perform less than K different functions; however, their function must be fully defined for each mode. For example, the conventional NAND gate considered for polymorphic circuits will perform the same function in all modes (denoted as NAND/NAND in Section 4).

A polymorphic circuit can formally be represented by the graph $G = (V, E, \varphi)$, where V is a set of vertices, and E is a set of edges between the vertices $E = \{(a, b)|a, b \in V\}$ and φ is a mapping assigning a function (polymorphic gate) to each vertex, $\varphi : V \rightarrow P$. As usually, V models the gates and E models the connections of the gates. A circuit (and also its graph) is in the *mode j* in the case that all gates are in the *mode j*.

Given P and logic functions $f_1 \ldots f_K$ required in different modes, the problem of the polymorphic circuit design at the gate level is formulated as follows: Find a graph G representing the digital circuit which performs functions $f_1 \ldots f_K$ in its modes $1 \ldots K$. Additional requirements can be specified, e.g. to minimize delay, area, power consumption etc.

3.2 Cartesian Genetic Programming

In other words, we are looking for a single circuit topology which will work for all functions of the circuit. The topology will be sought by CGP which is defined over graphs while the standard genetic programming operates with trees [3].

CGP models a reconfigurable circuit, in which digital circuits are evolved, as an array of u (columns) $\times$ v (rows) of programmable elements (gates). The number of the circuit inputs and outputs is fixed. Feedback is not allowed. Each gate input can be connected to the output of some gate placed in the previous columns or to some of the circuit inputs. L-back parameter defines the level of connectivity and thus reduces/extends the search space. For example, if L=1 only neighboring columns may be connected; if L=u, the full connectivity is enabled. We have to define for a given application the following: the number of inputs and outputs, L, u, v and the set of functions performed by programmable elements. Figure 1 shows an example and the corresponding chromosome. Similarly to this example, we will use $v = 1$ and $u = L$ in the proposed experiments.

Miller and Thomson originally used a very simple variant of evolutionary algorithm to produce configurations for the programmable circuit [3]. Our algorithm is based on their evolutionary technique. It operates with the population of 128 individuals; every new population consists of mutants of the best four individuals. Only the mutation operator has been utilized that modifies one randomly selected gene of an individual. In case that evolution has found a solution which produces the correct outputs for all possible input combinations, the number of gates is getting to minimize. Delay is not optimized. The computation is terminated in case that no improvement of the best fitness value has been reported in a given number of last generations (typically in 50000 generations).

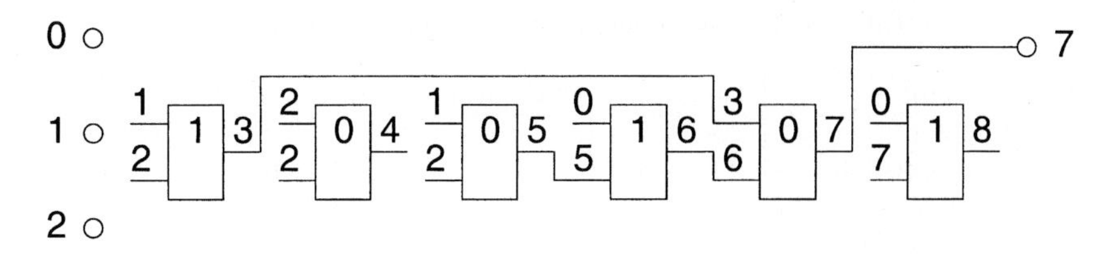

Fig. 1. An example of a 3-input circuit. CGP parameters are as follows: $L = 6$, $u = 6$, $v = 1$, functions AND (0) and OR (1). Gates 4 and 8 are not utilized. Chromosome: 1,2,1, 2,2,0, 1,2,0, 0,5,1 3,6,0, 0,7,1, 7. The last integer indicates the output of the circuit

As this paper deals with bi-functional gates, i.e. with bi-functional polymorphic circuits, the fitness value is obtained as follows:

1. Set all gates of a candidate circuit into *mode 1*.
2. Apply all possible input combinations at the circuit inputs and calculate the number of correct output bits B_1 obtained as response for these input values for the required function f_1.
3. Set all gates into *mode 2*.
4. Apply all possible input combinations at the circuit inputs and calculate the number of correct output bits B_2 obtained as response for these input values for the required function f_2.
5. Calculate $F_1 = B_1 + B_2$.

After achieving the required behavior for f_1 and f_2, the number of gates is being minimized and the fitness value is defined as:

$$F_2 = F_1 + u - g \tag{1}$$

where g denotes the number of gates utilized in a particular candidate circuit and u is the total number of gates available.

4 Experimental Results

This section presents some interesting examples of polymorphic combinational modules that we evolved. We dealt with small combinational circuits (up to 5 inputs and 4 outputs) with various types of bi-functional polymorphic gates. In particular the following circuits will be presented:

- 5-bit parity circuit vs 5-bit median circuit (denoted as 5b-parity-median)
- 2-bit multiplier vs 4-input sorting network (mult2b-sn4b)
- 2-bit multiplier vs 2-bit adder (2b-mult-add)

The following list shows the target circuits (considered as single circuits) and their conventional implementation costs measured in the number of two-input combinational gates.

Table 2. Evolved polymorphic combinational circuits

Circuit	u	run	corr.	opt. gates	gate1	gate2
5b-parity-median	24	200	16	14	NAND/NOR	XOR/XOR
5b-parity-median	20	900	48	13	NAND/XOR	XOR/NOR
Mult2b-sn4b	40	1000	8	25	NAND/NOR	AND/AND
Mult2b-sn4b	40	50	1	27	$(a \vee \bar{b})$/XOR	XOR/$(a \wedge \bar{b})$
2b-mult-add	40	200	11	20	NAND/NOR	OR/XOR
2b-mult-add	40	200	5	23	NAND/NOR	AND/AND

- 4-bit sorting network (sorts a 4-bit input vector) – 18 gates (9 AND, 9 OR)
- 2-bit adder (adds two 2-bit operands, 3-bit output) – 10 gates (a 1b full adder requires 2 XOR, 2 AND, 1 OR)
- 2-bit multiplier (multiplies two 2-bit operands, 4-bit output) – 7 gates (5 AND, 2 XOR)
- 5-bit parity (calculates even parity) – 5 gates (4 XOR, 1 NOT)
- 5-bit median circuit (returns the middle bit of a sorted 5-bit input vector) – 10 gates (5 AND, 5 OR)

Table 2 summarizes the circuits obtained using the algorithm described in Section 3.2. The symbol u denotes the number of gates used in CGP. The column *corr.* gives the number of correct circuits (i.e. perfectly working ones in both modes) out of *run* runs. The minimal number of gates we obtained for a particular circuit is given in the column denoted as *opt. gates. Gate1* and *gate2* are the polymorphic gates utilized in the design process. On average, tens of thousands of generations are needed to find a solution. We learned that the correct circuits represent only a fraction of all circuits generated by evolution. In the following figures Fig. 2, 3, 4 and 5, the label **0** denotes the first polymorphic gate and label **1** is the second gate of the two used. The bit 0 is LSB. These figures are taken from our design tool that we have developed.

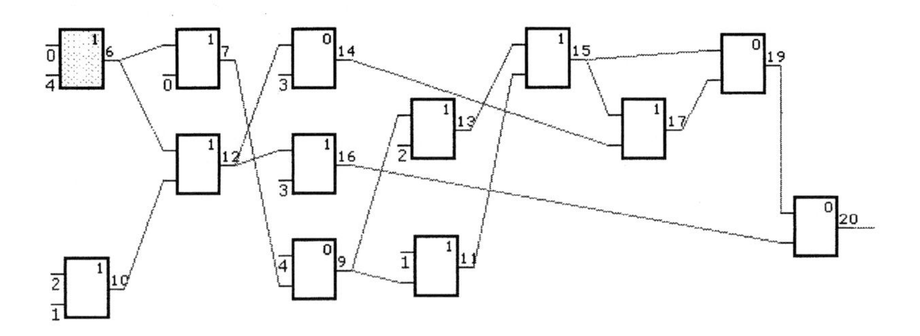

Fig. 2. Polymorphic median – parity circuit (the inputs: 0–4; gates: 0 – NAND/XOR, 1 – XOR/NOR)

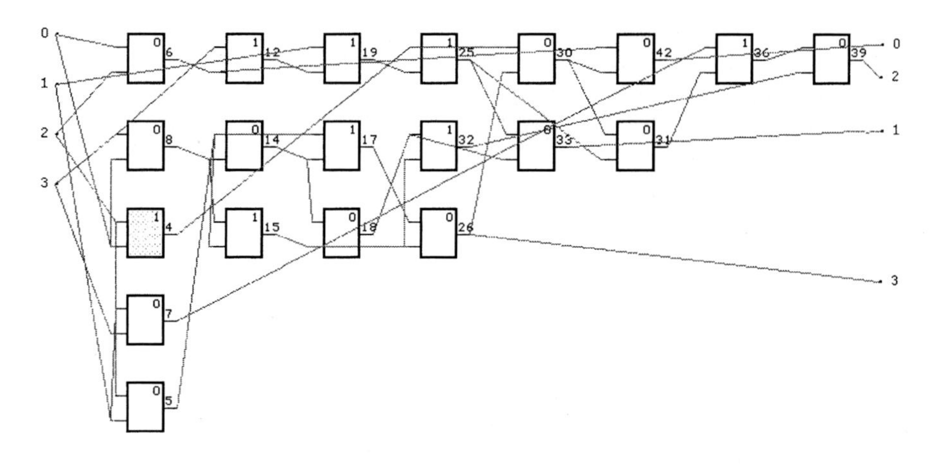

Fig. 3. Polymorphic multiplier – adder circuit (the inputs: A (0–1), B(2–3); the outputs: 0–3 (0–2 in case of adder); gates: 0 – NAND/NOR, 1 – OR/XOR)

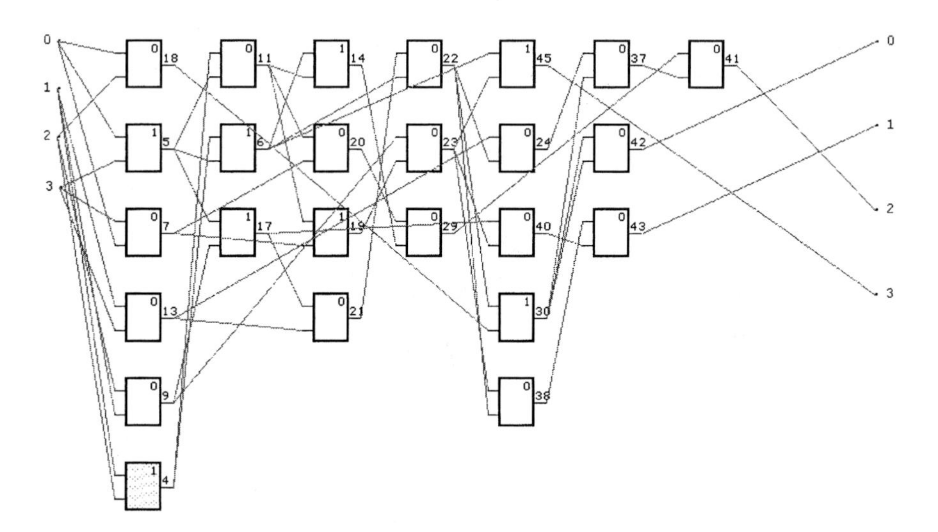

Fig. 4. Polymorphic multiplier – sorting network circuit (the inputs: A (0–1), B(2–3) in case of multiplier; the outputs: 0–3; gates: 0 – NAND/NOR, 1 – AND/AND)

5 Discussion

No useful circuits utilizing only a single type of a polymorphic gate were obtained. For instance, in case of circuits composed only of JPL's NAND/NOR gate, the same functionality of the circuit is obtained by exchanging NAND with NOR or vice versa; however, the circuit operates in the negative logic and with reordered signals. Therefore, no useful additional functionality can be achieved

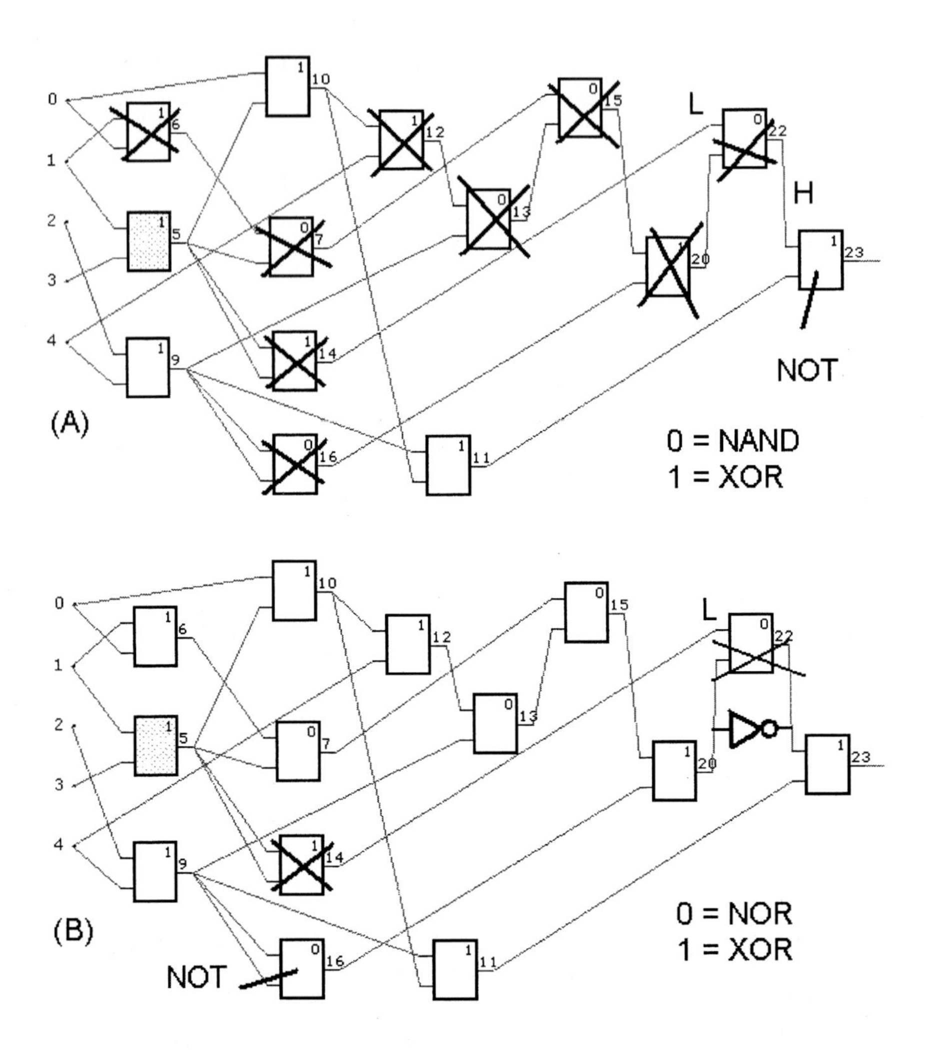

Fig. 5. Analysis of the polymorphic median – parity circuit (gates: 0 – NAND/NOR, 1 – XOR/XOR)

and so more than one type of (polymorphic) gate is needed. As Table 2 indicates, it seems useful to combine polymorphic gates with conventional gates.

The selection of suitable gates is also important for a particular polymorphic circuit. After our experimental work we have identified suitable polymorphic gates for the presented circuits (see Table 2 for concrete circuits). We performed the selection manually; next research will be conducted to use the evolution to accomplish this task. We recognized that only a few combinations of gates are suitable for a given problem; most combinations do not lead to a solution. As an example, we can mention the Mult2b-sn4b problem. We run 100 experiments for each combination of polymorphic gates taken from the following list:

- 1: NAND/NOR and XOR/XOR
- 2: NAND/NOR and OR/OR
- 3: NAND/NOR and $(a \vee \bar{b})/(a \vee \bar{b})$
- 4: NAND/NOR and AND/AND
- 5: NAND/NOR and $(a \wedge \bar{b})/(a \wedge \bar{b})$
- 6: NAND/NOR and OR/XOR
- 7: NAND/NOR and XOR/AND
- 8: $(a \vee \bar{b})$/XOR and XOR/$(a \wedge \bar{b})$
- 9: $(a \vee \bar{b})$/OR and XOR/$(a \wedge \bar{b})$
- 10: $(a \vee \bar{b})$/XOR and NXOR/$(a \wedge \bar{b})$

and obtained the perfect solution only for combinations (4) and (8). An open theoretical issue remains which combinations of polymorphic gates are sufficient to implement the required multifunctional behavior.

In some cases the evolved circuits have shown the lower number of gates when compared to a naive implementation multiplexing of two conventional modules. Note that the conventional two-input multiplexer requires 2 AND, OR and NOT gates. Hence the evolutionary approach seems to be a promising method for the design of polymorphic digital circuits. It is important to mention again that a typical polymorphic electronic device does not have a classical digital control signal. Hence the performed comparison is illustrative only.

In most cases we are not able to understand the topology of the evolved circuits since implementations of required behaviors are entangled. Fig 5 shows the analysis performed on the first circuit listed in Table 2. In the mode 1 (Fig. 5A) the realization is based on a classical implementation of parity circuits. In mode 2 (Fig. 5B) an inefficient implementation of the 5-input median circuit was evolved. For comparison, the implementation of the same behavior depicted in Fig. 2 is much more compact and difficult to decode.

The CGP is not scalable in its basic form, which means that neither our approach can be scaled. In order to illustrate the computational effort of the proposed method, we measured the average time of evolution for 500 runs of the multiplier – sorting network problem. In average 118,985 generations were produced in a single run, which corresponds to 143 seconds of the computational time at Pentium IV (2.6 GHz, 512MB RAM).

6 Conclusions

An evolutionary approach to the design of polymorphic combinational modules has been introduced. Area-efficient implementations have been discovered for various polymorphic gate-level circuits. We learned that it is useful to combine polymorphic gates and conventional gates in order to obtain the required functionality. The results of experiments allow us to predict that the approach could be useful for the design of real-world applications of polymorphic electronics. However, we have utilized hypothetic polymorphic gates. Hence a new development is needed in the basic polymorphic gate design.

Acknowledgment

The research was performed with the Grant Agency of the Czech Republic under No. 102/03/P004 *Evolvable hardware based application design methods.*

References

1. Higuchi, T. et al.: Evolving Hardware with Genetic Learning: A First Step Towards Building a Darwin Machine. In: Proc. of the 2nd International Conference on Simulated Adaptive Behaviour, MIT Press, Cambridge MA 1993, p. 417–424
2. Koza, J. R., Keane, M. A., Streeter, M. J.: What's AI Done for Me Lately? Genetic Programming's Human-Competitive Results. IEEE Intelligent Systems, May/June 2003, p. 25–31
3. Miller, J., Thomson, P.: Cartesian Genetic Programming. In: Proc. of the 3rd European Conference on Genetic Programming, LNCS 1802, Springer Verlag, Berlin 2000, p. 121–132
4. Miller, J., Job, D., Vassilev, V.: Principles in the Evolutionary Design of Digital Circuits – Part I. In: Genetic Programming and Evolvable Machines, Vol. 1(1), 2000, p. 8–35
5. Stoica, A., Zebulum, R. S., Guo X., Keymeulen, D., Ferguson, I., Duong, V.: Taking Evolutionary Circuit Design From Experimentation to Implementation: Some Useful Techniques and a Silicon Demonstration. IEE Proc.-Comp. Digit. Tech. Vol. 151(4) (2004) 295–300
6. Stoica, A., Zebulum, R. S., Keymeulen, D., Lohn, J.: On Polymorphic Circuits and Their Design Using Evolutionary Algorithms. In Proc. of IASTED International Conference on Applied Informatics (AI2002), Innsbruck, Austria 2002
7. Stoica, A., Zebulum, R., Keymeulen, D.: Polymorphic Electronics. In Proc. of International Conference on Evolvable Systems: From Biology to Hardware, LNCS 2210, Springer Verlag, 2001, p. 291–302
8. Thompson, A., Layzell, P., Zebulum, R., S.: Explorations in Design Space: Unconventional Electronics Design Through Artificial Evolution. IEEE Transactions on Evolutionary Computation. Vol 3(3), 1999, p. 167–196

A Biological Development Model for the Design of Robust Multiplier

Heng Liu, Julian F. Miller, and Andy M. Tyrrell

University of York, Department of Electronics, Bio-inspired Architectures Lab,
Heslington, York YO10 5DD, UK
{hl142, jfm, amt}@ohm.york.ac.uk

Abstract. A biologically inspired developmental model targeted at hardware implementation (off-shelf FPGA) is proposed which exhibits extremely robust transient fault-tolerant capability. All cells in this model have identical genotype (physical structures), and only differ in internal states. In a 3x3 cell digital organism, some individuals which implement a 2-bit multiplier were discovered using evolution that have the ability to "recover" themselves from almost any kinds of transient faults. An intrinsic evolvable hardware platform based on FPGA was realized to speed up the evolution process.

1 Introduction

Living multi-cellular biological organisms exhibit several intrinsic characteristics electronic engineers earnestly long for, such as growth and fault-tolerance. Multicellular living organisms achieved these traits through millions of years of evolution by means of cells which have identical genotypes. All of them come from a single special cell (zygote): this process is called development. The entire process of development is controlled by the interaction of cells rather than by a centralized process.

The development of an embryo is determined by genes, which control where, when and how many proteins are synthesized [1]. Complex interactions between various proteins and between proteins and genes within cells and hence interactions between cells are set up by activities of genes. It is these interactions that control how the embryo develops.

Development involves cell division, the emergence of pattern, change in form, cell differentiation and growth. The model proposed in this paper contains only two of these aspects, cell division and differentiation. Since in hardware, no new resources can be created as cells are pre-formatted and their number can not be increased, "growth" is used in this report to refer to "cell division".

Cell differentiation emerges as a result of differences in gene activities which lead to the synthesis of different proteins. As development is progressive, the fate of cells becomes determined at different times. Inductive interactions by means of chemicals or proteins between cells can make cells different from each other and the response to these inductive signals depends on the state of this cell.

Built-in redundancies and error handling capabilities are the most widely used conventional fault-tolerant technologies. Redundancies can be employed either

F. Rothlauf et al. (Eds.): EvoWorkshops 2005, LNCS 3449, pp. 195–204, 2005.

spatially or temporally. Spatial (area) redundancy can be applied using Dual Modular Redundancy or Triple Modular Redundancy, both of which are based on the majority vote of individual modules. In temporal (time) redundancy techniques, after an error output is detected, it is recomputed in an attempt to recover from the transient fault. Although time redundancy in general requires fewer resources than area redundancy, it demands error handling capability which will incontrovertibly increase the complexity of the system and its design cost. What's more, it is difficult to design such an error handling circuit which stores adequate information for recovery so that it can discover most transient faults.

Transient faults account most system failures [11], so at this stage we only concentrate on this kind of faults.

Development has been used as a bio-inspired technique in the past [6, 7, 8]. However, this paper considers a new development-inspired technique that makes use of a chemical signal which gives the system high tolerance to transient faults.

2 Development Cellular Model for Digital System

One of the most fundamental features of the development principle is the universal cell structure: each of the cells in a multi-cellular organism contains the entire genetic material, the genome.

Every cell only has direct access to the information of its four adjacent cells: No direct long term interaction between non-adjoining cells is permitted in this model.

In the digital hardware model proposed here (as shown in Fig. 1), the internal structure of digital cells is shown in Fig. 2. A digital cell is composed of three main components: Control Unit (CU), Execution Unit (EU) and Chemical Diffusion module (CD).

The Control Unit (CU) has a States Register, which stores the internal states of the cell, including the cell state (type) and chemicals. Each CU connects to its 4 immediate neighbors (shown in Fig. 1) and a Next States & Chemical Generator determines its own next state/chemicals according to the current states and chemicals of the neighbors, its own state and its own chemical (illustrated in Fig. 2). The NSCG contains two components: Next States Generator (NSG) and Next Chemical Generator (NCG), both of which are built from combinational circuits.

The EU Function Selection signal (the state of a cell) is 2-bit wide: 0 means this is a dead cell, and the EU will simply propagate its west (left) inputs to its south and east neighbors, otherwise this is a living cell (it can be in any type among 1, 2 or 3), and the EU will execute and propagate its calculated output to the south and east.

The Execution Unit (EU) is the circuit incorporated to do the real calculation of the target application. The inputs to each EU come from its immediate west and north neighbors, and the state of this cell (refer to Fig. 2). Every EU also propagates its output (Executing Signals) to its immediate north and east neighbors. The Execution Unit Core (EUC) is the evolvable core logic circuit, which determines how to process the input signals in the EU.

At present only combinational applications are considered, hence the EUs are purely combinational circuits. The state and chemical signals are 2-bit and 4-bit wide respectively, while the width of Executing Signal is 3-bit. Both the internal core logical structures of EU (EUC) and CU (NSG and NCG) are determined through evolution. As a result, the genotype encodes the EU and CU internal structures. The representation of the internal structure of EU and CU are based upon Cartesian Genetic Programming [4] (CGP): a program is expressed as an indexed graph which is encoded in a linear string of integers. So the genotype just contains a list of node connections and functions.

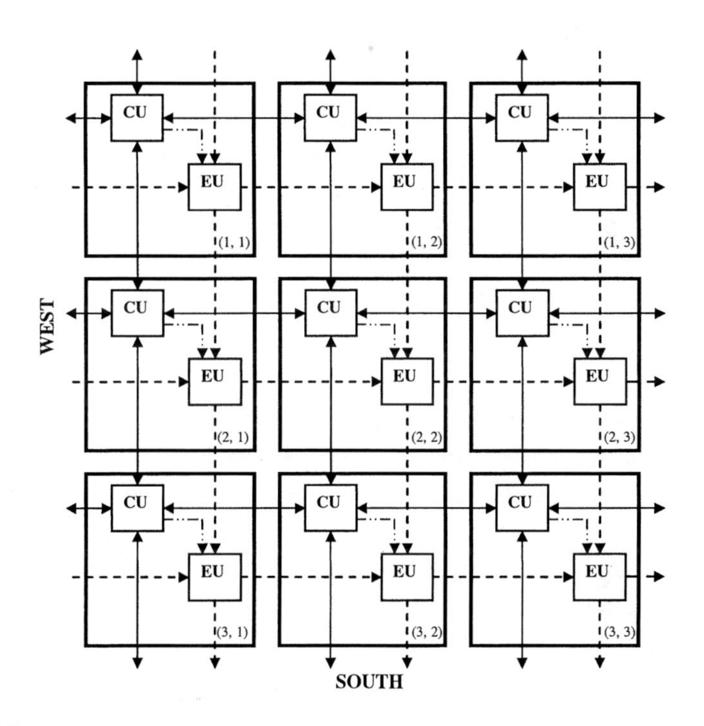

LEGENDS:

Fig. 1. Inter-connection of Cells

The Chemical Diffusion module (CD) mimics aspects of the real environment where biological cells live. In principle, CD should not be a component of a digital cell. However, this design decision makes it more convenient practically, so it is merged into the cell internal structure.

The chemical signal is introduced to transmit information between cells. Another function of the chemical is to serve as a resource which is required for a dead cell to transform to a living one.

Previous experiments [3, 5] suggest that chemicals are indispensable in order to achieve a robust solution: without chemicals, evolved individuals have poor stability and much lower fitness. The chemical diffusion regulation is the key mechanism which makes it such a significant aspect of this model: cells have a means to send long-distance messages.

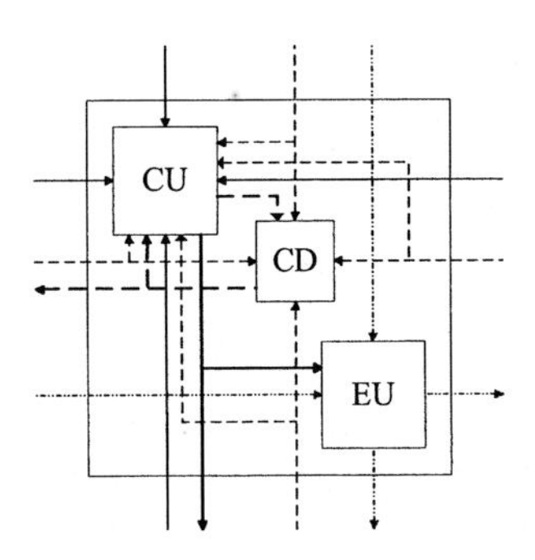

LEGENDS:

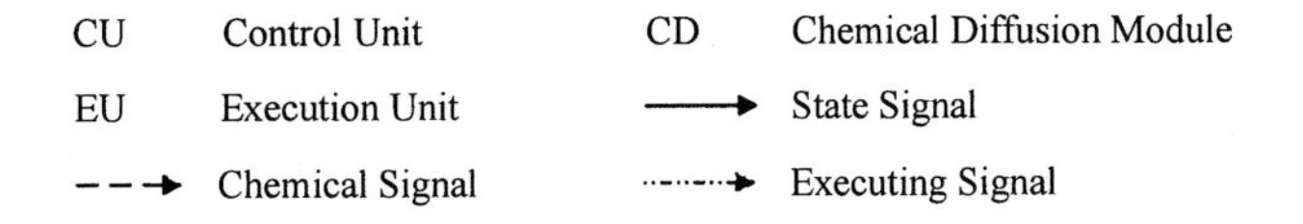

Fig. 2. Digital Cell Structure

The chemical diffusion rule employed in this work is similar to that in [3], except that there are only 4 immediate neighbors in this case. So the rule is:

$$(C_{ij})_{t+1} = \frac{1}{2}(C_{ij})_t + \frac{1}{8}\sum_{k,l \in N}(C_{kl})_t \tag{1}$$

Let N denote all the 4 immediate neighbors of a cell at (i, j) with neighboring position (k, l), the chemical at this position at the new time step is given by (1). The meaning of this equation is that each cell retains half of its previous chemical and distributes the other half equally to its four adjacent cells and receives the diffused chemical from them. It is evident the rule makes sure that chemicals are conserved (apart from the unavoidable loss when the level falls below one) in the diffusion procedure.

Calculating the diffused chemical in each grow step based on the chemicals from the four immediate neighbors and the cell's own chemical value is the main task of the Chemical Diffusion module (CD) (in Fig. 2). The CD also propagates the calculated value to the four adjacent cells.

Given a genotype, the inner-structure of the cells is determined and a zygote which is located at the centre of an 'artificial environment' with x rows and y columns of cells can be initiated and 'duplicate' itself. The position of the zygote was selected to speed up the growth: it takes least time for the digital organism to "cover" the entire area if the zygote is arranged in the centre. The inputs to the cells on the border of this environment are fixed to 0. Without chemicals no cells can live. This means that initially some chemicals must be injected at the position of the zygote.

Given a genotype, the growth procedure is described as follows:

1. Initialize chemical and the zygote;
2. Chemical diffusion;
3. All cells update their state simultaneously:
4. If no chemical at a position or all the cell's four neighbors and itself are dead, then this cell's internal program will not be executed;
5. Otherwise, it executes the program that is encoded by the genotype, to generate its next time chemical and state based on current states and chemicals;
6. If next state generated is alive, then overwrite chemical at this position with its own generated chemical;
7. Otherwise, do not touch the chemical at this position;
8. Unless stopping criterion reached go to 2.

3 2-Bit Multiplier: The First Real-World Application

A 2-bit multiplier was selected as an ideal test-case for this model to verify the feasibility and applicability of this model.

The digital organism employed in this application is made up of 3x3 identical digital cells. The maturity of the digital organism means that all the cells are grown to the pattern which implements a 2-bit multiplier.

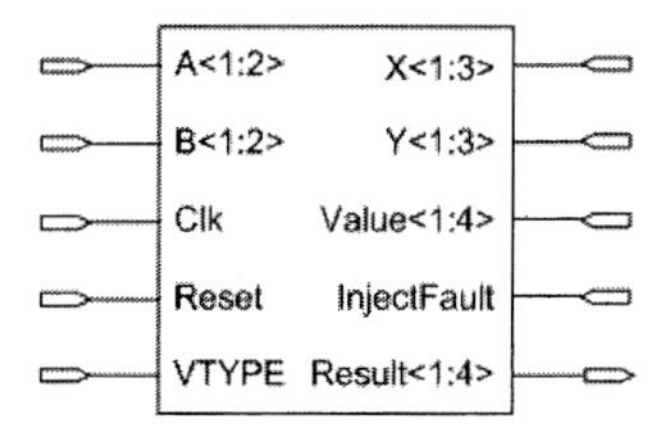

Fig. 3. Digital Organism External Interface

The external interface of the digital organism is shown in Fig. 3: Pin A, B and Result are the inputs and output of the 2-bit multiplier. Clk is the global clock signal; if the Reset pin is high, all the internal registers will be set to their initial values. All the remaining pins are dedicated to injecting transient fault(s) into the digital organism: when InjectFault pin is high, Value will be written into the chemical of cell at coordinate (X, Y) if VTYPE is low, otherwise the lowest 2-bit of Value is written into the state of the cell. Meanwhile the whole organism stops its growth process. Pin A and B are connected to the cells (1, 1) and (2, 1) in the digital organism, while the Result is driven by the cells (2, 3) and (3, 3).

Every cell has an identical structure: Pin InjectFault, VTYPE and Value are connected to their global counterparts. If this cell is at the coordinate (X, Y) and InjectFault is active (high), the CS pin of this cell will be driven to high and the cell will overwrite its own chemical or state with Value.

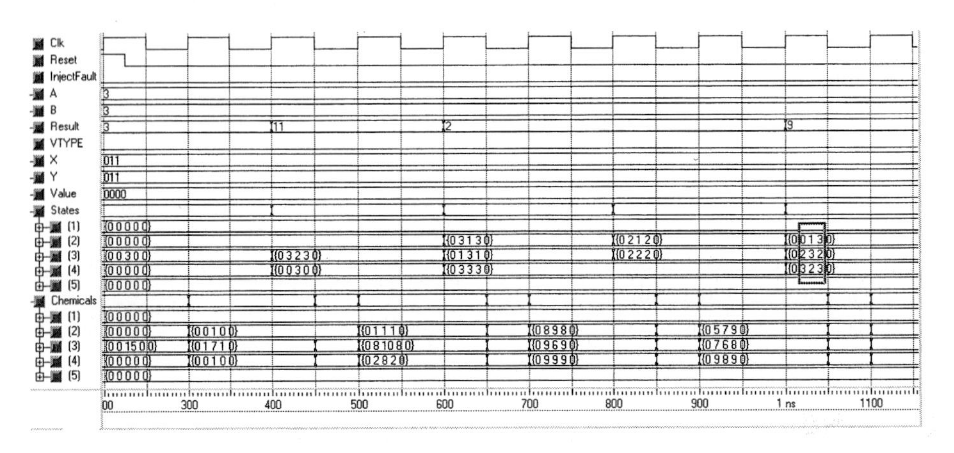

Fig. 4. Developmental Growth Procedure (the white rectangle circle the mature pattern)

A "growth step" lasts two clock cycles: at the falling edge of the first clock cycle a live cell (its state is not 0) will overwrite the existing chemical with its generated one; at the rising edge of the second clock, the chemicals diffuse according to the diffusion rule. At the rising edge of the first clock cycle in the next "growth step", the state will be updated.

The structures of the evolvable sub-circuits were evolved in software and a robust solution found was transformed into VHDL. The FPGA implementation was synthesized by ISE 6.1i from Xilinx, downloaded into the hardware. The detail of the waveform is demonstrated in Fig. 4. It can be seen that the organism matures at 1ns, when the state pattern is identical to that obtained in the software simulation. The following experiment was carried out: enough time was allocated to let the organism grow and mature (see Fig. 4). Subsequently, two sets of transient faults were injected: the first set composed of 4 transient errors in the chemicals of cell (2, 1), (2, 2), (2, 3) and (3, 3); the other set of faults were injected into the states of cell (2, 1), (2, 3) and (3, 1). Every fault was chosen to make the corresponding value 0. The time between the injections of the two sets of transient faults was more than enough for the organism to recover completely and stabilize itself again in terms of chemicals and states of the cells (see Fig. 5 and Fig. 6).

The recovery from of the first set of transient chemical faults is illustrated in Fig. 5. At the beginning, the chemical of some of the cells are modified, and then the organism resumes growing. It recovers flawlessly at 2.4ns and the result output regains the correct value.

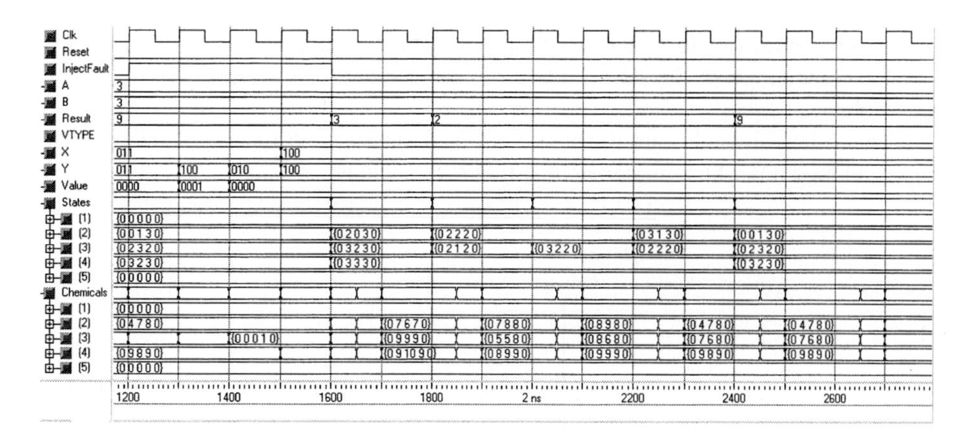

Fig. 5. Injection of the first set of faults and the recovery procedure

Fig. 6 demonstrates the recovery procedure from the second set of transient state faults. The states of the 3 selected "victim" cells are killed (state 0) at the beginning of this period. The organism again recovers completely to the correct pattern at 4ns.

The FPGA on the RC1000 board [9] connects to host PC with very limited data width: only 8-bit read and 8-bit write. So a further FPGA module "IOControler" was implemented to latch all the required inputs and feed them to the digital organism. Another function of IOControler is to cache the result output of the digital organism.

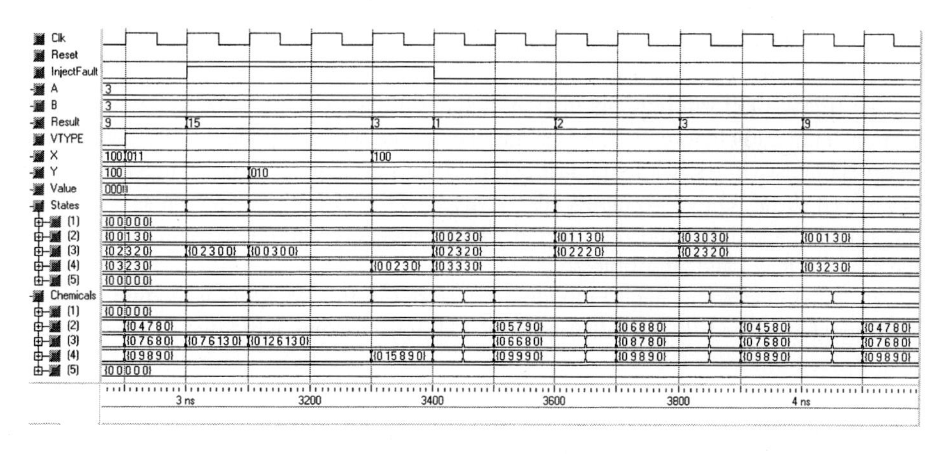

Fig. 6. Injection of the second set of faults and the recovery procedure

4 Intrinsic Evolvable Hardware (IEHW) Employing FPGA

An intrinsic evolvable hardware (IEHW) platform to accelerate the evolution progress was constructed.

The molecules (nodes in the CGP) are the most fundamental elements of the evolvable components of the model. Each evolvable sub-circuit is composed of several molecules.

The top level modules are illustrated in Fig. 7. There are 5 functional independent top-level modules which implement the IEHW as a whole.

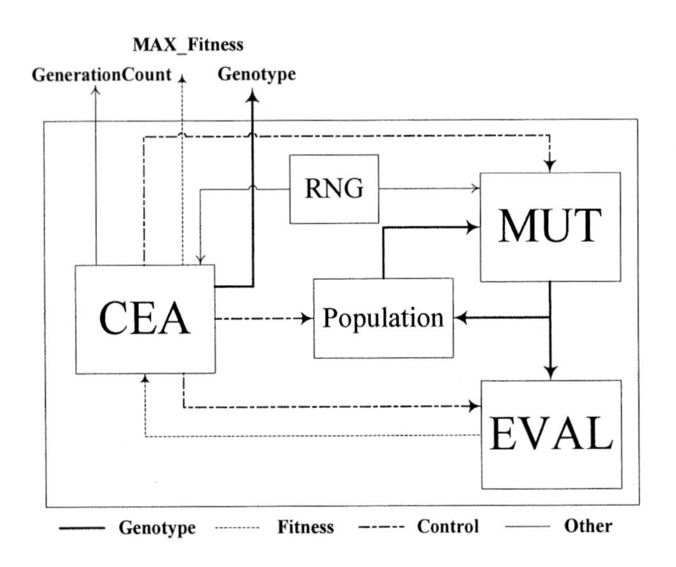

Fig. 7. Top-level Overview of the Intrinsic Evolvable Hardware Platform

The IEHW platform implementation includes 3 main outputs: MAX_Fitness, GenerationCount and Genotype. The first and the second will be updated every generation to reflect latest values, while the last output always propagates the best individual that is evolved so far. Only two inputs are required for this IEHW to function as expected: the global clock signal and reset signal. Other inputs are optional parameters, such as the seed for RNG and stop fitness.

All the genotypes of each individual are stored in the *Population* module. This is implemented in the FPGA as distributed RAM, for only one individual is manipulated at any given time.

The Controller of EA (*CEA*) supervises the entire evolution process and all the other modules. The fitness for all the individuals in the population is also stored in this module. The EA employed is an extended version of D. Levi's HereBoy Algorithm [10]: several parallel HereBoy are executed simultaneously. The CEA module is realized as a finite state machine (FSM).

RNG is a 64 bits Linear Feed-back Shift Register (LFSR), which is employed as a pseudo-Random Number Generator. If supplied, the seed of RNG will also be saved in this module.

The main function of Mutation Module (*MUT*) is to mutate a given genotype and latch the mutated genotype to be used by the *EVAL*. This module reads in the mutation rate and mutates molecules one by one until the specified mutation number is met. This module is also implemented as an FSM.

The core component of this IEHW is the Evaluation Module (*EVAL*), where the Digital Organism resides. Its main function is to evaluate the fitness of every individual. This module feeds every possible input to the 2-bit multiplier implemented by the evolved digital organism and sums up the total correct bits. Finally, the result of the subtraction of the total correct bits from the maximum possible correct bits (which is 64 in this case) is the fitness of this individual. The *EVAL* module is made up from 2 FSM: one is used to manage the digital organism and the other is in charge of feeding inputs, calculating correct output bits, the summary and the final subtraction to generate the fitness output of this module.

After reset signal is pulsed (low) for one clock cycle, all the modules, including all FSM and internal registers, are all cleared to their initial states. In this state, the IEHW will receive and latch any input parameters if provided, otherwise the default parameters are used. When the start signal is activated by the host PC, the *CEA* module will take all the responsibilities of the IEHW.

First, the population are initiated one by one, evaluated and saved into *Population*: *CEA* signals the *MUT* to mutate at the highest possible rate so all the molecules in the genotype are randomly generated, then *EVAL* evaluates it and propagates the fitness to *CEA*, finally the *CEA* saves the fitness and signal the *Population* module to store the new generated individual. These individuals make up the 0 generation.

Table 1. Synthesis Report

	Mole.	NCG	NSG	EUC	EVAL	RNG	CEA	MUT	All
LUTs	14	320	196	84	5507	64	164	106	7833
FFs					360	3	115	86	926

Second, after the initial population is ready, the main loop of evolution process begins: in each generation, the *CEA* selects each of the individuals in the *Population* and feeds it to the MUT. The mutated genotype is then evaluated by the *EVAl* module, and the fitness is again propagated back to *CEA*. If the mutated one (offspring) is better than the original one (parent), or with a probability p_r it substitutes the parent, which means the *CEA* asks the *Population* to store the mutated genotype; otherwise the content of *Population* module is untouched. After all the individuals have undertaken this procedure, a new generation is created. The evolution will continue to process the next generation unless the stop criterion, the specified fitness has been reached, is fulfilled. So no elitism is deployed in the IEHW.

When the main loop of the evolution process terminates, the best individual evolved is presented through the *Genotype* pin, while its fitness and the generation where this evolution stops are propagated out via MAX_Fitness and GenerationCount respectively.

Table 1 demonstrates how much hardware resources the various modules consume after the synthesis. The entire design occupies 31.9% LUTs, 3.77% Flip Flops and 58.6% IOBs totally.

5 Conclusion and Future Work

It was demonstrated that the biological development model proposed in this work can be applied to real world application and the solution discovered through evolution exhibits the intrinsic highly fault-tolerance feature similar to its living organism counterpart: the best solution found can virtually tolerate any transient damages.

Although this model may consume more resources for the 2-bit multiplier if compared with conventional majority voting systems, it does not require any voter systems and is not dependent on any single resource, so no single point fault. In the meantime, as this is a development model, we can apply it to more complex systems without fundamental modification.

In future, the module will be extended to explore more possibilities, such as making full use of chemical signals when dealing with state signals and unconstrained growth world. In addition, the hardware implementation will receive more improvement, including incorporating adaptive mutation. With the IEHW, we can also carry out more researches about the impacts of different parameters to the evolution outcome.

References

1. Lewis Wolpert: Principles of Development 2nd (2002), Chapter 1, Oxford University Press, c2002.
2. H. Ball and F. Hardy, "Effects and detection of intermittent failures in digital systems" 1969 FJCC, AFIPS Conf. Proc., vol. 35, pp.329-335
3. Julian F. Miller, "Evolving developmental programs for adaptation, morphogenesis, and self-repair" (2003), Seventh European Conference on Artificial Life, Lecture Notes in Artificial Life, Vol. 2801, pp. 256-265
4. 4. J. Miller and P. Thomson: Cartesian genetic programming. Lecture Notes in Computer Science, Vol. 1802, pp. 121-132, Poli, R., Banzhaf, W., Langdon, W.B., Miller, J. F., Nordin, P., Fogarty, T.C., (Eds.)
5. Julian F. Miller, "Evolving a self-repairing, self-regulating, French flag organism", GECCO 2004: Genetic and Evolutionary Computation Conference, Seattle, WA, USA, June 26-30, 2004. Proceedings, Part I
6. Ortega, C., Mange, D., Smith, S.L. and Tyrrell, A.M. 'Embryonics: A Bio-Inspired Cellular Architecture with Fault-Tolerant Properties' Journal of Genetic Programming and Evolvable Machines, Vol 1, No 3, pp 187-215, July 2000.
7. Jackson, A.H. and Tyrrell, A.M. 'Implementing Asynchronous Embryonic Circuits using AARDVArc', 4th NASA Workshop on Evolvable Hardware, Washington, pp 231-240, July 2002.
8. Canham, R. and Tyrrell, A.M. 'An Embryonic Array with Improved Efficiency and Fault Tolerance', 5th NASA Conference on Evolvable Hardware, Chicago, pp 265-272, July 2003.
9. http://www.celoxica.com/
10. D. Levi, "HereBoy: A Fast Evolutionary Algorithm", Proceedings of the 2nd NASA/DoD Evolvable Hardware Workshop, IEEE Computer Society, Los Alamitos, Ca, July 2000
11. H. Ball and F. Hardy, "Effects and detection of intermittent failures in digital systems" 1969 FJCC, AFIPS Conf. Proc., vol. 35, pp.329-335

Automatic Completion and Refinement of Verification Sets for Microprocessor Cores

Ernesto Sánchez, Matteo Sonza Reorda, and Giovanni Squillero

Politecnico di Torino – Dauin, C.so Duca degli Abruzzi 24, 10129 Torino – Italy
Tel: +39-0115647055, Fax: +39-0115647099
{edgar.sanchez, giovanni.squillero, matteo.sonzareorda}@polito.it

Abstract. In the design cycle of a microprocessor core, the unit is usually refined through a series of subsequent steps. To deliver a flaw free unit at the end of the process, in each stage a verification step is required. While it would be useful to automatically develop the set of test programs for verification concurrently to the design, in most of the existing approach verification is performed manually and starting from scratch. This paper presented a methodology for the automatic completion and refinement of existing verification programs. It shows a new technique for allowing a Genetic Programming-based framework to import an existing test-program set and assimilate it for further test generation. A case study is considered, in which a sample pipelined processor is used, and new test programs are generated starting from existing functional ones. Different metrics are targeted, and preliminary results are reported, showing the effectiveness of the method with respect to a pure random approach.

1 Introduction

Nowadays, the market is demanding more and more improved technological products in incredibly reduced times; this fact is leading manufactures to an ever shrinking time-to-market to be able to conquer their specific sell sector. It is true that the current VLSI densities offer a huge quantity of resources to design engineers for implementing new ideas; synthesis technologies, such as 90 nm, allow logic transistors densities of 90 millions of transistors per square centimeter, and additionally, the high clock rates and the low power consumption reachable by today's VLSI circuits permit high performance design levels. However, the verification and test methodologies for digital circuits are not progressing at the same pace. In fact, the roadmaps in the semiconductor association indicate the test and verification process as the main cost factor on a new design, reaching levels up to the 70% of the total cost [1].

In this scenario, the System-on-a-Chip (SoC) paradigm has a high acceptance because it is focused on reuse concepts looking for time-to-market reductions. The use of SoCs allows building more complex and complete systems in less time. A SoC complexity may greatly differ depending on the system goal, but typically SoCs contain at least one custom microprocessor core.

Designing a microprocessor core for a SoC is still a challenging task. Despite its limited size and reduced complexity, releasing an operational unit at the end of the

F. Rothlauf et al. (Eds.): EvoWorkshops 2005, LNCS 3449, pp. 205–214, 2005.

design cycle is made more difficult by time and resource constrains. To avoid implementation flaws, it is common practice to perform audits among different stages of the design cycle to guarantee the final system correctness [2] (Figure 1 shows a typical design cycle of *n* stages).

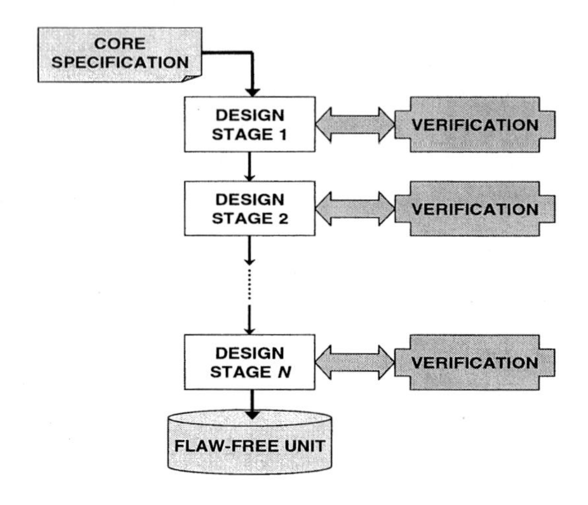

Fig. 1. Design Cycle

Design verification is usually performed by means of simulation tools able to simulate different design description levels and carry out a comparison to assert that the new stage implementation fits all device specifications. Unfortunately, verification technologies are not mature enough to fully automate the generation of verification patterns through all design stages.

Several authors propose automatic methodologies to generate a set of test programs for verifying a core. As shown in [3], the program VERTIS is able to generate verification programs based only on the processor instruction set architecture (ISA). In [4] the branch prediction mechanism of the PowerPC604 was considered. Even though the method is effective, profound processor knowledge is necessary to implement it, making the method extremely difficult to generalize. On the other hand, evolutionary computation techniques had been also used to automatically generate test programs, an almost completely automatic approach is described in [5]. Here the authors tackled a pipelined microprocessor attaining a successfully set of program maximizing the RTL (Register Transfer Level) statement coverage.

As shown in the figure 1, the microprocessor design cycle is constituted by several stages that represent different abstraction levels of the microprocessor model. In each stage verification teams may exploit the set of functional programs, often manually developed by designers. Since depending on the design phase different reference models are available for design verification, like an instruction set simulator (ISS) or RTL descriptions, the reuse of previously generated programs may not be straightforward. However, a set of test programs developed for a specific stage should be considered a good starting point for developing the new set. Additionally, the above methodologies for test program generation do not take advantage of pre-made test programs.

This paper presents a new methodology for the automatic completion and refinement of verification programs for microprocessor cores based on an evolutionary algorithm, which has been improved by an assimilation tool. The mechanism is able to *enhance* initial sets of functional programs. Previous sets are not merely *included* in the new set of programs for the verification, but *assimilated* and used as a starting point for the new test-program generation task. This approach allows to improve the quality of the initially generated test program set, and to significantly reduce the computational effort to generate the new test programs. Preliminary experimental results are reported on a simple pipelined processor, showing the results achievable in this way by resorting to different validation metrics. Currently, experiments are still running looking for additional improvements on the set of verification programs.

This paper is organized as follows: section 2 describes the proposed methodology. In section 3, a case study is shown. The experimental results are presented in section 4, and finally section 5 concludes the paper.

2 Proposed Approach

Each time that a new step into the design cycle is reached, the new design must be verified. Comparing the behavior of the new model with the previous one is a required step, but the new features should also be all checked.

To unveil design errors, the verification process should use a set of test programs able to excite all possible functionalities. Generally, at the earlier design stages, the set of test programs is composed of hand-written programs. These programs are usually targeted at specific corner cases and test specific functionalities, but they are not sufficient to reveal all possible design flaws. Then, verification engineers devise additional content to the test set to evaluate based on their experience, for example, targeting specific functional blocks, a complete set of instructions or a special operation mode.

As literature reveals, several and quite different approaches have been developed targeting automatic test program generation, but no one use previous test programs (some examples in section 1). If we consider a test program as an information block, containing data structures able to give us information about the processor testing. It must be useful to pick up this information to improve new test program generation processes. This kind of recovering information process has been called Data Reverse Engineering [6]. In this paper, a program "assimilator" able to take useful information in previously written test-programs will be presented.

The complete approach is composed of two blocks: the test-program generator and the test-program assimilator. The former is based on the μGP, an evolutionary algorithm described in [5]; the latter is a new module able to translate existing programs and fragments of code to graphs, usable by the μGP. The test-program assimilator reads the description of a specific assembly language and translates each line of an existing program to the appropriate reference in the library. Next sections better detail the two blocks.

Finally, The proposed methodology automatically refines and completes verification sets of test programs, maximizing several metrics. As supported by many authors, e.g. [8], it is not possible to accept a single coverage metric as the most reliable and complete one, thus 100% coverage on any particular metric cannot guarantee a 100% flaw free design. Therefore, the verification trend is to combine multiple coverage metrics together to obtain better results. However, not all metrics could be sensibly exploited in the earlier stage of the design flows. Therefore, it is extremely useful to automatically complete a verification set to reach complete coverage on different metrics.

2.1 Program Generator

An evolutionary tool called µGP has been used as program generator. Mainly, the µGP loop requires an Instruction Library (IL) and an external simulator that simulates the program execution on the processor under verification. The IL is composed of a set of macros that represent possible instructions and their operands. The evolutionary core evolves programs using the genetic programming paradigm and exploits the simulator as external evaluator.

Being a flexible tool, one can ask µGP to maximize or minimize diverse functions. Conceptually, the core evolves a population of verification programs (µGP Individuals) performing genetic operators such as mutation and crossover over some of them. Once generated, the new programs are evaluated based on a predefined fitness function and finally the complete set of individuals is sorted, discarding the worst. The process cycles until an optimal solution is found or a steady state is reached. Further information on the µGP can be found in [7]; however, since the assimilation process refers directly to the IL, a short description of the IL is also presented in the following.

2.2 The µGP Instruction Library

The instruction library defines the assembly language syntax. It enumerates all different sections and defines a set of macros. Macros are fragments of code of arbitrary length, with an arbitrary number of parameters. A very simple macro is shown in Figure 2. It encodes a single instruction, an "add" instruction, between a register and an 8-bit constant. The register is the first parameter ($1) and may be either R1 or R2. The 8-bit constant is the second parameter ($2) and takes values between -128 and 127. The instruction library is designed to be easily set up and understood by a human operator.

```
.macro
     add $1, $2
.parameter constant R1 R2
.parameter integer -128 127
.endmacro
```

Fig. 2. A Simple Macro

The instruction library supports different types of parameters, among them for instance, integer intervals, constants and program labels can be defined. The instruction library also allows specifying all syntactic details of the assembly language, like the format of label and subroutines, the format of comments, etc…

2.3 Assimilation Process

To assimilate information from the existing programs, a specific application was developed. The tool receives the original IL and the initial set of programs, then analyzes each program building an initial population for μGP containing these programs and possibly modifies the IL.

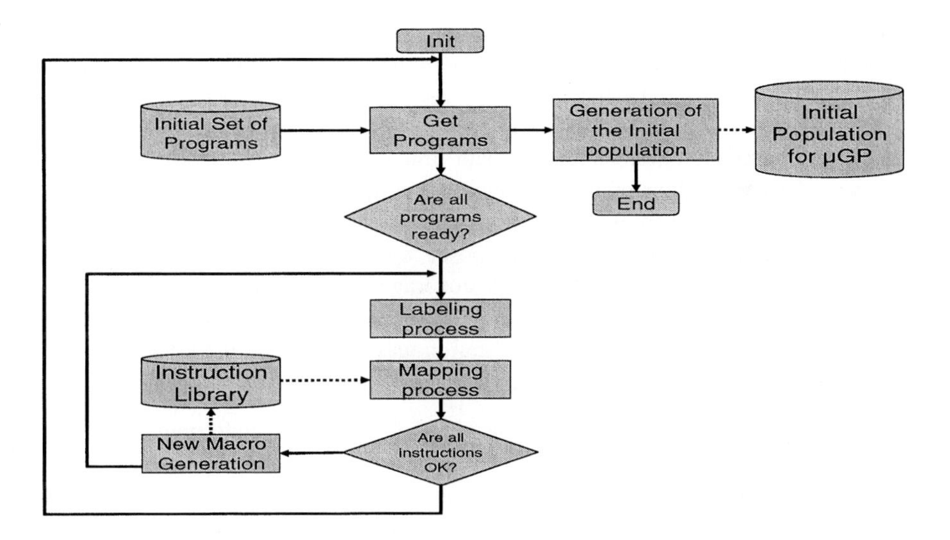

Fig. 3. Assimilation process

The general extraction process is shown in figure 3. An initial program analysis is performed to identify all program labels, and then the assimilation approach tries to map each instruction with a corresponding IL macro. Occasionally the macros belonging the IL cannot map a particular instruction because it is not implemented or the specific operators do not fit the established ranges, then a new macro is built containing the instruction using the appropriate operators. At this point the new macro is added to the IL and the mapping process can continue.

The assimilation process is performed by analyzing one code line each time; for each analyzed line, the tool recognizes three components: *label*, *instruction* and *comments*. The *label* importance leads on the loop and jump management of the assembler program. The *instruction* is split in operators and operands. Essentially, each operator could correspond a macro, however the initial instruction library may not be complete. Thus, the new instruction library will be improved by adding a new macro. On the other hand, the operands range is also analyzed to guarantee that, each one corresponds to a predefined type, on the contrary the new macro is build using as operators fix values. Even though the character played by the *comments* is not indispensable by the program execution, these are important to reach programs readability.

The assimilation process guarantees the acquisition of singular values implemented by design engineers at previous stages aimed to cover corner cases. Also, if the tool is specially configured to assimilate more complex structures, complete instruction sequences can be acquired.

Once all instructions have been mapped, the application generates a µGP initial population containing the initial programs and the IL is improved with the new macros.

2.4 Validation Metrics

Coverage analysis provides an easy, fast and objective way of measuring simulation effectiveness, and coverage metrics are the required starting point for every verification process.

As stated before, it is not possible to accept a single coverage metric as the most reliable and complete one. One of the main limitations of the previous usage of the µGP was the exploitation of the statement coverage verification metric alone. The work presented here overcomes this limitation: beside the statement coverage, the verification metrics supported in the proposed approach are: branch, condition and expression coverage. Moreover, the toggle coverage is also used.

The approach we propose can support different validation metrics. In the experimental evaluation, the design under verification was described at the RT level and the metrics that have been exploited are summarized in the following.

- **Statement coverage is** the most basic form of code coverage: statement coverage is a measure of the number of executable statements within the model that have been exercised during the simulation run. Executable statements are those that have a definite action during runtime and do not include comments, compile directives or declarations. Statement coverage counts the execution of each statement on a line individually, even if there are multiple statements on that line.
- **Branch coverage** reports whether Boolean expressions tested in control structures (such as the `if`-statement and `while`-statement) evaluated to both *true* and *false*. The entire Boolean expression is considered one true-or-false predicate regardless of whether it contains logical-and or logical-or operators. Branch coverage is sometimes called *decision coverage*.
- **Condition coverage** can be considered as an extension of branch coverage: it reports the *true* or *false* outcome of each Boolean sub-expression, separated by logical-and and logical-or if they occur. Condition coverage measures the sub-expressions independently of each other.
- **Expression coverage** is the same as condition coverage, but instead of covering branch decisions it covers concurrent signal assignments. It builds a focused truth table based on the inputs to a signal assignment using the same technique as condition coverage.

- **Toggle coverage** reports the number of bits that toggle at least once from 0 to 1 and at least once from 1 to 0 during the execution of a program. At the RT level registers are targeted and, since RT-level registers correspond to memory elements with an acceptable degree of approximation, the toggle coverage is an objective measure of the *activity* of the design. Indeed, this is a very peculiar metric and can be sensibly used in all late stages of the design cycle.

3 Case Study

The proposed methodology was implemented. The test-program assimilator consists of about 6,000 lines in ANSI C. To analyze the programs it exploits a publicly available library that implements regular expression pattern matching. The μGP consists in about 10,000 lines of ANSI C, but it did not require any modification for this work. To simulate the microprocessor, *Modelsim* v5.8b by (Model Technology) Mentor Graphics was used. A few scripts for accessing the different metrics were required. All experiments were run on a Sun Enterprise 250 with two UltraSPARC-II CPU at 400MHz and 2Gb of RAM.

As a case study the DLX/pII processor [9] was tackled. The DLX/pII is a small microprocessor with a 5-stage pipeline that implements the instruction set described in [10]. The core under verification is described in VHDL at RT-level with 4,558 statements. The description contains 3,695 branches that can be activated, 193 conditional statements based on 1,764 different expressions. A total of 8,283 logic bits could be toggled during the simulation. Due to its size and complexity, the DLX/pII can be considered a typical example of a small microprocessor core designed for a SoC.

The initial instruction library for the DLX consists of 98 macros, containing the information to map a large amount of the possible microprocessor instructions and operands. The ranges of most of the operands are integer intervals from 0 to $2^{16} - 1$.

A reference set of programs was chosen from the set of functional and application test programs performed by designer engineers. The set contains programs that stimulate mainly the LOAD/STORE instructions and the multiplication unit.

4 Experimental Results

As a starting set for the completion and enhancement process a set composed of 12 functional programs was considered. The functional programs were devised by the designer to validate the main functionalities of the core.

After assimilating the initial set, the μGP was asked to complete it by maximizing the 5 coverage metrics described above. The final verification set was composed of 31 programs, including the 5 initial programs and all the programs generated by the μGP. In more details:

- Targeting the statement coverage, the μGP selected 3 programs of cumulatively 297 instructions, able to obtain the 99.65% of coverage. The generation required 2 days.

- Targeting the branch coverage, the μGP selected 4 programs of cumulatively 170 instructions, able to obtain the 98.59% of coverage. The generation required 4 days.
- Targeting conditional coverage metric, the μGP devised 7 programs of cumulatively 135 instructions, reaching the 79.90% of coverage. The generation required about 2 days
- Targeting the expression coverage, the μGP selected 6 programs of cumulatively 315 instructions, reaching 48.71% of coverage. The generation required about 3 days.
- Targeting toggle coverage metric, the μGP devised 6 programs of cumulatively 327 instructions, reaching 78.28% of coverage. The generation required about 3 days.

Table 1. *Experimental Results*

Set	Statement	Branch	Condition	Expression	Toggle
Initial	68.91%	52.40%	43.01%	26.93%	30.34%
Statement	**99.65%**	98.43%	77.20%	36.96%	43.28%
Branch	99.23%	**98.59%**	76.17%	33.50%	39.39%
Condition	77.29%	68.69%	**79.90%**	25.11%	26.33%
Expression	93.07%	88.09%	55.44%	**48.71%**	44.02%
Toggle	96.31%	92.29%	62.18%	38.72%	**78.28%**
Total	99.65%	98.81%	79.79%	46.32%	80.31%

Results are summarized in Table 1. The different rows contain the results for the different set of test programs: the initial one and the 5 generated maximizing specific metrics. The values attained by each set on all possible metrics are also reported. The grayed cells represent the value on the metric that the set was intended to maximize. The last row contains the cumulative results of the complete set.

Significantly, the μGP was always able to maximize the coverage it targeted (the grayed cells), but it is interesting to notice the relationship between the different verification metrics. For instance, maximizing the condition coverage does not yield impressive results compared with the branch coverage, while, in theory, the former metric is an extension of the latter. On the other hand, toggle coverage is a very peculiar metric, while all sets attain very low results on this one, only the experiment targeting this metric was able to reach acceptable results.

It should be remarked than, while the first four metrics are similarly fast to be calculated, the toggle coverage requires a considerable overhead. However, the toggle coverage is an interesting metric in subsequent stage in the design cycle, when the design is eventually synthesized to logic gates.

For the sake of completeness, a verification set was devised using a pure random approach. The random instruction generator exploits the µGP instruction library to devise syntactically correct fragments of code and the µGP external evaluator to simulate them.

Differently from the evolutionary experiments, however, the approach is merely cumulative: coverage figures are not used as a feedback to optimize candidate programs, but a program is added to the test set if it increases at least one coverage figure.

About 20,000 random programs were generated and evaluated in about two weeks. The final test set contains about 230K lines of code in 2,000 random programs.

The table 2 shows the cumulative results obtained by the random experiment related to each code coverage metric. The row *size* reports the total instruction lines that compose the set of random programs.

Table 2. Pure Random Approach verification set

	Random
Statement coverage [%]	99.30
Branch coverage [%]	98.59
Condition coverage [%]	79.79
Expression coverage [%]	48.75
Toggle coverage [%]	56.84
Size [lines]	230K

As already remarked, the statement, branch and condition coverage metrics can be saturated easily, although the µGP shows slightly higher performances. Reaching high toggle coverage is more difficult and the random approach is significantly worse than µGP. Additionally, µGP speeds up by two the

5 Conclusions

This paper presented a methodology for the automatic completion and refinement of verification programs based on an evolutionary algorithm. A new technique is described, allowing an existing Genetic Programming-based framework for test-program generation to import an existing test-program set and assimilate it for further test-program generation.

Since in the design cycle of a microprocessor core the unit is usually refined through subsequent steps, it would be reasonable to develop the set of test programs used for verification in a similar way.

In each step, after modifying the design, the existing set of test programs and the test programs developed by designers can be automatically enhanced and completed to build a set able to maximize a given verification metric. The proposed process requires both less human intervention and less computational resources than starting from scratch. Moreover, it allows completing the work of verification engineers, modifying a set of test programs targeting new verification metrics.

A case study is considered, in which a sample pipelined processor is used, and new test programs are generated starting from existing functional ones. Different metrics are targeted, and preliminary results are reported, showing the effectiveness of the method with respect to a pure random approach. Additional experiments are still running, looking for further improvements on the set of verification programs.

Acknowledgements

Authors whish to thanks Giuseppe Trovato for his help and for implementing the assimilation tool.

References

[1] Semiconductor Industry Association, *International Technology Roadmap for Semiconductors 2002 Update*, `http://www.semichips.org/pre_stat.cfm?ID=153`
[2] V. Agrawal and M. Bushnell, *Essentials of Electronic Testing for Digital, Memory and Mixed-Signal VLSI Circuits*, Norwell: Kluwer Academic Publishers, 2000.
[3] J. Shen, J. A. Abraham, "Native Mode Functional Test Generation for processors with application to self test and design validation", *International Test Conference*, 1998, pp. 990-999
[4] N. Utamaphethai, R.D. Blanton and J.P. Shen, "Superscalar Processor Validation at the Microarchitecture Level", *12th IEEE International Conference on VLSI Design*, 1999, pp. 300-305.
[5] F. Corno, E. Sanchez, M. Sonza Reorda, G. Squillero, "Automatic Test Program Generation - a Case Study", *IEEE Design & Test, Special issue on Benchmarking for Design and Test*, Volume: 21, Issue 2, March-April 2004, pp. 102 – 109
[6] K. H. Davis, "Lessons Learned in Data Reverse Engineering", IEEE Proceedings, Eighth Working Conference on Reverse Engineering, October 2001, pp. 323 – 327
[7] G. Squillero, "MicroGP — An Evolutionary Assembly Program Generator", to appear on: *Genetic Programming and Evolvable Machines*, 2005
[8] Chien-Nan Jimmy Liu, Chen-Yi Chang, Jing-Yang Jou, Ming-Chih Lai, Hsing-Ming Juan, "A novel approach for functional coverage measurement in HDL Circuits and Systems", *ISCAS2000: The 2000 IEEE International Symposium on Circuits and Systems*, 2000, pp. 217-220.
[9] D. A. Patterson and J. L. Hennessy, *Computer Architecture - A Quantitative Approach,* (second edition), Morgan Kaufmann, 1996.
[10] P. M. Sailer, P. M. Sler, *DLX Instruction Set Architecture Handbook*, Morgan Kaufmann Publishers, 1996.

A Genetic Algorithm for VLSI Floorplanning Using O-Tree Representation

Maolin Tang and Alvin Sebastian

School of Software Engineering and Data Communications,
Queensland University of Technology,
2 George Street, Brisbane, Australia
{m.tang, a.sebastian}@qut.edu.au

Abstract. Floorplanning is one of the most important problems in VLSI physical design automation. A fundamental research problem in the VLSI floorplanning is representation because it determines the size of search space and the complexity of the transformation between a representation and its corresponding floorplan. O-tree representation is one of the most efficient floorplan representations as it has the smallest search space among all the admissible floorplan representations and the computational complexity of transformation between a representation and its corresponding floorplan is only $O(n)$. The efficiency of O-tree representation was demonstrated by a deterministic algorithm proposed by Guo *et al.*. The deterministic algorithm can quickly find a reasonably good floorplan. However, the deterministic floorplanning algorithm, by its nature, is a local search algorithm, and thereby may not be able to find an optimal or near-optimal solution sometimes. This paper presents a genetic algorithm for the VLSI floorplanning problem using O-tree representation. Experimental results show that the GA can consistently produce better results than the deterministic algorithm.

1 Introduction

Floorplanning is one of the most important problems in VLSI physical design automation. It determines the extent to which the performance of a VLSI chip will be. Given a set of rectangular modules of arbitrary sizes, floorplanning finds a placement of the modules such that no module overlaps and a predefined cost function is minimized.

A fundamental research problem in floorplanning is representation because it determines the size of search space and the complexity of the transformation between a representation and its corresponding floorplan. The representation of floorplans has been intensively studied for decades. For a floorplan with a slicing structure, the most popular floorplan representation is *slicing tree* [1] (or equivalent *Polish expression*). For a non-slicing floorplan, there are several efficient representations. One of the most efficient representations is the ordered tree (O-tree) representation proposed by Guo *et al.* [2]. The O-tree representation not only covers all optimal floorplans, but also has a small search space. For

F. Rothlauf et al. (Eds.): EvoWorkshops 2005, LNCS 3449, pp. 215–224, 2005.

a floorplanning problem of n modules, the search space is $n!c_n$, where $c_n = \frac{1}{2n+1}\binom{2n+1}{n}$. In addition, it only takes $O(n)$ time to transform between an O-tree representation and its corresponding floorplan.

In order to demonstrate the efficiency of the O-tree representation, Guo *et al.* further proposed a deterministic floorplanning algorithm based on the O-tree representation. Their experimental results showed that the algorithm could quickly produce reasonably good results. However, the deterministic floorplanning algorithm, by its nature, is a local search algorithm, and thereby it may not be able to find an optimal or near-optimal solution sometimes.

Genetic algorithm (GA) is a global search technique inspired by evolution [3]. The crux of GA lies in the "survival of the fittest" strategy. It takes an initial set of random *individuals*, termed as the *initial population*. Each individual in the population is a chromosome that represents a solution to the given problem in an encoded form. Using well-defined genetic operators, GA evolves the individuals in the population generation by generation until an optimal or near-optimal solution is found. The fitness of the individuals is evaluated in a *fitness function*. GAs have been successfully applied on VLSI floorplanning problems [4, 5, 6, 7, 8, 9]. However, all the existing GAs were proposed for solving slicing floorplan problems. This paper presents a GA for non-slicing floorplan problems using O-tree representation. Each individual in the population of the GA is an O-tree. The crossover operator generates a good floorplan structure by extracting meaningful structural components from two parents. The mutation operator keeps the population diverse by re-permutating the modules without changing the topology of the ordered tree. To make the search more efficient, the GA uses the basic idea behind the deterministic algorithm to generate the initial population and uses the deterministic algorithm to optimize the individuals during the evolution of the population. Preliminary experimental results show that this GA can consistently produce near-optimal solutions for all the benchmark problems.

The remaining paper is organized as follows. Section 2 gives a formal definition of the floorplanning problem. Section 3 briefly reviews the O-tree representation and the deterministic algorithm. Section 4 details the design of the GA. The experimental results are presented in Section 5. Finally, this research work is concluded in Section 6.

2 Problem Statement

A module m_i is a rectangular block with fixed height h_i and width w_i, and $M = \{m_1, m_2, \cdots, m_n\}$ is a set of modules that need to be placed on a floorplan.

A placement $P = \{(x_i, y_i) : 1 \leq i \leq n\}$ is an assignment of the coordinates of the bottom-left corners of the modules such that no two modules overlap. In a placement, a module has only two optional orientations, *portrait* or *landscape*. A placement is evaluated in a cost function consisting of two parts: one is the area of the smallest rectangle that encloses the modules and the other is the interconnection cost between the modules. The cost function is defined in Equation 1.

$$Cost(P) = w_1 \times Area(P) + w_2 \times Wirelength(P) \tag{1}$$

In Equation 1, P is the corresponding placement of (T, π), $Area(P)$ represents the normalized area of the minimal bounding rectangle of P, $Wirelength(P)$ is the normalized total wire length of P. The wire length of a net is the sum of half the perimeter of the bounding rectangle of all the net terminals in the circuit.

Given a set of modules M, the floorplanning problem is to find a placement for the modules such that the cost is minimized.

3 Related Work

This section briefly reviews the O-tree representation and the deterministic algorithm for the floorplanning problem, both of which were proposed by Guo *et al.* [2].

3.1 O-Tree Representation and Encoding

In the O-tree representation [2], a floorplan of n modules is represented in a horizontal (vertical) ordered tree of $(n+1)$ nodes, of which n nodes correspond to n modules $m_1, m_2, \cdots, m_n$, and one node corresponds to the left (bottom) boundary of the floorplan. The left (bottom) boundary is a dummy module with zero width (height) placed at $x = 0$ $(y = 0)$. In a horizontal ordered tree, there exists a directed edge from module m_i to module m_j if and only if $x_j = x_i + w_i$, where x_i is the x coordinate of the left-bottom position of m_i, x_j is the x coordinate of the left-bottom position of m_j, and w_i is the width of m_i. In a vertical ordered tree, there exists a directed edge from module m_i to module m_j if and only if $y_j = y_i + h_i$, where y_i is the y coordinate of the left-bottom position of m_i, y_j is the y coordinate of the left-bottom position of m_j, and h_i is the height of m_i. Figure 1 shows a floorplan and its horizontal ordered tree representation.

An ordered tree of n nodes can be encoded in a tuple (T, π), where T is a $2(n-1)$ bit string identifying the structure of the ordered tree and π is a permutation of the $(n-1)$ non-root nodes. For a horizontal O-tree, the tuple is obtained by DFS (Depth-First Search) traversing the non-root nodes and edges of the O-tree. When visiting a non-root node, we append it to π. When visiting an edge in descending direction we append an 0 to T and when visiting an edge in ascending direction we append a 1 to T. The horizontal ordered tree shown in Figure 1 is encoded into $(00110100011011, adbcegf)$. We can use the same idea to encode a vertical O-tree.

An O-tree is *admissible* if no module can be shifted left or bottom without moving other modules in its corresponding placement. Given any O-tree, we can construct an admissible O-tree using a so-called AOT (Admissible O-Tree) algorithm. Details about the AOT algorithm can be found in [2].

3.2 The Deterministic Algorithm

Given an initial placement encoded in an O-tree (T, π), the deterministic algorithm finds a local optimal solution through systematically permutating those

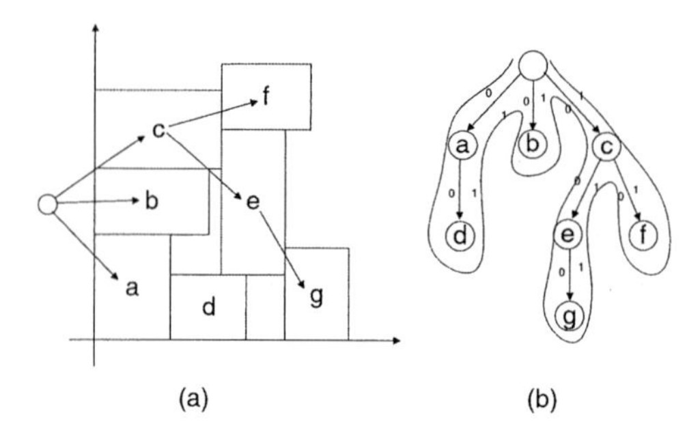

Fig. 1. A horizontal O-tree representation and its encoding. In this figure (a) shows a horizontal O-tree and (b) illustrates how to encode the O-tree

O-trees which can be obtained by changing a module's position in the given O-tree. Below is the algorithm description:

1. for each node in (T, π):
   (a) remove node m_i from (T, π);
   (b) insert node m_i in the position where we can get the best value of a predefined cost function among all possible inserting positions in (T, π) as an external node;
   (c) perform (a)-(b) on its orthogonal O-tree.
2. output (T, π).

4 The Genetic Algorithm

This GA is a steady-state GA. The crossover, which will be introduced later, produces only one child. This GA runs a fixed number of generations, $MaxGen$, and the size of population is *PopSize*. Below is the outline of the GA.

1. $t := 0$;
2. generate an initial population $P(t)$ of size *PopSize*;
3. evaluate the initial population $P(t)$ and find the best individual *best*;
4. while $t < MaxGen$
   (a) $t := t + 1$;
   (b) for each individual in $P(t)$
      i. this individual becomes the first parent p_1;
      ii. select a second parent using roulette wheel selection p_2;
      iii. probabilistically apply crossover and mutation to produce a child c_1;
      iv. use the deterministic floorplanning algorithm to optimize c_1;
      v. evaluate c_1;
      vi. if c_1 is better than p_1 then use $c_1$to replace p_1 in $P(t)$;
      vii. if c_1 is better than *best* then *best* $:= c_1$;
5. output *best*.

In the following we elaborate the genetic representation, the initial population generation, the fitness function, as well as the genetic operators.

4.1 Genetic Representation

Different from most GAs that use a binary string to present an individual, this GA uses a tuple (T, π) to present an individual. Each tuple is an encoded O-tree representing an admissible placement. In a tuple (T, π), T is a binary string of $2n$ bits and π is a sequence of n modules.

4.2 Fitness Function

The fitness of an individual (T, π) in the population is defined in Equation 2.

$$f((T, \pi)) = -Cost(P) \tag{2}$$

where P is the corresponding placement of (T, π), and $Cost(P)$ is the cost of P given by Equation 1.

4.3 Initial Generation

Instead of randomly generating the initial generation, the GA uses the idea that the deterministic algorithm [2] generates its initial O-tree to generate the initial population. The benefit of doing so is to make the search of the GA more efficiently. Below is the algorithm used to generate the initial population by the GA:

1. for $i := 1$ to $PopSize$
   (a) randomly generate a sequence of the modules π;
   (b) $T := \phi$;
   (c) $min\_fitness = infinite$;
   (d) for each module m
      i. insert m to T in an external node position;
      ii. $min\_T := T$;
      iii. $min\_fitness :=$ fitness (min_T);
      iv. for each of the other possible external node positions, p, in T
         A. $T_1 := T$;
         B. insert m to T_1 in position p;
         C. get admissible T_1 using AOT;
         D. $fitness := fitness(T_1)$;
         E. if $fitness < min\_fitness$ then
            $min\_fitness := fitness$;
            $min\_T := T_1$;
      v. $T := min\_T$;
   (e) $P[i] := (T, \pi)$;
2. output P.

4.4 Crossover

Given two parents, both of which are O-trees, the crossover generates one child by recombining meaningful structural components from the two parents.

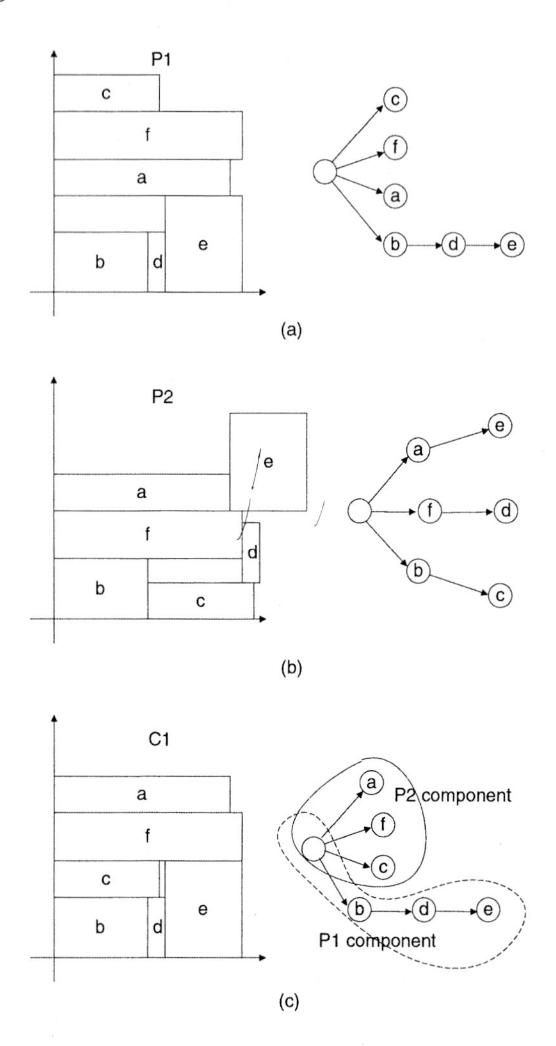

Fig. 2. Crossover. In this figure (a) is a parent p_1, which has four branches $\{b, d, e\}$, $\{a\}$, $\{f\}$ and $\{c\}$; (b) is another parent p_2, and (c) is the only generated child c_1. In this crossover, the branch $\{b, d, e\}$ of p_1 is randomly selected and duplicated to c_1. This part of c_1 is marked as "p_1 component" in (c). The other part of c_1 comes from p_2, and is obtained by removing nodes b, d and e, which are already present in c_1, from p_2. This part is marked as "p_2 component" in (c)

It is observed that branches of an O-tree are meaningful structural components because a branch represents a potential compact placement for those modules in the branch. Hence, the crossover uses branches of an O-tree as basic building blocks to generate an offspring.

When generating an offspring c_1 from two parents p_1 and p_2, the crossover randomly selects some branches from p_1, duplicates them and puts them in c_1. Then, the crossover operator takes a copy of p_2 and removes those modules

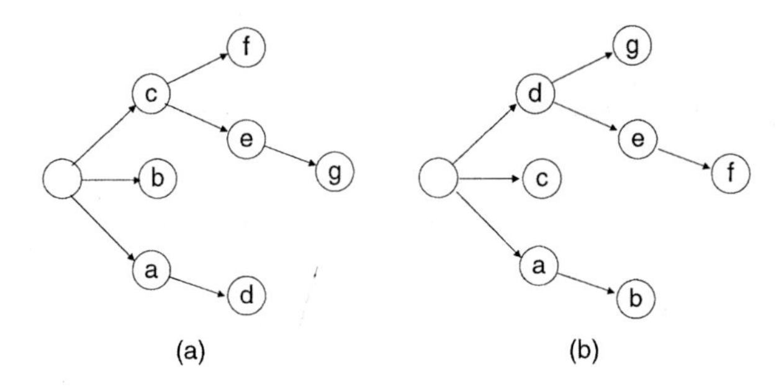

Fig. 3. Mutation. In this figure (a) shows the individual selected for mutation and (b) displays the mutated individual

that have been already present in c_1 from it and then adds it to c_1. Figure 2 illustrates the basic idea behind the crossover operator. In the figure, (a) and (b) are two O-trees, p_1 and p_2 respectively, and (c) is the offspring generated by the crossover operator. The corresponding placements are shown on the left hand side of the figure.

It takes $O(n)$ time to identify all branches of p_1, $O(1)$ time to randomly select some branches, and $O(n_1)$ time to duplicate and put them in c_1, where n is the number of modules and n_1 is the number of nodes in the selected branches of p_1. Since it takes $O(n)$ time to remove a node from an O-tree containing n nodes, it takes $O((n-n_1)*n)$ time to remove n_1 nodes from the copy of p_2. The time needed to combine the selected branches from p_1 and the branches after removing n_1 nodes from the copy of p_2 is $O(n-n_1)$. Hence, the computational complexity of this crossover is $O(n^2)$.

4.5 Mutation

Given an individual, or an O-tree encoded in a tuple (T, π), the mutation randomly re-permutates the sequence of the modules. The mutation does not change the topology of the O-tree, but generates a different placement.

Suppose that $(0011010001101, adbcegf)$ is an initial individual, and that $abcdefg$ is the randomly generated sequence of the module labels. The mutated individual is $(0011010001101, abcdefg)$. The mutation is illustrated in Figure 3 in which (a) shows the initial O-tree and (b) shows the mutated O-tree.

The computational complexity of this mutation is $O(n)$, where n is the number of modules.

5 Experimental Results

Since all the existing GAs are designed for slicing floorplanning problems, while our GA is designed for more challenging non-slicing problems, we could not compare our GA with the existing GAs. Hence, the experiment was to compare our GA with the deterministic algorithm [2]. In order to make them comparable, we implemented our GA and re-implemented the deterministic algorithm using the same programming language and in the same development environment. The programming language was C# and the development environment was Microsoft Visual Studio .NET 2003.

The experiments were carried out for five most popular MCNC benchmarks: *apte*, *xerox*, *hp*, *ami33* and *ami49*. The number of modules in these benchmarks ranges between nine and 49.

We compared our GA with the deterministic algorithm in terms of area and wire length. In our GA, the population size $PopSize$ was set to $20 * n$, where n is the number of modules in the corresponding benchmark problem. The probabilities of crossover and mutation were 0.95 and 0.05 respectively. The number of generations $MaxGen$ was set to 100. We ran our GA and the deterministic algorithm 10 times for each of the benchmark problems. When we ran the deterministic algorithm [2] we randomly generated the sequence of modules and their rotations.

For both the algorithms, we used the same cost function $w_1 \times Area(P) + w_2 \times Wirelength(P)$, where P is a placement, $Area(P)$ is normalized area of P, and $Wirelength(P)$ is normalized total wire length of P. Three sets of weights for w_1 and w_2 were used in the experiment: $\{w_1 = 0, w_2 = 1\}$, $\{w_1 = 0.5, w_2 = 0.5\}$, and $\{w_1 = 1, w_2 = 0\}$.

When using the first set of weights to test the two algorithms, we recorded the best and average results in terms of wire length only as the weight for area is 0. When using the second set of weights, we recorded the best and average results for both area and wire length. When using the third set of weights, we recorded the best and average results regarding area only as the weight for wire length is 0. The computation time for each set of weights and for each benchmark was also recorded. The experimental results for the deterministic algorithm and our GA are presented in Table 1 and Table 2 respectively. It should be pointed out that the experimental results for the deterministic algorithm are from our re-implemented program, rather than the original program, in order to make the experimental results are comparable with our GA experiential results.

For the first set of weights, our GA had 2.21% to 16.96% improvement in the best wire length and 8.97% to 26.62% improvement in the average wire length. On average, our GA had 8.88% improvement in the best wire length and 16.57% improvement in the average wire length. For the second set of weights, our GA

Table 1. Best and average results of the deterministic floorplanning algorithm with different weights for area and wire length

MCNC	$w_1 = 0, w_2 = 1$		$w_1 = 0.5, w_2 = 0.5$			$w_1 = 1, w_2 = 0$	
benchmarks	wire	time	area	wire	time	area	time
apte	214/234	1.1	49.7/51.3	228/253	1.1	48.6/49.2	1.1
xerox	535/559	2.6	20.8/21.8	565/636	2.4	20.4/21.3	2.7
hp	78.6/87.7	2.0	9.35/9.96	83.3/95.2	2.1	9.33/9.89	2.0
ami33	43.0/50.0	58.0	1.31/1.39	47.1/59.8	55.8	1.30/1.32	57.3
ami49	833/956	236.3	38.7/41.7	1003/1260	232.8	38.6/39.6	234.3

Table 2. Best and average results of our GA with different weights for area and wire length

MCNC	$w_1 = 0, w_2 = 1$		$w_1 = 0.5, w_2 = 0.5$			$w_1 = 1, w_2 = 0$	
benchmarks	wire	time	area	wire	time	area	time
apte	197/206	9.1	46.9/48.5	191/212	9.0	46.9/46.9	8.8
xerox	499/513	21.4	20.2/21.2	500/529	21.1	20.0/20.1	20.8
hp	67.2/69.2	16.4	9.85/9.91	68.3/69.8	16.8	9.03/9.13	16.8
ami33	39.3/41.8	1017.3	1.29/1.35	46.2/48.2	1009.2	1.22/1.23	1001.1
ami49	815/839	6277.3	39.5/40.7	912/976	6215.3	37.5/37.7	6222.2

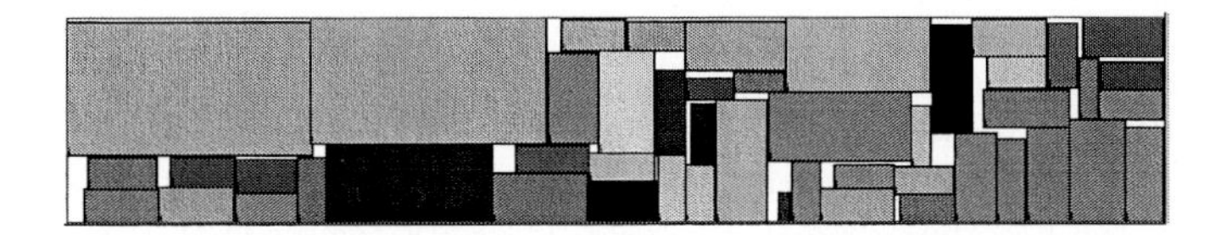

Fig. 4. A solution to the ami49 benchmark

had −5.08% to 5.97% improvement in the best wire length and 1.95% to 19.37% in the best area respectively. For all the five benchmarks, on average our GA improved the best and the average wire length by 0.74% and 7.46% respectively, and improved the best and average area by 13.25% and 25.83% respectively. It is pointed out that the area of the best floorplan for *hp* and *ami*49 obtained by our GA was not as small as that of the deterministic algorithm. However, the wire length of the results was significantly shorter than that of the deterministic algorithm. Hence, the results for the two benchmark problems were still better than those obtained by the deterministic algorithm overall. For the third set of weights, our GA had 2.00% to 6.56% improvement in the best area, and 4.90% to 8.32% improvement in the average area. On average, our GA improved the best area and the average area by 3.67% and 6.31% respectively. Our GA, however, used significantly longer computation time than the deterministic algorithm for the experiments. Figure 4 is the screen shot for the best solution to *ami*49 found by our GA.

6 Conclusion

This paper has presented a GA for non-slicing floorplanning problem using the O-tree representation. The crossover operator generates good floorplan structure by extracting meaningful structural components from two parents and combining the components to produce a highly fit child. The mutation operator keeps the population diverse. In order to increase the efficiency of this GA, the basic ideas behind the deterministic algorithm is used to generate the initial population. We also use the deterministic algorithm as a local optimizer in the GA to further improve the efficiency of the GA.

Preliminary experimental results showed that our GA can consistently produce better results than the deterministic algorithm although our GA had significantly longer computation time than the deterministic algorithm. We are going to use parallel implementation to the GA to speed up the computation.

References

1. Otten R.H.J.M.: Automatic Floorplan Design. Proc. ACM/IEEE Design Automation Conference. (1982) 261–267
2. Guo P.N., Takahashi T., Cheng C.-K., Yoshimura T.: Floorplanning Using a Tree Representation. IEEE Transactions on Computer-Aided Design of Integrated Circuits and Systems. **20** (2001) 281–289
3. Goldberg D.E.: Genetic Algorithms in Search, Optimization, and Machine Learning. Addison-Wesley, Reading. (1989)
4. Cohoon J.P., Hegde S.U., Martin W.N., Richards D.S.: Distributed Genetic Algorithms for the Floorplan Design Problem. IEEE Transactions on Computer-Aided Design of Integrated Circuits and Systems. **10** (1991) 483–492
5. Rebaudengo M., Reorda M.S.: GALLO: A Genetic Algorithm for Floorplan Area Optimization. IEEE Transactions on Computer-Aided Design of Integrated Circuits and Systems. **15** (1996) 943–951
6. Tazawa I., Koakutsu S., Hirata H.: An Immunity based Genetic Algorithm and Its Application to the VLSI Floorplan Design Problem. Proc. Int. Conf. on Evolutionary Computation. (1996) 417–421
7. Nakaya S., Koide T., Wakabayashi S.: An Adaptive Genetic Algorithm for VLSI Floorplanning based on Sequence-Pair. Proc. IEEE Int. Symposium on Circuits and Systems. (2000) 65–68
8. Liu C.-T., Chen D.-S., Wang Y.-W: An Effienct Genetic Algorithm for Slicing Floorplan Area Optimization. Proc. IEEE Int. Symposium on Circuits and Systems. (2002) 879–882
9. Valenzuela C.L., Wang P.Y.: VLSI Placement and Area Optimization Using a Genetic Algorithm to Bread Normalized Postfix Expressions. IEEE Transactions on Evolutionary Computation. **6** (2002) 390–401

Evolving Reversible Circuits for the Even-Parity Problem

Mihai Oltean

Department of Computer Science,
Faculty of Mathematics and Computer Science,
Babeş-Bolyai University, Kogălniceanu 1, Cluj-Napoca, 3400, Romania
moltean@cs.ubbcluj.ro

Abstract. Reversible computing basically means computation with less or not at all electrical power. Since the standard binary gates are not usually reversible we use the Fredkin gate in order to achieve reversibility. An algorithm for designing reversible digital circuits is described in this paper. The algorithm is based on Multi Expression Programming (MEP), a Genetic Programming variant with a linear representation of individuals. The case of digital circuits for the even-parity problem is investigated. Numerical experiments show that the MEP-based algorithm is able to easily design reversible digital circuits for up to the even-8-parity problem.

1 Introduction

The ultimate purpose of reversible computing is to perform computations less or not at all electrical power. Logically reversible operations occupy a central role in considerations of the fundamental physical limits of information handling [7]. The early work of Landauer showed that energy dissipation occurs during the destruction of information of the previous state of the system rather than the acquisition of information during the computational process. Subsequently, Bennett showed that computation could be carried out completely with operations that are logically reversible, i.e., operations in which the output uniquely defines the input [2].

One such reversible logic element is the Fredkin gate (FG) [3, 8] which contains 3 inputs and 3 outputs. Fredkin gate constitute a complete set of operators in that any logic operation (e.g., AND, OR, NOT) can be constructed from a combination of FGs.

In this paper, we propose a variant of the Multi Expression Programming (MEP) [11, 12] for designing reversible digital circuits for the even-parity problem. We choose to apply the MEP-based technique to the even-parity problems because according to Koza [6] these problems appear to be the most difficult Boolean functions to be detected via a blind random search.

Standard GP was able to solve up to even-5 parity when the set of gates F={AND, OR, NAND, NOR} is used [5]. Improvements, such as Automatically

F. Rothlauf et al. (Eds.): EvoWorkshops 2005, LNCS 3449, pp. 225–234, 2005.

Defined Functions [6] and Sub-symbolic node representation [14], allows GP programs to solve larger instances of the even-parity problem. Using MEP and reversible gates we are able to evolve a solution up to even-8-parity function using a reasonable population size.

The paper is organized as follows. MEP technique is briefly described in section 2. The way in which MEP can be applied for reversible circuits is introduced in section 3.1. Several numerical experiments for designing reversible digital circuits are performed in section 4. A comparison with standard digital circuits is described in section 4.3. Further research directions are indicated in section 5.

2 Basic on MEP

The *Multi Expression Programming* (MEP) [10, 11, 12] technique is briefly described in this section.

2.1 Individual Representation

MEP genes are represented by substrings of a variable length. The number of genes per chromosome is constant and it defines the length of the chromosome. Each gene encodes a terminal or a function symbol. A gene encoding a function includes references towards the function arguments. Function arguments always have indices of lower values than the position of that function in the chromosome.

This representation is similar to the way in which `C` and `Pascal` compilers translate mathematical expressions into machine code.

MEP representation ensures that no cycle arises while the chromosome is decoded (phenotypically transcripted). According to the representation scheme the first symbol of the chromosome must be a terminal symbol. In this way only syntactically correct programs (MEP individuals) are obtained.

Example

We employ a representation where the numbers on the left positions stand for gene labels (or memory addresses). Labels do not belong to the chromosome, they are provided here only for explanation purposes.

For this example, we use the set of functions $F = \{+, *\}$ and the set of terminals $T = \{a, b, c, d\}$. An example of chromosome using the sets F and T is given below:

1: a
2: b
3: + 1, 2
4: c
5: d
6: + 4, 5
7: * 3, 6

2.2 Decoding MEP Chromosome and Fitness Assignment Process

In this section we described the way in which MEP individuals are translated into computer programs and the way in which the fitness of these programs is computed.

This translation is achieved by reading the chromosome top-down. A terminal symbol specifies a simple expression. A function symbol specifies a complex expression obtained by connecting the operands specified by the argument positions with the current function symbol.

For instance, genes 1, 2, 4 and 5 in the previous example encode simple expressions formed by a single terminal symbol. These expressions are:

$E_1 = a,$
$E_2 = b,$
$E_4 = c,$
$E_5 = d,$

Gene 3 indicates the operation + on the operands located at positions 1 and 2 of the chromosome. Therefore gene 3 encodes the expression:

$$E_3 = a + b.$$

Gene 6 indicates the operation + on the operands located at positions 4 and 5. Therefore gene 6 encodes the expression:

$$E_6 = c + d.$$

Gene 7 indicates the operation * on the operands located at position 3 and 6. Therefore gene 7 encodes the expression:

$$E_7 = (a + b) * (c + d).$$

E_7 is the expression encoded by the whole chromosome.

There is neither practical nor theoretical evidence that one of these expressions is better than the others. Moreover Wolpert and McReady [17] proved that we cannot use the search algorithm's behavior so far for a particular test function to predict its future behavior on that function. Thus we cannot choose one of the expressions (let us say expression E_7) to store the output of the chromosome. Even this expression proves to be useful for the first 10 generations we cannot guarantee that it will be the best option for all generations.

This is why each MEP chromosome is allowed to encode a number of expressions equal to the chromosome length. Each of these expressions is considered as being a potential solution of the problem.

This is very important because we can get many solutions within the same running time as in the case of one solution/chromosome.

The value of these expressions may be computed by reading the chromosome top down. Partial results are computed by Dynamic Programming [1] and are stored in a conventional manner.

As MEP chromosome encodes more than one problem solution, it is interesting to see how the fitness is assigned. Usually the chromosome fitness is defined as the fitness of the best expression encoded by that chromosome. For instance, if we want to solve symbolic regression problems the fitness of each sub-expression E_i may be computed using the formula:

$$f(E_i) = \sum_{k=1}^{n} |o_{k,i} - w_k|,$$

where $o_{k,i}$ is the obtained result by the expression E_i for the fitness case k and w_k is the targeted result for the fitness case k. In this case the fitness needs to be minimized.

The fitness of an individual is set to be equal to the lowest fitness of the expressions encoded in chromosome:

$$f(C) = \min_i f(E_i).$$

When we have to deal with other problems we compute the fitness of each sub-expression encoded in the MEP chromosome and the fitness of the entire individual is given by the fitness of the best expression encoded in that chromosome.

2.3 Genetic Operators

Search operators used within MEP algorithm are crossover and mutation. These operators preserve the chromosome structure. All offspring are syntactically correct expressions.

Crossover. By crossover two parents are selected and recombined. For instance, within the uniform recombination the offspring genes are taken randomly from one parent or another.

Example

Let us consider the two parents $Parent_1$ and $Parent_2$ given in Table 1. The two offspring $Offspring_1$ and $Offspring_2$ are obtained by uniform recombination as shown in Table 1.

Table 1. MEP uniform recombination

Parents		Offspring	
$Parent_1$	$Parent_2$	$Offspring_1$	$Offspring_2$
1: b	1: a	1: a	1: b
2: *** 1, 1**	2: b	2: *** 1, 1**	2: b
3: **+ 2, 1**	3: + 1, 2	3: **+ 2, 1**	3: + 1, 2
4: a	4: c	4: c	4: a
5: *** 3, 2**	5: d	5: *** 3, 2**	5: d
6: a	6: + 4, 5	6: + 4, 5	6: a
7: - **1, 4**	7: * 3, 6	7: - **1, 4**	7: * 3, 6

2.4 Mutation

Each symbol (terminal, function or function pointer) in the chromosome may be the target of mutation operator. By mutation some symbols in the chromosome are changed with a fixed mutation probability p_m. To preserve the consistency of the chromosome its first gene must encode a terminal symbol.

2.5 MEP Algorithm

Standard MEP algorithm uses steady state [15] as its underlying mechanism. MEP algorithm starts by creating a random population of individuals. The following steps are repeated until a given number of generations [1] is reached. Two parents are selected using a selection procedure. The parents are recombined in order to obtain two offspring. The offspring are considered for mutation. The best offspring replaces the worst individual in the current population if the offspring is better than the worst individual.

The algorithm returns as its answer the best expression evolved along a fixed number of generations.

3 Reversible Computing

The ultimate purpose of reversible computing is to perform computations less or not at all electrical power. Logically reversible operations occupy a central role in considerations of the fundamental physical limits of information handling [7]. The early work of Landauer showed that energy dissipation occurs during the destruction of information of the previous state of the system rather than the acquisition of information during the computational process. Subsequently, Bennett showed that computation could be carried out completely with operations that are logically reversible, i.e., operations in which the output uniquely defines the input [2].

One such reversible logic element is the Fredkin gate (FG) [3] which contains an input control channel A, and two additional input channels, B and C, which exchange values if A is set at 1 or will go through the gate unchanged if A is set at 0. Fredkin gates constitute a complete set of operators in that any logic operation (e.g., AND, OR, NOT) can be constructed from a combination of FGs [3].

The Fredkin gate is depicted in Figure 1.

3.1 MEP for Reversible Circuits

The interpretation for a MEP chromosome needs to be modified because reversible gates have more than one output. Thus an MEP chromosome containing N Fredkin gates actually provides $3 * N$ outputs (plus the outputs provided

[1] In a steady-state algorithm, a generation is considered when the number of newly created individuals is equal to the population size.

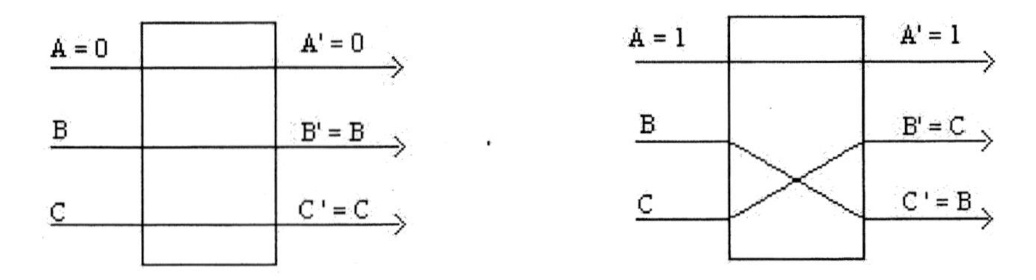

Fig. 1. Fredkin gate has 3 inputs and 3 outputs. If A = 0 the outputs are identical with the inputs. If A = 1 the inputs B and C are swapped. We can easily reconstruct the input from the output

directly from the inputs). MEP representation will be unchanged, but during the fitness evaluation we will have to handle more circuits than the case of standard gates.

Another modification is related to the number of inputs. Two constant inputs 0 (always-OFF) and 1 (always-ON) have been added. These 2 inputs are very important in simulating the standard gates (such as NOT, AND) [3]. Moreover, without these 2 inputs we are not able to build a circuit for the even-parity problems. For instance, in the case of even-3-parity problem our circuits must signal 0 when all inputs are 1. But, the Fredkin gate can never generate a 0 value when all inputs are 1 (see Figure 1).

4 Numerical Experiments

Several numerical experiments for evolving reversible digital circuits are performed in this section.

4.1 Test Problem

Our aim is to find a Boolean function that satisfies a set of fitness cases. The particular function that we want to find is the Boolean even-parity function. This function has k Boolean arguments and it returns **T** (**True**) if an even number of its arguments are **T**. Otherwise the even-parity function returns **F** (**False**) [6]. According to [6] the Boolean even-parity functions appear to be the most difficult Boolean functions to detect via a blind random search.

The terminal set T consists of the $k + 2$ Boolean arguments d_0, d_1, d_2, ... d_{k-1}, 0, 1.

The function set F consists of one three-argument gate: the Fredkin gate.

The set of fitness cases for this problem consists of the 2^k combinations of the k Boolean arguments. We have also added two constants inputs which are always signals 0 (respectively 1). These 2 fixed inputs are very important in simulating standard gates (such as NOT, AND, see [3] for more details). Thus each fitness case will have $k + 2$ inputs and one output.

Table 2. General parameters of the MEP algorithm for designing reversible circuits for the even-parity problem

Parameter	Value
Mutation probability	0.2
Crossover type	Uniform
Crossover probability	0.9
Selection	q-tournament ($q = 1\%$ of the population size)
Function set	$F = \{\text{Fredkin gate}\}$

Table 3. Success rate of the MEP-based algorithm for evolving reversible digital circuits. Success rate is computed over 100 independent runs. Circuit size is the minimum number of gates obtained in one of the successfull runs

Problem	Pop size	Number of generations	Chromosome length	Success rate %	Circuit size
even-3-parity	1000	50	10	95	3
even-4-parity	1000	50	15	35	4
even-5-parity	1000	100	20	15	5
even-6-parity	2000	200	30	18	6
even-7-parity	3000	500	30	29	8
even-8-parity	5000	500	30	11	12

4.2 Results

In this section we perform several experiments with MEP for solving several instances of the even-parity problem. General parameter settings for MEP are given in Table 2.

For reducing the chromosome length we keep all the terminals on the first positions of the MEP chromosomes.

The results along with the particular parameters used for obtaining them are given in Table 3. Success rate is computed as the number of successful runs over the total number of runs.

Table 3 shows that MEP algorithm is able to evolve reversible circuits for the even-parity problem. The shortest (regarding the number of gates) evolved reversible circuits for the even-3-parity and even-4-parity problem are depicted in Figures 2 and 3.

4.3 Comparison with Standard Approaches

Multi Expression Programming has been used [12] for designing standard digital circuits for the even-parity problem. Using the gates AND, OR, NAND, NOR we have been able to evolve up to even-5-parity problem using a population of 4000 individuals with 600 genes each evolved for 50 generations. The shortest evolved standard digital circuit has 6 gates for the even-3-parity problem, and 9 gates for the even-4-parity problem, whereas the reversible ones requires 4

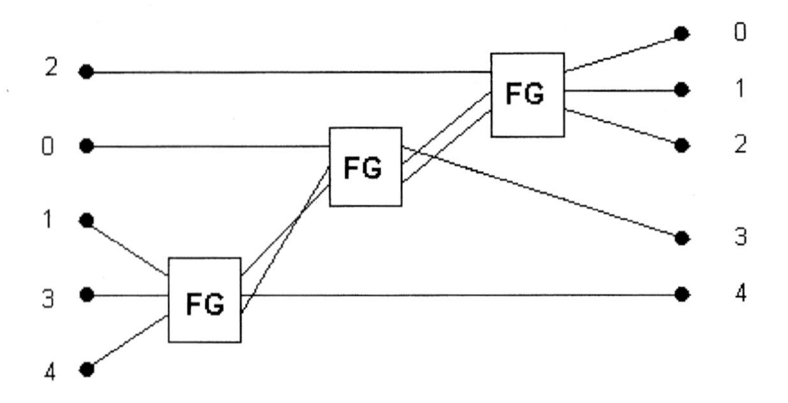

Fig. 2. The shortest evolved reversible digital circuit for the even-3-parity problem. Input 3 always signals 0 and input 4 always signals 1. Output 1 provides the result for the even-parity problem. The other outputs are used only for achieving the reversibility. FG stands for the Fredkin gate

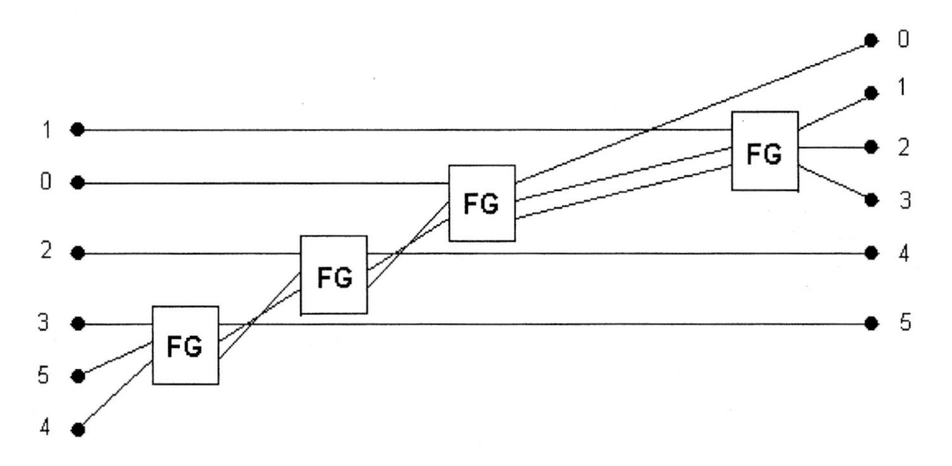

Fig. 3. The shortest evolved evolved reversible digital circuit for the even-4-parity problem. Output 3 provides the result for the even-parity problem. The other outputs are used only for achieving the reversibility. FG stands for the Fredkin gate

(even-3-parity) and 5 (even-4-parity) gates. The first remark is that reversible circuits might require less gates than the standard circuits.

However, when the entire set of 16 binary gates (including EQ, NOT, etc) was employed [12] the length of the evolved standard circuit is considerable shorter. Only 4 gates are required for a circuit implementing the even-5-parity problem and 5 standard gates are required for the even-6-parity problem [12]. The results obtained by using the Fredkin gate are similar (regarding the number of gates) to those obtained using the entire set of 16 gates with 2 binary inputs.

5 Conclusions and Further Work

An algorithm based on Multi Expression Programming has been used for designing reversible digital circuits. Numerical experiments have shown the ability of this algorithm to design reversible digital circuits. When compare to the standard circuits, we can see that the number of outputs of the reversible ones is larger than the case of the standard circuits. This is in full concordance with other studies [3] which have shown that reversible computing requires addition storage space. Further experiments will try to minimize the number of outputs required by the reversible digital circuit. However, this number cannot be less than 3 (the number of outputs of the Fredkin gate).

We will also be interested in extracting general principles from the evolved circuits in order to quickly build larger size reversible circuits. For instance, Cartesian Genetic Programming was used [13] for discovering of ripple-carry adder which is widely used for building large scale multipliers and adders. The evolution of Automatically Defined Functions [6] will also be an interesting aspect for reversible digital circuits.

The method will be used for designing other interesting digital circuits such as reversible adders and multipliers. Other reversible gates, such as CCNOT, will be considered in further experiments [8].

References

1. Bellman, R.: Dynamic Programming, Princeton University Press, New Jersey, (1957)
2. Bennett, C. H., and Landauer, R.: Fundamental physical limits of computation. Scientific American, 253, (1985), 48-56
3. Fredkin, E., Toffoli, T.: Conservative logic, International Journal of Theoretical Physics, Vol. 21, (1982) 219-253
4. Klein, JP., Leete, TH., Rubin, H.: A Biomolecular Implementation of Logically Reversible Computation with Minimal Energy Dissipation. BioSystems 52, (1999) 15-23
5. Koza, J. R.: Genetic Programming: On the Programming of Computers by Means of Natural Selection, MIT Press, Cambridge, MA, (1992)
6. Koza, J. R.: Genetic Programming II: Automatic Discovery of Reusable Subprograms, MIT Press, Cambridge, MA, (1994)
7. Landauer, R.: Irreversibility and heat generation in the computing process, IBM Journal of Research and Development, Vol 5, (1961), 183-191
8. Langdon, W. B.: THE DISTRIBUTION OF REVERSIBLE FUNCTIONS IS NORMAL, in Genetic Programming Theory and Practise, Rick L. Riolo and Bill Worzel (editors), Kluwer Academic Publishers, (2003), 173-188
9. Merkle, R. C.: Reversible Electronic Logic Using Switches, Nanotechnology, Vol 4, (1993), 21-40
10. Oltean, M.: Solving Even-parity problems with Multi Expression Programming, the 8th International Conference on Computation Sciences, North Carolina, Chen, K. et al. (Editors), (2003), 315-318

11. Oltean, M., Grosan, C.: Evolving Digital Circuits using Multi Expression Programming, NASA/DoD Conference on Evolvable Hardware, 24-26 June, Seattle, Edited by R. Zebulum, D. Gwaltney, G. Horbny, D. Keymeulen, J. Lohn, A. Stoica, IEEE Press, NJ, (2004), 87-90
12. Oltean, M.: Improving Multi Expression Programming: an Ascending Trail from Sea-level Even-3-parity Problem to Alpine Even-18-Parity Problem, contributed chapter 15, Evolutionary Machine Design, edited by Nadia Nedjah (et. al), Studies in Soft Computing and Fuzziness, Vol 161, Springer-Verlag, (2004), 229-255.
13. Miller, J. F., Job, D., Vassilev, V.K.: Principles in the Evolutionary Design of Digital Circuits - Part I, Genetic Programming and Evolvable Machines, Vol. 1(1), Kluwer Academic Publishers, (2000), 7-35.
14. Poli, R., Page, J.: Solving High-Order Boolean Parity Problems with Smooth Uniform Crossover, Sub-Machine Code GP and Demes, Journal of Genetic Programming and Evolvable Machines, Kluwer, (2000), 1-21
15. Syswerda, G.: Uniform Crossover in Genetic Algorithms, Proceedings of the 3rd International Conference on Genetic Algorithms, Schaffer, J.D. (editor), Morgan Kaufmann Publishers, San Mateo, CA, (1989), 2-9
16. Toffoli, T.: Reversible computing, In de Bakker, J. W. and van Leeuwen, Jan, editors, Automata, Languages and Programming, 7th Colloquium, LNCS 75, Springer-Verlag, (1980), 632-644
17. Wolpert D.H. and McReady W.G.: No Free Lunch Theorems for Search, Technical Report, SFI-TR-05-010, Santa Fe Institute, (1995)

Counter-Based Ant Colony Optimization as a Hardware-Oriented Meta-heuristic

Bernd Scheuermann[1] and Martin Middendorf[2]

[1] Institute AIFB, University of Karlsruhe, Germany
scheuermann@aifb.uni-karlsruhe.de
http://www.aifb.uni-karlsruhe.de

[2] Department of Computer Science, University of Leipzig, Germany
middendorf@informatik.uni-leipzig.de
http://pacosy.informatik.uni-leipzig.de

Abstract. In this paper, we present the Counter-based Ant Colony Optimization (C-ACO) algorithm as a meta-heuristic, which allows for a resource-efficient implementation on Field Programmable Gate Arrays. In comparison to the standard ACO approach in software on a sequential machine, the implementation of C-ACO in hardware leads to significant asymptotic speed-ups. In experimental studies, we investigate the performance of the proposed C-ACO algorithm. Furthermore, we introduce and examine alternative means of integrating heuristic information into the optimization process, thereby considering the requirements of the hardware architecture.

1 Introduction

Ant Colony Optimization (ACO) is a meta-heuristic which has been applied to a wide range of hard optimization problems [3]. Inspired by the foraging behavior of real ants, ACO uses a population of computational ants that iteratively construct solutions to a given combinatorial optimization problem (e.g., in each step they select the next city to be visited when constructing the solution for a TSP problem). Ants are thereby guided by so called pheromone information that previous ants which have found good solutions have disposed to mark their decisions in the solution construction process.

Usually, ACO algorithms are implemented in software on sequential machines. However, if short computation times become essential, there exist mainly two options to speed-up the execution. One option is to develop parallel variants of the algorithm to be executed on multi-processors machines (see [8] for an overview of such ACO algorithms). The other very promising approach is to directly map the ACO algorithm in hardware, thereby exploiting the parallelism and pipelining capabilities of the target architecture. Furthermore, ACO algorithms possess a core of iteratively repeated instructions and are therefore very attractive for an implementation in hardware.

As implementation platform we consider Field Programmable Gate Arrays (FPGAs). Since FPGAs can be re-configured by the user, different hardware

F. Rothlauf et al. (Eds.): EvoWorkshops 2005, LNCS 3449, pp. 235–244, 2005.

variants of ACO can be tested on the same chip. However, during hardware design one has to consider the constraints imposed by the available resources on chip. Various operations (e.g. multiplications, exponentiations) and data types (like floating point numbers) that are required by the standard ACO algorithm would demand a very large amount of chip resources. Therefore, we are interested in exploring alternative variants of ACO, which better fit the architectural constraints of FPGAs. In [10], we have presented an FPGA implementation of the Population-based ACO (P-ACO), which can be executed efficiently on an FPGA and leads to a significant speed-up compared to a software version on a PC. Using a different abstract computing model, we proposed an efficient implementation of a variant of the standard ACO on reconfigurable processor arrays ([7]).

Since P-ACO is not equally well suited for every combinatorial optimization problem, in this paper, we present the Counter-based ACO algorithm (C-ACO) as a variant of standard ACO which is suitable for FPGAs. In contrast to P-ACO, the counter-based variant allows to systolically pipe a sequence of artificial ants through a grid of processing cells, which promises a very efficient hardware realization. Furthermore, we present two different approaches to complement C-ACO with heuristic knowledge. Experiments are conducted on various instances of the Traveling Salesperson Problem (TSP). Note that the proposed algorithm does not aim to compete with other fast TSP solvers (e.g. Lin-Kernighan [6]). TSP is chosen as a test problem due to its close relation to the natural paradigm of real ants, which are capable of finding shortest paths between their nest and food sources. Furthermore, TSP can be considered as the standard test bed for ACO. New algorithmic ideas, which have shown a good performance on TSP, could often successfully be adapted to other optimization problems, for which ACO algorithms belong to the best known approaches. The proposed C-ACO algorithm was created with an FPGA implementation in mind, though it may also be interesting as alternative ACO implementation in software.

2 Standard Ant Colony Optimization for TSP

In this section, we briefly introduce the standard ACO approach for the TSP and describe the problems when mapping it to FPGAs. For a detailed introduction to ACO see for example [3].

2.1 Algorithm

We apply ACO to search for short tours that connect all n given cities of an instance of TSP such that each city is visited exactly once. The pheromone information is encoded in an $n \times n$ pheromone matrix $[\tau_{ij}]$. Pheromone value τ_{ij} expresses how beneficial it was for preceding ants to visit city j directly after city i. Typically, when constructing a solution, ants do not solely rely on pheromone information, but also use heuristic information η_{ij}. A possible heuristic for the TSP is to set $\eta_{ij} = 1/d_{ij}$ where d_{ij} is the distance between city i and city j.

An ant builds a tour by making a sequence of local decisions, i.e. successive selections of cities. Every decision is made randomly according to a probability distribution over the so far unchosen cities in selection set S:

$$\forall j \in S : \; p_{ij} = \frac{\tau_{ij}^{\alpha} \eta_{ij}^{\beta}}{\sum_{h \in S} \tau_{ih}^{\alpha} \eta_{ih}^{\beta}} \tag{1}$$

where parameters α and β determine the relative influence of pheromone values and heuristic values. Initially, the selection set S contains all cities and after each decision, the selected city is removed from S. At the end of an iteration, when m tours have been generated (m being the number of ants per iteration), the shortest tour π^* of the current iteration is determined. The pheromone matrix is then updated in two steps:

1. Evaporation: All pheromone values in the matrix are reduced by a certain percentage ρ: $\forall\, i, j \in [1 : n] : \tau_{ij} := (1 - \rho)\tau_{ij}$.
2. Intensification: The pheromone values along the best solution π^* are increased by a fixed amount Δ: $\forall\, i \in [1 : n] : \tau_{i\pi^*(i)} := \tau_{i\pi^*(i)} + \Delta$.

The ACO algorithm executes a number of iterations until a specified stopping criterion has been met, e.g. a predefined maximum number of iterations has been executed.

2.2 Problems Mapping Standard ACO onto FPGA

When designing ACO for FPGAs, the typical characteristics of the algorithm make a hardware realization difficult: (i) pheromones, heuristic values and random numbers require a floating point representation, (ii) evaporation and the integration of heuristic information requires a large number of multiplication operations, (iii) the integration of weights α and β into the probabilities p_{ij} in Eq. 1 demands exponentiation operations. The realization of the required floating point numbers, multiplication operations, and exponentiation operations in hardware is possible, but would afford a high amount of chip resources on fine-grained programmable logic devices like FPGAs. Thus, we suggest an alternative approach that is described in the following section.

3 Counter-Based Ant Colony Optimization for TSP

In this section we describe the C-ACO algorithm for the TSP and sketch how it can be implemented on an FPGA.

3.1 Algorithm

In the following, we explain the characteristics of C-ACO for asymmetric TSP (see Algorithm 1).

Pheromone Representation: In contrast to standard ACO, pheromone values $\tau_{ij} \geq \tau_{min} > 0$ are represented by integer numbers, which demand less chip

Algorithm 1. C-ACO for Asymmetric TSP

```
1: for i := 1 to n do
2:     for j := 1 to n do
3:         if i = j then                                   /* initialize pheromone matrix */
4:             τ_ij := 0                                   /* exception: diagonal elements */
5:         else
6:             τ_ij := τ_min + τ_init                      /* regular initialization */
7:         end if
8:     end for
9:     U_i := 0                                            /* initialize update counters */
10: end for
11: while stopping condition not met do                    /* begin iterations */
12:     for a = 1 to m do                                  /* construct m solution */
13:         S := [1 : n]                                   /* initialize selection set */
14:         select start city c ∈ S
15:         S := S \ {c}; i := c
16:         for h := 1 to n do                             /* n ant decisions per tour */
17:             if h < n then
18:                 randomly select item j ∈ S according to Eq. 1
19:                 S := S \ {j}
20:             else
21:                 j := c                                 /* finally return to start city */
22:             end if
23:             π^a(i) := j                                /* insert selected city into tour */
24:             Δ^ρ := min{τ_ij − τ_min, 1}
25:             τ_ij := τ_ij − Δ^ρ                         /* evaporation */
26:             U_i := U_i + Δ^ρ                           /* increment update counter */
27:             i := j                                     /* move to selected city */
28:         end for
29:     end for
30:     a* := arg min_{a∈{1,...,m}} F(π^a); π* := π^{a*}   /* determine best solution */
31:     for i = 1 to n do
32:         τ_{iπ*(i)} := τ_{iπ*(i)} + U_i                 /* pheromone update */
33:         U_i := 0                                       /* reset update counters */
34:     end for
35: end while                                              /* end iterations */
```

resources than floating point numbers. During initialization all pheromone values τ_{ij} with $i \neq j$ receive the same start value $\tau_{min} + \tau_{init}$ (line 6).

Selection: Instead of evaporation by multiplying with $1 - \rho$, the pheromones evaporate during the selection process: When a city j is selected, the respective pheromone value is decremented by $\Delta^\rho = \min\{\tau_{ij} - \tau_{min}, 1\}$, i.e. pheromones behave like decremental counters (lines 24 and 25). Each time a city is selected, the pheromone value is decreased, thereby reducing the attractiveness of the respective arc for following ants. Consequently, the exploration of not yet visited arcs is supported and ants are less likely to converge to a common path. Dorigo and Gambardella [4] have also investigated a similar form of on-line evaporation (local pheromone update) where they reduced the pheromone values by a certain percentage and performed a partial reset to the initial value: $\tau_{ij} := (1-\rho)\tau_{ij} + \rho\tau_0$. The selection process in C-ACO, which results in a subtraction of a constant integer value, is better suited for an FPGA implementation.

Update Counters: Per row of the pheromone matrix there exists an update counter U_i which accumulates the total amount of pheromone evaporation in that row during an iteration (line 26). At the end of an iteration, when m tours have been constructed, each update counter holds a value $0 \leq U_i \leq m$.

Pheromone Update: When the best tour π^* of the current iteration has been determined, the pheromone update is done according to the respective solution as follows: The pheromone value $\pi^*(i)$ in row i is incremented by the value of the update counter in the same row, i.e. $\tau_{i\pi^*(i)} := \tau_{i\pi^*(i)} + U_i$ (line 32), instead of value Δ as stated in Sec. 2.1. Thus, the total amount of pheromone in each row remains constant.

The algorithm described before has to be modified to handle symmetric TSP instances. If an item j is selected in row i, then pheromone values τ_{ij} and τ_{ji} have to be decremented and the respective update counters U_i and U_j are incremented. Accordingly, the update process has to be adapted. In order to maintain a symmetric matrix, pheromone values τ_{ij} and τ_{ji} are increased by the same amount. Since in every row i, two updates are performed, the pheromone values are only increased by at most $\lfloor U_i/2 \rfloor$. If in a row the amount of update is smaller than the update counter value, the remainder is transferred into the next iteration.

3.2 Mapping C-ACO onto FPGA

For an implementation of C-ACO on an FPGA the pheromone matrix is mapped directly onto the chip area. This design comprises all processing and memory resources for each element of the statically allocated matrix. The ants are piped through the matrix in a systolic fashion. Thus, the index of ant a and its selection set is propagated top-down through the cells of the matrix circuitry. Such an individual cell is depicted in Fig. 1.

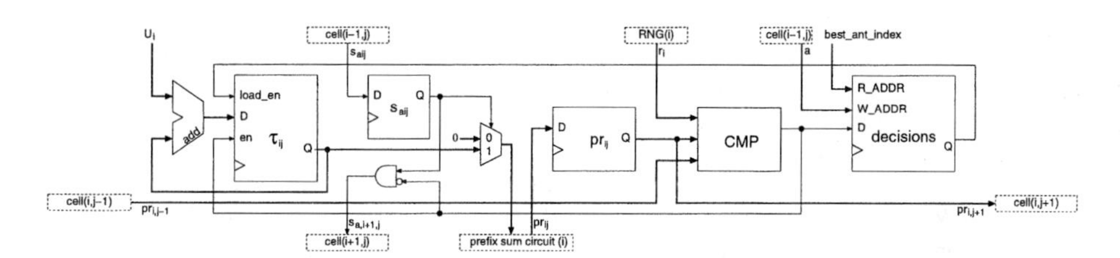

Fig. 1. Circuit of cell (i, j) of the pheromone matrix. For reasons of clarity, signals for clock, reset and control are omitted

The current selection set of ant a in row i is denoted by $S_{ai} = \{s_{ai1}, \ldots, s_{ain}\}$ with $s_{aij} = 1$ if j has not yet been visited as the next city after city i by ant a, else $s_{aij} = 0$. Pheromone value τ_{ij} is stored in a loadable decremental counter with minimum value τ_{min}. In this design, heuristic knowledge is disregarded. However, the integration of heuristics will be discussed in Sec. 6. We set pheromone weight $\alpha := 1$ (common choice in standard ACO) to avoid exponentiations. Therefore, the calculation of selection probabilities (cmp. Eq. 1) can be simplified to $p_{ij} = \tau_{ij} / \sum_{h \in S} \tau_{ih}$. In order to select a city, it is not necessary to calculate the denominator in this equation and to perform the division. It is sufficient to calculate the prefix sum over the numerators of the yet unselected cities. Hence, in every row of the pheromone matrix, the design contains a circuit to calculate the prefix sum $pr_{ij} = \sum_{k=1}^{j} s_{aik}\tau_{ik}$. Which city is visited next is decided probabilistically by a random number r_i (see e.g. [1] for a random number generator on FPGA). All decisions d_{aij} in cell (i,j) are stored in a decision memory with $d_{aij} = 1$ if $pr_{i,j-1} \leq r_i < pr_{ij}$ (CMP), else $d_{aij} = 0$.

Every ant steps through the n rows of the cell matrix and requires $\Theta(\log n)$ time (due to the calculation of the prefix sums) in every row. So the time to construct a solution is $\Theta(n \log n)$. Since ants are piped in a systolic fashion, the solution constructions are started with a period of $\Theta(\log n)$ and the total runtime to compute z solutions is $\Theta((z+n)\log n)$ if a non-generational approach with solution evaluation, comparison, and update processes running in parallel to the solution construction (cmp. [7]). The standard ACO on a single sequential CPU needs time $O(zn^2)$. Hence, we obtain a speed-up of $O(zn^2/((z+n)\log n))$.

4 Effect of Pheromone Intensification

In this section, the pheromone update rule of C-ACO motivated and its relation to the standard pheromone update rule is discussed. In the standard ACO pheromone update, each pheromone value $\tau_{i\pi^*(i)}$ that belongs to the best tour π^* is increased by a fixed value $\Delta > 0$. In the following, we consider how much such an individual portion of pheromone affects the decisions of ants in the succeeding iterations and demonstrate that pheromone intensification in C-ACO and standard ACO have a similar stochastic effect. We assume that for standard ACO the pheromone values are normalized, such that the sum over all values in a row is $T = \sum_{j=1}^{n} \tau_{ij} = 1$. Therefore, the total amount of pheromone evaporated per iteration is equal to ρ. In order to maintain T constant, we set $\Delta := \rho$ (if only one ant per iteration is allowed to update). This version of pheromone update for standard ACO has been used by several authors (e.g. [2, 5]).

Standard ACO: Consider an amount of pheromone Δ which has been added to τ_{ij} in an arbitrary iteration, and define an iteration counter starting from this specific iteration $t := 0$. As an idealized situation, it is assumed that all cities (except city i) are contained in selection set S. Then $p_1 := \Delta/T = \rho$ denotes the probability, that in the following iteration $t = 1$ an ant selects city j after city i on accounts of value Δ. Since pheromone is evaporated in every iteration, the probability that city j is selected in iteration $t = 2$, is $p_2 = (1-\rho)\rho = (1-p_1)p_1$, or more general: $p_t = (1-p_1)^{t-1}p_1$. Let X_t be a random variable which expresses the number of ants selecting j as the next city (due to value Δ) in iteration t. Obviously, X_t is distributed binomially according to $B(m, p_t)$ with expected value:

$$E(X_t) = mp_t = m(1-p_1)^{t-1}p_1. \tag{2}$$

C-ACO: In C-ACO, the intensification received by τ_{ij} is $\Delta = U_i$. Let $p_1 = U_i/T$ (with $T = (n-1)(\tau_{min} + \tau_{init})$ denote the probability, that the first ant in iteration $t = 1$ selects city j after city i on accounts of value Δ. Since ants remove pheromone when selecting a city, random variables X_t are distributed hyper-geometrically $H(T, U_i, m)$. Accordingly, the expected value in iteration $t = 1$ is $E(X_1) = mU_i/T$, and in the succeeding iterations $t > 1$: $E(X_t) =$

$\frac{m}{T}(U_i - \sum_{t'=1}^{t-1} E(X_{t'}))$. Assuming that arc (i,j) received the maximum update $U_i = m$, then the following equation for the expected value can be proved by complete induction:

$$E(X_t) = m(1 - \frac{m}{T})^{t-1}\frac{U_i}{T} = m(1-p_1)^{t-1}p_1, \tag{3}$$

which is equal to Eq. 2. Hence, for standard ACO and C-ACO an update of τ_{ij} causes arc (i,j) to be selected in average $m(1-p_1)^{t-1}p_1$ times in iterations $t > 0$. Let random variable $Y_N = \sum_{t=1}^{N} X_t$ denote the frequency of selecting city j as the next city in the next N iterations. Then the expected value $E(Y_N) = \sum_{t=1}^{N} E(X_t)$ is determined by $E(Y_N) = mp_1 \sum_{t=1}^{N}(1-p_1)^{t-1} \rightarrow m$. Hence, city j is selected approximately m times. For these similarities, we expect that C-ACO shows a comparable optimization behavior as standard ACO.

5 Comparison Standard ACO Versus C-ACO

Experimental studies to compare standard ACO and C-ACO algorithm are described in this Section. Since C-ACO is not yet implemented in hardware it cannot be evaluated in terms of exact computation time. In the following we compare its optimization behavior with standard ACO.

The first experiment was conducted on a range of symmetric TSP instances `gr48, eil101, d198` with 48, 101, 198 cities and asymmetric instances `ry48p, kro124p, ftv170` with 48, 100, 171 cities from the TSPLIB benchmark [9]. For both algorithms, we set $\alpha = 1$, and $m = 8$ ants per iteration (generational approach, i.e. no systolic piping of ants). The results were computed for $\beta = 0$ (no heuristic) and $\beta = 5$ (standard heuristic $\eta_{ij} = 1/d_{ij}$). With probability q_0 an ant deterministically selected the next city j, which had the maximum product $\tau_{ij}^{\alpha}\eta_{ij}^{\beta}$ (exploitation). With probability $1-q_0$ the next city was selected according to Eq. 1 (exploration). The probabilities were chosen from $q_0 \in \{0, 0.5, 0.9\}$. Experiments were run with and without elitism (i.e., not only the best ant of the current iteration, but also the best ant so far was allowed to update).

Standard ACO was implemented such that $T = 1$ and $\Delta = \rho$ (see Sec. 4) with $\rho \in \{0.005, 0.01, 0.02, 0.05, 0.1\}$. We introduced a minimum pheromone value τ_{min} to allow for a fair comparison with C-ACO, which was also equipped with a lower pheromone bound. Hence, cities were selected with probability $p_{ij} = (\tau_{ij} + \tau_{min})^{\alpha}\eta_{ij}^{\beta} / \sum_{h \in S}(\tau_{ih} + \tau_{min})^{\alpha}\eta_{ih}^{\beta}$. The minimum pheromone value was determined by $\tau_{min} = \gamma_s \tau_{init} = \gamma_s/(n-1)$ with $\gamma_s \in \{0, 0.001, 0.01, 0.1, 0.5, 1\}$. For C-ACO, we chose $\tau_{init} \in \{1, 2, 3, 4, 5, 8, 10, 50, 100, 1000\}$. The minimum pheromone value was set to $\tau_{min} = \lceil \gamma_c \tau_{init} \rceil$, $\gamma_c \in \{0.001, 0.01, 0.1, 0.5, 1\}$.

Each run was terminated after $t = 50000 + 1000n$ iterations (when practically all runs have converged). Solutions qualities were calculated as the average tour lengths over 10 runs. The average tour lengths were measured in iterations $t_i = \lfloor \frac{1}{7} it \rfloor$, $i = 1, \ldots, 7$. For a specific TSP instance and fixed pair (β, q_0), the solution qualities for all 122 parameter combinations (60 for standard ACO and 62 for C-ACO after removing duplicate pairs $(\tau_{min}, \tau_{init})$) were ranked. The obtained

Table 1. Average ranks of standard ACO and C-ACO in iterations $t_1, \ldots, t_7$

q_0	Algorithm	$\beta = 0$							$\beta = 5$						
		t_1	t_2	t_3	t_4	t_5	t_6	t_7	t_1	t_2	t_3	t_4	t_5	t_6	t_7
0	Standard	54.57	56.49	57.46	57.93	58.32	58.59	58.80	57.10	59.12	59.55	60.31	59.15	60.18	59.00
	C-ACO	68.20	66.34	65.41	64.93	64.58	64.05	63.81	59.78	58.01	56.72	56.54	55.13	55.81	54.80
0.5	Standard	53.66	55.01	55.67	56.06	56.23	56.74	56.33	68.48	68.53	68.79	68.10	68.39	68.39	68.77
	C-ACO	68.94	67.61	67.09	66.50	66.03	66.10	65.41	50.85	50.10	50.14	49.22	49.58	49.26	49.32
0.9	Standard	73.94	73.73	73.49	73.49	73.31	73.56	73.22	83.85	83.31	82.44	82.34	82.56	82.57	81.03
	C-ACO	49.23	49.16	49.17	49.38	49.23	49.34	48.81	37.64	38.10	38.12	38.08	38.34	38.76	37.99

instance-independent average ranks for standard ACO and C-ACO are given in Table 1.

For $\beta = 0$ and $q_0 = 0$ or $q_0 = 0.5$, standard ACO performs better than C-ACO. For $q_0 = 0.9$, C-ACO has significantly better ranks than the standard ACO. Presumably, C-ACO benefits from a higher degree of exploitation. In runs with heuristic guidance ($\beta = 5$), C-ACO performed consistently better than standard ACO with only one exception (t_1, $q_0 = 0$). The simulation results indicate that in average C-ACO shows a competitive optimization behavior. Overall the average ranks were 66.16 (standard ACO) and 54.32 (C-ACO).

6 Heuristics

Optimization performance usually benefits from the integration of heuristic information. Generally, including heuristic information into the calculation of selection probabilities (see Eq. 1) requires multiplication and exponentiation operations with floating point numbers. To save computation time and FPGA resources we propose the following 2 steps: 1) Weighing heuristic values by β and scaling the resulting values is pre-computed on an exterior processor (if the heuristic allows this), 2) applying one of the following integer-based types of η-heuristics or τ-heuristic.

η-Heuristics. The η-heuristics represent the standard way of integrating heuristic information into the ant decision process by multiplying transformed heuristic values η'_{ij} with pheromone values τ_{ij}. We consider three variants:

- **REALVAL**: Heuristic values are processed unchanged: $\eta'_{ij} = \eta^{\beta}_{ij}$.
- **INTVAL**: Heuristic values are transformed into integer numbers $\eta'_{min} \leq \eta'_{ij} \leq \eta'_{max}$, where η'_{ij} is determined by $\eta'_{ij} = \lfloor f(\eta^{\beta}_{ij}) \rfloor \in I\!N$ with:

$$f(\eta^{\beta}_{ij}) = \frac{\eta'_{max} - \eta'_{min}}{\eta^{\beta}_{max} - \eta^{\beta}_{min}} (\eta^{\beta}_{ij} - \eta^{\beta}_{min}) + \eta'_{min}$$

  where η_{min} and η_{max} denote the minimum and resp. the maximum of all values η_{kl} in the heuristic matrix with $k \neq l$. Multiplications by integer heuristic values save FPGA resources.
- **POTVAL**: Heuristic information is transformed to an interval of integer numbers $\eta'_{min} \leq \eta'_{ij} \leq \eta'_{max}$, where $\eta'_{ij} = 2^k$ with $k = \lfloor \log_2 f(\eta^{\beta}_{ij}) \rfloor$ and f

being the same scaling function as in INTVAL. Multiplications by numbers $\eta'_{ij} = 2^k$ can be substituted by shifting the respective bit representation of the pheromone value by k digits.

τ-Heuristic. In the τ-heuristic, η_{ij}^{β} values are transformed (as in η-INTVAL) into integer values $\eta'_{min} \leq \eta'_{ij} \leq \eta'_{max}$ with $\eta'_{ij} \in \mathbb{N}$ and then included into the pheromone matrix as lower thresholds $\tau_{ij}^{t} = \tau_{min} + \eta'_{ij}$. Pheromone values are not allowed to fall below the threshold, i.e. $\tau_{ij} \geq \tau_{ij}^{t}$. Initial pheromone values are calculated as $\tau_{ij} := \tau_{ij}^{t} + \tau_{init}$, where $\tau_{init} = \lceil v\bar{\eta} \rceil$ with v a parameter and $\bar{\eta}$ is the average over all values η'_{kl} with $k \neq l$. Different to Alg. 1, pheromone values of selected cities are drecremented by $\delta = \lceil w\tau_{init} \rceil$ with w a parameter. Selection probabilities are computed as $p_{ij} = \tau_{ij} / \sum_{h \in S} \tau_{ih}$. Thus, no multiplications with heuristic values are required.

6.1 Experimental Results

To compare these two variants of heuristics they were tested on the `kro124p` TSP instance using parameter values: $q_0 = 0.3$, $\alpha = 1$, $\beta = 5$, $m = 8$, $\tau_{min} = 1$, $\eta'_{min} = 1$, and $\eta'_{max} = 2^k$ with $k \in [0 : 25]$. The η-heuristics were run with $\tau_{init} \in \{1, 5, 10, 50, 100\}$, the τ-heuristic with $v \in \{0.5, 1.0, 5.0, 50.0, 100.0\}$ and $w \in \{0.00001, 0.001, 0.1, 0.5, 1.0\}$. Fig. 2 shows the average solution qualities (10 repetitions) after 150000 iterations. For every value of k, the corresponding solution quality was determined as the minimum average tour length over the input parameter combinations. The best tour length reached by the standard heuristic (η-REALVAL) is drawn as a horizontal line in Fig. 2. With an increasing exponent k the alternative heuristics achieve solution qualities comparable to the standard heuristic. With a value of $\eta'_{max} = 2^8$ the η-POTVAL-heuristic can reasonably approximate the standard heuristic, i.e. multiplications can be substituted by shifts of at most $k = 8$ steps. For the η-INTVAL-heuristic, heuristic values can be approximated by multiplications with integer numbers with a size of at most $k = 10$ bits. Both alternative heuristics perform slightly better than the τ-heuristic, for which we would have to provide pheromone counters with a lower bound of size $k = 12$ bits. All three alternative heuristics allow to approximate the standard heuristic with a reasonable amount of hardware resources. The choice

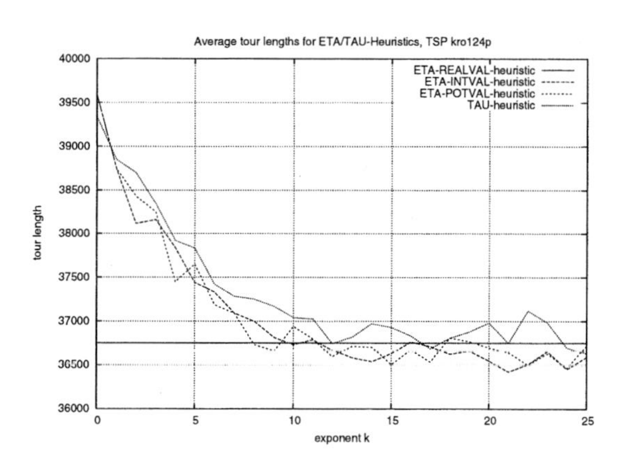

Fig. 2. Comparison of the η and τ heuristics

of the appropriate heuristic type depends on the resources which are available on the target device.

7 Conclusion

We presented the C-ACO algorithm, which is suitable for a resource-efficient implementation on FPGAs. Compared to the standard ACO approach in software on a sequential machine, the implementation of C-ACO in hardware attains significant asymptotic speed-ups. Software simulations demonstrated that C-ACO is a competitive approach in comparison to the standard ACO approach. Finally, we proposed alternative ways of integrating heuristic information to save further hardware resources. Our future work includes the examination of systolic solution construction, implementing C-ACO on FPGAs and considering other optimization problems.

References

1. J. Ackermann, U. Tangen, B. Bödekker, J. Breyer, E. Stoll, and J.S. McCaskill. Parallel random number generator for inexpensive configurable hardware cells. *Computer Physics Communications*, 140(3):293–302, 2001.
2. C. Blum, A. Roli, and M. Dorigo. The hyper-cube framework for ant colony optimization. In *Proceedings of MIC'2001 – Metaheuristics International Conference*, volume 2, pages 399–403, 2001.
3. M. Dorigo and G. Di Caro. The ant colony optimization meta-heuristic. In D. Corne, M. Dorigo, and F. Glover, editors, *New Ideas in Optimization*, pages 11–32. McGraw-Hill, 1999.
4. M. Dorigo and L.M. Gambardella. Ant colony system: A cooperative learning approach to the traveling salesman problem. *IEEE Transactions on Evolutionary Computation*, 1:53–66, 1997.
5. M. Guntsch and M. Middendorf. Pheromone modification strategies for ant algorithms applied to dynamic TSP. In E.J.W. Boers et al., editor, *Applications of Evolutionary Computing: Proceedings of EvoWorkshops 2001*, number 2037 in Lecture Notes in Computer Science, pages 213–222. Springer Verlag, 2000.
6. S. Lin and B. W. Kernighan. An effectice heuristic algorithm for the traveling salesman problem. *Operations Research*, 21, 1973.
7. D. Merkle and M. Middendorf. Fast ant colony optimization on runtime reconfigurable processor arrays. *Genetic Programming and Evolvable Machines*, 3(4):345–361, 2002.
8. M. Randall and A. Lewis. A parallel implementation of ant colony optimization. *Journal of Parallel and Distributed Computing*, 62(9):1421–1432, 2002.
9. G. Reinelt. TSPLIB - a traveling salesman problem library. *ORSA Journal on Computing*, 3:376–384, 1991.
10. B. Scheuermann, K. So, M. Guntsch, M. Middendorf, O. Diessel, H. ElGindy, and H. Schmeck. FPGA implementation of population-based ant colony optimization. *Applied Soft Computing*, 4:303–322, 2004.

Use of an Evolutionary Tool for Antenna Array Synthesis

Luca Manetta, Laura Ollino, and Massimiliano Schillaci

Politecnico di Torino, Dipartimento di Elettronica,
C.so Duca degli Abruzzi 24,10129 Torino, Italy
{s86295, s84066, s63849}@studenti.polito.it

Abstract. This paper describes an evolutionary approach to the optimization of element antenna arrays. Classic manual or automatic optimization methods do not always yield satisfactory results, being either too labour-intensive or unsuitable for some specific class of problems. The advantage of using an evolutionary approach is twofold: on the one hand it does not introduce any arbitrary assumptions about what kind of solution shows the best promise; on the other hand, being intrinsically non-deterministic, it allows the whole process to be repeated in search of better solutions. A generic evolutionary tool originally developed for a totally different application area, namely test program generation for microprocessors, is employed for the optimization process. The results show both the versatility of the tool (it's able to autonomously choose the number of array elements) and the validity of the evolutionary approach for this specific problem.

1 Introduction

Antenna arrays have long been used to achieve performance impossible to obtain from a single antenna. High-directivity antennas and shaped beam arrays are examples of products that take advantage of the array concept. Uniform arrays, however, may be unsuitable for a given specification. This drives us to the need for array synthesis and optimization, in order to obtain a given functional specification at a reduced cost. Numerous manual or automatic methods exist to achieve this goal: Conjugate Gradient [6], Fourier series and Woodward-Lawson methods [7] first explored the concept of automatic array synthesis; Monte Carlo method follow as a statistical approach [8] and finally genetic algorithms are used.

Previous work in this field includes the use of GAs [4], evolutionary programming [3] and hybrid methods [5].

Marcano and Duran [4] introduce the use of GAs for the optimization of linear and planar arrays. However, the problems presented do not seem to be particularly stressful to the method employed. Chellapilla and Hoorfar [3] present an EP method for the generation of optimally thinned linear arrays, showing increased performance with respect to GAs. Hollapilla and Zhu [5], finally, show that hybrid methods perform better than pure GA or EP alorithms on some problems.

F. Rothlauf et al. (Eds.): EvoWorkshops 2005, LNCS 3449, pp. 245–253, 2005.

We employed a rather generic evolutionary tool to address the problem of array synthesis and optimization. One peculiarity of our work is indeed the use of a tool developed for a totally different application area, namely test program generation for microprocessors. This not only allows us to critically assess the validity of the evolutionary approach to array synthesis, but also helps the development of the tool itself. Some of its new features, in fact, have been added on the consideration of their usefulness for this specific application, and are also being used in the original context.It's interesting to note that the used tool shows hybrid GA/EP properties since it employs both mutation and crossover. The tool itself will be described later. The paper is organized as follows: a brief introduction on antenna arrays is given in section 2; section 3 introduces the evolutionary computation paradigm and describes in more detail the evolutionary tool used; in section 4 we describe our workflow and the performed numerical experiments; section 5 reports the obtained results; finally, the conclusions are reported in section 6.

2 Antenna Arrays

To convert an electrical signal into electromagnetic waves and vice versa we need particular actuators and sensors: the antennas. High gain applications require high directivity antennas; this can be achieved by arranging them in an array: more antennas are placed near each other to fuse their individual irradiation diagrams to obtain a collective diagram more fitted to specified application. Also it's possible to design antenna arrays with a shaped beam; these arrays irradiate in a particular space zone according to a pre-arranged form (for example: for a satellite which must irradiate a country one must design an antenna that has a shaped beam which covers only the desired territory). However the design of this type of antennas presents, unfortunately, various problems.

The problems which one meets during the design of a shaped beam antenna are substantially due to the fact that the design operation is of inverse type: from the normalized array factor we must pass to the position of radiators and to their feeding phase. To represent the array factor rigorously it's possible to express it as a polynomial whose roots represent the feed coefficients of the radiators. Changing the modulus or the phase of a root we change the overall shape. Another important problem is that the various radiators must furthermore be in such positions that their mutual coupling be minimum. With all these constraints the problem becomes quickly intractable. Also in the past the problem was relegated to the most expert designers; they started with different mathematical methods to do the synthesis of the antenna and, with little shifting of the various radiators, were able to obtain good approximate results; however the cost in terms of time was huge. The development of the computer technology gives us, today, various methods with which it is possible to automatically design this type of antennas, and with good results.

In the past the growth of the antennas was of evolutionary type: from the first systems we passed to the sophisticated ones, we can think about the example of the ground plane antenna which presents an impedance of 38 Ω, to pass to the ground plane with the folded arms with a 50 Ω impedance, to finish with the skirt dipole. Imitating this process, we will show it is possible to get working and well approximated solutions, beginning from inefficient ones, with an evolutionary method.

3 Evolutionary Method

Evolutionary computation is a computer-based problem solving paradigm based on Darwin's evolution theory [1]. In this paradigm possible solutions to a given problem are seen either as individuals inside a larger population or as species within an environment. These compete against each other and periodically undergo a selection process. The best solutions, i.e. the 'fittest' ones, survive the selection and are allowed to reproduce, that is to produce other solutions similar, but not completely identical, to themselves. These offspring are in turn subjected to the same selection process as their ancestors. This process leads, in turn, to an increment in the average fitness. The term fitness is historically used to denote a measure of the compliance of a candidate solution with its goals. An increment in the average fitness usually goes together with an increase in its maximum value. Evolutionary computation itself has evolved over time, producing many different kinds of evolutionary algorithms. The best-known ones are Genetic Algorithms, Evolutionary Programming, Evolution Strategies, Classifier Systems and Genetic Programming. None of these methods is perfect for all problems, but they offer a large choice of approaches for the user to try. Evolutionary methods are particularly suited to solve computationally hard problems for which no good heuristic is known.

The main goal of an evolutionary method is to make a computer obtain an exact or, more often, approximate solution to a problem without being explicitly told how to do so.

In our work we use a tool named μGP (MicroGP). MicroGP [2] is an evolutionary approach to generic optimization problems with a focus on the generation of test programs for microprocessors, similar to both Evolutionary Programming and Evolution Strategies. It is not strictly a genetic algorithm since it does not employ a fixed-size chromosome setting, but a graph structure, to describe the individuals it cultivates. In Evolutionary Algorithms parlance, it is a steady-state evolutionary method that implements a variation of the $(\mu + \lambda)$ strategy on a single population of individuals. This means that, given an initial population of μ individuals, λ genetic operators are applied on it to produce a variable number of offspring; the parents and offspring are then merged into a single population, which undergoes selection: the μ individuals with the highest fitness are selected for survival, and the rest are discarded. Individuals with high fitness may remain indefinitely in the population.

It is different from Evolutionary Programming mainly because it employs crossover, currently in two forms; additionally, mutation operators are not implemented in many forms for strength selection, but rather a great variety of operators is implemented. Additionally, population selection is always deterministic. In common with Evolutionary Programming there is no requirement that a single offspring be generated from each parent. It is also different from Evolution Strategies in that it (currently) only employs the $(\mu + \lambda)$ strategy. It is, however, conceptually similar since its evolutionary basis is the individual, not the species. It finally differs from both since it dynamically self-adapts many of its parameters.

One of the main peculiarities of MicroGP is the fact that the focus during the reproduction process is not so much on the reproducing individual as on the genetic operator employed. In fact, the λ in the $(\mu + \lambda)$ expression is not the number of generated offspring as the standard terminology dictates, but the number of genetic operators used.

Although its original focus is the generation of test programs, MicroGP is a very versatile tool that can be employed to successfully approach a number of other problems, on the only condition that a solution can be expressed with the syntactical constraints as an assembly program. So, for example, any problem whose solution can be represented with a table, a tree or a directed graph is eligible for approach.

The evolutionary core is continually being developed, and many features have been added to it over time, many of which may seem somewhat odd, to improve its performance: clone detection and optional extermination to avoid the evaluation of identical individuals and to improve genetic variety; a fitness hole in tournament selection, that is a small but nonzero probability that the tournament selection criterion is not the fitness but the entropy value of the individuals, again to improve genetic variability; parallel fitness evaluations; an initial population size optionally greater than μ, to better exploit the initial random search phase. In this paper the support for real numbers in the individuals and a new form of mutation for MicroGP have been developed, and new features can be expected to appear in the near future.

4 Numerical Experiments

The main goal of our numerical experiments was to obtain a working environment through which we could perform an automatic process of array synthesis and optimization. One of our objectives is to reduce as much as possible the manual effort of the human designer, while still obtaining an acceptable solution. To set up our environment we wrote a very simple instruction library for MicroGP, specifying the allowed range for the roots of the array factor. The only thing the designer is left to do is specifying a wanted array factor, and optionally a desired maximum number of elements.

In our experimental setting we used MicroGP to minimize a measure of distance between an objective array factor and the synthesized antenna's own array factor. We used three different measures of distance, to evaluate the effect of various evaluation criteria on the quality of the result. The objective array factor is passed directly to the fitness evaluator, in the form of a series of $(\psi, F(\psi))$

values. The evaluator reads and normalizes this series, builds a second series containing the corresponding values of the current array factor, with the same normalization, then computes one of the three distances between the two series, as configured in a parameter file. The distances implemented so far are the classic sum of absolute differences, root-mean-square and maximum absolute difference between the two series.

The obtained results are rather different from each other, as will be shown in the next section. This shows the importance of a careful selection of the fitness function.

As the parameter that most critically influences the quality of the solution found by an evolutionary method is the population size we used very big populations in our numerical experiments. Also, we did not employ a hard selection scheme, to let the evolutionary core explore a greater portion of the search space.

To test the suitability of our approach we performed two types of experiments: in the first one we tried to approximate the array factor of an uniform array, while the second was concerned with the synthesis of a rectangular array factor. Approximating the uniform array is seemingly trivial, but, since the evolutionary tool starts from random solutions, we should not take it for granted that it will quickly converge to the exact solution. The approximation of a well-known array type, moreover, gives us confidence in the employed methodology and lets us assess the quality of the obtained solutions. The rectangular array factor, on the other hand, allows us to push the method used to its limits, evidences the differences in performance between the various measures of distance and provides us further insight on the best ways to improve the fitness evaluator.

While performing the optimization, we noticed that the choice of the initial number of roots has a noticeable effect on the achieved quality of the solution. This is due to the fact that the initial phase of the evolutionary method consists of a random search: giving the right number of roots allows the algorithm to randomly hit promising regions of the search space that would remain hidden during a normal search process that starts from a low number of roots. In this latter case, in fact, the evolutionary algorithm may generate solutions with the right number of roots when it is already in the exploitation phase, with a very uniform population, and thus unable to broadly explore the resulting higher-dimensional search space. Only the most significant results are therefore provided.

5 Results

The optimization on the uniform array approximation were conducted with a population of 300 individuals, applying 200 genetic operators per generation and carried on for 100 generations. The obtained results clearly show that even the approximation of an array factor is not a trivial operation. The best fit is obtained using the root-mean-square measure of difference between the objective function and the approximating function. The sum of absolute values yields a somewhat worse performance since it does not discriminate between small and large deviations from the objective, but lumps everything together with

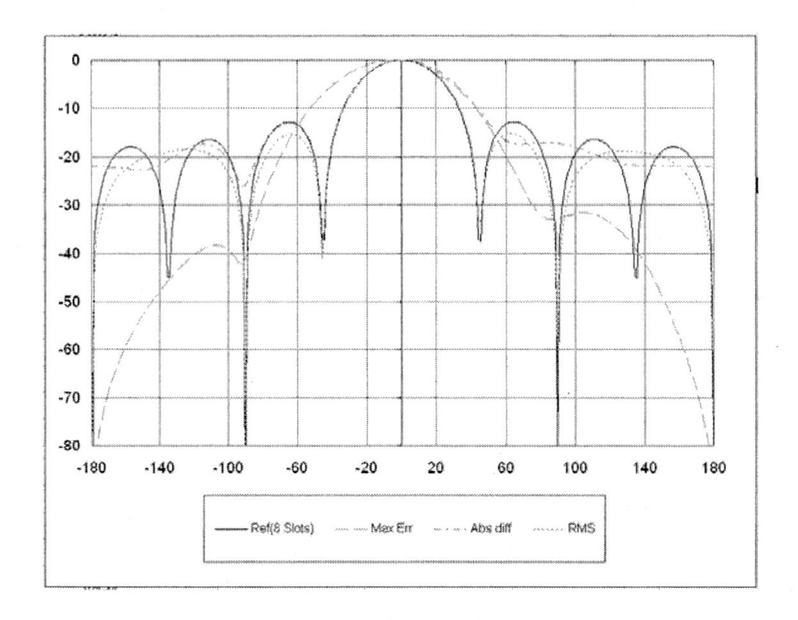

Fig. 1. Approximations of the uniform array factor

the final sum. The worst result of all is obtained using the maximum absolute difference between the two functions; this happens because the fitness landscape has large flat regions in it. To give a hint of why it is so, consider an objective function and a given candidate solution that has a specific value in point ψ_0 (named F_C), where the difference between F_C and the corresponding value of the objective function (named F_O) is maximum; call M this maximum; there obviously exists an infinite number of functions that pass the (ψ_0, F_C) and remain within distance M from the objective function and therefore exhibit the same fitness as the first one. This makes it extremely difficult to find a path even to local maxima. Figure 1 shows the results obtained with the three fitness measures. For the case of the rectangular objective function we used a really large population of 3000 individuals, applying 2000 genetic operators per generation and allowing the evolution to proceed for 1000 generations. The rectangular array factor proves a much harder problem to solve than the uniform array factor, not only needing more elements for an acceptable approximation, but also showing a poorer quality of the solution (Figure 2).

For a comparison, a similar numerical experiment performed approximating the uniform array leads to a result visually indistinguishable from the objective. Again the performance of the three fitness measures shows the same order. The root-mean-square difference measure leads to an imperfect approximation of the low level of the objective function, but to the overall better approximation of the high level; the sum of absolute differences yields the best approximation for the low level but a slightly worse aproximation of the high level; finally, for the same reasons outlined above, the maximum absolute difference gives us the worst

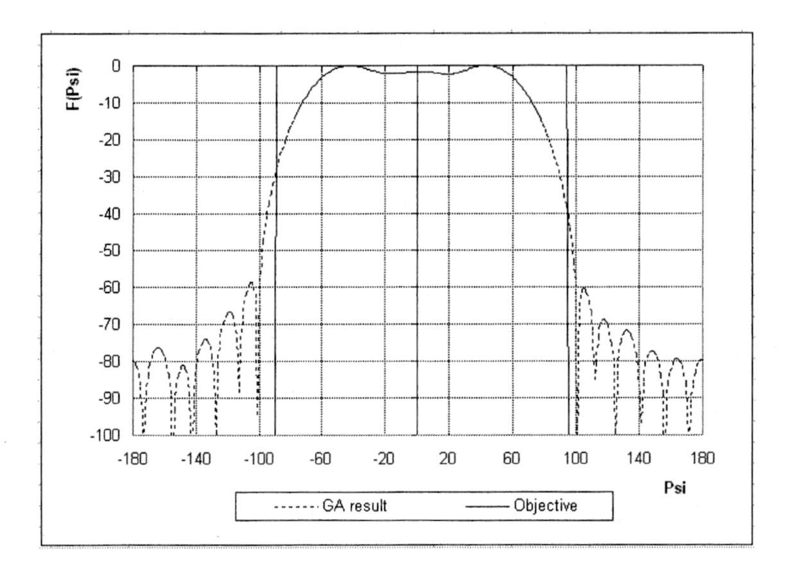

Fig. 2. Approximation of the constant array factor with a 15 roots polynomial

performance, and the resulting evolutionary process is unable to satisfactorily approximate the objective.

It is noteworthy that we let the evolutionary method autonomously choose the number of roots used to approximate the objective function: while this is meant to increase the quality of the obtained solution, it also greatly increases the size of the search space, making it more difficult to find an exact solution.

One significant advantage of an evolutionary method over the deterministic ones is that the latter ones generate very critical solutions, that is, solutions

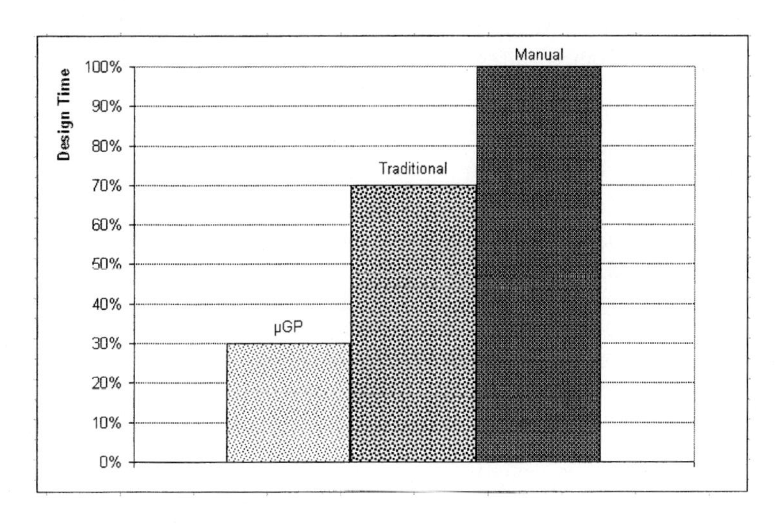

Fig. 3. Comparison between different approachs

that cannot be modified, even slightly, without degrading their quality. The evolutionarily generated ones, instead, can undergo greater modifications before losing as much quality as the deterministic ones. This is most probably the effect of these solutions belonging to a population of similar candidate solutions which, during the search process, are selected and mutated: the evolutionary core has a natural tendency to concentrate its population around local maxima which cover large parts of the search space, while very narrow peaks in the fitness function are harder to be detected. The solutions generated with the evolutionary method may undergo further manual optimization. While this is not a desired situation, it may be necessary for some particularly critical problem, anyway comparing this evolutive approach versus classical methods we can observe that the evolutive approach allows the designer to minimize the design time (figure 3).

6 Conclusions

We set up a working environment to perform array antenna synthesis and optimization using an evolutionary approach. We performed a series of experiments trying to approximate two different objective array factors using different performance measures. The obtained results clearly indicate the need for careful selection of the fitness function within the evolutionary process. They also show that acceptable solutions can be obtained rather quickly and, most importantly, with little human intervention.

The evolutionary tool itself proved very versatile, being able to successfully cope with a problem totally outside of its original application area. This encourages both further investment in the application of evolutionary methods to antenna array syntesis and optimization and development of the tool itself.

In the near future we expect to be able to add support for mask specification as well as new fitness measures in the quest for higher-quality solutions. Later on, we plan to integrate it under a graphic interface for simplified usage.

References

1. T. Baeck, D. B. Fogel, and Z. Michalewicz (eds.): Handbook on Evolutionary Computation. IOS Press, (1997)
2. F. Corno, E. Sanchez, M.S. Reorda, G. Squillero: Automatic test program generation: a case study Design & Test of Computers, IEEE (2004)
3. Chellapilla, K.; Hoorfar, A.: Evolutionary programming: an efficient alternative to genetic algorithms for electromagnetic optimization problems Antennas and Propagation Society International Symposium, 1998. IEEE , Volume: 1 , 21-26 June 1998 Pages:42 - 45 vol.1
4. Marcano, D.; Duran, F.: Synthesis of antenna arrays using genetic algorithms Antennas and Propagation Magazine, IEEE , Volume: 42 , Issue: 3 , June 2000 Pages:12 - 20

5. Hoorfar, A.; Jinhui Zhu: A novel hybrid EP-GA method for efficient electromagnetics optimization Antennas and Propagation Society International Symposium, 2002. IEEE , Volume: 1 , 16-21 June 2002 Pages:310 - 313 vol.1
6. S. Choi, Tapan K. Sarkar; J. Choi Adaptive antenna array for direction-of-arrival estimation utilizing the conjugate gradient method Signal Processing Volume: 45 (1995)
7. Robert S.Elliot: Antenna theory and design Prentice-Hall, Inc., Englewood Cliffs, New Jersey 07632 (1981)
8. R.Y. Rubinstein: Simulation and the Montecarlo Method John Wiley and Sons, New York. (1981)

A Coevolutionary Approach for Clustering with Feature Weighting Application to Image Analysis

Alexandre Blansché, Pierre Gançarski, and Jerzy J. Korczak

LSIIT, UMR 7005 CNRS-ULP, Parc d'Innovation,
Boulevard Sébastien Brant, 67412 Illkirch, France
{blansche, gancarski, jjk}@lsiit.u-strasbg.fr
http ://lsiit.u-strasbg.fr/afd/

Abstract. This paper presents a new process for modular clustering of complex data, such as that used in remote sensing images. This method performs feature weighting in a wrapper approach. The proposed method combines several local specialists, each one extracting one cluster only and using different feature weights. A new clustering quality criterion, adapted to independant clusters, is defined. The weight learning is performed through a cooperative coevolution algorithm, where each species represents one of the clusters to be extracted.

1 Introduction

Data mining methods have been used for many years on less and less elementary data : intervals, distributions, histograms, fuzzy data, temporal data, images, etc. In general, when objects are described by a large set of features, many features are correlated, some of them are noisy or irrelevent. In image analysis, there are two main problems. In hyperspectral data, the superabondance of noisy, correlated and irrelevent bands disturbs the classical procedures of extraction (per pixel clustering). In object-oriented approach, regions are described by a large set of features of heretogeneous types.

Many methods have been proposed for feature weighting or feature selection [1, 2, 3], but almost all these methods are supervised and many of them use only one set of feature weights for clustering the entire data set. In [3] it is shown that a wrapper approch for feature weighting provides better results, because of the feedback from the classification algorithm ; and in [4] it is shown that continuous weights provide better results than binary weights (feature selection). Moreover, in agreement with [5, 6], we believe that even if all features are relevant, their relative importance depends on the classes to extract.

Few methods exist for unsupervised feature weighting [6, 7]. In fact, these methods are based on weighted (dis)similarity measures and use a K-means or prototype-based clustering paradigm.

Our approach is different. A set of extractors (of individual clusters) are defined. Each extractor uses an algorithm to discriminate one cluster and a

F. Rothlauf et al. (Eds.): EvoWorkshops 2005, LNCS 3449, pp. 254–263, 2005.

(local) feature weight vector to optimize the discrimination of this cluster. The global result is obtained by the union of these extracted clusters.

The quality of this global clustering must be estimated with two criteria. A first criterion tests if the clustering is a partition or if there are many unclassified objects and many clusters overlapping. A second one depends on the quality of its extracted cluster.

Consequently, the global clustering quality criterion value is high if, on the one hand, each cluster has a good quality and, on the other hand, each object belongs to one and only one cluster (i.e. each cluster is different from all the others without overlapping).

To find the best set of feature weights according to this global clustering quality criterion, we have defined a cooperative coevolution method.

In this paper, we first present the proposed method, that we call MACLAW (Modular Approach for Clustering with Local Attribute Weighting). Then we validate it on two datasets, the segment dataset from the UCI repository (regions clustering) and a hyperspectral remote sensing image (per pixel clustering).

2 The Proposed Method: MACLAW

The proposed method consists of a cooperative coevolution algorithm in which individuals are extractors, represented by the feature weights they use. K populations of extractors are defined where K is the number of clusters in the global result. The final classification is obtained by the union of one cluster of each population. Each population evolves to extract the best possible cluster according to the global clustering quality criterion. But individuals evolve (by crossover, mutation, ...) inside their population only. The evaluation of an extractor X_i^j (the j-th extractor of the i-th population) is carried out by the use of a set of representative individuals of the other populations, called reference PSC, and updated every generation.

Each generation can be divided into 3 main steps (as shown in Fig. 1) : clusters extraction and quality evaluation, genetic evolution (selection and reproduction) and reference PSC update. The two last steps can be performed in parallel.

Consequently, to define our modular clustering method and in particular the feature weighting process, four main questions have to be addressed :

- how is extracted a cluster (2.1) ?
- how is evaluated the global clustering quality criterion (2.2) ?
- how evolve extractors (2.3) ?
- how are combined extracted clusters (2.4) ?

2.1 Cluster Extraction

Formally, an *extractor* is a function X, such that $X(S) \subset S$ where S is a set of objects to be classified. Let an extractor be a triplet $X = (M, w, r)$, where M is a clustering method, w a set of weights and r a cluster quality criterion. To extract the cluster $X(S)$, first the dataset is classified, using method M and

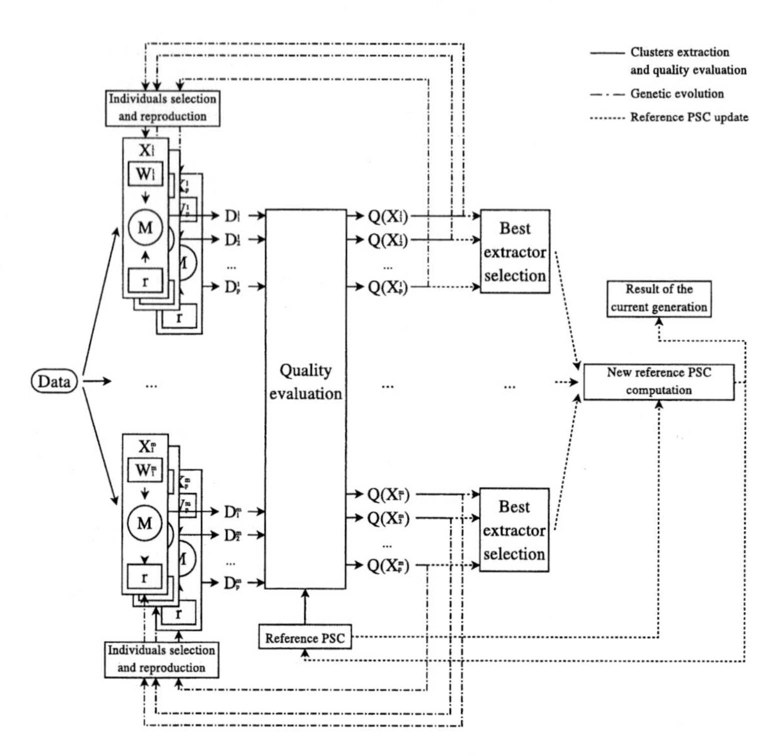

Fig. 1. Schematic diagram of MACLAW

weights w on the attributes of the elements of S, to obtain a set of clusters $\{C_1, \ldots, C_{K_M^w}\}$: for example, for distance-based methods, the weights are used to compute the global distance between objects and compactness of clusters.

Then the cluster C_e such that $r(C_e) = max\{r(C_i), i = 1, \ldots, K_M^w\}$ is selected.

Let $D = \{D_1, D_2, \ldots, D_K\}$ be the *Partial Soft Clustering* (PSC) of the data composed of K extracted clusters $D_i = X_i(S)$, where the X_i are extractors.

2.2 Classification Quality

Many criteria have been proposed to define quality of an unsupervised classification, such as compactness or inertia [8, 9, 10, 11]. But, in our case, it is difficult to use typical criteria because the global classification is a unification of clusters obtained by independant extractors:

- it is not possible to use criteria based on the distance between clusters (e.g. interclass inertia), because each extractor uses a different metric;
- internal criteria (e.g. intraclass inertia) cannot be used because objects may belong to zero, one or several clusters. Consequently, some clusters may all have a good quality, but do not produce a partition of the data.

It is necessary to define a new quality criterion which depends on the quality of the PSC of the extracted clusters.

We first define the *quality of unambiguity* of an object o in a PSC D by $Q_o(o, D) = 1 - \left|Card\{D_i \mid o \in D_i, D_i \in D\} - 1\right|$.

Thus, if o is extracted by one and only one extractor, $Q_o(o, D) = 1$. If o is unclassified or if it is extracted by two extractors, $Q_o(o, D) = 0$, if it is extracted by three extractors, $Q_o(o, D) = -1$, and so on.

Then, the *partition quality* of a PSC D can be defined using the quality of unambiguity by $Q_P(D) = \max\left(0, \frac{1}{N}\sum_{k=1}^{N} Q_o(o, D)\right)$ where N is the number of objects to be classified, i.e. $N = |S|$.

One can notice that $Q_P(D) = 1$ if and only if D is a partition of S. Indeed, if D is a partition, $Q_o(o, D) = 1$ for all objects o of S, thus $\sum_{k=1}^{N} Q_o(o, D) = N$ and $Q_P(D) = 1$. If D is not a partition, there is an object o such that $Q_o(o, D) < 1$, so $\sum_{k=1}^{N} Q_o(o, D) < N$ and $Q_P(D) < 1$.

However, if this criterion is used alone, a cluster can include almost all the objects. The PSC may be a good partition. Nevertheless, it is probably semantically incorrect.

A second criterion is added to take into account the quality of the extracted clusters. The *cluster quality* of a PSC D can be defined by $Q_C(D) = \prod_{D_i \in D} q(D_i)$, where $q(D_i)$ is a quality criterion (e.g. compactness) for one single cluster, which takes values in $[0; 1]$.

Finally, the *quality of a PSC* can be defined, using the partition quality and the cluster quality, by $Q(D) = Q_P(D) \times Q_C(D)$.

First experiments, with this first definition of quality, have shown encouraging results [12], but improvements can be obtained by the use of a fuzzy definition of cluster membership.

The quality of unambiguity of an object o in a PSC D can be extended to a fuzzy definition of cluster membership by $Q_o(o, D) = p(o, D_m) - \sum_{D_i \neq D_m} p(o, D_i)$ where D_i is the i-th extracted cluster and $p(o, D_m) = \max_{D_i \in D} p(o, D_i)$ and $p(o, D_i)$ is the membership degree of the object o for the cluster D_i.

2.3 Evolution of Extractors

The weights are learned through an evolutive algorithm. Each individual is a single extractor. A *chromosome* is the set of weights $w_i \in [0, 1]$ on each attribute for the extractor that it represents. The goal is to evolve several individuals and make them cooperate, so that the PSC built using their extracted clusters is a good classification.

A cooperative coevolution algorithm is used. Cooperative coevolution has been defined in [13, 14]. It is an evolutionary algorithm which uses several populations. A population evolves in an environment which depends on the other populations and evolves with them.

One population for each cluster to extract is used. A PSC is built using one extracted cluster from each population. Because it is an unsupervised method, the cluster of each population is not known at the beginning, but is to be discovered during the learning.

We decide to calculate individual quality only in a single environment as defined in [15]. In the case of modular clustering, we call the environment reference PSC. At a given generation g ($g \neq 1$), the *reference PSC* is the best PSC found during previous generations. It is defined by $\Delta(g) = \{\Delta_1(g), \ldots, \Delta_i(g), \ldots, \Delta_K(g)\}$ where $\Delta_i(g)$ is a representative cluster of the i-th population.

The quality of an extractor X_i^j (j-th individual in the i-th population) is defined by the quality of the PSC obtained by replacing, in the reference PSC $\Delta(g)$, the i-th cluster by $X_i^j(S)$. Thus, $Q\left(X_i^j\right) = Q\left(D_i^j\right)$, where $D_i^j = \{\Delta_1(g), \ldots, X_i^j(S), \ldots, \Delta_K(g)\}$.

A roulette-wheel method (fitness proportionate selection) is used to select individuals and classic genetic operators (crossovers, mutations and new individuals) are used for reproduction.

2.4 Extracted Clusters Combination

At each generation, the PSC obtained is a set of K independant clusters. Thus there may be unclassified objects or objects classified in more than one cluster. To obtain partitioning, a first method (rough combination) for combining clusters simply consists of grouping objects that are not classified in one cluster, in a new cluster of rejected objects.

A better method (conflictless combination) is to affect each unclassified object to one single cluster :

- for each object o, for each extractor $X_i = (M_i, w_i, r_i)$ relative distance to the centre $d_r(o, g_e) = \frac{d_{w_i}(o, g_e)}{d_{w_i}(o, g)}$ is computed, where g_e is the centre of $X_i(S)$ and g the nearest cluster centre (obtained by using M_i with w_i on S) to o, different to g_e ;
- the object is added to the extracted cluster which provides the lowest relative distance.

It is easy to see that if an object is extracted by one extractor X only, it will be added to the cluster $X(S)$.

3 Experiments

3.1 Segment Database from UCI Repository

The new method has been tested on the segment database from the UCI repository [16]. This dataset consists of 7 classes of 330 objects each. The objects were drawn randomly from a database of 7 outdoor images. The images were hand-segmented to create a classification for every pixel. The objects are described by

19 continuous features. All features are defined on different scales. As proposed in [1], we normalized each feature so that they had an expected difference of 1.

We carried out two series of tests with 3 methods, namely K-means, Weighting K-means [7] and MACLAW. We have applied each method 100 times to the raw data and the normalized data. Each method was configured to find 7 clusters. MACLAW was configured as follows :

- Clustering method used by the extractors : K-means;
- Cluster quality criterion : compactness, as defined in [17];
- Membership degree (for quality of unambiguity definition) :

$$p_{\alpha,\varphi}(o, D) = \begin{cases} \exp\left(-\ln\left(\frac{1}{\varphi}\right) \times \left(\frac{d(o,g_C)}{d(o,g)}\right)^{\alpha}\right), \text{ if } d(o,g) \neq 0 \\ 0, \text{ otherwise} \end{cases}$$

  where g_D is the centre of D and g is the nearest cluster centre from o different from g_D, and whith $\alpha > 0$ et $0 < \varphi < 1$;
- Populations sizes : 20 individuals;
- Number of generations : 250 generations;
- Ratios to create new populations :
  - new individuals : 20 %;
  - mutated indivuduals : 20 %;
  - individuals obtained by crossover : 20 %;
  - surviving individuals : 40 %.

Table 1 shows that, for all the methods, normalized data provide better results than raw data. It also shows that with normalized data, methods with feature weighting provides better results than simple K-means and Weighting K-means outperforms MACLAW. With raw data Weighting K-means fails, but MACLAW still provides better results than K-means and largely outperforms Weighting K-means.

These results show that MACLAW seems more robust. The Weighting K-means can provide better results, but is strongly dependent on the initial scales of the features because the weights are computed iteratively. On the other hand, MACLAW searches weights stochastically. The results are lower with raw data only because the search space is not accorded to the data : when a feature has a high expected difference, it is not necessary to test high valued weights.

Table 1. Average, minimum and maximum accuracy over 100 tests

Method	Data	Av. Acc. (%)	Min. acc. (%)	Max. acc. (%)
K-means	raw	41.2 ± 3.53	39.19	47.7
	norm.	44.45 ± 6.97	30.39	56.96
W. K-means	raw	14.15 ± 2.89	13.25	23.33
	norm.	55.01 ± 6.6	44.29	70.91
MACLAW	raw	43.36 ± 7.32	30.13	65.15
	norm.	47.86 ± 6.83	32.34	63.29

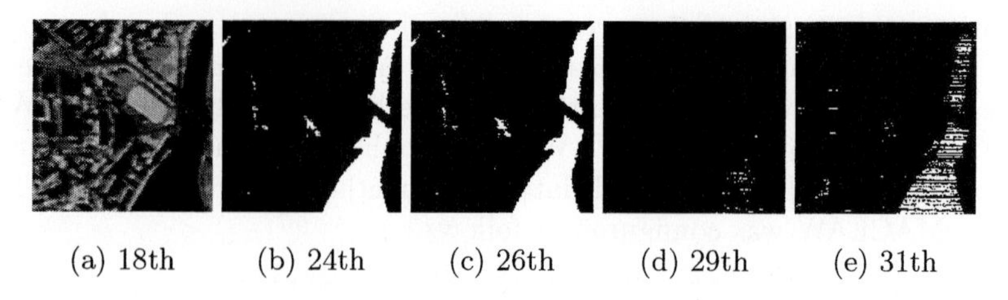

(a) 18th (b) 24th (c) 26th (d) 29th (e) 31th

Fig. 2. Examples of radiometric bands

3.2 Application to Radiometric Bands Selection in Hyperspectral Remote Sensing

We have applied our method to a part of a hyperspectral remote sensing image (DAIS image) from the city of Strasbourg (France), with 44 channels and a standard resolution. The image contains 152 × 156 pixels. In figure 2, one can see 5 different bands :

- the 18th band (Fig. 2(a)) seems relevant and not noisy ;
- the 24th and 26th (Fig. 2(b) Fig. 2(c)) are correlated ;
- the 29th (Fig. 2(d)) does not seem to be relevante ;
- the 31th band (Fig. 2(e)) is strongly noisy.

It is clear that it is necessary to carry out a selection of these bands to obtain a relevant classification.

We carried out some tests with 5 clusters expected, 150 individuals by population and 50 generations. Figure 3 shows the evolution of the quality criterion :

- from generation 0 to 12, a strong improvement of the quality of classification ;
- from generation 13 to 28, a very slow improvement, even a stagnation of this quality ;
- from generation 29 to 37, a new very strong improvement of the quality ;
- from generation 38 to 50, again a very slow increase.

We explain this evolution by :

- during the first generations (from 0 to 12), the improvement comes mainly from the specialization of each population of extractors ;
- during the second period (13 to 28), the specializations found in the preceding stage did not allow any more significant evolution of quality, probably due to the presence of a local maximum ;
- at the 31st generation, the first extractor has specialized in a radically new cluster compared to the preceding generation, but also, and especially, compared to the other extractors of generation 31. This specialization removed a conflict between extractor 1 and extractor 4 ;
- the extractors preserved these new specializations until the end of the training. Only small improvements have been observed. For example an extractor has grouped all pixels of water.

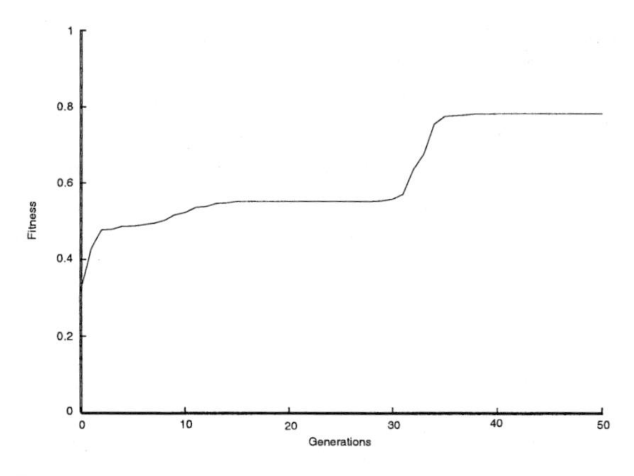

Fig. 3. Fitness evolution during the learning on a hyperspectral image

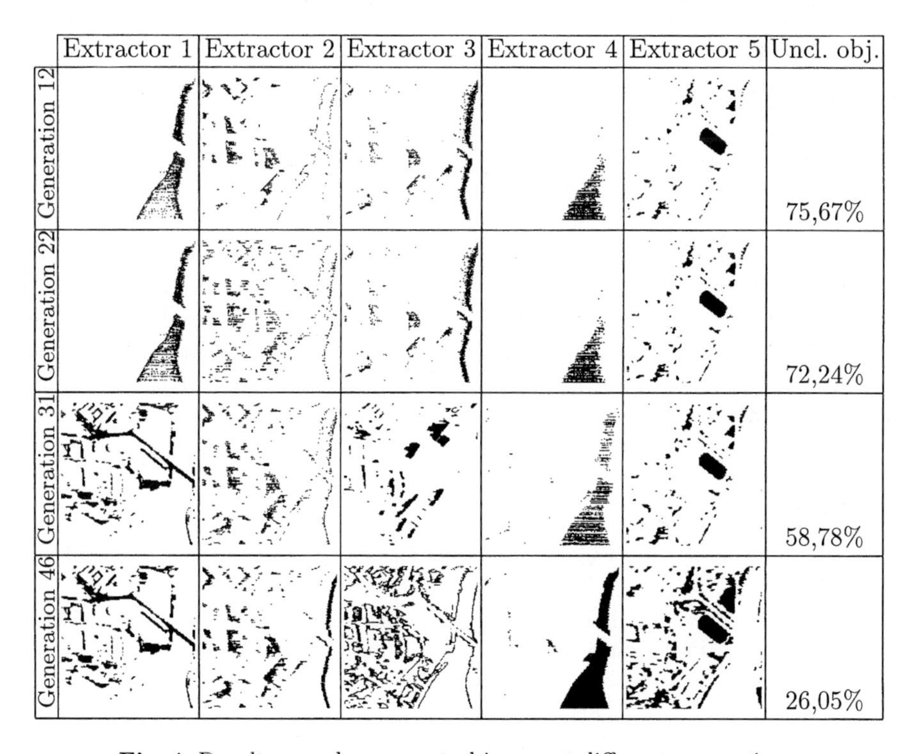

Fig. 4. Results on a hyperspectral image at different generations

Figure 4 shows the reference PSC (as defined in 2.3) corresponding to points of inflections.

Overall, we observe at the 46th generation that interesting clusters seem to have been discovered. Thus, the first extractor "identified" the roads, the 2nd, the shadows, the 4th the water, and the 5th, the vegetation (a stadium and parks). The third extractor has extracted the pixels of edges, often difficult to classify.

4 Conclusion

We have presented MACLAW, a new process of collaborative clustering which is able to classify complex objects described by a large set of features, which might be noisy, correlated, irrelevant and eventually of heterogeneous types. MACLAW decomposes this problem into several simpler problems : rather than trying to classify the set of data in a monolithic way, several extractors are used, each one specialized in one class. The extractors learn in collaboration, through a genetic algorithm, to improve accuracy. It has been necessary to define a new quality criterion, because each cluster is defined independantly. We have shown that MACLAW is efficient in image analysis, in both per pixel clustering or region clustering cases.

We define a general method which has been tested with a distance-based method, cluster quality criterion and membership degree definition, but can be completely independant of the notion of distance or similarity by using some other methods.

However, our method shows some limits. First, genetic learning methods, and especially coevolution methods, have well-known shortcomings. Then, the proposed method needs to know how many clusters there are. This is a common problem for many clustering methods. In our case, we propose that, in the future, the method will be able to add or remove populations dynamically, during the learning, to find the correct number of clusters.

Further research will focus first on the possibility of the collaboration of various classification methods, where each one uses a different model from the data. In the case of remote sensing images for example, we could use various data sources on the same zone (radar, radiometry, photo...). Secondly, we are interested in using domain knowledge to improve each step of the collaborative mining.

References

1. Howe, N., Cardie, C. : Weighting unusual feature types. Technical Report TR99-1735, Ithaca (1999)
2. John, G., Kohavi, R., Pfleger, K. : Irrelevant features and the subset selection problem. In : Proceedings of the Eleventh International Conference on Machine Learning. (1994) 121–129
3. Wettschereck, D., Aha, D. : Weighting features. In Veloso, M., Aamodt, A., eds. : Case-Based Reasoning, Research and Development, First International Conference, Berlin, Springer Verlag (1995) 347–358
4. Wettschereck, D., Aha, D., Mohri, T. : A review and empirical evaluation of feature weighting methods for a class of lazy learning algorithms. Artificial Intelligence Review **11** (1997) 273–314
5. Howe, N., Cardie, C. : Examining locally varying weights for nearest neighbor algorithms. In : ICCBR. (1997) 455–466
6. Frigui, H., Nasraoui, O. : Unsupervised learning of prototypes and attribute weights. Pattern Recognition **34** (2004) 567–581

7. Chan, E., Ching, W., Ng, M., Huang, J. : An optimization algorithm for clustering using weighted dissimilarity measures. Pattern Recognition **37** (2004) 943–952
8. Bolshakova, N., Azuaje, F. : Cluster validation techniques for genome expression data. Signal processing **83** (2003) 825–833
9. Günter, S., Burke, H. : Validation indices for graph clustering. In : Proc. 3rd IAPR- TC15 Workshop on Graph-based Representations in Pattern Recognition. J.-M. Jolion, W. Kropatsch, M. Vento (2001) 229–238
10. Halkidi, M., Batistakis, Y., Vazirgiannis, M. : On clustering validation techniques. Journal of Intelligent Information Systems **17** (2001) 107–145
11. Levine, E., Domany, E. : Resampling method for unsupervised estimation of cluster validity. Neural Computation **13** (2001) 2573–2593
12. Blansché, A., Gançarski, P. : Application aux images hyperspectrales d'une nouvelle méthode de sélection d'attributs pour la classification d'objets complexes. In : Proc. of workshop "Fouille de Données Complexes dans un processus d'extraction de connaissances" in EGC'04, Clermont-Ferrand (2004) 103–114
13. Potter, M., De Jong, K. : A cooperative coevolutionary approach to function optimization. In : Proceedings of the Third Conference on Parallel Problem Solving from Nature. (1994) 249–257
14. Potter, M., De Jong, K., Grefenstette, J. : A coevolutionary approach to learning sequential decision rules. In : Proceedings of the Sixth International Conference on Genetic Algorithms. (1995) 366–372
15. Mayer, H. : Symbiotic coevolution of artificial neural networks and training data sets. In : Proceedings of the Fifth International Conference on Parallel Problem Solving from Nature. (1998) 511–520
16. Blake, C., Merz, C. : UCI repository of machine learning databases (1998)
17. Wemmert, C., Gançarski, P., Korczak, J. : An unsupervised collaborative learning method to refine classification hierarchies. In : Proceedings of the IEEE 11th International Conference on Tools with Artificial Intelligence. (1999) 263–270

A New Evolutionary Algorithm for Image Segmentation

Leonardo Bocchi[1], Lucia Ballerini[2], and Signe Hässler[3]

[1] Dept. of Electronics and Telecommunications,
University of Florence, Via S.Marta 3, 50139 Firenze, Italy
leo@asp.det.unifi.it
[2] lucia@cb.uu.se
[3] Department of Medical Sciences, Uppsala University,
Uppsala University Hospital, 751 85 Uppsala
signe.hassler@medsci.uu.se

Abstract. This paper describes a new evolutionary algorithm for image segmentation. The evolution involves the colonization of a bidimensional world by a number of populations. The individuals, belonging to different populations, compete to occupy all the available space and adapt to the local environmental characteristics of the world. We present experiments with synthetic images, where we show the efficiency of the proposed method and compare it to other segmentation algorithm, and an application to medical images. Reported results indicate that the segmentation of noise images is effectively improved. Moreover, the proposed method can be applied to a wide variety of images.

1 Introduction

Image segmentation is typically the most difficult task in image processing and it is usually the starting point for any subsequent analysis. Image segmentation has been the subject of intensive research, and a wide variety of techniques have been reported in literature. A good review of these methods can be found in [1].

In many applications, clustering algorithms can be used for image segmentation [2]. Among them, the fuzzy c-means clustering algorithm (FCM) is one of the best known and the most widely used clustering technique [3, 4]. However, FCM exploits the homogeneity of data only in the feature space and does not adapt to their local characteristics. This is a major drawback of the use of FCM in image segmentation, because it does not take into account the spatial distribution of pixels in images. Many optimization methods have been reported to improve and further automate the fuzzy clustering. Among them, various authors proposed the use of genetic algorithms [5] with promising results [6, 7, 8].

Alternative approaches to exploit the metaphor of natural evolution in the context of image segmentation have been proposed. The genetic learning system proposed by Bhanu et al. [9] allows the segmentation process to adapt to image characteristics, which are affected by varying environmental factors such as the

F. Rothlauf et al. (Eds.): EvoWorkshops 2005, LNCS 3449, pp. 264–273, 2005.

time of the day, condition on cloudiness, etc. Bhandarkar and Zhang [10] use the genetic algorithm to minimize the cost function that is used to evaluate the segmentation results. Andrey [11] describe a selectionist relaxation algorithm, whereby the segmentation of an input image is achieved by a population of elementary units iteratively evolving through a fine-grained distributed genetic algorithm. Liu and Tang [12] present an autonomous agent-based approach, where a digital image is viewed as a two-dimensional cellular environment in which the agents inhabit and attempt to label homogeneous segments. Veenman et al. [13] use a similar image segmentation model and propose a cellular coevolutionary algorithm to optimize the model in a distributed way. Methods based on ant colonies and artificial life algorithms are also investigated for image segmentation and clustering problems [14].

The application of heuristic methods on image segmentation looks very promising, since segmentation can be seen as a clustering and combinatorial problem. Throughout this paper, we will consider the clustering problem and the segmentation problem as being similar. Accordingly, we consider solution methods for both problems interchangeably.

In this paper, a system based on an evolutionary algorithm, which is reminiscent of the well-know 'Life' game, invented by John Horton Conway [15], is presented. The evolution involves the colonization of a bidimensional world by a number of populations, which represent the different regions which are present in the image. The individuals, belonging to different populations, compete to occupy all the available space and adapt to the local environmental characteristics of the world.

The paper is organized as follow: section 2 describes the standard FCM algorithm we use as a reference, section 3 reports the details of the proposed method, section 4 reports numerical results obtained on synthetic images and section 5 shows the results obtained on a set of medical images.

2 Fuzzy c-Means Clustering Algorithm

We choose to use a standard FCM algorithm as a reference for the evaluation on the proposed algorithm. The standard FCM algorithm is based on the minimization of the following objective function:

$$J_m(U,V) = \sum_{i=1}^{c}\sum_{k=1}^{n} u_{ik}^m \|\mathbf{x}_k - \mathbf{v}_i\|^2 \tag{1}$$

where:

- $\mathbf{x}_1, \mathbf{x}_2, ..., \mathbf{x}_n$ are n data sample vectors;
- $V = \{\mathbf{v}_1, \mathbf{v}_2, ..., \mathbf{v}_c\}$ are cluster centers;
- $U = [u_{ik}]$ is a $c \times n$ matrix, where u_{ik} is the ith membership value of the kth input sample $\mathbf{x}_k$, such that $\sum_{i=1}^{c} u_{ik} = 1$
- $m \in [1, \infty)$ is an exponent weight factor that determines the amount of "fuzziness" of the resulting classification.

If we assume that the norm operator $\|\cdot\|$ represents the standard Euclidean distance, the objective function is the sum of the squared Euclidean distances between each input sample and its corresponding cluster center, with the distance weighted by the fuzzy membership.

As pointed out by several authors [16, 17], the FCM algorithm always converges to strict local minima of J_m starting from an initial guess of u_{ik}, but different choices of initial u_{ik} might lead to different local minima.

Bezdek et al. [6, 7] introduces a general approach based on GA for optimizing a broad class of clustering criteria. Since the cluster centers are the only variable used by the GA, they reformulate the FCM functional as:

$$R_m(V) = \sum_{k=1}^{n} \left(\sum_{i=1}^{c} \|\mathbf{x}_k - \mathbf{v}_i\|^{1/(1-m)} \right)^{(1-m)} \quad (2)$$

They demonstrate [18] that this functional is fully equivalent to the original one for each clustering criterion (hard, fuzzy and probabilistic).

3 System Architecture

The system is based on an evolutionary algorithm which simulates the colonization of a bidimensional world by a number of populations. The world is organized in a bidimensional array of locations, or cells, where each cell is always occupied by an individual.

The world is represented by a matrix, associated with a vector of input images I_z (i.e. RGB components, textural parameters, or whatever), which are stacked one above the other. Each cell of the matrix corresponds to a pixel of the image stack, and therefore, the cell having coordinates $P = (x, y)$ is associated to a vector of features $e(x, y) = \{I_z(x, y)\}$. In our simulation, this feature vector is assumed to represent the environmental conditions at point P of our world.

During each generation, each individual has a variable probability S_r, depending both on the environmental conditions and on the local neighborhood, to survive to the next generation. When the individual fails to survive, the empty cell is immediately occupied by a newly generated individual.

3.1 Environmental Constraints

The environmental conditions in a cell influence the probability of the individual surviving in that location. If the population (which the individual belongs to) is well suited to the proposed environment, the survival chances of that individual are very high. On the other hand, if the population is suited to an environment which is very different from the local one, the possibilities for that individual to survive to the next generation are very low.

This requires us to define an ideal environment which maximises the chances of survival of an individual of a given population. This ideal environment has been obtained by averaging, in each iteration, the environment in all the cells occupied by individual of the population.

For instance, if the population A is composed of individuals mainly located in dark zones of the input image, the few individuals belonging to the population A, which are situated in bright zones of the input image have a low survival rate. After a few iterations, the percentage of individual situated in dark areas will be increased.

A second parameter used to increase the selective pressure over the population is the variance of the feature vector. This is used to normalize the evaluation of the similarity between the ideal environment and the local environmental conditions.

The environmental factor described above has been modeled in our system by means of a survival factor S_e which is represented, for an individual belonging to the population i and situated in the point (x, y), as:

$$S_e = \frac{1}{1 + \exp\left(c_i(x, y)/c_0 - c_t\right)} + m_e \tag{3}$$

The expression above represents a sigmoid-like function, centered in c_t. Parameters c_t and c_0 describe the position and the steepness of the sigmoid function, while the constant m_e represents a minimal survival rate. The variable $c_i(x, y)$ represents the similarity between the local environment $e(x, y)$ and the ideal environment e_i for the population i, evaluated as:

$$c_i(x, y) = \left| \frac{e(x, y) - e_i}{\sigma_i} \right| \tag{4}$$

where, as described above, e_i and σ_i are, respectively, the mean and the standard deviation of $e(x, y)$ over all points of the image occupied by individuals belonging to the population i.

3.2 Neighborhood Constraints

The presence of individuals of the same population in a neighborhood is known to increase the survival rate of them. In our simulation, this has been taken into account by including in the model a survival factor S_n which depends on the number of individuals n_i in a 3×3 neighborhood which belong to the same population of the individual located in the position (x, y).

The neighbor factor associated, named S_n, is evaluated as:

$$S_n = \frac{1}{1 + \exp\left(n_t - n_i(x, y)/n_0\right)} + m_s \tag{5}$$

where, as above, parameters n_t and n_0 describe the position and the steepness of the sigmoid function, while the constant m_s represents a minimal survival rate. It is worth noting the difference between the two survival rates is in the sign: in this case the survival rate increases when n_i increases, while S_e decreases when c_i increases.

3.3 Splitting and Merging

As presented above, the method does not prevent the situation were two populations are competing to colonize regions having similar environmental constraints. We overcome this problem by including, once over a predefined number of iterations, a split and merge step. In this step, we evaluate how different the separation are from each other by means of a statistical analysis of the populations descriptors. For each pair (i, j) of populations we evaluate a separation coefficient s_f as:

$$s_f = \sum_z \frac{(e_i - e_j)^2}{\sigma_i \sigma_j} \tag{6}$$

when this coefficient is too small, we assume that the two population are statistically equivalent, and we merge them in a single one. At the same time, in order to preserve the total number of populations, the population having the highest dishomogeneity, measured as the largest value of $|\sigma_i|$ is split in two new populations.

3.4 Algorithm

The algorithm can be described according to the following steps:

1. On each point in the image is placed a random individual
2. For each generation:
   (a) The average feature vector e_i and its standard deviation σ_i are computed for each population
   (b) For each individual:
      i. The survival probability is computed as $S_r = S_e * S_n$.
      ii. If the individual does not survive, a new one replaces it. The new individual is assigned to a population randomly selected with probabilities proportional to the survival factor S_e of an individual of each population.
3. The separation s_f among populations is evaluated, and split and merge operation are performed

4 Experiments with Synthetic Images

This experiment uses synthetic images containing geometric objects (a circle, square and a star-shaped object). The intensity image is generated by assigning grey level value 100 to pixels belonging to the background, and 70, 130, and 160 respectively to the objects.

A zero mean, white Gaussian noise is added to this image. Three different noise levels (corresponding to the standard deviation values: 10, 20, 30) are considered, as shown in Figure 1. This allow to study the robustness of our segmentation technique with respect to noise variance and to determine an adequate set of parameters.

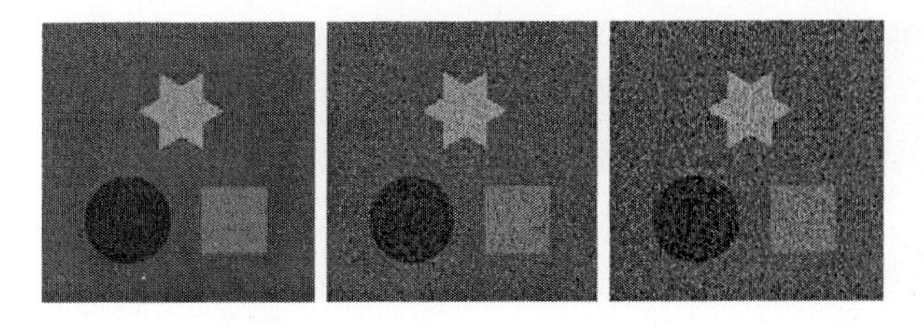

Fig. 1. Examples of synthetic test images with different values of Gaussian noise (from left to right: σ_{noise} =10, 20, 30)

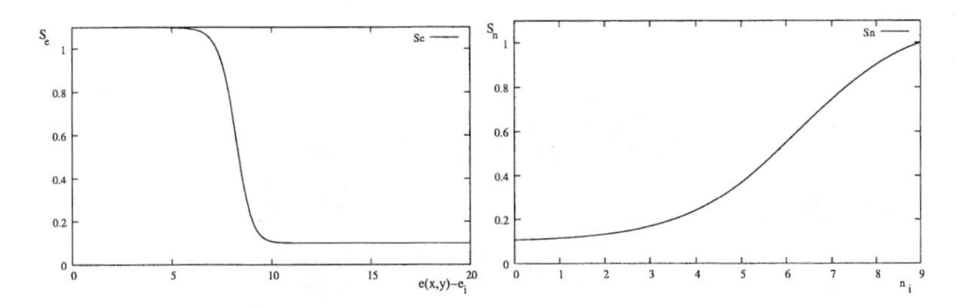

Fig. 2. Plot of S_e (left) and S_n (right) obtained using $n_t = 5$, $n_0 = 1.25$, $n_t = 10$, $c_0 = 20$, $\min(s_f) = 0.2$, $m_e = 0.1$, $m_s = 0.3$

On these images we perform experiments using the following parameters: populations = 4 (circle, square, star, background), $n_t = 5$, $n_0 = 1.25$, $n_t = 10$, $c_0 = 20$, $\min(s_f) = 0.2$, $m_e = 0.1$, $m_s = 0.3$. Using this parameter set, we obtain the plot reported in Figure 2 for S_n and S_e. We can observe that an individual which is located in a cell where $e(x, y)$ differs from e_i more than 8, has practically no probability to survive. As concerns the neighborhood factor S_n, the probability of surviving is almost proportional to the number of neighbors belonging to the same population, when this number is larger than four.

The effect of the split and merge operations is depicted in Figure 3. The image on the left shows an intermediate evolution step, where two different populations are competing to colonize the round object. This situation is a sort of local minimum which traps the evolutionary algorithm. However, when the two competing populations are restricted to the same part of the image, their mean and standard deviations indicate there is no significative difference between them, and in the next generation they will be merged together (center image). At the same time, to maintain the total number of population constant, one of the remaining populations is split in two parts. In this case, the population having a larger variance between its individual is the population which colonizes the other objects. After a few iterations (right image), the two populations adapt to the different objects, achieving a correct discrimination among them.

The final segmentation results, see Figure 4, show that the algorithm is able to give a sensible segmentation also with the highest value of noise in the image.

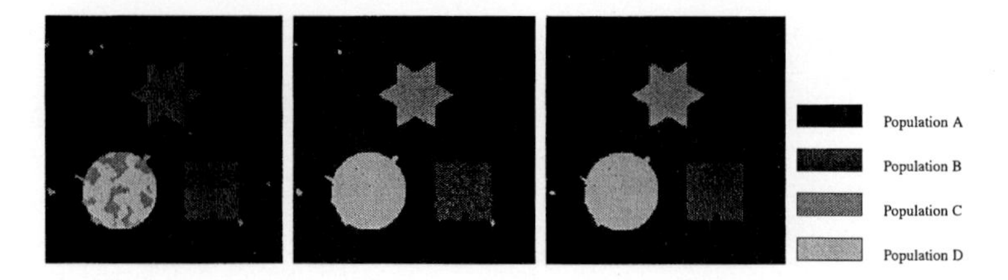

Fig. 3. Intermediate evolution steps. Left image: two populations which are competing for the round object. Center image: a split and merge is occurred: the round object is colonized by a single population, while the two populations generated by split are adapting to the different objects

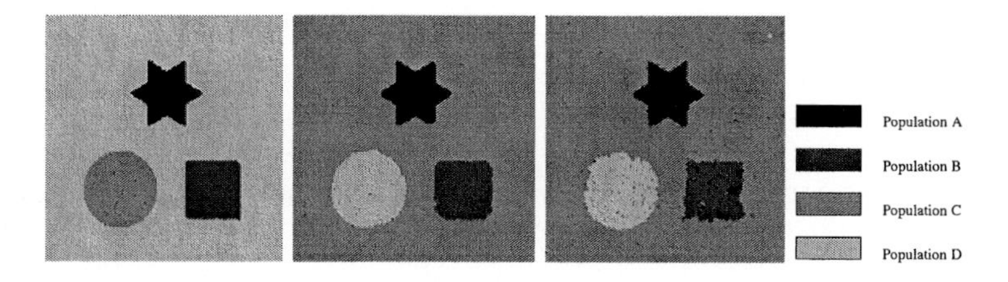

Fig. 4. Simulation results on synthetic test images with different values Gaussian noise

The experimental results have been compared with the segmentation results of a standard fuzzy c-means algorithm.

A quantitative evaluation of the segmentation results, reported in table 1, can be obtained measuring the classification errors A^+ and A^-. The parameter A^+ represents the percentage of pixel which do not belong to the object, but which have been assigned to it. On the other hand, A^- represents the percentage of pixels which belong to the object, but which have been classified as background.

Table 1. Average classification results on synthetic images

Object	circle		square		star	
	A^-(%)	A^+(%)	A^- (%)	A^+(%)	A^- (%)	A^+ (%)
$\sigma = 10$	0.5	0.3	0.9	0.3	0.5	1.7
$\sigma = 20$	2.7	1.3	6.7	0.7	2.3	4.3
$\sigma = 30$	6.2	3.7	14.7	4.6	1.8	18.7

Table 2. Average classification results on synthetic images using the standard FCM algorithm

Object	circle		square		star	
	A^-(%)	A^+(%)	A^- (%)	A^+(%)	A^- (%)	A^+ (%)
$\sigma = 10$	7.2	62.9	39.9	> 100	0.3	49.0
$\sigma = 20$	30.2	> 100	51.8	> 100	9.8	> 100
$\sigma = 30$	47.8	> 100	57.7	> 100	24.1	> 100

The use of local information allows the proposed method to obtain results significatively better than the FCM algorithm. As it can seen from table 2, the variance of the noise is large enough to disallow the FCM algorithm to successfully discriminate objects from background in most cases. The notation > 100 actually means that a large part of the background has been assigned to the same class of the examined object.

5 Application to Medical Images

The pancreas is composed of two different types of tissues: an exocrine parenchima which produces different digestive enzymes and the islets of Langherans, which are found scattered throughout the tissue [19]. The islets have an endocrine function and produce hormones important in metabolism like insulin and glucagon. The islets of Langherans are attacked and destroyed by the immune system in type I diabetes and have a decreased ability to produce insulin in type II diabetes. By analysing the number of islets and the surface they occupy in proportion to the total surface of the pancreatic tissue section we can define the normal values for different mouse strains and later on we can compare these numbers with genetically modified mice in order to identify genes that are important in the embryonic/postnatal development of the islets of Langerhans, which might be of importance for the development of diabetes later on in life. The picture was taken on the pancreas of a 5 months old mouse, C57Bl/6 strain. The pancreas was fixed in formalin, paraffin embedded and sectioned into 14 mm slices. The sections were stained with hematoxilin and eosin. The image acquisition were performed with a LEICA DMRB, Camera LEICA DC 200, acquisition software LEICA QWin, using a 20 × magnification.

Two images of a pancreas, containing the islets of Langherans are shown in Figure 5.

The proposed method has been applied for the segmentation of these images, using the following parameters: populations = 2 (islets of Langherans, parenchima), $n_t = 4$, $n_O = 1$, $n_t = 10$, $c_O = 20$, $\min(s_f) = 0.2$, $m_e = 0.1$, $m_s = 0.3$.

Examples of the obtained segmentation are shown in figure 6.

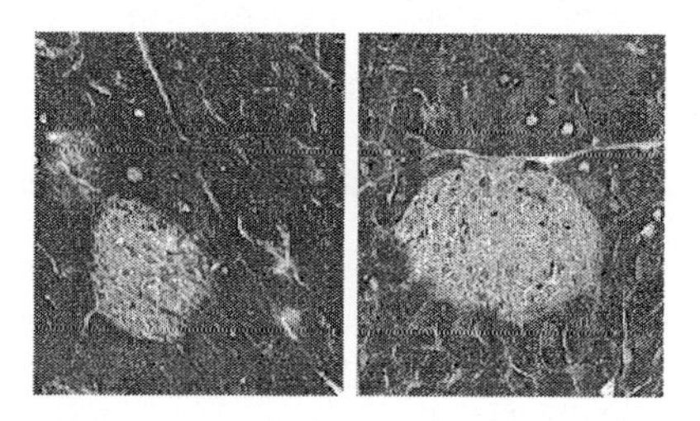

Fig. 5. Two sections of a pancreas, including the islets of Langherans

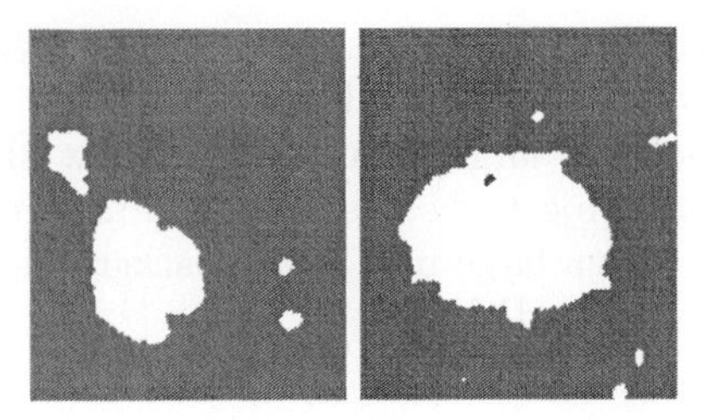

Fig. 6. Segmentation of the images in Figure 5

A quantitative evaluation of these results is currently under study. There is not a standard segmentation technique in this context to compare with our method, in fact medical doctors still perform it manually. We are therefore asking to more than one clinician to segment all the islets blindly and compare their agreement with the described approach.

6 Conclusions

In this paper, we presented an evolutionary algorithm for image segmentation. In the experiments, we showed the effectiveness of the method and compared it to a well-known clustering method. The proposed algorithm can be used for the segmentation of gray-scale, color and textural images. The representation of the environmental constraints as a feature vector allows us to easily extend the method to any vector-valued parametric images, independently on the number of components. Moreover, the normalization of each component of the similarity term c_i enables to use paramedic images having different ranges of values.

We are planning to extend the method in order to include local properties on each population in the evaluation of survival rates. In this way we will enable the system to better adapt to slow variations present in the image (for instance, uneven illumination) which cannot be captured by the overall mean on the population. This extension could also allow for the introduction of a new split procedure based on the differentiation between local properties and the overall mean.

References

1. Pal, N.R., Pal, S.K.: A review on image segmentation techniques. Pattern Recognition **26** (1993) 1277–1294
2. Jain, A.K.: Cluster analysis. In Young, T.Y., Fu, K.S., eds.: Handbook of Pattern Recognition and Image Processing. Academic Press (1986) 33–57
3. Cannon, R.L., Dave, J.V., Bezdek, J.C.: Efficient implementation of the fuzzy c-means clustering algorithms. IEEE Transactions on Pattern Analysis and Machine Intelligence **PAMI-8** (1986) 248–255
4. Chi, Z., Yan, H., Pham, T.: Fuzzy algorithms: with application to image processing and pattern recognition. World Scientific, Singapore (1996)

5. Goldberg, D.E.: Genetic Algorithms in Search, Optimization, and Machine Learning. Addison-Wesley, Reading, MA (1989)
6. Bezdek, J.C., Hathaway, R.J.: Optimization of fuzzy clustering criteria using genetic algorithms. In: Proc. 1st IEEE Conf. Evolutionary Computation. (1994) 589–549
7. Hall, L.O., Ozyurt, I.B., Bezdek, J.C.: Clustering with a genetically optimized approach. IEEE Transactions on Evolutionary Computation **3** (1999) 103–112
8. Ballerini, L., Bocchi, L., Johansson, C.B.: Image segmentation by a genetic fuzzy c-means algorithm using color and spatial information. In: Application of Evolutionary Computation. Number 3005 in Lectures Notes in Computer Science, Coimbra, Portugal (2004) 260–269
9. Bhanu, B., Lee, S., Ming, J.: Adaptive image segmentation using a genetic algorithm. IEEE Transactions on Systems, Man and Cybernetics **25** (1995) 1543–1567
10. Bhandarkar, S.M., Zhang, H.: Image segmentation using evolutionary computation. IEEE Transactions on Evolutionary Computation **3** (1999) 1–21
11. Andrey, P.: Selectionist relaxation: Genetic algorithms applied to image segmentation. Image and Vision Computing **17** (1999) 175–187
12. Liu, J., Tang, Y.Y.: Adaptive image segmentation with distributed behavior-based agents. IEEE Transactions on Pattern Analysis and Machine Intelligence **21** (1999) 544–551
13. Veenman, C.J., Reinders, M.J.T., Backer, E.: A cellular coevolutionary algorithm for image segmentation. IEEE Transactions on Image Processing **12** (2003) 304–313
14. Ramos, V., Almeida, F.: Artificial ant colonies in digital image habitats - a mass behaviour effect study on pattern recognition. In: Proc. of ANTS'2000 - 2nd Int. Workshop on Ant Algorithms (From Ant Colonies to Artificial Ants), Brussels, Belgium (2000) 113–116
15. Gardner, M.: The fantastic combinations of John Conway's new solitaire game "life". Scientifican American **223** (1970) 120–123
16. Lim, Y.W., Lee, S.U.: On the color image segmentation algorithm based on the thresholding and the fuzzy c-means techniques. Pattern Recognition **23** (1990) 1935–952
17. Xie, X.L., Beni, G.: A validity measure for fuzzy clustering. IEEE Transactions on Pattern Analysis and Machine Intelligence **13** (1991) 841–847
18. Hathaway, R.J., Bezdek, J.C.: Optimization of clustering criteria by reformulation. IEEE Transactions on Fuzzy Systems **3** (1995) 241–254
19. Stevens, A., Lowe, J.: Human Histology. C.V. Mosby, 2nd edition (1997)

An Interactive EA for Multifractal Bayesian Denoising

Evelyne Lutton, Pierre Grenier, and Jacques Levy Vehel

INRIA - COMPLEX Team,
B.P. 105, 78153 Le Chenay cedex, France
http://fractales.inria.fr

Abstract. We present in this paper a multifractal bayesian denoising technique based on an interactive EA. The multifractal denoising algorithm that serves as a basis for this technique is adapted to complex images and signals, and depends on a set of parameters. As the tuning of these parameters is a difficult task, highly dependent on psychovisual and subjective factors, we propose to use an interactive EA to drive this process. Comparative denoising results are presented with automatic and interactive EA optimisation. The proposed technique yield efficient denoising in many cases, comparable to classical denoising techniques. The versatility of the interactive implementation is however a major advantage to handle difficult images like IR or medical images.

1 Introduction

Interactive Evolutionary Algorithms (IEA) have now many applications in various domains, where quantities to be optimised are related to subjective rating (visual or auditive interpretation). Following founding works, [1, 2, 3, 4] oriented towards artistic applications, characteristic examples are [5] for Hearing Aids fitting, [6] for smooth, human-like, control rules design for a robot arm, or [7] for the design of HTML style sheets. An overview of this vast topic can be found in [8].

The specific context of human interaction constrains the evolutionary engine a different way as classical EA approaches. The "user bottleneck" [9], i.e. the human fatigue, is a major fact. Solutions have to be found in order to efficiently drive the evolution of the system while avoiding systematic and boring interactions [9, 8, 10]. Usually, the IEA populations are small, and interfaces are designed in order to let the user interact in various ways with evolution (initialisations, solutions rating, and if possible direct modifications on genomes [11]).

The present work deals with complex image analysis techniques that depend from a set of control parameters. These techniques are precise and efficient, but depending on applications, the aim of the end-used may be very different (medical practicionner, teledetection, stereophotogrametry), and some judgment criteria are fully user dependent.

We explore the idea of developing an interactive EA to cope with unpredictibility of the exact aim of the user. As an example, for denoising applications, computed image distance is not sufficient to decide which algorithm is the best. Criteria related to details

F. Rothlauf et al. (Eds.): EvoWorkshops 2005, LNCS 3449, pp. 274–283, 2005.

preservation, and depending on psychovisual factors are extremely important. Moreover, the end-user judgment depends on the way he will use the denoised image ...

The paper is organised as follows. In section 2 the principle of the multifractal bayesian denoising technique is presented, and the free parameters are identified. These free parameters are optimised with the help of an interactive evolutionary algorithm, see section 3. Results of interactive and automatic optimisations are presented in section 4, with comparisons with another efficient denoising technique based on wavelet coefficients. Conclusions and future work are presented in section 5.

2 Multifractal Bayesian Denoising

2.1 Multifractal Analysis

The multifractal analysis of a signal consists in measuring its regularity at each sample point, in grouping the points having the same irregularity, and then in estimating the Hausdorff dimension (i.e. the "fractal dimension") of each iso-regularity set. Irregularity is measured via the local Hölder exponent [12] defined for a continuous function f at x_0 as the largest real α such that:

$$\exists C, \rho_0 > 0 : \forall \rho < \rho_0 \qquad sup_{x,y \in B(x_0,\rho)} \frac{|f(x) - f(y)|}{|x - y|^\alpha} \leq C$$

Since α is defined at each point, we may associate to f the function $x \rightarrow \alpha(x)$ which measures the evolution of its regularity.

A multifractal spectrum f_H is a representation of the irregularity of the signal over its definition domain. For each irregularity value, i.e. for each α, one estimates $f_H(\alpha)$ as the Hausdorff dimension (the "fractal dimension," intuitively related to a frequency/geometrical distribution) of the corresponding iso-α set. As an example, for image data, a $f_H(\alpha) \simeq 1$ corresponds to a linear and smooth structure, while $f(\alpha) \simeq 0$ is a set of scattered points (singular points), or $f(\alpha) \simeq 2$ is a uniformly textured area.

The multifractal spectrum is a central notion exploited in multifractal image and signal analysis, as it provides in the same time a local (α) and a global ($f_H(\alpha)$) viewpoint on data. It has been exploited with success in many applications where irregularity bears some important informations (image segmentation [13], signal and image denoising [12], etc ...)

Wavelet transforms are convenient tools for the estimation of the Hölder exponents. Our method is thus based on a discrete wavelet transform, and has been compared to another denoising technique based on wavelets (soft thresholding), know as very efficient in many cases, see section 4.

2.2 Bayesian Denoising

The principle of the method is the following: For a noisy image I_1, we search for a denoised image I_2 that satisfies two conditions:

- I_2 has a given multifractal spectrum,
- the probability that the addition of a white gaussian noise (with variance σ) to I_2 produces an observed image I_1, is maximal.

If the wavelet coefficient of the noisy image at scale j is y, we get the following coefficient at the same scale for the denoised image (for details, see [14]):

$$\widehat{x} = argmax_{x>0}\left(jg\left(\frac{log_2(\widehat{K}x)}{-j}\right) - \frac{(|y|-x)^2}{2\sigma^2}\right) sgn(y) \tag{1}$$

where

- $\widehat{K}$ is a constant (that can depend from the scale j) such that $\widehat{K}|y| < 1$ for every coeficients y at scale j of the noisy image. In what follows, $\widehat{K}$ is taken as the inverse of the maximal coefficient of each scale.
- g is the function that defines the a priori spectrum of the denoised image.

  We have chosen to use functions that verify the following properties:

  - g is defined on a interval $[\alpha_{min}, \alpha_{max}]$,
  - $g(x) \in [0, 1]$,
  - there exists an $\alpha_{mod} \in [\alpha_{min}, \alpha_{max}]$ such that $g(\alpha_{mod}) = 1$,
  - g is affine on $[\alpha_{min}; \alpha_{mod}]$ and on $[\alpha_{mod}; \alpha_{max}]$.

  The g functions are thus fully determined by 5 values: α_{min}, α_{mod}, α_{max}, $g(\alpha_{min})$ and $g(\alpha_{max})$.

As far as the previous set of values is chosen, the computation of the optimal wavelet coefficients of equation (1) is a simple deterministic calculation (as the a priori spectrum g is affine by parts). The results provided by the method are fully determined.

2.3 Free Parameters

A full set of parameters needs to be tuned in order to produce an efficient denoising with the previous method. Among them[1], the most important ones are the values α_{min}, α_{mod}, α_{max}, $g(\alpha_{min})$ and $g(\alpha_{max})$, that define the profile of the a priori spectrum. For image denoising, these values have to be defined for the horizontal/vertical coefficients and for the diagonal ones.

Actually, we have chosen to distinguish the a priori hypotheses made on the horizontal/vertical coefficients and on the diagonal ones. Usually, on "non-noisy" images, the diagonal wavelet coefficients are significantly lower than the horizontal and vertical ones, while if these images are perturbed by an additive gaussian noise, this discrepancy

[1] The choice of the wavelet basis is of course important, and visually influences the results. For the present work we have chosen to optimise this parameter offline, independently from the interactive evolution process. Inded, experiments have been performed whith a genome including a symbolic component identifying the wavelet basis. It has been noticed that experimentally the wavelet coefficients that yield best results where Daubechies 8 to 12. It has however been noticed that as far as a "correct" wavelet has been chosen, the shape of the a priori spectrum is determining for the quality of results.

vanishes[2]. Diagonal wavelet coefficients seems to be more sensitive to additive noise, and a "stronger" denoising of them often yield better results.

We have thus chosen to use a different g function for the diagonal coefficients: g_{Diag} is similar to g, but translated with respect to its abcissa. This translation is an additional parameter of the method.

3 Interactive Optimisation of Free Parameters

In [12], the multifractal denoising technique was based on another hypothesis with respect to the multifractal spectrum. We supposed that the noise was resulting into a translation of the multifractal spectrum of the initial image. In this paper, we relax this hypothesis and do not suppose that the spectrum of the initial image can be deduced from the degraded one. Moreover we do not suppose we know the variance of the noise (that we still suppose to be a white Gaussian noise however).

The resulting method is thus more versatile, but in the same time the quality of results is stongly dependent from the choice of the parameter set. The solution we propose is to let a human user choose among the various parameters combinations. This problem is of course strongly combinatorial and an interactive evolutionary algorithm has been designed in order to focus the search.

The population is made of a small number of individuals. Each individual is a parameter setting. It is presented to the user as an image, result of the corresponding denoising algorithm. The initial image (noisy) is simulatneously presented in the interface, so that the user can easily compare the results, see figure 1.

3.1 Genome

The genome that will be evolved by the IEA is made of 7 real genes:

- 5 values to define the g function used in formula (1) for the horizontal/vertical wavelet coefficients: ranges are chosen in order to ensure that the general shape of the spectrum is respected, i.e. $\alpha_{min} \in [0, 0.5]$, $g(\alpha_{min}) \in [0, 0.1]$, $\alpha_{mod} > \alpha_{min}$, $\alpha_{mod} \in [0, 1]$, $\alpha_{max} > \alpha_{mod}$, $\alpha_{max} \in [0.5, 5]$, $g(\alpha_{max}) \in [0.9, 1]$
- the shift of the g function for the diagonal coefficients (range $[0, 0.5]$),
- the noise variance σ (range $[3.0, 40.0]$).

3.2 Fitness and User Interaction

The fitness function is given by the user via a cursor attached to each denoising result of the window. The range of values is $[-10, 10]$. A default value of "0" (indifferent) is set

[2] Diagonal coefficients are roughly related to a second derivative of the signal while the horizontal/vertical ones are related to a first derivative. Additionally, this behaviour has been verified in a simple experiment on a set of 80 sample images: the mean values of horizontal/vertical wavelet coefficients have been computed and compared to the mean value of the diagonal ones. In average the quotient (mean diagonal values) over (mean vertical/horizontal values) is 0.52 (standard deviation 0.16).

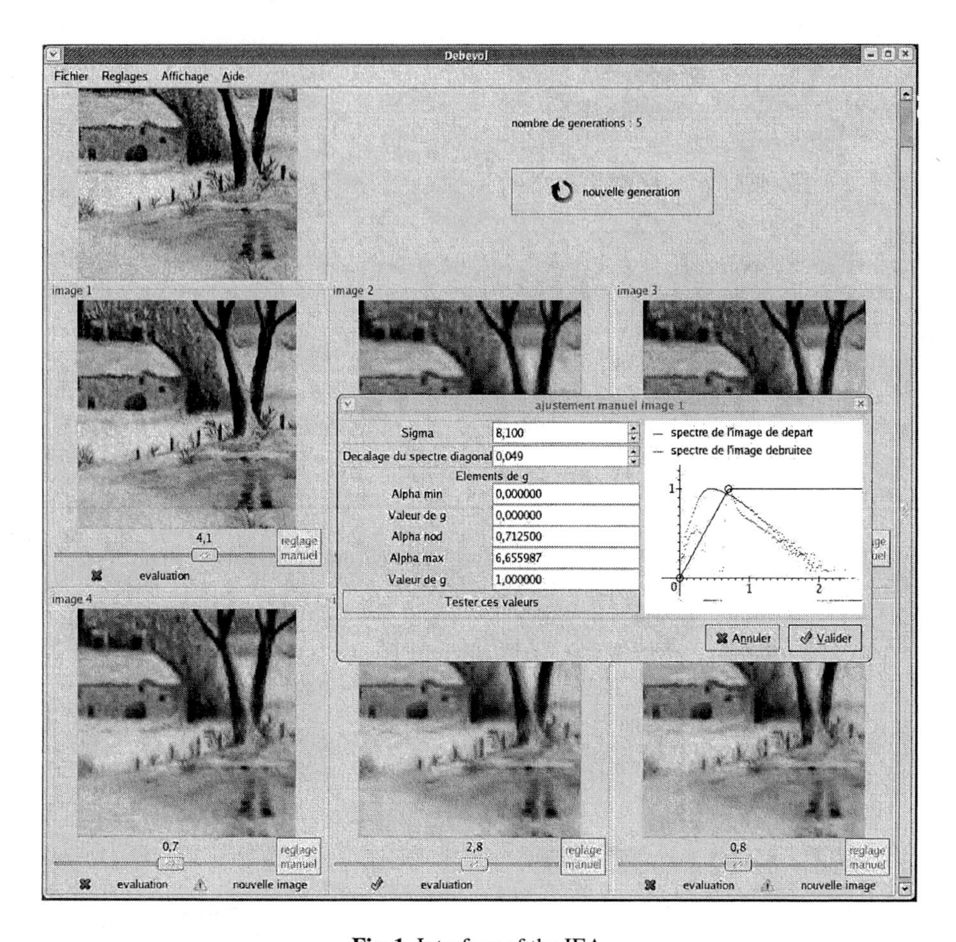

Fig. 1. Interface of the IEA

for each new genome (each image in the interface), the user can either increase, decrease or accept this value.

A sharing function is used in the selection process, in order to maintain diversity in the small population. The sharing is based on a weigthed L_2 distance computed on the real genes (parameters).

3.3 Genetic Engine

The replacement step of the algorithms consists in replacing the 3 worst individuals of the population by new ones.

- **Selection** is performed by tournament of size 3.
- **Crossover** is a barycentric crossover (the new individual is a weighted combination of his parents with a randomly chosen weight in $[0, 1]$).
- **Mutation** is an independent uniform perturbation of each gene value within a given range.

3.4 Interaction Interface

When a new image to be denoised is loaded in the interface, 6 images are displayed that correspond to 6 initial random genomes with values within the range of admissible values. The user interacts with the system either by affecting a notation to some images of the current popUlation (cursor : 10 is good, -10 is bad), or by directly modifying the values of the parameter via a specialised window that appear when clicking on "manual interaction": results are directly observable inside the window and on the associated spectra. The result can be included in the current population and evolve the same way as automatically generated individuals.

This direct interaction is thus to be considered as an additionnal genetic operator, fully driven by user interaction (this is a factor that reduces the "user fatigue", by letting him the possibility to be more or less directive in the evolution process). The production of a new generation is triggered by a "next generation" button. Experimentally, this direct interaction appears as a important component for the efficiency of the evolution, and it allows the user to gain intuition –to some extent– on the influence of some of the genome parameters.

The whole set of EA parameters (genome values ranges, probabilities and various parameters associated to the genetic operators) are tunable via a specialised window. The image display can also be tuned for large images, in order to be able to have a global view on the whole population with reduced resolution, and a precise view to look at the details of each denoised image.

4 Results

Figure 2 shows a result of the interactive denoising on a radar image, figures 3 and 4 present comparative results with a soft thresholding technique.

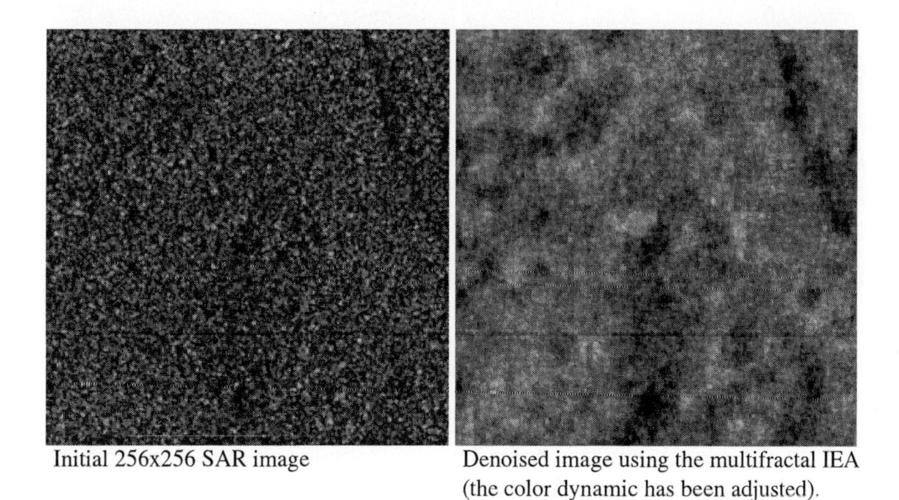

Initial 256x256 SAR image

Denoised image using the multifractal IEA (the color dynamic has been adjusted).

Fig. 2. Results on a radar image in 14 generations

Wavelet thresholding techniques consist in supressing the too small coefficients of the wavelet transform. There exist various thresholding procedures, the two most known ones are soft thresholding and hard thresholding. Hard thresholding consists in setting to 0 all the coefficients whose value are under a given threshold. Soft thresholding (also called wavelet shrinkage) lowers all the wavelet coefficient by a given quantity (the threshold), coefficients that are then negative are set to 0. These techniques were proposed by D.Donoho and I.Johnstone in the beginning of 1990-ies [15].

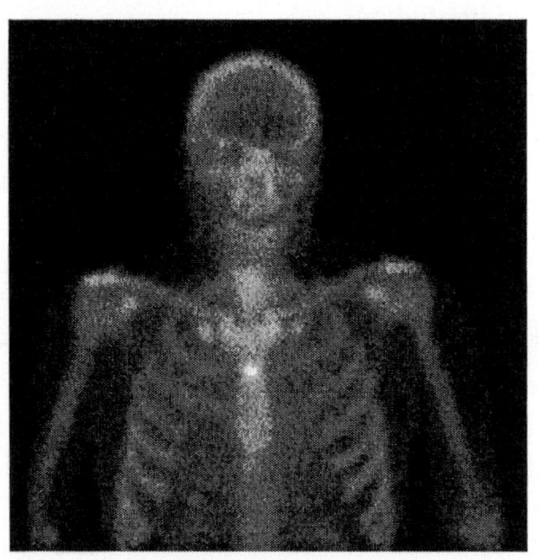

Initial noisy image

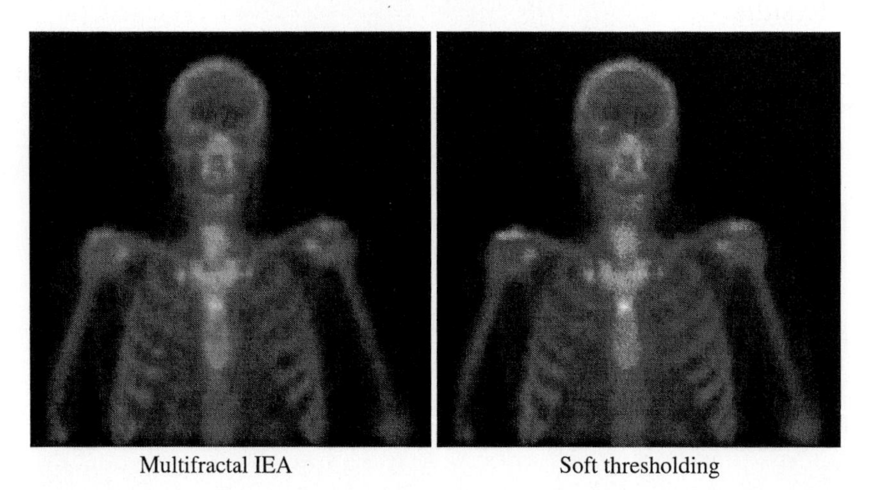

Multifractal IEA Soft thresholding

Fig. 3. Bones scintigraphy : 512x512 image

For fair comparison purpose, a non interactive version of the algorithm has been developed in order to test the multifractal denoising method: If the original noise-free

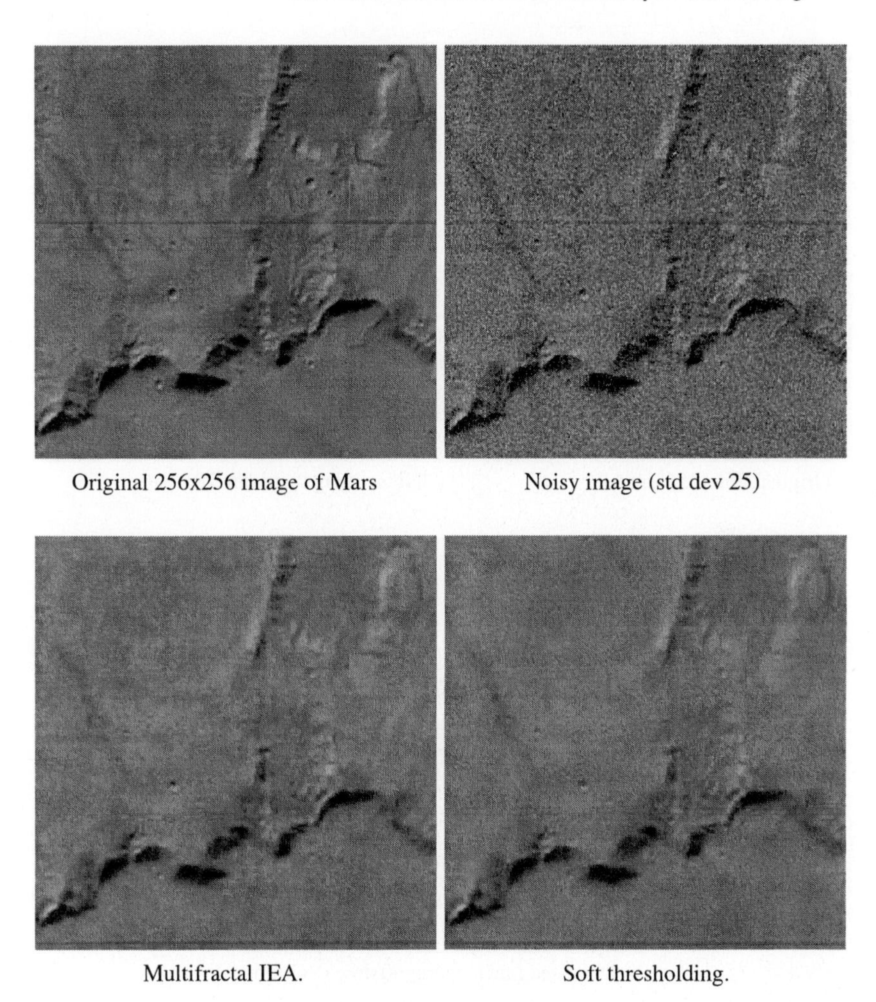

Original 256x256 image of Mars | Noisy image (std dev 25)

Multifractal IEA. | Soft thresholding.

Fig. 4. Comparative test with the interactive method

image is available, an automatic fitness can be computed as the L_2 distance between denoised and original images. Successive populations of parameter setting can thus be evolved without user interaction. This process allows to obtain objective comparison data, even if the L_2 distance does not always reflect the visual impression of denoising quality.

Figure 5 shows an automatic experiment on the Lena image with a white Gaussian noise of variance 20. The non-interactive EA has run during 300 generation. The initial distance between original and noisy image is 9119. The distance obtained with the off-line evolution after 300 generation was 4809. A soft thresholding with optimal threshold yield a distance of 4935, for a result that is visually very similar.

Original 256x256 image Noisy image (std dev 20)

Denoised image using off-line evolution.

Fig. 5. Automatic tests on Lena

5 Conclusions and Future Work

Complex image processing techniques are often necessary for specialized purpose and/or "difficult" images. However their usage often necessitates a parameter tuning stage, that may be very combinatorial. The solution we propose is to assist the user in the search of its optimal parameter setting via IEA. Experiments have been produced here for a denoising application, that prove the versatility of the approach, and the efficiency of results in comparison to other denoising techniques.

We intend to continue in this direction for other multifractal image analysis methods. A version of the presented software will be soon availble in the fraclab toolbox (see http://fractales.inria.fr).

References

1. Karl Sims. Artificial evolution for computer graphics. *Computer Graphics*, 25(4):319–328, July 1991.
2. K. Sims. Interactive evolution of dynamical systems. In *First European Conference on Artificial Life*, pages 171–178, 1991. Paris, December.
3. S.J.P. Todd and W. Latham. *Evolutionary Art and Computers*. Academic Press, 1992.
4. P. J. Angeline. Evolving fractal movies. In *Genetic Programming 1996: Proceedings of the First Annual Conference, John R. Koza and David E. Goldberg and David B. Fogel and Rick L. Riolo (Eds)*, pages 503–511, 1996.
5. Hideyuki Takagi and Miho Ohsaki. Iec-based hearing aids fitting. In *IEEE Int. Conf. on System, Man and Cybernetics (SMC'99)*, volume 3, Tokyo, Japan, Oct. 12-15 1999.
6. S. Kamohara, Hideyuki Takagi, and T. Takeda. Control rule acquisition for an arm wrestling robot. In *IEEE Int. Conf. on System, Man and Cybernetics (SMC'97)*, volume 5, Orlando, FL, USA, 1997.
7. Nicolas Monmarche, G Nocent, Gilles Venturini, and P. Santini. *Artificial Evolution, European Conference, AE 99, Dunkerque, France, November 1999, Selected papers,*, volume Lecture Notes in Computer Science 1829, chapter On Generating HTML Style Sheets with an Interactive Genetic Algorithm Based on Gene Frequencies. Springer Verlag, 1999.
8. Hideyuki Takagi. Interactive evolutionary computation : System optimisation based on human subjective evaluation. In *IEEE Int. Conf. on Intelligent Engineering Systems (INES'98)*, Vienna, Austria, Sept 17-19 1998.
9. Riccardo Poli and Stefano Cagnoni. Genetic programming with user-driven selection : Experiments on the evolution of algorithms for image enhancement. In *2nd Annual Conf. on Genetic Programming*, 1997.
10. Wolfgang Banzhaf. *Handbook of Evolutionary Computation*, chapter Interactive Evolution. Oxford University Press, 1997.
11. J. Chapuis and E. Lutton. Artie-fract : Interactive evolution of fractals. In *4th International Conference on Generative Art*, Milano, Italy, December 12-14 2001.
12. Jacques Lévy Véhel and Evelyne Lutton. Evolutionary signal enhancement based on hölder regularity analysis. In *EVOIASP2001 Workshop, Como Lake, Italy, Springer Verlag, LNCS 2038*, 2001.
13. J. Lévy Véhel. Introduction to the multifractal analysis of images. In Y. Fisher, editor, *Fractal Image Encoding and Analysis*. Springer Verlag, 1997.
14. Jacques Lévy Véhel and Pierrick Legrand. Bayesian multifractal signal denoising. In *IEEE ICASSP Conference*, 2003.
15. D.L Donoho. De-noising by soft-thresholding. *IEEE, Trans. on Inf. Theory*, 3(41):613–627, 1995.

Object Detection for Computer Vision Using a Robust Genetic Algorithm

Tania Mezzadri Centeno, Heitor Silvério Lopes,
Marcelo Kleber Felisberto, and Lúcia Valéria Ramos de Arruda

CPGEI/CEFET-PR, Av. 7 de setembro, 3165, CEP: 80230-000 Curitiba-PR-Brazil
{mezzadri, hslopes, mkf, arruda}@cpgei.cefetpr.br

Abstract. This work is concerned with the development and implementation of an image pattern recognition approach to support computational vision systems when it is necessary to automatically check the presence of specific objects on a scene, and, besides, to describe their position, orientation and scale. The developed methodology involves the use of a genetic algorithm to find known 2D object views in the image. The proposed approach is fast and presented a robust performance in several test instances including multiobject scenes, with or without partial occlusion.

1 Introduction

Detecting and describing how a specific object appears in an image using traditional matching procedures usually involves hard computational effort, particularly when rotation, translation and scale factors are necessary. In addiction, the complexity of the object recognition problem increases when it is possible to have the target object partially occluded [1].

This work reports the development and implementation of an object detection technique, using a robust genetic algorithm, in order to support a computational vision system for object detection and recognition. The final goal is the development of an automatic vision system for checking the presence of specific objects in a scene, describing their position, orientation and scale.

In the proposed approach, image processing techniques are used to extract properties from an object image in order to construct a compact object view representation. Then, a genetic algorithm manages the search for occurrences of the target object in an input image. Next, using the results found by the genetic search, the recognized objects are correctly extracted from the tested images. Thanks to the compact object representation form in our approach, few amounts of data are processed and, consequently, less computational effort is spent in the search process.

1.1 Related Work

Detecting and describing how a target object appears on images is addressed as an object verification problem by [1]. Eventually, matching procedures, such as template matching, morphological approaches and analogical methods offer feasible solutions

F. Rothlauf et al. (Eds.): EvoWorkshops 2005, LNCS 3449, pp. 284–293, 2005.

for this problem, but only if rotation and changes in the scales, as well as others geometrical transformations, are not considered. Otherwise, the matching procedure can easily fall into an exhaustive search [1].

Nevertheless, according to [2], genetic algorithms are especially appropriate for optimization in large search spaces, where exhaustive search procedures are not feasible. And a review in more recent works [3] [4] [5] [6] also reveals a tendency towards the employment of heuristic based search and optimization techniques, such as genetic algorithms, in order to improve image matching procedures for shape and object recognition tasks, with many advantages.

2 Methodology

The first steps in the proposed approach are the definition of the object model representation and the search objective. Next, the genetic algorithm parameters shall be detailed, with special emphasis in the encoding and the computation of the fitness function.

2.1 Object Model Representation

Here we describe the steps of the procedure to encode the image model (object view), representing the pattern, in such a way that a reference matrix ($\boldsymbol{M}_{ref}$) and two distances, dx and dy, fully represents the object model.

a. The image (Fig. 1a) is sliced by n horizontal lines evenly spaced by dy pixels, where $dy = number\ of\ image\ lines\,/\,(n+1)$.
b. Similarly as before, the image is sliced by n vertical lines evenly spaced by dx pixels, where $dx = number\ of\ image\ columns\,/\,(n+1)$.
c. Crossing lines define points that are named reference points, and are represented by P_{ij}, where $i = 0, 1, \ldots, n-1$, and $j = 0, 1, \ldots, n-1$.
d. The point P_0 at coordinates (x_0, y_0), with $x_0 = dx \times (n+1)/2$ and $y_0 = dy \times (n+1)/2$, is named the central reference point.
e. A function $f(P_{ij})$ assigns to each reference point the mean value of the pixels in the P_{ij} neighborhood as defined by the delimited region shown in Fig. 1b.

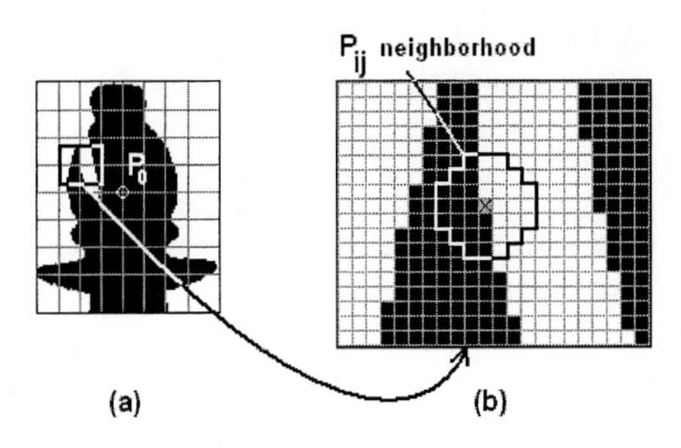

Fig. 1. (a) The image model sliced by horizontal and vertical lines with the central reference point (P_o); (b) Neighborhood of a generic point P_{ij}

f. All the computed $f(P_{ij})$ values are archived as a $n \times n$ matrix, called reference matrix ($\boldsymbol{M}_{ref}$), given by Equation 1.

$$M_{ref} = \begin{bmatrix} f(P_{11}) & \dots & f(P_{1n}) \\ \dots & f(P_{ij}) & \dots \\ f(P_{n1}) & \dots & f(P_{nn}) \end{bmatrix} \quad (1)$$

2.2 Genetic Algorithms

2.2.1 Individual Encoding

In our approach, five parameters are necessary to describe a simple individual: a threshold value t for the input image, a scale factor s, a rotation angle θ, and the pair of coordinates (x_0'; y_0') for the central reference point (P_0') in the input image. Hence, the k-th individual of the population will be represented by the 5-tuple ($x_0'{}_k$, $y_0'{}_k$, s_k , θ_k , t_k), whose ranges are shown in the Table 1.

Table 1. Ranges of parameters encoded in an individual k

Parameter	**Range**
Po' column	$0 \leq x_0'{}_k \leq 2047$
Po' lines	$0 \leq y_0'{}_k \leq 2047$
scale factor	$0.5 \leq s_k \leq 2.0$
rotation angle	$0 \leq \theta_k \leq 2\pi$ rad
threshold	$0 < t_k < 255$

For the tests, detailed later, we shall use images with no more than 2047 lines or columns. Therefore, 11 bits will be enough to represent the central reference point coordinates. For the sake of simplicity, we used the same length for the remaining parameters, t, s and θ. Therefore, an individual will have 55 bits long, leading to a search space $>10^{16}$.

2.2.2 Objective and Fitness Functions

Firstly, the input image is binarized, based on the threshold value t_k. For the pixels of the image with value less than t_k is assigned the value 0 (black) and, for the remainder pixels, 255 (white).

Based on the parameters encoded in an individual, the coordinates of a generic point (P_{ij}') in the input image is given by Equations 2 and 3, considering translation and rotation [7], respectively:

$$x_i' = x_0' + s.[(x_i - x_0).\cos\theta + (y_i - y_0).\sin\theta] \quad (2)$$

$$y_i' = y_0' + s.[(y_i - y_0).\cos\theta + (x_i - x_0).\sin\theta] \quad (3)$$

where: x_i and y_i are the coordinates of point P_{ij} relative to point P_O (in the object model image), and the x_i' and y_i' are the coordinates of the point P_{ij}' (projection of point P_{ij} in the input image). For the central reference point P_O', Equations 2 and 3 give $P_O' = (x_O'; y_O')$.

For instance, suppose that the individual represented by vector (23; 311; 6.28; 1, 56) has been generated as a possible solution for the search of the object model in Fig. 1. The reference points would be located as shown in Fig. 2a, for that input image. For better visualizing the result, Fig. 2b shows the object model projection over the input image matched with the proposed solution. Also, in Fig. 2a, it is shown that it is possible some reference points fall off the image limits. Such points are called invalid points, and the result of $f(P_{ij}')$ is represented by an asterisk. The total of invalid points is denoted by n*.

Once reference points have been located, a new reference matrix can be generated for the proposed solution, by following steps (e) and (f) of section 2.1. Such matrix, for the k-th individual ($x_O'_k$; $y_O'_k$; s_k ; θ_k ; t_k), is denoted by $\boldsymbol{M}_{ref}'$ (k).

Equation 4 is the objective function that measures the distance between $\boldsymbol{M}_{ref}'$ and $\boldsymbol{M}_{ref}$. It is based on the sum of the quadratic errors and, the small the distance $S_{SQE}(k)$, the better is a given solution k.

$$\mathrm{S_{SQE}}(k) = \sum_{i=0}^{n-1}\sum_{j=0}^{m-1}(g(M_{ref}i,j - M_{ref}^{'}(k)i,j))^2 \tag{4}$$

where:

$$\begin{cases} if \;\; f(P_{ij}^{'}) = * \Rightarrow g(Mi,j - M'(k)i,j) = 0 \\ Otherwise \Rightarrow g(Mi,j - M'(k)i,j) = Mi,j - M'(k)i,j \end{cases} \tag{5}$$

By default, the fitness function of a genetic algorithm deals with a maximization problem. Since it is searched for a given solution k ($x_O'_k$; $y_O'_k$; s_k ; θ_k ; t_k) that minimizes S_{SQE} (k), the fitness function can be defined by Equation 6:

$$fit(k) = \frac{S_{SQE-MAX} - S_{SQE} \cdot (k)}{S_{SQE-MAX}} \tag{6}$$

where:

$$S_{SQE-MAX} = (99 \cdot (n \times n - n^*))^2 \tag{7}$$

Here, nxn is the $\boldsymbol{M}_{ref}$ dimensions and n^* is the number of invalid points of reference. Since $S_{SQE\text{-}MAX} \geq S_{SQE}$ (k) for any feasible solution k, the fitness function values will be defined within the range [0...1].

A restriction to the maximum number of invalid points was incorporated in the fitness function definition to limit the maximum number of $n*$ occurrences. Therefore, the fitness function is redefined, as follows, considering the index "*WR*" as the fitness value with restriction:

$$\begin{cases} if \;\; n^* > \dfrac{n \times n}{2} \;\; \Rightarrow fit_{WR}(k) = fit(k) \\ \\ Otherwise \;\; \Rightarrow fit_{WR}(k) = 0 \end{cases} \tag{8}$$

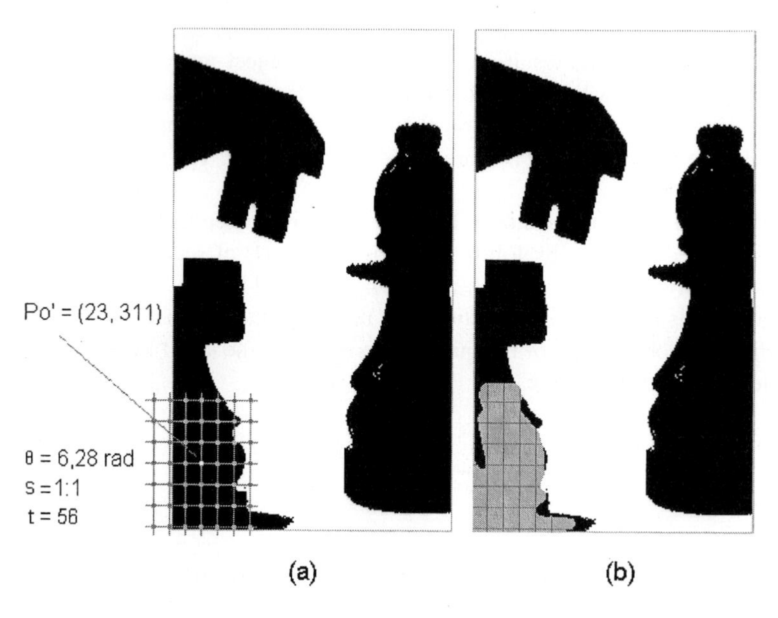

Fig. 2. (a) Reference points plotted in a test image, for individual (23; 311; 6.28; 1, 56); (b) The object model projection in the input image matched with the proposed solution

2.2.3 The Genetic Search

The genetic search starts with the random generation of the initial population of z individuals: $(x_0{'}_1 ; y_0{'}_1 ; s_1 ; \theta_1; t_1)$, $(x_0{'}_2 ; y_0{'}_2 ; s_2 ; \theta_2; t_2)$, … , $(x_0{'}_z ; y_0{'}_z ; s_z ; \theta_z; t_z)$. Each individual of the population is evaluated by the fitness function, and the probability of each individual to be selected for reproduction increases proportionally to its fitness value. An appropriated selection method [8] is used to select candidates for crossover and mutation. Such genetic operators will generate new individuals to form a new population. Some population generating strategies include elitism that means to copy some of the fittest individuals of the current population to the next one. Basically these same procedures are used to generate the following populations until some stop criterion is met. Usually a maximum number of generations or a satisfactory fitness value reached is used as stop criterion.

3 Implementation

The system was implemented in C++ object-oriented programming language on Microsoft Windows 2000 platform. For the GA implementation we used the GAlib genetic algorithm package [9]. Fig. 3 shows a block diagram that illustrates the information flow through the system components, which are described as follows:

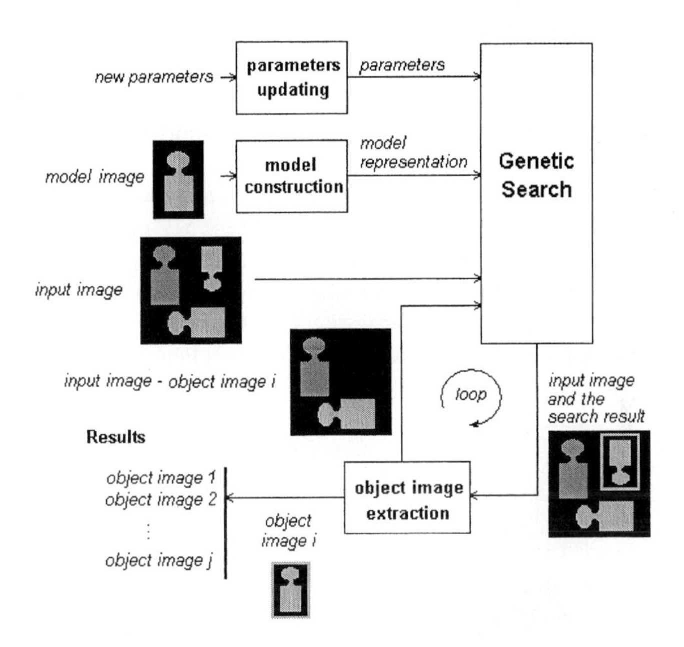

Fig. 3. Block diagram of the proposed system showing its components

a) The parameters updating block is used to modify running parameters of the GA.
b) The model construction block applies operators to the model image in order to construct the object model representation (M_{ref}, dx, dy), as explained before.
c) The genetic search block is the main part of this system. It receives 3 kinds of input data: (1) the GA parameters (population size, maximum number of generation, crossover and mutation probabilities and the selection method); (2) the model representation ($\boldsymbol{M}_{ref}$, dx, dy); and (3) the input image. As an output, the genetic search block gives the set of parameters (x_0' ; y_0' ; s ; θ; t) found by the search process, as well as its fitness value (fit_{WR}).
d) The object image extraction block makes a decision based on the fitness value of the current solution. For a fitness value less than a fixed fitness threshold (t_{FIT}), the solution is just discarded and the search stops. Otherwise, the object is extracted from the input image and saved as a new image denominated *object image i* (i=1… j). Besides, the result of the image subtraction (*input image – object image i*) is feed-backed to the genetic search block, for a new search.

The loop between the blocks genetic search and object image extraction allows the system to find further copies of the same object in the input image.

4 Experiments and Results

Several experiments were done using 25 images of chess pieces. An image of a model object, showing a single object view, was used as model. Fig. 4 shows the object model image used in the experiments and its reference matrix (*Mref*), generated by the model construction block (Fig. 3). In the 25 images used in the tests, the same object appears in different orientations, scales and positions (in the image plan), in multiobject scenes and partial occlusion occurrences.

During the GA run, the number of solutions generated and evaluated is $z \times g = 100 \times 1000 = 10^5$. The search space, considering a typical 512 × 512 pixels image, 3600 possible values for the rotation angle (0.0°, 0.1°, 0.2°,..., 359.9°), and 101 scale factors (0.500, 0.515, 0.530, ..., 2.000), is larger than 9.53×10^{10}. Consequently, the genetic algorithm can find an acceptable solution testing less than 0.0002 % of that huge search space, spending no more than 15 second by search (using a Pentium-IV 2.0 GHz).

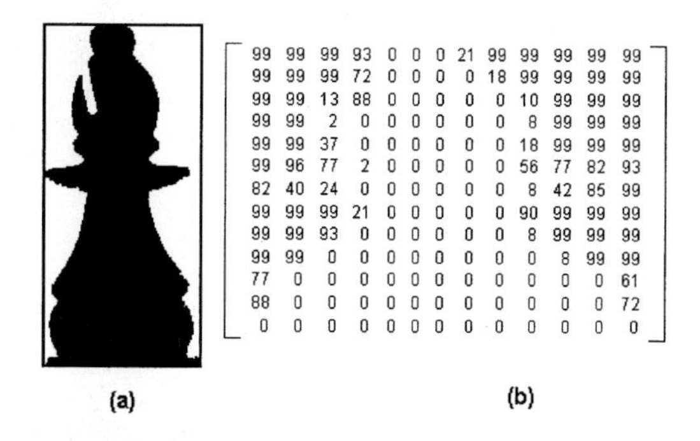

99 99 99 93 0 0 0 21 99 99 99 99 99
99 99 99 72 0 0 0 0 18 99 99 99 99
99 99 13 88 0 0 0 0 0 10 99 99 99
99 99 2 0 0 0 0 0 0 8 99 99 99
99 99 37 0 0 0 0 0 0 18 99 99 99
99 96 77 2 0 0 0 0 0 56 77 82 93
82 40 24 0 0 0 0 0 0 8 42 85 99
99 99 99 21 0 0 0 0 0 90 99 99 99
99 99 93 0 0 0 0 0 0 8 99 99 99
99 99 0 0 0 0 0 0 0 0 8 99 99
77 0 0 0 0 0 0 0 0 0 0 0 61
88 0 0 0 0 0 0 0 0 0 0 0 72
0 0 0 0 0 0 0 0 0 0 0 0 0
(a)
(b)

Fig. 4. (a) Object model image used in the experiments. (b) Corresponding reference matrix

Fig. 5. Images used in the experiments

Fig. 5 shows one of the 25 images used in the experiments, where the target object appears in three different locations and positions in the same image. Table 2 shows the search results for the experiments using the image of Fig. 5 as input. Each table column shows the object extracted from the original input image, followed by the parameters found in the genetic search, and the fitness value for the current solution.

These results show that all occurrences of the target object in Fig. 5 were found. Note that the partially occluded object (*object image #1*), in the right side of the image (Fig. 5), was also detected and correctly extracted from the image.

Table 2. Results obtained using the image of the Fig. 5 as an input

object image #1	object image #2	object image #3
P_0' = (528 , 150)	P_0' = (80, 113)	P_0' = (245 , 144)
S = 1.523926	s = 0.934570	s = 0.747803
Θ = 1.386719 rad	θ = 6.273985 rad	θ = 0.033748 rad
T = 140	t = 160	t = 118
fitness = 0.902464	fitness = 0.894589	fitness = 0.907113

Fig. 6. Multiobject image used in the experiments

Fig. 7. Matching of the first solution proposed by the GA, using as input the image in Fig.6

Table 3. Results obtained using as input the image shown in Fig. 7

object image #1	Discarded solution #1	discarded solution #2
P_0' = (555 , 152)	P_0' = (433 , 169)	P_0' = (321 , 138)
s = 0.983545	s = 0.983545	s = 1.222607
θ = 0.000000 rad	θ = 0.012272 rad	θ = 6.258645 rad
t = 128	t = 125	t = 238
fitness = 0.881313	fitness = 0.832775	fitness = 0.844708

Fig. 6 shows another test image used, where the target object appears in a multiobject scene. Fig. 7 shows the result of the object model matching with the solution proposed by the system.

Table 3 shows the results for the experiments using as input the image shown in Fig. 6. Note that the Table 3 shows the two next solutions that would be found by the system if the fitness threshold value t_{FIT} would decrease from 0.85 to 0.80. For the other tested images, the results presented the same accuracy without false alarms e no-detections.

5 Conclusions

This work proposed an object detection approach based on a genetic algorithm. The implemented system is the core of an upcoming computer vision system. This approach is useful in many cases where it is necessary to check the presence of specific 2D shapes, or a specific view of a 3D object, in a scene, and, further, to describe their position, orientation and scale.

During the reported experiments, the system displayed a good performance (without false alarms and no-detections) for all test sets. However, it was observed that, for most cases, the solutions presented by the system were not the optima, but something very close to the optimum value (see, for instance, Fig. 7).

The developed system is also computationally efficient, obtaining good solutions spending no more than 15 seconds by search (using a Pentium-IV 2.0 GHz). It is also important to point out that, it is possible to extend the system performance for reliably 3D object detection, by using several object views as object models.

Despite the good results, future work will focus accuracy, including the implementation of some local search strategy in order to fine-tune results from the genetic search, leading to even more accurate results.

Acknowledgements

This work has been partially supported by Agência Nacional do Petróleo (ANP) and Financiadora de Estudos e Projetos (FINEP) - ANP/MCT (PRH10-CEFET-PR).

References

1. Jain R, Kasturi R and Schunck BG (1995) *Machine Vision*. McGraw-Hill, New York
2. Simunic KS and Loncaric S (1998) A genetic search-based partial image matching. In: *Proc 2nd IEEE Int Conf on Intelligent Processing Systems*, pp. 119-122
3. Cordon O, Damas S. and Bardinet E (2003) 2D Image registration with iterated local search. In: Advances in Soft Computing – Engineering, Design and Manufacturing. Proc. of NSC7. pp. 1-10.
4. Han K P, Song K W, Chung E Y, Cho S J, Ha Y H (2001) Stereo matching using genetic algorithm with adaptive chromosomes. In: *Pattern Recognition*, vol. 34, pp. 1729-1740.
5. Yamany S M, Ahmed M N, Farag A A (1999) A new genetic based technique for matching 3D curves and surfaces. In: *Pattern Recognition*, vol. 32, pp. 1817-1820.
6. Bhanu B and Peng J (2000) Adaptive integrated image segmentation and object recognition. *IEEE T Syst Man Cy C* 30(4):427–441
7. Gonzalez RC and Wintz P (1987) *Digital Image Processing*, Addison-Wesley, Boston
8. Bäck T and Hoffmeister F (1991) Extended selection mechanisms in genetic algorithms. In: *Proc 4th Int Conf on Genetic Algorithms*, pp. 92–99
9. Wall M (2003) *GAlib A C++ Library of Genetic Algorithm Components* vs. 2.4.5. http://lancet.mit.edu/ga/.

An Evolutionary Infection Algorithm for Dense Stereo Correspondence

Cynthia B. Pérez[1], Gustavo Olague[1],
Francisco Fernandez[2], and Evelyne Lutton[3]

[1] CICESE, Research Center, Applied Physics Division,
Centro de Investigación Científica y de Educación Superior de Ensenada,
B.C., Km. 107 Carretera Tijuana-Ensenada, 22860, Ensenada, B.C., México
{cbperez, olague}@cicese.mx
[2] University of Extremadura, Computer Science Department,
Centro Universitario de Merida, C/Sta Teresa de Jornet, 38. 06800 Mérida, Spain
fcofdez@unex.es
[3] INRIA Rocquencourt, Complex Team,
Domaine de Voluceau BP 105. 78153 Le Chesnay Cedex, France
Evelyne.Lutton@inria.fr

Abstract. This work presents an evolutionary approach to improve the infection algorithm to solve the problem of dense stereo matching. Dense stereo matching is used for 3D reconstruction in stereo vision in order to achieve fine texture detail about a scene. The algorithm presented in this paper incorporates two different epidemic automata applied to the correspondence of two images. These two epidemic automata provide two different behaviours which construct a different matching. Our aim is to provide with a new strategy inspired on evolutionary computation, which combines the behaviours of both automata into a single correspondence process. The new algorithm will decide which epidemic automata to use based on inheritance and mutation, as well as the attributes, texture and geometry, of the input images. Finally, we show experiments in a real stereo pair to show how the new algorithm works.

1 Introduction

Dense stereo matching is one of the main and most interesting problems to solve in stereo vision. It consists in determining which pair of pixels projected on at least two images, belongs to the same physical 3D point [1]. The correspondence problem has been one of the main subjects in computer vision and is clear that these matching tasks are to be solved by computer algorithms. Currently, there is no general solution to the problem and it is also clear that successful matching by computer can have a large impact on computer vision [2]. The matching problem has been considered the hardest and most significant problem in computational stereo [3, 4]. The difficulty is related to the inherent ambiguities being produced during the image acquisition concerning the stereo pair; i.e., geometry, noise, lack of texture, occlusions, and saturation. Occlusion is the cause of complicated

F. Rothlauf et al. (Eds.): EvoWorkshops 2005, LNCS 3449, pp. 294–303, 2005.

problems in stereo matching, especially when there are narrow objects with large disparity and optical illusion in the scene [5, 6, 7]. Moreover, classical techniques are limited and fail in the general case, due to the complexity and nature of the problem. This is the main reason why we are considering a new approach based on artificial life and evolutionary methods that we called evolutionary infection algorithm. The infection algorithm presented in [8] uses an epidemic automaton that propagates the pixel matches as an infection over the whole image with the purpose of matching the contents of two images. Finally, the algorithm provides the rendering of 3D information allowing the visualization of the same scene from novel viewpoints. In our past work, we had four different epidemic automata in order to observe and analyse the behaviour of the matching process. The better results we have obtained were those related to the case of 47% and 99% of effort savings in the correspondence process. The first case represents geometrically a good image with a moderate percentage of computational effort saving, see Fig. 7a. The second case represents a high percentage of automatically allocated pixels producing an excellent percentage of computational effort saving, with an acceptable image quality, see Fig. 7c. Our current work aims to improve the results based on a new algorithm that uses concepts from evolution; such as: inheritance and mutation. We want to combine the best of both epidemic automata in order to obtain a high computational effort saving with an excellent image quality.

This paper is organized as follows. The next section describes the nature of the correspondence problem. Section 2 introduces the new algorithm giving emphasis to the explanation on how the evolution was applied to the epidemic

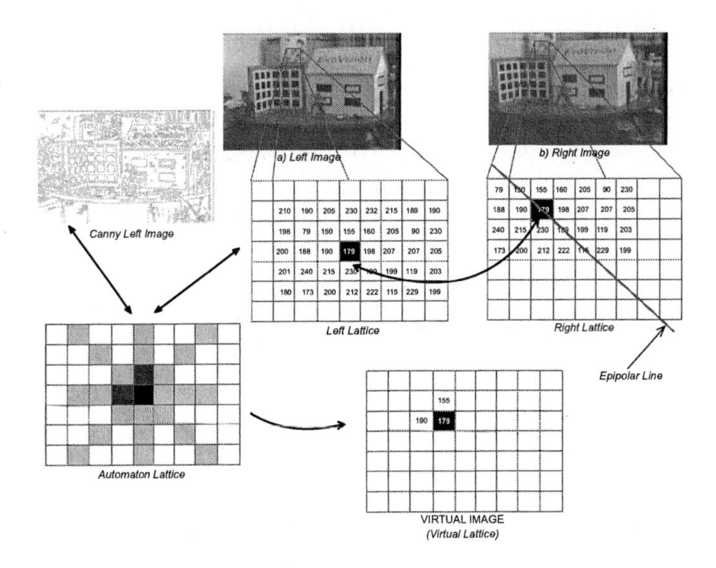

Fig. 1. Representation of the interactions among the five lattices used by the infection algorithm. A pixel in the left image is related to the right image, using several lattices such as: automaton lattice, canny image and virtual image lattice during the correspondence process

automata. Finally, the section 3 shows the results of the algorithm illustrating the behaviour, performance, and quality of the evolutionary infection algorithm.

1.1 Problem Statement

Computational stereo studies how to recover the three dimensional characteristics of a scene from multiple images taken from different viewpoints. A major problem in computational stereo is how to find the corresponding points between a pair of images, which is known as the correspondence problem or stereo matching. In this work the images are taken by a moving camera, with the hypothesis that a unique three-dimensional physical point is projected into a unique pair of image points, see Fig. 1. In order to solve the problem in a more efficient manner, several constraints and assumptions regarding occlusions, lack of texture, saturation, or field of view are exploited. The movement between the images is a translation with a small rotation along the x, y and z axes respectively: $T_x = 4.91\ mm$, $T_y = 114.17\ mm$, $T_z = 69.95\ mm$, $R_x = 0.84°$, $R_y = 0.16°$, $R_z = 0.55°$. Fig. 1 shows also five lattices that we have used in the evolutionary infection algorithm. The first two lattices correspond to the images acquired by the stereo rig. The third lattice is used by the epidemic cellular automata in order to process the information that is being computed. The fourth lattice corresponds to the reprojected image, while the fifth lattice is used as a database in which we save information related to contours and texture.

2 The Evolutionary Infection Algorithm

The infection algorithm is based on the concept of natural virus for searching the correspondences between real stereo images. The purpose is to find all existing corresponding points in stereo images while saving the maximum number of calculations and maintaining the quality of the reconstructed data.

The motivation to use what we called an infection algorithm is based on the following: when we observe a scene, we do not observe everything in front of us. Instead, we focus our attention in some regions which keeps our interest on the scene. As a result, it is not necessary to analyse each part of the scene in detail. The infection algorithm attempts to *"guess"* some parts of the scene through a process of propagation based on artificial epidemics with the purpose of saving computational time. The mathematical description of the infection algorithm is presented in [9]. This paper introduces the idea of evolution within the infection algorithm using the concepts of inheritance and mutation in order to achieve a balance between exploration and exploitation. As we could see in the paper, the idea of evolution is rather different from the traditional genetic algorithm. Concepts like an evolving population are not considered in the evolutionary infection algorithm. Instead, we incorporate aspects such as inheritance and mutation to develop a dynamical matching process. In order to introduce the new algorithm let us define some notations:

```
1. BEGIN
2. Corner detection of the left image with O&H or K&R detectors.
3. Calculate Projection Matrices MM1 and MM2 of the Left and Right image.
4. Calculate Fundamental Matrix (F) Left to Right.
5. Coding of rules matrices 47% and 99%.
6. Process the left image with the canny edge detector.
7. Instantiate epidemic automata to Healthy(Not-Explored) state.
8. Obtain the left image with canny edge detector.
9. While the number of Immune (Automatic) and Sick (Explored) states is
   different at time t and t+1 step.
10.DO
11. {  FOR col=6 to N
12.      { FOR row=6 to M
13.          neighbors=countKindNeighbor(row,col)
14.          pos=statesPosition(row,col)
15.          stateCell=automata[row][col].cell
16.          actionFam= critFam(stateCell,neighbors,pos,probFam)
17.          actionCont= critCont(stateCell,probContour)
18.          actionMut= critMut(stateCell, probMutation)
19.          IF(( probFam > probContour && probFam>probMutation)
20.               action=actionFam
21.          IF(( probContour > probFam && probContour>probMutation)
22.               action=actionCont
23.          IF(( probMutation > probFam && probMutation>probContour)
24.               action=actionMut
25.          SWITCH(action)
26.          { 1: EXPLORED
27.            2: PROPOSED
28.            3: Analyse the inclination of the epipolar line
29.               IF the inclination = to the inclination of at least three
                     neighbors
30.               {  AUTOMATICALLY ALLOCATED }
31.               ELSE
32.               {  EXPLORED } } } } }
33. Virtual image obtained.
34. END
```

Fig. 2. Pseudo-code for the evolutionary infection algorithm

Definition 1. Epidemic cellular automata: Our epidemic cellular automata could be formally represented as a quadruple $E = (S, d, N, f)$ where $|S| = 5$ is a finite set composed of 4 states and the wild card $(*)$, $d = 2$ a positive integer, $N \subset Z^d$ a finite set, and $f_i : S^N \rightarrow S$ an arbitrary set of (local) functions, where $i = \{1, \ldots, 14\}$. The global function $G_f : S^L \rightarrow S^L$ is defined by $G_f(c_v) = f(c_v + N)$. Also, it is useful to mention that S is defined by the following sets:

- $S = \{\alpha_1, \varphi_2, \beta_3, \varepsilon_0, *\}$ a finite alphabet,
- $S_f = \{\alpha_1, \beta_3\}$ is the set of final output states,
- $S_0 = \{\varepsilon_0\}$ is called the initial input state.

Each epidemic cellular automaton has 4 states that are defined as follows:

- *Explored*, representing the cells that have been infected by the virus (it refers to the pixels that have been computed in order to find their matches).
- *Not-explored*, representing the cells which have not been infected by the virus (it refers to the pixels which remain in the initial state).
- *Automatically allocated*, representing the cells which cannot been infected by the virus. This state represents the cells which are immune to the disease (it refers to the pixels which have been confirmed by the algorithm in order to automatically allocate a pixel match).
- *Proposed*, representing the cells which have acquired the virus with a probability of recovering from the disease (it refers to the pixels which have been guessed by the algorithm in order to decide later the better match based on local information).

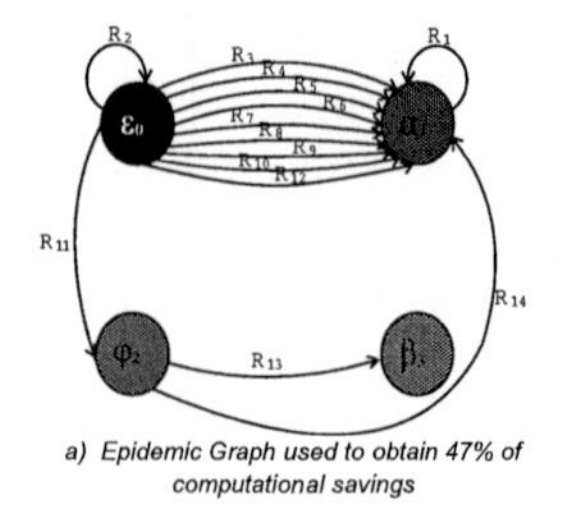

a) Epidemic Graph used to obtain 47% of computational savings

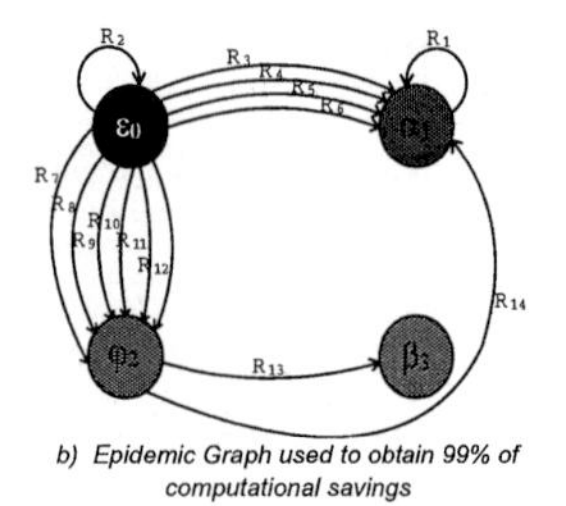

b) Epidemic Graph used to obtain 99% of computational savings

Fig. 3. Evolutionary epidemic graphs used in the infection algorithm

2.1 Infection Algorithm with an Evolutionary Approach

In previous work, we defined four different epidemic cellular automata from which we detect two epidemic graphs that provide singular results in our experiments, see Fig. 3. One epidemic cellular automaton produces a 47% of effort saving while the other a 99% of effort saving. These automata use a set of transformations expresed by a set of rules grouped within a single graph. Each automaton transforms a pattern of discrete values over a spatial lattice. A whole different behaviour is achieved by changing the relationships between the four states using the same set of rules. Each rule represents a relationship which produces a transition based on local information. These rules are used by the epidemic graph to control the global behaviour of the algorithm. In fact, the evolution of cellular automata is governed typically not by a function expressed in closed-form, but by a "rule table" consisting of a list of the discrete states that occur in an automata together with the values to which these states are to be mapped in one iteration of the algorithm.

The goal of the search process is to achieve a good balance between two rather different epidemic cellular automata in order to combine the benefit of each automaton. Hence, our algorithm not only finds a match within the stereo pair, but it provides an efficient and general process using geometric and texture information. It is efficient because the final image combine the best of each partial image within the same amount of time. It is also general because the algorithm could be used with any pair of image with little additional effort to adapt it.

Our algorithm attempts to provide a remarkable balance between the exploration and exploitation of the matching process. Two cellular automata were selected because each one provides a particular characteristic from the exploration and exploitation standpoint. The 47% epidemic cellular automaton, we called A, provides a strategy which exploits the best solution. Here, best solution refers to areas where matching is easier to find. The 99% epidemic cellular automaton, we called B, provides a strategy which explores the search space when the matching is hard to achieve.

The pseudo-code for the evolutionary infection algorithm is depicted in figure 2. First, we calculate the projection matrices MM1 and MM2 related to the left and right image, through a process of calibration. Then, two sets of rules are defined A=46% and B=99%. The sets of rules contain information about the configuration of the pixels in the neighborhood. Next, we built a lattice with the contour and texture information, that we called canny left image. Thus, we iterate the algorithm while the number of pixels with immune and sick states is different between time t and $t+1$. Each pixel is evaluated according to a decision that is made based on three criteria as follows:

1. The decision of using A or B is weighted considering the current evaluated pixels in the neighborhood. In this way inheritance is incorporated within the algorithm.
2. The decision is also made based on the current local information (texture and contour). Thus, enviromental adaptation is contemplated as a driving force to the dynamical matching process.
3. A probability of mutation which could change the decision of using A or B is computed. This provides the system with the capability of stochastically adapting the dynamic search process.

In this way, an action is activated producing a path and sequence around the initial cells. When the algorithm needs to execute a rule to evaluate a pixel, it calculates the corresponding epipolar line using the fundamental matrix information. The correlation window is defined and centered with respect to the epipolar line when the search process is started. Thus during the search process, o sort of inheritance is used to decide which cellular automata to apply. Finally, our algorithm produces an image that is generated combining the two epidemic cellular automata.

2.2 Transitions of Our Epidemic Automata

The transitions of our epidemic automata are based on a set of rules. Each relationship between the four states is represented as a transition based on a local rule, which as a set is able to control the global behaviour of the algorithm. Each epidemic graph has 14 transition rules that we divide in three classes: basic rules, initial structure rules, and complex structure rules. Each rule could be represented as a predicate that encapsulates an action allowing a change of state on the current cell based on their neighborhood information. The basic rules relate the obvious information between the initial and explored states. The initial structure rules consider only the spatial set of relationships between the close neighborhood. The complex structure rules consider not only the spatial set of relationships between the close neighborhood, but also those within the external neighborhood. The next section explains how each rule works according to the above classification. The basic rules correspond to rules 1 and 2. The initial structure is formed by the rules 3,4,5 and 6. Finally, the complex structure rules correspond to the rest of the rules. The basic and initial structure rules until this moment have not been changed. The rest of the rules are easily modified in order to produce different behaviours and a certain percentage of computational effort saving, see Fig. 3. The most important epidemic rules related to the case of 47% are explained next.

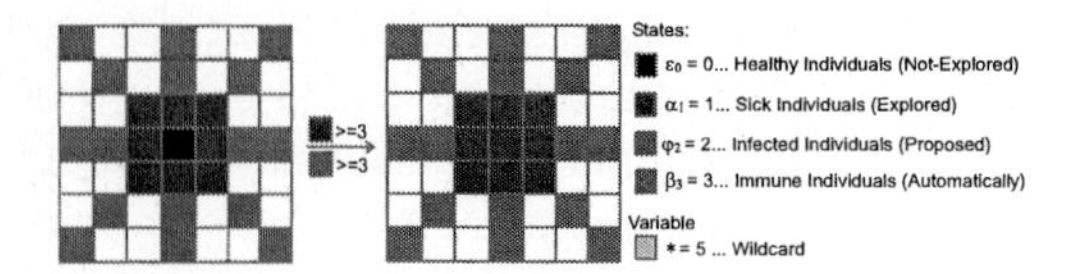

Fig. 4. Rule 7. The epidemic transition indicates that it is necessary to have at least three Sick (Explored) individuals on the closed neighborhood and three Immune (Automatically allocated) individuals on the external neighborhood in order to change the central cell

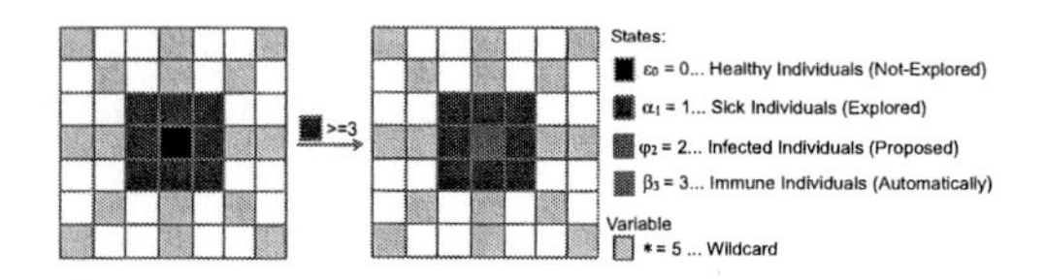

Fig. 5. Rule 11. This epidemic transition represents the quantity of Infected (Proposed) individuals in order to obtain the Immune (Automatically Allocated) individuals. In this case, if there are three Sick (Explored) individuals in the closed neighborhood, then the central cell changes to an Infected (Proposed) state

- **Rules 3,4,5 and 6**. The infection algorithm begins the process with the nucleus of infection around the whole image. The purpose of creating an initial structure in the matching process is to explore the search space in such way that the information is distributed in several process. Thus, the propagation of the matching is realized in a higher number of directions from the central cell. We use these rules at the initialisation of the process only.
- **Rule 7 and 8**. These rules ensure the evaluation of the pixels in a region where exist Immune (Automatically Allocated) individuals, see Fig. 4. The figure of the rule 8 is similar to 7. The main purpose of these rules is to control the quantity of immune individuals within a set of regions.
- **Rule 11**. This rule generates the Infected (Proposed) individuals in order to obtain later a higher number of Immune (Automatically Allocated) individuals. If the central cell is on Healthy state (Not-Explored) and there are at least three Sick individuals (Explored) in the closed neighborhood, then, the central cell is Infected (Proposed).
- **Rules 12 and 14**. The reason for these transitions is to control the Infected (Proposed) individuals. If we have at least three Infected (Proposed) individuals in the closed neighborhood, the central cell is evaluated.
- **Rule 13**. This rule is one of the most important epidemic transition rules because it indicates the computational effort saving without any computation of individual pixels during the matching process. This is the reason why we called it immune or automatically allocated. Rule 13 can be summarized as follows: if the central cell is Infected (Proposed) and if there are at least three Sick (Explored) individuals in the closed neighborhood; then, we know automatically that the corresponding pixel in the right image is made

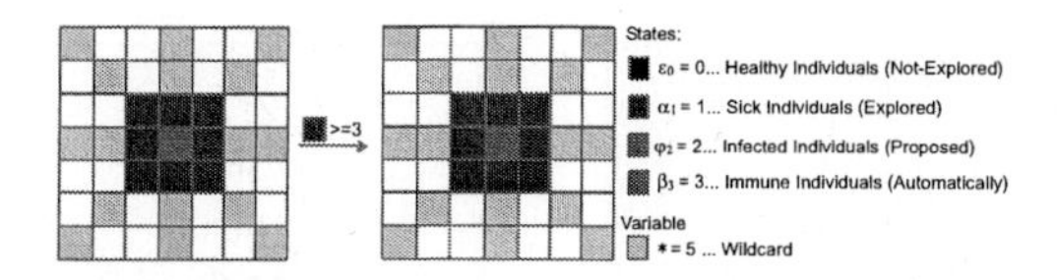

Fig. 6. Rule 13. This transition indicates when the Infected (Proposed) individuals will change to Immune (Automatically Allocated) individuals

without any computation. The number of Sick (Explored) individuals can be changed depending on the percentage of computational savings that we need.

Obviously, if we modified some of the complex structure rules from the case 47% we could obtain up to 99% of computational effort saving. The basic and initial structure rules do not change the coding, implying that the central cell remains without changes. The complex structure rules, except rule 14, are modified in two ways: the central cell state is modified according to the information within the neighborhood and the number of cells involved.

3 Experimental Results and Conclusions

We have tested the infection algorithm with an evolutionary approach on a real stereo pair of images. The infection algorithm was implemented under the Linux operating system on an Intel Pentium 4 at 2.0Ghz with 256Mb of RAM. We have used libraries programmed in C++, designed specially for computer vision, called V*X*L (Vision *X* Libraries). We have proposed to improve the results obtained by the infection algorithm through the implementation of an evolutionary approach using inheritance and mutation operations. The idea was to combine the best of both epidemic automata 47% and 99%, in order to obtain a high computational effort saving together with an excellent image quality. We used knowledge based on geometry and texture in order to decide during the correspondence process, which epidemic automata is better to apply during the evolution of the algorithm.

Figure 7 shows a set of experiments where the epidemic cellular automata was changed in order to modify the behaviour of the algorithm and to obtain a better virtual image. Fig. 7a is the result of obtaining 47% of computational effort savings, while Fig. 7b is the result of obtaining 70% of computational effort savings and Fig. 7c shows the result to obtain 99% of computational effort savings. Fig. 7d presents the result that we obtain with the new algorithm. Clearly, the final image shows how the algorithm combines both epidemic cellular automata. However, the quality of the final image depends on the texture and contours of the stereo pair. We observe that the geometry is preserved with a nice texture reconstruction. We also observe that the new algorithm spends about the same time employed by the 70% epidemic cellular automaton with a slightly better texture result. Figure 8a shows the behaviour of the evolution-

a) *Final view with 47% savings*

b) *Final view with 70% savings*

c) *Final view with 99% savings*

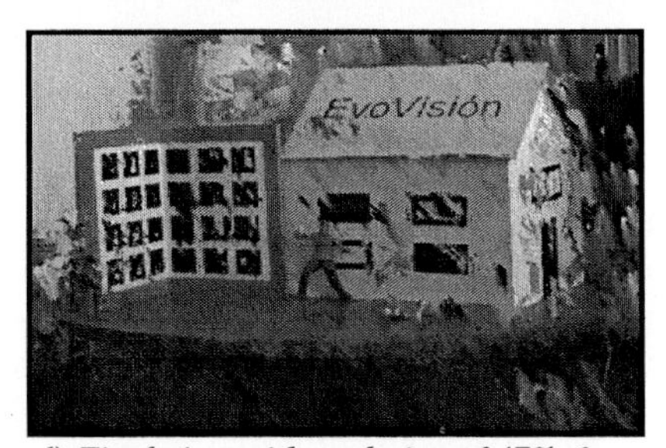

d) *Final view with evolution of 47% & 99% epidemic automata*

Fig. 7. Set of different experiments where epidemic cellular automata were change

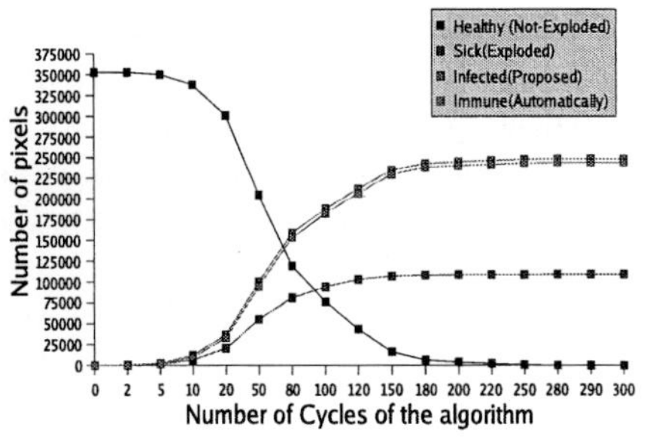

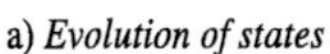

a) *Evolution of states*

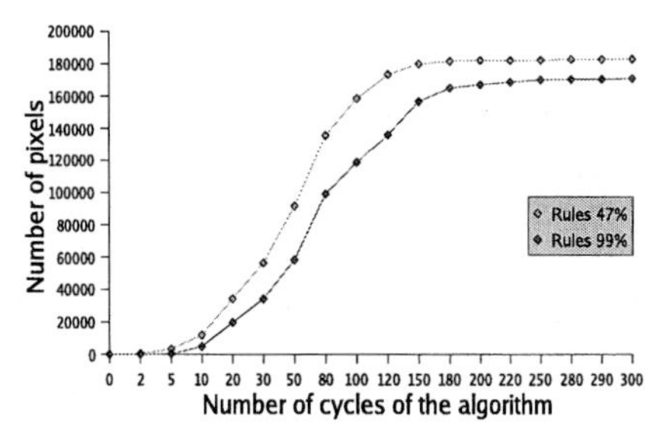

b) *Evolution of epidemic automatas*

Fig. 8. Evolution of the epidemic cellular automata during the dense correspondence process

ary infection algorithm that corresponds to the final result of Fig. 7d. Figure 8b describes the behaviour of the two epidemic cellular automata during the correspondence process. In the near future, we expect to use the Evolutionary Infection Algorithm in the search of novel viewpoints.

Acknowledgments. This research was funded by Ministerio de Ciencia y Tecnología, SPAIN, research project number TIC2002-04498-C05-01, and by CONACyT and INRIA through the LAFMI project. First author supported by scholarship 171325 from CONACyT.

References

1. Lionel Moisan and Bérenger Stival. A Probabilistic Criterion to Detect Rigid Point Matches Between Two Images and Estimate the Fundamental Matrix. International Journal of Computer Vision. Vol.57(3):201-218. 2004.
2. Myron Brown, Darius Burschka and Gregory Hager. Advances in Computational Stereo. IEEE Trans. on Pattern Analysis and Machine Intelligence. Vol.25(8):993-1008. 2003.
3. Umesh Dhond and K. Aggarwal. Structure from Stereo. IEEE Transactions on Systems and Man and Cybernetics. Vol.19(6):1489-1509. 1989.
4. Barnard S, Fischler M."Computational stereo". ACM, Computer Surveys, Vol. 14(4):553-572. 1982.
5. Quiming Luo, Jingli Zhou, Shengsheng Yu, Degui Xiao. "Stereo Matching and occlusion detection with integrity and illusion sensitivity". Pattern Recognition Letters. 24(2003):1143-1149.
6. Stan Birchfield and Carlo Tomasi. Depth Discontinuities by Pixel-to-Pixel Stereo. International Journal of Computer Vision. Vol.35(3):269-293. 1999.
7. C.Lawrence Zitnick and Takeo Kanade. A Cooperative Algorithm for Stereo Matching and Occlusion Detection. IEEE Trans. on Pattern Analysis and Machine Intelligence. Vol.22(7):675-684. July 2000.
8. Gustavo Olague, Francisco Fernández, Cynthia B. Pérez, and Evelyne Lutton. "The infection Algorithm: An Artificial Epidemic Approach to Dense Stereo Matching. Parallel Problem Solving from Nature VIII. X. Yao et al.(Eds.): LNCS 3242, pp.622-632, Springer-Verlag, Birmingham, UK, September 18-22, 2004.
9. Gustavo Olague, Francisco Fernández, Cynthia B. Pérez, and Evelyne Lutton. "The infection Algorithm: An Artificial Epidemic Approach for Dense Stereo Correspondence." Available as CICESE research report No.25260

Automatic Image Enhancement Driven by Evolution Based on Ridgelet Frame in the Presence of Noise

Tan Shan, Shuang Wang, Xiangrong Zhang, and Licheng Jiao

National Key Lab for Radar Signal Processing and Institute of Intelligent, Information Processing, Xidian University, 710071 Xi'an, China
tanshan5989@yahoo.com.cn

Abstract. Many conventional and well-known image enhancement methods suffer from a tendency to increase the visibility of noise when they enhance the underlying details. In this paper, a new kind of image analysis tool — ridgelet frame is introduced into the arena of image enhancement. We design an enhancement operator with the advantages that it not only enhance image details but also avoid the amplification of noise within source image. Different from those published previously, our operator has more parameters, which results in more flexibility for different category images. Based on an objective criterion, we search the optimal parameters for each special image using Immune Clone Algorithm (ICA). Experimental results show the superiority of our method in terms of both subjective and objective evaluation.

1 Introduction

Human visual system has certain limitations in perceiving information carried by images. For example, it cannot detect brightness contrasts that are lower than a certain contrast sensitivity threshold. Image enhancement is intended to convert images into a form that makes use of capabilities of human visual system to perceive information to their highest degree.

The conventional image enhancement methods such as histogram equalization usually suffer from a tendency to increase the visibility of noise at the same time as they enhance the visibility of the underlying signal. Practically, the imaging techniques used in some important application areas such as medical imaging tend to result in images with poor contrast along with relatively high noise levels. It is in such application areas that there is the greatest need for improved enhancement methods, capable of enhancing often-weak signal features against a background of relatively high noise [1].

Considerable success has already been achieved in the development of transform-domain based image enhancement methods with noise suppression [1-5]. The transform-domain image enhancement methods commonly begin with taking a transform to the image. Then, a non-linear operator is applied to the transform domain coefficients. Finally, the enhanced image is obtained when the inverse transform is applied. A typical non-linear operator for image enhancement is shown in Fig.1. The transform-domain coefficients of absolute magnitude below a lower threshold $\mathbf{T}_1$ are suppressed. The coefficients whose absolute magnitude exceeds $\mathbf{T}_1$, but fall below a

F. Rothlauf et al. (Eds.): EvoWorkshops 2005, LNCS 3449, pp. 304–313, 2005.

second threshold $\mathbf{T}_2$ are subject to a uniform gain. For coefficients whose absolute magnitude exceeds $\mathbf{T}_2$ the gain diminished with increasing coefficients magnitude.

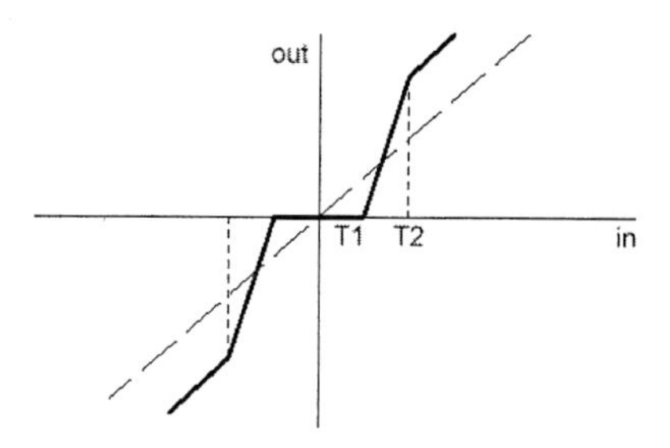

Fig. 1. A typical non-linear operator for transform-domain image enhancement

Commonly, the used transformation compresses the essential information into relatively few, large coefficients that capture important features such as points and edges, whereas the input noise is spread out evenly over all coefficients. Practically, the threshold $\mathbf{T}_1$ corresponds to the noise level in the source image. As a result, image enhancement methods using the non-linear operator such as that in Fig.1 have the advantages that they not only enhance image details but also avoid the amplification of noise within source images.

When the source images were contaminated with additive Gaussian noise, the quality of enhanced images resulting from transform-domain based enhancement methods is essentially determined by the nonlinear approximation ability of the used transformation. In this paper, we introduce a new kind of image analysis tool, namely, ridgelet frame into the arena of image enhancement. Ridgelet frame can effectively represent edges in images and recover well the faint edges in the presence of noise [6]. The main purpose of image enhancement is to increase the visibility of faint details including point and edges in source image. Hence, ridgelet frame is a good candidate for the task of image enhancement especially in the presence of noise. We design a nonlinear operator in the ridgelet frame domain for image enhancement. The proposed operator has *six* parameters that control the quality of the enhanced images. Compared with those previously published, our operator has more parameters which results in more flexibility, considering that the optimal parameters of any enhancement operator are image dependent and no one with fixed parameters is suitable for all category of images. We introduce an objective evaluation criterion to estimate the quality of enhanced images, and then, Immune Clone Algorithm (ICA) [7] [8], which is an optimization algorithm inspired by Artificial Immune System is used to obtain the optimal parameters for a special image. The whole process is completed automatically without the human interpreter.

This paper is organized as follows. In section 2, the ridgelet frame is reviewed briefly and a nonlinear operator in ridgelet frame domain for image enhancement is designed, then, an objective criterion is present also. In section 3, the ICA is used to obtain the optimal parameters of the proposed operator. Results are shown in section 4. Finally, concluding remarks are given in section 5.

2 Ridgelet Frame Based Image Enhancement

2.1 Ridgelet Frame

In the area of computed tomography, it is well known that there exists an isometric map from Radon domain $\Re$ to spatial domain $L^2(R^2)$. Hence, one can construct a tight frame in Radon domain first, then, the image of the tight frame under the isometric map constitutes a tight frame also in $L^2(R^2)$.

The construction of tight frame in Radon domain can start with an orthonormal basis obtained from tensor product of two 1-D wavelet orthonormal systems respectively for $L^2(R)$ and $L^2([0,2\pi))$. Let w''_λ ($\lambda \in \Lambda$) denote an orthonormal system of this kind in $L^2(R \otimes [0,2\pi))$, where Λ is the collection of index λ. Let $w'_\lambda := 2\sqrt{\pi} w''_\lambda$ and define orthoprojector $P_\Re$ from $L^2(R \otimes [0,2\pi))$ to Radon domain by

$$(P_\Re F)(t,\theta) = (F(t,\theta) + F(-t,\theta+\pi))/2 , \tag{1}$$

where $F \in L^2(R \otimes [0,2\pi))$ .Then, applying $P_\Re$ on w'_λ, we obtain

$$w_\lambda := P_\Re(w'_\lambda) = (\frac{I + T \otimes S}{2}) w'_\lambda = 2\sqrt{\pi} P_\Re(w''_\lambda) , \tag{2}$$

where operator T is defined by $(Tf)(t) = f(-t)$ and S is defined by $(Sg)(\theta) = g(\theta+\pi)$. Then, it turned out that w_λ is a tight frame with frame bound 1 in Radon domain. As mentioned above, one exactly obtains a tight frame by mapping the one in Radon domain $\Re$ to spatial domain $L^2(R^2)$. And the resulting tight frame in $L^2(R^2)$ is with the same frame bound 1 as its counterpart in Radon domain $\Re$. Tan etc. called the tight frame in $L^2(R^2)$ ridgelet frame [6].

Ridgelet frame can be viewed as an extension of orthonormal ridgelet [9]. It is worth emphasizing that one can obtain the orthonormal ridgelet when Meyer wavelet is used in the above construction and also the redundancy of resulting tight frame is removed. Elements of orthonormal ridgelet and of ridgelet frame constructed using Symlet-8 wavelet are displayed in Fig. 2.

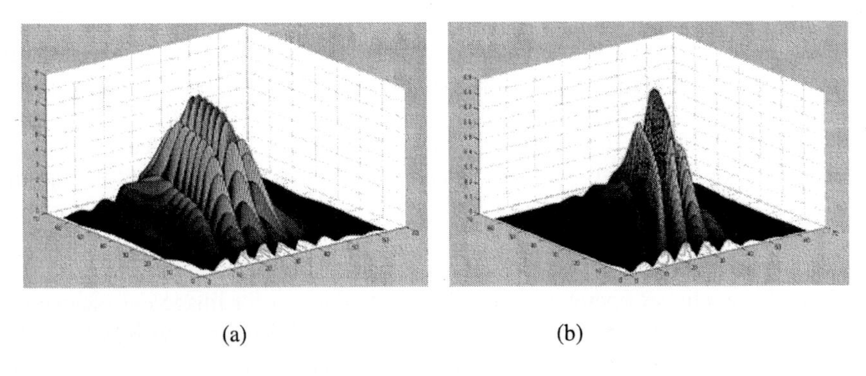

(a) (b)

Fig. 2. (a) An element of orthonormal ridgelet constructed using Meyer wavelet (b) An element of ridgelet frame constructed using Symlet-8 wavelet

As its forerunner—orthonormal ridgelet, ridgelet frame inherits the key idea that deals with straight singularities by transferring it to point singularities. As a result, the ridgelet frame retains the ability of orthonormal ridgelet to effectively representing function smooth away from straight singularity. And it is good at recovering the line structures in images at the presence of noise.

2.2 Nonlinear Mapping Function in Ridgelet Frame Domain

Commonly, the transform-domain based image enhancement methods apply a non-linear operator G to individual transform coefficients. The operator G maps a coefficient, x, to a new one, y, namely, $y=G(x)$.

The operator G we propose, which allows for enhancing image details at the same time suppress noise can be expressed as

$$y = G(x;M,\sigma,a,b,c,d,e,f) = \begin{cases} x & \text{if } |x| < e\sigma \\ \left(\dfrac{|x|-e\sigma}{e\sigma}\left(\dfrac{dM}{c\sigma}\right)^{a} + \dfrac{2e\sigma-|x|}{e\sigma}\right)x & \text{if } |x| < fe\sigma \\ \left(\dfrac{dM}{|x|}\right)^{b} x & \text{if } fe\sigma \le |x| < dM \\ \left(\dfrac{dM}{|x|}\right)^{c} x & \text{if } |x| \ge dM \end{cases}, \tag{3}$$

where, the σ is the noise standard deviation, M is the maximum transform coefficient, a,b,c,d,e and f are parameters that control the quality of enhanced image. Specially, a, b determines the degree of nonlinearity and c introduces dynamic range compression. Using a nonzero c will diminish the large coefficients. The coefficients of absolute magnitude below dM are amplified. An e value larger than 3 guaranties that the noise will not be amplified. f controls the partition of coefficients where a different gain is applied to. Our operator is an extended version of that used in paper [5], where $a=b$, $d=1$ and $f=2$. Our operator in (1) allows for more broad range of the changed coefficients which accord to more flexible enhancement. An operator G defined in (1) is shown in Fig.3, where the parameters are $M=30$, $d=1$, $a=0.5$, $b=0.5$, $c=0$, $e=3$ and $f=2$.

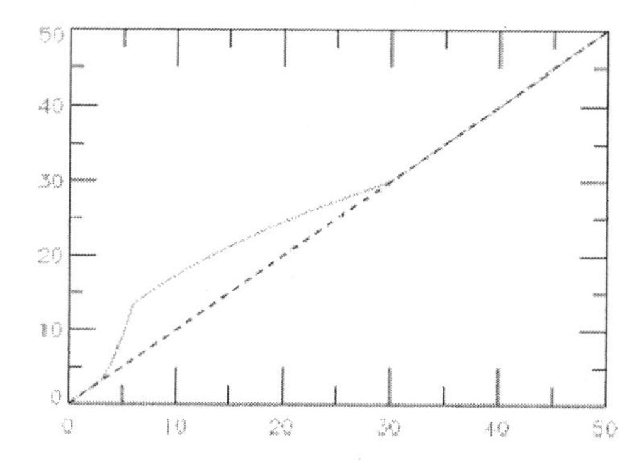

Fig. 3. An non-linear operator G in (1)

2.3 Enhancement Evaluation Criterion

In order to apply an automatic image enhancement technique, which does not require human intervention, an objective criterion for quality evaluation of the enhancement method should be chosen. However, it is a notoriously difficult problem to define an objective measure for images. In the literature of image enhancement, different measurements such as entropy, signal to noise ration, or average contrast are used as an indicator for the quality of enhanced image, but the extent to which they correlate with the results of human perception is open to debate. In paper [10], an evaluation criterion is proposed taking account several measurements simultaneously. The enhancement evaluation criterion is composed of three basic elements.

The first is the entropy measure. For image *I*, the entropy can be defined as

$$H(I) = -\sum_i p_i \bullet \ln p_i \ , \tag{4}$$

where p_i is the frequency of pixels having the *i*th gray-levels in the histogram of image *I*. The measure *H(I)* is a quantification of the number of gray-levels present in the image. When *H(I)* increases, the histogram of the image has the tendency to approach the uniform distribution.

The second is to measure the intensity of edges in image *I*. It can be expressed as

$$E(I) = \sum_x \sum_y \sqrt{\delta h_I(x,y)^2 + \delta v_I(x,y)^2} \ , \tag{5}$$

where

$$\begin{aligned}\delta h_I(x,y) = {} & I(x-1,y+1) + 2I(x,y+1) + I(x+1,y+1) \\ & -I(x-1,y-1) - 2I(x,y-1) - I(x+1,y-1) \\ \delta v_I(x,y) = {} & I(x+1,y+1) + 2I(x+1,y) + I(x+1,y-1) \\ & -I(x-1,y+1) - 2I(x-1,y) - I(x-1,y-1)\end{aligned}$$

and *I(x,y)* denotes the gray-level intensity of the pixel at location (*x*, *y*) in image *I*. In fact, (5) accords with the Sobel edge detector. And this measurement is introduced by taking an assumption that the enhanced image should have a higher intensity of the edges.

The third measurement $\eta(I)$ takes the assumption that the enhanced image should have high number of edgels, namely, the pixels belonging to an edge. It is realized by accounting how many pixels have bigger intensity than a threshold in the output of Sobel detector, that is an image with pixels $\delta h_I(x,y)^2 + \delta v_I(x,y)^2$.

Then, an enhancement evaluation criterion *Eval(I)* for image *I* is proportional to the three measurem

$$Eval(I) \sim \begin{cases} H(I) \\ \eta(I) \ . \\ E(I) \end{cases} \tag{6}$$

A maximal *Eval(I)* will correspond to an image with maximal number of edgels $\eta(I)$, having sharp edges (e.g. *E(I)* maximal), and a uniform histogram equivalent to a maximal entropy measurement $H(I)$.

The task for our automatic image enhancement based on ridgelet frame is to find the best combination of the parameters a,b,c,d,e and f according to the objective criterion (6), which may be realized using Immune Clone Algorithm (ICA) presented below.

3 Automatic Image Enhancement Algorithm

3.1 Immune Clonal Algorithm

Immune Clone Algorithm introduced in paper [7][8], is an evolutionary strategy capable of solving complex function optimization problem by imitating the main mechanisms of clone in biology immune system [11] [12] [13].

Note that each individual in ICA is called antibody, which is called gene in Genetic Algorithm (GA). ICA includes three operators, i.e., clone, mutation and selection that can be described as follows. Let the antibody population be $A=\{a_1, a_2 ... a_N\}$, here N is the population size and $a_i \in B, i=1,...,N$. B is the range of variables of the objective function.

Clone operator Θ: For $\forall a_i$ in A, we define the clone operator as

$$\Theta(a_i) = I_i \times a_i, \tag{7}$$

where $I_i = [1\,1 ... 1]$ is the q_i dimension vector whose each element is 1, and the q_i is given by

$$q_i = N' \times \frac{aff(a_i)}{\sum_{j=1}^{N} aff(a_j)}, \quad j=1,2,...,N \tag{8}$$

where $N' > N$ is a given integer called clone size , $aff(a_i)$ is the value of objective function for an individual a_i, called affinity coming from the terminology in the biology immune system.

Applied the Θ to antibody population A, we obtain a new population

$$A' = \{A, \Theta(a_1), \Theta(a_2), \ ..., \Theta(a_N)\} = \{A, A_1', A_2', ..., A_N'\} \tag{9}$$

Mutation Operator T_m^C : Different from GA, ICA only apply mutation operator to the cloned individuals, namely, $A_1', A_2', ..., A_N'$. In our automatic image enhancement algorithm, we use the non-uniform mutation in which the Gaussian distribution starts wide, and narrows to a point distribution as the current generation approaches the maximum generation.

Selection Operator T_s^C : For $\forall i = 1,2,\cdots N$, let $b = \max\{f(a_{ij}) \mid j = 2,3,\cdots q_i - 1\}$, if $f(a_i) < f(b)$, $a_i \in A$,Then b replaces the antibody a_i in the aboriginal population. So a new antibody population with the same size as the original one is realized

3.2 Automatic Image Enhancement Driven by ICA

A real-code ICA is used in our automatic image enhancement task, where the ICA find the best combination of parameters *a,b,c,d,e* and *f*, that gives the best enhancement for a given image. The parameters have real values, therefore the simplest coding of the ICA individual is a direct one to one coding: the ICA individual is as a string of 6 real numbers denoting the 6 parameters.

The affinity of the ICA allocated to each antibody a_i is computed as follows. First, we enhance the given image I by

$$I' = T^{-1}G(T(I); M, \sigma, a_i) , \tag{10}$$

where T and T^{-1} denote the ridgelet frame transform and its reversion respectively. Then, according to (6), the affinity of each antibody a_i is obtained by

$$aff(a_i) = \ln(\ln(E(I') + e)) \times \eta(I') \times e^{H(I')} \tag{11}$$

We summarize our automatic image enhancement algorithm as follows.

Step1: Estimate the standard variance

The standard variance σ of noise in the source image I can be estimated using the median estimator in the highest *HH* subband of wavelet transform of I.

Step2: Initialize Population

We only need to optimize five parameters: *a, b, c, d* and *f* since the *e* is fixed to 3. And we found that suitable intervals in (1) is [0.1 0.3] for *a* and *b,* [0 1] for *c,* [0.5 0.8] for *d,* [1 3] for *f.* The initial population is generated randomly within these specific bounds. We have chosen a population size N=10.

Step3: Determine the Iterative Condition

The ICA's termination criterion is triggered whenever the maximum number of generations is attained. For a large category of images, experiments performed show that a maximum number of generations equal to 10 suffice for the ICA to find good solutions.

Step4: Evaluate Affinity

To compute the affinity of each antibody in population using (10) and (11).

Step5: Clone, Mutation and Selection

Here, the clone size $N' = 40$. All three operators have been defined in the section 3.1.

Step6: Return to Step 3

4 Experimental Results

We compare our automatic image enhancement algorithm for a large category of images with the histogram equalization and that proposed in [10].

The first experiment is carried out on some test images widely used in the literatures of image enhancement. And these images have low noise level, for example, 1-3dB. Then, we contaminate these images using additive Gaussian white noise with standard variance 10. And we carried out the second experiment on these noisy images to test the ability of these enhancement methods to suppress noise.

Some images used in our experiment are shown in Fig. 4.

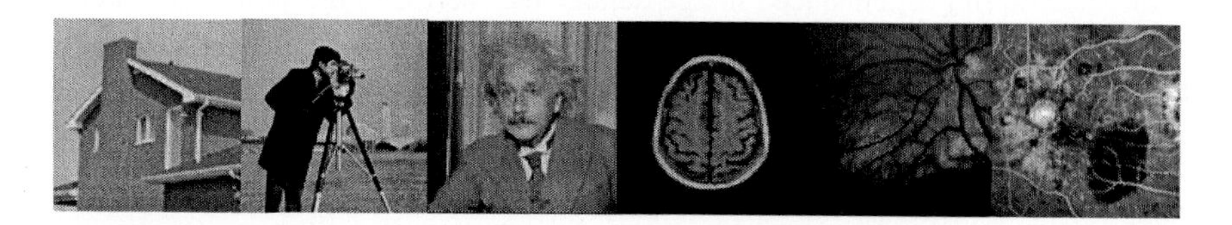

Fig. 4. Some images used in our experiment. From left to right: Image-1, Image-2, Image-3, Image-4, Image-5, Image-6

Table 1. Results in terms of quality evaluation using (11)

Image/quality	Original	Histogram Equalization	Algorithm in [10]	Our Algorithm
Image-1	147.85	52.01	267.48	**489.53**
Image-2	254.97	78.28	**468.32**	397.41
Image-3	149.10	63.49	281.60	**321.72**
Image-4	41.69	23.84	51.62	**58.19**
Image-5	93.77	39.16	**183.65**	176.05
Image-6	308.43	72.11	279.49	**380.87**

Fig. 5. Enhancement results for Image-1 and Image-6. From left to right: result using histogram equalization, result using the method in [10], result using our method

The results of the first experiment are listed in Table 1 and Fig. 5. The quality evaluations of the original and the enhanced images using (11) are listed in Table 1. For the limited space of this paper, we only give two images for visual comparison in Fig 5.

From Table 1, it is clear that our method scores much better than the other methods for most of the test images. From Fig. 5, it is observed that the histogram equalization and the method in [10] enhanced image details and increased the visibility of noise at the same time. For example, for Image-1, the results using both of the methods have much higher noise level in the sky than the original, though the noise standard variance in original images is little, 1.15.

Fig. 6. Enhancement results for images contaminated by additive Gaussian white noise with standard variance 10. From left to right: original image, result using histogram equalization, result using the method in [10], result using our method

One result of our second experiment is shown in Fig. 6. Obviously, only our method works in this case.

5 Conclusions

We have developed an automatic image enhancement algorithm based on ridgelet frame and driven by ICA. By taking into account the existence of noise in the source image, we introduce a non-linear operator, which make the enhancement method suppress noise and increase the visibility of the underlying signal at the same time. Also, through introducing an objective enhancement criterion, ICA can find the optimal parameters of the non-linear operator for a special image without the human interpreter.

References

1. Brown, T.J.: An Adaptive Strategy for Wavelet Based Image Enhancement. Proceedings IMVIP (2000) 67–81
2. Lu., Healy, D.M.: Contrast Enhancement of Medical Images Using Multiscale Edge Representation. Proceeding of SPIE. Wavelet Applications, Orlando Florida (1994)
3. Zong, X., Laine, A.F., Geiser, E.A., Wilson, D.C.: Denoising And Contrast Enhancement via Wavelet Shrinkage and Non-linear Adaptive Gain. Wavelet Applications 3: Proceeding of SPIE Vol. 2762, (1996) 566–574

4. Fan, J., Laine, A.: Contrast Enhancement by Multiscale and Non-linear Operators. (1995)
5. Starck, J. L., Candes, E. J., Donoho, D. L.: Gray and Color Image Constrast Enhancement by the Curvelet Transform. IEEE Trans. on Image Processing, Vol. 12, (2003) 706–716
6. Tan Shan, Jiao, L.C., Feng, X.C.: Ridgelet Frame. In: Proc. Int. Conf. Image Analysis and Recognition, Porto, (2004) 479–486
7. Du, H.F., Jiao, L.C., Wang Sun'an: Clonal Operator and Antibody Clone Algorithms. In: Shichao, Z., Qiang, Y., Chengqi, Z. (eds.): Proceedings of the First International Conference on Machine Learning and Cybernetics, Beijing (2002) 506–510
8. Du, H.F., Jiao, L.C., Gong, M.G., Liu, R.H.: Adaptive Dynamic Clone Selection Algorithms. In: Zdzislaw, P., Lotfi Z. (eds): Proceedings of the Fourth International Conference on Rough Sets and Current Trends in Computing (RSCTC'2004). Uppsala, Sweden (2004)
9. Donoho, D. L.: Orthonormal Ridgelet and Linear Singularities. SIAM J. Math Anal., Vol. 31, (2000) 1062–1099
10. Munteanu, C., Rosa, A.: Gray-Scale Image Enhancement As An Automatic Process Driven by Evolution. IEEE Trans. on SMC- B, Vol 34, (2004) 1292–1298
11. Jiao, L.C., Wang. L.: A Novel Genetic Algorithm Based on Immunity. IEEE Trans. on SMC-A, Vol. 30, (2000) 552–561
12. Zhong, W.C., Liu, J., Xue, M.Z., Jiao, L.C.: A Multiagent Genetic Algorithm for Global Numerical Optimization. IEEE Trans. System, Man, and Cybernetics-Part B. Vol. 34, (2004) 1128–1141
13. Zhang, X.R., Shan, T., Jiao, L.C.: SAR Image Classification Based on Immune Clonal Feature Selection. In: Proceedings of Image Analysis and Recognition, Lecture Notes in Computer Science, Vol. 3212. Springer-Verlag, Berlin Heidelberg New York (2004) 504-511

Practical Evaluation of Efficient Fitness Functions for Binary Images

Róbert Ványi

Institute of Informatics, University of Szeged,
Árpád tér 2, H-6720 Szeged, Hungary
robert@vanyi.org

Abstract. Genetic Programming can be used to evolve complex objects. One field, where GP may be used is image analysis. There are several works using evolutionary methods to process, analyze or classify images. All these procedures need an appropriate fitness function, that is a similarity measure. However, computing such measures usually needs a lot of computational time. To solve this problem, the notion of efficiently computable fitness functions was introduced, and their theory was already examined in detail. In contrast to that work, in this paper the practical aspects of these fitness functions are discussed.

1 Introduction

In this paper the *efficiently computable fitness functions* [1] are discussed in a practical aspect. First the theoretical results are converted into algorithms, then some practical problems are examined, and solutions for them are given.

These fitness functions have a great practical importance, since they can be used in many applications, like hand tracking [2], edge detection [3] or road identification in images [4]. The only assumption is that a structural description is evolved to describe an image as closely as possible. Such structures can be for example Lindenmayer systems [5][6], or iterated function systems [7][8].

During the following sections it is assumed that some kind of structural description is evolved. To calculate the fitness, this description has to be interpreted, or converted to an image, that is the set of the foreground pixels has to be determined. Using the theoretical results a fast fitness functions calculation algorithm can be built into the conversion algorithm. During this conversion several points may be generated more than once, this is called *overwriting*. Furthermore, several description forms may use sets of lines, which may be thick. Therefore it is possible that parts of these lines are *overlapping*. This causes a similar problem as overwriting. Solution for these problems are also given in this paper.

First, however, several basic notions are explained, and the theoretical results on efficiently computable fitness functions are summarized. In Section 2 the algorithm for efficient fitness calculation is given. In Section 3 the algorithm is extended to solve the previously mentioned problems. These problems themselves are also described in detail. In Section 4 several experiments are carried

F. Rothlauf et al. (Eds.): EvoWorkshops 2005, LNCS 3449, pp. 314–324, 2005.

out using a simple test environment. Finally in Section 5 some conclusions are drawn, and future plans are mentioned.

1.1 Binary Images

An image is usually a function that maps colors or gray level values to a set of pixels. In general a two-dimensional image is a function $f : \mathbb{R} \times \mathbb{R} \to \mathbb{R}$ with some restrictions. These restrictions are not of interest here, since they are automatically satisfied in computer imaging, where only finite discrete images are used. This means the function is defined over $\mathbb{Z} \times \mathbb{Z}$ and the possible values come from $\mathbb{Z}_k = \{0, 1, \ldots, k-1\}$ $(k \geq 2)$. Also the domain of this functions is finite. Later it will be easier to have a definition for this domain, which will be referred to as *container.*

Definition 1. *(Image container)*
An image container of size $n \times m$ is the set $C_{n,m} = \mathbb{Z}_n \times \mathbb{Z}_m$, where $n, m \in \mathbb{Z}^+$. An element (x, y) of the container is called *pixel.*

Definition 2. *(Digital image)*
A digital image over a given container C is a function $I : C \to \mathbb{Z}_k$, where $k \in \mathbb{Z}^+$. C is called the *container of the image* I. For a pixel $(x, y) \in C$ $I(x, y)$ is called the *value* or the *color* of the pixel. Pixels with value 0 are called *background pixels.* The number of pixels that are not background pixels is denoted by $|I|$.

Integer k is called the *color depth* of the image I, and usually $k \geq 2$. When $k = 2$ the image is called *bi-level or binary image.* In this case $|I|$ denotes the number of the *foreground pixels*, that is pixels with value 1.

The set of digital images with color depth of k over a given container $C_{n,m}$ is denoted by $\mathcal{I}^k_{n,m}$ Formally: $\mathcal{I}^k_{n,m} = \mathbb{Z}_k^{C_{n,m}}$. The set of digital images with color depth of k is denoted by $\mathcal{I}^k$. That is $\mathcal{I}^k = \bigcup_{n=1}^{\infty} \bigcup_{m=1}^{\infty} \mathcal{I}^k_{n,m}$.

1.2 Efficient Fitness Functions

In this section the most important theoretical results on efficient fitnesses are summarized briefly. The details can be found in the previously mentioned work.

Definition 3. *(Fitness function)*
A fitness function for digital images is a function $\Phi^{k,l} : \mathcal{I}^k \times \mathcal{I}^l \to \mathbb{R}$, where $k, l \in \mathbb{Z}^+$ are positive integers. That is Φ orders a real value to a pair of images with color depths at most k and l respectively. One of the images is the target image. The other image is generated by the evolutionary algorithm. Therefore they are denoted by T and G respectively.

When the color depths are unambiguous the index is omitted, and only Φ is used. When it is also clear from the context, fitness can have the meaning of either fitness function (Φ) or fitness value ($\Phi(T, G)$).

Definition 4. *(Efficient fitness)*
A fitness function is called *efficiently computable* or *efficient* for short when for any given destination image T, and any finite sequence of generated images $G_0, G_1, \ldots, G_{s-1}$ the following holds:

$$t_\Phi(T, G_0, G_1, \ldots, G_{s-1}) \leq t_{\Phi,p}(T) + t_{\Phi,c} \sum_{i=0}^{s-1} |G_i|,$$

where $t_\Phi(T, G_0, G_1, \ldots, G_{s-1})$ is the shortest time within $\Phi(T, G_i)$ can be computed for each $i \in \{0, 1, \ldots, s-1\}$, $t_{\Phi,p} : \mathcal{I} \to \mathbb{R}$ is a function, usually the preprocessing time, and $t_{\Phi,c}$ is a constant, called *the comparison coefficient* of Φ.

In words, the computation time is less then a sum of two values, where the first value depends only on the target image, and is independent from the number of the generated images and from the generated images themselves. The second value is independent form the target image and is linearly dependent on the number of non-zero pixels in the generated images. The motivation is that the positions of the pixels have to be computed anyway, and this needs at least linear time.

1.3 Theoretical Results on Efficient Fitness Functions

The first result is on simple fitness functions, which are defined using a so-called loop operator as follows.

Definition 5. *(Loop operator)*
Given an associative binary operator $\omega : M \times M \to M$, with an identity $\varepsilon_\omega \in M$ ($\varepsilon_\omega m = m\varepsilon_\omega = m \; \forall m \in M$). That is $(M, \omega, \varepsilon_\omega)$ is a monoid. A loop operator Ω belonging to ω is defined as follows:

$$\Omega_X(f) = \begin{cases} \varepsilon_\omega & \text{, if } X = \emptyset \\ \omega(f(p), \Omega_{X \setminus \{p\}}(f)) \text{ for an arbitrary } p \in X & \text{, otherwise,} \end{cases}$$

where $X \subseteq P$ is a finite set and $f : P \to M$ is an arbitrary computable function. Note that the loop operator is unambiguous only if the operator is commutative or the set X is ordered, and p is not an arbitrary, but for example the first element of the set.

Definition 6. *(Simple fitness functions)*
A fitness function is called *simple*, if it can be written in the following form:

$$\Phi(T, G) = f'(T)\sigma\Omega_{G(x,y)\neq 0}(f''(T(x,y), G(x,y), x, y)),$$

where σ is a binary operator, Ω is a loop operator belonging to a binary operator ω, and the computation times of σ, ω and f'' are $\mathcal{O}(1)$.

Theorem 1. *(Efficiency of simple fitnesses)*
Every simple fitness function can be computed efficiently.

An example for simple fitness is the generalized quadratic error, where the fitness is defined as follows:

$$\Phi(T,G) = \sum_{(x,y)\in C} f(T(x,y),G(x,y)).$$

$$f(i,j) = \begin{cases} W_0 & \text{if } i = j = 0, \\ W_1 & \text{if } i = j = 1, \\ W_2 & \text{if } i = 0, \text{ and } j = 1, \\ W_3 & \text{if } i = 1, \text{ and } j = 0, \end{cases}$$

where the W_i $(i = 0,1,2,3)$ values are predefined constants. These fitness functions are not powerful enough, therefore another class of fitness functions is also defined, which is also efficient. These are the semi-local fitness functions.

Definition 7. *(Semi-local fitness function)*
A fitness function is called *local (additive) in* G, when it has the following form:

$$\Phi(T,G) = \sum_{(x,y)\in C} \varphi(T,G,x,y) = \sum_{(x,y)\in C} f(T,G(x,y),x,y).$$

Similarly fitness functions local in T can also be defined. A fitness function is called *semi-local*, when it is local in either T, or G.

Theorem 2.
Any fitness function $\Phi^{k,2}$ that is local in G can be converted into a form

$$\Phi(T,G) = f'(T) + \sum_{G(x,y)\neq 0} T'(x,y),$$

where $f'(T) = \sum_{(x,y)\in C} f(T,0,x,y)$ and $T'(x,y) = f(T,1,x,y) - f(T,0,x,y)$.

Corollary 1. *(Efficiency of semi-local fitnesses)*
Given a fitness function $\Phi^{k,2}$. If Φ is local in G and T can be preprocessed then Φ can be computed efficiently. By definition T *can be preprocessed*, and converted to T' if there exists a function T_o, and two integers x_b and y_b, such that $(x \geq x_b \wedge y \geq y_b) \Rightarrow T'(x,y) = T_o(x,y)$, and the calculation time of T_o is $\mathcal{O}(1)$.

Semi-local fitnesses are for example the distance based fitness functions, where the fitness for a misplaced pixel is a function of the distance of the given pixel and the target image as a set of pixels.

$$\varphi(T,G,x,y) = \begin{cases} W_0 & \text{if } T(x,y) = G(x,y) = 0, \\ W_1 & \text{if } T(x,y) = G(x,y) = 1, \\ W(h(T,x,y)) & \text{if } T(x,y) = 0, \text{ and } G(x,y) = 1, \\ W_3 & \text{if } T(x,y) = 1, \text{ and } G(x,y) = 0. \end{cases}$$

where $h(T,x,y)$ is the distance of (x,y) and the set of 1 pixels in T.

2 Fitness Calculation

Theoretical results can be useless toys, unless one can convert them into algorithms. In this section the algorithms are given for the efficient fitness calculation. There are two ways to calculate the fitness. The first method is the *off-line* way: first the generated image is drawn then in the second step it is compared with the target image. The second way is to do this *on-line*: draw the image and calculate the fitness at the same time. Usually the on-line method is better, since the image does not have to be stored. When the off-line method is used, the non-zero pixels have to be stored for example in a list, to allow a fast loop over them. Since the main part of the algorithm is almost the same in both cases, only the on-line version is considered here.

The algorithm can be designed using Theorem 2. It gives a preprocessing algorithm for calculating f' (`fitc`) and T' (`Tpre`), and the main algorithm. The preprocessing algorithm can be seen in Algorithm 2.1 together with function `Target`. The latter one is used to return the value of the preprocessed image for a given pixel. `Tpre` cannot be used directly, since x and y may be outside of this image. In this case the return value is determined by `Tout`, which returns the value of T_o.

```
PREPROCESS()
1  fitc ← 0
2  for each (x, y) in image T
3  do f1 ← f(T, 1, x, y); f0 ← f(T, 0, x, y)
4       Tpre[x, y] ← f1 − f0; fitc ← fitc + f0

TARGET(x, y : integer)
1  if (x, y) is outside of Tpre
2     then return Tout(x, y)
3     else return Tpre[x, y]
```

Algorithm 2.1. Preprocessing

The main part of the algorithm is nothing else but summing the pixel fitness values for each newly generated pixel. The pixel fitness here is the value of T', but instead of using function `Target`, a new function is introduced, to be able to easily extend the algorithm later.

```
FITNESS(Gdesc : image_description)
1  result ← fitc
2  while the image is not completely generated
3  do calculate one or more pixels of the image
4       for each new pixel (x, y)
5       do result ← result + pixel_fitness(x, y)
6            after_draw_commands
7  return result

PIXEL_FITNESS(x, y)
1  return Target(x, y)

MAIN()
1  preprocess
2  start_EA
```

Algorithm 2.2. Main part of the fitness calculation

3 Fitness Extensions

In some cases the simple fitness functions have to be extended. It is, however, important not to loose the possibility of the fast calculation, therefore the new functions are designed with this requirement in mind.

3.1 Overwriting Pixels

When a set of graphical primitives is used as a representation, it can happen during the interpretation that some pixels are drawn more than once. For example when they are covered by more than one primitive. With the previous algorithm the more times a pixel is covered the higher the fitness increment it causes. It is usually not what is expected. To handle this problem an additional *memorizing image* $M: C \to \{f, v\}$ is introduced. This image is initially filled with the value f (free). After a pixel is drawn, it is changed to v (visited). Thus the comparator function $f(i, j)$ must be extended to $f_o(i, j, k)$, where $k \in \{f, v\}$. A straight implementation of this method can be seen in Algorithm 3.1.

```
PIXEL_FITNESS(x, y)
1  if M[x, y] = f
2     then return Target(x, y)
3     else return penalty

AFTER_DRAW_COMMANDS()
1  for each (x, y) in C
2  do M[x, y] ← f
```

Algorithm 3.1. Dummy algorithm for handling overwrites

A problem with this image M is that it has to be cleared (set to f) before each fitness calculation. That needs time proportional to the image size, which is usually much greater than the time needed to calculate the fitness. In an experiment the CPU spent more than 94% of the running time with clearing this image. Therefore a greater set is used instead of $\{f, v\}$, for example $\mathbb{Z}_{256}$, or $\mathbb{Z}_{65536}$ and the actual value showing the visited state is increased after each step. Image M is cleared only after the maximal value is reached. Using this modification, the fitness calculation is presented by Algorithm 3.2.

```
PIXEL_FITNESS((x, y))
1  if M[x, y] < overwrite
2     then result ← Target(x, y)
3          M[x, y] ← overwrite
4     else result ← penalty
5  return result

AFTER_DRAW_COMMANDS()
1  if overwrite ≥ MAX
2     then overwrite ← 0
3          for each (x, y) in I
4          do M[x, y] ← 0
5  overwrite ← overwrite + 1
```

Algorithm 3.2. Calculating fitness with overwriting

Note that this method breaks the linearity of the fitness computation. However, one can avoid the usage of the additional image by using a sorted list of the drawn pixels. Though the time will not be constant either, but logarithmic, for each pixel. The implementation of this method can be seen in Algorithm 3.3.

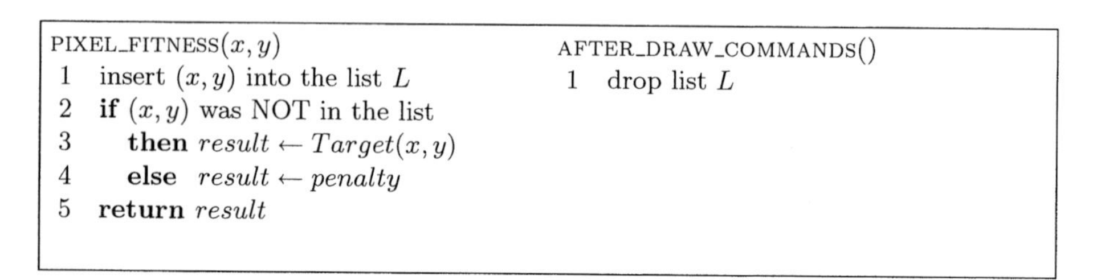

```
PIXEL_FITNESS(x, y)
1  insert (x, y) into the list L
2  if (x, y) was NOT in the list
3     then result ← Target(x, y)
4     else result ← penalty
5  return result

AFTER_DRAW_COMMANDS()
1  drop list L
```

Algorithm 3.3. Calculating fitness with overwriting with list

3.2 Handling Thick Lines

In Figure 1a a possible problem of the thick line based representations can be seen. A different line segment is described, but because of the thickness of the line, and the non-zero angle, some overlapping may occur. However, this should not be considered as a normal overwriting.

To solve the problem two classes of pixels are introduced: the *main line*, which is more important, and the *extension*. For the fitness calculation the memorizing image has three values. These are f (free), e (exhausted) and p (prohibited). Value f means that this pixel has not been covered yet. Value e means that this pixel has been covered by an extension pixel, covering it again by an extension pixel will give no penalty, but the fitness will not be increased either. If a pixel is p, then it has been covered by the main-line, thus covering it again gives penalty. The calculation scheme can be seen in Figure 1b.

The algorithm defined for simple overwriting can easily be modified to handle this case. Only a new function should be introduced to draw extension pixels and calculate fitness, and the value of *overwrite* has to be increased by 2. An algorithm using lists can also be given, where the pixels of the main line and the pixels of the extension are stored in two separate lists.

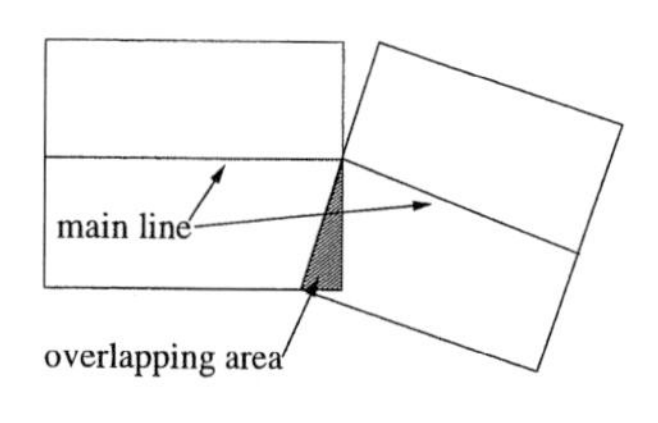

a) Overlapping thick lines

pixel	old	new	fitness
main line	f	p	full gain
main line	e	p	some error
main line	p	p	huge error
extension	f	e	some gain
extension	e	e	nothing
extension	p	p	some error

b) Fitness for thick lines with overwriting

Fig. 1. Thick lines

4 Experimental Results

In this section some experimental results are discussed. For the tests a simple image was used as the target image. The generated image was the displaced version of the original. During the evolution the individuals were represented by (d_x, d_y) pairs, that is by the displacement vectors, this means the optimum is $(0, 0)$. Note that in another experiment the rotation angle and scale factor were used with similar results.

4.1 Computation Times

In the first experiment the computation times were compared. A simple image description was created and interpreted for different image sizes. Then the images were compared with a simple target image. To test calculation speed the distance based fitness was used, with different $W(x)$ functions. Using the algorithm presented in this paper, $W(x)$ does not influence the calculation speed. However, for testing purposes the $W(x)$ was also computed on-line. The results for $W(x) = 10x$, $W(x) = x^2$ and $W(x) = \sqrt{x}$, for drawing without fitness calculation and for the general method using the previously defined algorithm can be seen in Figure 2.

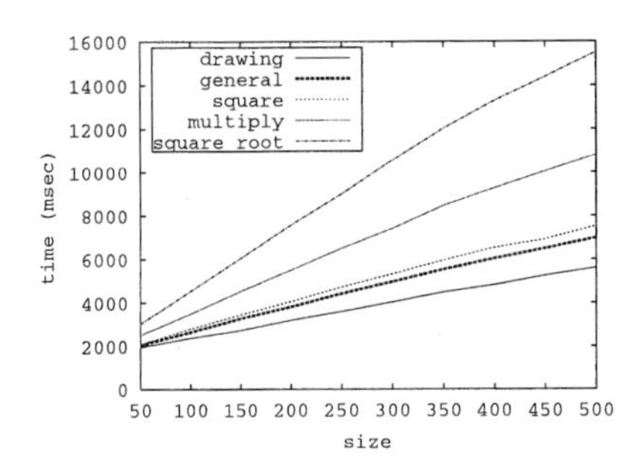

Fig. 2. Comparison of drawing and some efficient fitness calculations

4.2 Convergence Speed

In the second experiment the previously introduced simple test environment was used. The population size was 20 and 1000 steps were made. Two graphs were generated for the quadratic error (Figure 3), for the distance based fitness (Figure 4) and for the Hausdorff distance (Figure 5). The last one is unfortunately not efficient, and used here only for comparison. The first graphs show the deviation of the population in the solution space in the first 150 steps. The second images show the fitness values of the individuals, plotted against the generation. The individuals were initially placed at the perimeter of the solution space, whereas the optimum $(d_x = 0, d_y = 0)$ is at the center.

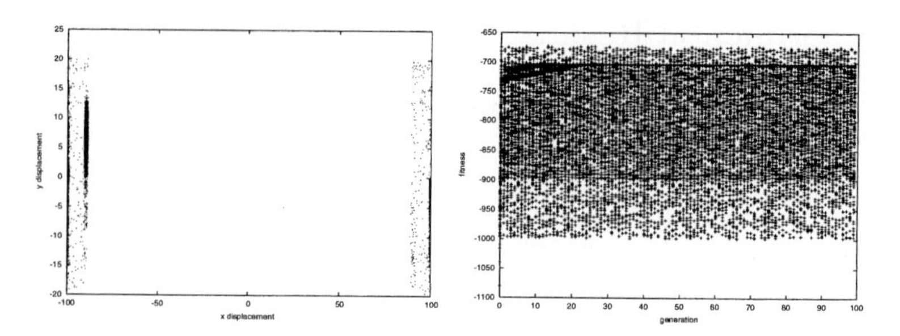

Fig. 3. Convergences for quadratic error

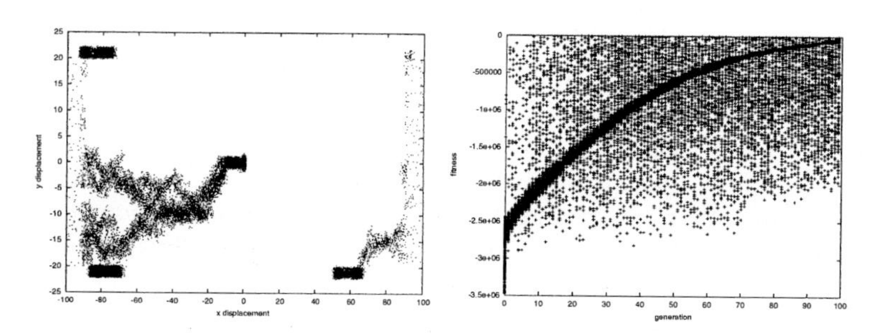

Fig. 4. Convergences for distance based fitness

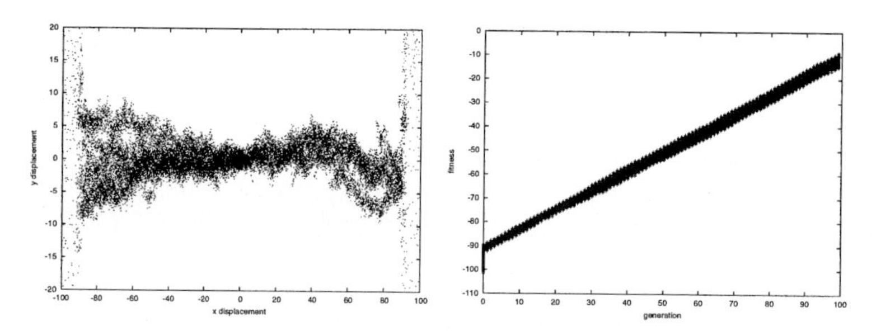

Fig. 5. Convergences for Hausdorff distance

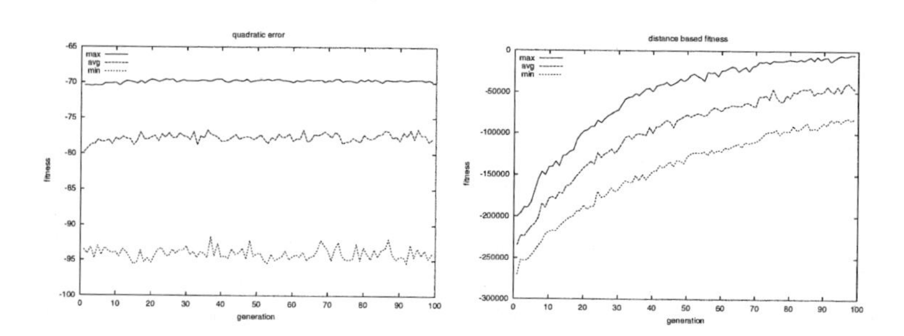

Fig. 6. Convergences for different fitness functions

In Figure 6 the results are summarized, and the minimal, maximal and average fitness values are plotted averaged over the 10 independent runs, but not for the Hausdorff distance, since the results are obvious in this case.

5 Conclusion

In this paper several algorithms were given for efficient fitness computation. With them the theoretical results on these fitnesses can be used in practice, making fitness calculation only a negligible additional extra work. Tests have shown that using the algorithms given in this paper, any kind of function of the distance can be used for fitness computation, without risking any loss of computational speed, meanwhile the calculation of these functions would normally need a considerable additional time. Some extensions were also given to solve two frequent practical problems.

Other tests were also made showing that the quadratic error converges poorly, but an appropriate distance based fitness may converge much better. The results show that the distance based fitness functions can combine the good convergence properties of the Hausdorff distance and the high calculation speed of the efficient fitness functions.

In the future more tests are also planned with more compilcated images. Hopefully completely new classes of fitness functions may also be found that can be proven to be efficient as well.

References

1. Ványi, R.: Efficiently computable fitness functions for binary image evolution. In: Applications of Evolutionary Computing EvoWorkshops 2002: EvoCOP, EvoIASP, EvoSTIM/EvoPLAN Proceedings. Volume 2279 of LNCS., Kinsale, Ireland, Springer-Verlag (2002) 280–291
2. Heap, T., Samaria, F.: Real-time hand tracking and gesture recognition using smart snakes. In: Proceedings of Interface to Human and Virtual Worlds, Montpellier, France (1995)
3. Harris, C., Buxton, B.: Evolving edge detectors with genetic programming. In Koza, J.R., Goldberg, D.E., Fogel, D.B., Riolo, R.L., eds.: Genetic Programming 1996: Proceedings of the First Annual Conference, Stanford University, CA, USA, MIT Press (1996) 309–315
4. Boggess, J.E.: A genetic algorithm approach to identifying roads in satellite images. In: Proceedings of the Ninth Florida Artificial Intelligence Research Symposium, FLAIRS-96, Eckerd Coll., St. Petersburg, FL, USA (1996) 142–146
5. Jacob, C.: Genetic L-system programming. In: Proceedings of the Parallel Problem Solving from Nature, International Conference on Evolutionary Computation, PPSN III. Volume 866 of Lecture Notes of Computer Science., Springer-Verlag, Berlin (1994) 334–343

6. Koza, J.R.: Discovery of rewrite rules in Lindenmayer systems and state transition rules in cellular automata via genetic programming. In: Symposium on Pattern Formation (SPF-93), Claremont, California, USA. (1993)
7. Collet, P., Lutton, E., Raynal, F., Schnoenauer, M.: Polar IFS + parisian genetic programming = efficient IFS iverse problem solving. Genetic Programming and Evolvable Machines **1** (2000) 339–362
8. Hart, J.C.: Fractal image compression and the inverse problem of recurrent iterated function systems. IEEE Computer Graphics and Applications **16** (1996)

Selective SVMs Ensemble Driven by Immune Clonal Algorithm

Xiangrong Zhang, Shuang Wang, Tan Shan, and Licheng Jiao

National Key Lab for Radar Signal Processing,
Institute of Intelligent Information Processing,
Xidian University, 710071 Xi'an, China
xrzhang@mail.xidian.edu.cn

Abstract. A selective ensemble of support vector machines (SVMs) based on immune clonal algorithm (ICA) is proposed for the case of classification. ICA, a new intelligent computation method simulating the natural immune system, characterized by rapid convergence to global optimal solutions, is employed to select a suitable subset of the trained component SVMs to make up of an ensemble with high generalization performance. The experimental results on some popular datasets from UCI database show that the selective SVMs ensemble outperforms a single SVM and traditional ensemble method that ensemble all the trained component SVMs.

1 Introduction

Two major recent advances in learning theory are SVM [1] and ensemble methods such as boosting and bagging [2][3]. SVM, proposed by Vapnik and his group, has been proved to be an effective learning machine by successful applications, such as face detection [4], hand-written digit recognition [5], as well as image retrieval [6]. In recent years, combining machines instead of using a single one for increasing learning accuracy is an active research area. And SVMs ensembles have been proposed to further improve the generalization performance also. In [1], a boosting technique is used to train each component SVM and another SVM is used to combine the SVMs trained. In [7] and [8], SVMs ensemble is realized based on bagging algorithm and applied to face detection and classification. These methods mentioned above make a collective decision of all the individual SVMs, which may restrain the improvement of the performance of the ensemble in the case of the existence of the same component SVMs.

Researches have shown that an effective ensemble should consist of a set of models that are not only highly correct, but ones that make their errors on different parts of the input space as well. [9][10]. In other words, an ideal ensemble consists of highly accurate classifiers that disagree as much as possible. In [11], a selective neural networks ensemble based on genetic algorithm (GA) is proposed to improve the generalization performance of the ensemble and a conclusion is drawn that ensemble many could be better than all.

F. Rothlauf et al. (Eds.): EvoWorkshops 2005, LNCS 3449, pp. 325–333, 2005.

Despite the good global searching capability and its wide applications in different areas, GA has the unavoidable disadvantages that the speed of convergence is low and the optimal solutions cannot be obtained in limited generations since it emphasizes the competition only and the communication between individuals is ignored. Immune clonal algorithm (ICA) overcomes the shortcoming to some degree and has been proved to be an effective optimization algorithm [12][13] since it imitates the artificial immune system and the competition and cooperation coexist in it. ICA demonstrates the self-adjustability function by accelerating or restraining the generation of antibodies, which enhances the diversity of the population.

In this paper, a selective SVMs ensemble is proposed and ICA is used to the selection of component SVMs. Each individual SVM is trained using a randomly chosen training set from the original training data via the bootstrap technique in order to reduce error correlation between individual SVMs and to improve the generalization performance of SVMs. Experimental results on some popular datasets from UCI database show the effectiveness of the new method.

This paper is organized as follows. In section 2, SVM for classification problem is reviewed briefly. Section 3 presents the principle of SVMs ensemble first, followed by the introduction of ICA, based on which, selective SVMs ensemble is developed. Section 4 gives the experimental results and discussions. Finally, conclusive remarks are given in section 5.

2 Support Vector Machine

As a member of many kernel methods, SVM is a relatively new learning algorithm, in which data points are non-linearly mapped to a high dimension feature space by replacing the dot product operation in the input space with a kernel function $K(\cdot,\cdot)$. It is to find the best decision hyperplane that separates the positive examples and negative examples with maximum margin. By defining the hyperplane this way, SVM can be generalized to unknown instances effectively.

Suppose training samples $\{(x_1, y_1),\cdots(x_l, y_l)\}$, in which $x \in R^N$, $y \in \{-1,+1\}$ and R^N denotes the input space. Then original input space is projected to high dimension feature space Ω by kernel projection, which ensures that the patterns can be recognized linearly in feature space. For pattern recognition problem, it turns to a programming question as follows.

$$\max Q(x) = \sum_{i=1}^{l} \alpha_i - \frac{1}{2}\sum_{i,j=1}^{l} \alpha_i \alpha_j y_i y_j K(x_i, x_j) \tag{1}$$

under constraints $\sum_{i=1}^{l} \alpha_i y_i = 0$ and $0 \le \alpha_i \le C, i = 1,...,l$. The according decision function is:

$$f(x) = sign(\sum_{i=1}^{l} y_i \alpha_i^* K(x \cdot x_i) + b^*) . \tag{2}$$

3 Selective SVMs Ensemble Based on ICA

3.1 SVMs Ensemble

The term ensemble of classifiers is used to identify a set of classifiers whose individual decisions are combined in some way (typically by voting) to classify new examples. It is known that an ensemble often exhibits a much better performance than the individual classifiers that compose it [9]. Bagging and boosting techniques, which use a particular learning algorithm trained with different distributions of the training set, are used popularly in SVMs ensembles.

Bagging is a particular ensemble architecture where a voting combination of learning machines is used and each component one is trained using a subset with replacement of the initial training data. The size of the subsamples set is equal to that of the original training set, but repetitions of points occur. In this way, we can get an ensemble $\{f_1, f_2, \cdots f_N\}$ of N classifiers. If all the classifiers are identical, their results are matched on the same data and an ensemble will exhibit the same performance as individual classifiers. Then we can draw one conclusion that combining identical classifiers is useless. However, if all the classifiers are different and their errors are uncorrelated, most of the other classifiers except for $f_i(x)$ may be correct when $f_i(x)$ is wrong. In this case, the result of majority voting can be correct.

Hansen and Salamon [9] first introduced the hypothesis that the ensemble of models is most useful when its member models make errors independently with respect to one another. Tumer and Ghosh [14] have shown how the error rate obtained by a combiner is related to that of a single classifier. The equation that relates both is:

$$Error_{ensemble} = \frac{1+\rho(N-1)}{N} Error + Error_{Optimal\ Bayes}, \tag{3}$$

where ρ denotes the correlation among classifier errors, and $Error_{Optimal\ Bayes}$ the error rate obtained using the Bayes rule on condition that all the conditional probabilities are known. If $\rho = 0$, the error of the ensemble decreases in proportion to the number of classifiers. When $\rho = 1$, the error of the ensemble is equal to the error of a single classifier.

From the above study, two main conclusions can be got. Firstly, the performance of individual classifiers must be better than that of a random guess, namely, the accuracy of individual classifier cannot be low seriously. Secondly, the diversity of individual classifiers consisted in ensemble is one of the requirements. In other word, we should combine the classifiers that make uncorrelated errors because it is evident that combining several identical predictors produces on gain.

For SVMs ensemble, the first condition is satisfied obviously. To satisfy the second condition, the idea of selection is introduced to the SVMs ensemble. Similar or redundant individual SVMs trained should be removed from ensemble and better generalization performance of the ensemble can be obtained in this way. Then the problem is transformed into an optimization problem that is finding the most suitable subset of component SVMs trained. ICA is a good choice for us since it has been shown to be a fast and effective global optimization technique.

3.2 A Brief Review of ICA

Derived from traditional evolutionary algorithm, ICA [15][16] introduces the mechanisms of avidity maturation, clone and memorization. Rapid convergence and good global search capability characterize the performance of the corresponding operators. The property of rapid convergence to global optimum of ICA is made use of to speed up the searching of the most suitable subset among a number of individual SVMs trained.

The clonal selection theory is used by the immune system to describe the basic features of an immune response to an antigenic stimulus; it establishes the idea that the cells are selected when they recognize the antigens and proliferate. When exposed to antigens, immune cells that may recognize and eliminate the antigens can be selected in the body and mount an effective response against them during the course of the clonal selection. The clonal operator is an antibody random map induced by the avidity including three steps: clone, clonal mutation and clonal selection. The state transfer of antibody population is denoted as follows:

$$C_{MA}: A(k) \xrightarrow{clone} A'(k) \xrightarrow{mutation} A''(k) \xrightarrow{selection} A(k+1) \tag{4}$$

According to the avidity function f, a point $a_i = \{x_1, x_2, \cdots, x_m\}$, $a_i(k) \in A(k)$ in the solution space will be divided into q_i same points $a_i^{'}(k) \in A'(k)$ by using clonal operator. A new antibody population is produced after the clonal mutation and clonal selection are performed. Here, an antibody corresponds to an individual in EA. The term of avidity denotes the fitness of an antibody to the objective function. The fundamental steps of ICA are summarized in Fig. 1.

1 $k=0$; Initial the antibody population $A(0)$, set the parameters, calculate the avidity of the initial population;
2 According to the avidity and the clonal size set of the antibody, perform operations of clone T_c^C, clonal mutation T_g^C and clonal selection T_s^C, then obtain the new antibody population $A(k)$;
3 Calculate the avidity of $A(k)$;
4 $k=k+1$; If satisfying the iterative termination condition, stop the iteration, otherwise, return to 2.

Fig. 1. Immune Clonal Algorithm

3.3 Selective SVMs Ensemble Based on ICA

The leading idea of our method is to select a part of the trained component SVMs to construct an ensemble with the lowest generalization error. Like traditional bagging ensemble method, the bootstrap technique is adopted to sample training data for individual SVMs in the procedure of training. For a binary classification, given a training set $S = \{(x_i, y_i) | i = 1,2,\cdots,l\}$ where $y_i \in \{-1,1\}$, suppose we want to get a ensemble selected from K individual SVMs, bootstrap technique is used to generate K training sets $\{S_k^B | k = 1,2,\cdots,K\}$ with different distributions by randomly re-sampling from the original training set repeatedly. For a special sample, it may appear repeatedly or may not appear at all in a certain training set. Then, K SVMs $\{C_k | k = 1,2,\cdots,K\}$ are trained respectively on the K training set.

For *N*-class problem, one-against-one scheme is applied to reduce it to $N(N-1)/2$ 2-class problems. Like binary classification, K training sets are generated also. It is different that $N(N-1)/2$ SVMs will be trained on each generated training set, and the $N(N-1)/2$ SVMs will be viewed as a whole in the selection and decision procedures since they are trained using the same training set. For convenience, by a component SVM of an ensemble we mean the $N(N-1)/2$ SVMs trained on the same training set for a *N*-class problem.

Input: training set S, SVM C, Number of SVMs trained K, test set $\{x_i | i = 1,2,\cdots,L\}$

Procedure:

Step1 for *k*=1 to *K*

S_k^B = bootstrap samples from S

$C_k = C(S_k^B)$

end for

Step2 Select an optimal subset from $\{C_k | k = 1,2,\cdots,K\}$ to make up of an ensemble C^* using ICA

Step3 for *i*=1 to L

Make predictions of a test sample x_i using each component SVMs included in the ensemble C^*

Give the final decision $C^*(x_i)$ via the majority voting

end for

Output: the final decisions of test samples

Fig. 2. The selective ensemble of SVMs based on ICAï

Then our focus is to create an ensemble with SVMs selected from the available individual SVMs with ICA. As mentioned in section 3.2, ICA is induced by the avidities of antibodies in the evolving population. The evaluation of the performance of an ensemble is related to the avidity function of ICA directly. Intuitively, the lower the generalization error is, the better the performance of the ensemble. Therefore, the generalization error $\hat{E}$ of the ensemble on a validation data set is used for evaluating the performance of the ensemble, which is obtained via the majority voting of the component SVMs of a given ensemble considered. For ICA, $1/\hat{E}$ is used as the avidity function. And now, the task of ICA is to find an optimal subset of individual SVMs to generate an ensemble with which a minimum of $\hat{E}$ on the validation set can be obtained. The used validation set is bootstrap sampled from the original training set here. Certainly, it is better that a validation set is a separate one from the training set.

After we get the ensemble of selected individual SVMs, the final decision for a test example is made using a majority-voting rule based on the ensemble whose output is the class label received the most number of votes in classification. The approach proposed is summarized in Fig. 2.

In step 2, a binary encoding scheme is used to represent the presence or absence of individual SVMs. An antibody is a binary string whose length D is determined by the number of total individual SVMs trained. Let $(a_{v_1}, a_{v_2}, \cdots, a_{v_D})$ denote an antibody, where $a_{v_i} = 0$ denotes the associated individual SVM is absent; $a_{v_i} = 1$ the associated individual SVM is present in the ensemble. When evaluating the avidity of a given antibody, the binary string is decoded to a combination of individual SVMs through removing the SVMs corresponding to $a_{v_i} = 0$. The initial antibody population is generated randomly, each antibody denotes a kind of combination of component SVMs. Evolving the population with ICA according to avidity $1/\hat{E}$, and one will get the optimal antibody whose avidity is the maximum. The final solution presents the appropriate ensemble of component SVMs with good generalization performance.

When implementing the clone operator on current parent population $A(k)$, the clonal size N_c of each antibody can be determined proportionally by the avidity between antibody and antigen or be a constant integer for convenience. Then we obtain $A(k)' = \{A(k), A_1'(k), A_2'(k), \cdots, A_{N_p}'(k)\}$. The clonal mutation operator is only implemented on $A'(k)$, the cloned part of $A(k)'$, which changes each of the bits based on the probability of mutation $p_m = 1/D$, and then $A''(k)$ is achieved. The clonal selection operator is carried out as follows. In subpopulation, if mutated antibody $b = \max\{f(a_{ij}) \mid j = 2, 3, \cdots, q_i - 1\}$ exists such that $f(a_i) < f(b)$, $a_i \in A(k)$, b replaces the antibody a_i and is added to the new parent population, namely, the antibodies are selected proportionally as the new population of next generation $A(k+1)$ based on the avidity. It is a map $I^{N_c(k)+n} \to I^n$, which realizes population compressing through selecting local optimum.

In our method, the ICA's termination criterion is triggered whenever the maximum number of generations is attained. Then the optimal antibody in current population is the final solution.

4 Experiment Results and Discussion

To evaluate the performance of the method proposed, seven data sets are selected for classification according to the data size from UCI machine learning repository [17].

Table 1 gives the characteristics of the data sets used in this paper. The missing values have been removed from Breast-cancer-W data set. Pen digits (Pen-based handwritten digits) data set has been normalized in advance.

Table 1. Data sets used for classification

Data set	Class	Attribute	Train size	Test size
Sonar	2	60	208	-
Glass	7	9	214	-
Breast-cancer-W	2	9	683	-
Segmentation	7	19	2310	-
Chess (kr-vs-kp)	2	35	3196	-
Waveform	3	22	5000	-
Pen digits	10	16	7494	3498

Except pen digits data, the available samples of each other dataset are randomly divided into two equivalent subsets, training set and test set. For each data, 20 SVMs are trained individually with 20 training sets that are bootstrap sampled from the original training set. The validation set for evaluating the generalization error of the ensemble in the selection is obtained from the original training set using bootstrap technique as well. For comparison, we use all SVMs trained (traditional bagging technique) and the selected SVMs using the proposed method to create an ensemble separately and make final prediction for test samples. For each data set, we perform 10 runs and record the average error rate and the standard deviation of the ensembles on the test set. The experimental results of two ensemble methods and a single SVM are listed in table 2.

In ICA, the size of the antibody population is 10 and the length of each antibody is the number of trained component SVMs, 20. The maximum evolutionary generation is 50 and the clonal size is 5. In SVMs, the RBF kernel function is used. The parameters of SVMs are experimentally chosen for each dataset.

Table 2. Comparison of the error rates and the standard deviations of single SVM, traditional bagging and selective SVMs ensemble

Data set	Single SVM	Traditional bagging	Selective ensemble	Num. of SVMs selected
Sonar	18.08 ± 4.02	18.08 ± 3.45	15.77 ± 2.49	8.8
Glass	36.17 ± 5.06	37.20 ± 6.37	31.40 ± 5.25	7.6
Breast-cancer-W	4.24 ± 0.85	4.15 ± 0.88	3.71 ± 0.85	8.4
Segmentation	3.84 ± 0.55	3.70 ± 0.59	2.99 ± 0. 49	10.2
Chess(kr-vs-kp)	1.80 ± 0.40	1.92 ± 0.35	1.48 ± 0.43	8.6
Waveform	16.73 ± 0.80	16.15 ± 0.50	15.00 ± 0.50	8.2
Pen digits	2.86 ± 0.00	2.50 ± 0.08	2.48 ± 0.11	10.5
Average	11.96 ± 1.67	11.96 ± 1.75	10.40 ± 1.45	8.9

Table 2 shows that the selective SVMs ensemble driven by ICA is better than a single SVM and traditional bagging algorithm. It is important that the average number of the SVMs comprised in the selective ensemble is approximately a half of the number of all trained SVMs, consequently the test time will be remarkably shorten compared to the traditional bagging algorithm. From the average result of all used data sets, we find that there is no significant difference between the performance of the traditional bagging algorithm and single SVM, which may be due to the fact that bagging algorithm works better when the individual classifiers are "unstable".

5 Conclusion

In this paper, we have proposed an algorithm for selective ensemble of SVMs driven by ICA. The characteristic of rapid convergence of ICA, which putts both the global and local searching into consideration, ensures that the satisfied ensemble of the selected SVMs trained can be achieved rapidly. Experiment results on some UCI data sets show that this approach outperforms single SVM and traditional bagging algorithm that ensembles all of the individual SVMs.

Further studies on the selective ensemble based on diversity measures of component classifiers and the applications of the method proposed in this paper will be made.

Acknowledgements. This work is supported by grants from the Key Program of National Natural Science Foundation of P.R.China (No. 60133010) and the National Defence Foundation of P.R.China for Key Lab (No. 51483050201DZ0102).

References

1. Vapnik, V.: The Nature of Statistical Learning Theory. Springer-Verlag, Berlin Heidelberg New York (1999)
2. Freund, Y.: Boosting a Weak Algorithm by Majority. Information and Computation. 121 (1995) 256–285
3. Breiman, L.: Bagging Predictors. Machine Learning. 24 (1996) 123–140
4. Osuna, E., Freund, R., Girosi, F.: Training Support Vector Machines: An Application to Face Detection. In Proceedings of IEEE Conference on Computer Vision and Pattern Recognition. (1997) 130–136
5. Scholkopf, B., Smola, A.J.: Learning with Kernels, The MIT Press, 2002
6. Chapelle, O., Haffner, P., Vapnik, V.: SVMs for Histogram-Based Image Classification. IEEE Trans. on Neural Networks. 10 (1999) 1055-1065
7. Je, H.M., Kim, D., Bang, S.Y.: Human Face Detection in Digital Video Using SVM Ensemble. Neural Processing Letters. 17 (2003) 239–252
8. Pang, S.N., Kim, D., Bang, S.Y.: Membership Authentication in the Dynamic Group by Face Classification Using SVM Ensemble. Pattern Recognition Letters. 24 (2003) 215–225
9. Hansen, L., Salamon, P.: Neural Network Ensembles. IEEE Trans. on Pattern Analysis and Machine Intelligence. 12 (1990) 993–1001

10. Krogh, A., Vedelsby, J.: Neural Network Ensembles, Cross Validation, and Active Learning. In Advances in Neural Information Processing Systems. Cambridge, MA: MIT Press. 7 (1995) 231–238
11. Zhou, Z.H., Wu, J.X., Tang, W.: Ensembling Neural Networks: Many Could Be Better Than All. Artificial Intelligence. 137 (2002) 239–263
12. Jiao, L.C., Du, H.F.: Development and Prospect of the Artificial Immune System. Acta Electronica Sinica. 31 (2003) 73–80
13. Zhang, X.R, Shan, T, Jiao, L.C.: SAR Image Classification Based on Immune Clonal Feature Selection. In: Aurélio, C.C, Mohamed, S.K. (Eds.): Proceedings of Image Analysis and Recognition. Lecture Notes in Computer Science, Vol. 3212. Springer-Verlag, Berlin Heidelberg New York (2004) 504-511
14. Tumer, K., Ghosh, J.: Error Correlation and Error Reduction in Ensemble Classifiers. Connection Science, Special Issue on Combining Artificial Neural Networks: Ensemble Approaches. 8 (1996) 385-404
15. Du, H.F., Jiao, L.C., Wang S.A.: Clonal Operator and Antibody Clone Algorithms. In: Proceedings of the First International Conference on Machine Learning and Cybernetics. Beijing. (2002) 506–510
16. Du, H.F., Jiao, L.C., Gong M.G., Liu R.C.: Adaptive Dynamic Clone Selection Algorithms. In: Pawlak, Z., Zadeh, L. (eds): Proceedings of the Fourth International Conference on Rough Sets and Current Trends in Computing. Uppsala, Sweden. (2004) 768-773
17. Blake, C.L., Merz, C.J.: UCI Repository of Machine Learning Databases [http://www.ics.uci.edu/~mlearn/MLRepository.html]. Irvine, CA: University of California, Department of Information and Computer Science. (1998)

Sensory-Motor Coordination in Gaze Control

G. de Croon, E.O. Postma, and H.J. van den Herik

IKAT, Universiteit Maastricht,
P.O. Box 616, 6200 MD, Maastricht, The Netherlands
{g.decroon, postma, herik}@cs.unimaas.nl
http://www.cs.unimaas.nl

Abstract. In the field of artificial intelligence, there is a considerable interest in the notion of sensory-motor coordination as an explanation for intelligent behaviour. However, there has been little research on sensory-motor coordination in tasks that go beyond low-level behavioural tasks. In this paper we show that sensory-motor coordination can also enhance performance on a high-level task: artificial gaze control for gender recognition in natural images. To investigate the advantage of sensory-motor coordination, we compare a non-situated model of gaze control (incapable of sensory-motor coordination) with a situated model of gaze control (capable of sensory-motor coordination). The non-situated model of gaze control shifts the gaze according to a fixed set of locations, optimised by an evolutionary algorithm. The situated model of gaze control determines gaze shifts on the basis of local inputs in a visual scene. An evolutionary algorithm optimises the model's gaze control policy. From the experiments performed, we may conclude that sensory-motor coordination contributes to artificial gaze control for the high-level task of gender recognition in natural images: the situated model outperforms the non-situated model. The mechanism of sensory-motor coordination establishes dependencies between multiple actions and observations that are exploited to optimise categorisation performance.

1 Introduction

In the field of artificial intelligence there is a considerable interest in situated models of intelligence that employ sensory-motor coordination to solve specific tasks [1, 2]. A situated model of intelligence is a model in which motor actions co-determine future sensory inputs. Together, the sensory inputs and the motor actions form a closed loop. Sensory-motor coordination exploits this closed loop in such a way that the performance on a particular task is optimised.

Several studies have investigated the mechanism of sensory-motor coordination [3, 4, 5, 6]. For instance, they show that sensory-motor coordination can simplify the execution of tasks, so that the performance is enhanced. However, until now, research on sensory-motor coordination has only examined low-level tasks, e.g., categorising geometrical forms [3, 4, 5, 6, 7]. It is unknown to what extent sensory-motor coordination can contribute to high-level tasks.

F. Rothlauf et al. (Eds.): EvoWorkshops 2005, LNCS 3449, pp. 334–344, 2005.

So, the research question in this subdomain of AI research reads: *Can sensory-motor coordination contribute to performance of situated models on high-level tasks?* In this paper we restrict ourselves to the analysis of two models both performing the same task, viz. gaze control for gender recognition in natural images. The motivation for the choice of this task is two-fold: (1) it is a challenging task, to which no situated gaze control models have been applied before; (2) it enables the comparison of two models that are identical, except for their capability to coordinate sensory inputs and motor actions. Thus, we will compare a situated with a non-situated model of gaze control. If the situated model's performance is better, we focus on a second research question: *How does the mechanism of sensory-motor coordination enhance the performance of the situated model on the task?* We explicitly state that we are interested in the relative performance of the models and the cause of an eventual difference in performance. It is not our intention to build the gender-recognition system with the best categorisation performance. Our only requirement is that the models perform above chance level (say 60% to 80%), so that a comparison is possible.

The rest of the paper is organised as follows. In Sect. 2 we describe the non-situated and the situated model of gaze control. In Sect. 3 we outline the experiment used to compare the two models of gaze control. In Sect. 4 we show the experimental results and analyse the gaze control policies involved. In Sect. 5 we discuss the relevance of the results. We draw our conclusions in Sect. 6.

2 Two Models of Gaze Control

Below, we describe the non-situated model of gaze control (Sect. 2.1) and the situated model of gaze control (Sect. 2.2). Then we discuss the adaptable parameters of both models (Sect. 2.3).

2.1 Non-situated Model of Gaze Control

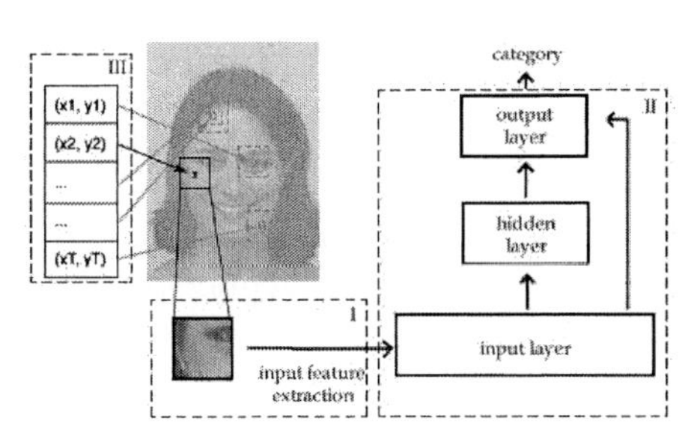

Fig. 1. Overview of the non-situated model of gaze control

The non-situated model consists of three modules. The first module receives the sensory input and extracts input features, given the current fixation location. The second module consists of a neural network that determines a categorisation based on the extracted input features. The third module controls the gaze shifts. Figure 1 shows an overview of the model. The three modules are illustrated by the dashed boxes, labelled 'I', 'II', and 'III'. The current fixation location is indicated by an 'x' in the face.

Module I receives the raw input from the window with centre x as sensory input. In Fig. 1 the raw input is shown on the left in box I; it contains a part of the face. From that window, input features are extracted (described later). These input features serve as input to module II, a neural network. The input layer of the neural network is illustrated by the box 'input layer'. Subsequently, the neural network calculates the activations of the hidden neurons in the 'hidden layer' and of the output neuron in the 'output layer'. There is one output neuron that indicates the category of the image. The third module determines the next fixation location, where the process is repeated. Below we describe the three modules of the non-situated model of gaze control in more detail.

Module I: Sensory Input. Here, we focus on the extraction procedure of the input features. For our research, we adopt the set of input features as introduced in [8], but we apply them differently.

An input feature represents the difference in mean light intensity between two areas in the raw input window. These areas are determined by the feature's type and location. Figure 2 shows eight different types of input features (top row) and nine differently sized locations in the raw input window from which the input features can be extracted (middle row, left). The sizes vary from the whole raw input window to a quarter of the raw input window. In total, there are $8 \times 9 = 72$ different input features. In the figure, two example input features are given (middle row, right). Example feature 'L' is a combination of the first type and the second location, example feature 'R' of the third type and the sixth location. The bottom row of the figure illustrates how an input feature is calculated. We calculate an input feature by subtracting the mean light intensity in the image covered by the grey surface from the mean light intensity in the image covered by the white surface. The result is a real number in the interval $[-1, 1]$. In the case of example feature L, only the left half of the raw input window is involved in the calculation. The mean light intensity in the raw input window of area 'A' is subtracted from the mean light intensity of area 'B'.

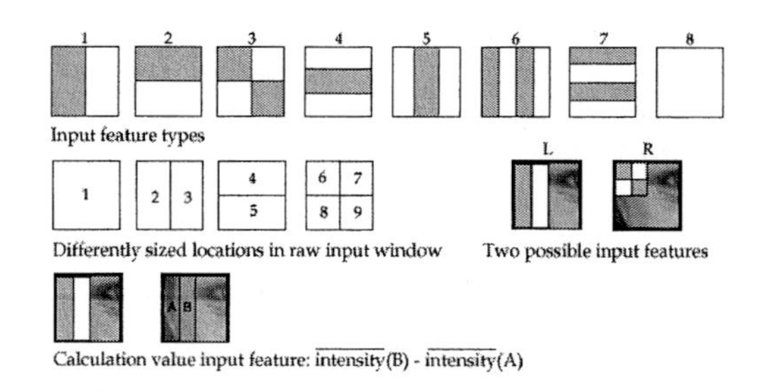

Fig. 2. An input feature consists of a type and a location

Module II: Neural Network. The second module is a neural network that takes the extracted input features as inputs. It is a fully-connected feedforward neural network with h hidden neurons and one output neuron. The hidden and output neurons all have sigmoid activation functions: $a(x) = \tanh(x)$, $a(x) \in \langle -1, 1 \rangle$. The activation of the output neuron, o_1, determines the categorisation (c) as follows.

$$c = \begin{cases} \text{Male} & \text{, if } o_1 > 0 \\ \text{Female} & \text{, if } o_1 \leq 0 \end{cases} \tag{1}$$

Module III: Fixation Locations. The third module controls the gaze in such a way, that for every image the same locations in the image are fixated. It contains coordinates that represent *all* locations that the non-situated model fixates. The model first shifts its gaze to location (x_1, y_1) and categorises the image. Then, it fixates the next location, (x_2, y_2), and again categorises the image. This process continues, so that the model fixates all locations from (x_1, y_1) to (x_T, y_T) in sequence, assigning a category to the image at every fixation. The performance is based on these categorisations (see Sect. 3.2). Out of all locations in an image, an evolutionary algorithm selects the T fixation locations. Selecting the fixation locations also implies selecting the order in which they are fixated.

2.2 Situated Model of Gaze Control

The situated model of gaze control (inspired by the model in [7]) is almost identical to the non-situated model of gaze control. The only difference is that the gaze shifts of the situated model are not determined by a third module, but by the neural network (Fig. 3). Therefore, the situated model has only two modules. Consequently, the current neural network has three output neurons. The first output neuron indicates the categorisation as in (1). The second and the third output neurons determine a gaze shift (Δx, Δy) as follows.

$$\Delta x = \lfloor m o_2 \rfloor \tag{2}$$

$$\Delta y = \lfloor m o_3 \rfloor , \tag{3}$$

where o_i, $i \in \{2, 3\}$, are the activations of the second and third output neurons. Moreover, m is the maximum number of pixels that the gaze can shift in the x- or y-direction. As a result, Δx and Δy are expressed in pixels. If a shift results in a fixation location outside of the image, the fixation location is repositioned to the nearest possible fixation location. In Fig. 3 'x' represents the current fixation location, and 'o' represents the new fixation location as determined by the neural network.

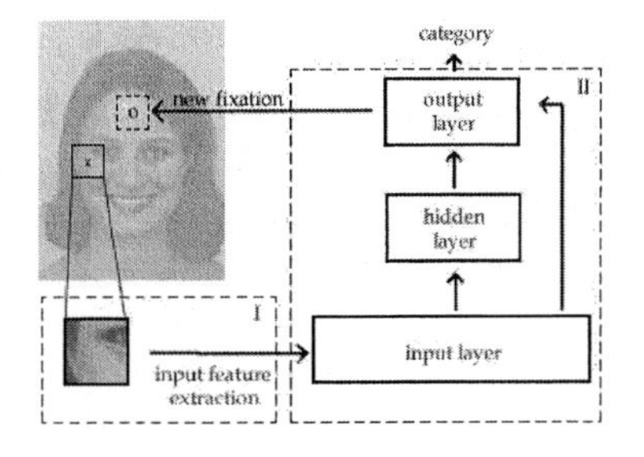

Fig. 3. Overview of the situated model of gaze control

2.3 Adaptable Parameters

In subsections 2.1 and 2.2 we described the non-situated and the situated model of gaze control. Four types of parameter values define specific instantiations of both models. We refer to these instantiations as agents. The four types of parameters are: the input features, the scale of the raw input window from which features are extracted, the neural network weights, and for the non-situated model the coordinates of all fixation locations. An evolutionary algorithm generates and optimises the agents (i.e., parameter values) by evaluating their performance on the gaze-control task.

3 Experimental Setup

In this section, we describe the gender-recognition task on which we compare the non-situated and the situated model of gaze control (Sect. 3.1). In addition, we discuss the evolutionary algorithm that optimises the models' adaptable parameters (Sect. 3.2). Finally, we mention the experimental settings (Sect. 3.3).

3.1 Gender-Recognition Task

Below, we motivate our choice for the task of gender recognition. Then we describe the data set used for the experiment. Finally, we outline the procedure of training and testing the two types of gaze-control models.

We choose the task of gender recognition in images containing photos of female or male faces, since it is a challenging and well-studied task [9]. There are many differences between male and female faces that can be exploited by gender-recognition algorithms [10, 11]. State-of-the-art algorithms use global features, extracted in a non-situated manner. So far, none of the current algorithms is based on gaze control with a local fixation window.

The data set for the experiment consists of images from J.E. Litton of the Karolinska Institutet in Sweden. It contains 278 images with angry-looking and happy-looking human subjects. These images are converted to gray-scale images and resized to 600×800 pixels.

One half of the image set serves as a training set for both the non-situated and the situated model of gaze control. Both models have to determine whether an image contains a photo of a male or female, based on the input features extracted from the gray-scale images. For the non-situated model, the sequence of T fixation locations is optimised by an evolutionary algorithm. For the situated model, the initial fixation location is defined to be the centre of the image and the subsequent $T-1$ fixation locations are determined by the gaze-shift output values of the neural network (outputs o_2 and o_3). At every fixation, the models have to assign a category to the image. After optimising categorisation on the training set, the remaining half of the image set is used as a test set to determine the performance of the optimised gaze-control models. Both training set and test set consist of 50% males and 50% females.

3.2 Evolutionary Algorithm

As stated in subsection 2.3, an evolutionary algorithm optimises the parameter values that define the non-situated and the situated agents, i.e., instantiations of the non-situated and situated model, respectively. We choose an evolutionary algorithm as our training paradigm, since it allows self-organisation of the closed loop of actions and inputs.

In our experiment, we perform 15 independent 'evolutionary runs' to obtain a reliable estimate of the average performance. Each evolutionary run starts by creating an initial population of M randomly initialised agents. Each agent operates on every image in the training set, and its performance is determined by the following fitness function:

$$f(a) = \frac{t_{c,I}}{IT} , \tag{4}$$

in which a represents the agent, $t_{c,I}$ is the number of time steps at which the agent correctly classified images from the training set, I is the number of images in the training set, and T is the total number of time steps (fixations) per image. We note that the product IT is a constant that normalises the performance measure. The $\frac{M}{2}$ agents with the highest performance are selected to form the population of the next generation. Their adaptable parameter sets are mutated with probability P_f for the input feature parameters and P_g for the other parameters, e.g., representing coordinates or network weights. If mutation occurs, a feature parameter is perturbed by adding a random number drawn from the interval $[-p_f, p_f]$. For other types of parameters, this interval is $[-p_g, p_g]$. For every evolutionary run, the selection and reproduction operations are performed for G generations.

3.3 Experimental Settings

In our experiment the models use ten input features. Furthermore, the neural networks of both models have 3 hidden neurons, $h = 3$. All weights of the neural networks are constrained to a fixed interval $[-r, r]$. Since preliminary experiments showed that evolved weights were often close to 0, we have chosen the weight range to be $[-1, 1]$, $r = 1$. The scale of the window from which the input features are extracted ranges from 50 to 150 pixels. Preliminary experiments showed that this range of scales is large enough to allow gender recognition, and small enough for local processing, which requires intelligent gaze control. The situated model's maximal gaze shift m is set to 500, so that the model can reach almost all locations in the image in one time step.

For the evolutionary algorithm we have chosen the following parameter settings: $M = 30$, $G = 300$, and $T = 5$. The choice of T turns out not to be critical to the results with respect to the difference in performance of the two models (see Sect. 1). The mutation parameters are: $P_f = 0.02$, $P_g = 0.10$, $p_f = 0.5$, and $p_g = 0.1$.

4 Results

In this section, we show the performances of both models (Sect. 4.1). Then we analyse the best situated agent to gain insight into the mechanism of sensory-motor coordination (Sect. 4.2).

4.1 Performance

Table 1 shows the mean performances on the test set (and standard deviation) of the best agents of the 15 evolutionary runs. Performance is expressed as the proportion of correct categorisations. The table shows that the mean performance of the best situated agents is 0.15 higher than that of the best non-situated agents. Figure 4 shows the histograms of the best performances obtained in the 15 runs for non-situated agents (white) and for situated agents (gray). Since both distributions of the performances are highly skewed, we applied a bootstrap method [12] to test the statistical significance of the results. It revealed that the difference between the mean performances of the two types of agents is significant ($p < 0.05$).

Table 1. Mean perf. ($\bar{f}$) and std. dev. (σ) of the performance on the test set of the best agents of the evolutionary runs

	$\bar{f}$	σ
Non-situated	0.60	0.057
Situated	0.75	0.055

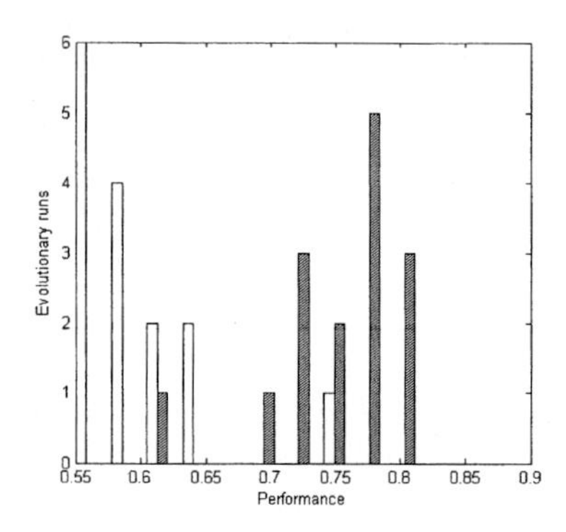

Fig. 4. Histograms of the best fitness of each evolutionary run. White bars are for non-situated agents, gray bars for situated agents

4.2 Analysis

In this subsection, we analyse the evolved gaze-control policy of the best situated agent of all evolutionary runs. The analysis clarifies how sensory-motor coordination optimises performance on the gender-recognition task.

The first part of the analysis shows that for each category the situated agent controls the gaze in a different way. This evolved behaviour aims at optimising performance by fixating suitable categorisation locations. The second part of the analysis shows that for individual images, too, the situated agent controls the gaze in different ways to fixate suitable categorisation locations.

Gaze Control per Category. Depending on the category, the situated agent fixates different locations. Below, we analyse per category the gaze path of the situated agent when it receives inputs that are typical of that category. The fixations take place at locations that are suitable for categorisation.

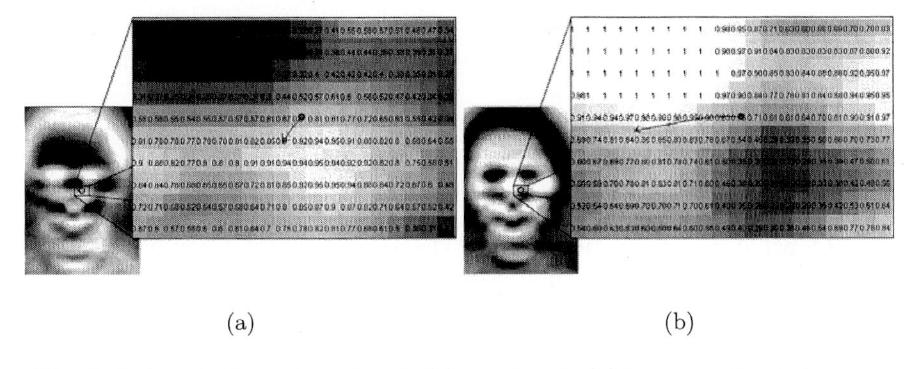

(a) (b)

Fig. 5. Categorisation ratios of male (a) and female (b) images in the training set

To find the suitable categorisation locations per category, we look at the situated agent's categorisation performance on the training set at all positions of a 100×100 grid superimposed on the image. At every position we determine the categorisation ratio for both classes. For the category 'male', the categorisation ratio is: $\frac{c_m(x,y)}{I_m}$, where $c_m(x,y)$ is the number of correctly categorised male images at (x,y), and I_m is the total number of male images in the training set. The left part of Fig. 5(a) shows a picture of the categorisation ratios represented as intensity for all locations. The highest intensity represents a categorisation ratio of 1. The left part of Fig. 5(b) shows the categorisation ratios for images containing females. The figures show that dark areas in Fig. 5(a) tend to have high intensity in Fig. 5(b) and vice versa. Hence, there is an obvious trade-off between good categorisation of males and good categorisation of females[1]. The presence of a trade-off implies that categorisation of males and females should ideally take place at different locations.

If we zoom into the area in which the agent fixates, we can see that it always moves its fixation location to an area in which it is better at categorising the presumed category. The right part of Fig. 5(a) zooms in on the categorisation ratios and shows the gaze path that results when the agent receives average male inputs at all fixation locations. The first fixation location is indicated by an 'o'-sign, the last fixation location by an arrow. Intermediate fixations are represented with the 'x'-sign. The black lines in Fig. 5(a) connect the fixation locations. The agent moves from a region with categorisation ratio 0.8 to a region with categorisation ratio 0.9. The right part of Fig. 5(b) shows the same information for images containing females, revealing a movement from a region with a categorisation ratio of 0.76 through a region with a ratio of 0.98. Both figures show that the situated agent takes misclassifications into account: it avoids areas in which the categorisation ratios for the other category are too low. For example, if we look at the right part of Fig. 5(a), we see that the agent fixates locations to the bottom left of the starting fixation, while the

[1] Note that the images are not inverted copies: in locations where male and female inputs are very different, good categorisation for both classes can be achieved.

categorisation ratios are even higher to the bottom right. The reason for this behaviour is that in that area, the categorisation ratios for female images are very low (Fig. 5(b)).

Non-situated agents cannot exploit the trade-off in categorisation ratios. They cannot select fixation locations depending on a presumed category, since the fixation locations are determined in advance for all images.

Gaze Control per Specific Image. The sensory-motor coordination of the situated agents goes further than selecting sensory inputs depending on the presumed category. The categorisation ratios do not explain the complete performance of situated agents. In this section we demonstrate that in specific images, situated agents often deviate from the exemplary gaze paths shown in Fig. 5(a) and 5(b) to search for facial properties that enhance categorisation performance.

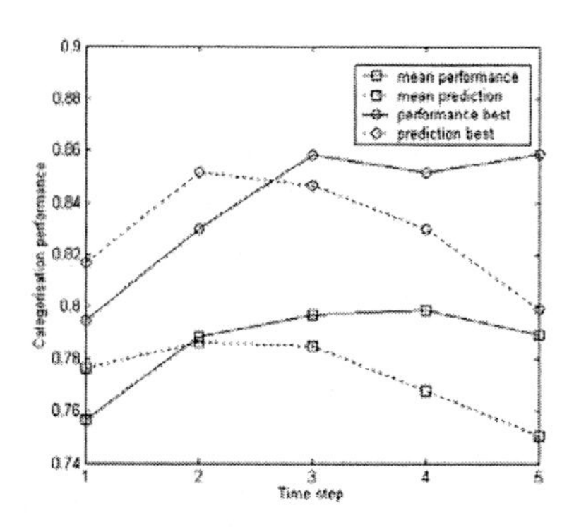

Fig. 6. Actual performance (solid lines) and predicted perf. (dotted lines) over time, averaged over all agents (squares) and for the best situated agent in particular (circles)

To see that the categorisation ratios do not explain the complete performance of situated agents, we compare the actual performance of situated agents over time with a predicted performance over time that is based on the categorisation ratios. For the prediction we assume that the categorisation ratios are conditional categorisation probabilities. The categorisation ratio $\frac{c_m(x,y)}{I_m}$ approximates $P_{x,y}(c = M \mid M)$ and $\frac{c_f(x,y)}{I_f}$ approximates $P_{x,y}(c = F \mid F)$. In addition, we assume that conditional probabilities at different locations are independent from each other. We determine the predicted performance of a situated agent by tracking its fixation locations over time for all images and averaging over the conditional categorisation probabilities at those locations. Figure 6 shows both the actual performance (solid lines) as the predicted performance (dotted lines) over time, averaged over all situated agents (squares) and for the best situated agent in particular (circles).

For the last three time steps the actual performances of the situated agents are higher than the predicted performances. The cause of this discrepancy is that the predicted performance is based on the assumption that the conditional categorisation probabilities at different positions are independent from each other. This assumption can be violated, for example, in the case of two adjacent locations.

The situated agent exploits the dependencies by using input features to shift gaze to fixation locations that are well suited for the task of gender recognition. For example, the best situated agent bases its categorisation partly on the eye-brows of a person. If the eye-brows of a male are lifted higher than usual, the agent occasionally fixates a location right and above of the starting fixation. This area is generally not good for male categorisation ($\frac{c_m(x,y)}{I_m} = 0.57$, see Fig. 5(a)), since eye-brows in our training set are usually not lifted. However, for some specific images it *is* a good area, because it contains a (part of a) lifted eye-brow.

The gaze control policy of the situated agents results in the optimisation of (actual) performance over time. Figure 6 shows that the actual performance augments after $t = 1$. The fact that performance generally increases over time reveals that sensory-motor coordination establishes dependencies between *multiple* actions and observations that are exploited to optimise categorisation performance. As mentioned in Sect. 3.3, other settings of T ($T > 1$) lead to similar results. Finally, we remark that for large T ($T > 20$), the performance of the non-situated model deteriorates due to the increased search space of fixation locations.

5 Discussion

We expect our results to generalise to other image classification tasks. Further analysis or empirical verification is necessary to confirm this expectation. Our results may be relevant to two research areas. First, the results may be relevant to the research area of computer vision. Most research on computer vision focuses on improving pre-processing (i.e., finding appropriate features) and on classification (i.e., mapping the features to an appropriate class) [13]. However, a few studies focus on a situated model (or 'closed-loop model') [14, 15]. Our study extends the application of a situated model using a local input window to the high-level task of gender recognition. Second, the results are related to research on human gaze control. Of course there is an enormous difference between the sensory-motor apparatus and neural apparatus of the situated model and that of a real human subject. Nonetheless, there might be parallels between the gaze-control policies of the situated model and that of human subjects. There are a few other studies that focus explicitly on the use of situated computational models in gaze control [16, 17], but they also rely on simplified visual environments. Our model may contribute to a better understanding of gaze control in realistic visual environments.

6 Conclusion

We may draw two conclusions as an answer to the research questions posed in the introduction. First, we conclude that sensory-motor coordination contributes to the performance of situated models on the high-level task of artificial gaze control for gender recognition in natural images. Second, we conclude that the mechanism of sensory-motor coordination optimises categorisation performance by establishing useful dependencies between multiple actions and observations; situated agents search adequate categorisation areas in the image by determining fixation locations that depend on the presumed image category and on specific image properties.

References

1. Pfeifer, R., Scheier, C.: Understanding Intelligence. MIT Press, Cambridge, MA (1999)
2. O'Regan, J.K., Noë, A.: A sensorimotor account of vision and visual consciousness. Behavioral and Brain Sciences **24:5** (2001) 883–917
3. Nolfi, S.: Power and the limits of reactive agents. Neurocomputing **42** (2002) 119–145
4. Nolfi, S., Marocco, D.: Evolving robots able to visually discriminate between objects with different size. International Journal of Robotics and Automation **17:4** (2002) 163–170
5. Beer, R.D.: The dynamics of active categorical perception in an evolved model agent. Adaptive Behavior **11:4** (2003) 209–243
6. van Dartel, M.F., Sprinkhuizen-Kuyper, I.G., Postma, E.O., van den Herik, H.J.: Reactive agents and perceptual ambiguity. Adaptive Behavior (in press)
7. Floreano, D., Kato, T., Marocco, D., Sauser, E.: Coevolution of active vision and feature selection. Biological Cybernetics **90:3** (2004) 218–228
8. Viola, P., Jones, M.J.: Robust real-time object detection. Cambridge Research Laboratory, Technical Report Series (2001)
9. Bruce, V., Young, A.: In the eye of the beholder. Oxford University Press (2000)
10. Moghaddam, B., Yang, M.H.: Learning gender with support faces. IEEE Trans. Pattern Analysis and Machine Intelligence **24:5** (2002) 707–711
11. Calder, A.J., Burton, A.M., Miller, P., Young, A.W., Akamatsu, S.: A principal component analysis of facial expressions. Vision research **41:9** (2001) 1179–1208
12. Cohen., P.: Empirical Methods for Artificial Intelligence. MIT Press, Cambridge, Massachusetts (1995)
13. Forsyth, D.A., Ponce, J.: Computer Vision: a Modern Approach. Prentice Hall, New Jersey (2003)
14. Köppen, M., Nickolay, B.: Design of image exploring agent using genetic programming. In: Proc. IIZUKA'96, Iizuka, Japan (1996) 549–552
15. Peng, J., Bhanu, B.: Closed-loop object recognition using reinforcement learning. IEEE Transactions on Pattern Analysis and Machine Intelligence **20:2** (1998) 139–154
16. Schlesinger, M., Parisi, D.: The agent-based approach: A new direction for computational models of development. Developmental review **21** (2001) 121–146
17. Sprague, N., Ballard, D.: Eye movements for reward maximization. In Thrun, S., Saul, L., Schölkopf, B., eds.: Advances in Neural Information Processing Systems 16. MIT Press, Cambridge, MA (2004)

Region Merging for Severe Oversegmented Images Using a Hierarchical Social Metaheuristic

Abraham Duarte[1], Ángel Sánchez[1], Felipe Fernández[2], and Antonio Sanz[1]

[1] Universidad Rey Juan Carlos, c/ Tulipán s/n,
28933 Móstoles, Madrid, Spain
{a.duarte, an.sanchez, a.sanz}@escet.urjc.es
[2] Dept. Tecnología Fotónica, FI-UPM, Campus de Montegancedo,
28860, Madrid, Spain
Felipe.Fernandez@es.bosch.com

Abstract. This paper proposes a new evolutionary region merging method to improve segmentation quality result on oversegmented images. The initial segmented image is described by a modified Region Adjacency Graph model. In a second phase, this graph is successively partitioned in a hierarchical fashion into two subgraphs, corresponding to the two most significant components of the actual image, until a termination condition is met. This graph-partitioning task is solved as a variant of the min-cut problem (normalized cut) using a Hierarchical Social (HS) metaheuristic. We applied the proposed approach on different standard test images, with high-quality visual and objective segmentation results.

1 Introduction

Image segmentation is one of the most complex stages in image analysis. It becomes essential for subsequent image description and recognition tasks. The problem consists in partitioning an image into its constituent regions or objects [1]. The level of division depends on the specific problem being solved. The segmentation result is the labelling of the image pixels that share any property (brightness, texture, colour…). The oversegmentation, which occurs when a single semantic object is divided into several regions, is a tendency of some segmentation methods like watersheds [2,3]. Therefore, some subsequent region merging process is needed to improve the segmentation results.

The proposed segmentation method can be considered as a region-based one and pursuits a high-level extraction of the image structures. After a required oversegmentation of the initial image, our method produces a hierarchical top-down region-based decomposition of the scene. The way to solve the segmentation problem is a pixel classification task, where each pixel is assigned to a class or region by considering only local information [1]. We take into account this pixel classification approach by representing the image as a simplified weighted graph, called Modified Region Adjacency Graph (MRAG). The application of a Hierarchical Social (HS) metaheuristic [4] to efficiently solve the normalized cut (NCut) problem for the image MRAG is the core of the proposed method. An evident computational advantage is

F. Rothlauf et al. (Eds.): EvoWorkshops 2005, LNCS 3449, pp. 345–355, 2005.

obtained describing the image by a set of regions instead of pixels in the MRAG structure. It enables a faster region merging in images with higher spatial resolution.

Today, the applications of evolutionary techniques to Image Processing and Computer Vision problems have increased mainly due to the robustness of these methods [5]. Evolutionary image segmentation [6,5,7] has reported a good performance in relation to more classical segmentation methods. Our approach of modelling and solving image segmentation as a graph-bipartitioning problem is related to Shi and Malik's work [8]. They use a computational technique based on a generalized eigenvalue problem for computing the segmentation regions. Instead, we found that high quality segmentation results can be obtained when applying an HS metaheuristic to image segmentation through a normalized cut solution.

2 Modified Region Adjacency Graph

Several techniques have been proposed to decrease the effect of oversegmentation on watershed-based approaches [2,3]. These usually involve a preprocessing of the original image. Many of them are based on the Region Adjacency Graph (RAG) which is a usual data structure for representing region neighbourhood relations in a segmented image [9].

As stated in [8,10] the image partitioning task is inherently hierarchical and it would be appropriate to develop a top-down segmentation strategy that returns a hierarchical partition of the image instead of a flat partition. Our approach shares this perspective and provides as segmentation result an adaptable tree-based image bipartition where the first levels of decomposition correspond to major areas or objects in the segmented image.

The MRAG structure takes advantage of both, region-based and pixel-based representations [8,11]. The MRAG structure is an undirected weighted graph $G=\{V,E,W\}$, where the set of nodes (V) represents the set of centres-of-gravity of each region. These regions result from the initial oversegmentated image. The set of edges (E) are the relationships between pairs of regions, and the edge weights (W) represent a similarity measure between pair of regions. In this context, the segmentation problem can be formulated as a graph bipartition problem, where the set V is partitioned into two subsets V_1 and V_2, with high similarity among vertices inside each subset and low similarity among vertices of different subsets.

As starting hypothesis, we suppose that each initial pre-segmented region must be small enough in size with respect to the original image and not having much semantic information. Some characteristics of the MRAG representation that yield to some advantages respect to RAG are:

1) It is defined once and it does not need from any dynamic updating when merging regions.

2) The number of pixels associated to each MRAG node (size of initial oversegmented regions) must be approximately the same.

3) MRAG-based segmentation approach is hierarchical and the number of final regions is controlled by the user according to the required segmentation precision.

4) The segmentation, formulated as a graph partition problem, leads to the fact that extracted objects are not necessarily connected.

The set of edge weights reflects the similarity between each pair of related regions (nodes) v_i and v_j. These connected components may or may not be adjacent, but if they are not adjacent, these components are close than a determined distance threshold r_x. The weights $w_{ij} \in W$ are computed by the conditional function:

$$if \left(\left|x_i - x_j\right| < r_x\right) \ then \ w_{ij} = e^{\frac{-C_{ij}(I_i - I_j)^2}{\sigma_I^2}} \cdot e^{\frac{-C_{ij}(x_i - x_j)^2}{\sigma_x^2}} \ else \ w_{ij} = 0 \tag{1}$$

where r_x, σ_x and σ_I are experimental values, I_i is the mean intensity of region i, and x_i is the spatial centre-of-gravity of that region. Finally, the factor C_{ij} takes into account the cardinality of the regions i and j. This value is given by:

$$C_{ij} = \frac{\| E_i \| \cdot \| E_j \|}{\| E_i \| + \| E_j \|} \tag{2}$$

where $\|E_i\|$, $\|E_j\|$ are, respectively, the number of pixels in regions v_i and v_j. Non-significant weighted edges, according to the defined similarity criteria, are removed from the image graph.

3 Image Partitioning via Graph Cuts

The recent literature has witnessed two popular image graph-based segmentation methods: the minimum cut (and their derivates) using graph cuts analysis [8,12,13] and the energy minimization, using the max flow algorithm [14,15]. More recently, it has been proposed a third major approach based on a generalization of Swendsen-Wang method [16]. In this paper, we focus on min-cut approach because they can be easily solved with an HS metaheuristic.

The min-cut optimization problem, defined for a weighted undirected graph $S=(V,E, W)$, consists in finding a bipartition G of the set of nodes of the graph: $G=(C,C')$ such that the sum of the weights of edges with endpoints in different subsets is minimized. Every partition of vertices V into C and C' is usually called a cut or cutset and the sum of the weights of the edges is called the weight of the cut or similarity (s) between C and C'. For the considered min-cut optimization problem, it is minimized the cut or similarity s, between C and C':

$$w(C,C') = s(C,C') = \sum_{v \in C, u \in C'} w_{vu} \tag{3}$$

In [17] is demonstrated that the decision version of Max-Cut (dual version of Min-Cut problem) is NP-Complete. This way, we need to use approximate algorithms for finding the solution in a reasonable time.

The Min-Cut approach has been used by Wu and Leahy [13] as a clustering method and applied to image segmentation. These authors look for a partition of the graph into k subgraphs such that the similarity (min-cut) among subgraphs is minimized. They pointed out that although in some images the segmentation is acceptable; in general, this method produces an oversegmentation because small regions are favoured. To avoid this fact, in [18] other functions that try to minimize the effect of this problem are proposed. The optimization function called min-max cut is:

$$cut(G) = \frac{\sum_{v\in C, u\in C'} w_{vu}}{\sum_{v\in C, u\in C} w_{vu}} + \frac{\sum_{v\in C, u\in C'} w_{vu}}{\sum_{v\in C', u\in C'} w_{vu}} = \frac{s(C,C')}{s(C,C)} + \frac{s(C,C')}{s(C',C')} \tag{4}$$

where the numerators of this expression are the similarity $s(C,C')$ and the denominators are the sum of the arc weights belonging to C or C', respectively. It is important to remark that in an image segmentation framework, it is necessary to minimize the similarity between C and C' (numerators of eq. 2) and maximize the similarity inside C, and inside C' (denominators of eq. 2). In this case, the sum of arcs between C and C' is minimized, and simultaneously the sums of weights inside of each subset are maximized. Other authors [2] propose an alternative cut value called *normalized cut (NCut)*, which, in general, gives better results in practical image segmentation problems. Mathematically this cut is defined as:

$$Ncut(G) = \frac{\sum_{v\in C, u\in C'} w_{vu}}{\sum_{v\in C, u\in C\cup C'} w_{vu}} + \frac{\sum_{v\in C, u\in C'} w_{vu}}{\sum_{v\in C', u\in C\cup C'} w_{vu}} = \frac{s(C,C')}{s(C,G)} + \frac{s(C,C')}{s(C',G)} \tag{5}$$

where $G = C \cup C'$.

4 Hierarchical Social (HS) Algorithms

This section shows general features of a new evolutionary metaheuristic called hierarchical social (HS) algorithm. In order to get a more general description of this metaheuristic, the reader is pointed to references [4,19,20,21,22]. This metaheuristic has been successfully applied to several problems such as: critical circuit computation [22], scheduling [4,21], MAX-CUT problem [19] and region-based segmentation[20].

HS algorithms are inspired in the hierarchical social behaviour observed in a great diversity of human organizations. The key idea of HS algorithms consists in a simultaneous optimization of a set of disjoint solutions. Each group of a society contains a feasible solution. These groups are initially randomly distributed to produce a disjoint partition of the solution space. Better solutions are obtained using intra-group cooperation and inter-group competition as evolution strategies. Through this process groups with lower quality tend to disappear. As a result, the objective functions of winners groups are optimized. The process typically ends with only one group that contains the best solution.

4.1 Metaheuristic Structure

For the image segmentation problem, the feasible society is modelled by the specified undirected weighted graph, also called feasible society graph. The set of individuals are modelled by nodes of the graph V and the set of feasible relations are modelled by edges E of the specified graph. The set of similarity relations are described by the weights W. Notice that when the graph also models an image, nodes represent initial watershed resulting regions and edges model the similarity between these regions.

Figure 1.a shows an example of a feasible society graph, which represents a simple synthetic image with two major dark and white squares. This image is a noisy and deformed chess board. In figure 1.b is shown the watershed segmentation of the image presented in figure 1.a. In this image there are 36 regions, 9 regions in each square. Figure 1.c shows the MRAG built from the watershed image. Obviously, the graph has 36 nodes.

The state of a society is modelled by a hierarchical policy graph [4,22]. This graph also specifies a society partition composed by a disjoint set of groups $\Pi=\{g_1,g_2,\ldots,g_g\}$, where each individual or node is assigned to a group. Each group $g_i \subset S$ is composed by a set of individuals and active relations, which are constrained by the feasible society. The individuals of all groups cover the individuals of the whole society. Notice that each group exactly contains one solution.

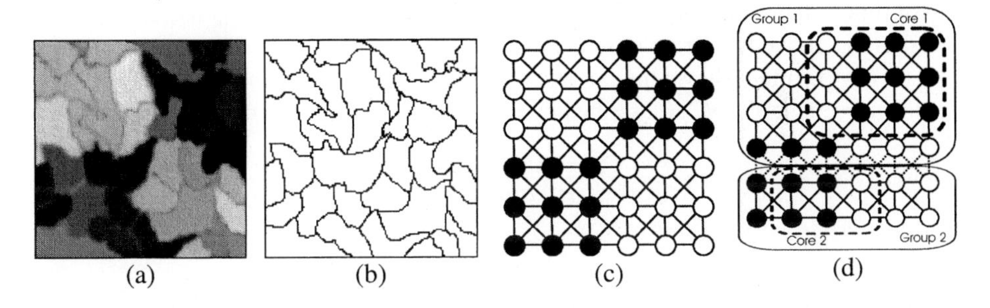

Fig. 1. (a) Synthetic chess board image. (b) Watershed segmentation. (c) Feasible society graph. (b) Society partition and groups partition

The specification of the hierarchical policy graph is problem dependent. The initial society partition determines an arbitrary number of groups and assigns individuals to groups. Figure 1.d shows a society partition example formed by two groups.

Each individual of a society has two objective functions: *individual objective function f1* and *group objective function f2* that is shared by all individuals in the same group. Furthermore each group g_i is divided into two disjoint parts: *core* and *periphery*. The core determines the value of the corresponding group objective function *f2* and the periphery defines the alternative search region of the group.

In the image segmentation framework, the set of nodes of each group g_i is divided into two disjoint parts: $g_i = (C_i, C'_i)$ where C_i is the core or group of nodes belonging to the considered cutset and C_i' is the complementary group of nodes. The core edges are the arcs that have their endpoints in C_i and C_i'. Figure 1.d also shows an example of core for the previous considered partition. The core nodes of each group are delimited by one dotted line. For each group of nodes $g_i = (C_i, C'_i)$, the group objective function $f2(i)$ is given by the corresponding normalized cut $Ncut(i)$ referred to the involved group g_i:

$$f2(i) = NCut\ (i) = \frac{\sum_{v\in C_i, u\in C_i'} w_{vu}}{\sum_{v\in C_i, u\in C_i\cup C_i'} w_{vu}} + \frac{\sum_{v\in C_i, u\in C_i'} w_{vu}}{\sum_{v\in C_i', u\in C_i\cup C_i'} w_{vu}} = \frac{s(C_i, C_i')}{s(C_i, g_i)} + \frac{s(C_i, C_i')}{s(C_i', g_i)}$$
$$\forall v \in g_i \quad f2(v,i) = f2(i) = NCut\ (i) \tag{6}$$

where $gi = Ci\cup Ci'$ and the weights w_{vu} are supposed to be null for the edges that do not belong to the specified graph.

For each individual or node v, the individual objective function $f1(v,i)$ relative to each group g_i is specified by a function that computes the increment in the group objective function when an individual makes a movement. There are two types of movements: *intra-group movement* and *inter-group movement*. In the intra-group movement there are two possibilities: the first one consists in a movement from $\boldsymbol{C_i}$ ***to*** $\boldsymbol{C_i'}$, the second one is the reverse movement ($\boldsymbol{C'_i}$ ***to*** $\boldsymbol{C_i}$).

The inter-group movement is accomplished by an individual v that belongs to a generic group g_x ($g_x = \Pi \setminus g_i$) that wants to move from g_x to g_i. There are two possibilities: the first one consists in a movement from $\boldsymbol{g_x}$ ***to*** $\boldsymbol{C_i}$, the second one consists in a movement from $\boldsymbol{g_x}$ ***to*** $\boldsymbol{C'_i}$.

The next formula shows the incremental computation of the individual function $f1$ for the movement $C_i \to C'_i$.(described by the function $C\_to\_C'(v,i)$).

$$f1(v,i) = C\_to\_C'(v,i) = \frac{s(C_i, C_i') - \sigma'(i) + \sigma(i)}{s(C_i, gi) - \sigma'(i)} + \frac{s(C_i, C_i') - \sigma'(v,i) + \sigma(v,i)}{s(C_i', g_i) + \sigma(v,i)}$$
$$\text{where} \quad \sigma(v,i) = \sum_{u\in C_i} w_{vu} \ \ and \ \ \sigma'(v,i) = \sum_{u\in C_i'} w_{vu} \tag{7}$$

The other movements ($C'_i \to C_i$, $X \to C_i$,$X \to C'_i$) have similar expressions. During a competitive strategy, function $f1$ allows selecting for each individual v, the group that achieves the corresponding minimum value.

The HS algorithms here considered, try to optimize one of their objective functions ($f1$ or $f2$) depending on the operation phase. During cooperative phase, each group g_i aims to improve independently the group objective function $f2$. During a competitive phase, each individual tries to improve the individual objective function $f1$, the original groups cohesion disappeared and the graph partition is modified in order to optimize the corresponding individual objective function.

4.2 Metaheuristic Process

The algorithm starts from a random set of feasible solutions. Additionally for each group, an initial random cutset is derived. The groups are successively transformed through a set of social evolution strategies. For each group, there are two main strategies: *intra-group cooperative strategy* and *inter-group competitive strategy*. The first strategy can be considered as a local search procedure in which the quality of the solution contained in each group is autonomously improved. This process is maintained during a determined number of iterations (autonomous iterations).

The intra-group competitive strategy can be considered as a constructive procedure and is oriented to let the interchange of individuals among groups. In this way the groups with lower quality tend to disappear because their individuals move from these groups to another ones with higher quality.

Cooperative and competitive strategies are the basic search tools of HS algorithms. These strategies produce a dynamical groups partition, where group annexations and extinctions are possible. A detailed description of HS algorithms, cooperative strategy and competitive strategy and their corresponding pseudo-codes can be found in [4].

5 Method Overview

Figure 2 outlines the three major stages considered in the proposed evolutionary segmentation approach. First, we create an over-segmented image applying a standard watershed segmentation to the initial brightness image.

In the second stage, the corresponding MRAG for the oversegmented image is built. This graph is defined by representing each resulting region by a unique node and defining the edges and corresponding edge weights as a measure of spatial location, grey level average difference and cardinality between the corresponding regions (see Eq.1 in Section 2).

The third major stage consists in iteratively applying the considered HS metaheuristic in a hierarchical fashion to the corresponding subgraph, resulting from the previous graph bipartition, until a termination condition is met. This stage itself constitutes an effective region merging for oversegmented images.

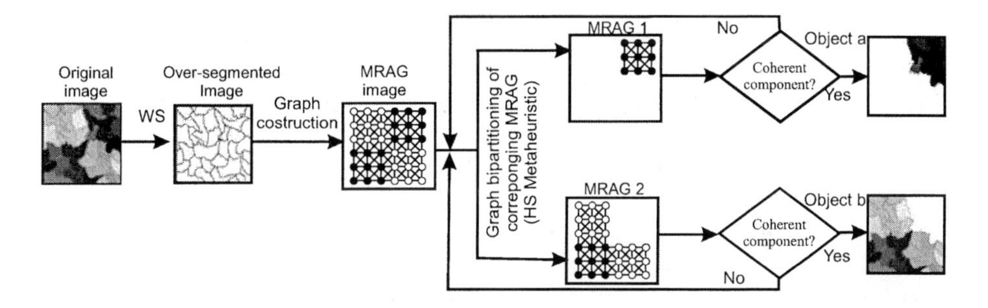

Fig. 2. Block diagram of the proposed method

6 Experimental Results

Table 1 shows the standard test images used, the characteristics of the corresponding MRAG and the value of the first *NCut* after the application of the HS metaheuristic. Considered characteristics of MRAG are: number of nodes (regions of the watershed oversegmented image), number of edges and parameters of the weight function $(\sigma_I, \sigma_x, r_x)$. The last column shows the first *NCut* value for the first MRAG bipartition, which can be considered as a quantitative measure of the segmentation quality [8].

Figure 3.a shows the input image (*Lenna*), its corresponding oversegmented image. Figure 3.b shows the obtained oversegmentation by means of a watershed algorithm.

Table 1. Image characteristics and quantitative results

	MRAG	*MRAG Parameters*				*HSA*
Image	*Nodes*	*Arcs*	σ_I	σ_X	r_X	*NCut*
Lenna256x256	7156	1812344	200	200	40	0.0550
Windsurf480x320	11155	1817351	200	200	35	0.0396
Cameraman256x256	4181	460178	100	200	35	0.0497

The resulting segmentation tree (Figure 3.c) gives a hierarchical view of the segmentation process. In the first phase of the algorithm, the original image is split into two parts (Figures 3.d and 3.h). Notice that the segmented objects are not connected. This property is especially interesting in images with partially occluded objects, noisy images, etc.

The most important part (Figure 3.d) is split again, obtaining the images presented in Figure 3.e and 3.i. As in the previous case, the most significant region (Figure 3.e) is split again, obtaining the images 3.f and 3.j. This process can be repeated until a determined minimum *NCut* value is obtained or the process is stopped by the user. The segmented image is given by the union of the final components. The resulting objects correspond to the tree segmentation leafs. For *Lenna* image, a high segmentation quality is achieved. Note that the images presented in the rest of figures (Figures 3.g, 3.h, 3.i, 3.j and 3.k) could be also bipartitioned, in order to achieve a more detailed segmentation.

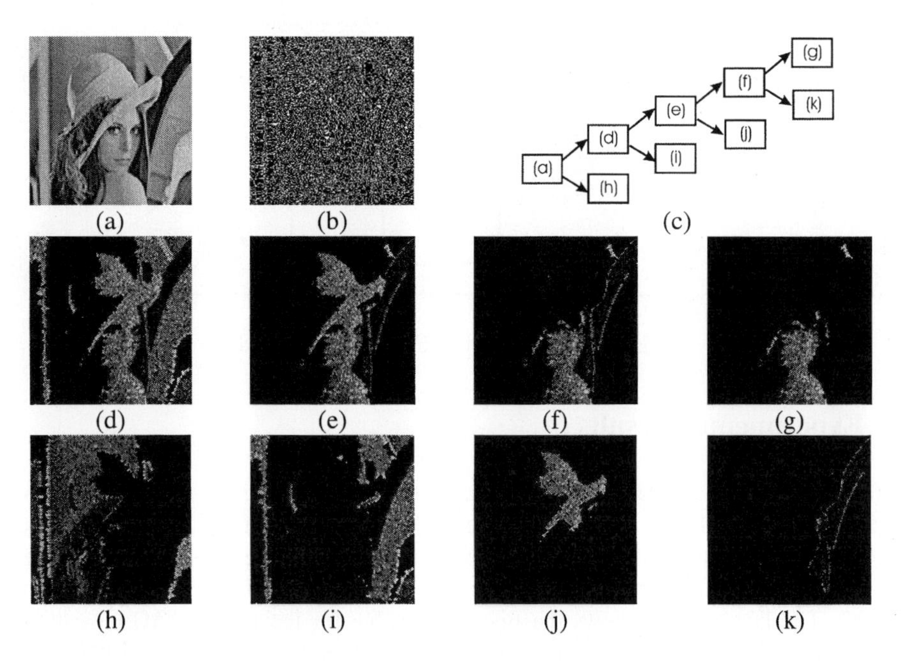

Fig. 3. Segmentation results for image *Lenna*: (a) Initial image. (b) Watershed segmentation (c) Structure of the segmentation tree. (d),...,(k) Resulting segmented regions according to the segmentation tree

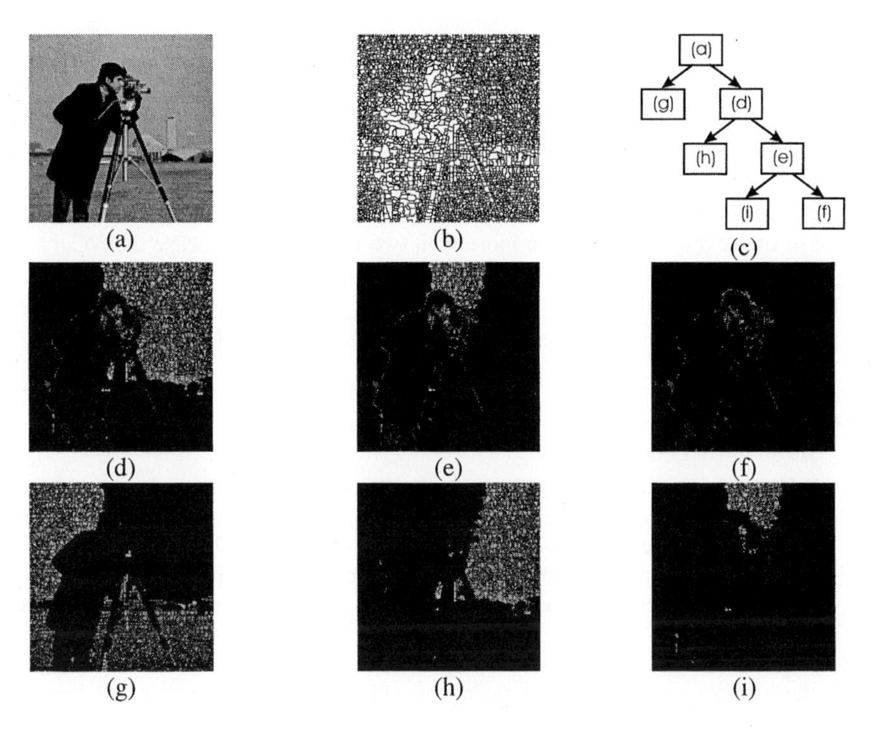

Fig. 4. Segmentation results for image *Cameraman*: (a) Initial image. (b) Watershed segmentation (c) Structure of the segmentation tree. (d),...,(i) Resulting segmented regions according to the segmentation tree

The segmentation process can have some peculiarities relative to the obtained *NCut*. Sometimes, the obtained segmentation contains spurious cuts that do not correspond to objects. An example of this phenomenon can be observed in Figure 4.h, where the background is segmented into two regions.

This fact occurs because *NCut* favours approximately equal size cuts. Sometimes, these spurious cuts do not affect the segmentation results, as in this case, because in the next cut it is extracted the rest of the main information. If an important object has been split, the algorithm can not correctly extract the corresponding object. In this case, a different choice of the edge weights (similarity measure) or metaheuristic parameters should be considered to improve the segmentation results.

7 Conclusions

This paper has introduced an HS metaheuristic, as a region merging technique, to efficiently improve the image segmentation quality results. Also, a new RAG is proposed, called MRAG. This representation considers neighbourhood relations between pair of regions that are not adjacent. This new model allows the processing of larger spatial resolution images than other typical graph-based segmentation methods [10, 8]. The image problem is now equivalent to minimize the *NCut* value in the corresponding MRAG. As we have experimentally shown, the HS algorithms provide an effective region merging method for achieving high quality segmentation. An important advantage of the approach is that MRAG structure does not need to be updated when merging regions. Moreover, the resulting hierarchical top-down segmentation is adaptable to the complexity of the considered image.

The capability of the method can be improved by decomposing the image at each level of the segmentation tree in more than two regions. In this case the *NCut* value is not an adequate group objective function, because it is not defined for several cuts. We propose as a future work the use of other group objective functions in order to exploit all the potential of HS metaheuristics for segmentation applications.

References

1. R.C. Gonzalez and R. Woods, *Digital Image Processing*, 2nd Edition, Prentice Hall, 2002.
2. K.Haris,et al, " Hybrid Image Segmentation Using Watersheds and Fast Region Merging", IEEE Trans. on Image Processing, v.7, n.12, pp. 1684-1699, 1998.
3. S. E. Hernández and K.E. Barner, "Joint Region Merging Criteria for Watershed-Based Image Segmentation", International Conference on Image Processing, v. 2, pp.108-111, 2000.
4. A. Duarte, "Algoritmos Sociales Jerárquicos: Una metaheurística basada en la hibridación entre métodos constructivos y evolutivos", PhD Thesis, Universidad Rey Juan Carlos, Spain, 2004.
5. R. Poli, "Genetic programming for image analysis", J. Koza (ed): *Genetic Progr.,* 1996.
6. S.Y. Ho and K.Z. Lee, "Design and Analysis of an Efficient Evolutionary Image Segmentation Algorithm, *J. VLSI* Signal Processing, Vol 35, pp. 29-42, 2003.
7. M. Yoshimura and S. Oe, "Evolutionary Segmentation of Texture Image using Genetic Algorithms", *Pattern Recognition*, Vol. 32, pp. 2041-2054, 1999.
8. J. Shi and J. Malik, "Normalized Cuts and Image Segmentation", *IEEE Trans. Pattern Analysis and Machine Intelligence*,V. 22, no. 8, pp. 888-905, Aug. 2000.
9. M. Sonka et al. *Image Processing, Analysis and Machine Vision*, 2nd Ed., PWS, 1999.
10. P. F. Felzenszwalb and D. P. Huttenlocher, " Efficient Graph-Based Image Segmentation", International Journal of Computer Vision, Kluwer 59(2), 167–181, 2004
11. A. Gothandaraman, "Hierarchical Image Segmentation using the Watershed Algorithm with a Streaming Implementation", PhD Thesis, University fo Tennessee, USA, 2004
12. O. Veskler "Image Segmentation by Nested Cuts", In Proc. of IEEE CVPR Conf, June 2000, p.339-344
13. Z. Wu et al, "*Optimal Graph Theoretic Approach to Data Clustering: Theory and its Application to Image Segmentation*", IEEE Trans. PAMI, V. 15, n. 11, pp. 1101-1113, 1993.
14. V. Kolmogorov and R. Zabih, "What energy functions can be minimized via graph cuts?", Proc. ECCV, pp. 65-81. vol. 3, Copenhagen, Denmark, 2002.
15. S. Roy, I. Cox, "A maximum-flow formulation of the n-camera stereo correspondence problem", Proc. ICCV Conference, 1998.
16. A.Barbu and S.Zhu, "Graph Partitioning by Swendsen-Wang Cuts", J. *of Pattern Recognition and Machine Intelligence*. to appear in 2005.
17. R.M. Karp, Reducibility among Combinatorial Problems, R. Miller and J. Thatcher (eds.): Complexity of Computer Computations, Plenum Press, pp. 85-103, 1972.

18. C. Ding, X. He, H. Zha, M. Gu and H. Simon, "A Min-Max Cut Algorithm for Graph Partitioning and Data Clustering", Proc. of ICDM Conference, 2001.
19. A. Duarte, F. Fernández, A. Sánchez and A. Sanz, "A Hierarchical Social Metaheuristic for the Max-Cut Problem", Lecture Notes in Compute Science, v. 3004, pp. 84-93, 2004.
20. A. Duarte, F. Fernández, A. Sánchez, A. Sanz, J.J. Pantrigo, "Top-Down Evolutionary Image Segmentation using a Hierarchical Social Metaheuristic ", LNCS, v. 3005, pp.301-310, 2004.
21. A. Duarte, F. Fernández, A. Sánchez, "Software Pipelining using Hierarchical Social Metaheuristic", In Proc. of RASC'04.
22. F. Fernández, A. Duarte and A. Sánchez, "A Software Pipelining Method based on a Hierarchical Social Algorithm", *Proc. 1st* MISTA'03 Conference, pp. 382-385, 2003.

Automated Photogrammetric Network Design Using the Parisian Approach

Enrique Dunn[1], Gustavo Olague[1], and Evelyne Lutton[2]

[1] Centro de Investigación Científica y Educación Superior de Ensenada,
División de Física Aplicada, EvoVisión Laboratory
{edunn, olague}@cicese.mx
[2] INRIA - COMPLEX Team,
Domaine de Voluceau BP 105 78153 Le Chesnay Cedex - France
Evelyne.Lutton@inria.fr

Abstract. We present a novel camera network design methodology based on the *Parisian* approach to evolutionary computation. The problem is partitioned into a set of homogeneous elements, whose individual contribution to the problem solution can be evaluated separately. These elements are allocated in a population with the goal of creating a single solution by a process of aggregation. Thus, the goal of the evolutionary process is to generate individuals that jointly form better solutions. Under the proposed paradigm, aspects such as problem decomposition and representation, as well as local and global fitness integration need to be addressed. Experimental results illustrate significant improvements, in terms of solution quality and computational cost, when compared to canonical evolutionary approaches.

1 Introduction

Automatic camera placement for artificial perception tasks consists on deciding the position of a set of sensors with respect to an observed scene. Depending on the selected task, the resulting problem offers an intricate combination of interactions between the sensor physical constraints, the mathematical modeling of the problem, as well as the numerical methods used to solve it [1], [2]. Accurate 3D reconstruction is a particularly difficult problem that needs to be automated. The complexity is mainly due to the stochastic nature of the uncertainty assessment process which requires multiple redundant image measurements [3]. Indeed, a difficult numerical adjustment problem for 3D reconstruction arises along with a highly discontinuous design search space for imaging geometry. The Bundle Method for optical triangulation is a fundamental aspect involved in the attainment of precise mensuration in photogrammetric projects. Indeed, the lack of a widespread utilization outside this community can be attributed to its expensive computational requirements and inherent design complexity. However, the development of an effective camera network configuration should also be based on a rigorous photogrammetric approach. This work presents the ongoing

F. Rothlauf et al. (Eds.): EvoWorkshops 2005, LNCS 3449, pp. 356–365, 2005.

development of the EPOCA [4],[5] sensor planning system and implements an evolutionary computation methodology based on the Parisian approach. This is done in order to efficiently search the space of possible camera configurations while maintaining high qualitative solutions of the photogrammetric adjustment process.

Photogrammetric Network Design is an active research field in photogrammetry, see [6], where recent works have provided important insights into the problem of determining an optimal imaging geometry. Mason [7] proposed an expert system approach based on the theory of generic networks in order to automate the viewpoint selection process. Thus, the decision making is carried out by heuristic means utilizing extensive expert prior knowledge. On the other hand, the work of Olague [8] uses an evolutionary computation approach, developing a criterion based on the error propagation phenomena. In this way, the design search space is explored by an stochastic meta-heuristic that yields human competitive results.

2 Problem Statement: Photogrammetric Network Design

Accuracy assessment of visual 3D reconstruction consists on attaining some characterization of the uncertainty of our results. The design of a photogrammetric network is the process of determining an imaging geometry that allows accurate 3D reconstruction. Rigorous photogrammetric approaches toward optical triangulation are based on the bundle adjustment method [9], which simultaneously refines scene structure and viewing parameters for multi-station camera networks. Under this nonlinear optimization procedure, the image forming process is described by separate functional and stochastic models. The functional model is based on the collinearity equations given by $s(\mathbf{p}-\mathbf{c}^p) = \mathbf{R}(\mathbf{P}-\mathbf{C}^o)$, where s is a scale factor, $\mathbf{p}$=$(x, y, -f)$ is the projection of an object feature into the image, $\mathbf{c}^p = (x^p, y^p, 0)$ is the principal point of the camera , $\mathbf{P} = (X, Y, Z)$ represents the position of the object feature, $\mathbf{C}^o = (X^o, Y^o, Z^o)$ denotes the optical center of the camera, while $\mathbf{R}$ is a rotation matrix expressing its orientation.

This formulation is readily extensible to multiple features across several images. For multiple observations a system of the form $\mathbf{l} = f(\mathbf{x})$ is obtained after rearranging and linearizing the collinearity equations, where $\mathbf{l} = (x_i, y_i)$ are the observations and $\mathbf{x}$ the viewing and scene parameters. Introducing a measurement error vector $\mathbf{e}$ we obtain a functional model of the form $\mathbf{l} - \mathbf{e} = \mathbf{A}\mathbf{x}$.

The design matrix $\mathbf{A}$ is of dimension $n \times u$, where n is the number of observations and u the number of unknown parameters. Assuming the expectancy $E(\mathbf{e}) = 0$ and the dispersion operator $D(\mathbf{e}) = E(\mathbf{e}\mathbf{e}^t) = \sigma_0^2\mathbf{W}^{-1}$, where $\mathbf{W}$ is the "weight coefficient" matrix of observations, we obtain the corresponding stochastic model: $E(\mathbf{l}) = \mathbf{A}\mathbf{x}\Sigma_{ll} = \Sigma_{ee} = \sigma_0^2\mathbf{W}^{-1}$. Here Σ is the covariance operator and σ_0^2 the variance factor. The estimation of $\mathbf{x}$ and σ_0^2 can be performed by least squares adjustment in the following form

$$\mathbf{x} = (\mathbf{A}^{\mathrm{T}}\mathbf{W}\mathbf{A})^{-1}\mathbf{A}^{\mathrm{T}}\mathbf{W}\mathbf{l} = \mathbf{Q}\mathbf{A}^{\mathrm{T}}\mathbf{W}\mathbf{l},$$

$$\mathbf{v} = \mathbf{A}\mathbf{x} - \mathbf{l} \qquad \sigma_0^2 = \frac{\mathbf{v}^t\mathbf{W}\mathbf{v}}{r}$$

where r is the number of redundant observations, $\mathbf{v}$ is the vector of residuals after least squares adjustment and $\mathbf{Q}$ is the corresponding cofactor matrix. Additionally, the covariance of the parameters is given by $\Sigma_{xx} = \sigma_0^2\mathbf{Q}$. The vector of parameters can be separated in the form $\mathbf{x} =< \mathbf{x_1}, \mathbf{x_2} >$, where $\mathbf{x_1}$ contains the viewing parameters while $\mathbf{x_2}$ expresses the scene structure correction parameters. Thus, we obtain a system of the form

$$\begin{pmatrix} \mathbf{x}_1 \\ \mathbf{x}_2 \end{pmatrix} = \begin{pmatrix} \mathbf{A}_1^T\mathbf{W}\mathbf{A}_1 & \mathbf{A}_1^T\mathbf{W}\mathbf{A}_2 \\ \mathbf{A}_2^T\mathbf{W}\mathbf{A}_1 & \mathbf{A}_2^T\mathbf{W}\mathbf{A}_2 \end{pmatrix}^{-1} \begin{pmatrix} \mathbf{A}^T\mathbf{W}\mathbf{l} \\ \mathbf{A}^T\mathbf{W}\mathbf{l} \end{pmatrix}$$

Accordingly, the cofactor matrix $\mathbf{Q}$ can be written as

$$\mathbf{Q} = \begin{pmatrix} \mathbf{Q}_1 & \mathbf{Q}_{1,2} \\ \mathbf{Q}_{2,1} & \mathbf{Q}_2 \end{pmatrix}$$

The matrix $\mathbf{Q}_2$ describes the covariance structure of scene coordinate corrections. Hence, an optimal form of this matrix is sought in order to obtain accurate scene reconstruction. The criterion we selected for minimization is the average variance along the covariance matrix, see [8],

$$f_1(\mathbf{x_1}, \mathbf{x_2}) = \sigma_{\mathbf{x}_2}^2 = \frac{\sigma_0^2 \, trace(\mathbf{Q}_2)}{3n}. \tag{1}$$

3 The Parisian Approach

The Parisian Approach differs from typical approaches to evolutionary computation in the sense that a single individual in the population represents only a part of the problem solution. In this respect, it is similar to the Michigan approach developed for Classifier Systems. Hence, an aggregation of multiple individuals must be considered in order to obtain a solution for the problem being studied. Thus, the evolution of the whole population is favored instead of the emergence of only a single dominant solution. The motivation for such an approach is to make an efficient use of the genetic search process. This can be achieved from two different perspectives. First, the algorithm discards less computational effort at the end of execution, while considering more than a single best individual as output. Second, the computational expense of the fitness function evaluation is considerably reduced for a single individual.

Successful examples of such an approach can be found in the image analysis and signal processing literature. The *Fly Algorithm* developed by Louchet et al. [10] is a real-time pattern recognition tool used in stereo vision systems. In such a work, the population is formed by individuals representing each a single 3D point. The evolutionary algorithm favors the positioning of each so called "fly" to a surface point in the observed scene using insightful problem modeling. The work of Raynal et al.[11] incorporates the Parisian approach to the solution

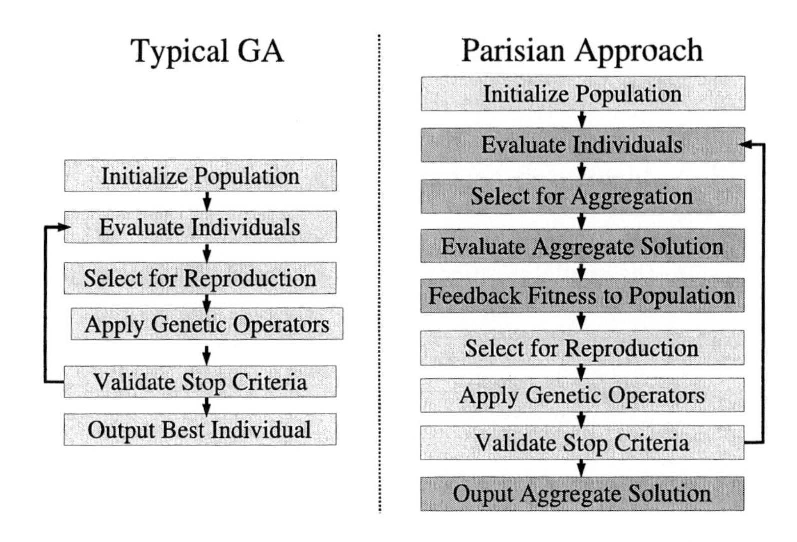

Fig. 1. Outline of our implementation of the *Parisian Sensor Planning* approach. Fitness evaluation is modified in order to consider the local and global contribution of an individual

of the inverse problem for Iterated Function Systems (IFS). In this instance a Genetic Programming methodology was adopted and experimentation on 2D images presented.

Many of the canonical aspects of evolutionary algorithms are retained under the Parisian approach, see Figure 1, allowing for a great flexibility in its deployment. However, the applicability of this paradigm is restricted to problems where the solution can be decomposed into homogeneous elements or components, whose individual contribution to the complete solution can be evaluated. Therefore, each implementation is necessarily application dependent, where the design of a suitable problem decomposition is determinant factor. Thus, the following implementation issues have been identified:

- **Partial Encoding.** The genetic representation used for a single individual encodes a partial solution.
- **Local Fitness.** A meaningful merit function must be designed for each partial solution. In this way, the worthiness of a single individual can be evaluated in order to estimate the potential contribution to an aggregate solution.
- **Global Fitness.** A method for the aggregation of multiple partial solutions must be determined. In turn, a problem defined fitness function can be evaluated from this complete solution. However, the worthiness of this composite solution should be reflected on each partial solution.
- **Evolutionary Engine.** The evolution of the complete population should promote the emergence of better aggregate solutions. The evolutionary engine requires a scheme for combining local and global fitness values. Also,

it requires a diversity preserving mechanism in order to maintain a set of complementary partial solutions.

4 Parisian Approach to Camera Network Design

Camera placement can be viewed as a geometric design problem where the control variables are the spatial positioning and orientation of a finite set of cameras. In order to state such design problem in optimization terms the criterion expressed in section 2 is adopted. However, due to the sensor characteristics and mathematical modeling of the problem a strongly constrained optimization problem emerges. In this section we will discuss the different implementation issues involved in our incorporation of the Parisian approach into the sensor planning problem.

4.1 Problem Partitioning and Representation

A viewing sphere model for camera placement is adopted in order to reduce the dimensionality of the search space. Therefore, each camera position is defined by its polar coordinates $[\alpha_i, \beta_i]$. A network of M cameras is represented by a real valued vector

$$\boldsymbol{\Psi} \in \mathbb{R}^{2M} \quad \textit{where} \quad \alpha_i = \Psi_{2i-1}, \beta_i = \Psi_{2i} \quad \textit{for} \quad i = 1, \ldots, M. \tag{2}$$

Our design problem allows the decomposition into individual elements since the complete camera network is formed by a set of homogeneous components. Nevertheless, a decision on the level of *granularity* of our decomposition is crucial. Here we have the choice of an individual representing a single camera or a camera subnetwork (i.e. a set of cameras). We have decided for the latter option since such an individual can be meaningfully evaluated in terms of its imaging geometry contribution to 3D reconstruction. Hence, each individual in the population represents a fixed size subnetwork of N cameras, denoted by a vector of the form

$$\psi^j \in \mathbb{R}^{2N} \quad \textit{where} \quad \alpha_i = \psi^j_{2i-1}, \beta_i = \psi^j_{2i} \quad \textit{for} \quad i = 1, \ldots, N, \tag{3}$$

where j is defined as the subnetwork population index. Accordingly, a complete camera network specification is given by the aggregation of J subnetworks

$$\boldsymbol{\Psi} \in \mathbb{R}^{2M} = \bigcup_{j=1}^{J} \psi^j, \quad \textit{where} \quad M = J \times N \tag{4}$$

4.2 Local Fitness Evaluation

Section 2 presented a photogrammetric approach for estimating the variance of 3D point reconstruction using redundant measurements, see Eq. (1). Such methodology is generally applied to the complete measured object considering all cameras concurrently. Since in our representation we are working with camera subnetworks, it is unlikely that any single individual successfully captures the complete 3D object denoted by the whole set of 3D points $\mathbf{P}$. Hence, the object is also partitioned into R disjoint regions or subsets of points, in such a way that $\mathbf{P} = \bigcup_{i=1}^{R} P_i$. The visibility of a camera subnetwork ψ^j is limited to a subset of the whole object, expressed by $\mathbf{V}(\psi^j) \subseteq \mathbf{P}$. Therefore, we define the field of view constraint in the form

$$C_{fov}(\psi^j, P_i) = \begin{cases} 1 & \text{if } P_i \subset \mathbf{V}(\psi^j) \\ 0 & \text{otherwise} . \end{cases}$$

Thus, the local fitness evaluation uses the idea of decomposing the problem in subnetworks which provide greater object coverage with higher precision in order to attain higher fitness values. The reconstruction uncertainty for each set P_i is evaluated for a single individual accordingly to Equation (1), discarding the portions of the object not sensed by such a subnetwork. Thus, we have

$$f_{loc}(\psi^j) = \sum_{i=1}^{R} \frac{1}{f_1(\psi^j, P_i)} \quad \forall P_i : C_{fov}(\psi^j, P_i) = 1. \tag{5}$$

4.3 Global Fitness Evaluation

Once the local fitness of each individual has been evaluated, a process of aggregation is needed to obtain a solution to our camera network design problem. In order to achieve this, a *selection* of a group of individuals from the population must be made. The selection should be based on the merit of each individual fitness and can be realized by deterministic (i.e. selecting the top J individuals in the population) or stochastic (i.e. roulette, tournament) means. In this way, at each generation t an aggregate solution $\boldsymbol{\Psi}(t)$ has been formed for global fitness evaluation. This global evaluation uses the same criterion as in the local fitness evaluation. Therefore, we obtain:

$$f_{global}(\boldsymbol{\Psi}(t)) = \sum_{i=1}^{R} \frac{1}{f_1(\boldsymbol{\Psi}(t), P_i)} \quad \forall P_i : C_{fov}(\boldsymbol{\Psi}(t), P_i) = 1. \tag{6}$$

Such value describes the aptitude of the aggregate solution obtained at generation t. Obviously, the goal of the algorithm is to foster the improvement of this global fitness along the course of successive generations. However, another purpose of this evaluation is to be able to reflect on the population the effect of the evolutionary process. The individuals that form part of the aggregate solution will be rewarded or punished based on its global fitness. Also, based on the complete solution characteristics, promising individuals not selected should be compensated so they might contribute in latter stages of the evolutionary process.

A valid solution to the network design problem is one that reconstructs accurately the complete object. This requires addressing the aspects of constraint satisfaction and function optimization. We shall now describe how we use global fitness evaluation to deal concurrently with both of these issues.

Function optimization will be addressed first. In order to reflect the quality of an aggregate solution $\boldsymbol{\Psi}(t)$ on each of the individuals ψ^j that compose it, we use the ratio of improvement in global fitness among successive generations. The magnitude of the adjustment of an individual's local fitness is proportional to this ratio as follows

$$g_1(\psi^j) = f_{loc}(\psi^j)\left[\frac{f_{global}(\boldsymbol{\Psi}(t))}{f_{global}(\boldsymbol{\Psi}(t-1))} - 1\right] \qquad \forall \psi^j \in \boldsymbol{\Psi}(t). \tag{7}$$

Now we shall consider constraint satisfaction. It is very likely that each individual subnetwork will only cover part of the object. It is also possible that a given aggregation of individuals will not provide full object coverage. In this respect, when a particular aggregate solution $\boldsymbol{\Psi}(t)$ does not cover some object region P_i (e.g. $C_{fov}(\boldsymbol{\Psi}(t), P_i) = 0$), it would be desirable to enhance the fitness value of those individuals on the population that indeed cover such region. The amount of enhancement of those individuals shall be proportional to their difference in fitness with respect to the best individual in the population. Hence, we have

$$g_2(\psi^j) = f_{loc}(\psi^{best}) - f_{loc}(\psi^j) \qquad \forall \psi^j : \mathbf{V}(\psi^j) \bigcap \overline{\mathbf{V}(\boldsymbol{\Psi}(t))} \neq \emptyset. \tag{8}$$

Note that this value is only calculated for those individuals that cover an object region not sensed by the aggregate solution formed at that generation.

In this way, the global fitness is "fedback" to the general population in the following manner:

$$F(\psi^j) = \begin{cases} f_{loc}(\psi^j) + \lambda_1 g_1(\psi^j) & \text{if } \psi^j \in \boldsymbol{\Psi}(t) \\ f_{loc}(\psi^j) + \lambda_2 g_2(\psi^j) & \text{if } \mathbf{V}(\psi^j) \bigcap \overline{\mathbf{V}(\boldsymbol{\Psi}(t))} \neq \emptyset \\ f_{loc}(\psi^j) & \text{otherwise .} \end{cases}$$

Here, λ_1 and λ_2 are user defined parameters that reflect the relative importance given to each of the aspects involved in the global fitness evaluation.

5 Experimental Results

The reconstruction of a complex 3D object is considered in our experimentation. The goal is to determine a viewing configuration that will offer optimal results in terms of reconstruction accuracy. Here, we shall consider the design of a fixed size camera network of $M = 9$ stations. According to our approach, the level of granularity of our problem decomposition needs to be established. For these series of experiments we will use camera subnetworks of $N = 3$ cameras. In this way, each of the individuals in the population will consist of a vector $\psi \in \mathbb{R}^6$. Hence, a total of $J = 3$ subnetworks will need to be aggregated in order to form a complete solution to our network design problem. The convex polyhedral

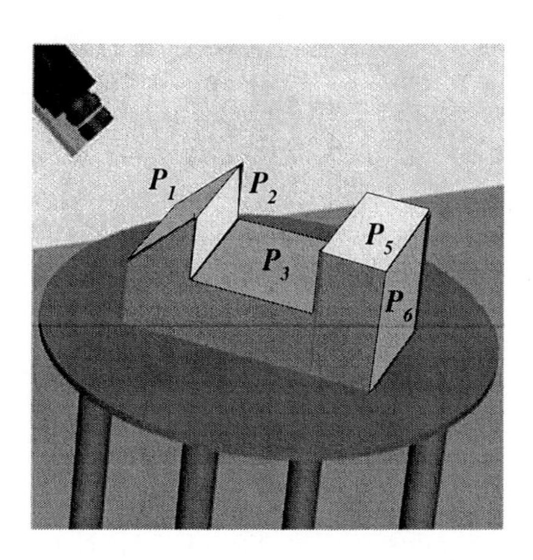

Fig. 2. The 3D object under observation. The concave object is partitioned into different regions in order to facilitate the fitness evaluation of sub-networks of small size. A photogrammetric network formed by 9 cameras is illustrated on the right

object under study, depicted in Figure 2, is partitioned into $R = 6$ regions. Elitist selection of individuals for solution aggregation is based on their fitness value. Finally, the user defined valued λ_1 and λ_2 are set to $\lambda_1 = \lambda_2 = 1.0$.

For all our experiments, SBX-crossover is utilized with a probability $P_c = 0.95$ along with polynomial mutation subject to probability $P_m = 0.05$. We have used tournament selection for reproduction under generational replacement. Alongside of our methodology, the same global fitness function was optimized by a typical genetic algorithm (e.g. each individual encodes a complete solution). This was done in order to have some reference point in the assessment of our proposed methodology. Both evolutionary algorithms were executed for 100 generations, using a population of 30 individuals.

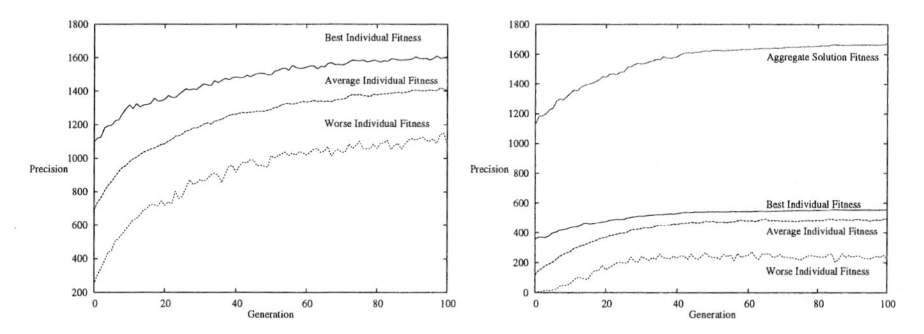

Fig. 3. Performance Comparison. On the left, the population evolution of a typical genetic algorithm is depicted. On the right, higher fitness values are consistently attained by the aggregate solutions of our proposed methodology. Plotted values reflect the averages over 20 executions

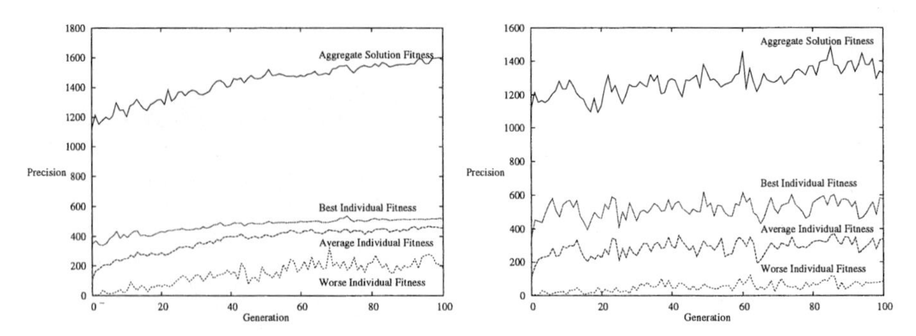

Fig. 4. Dependence on parameters λ_1, λ_2.The plot on the left corresponds to an execution with mixing values $[\lambda_1 = 0.8, \lambda_2 = 0.2]$. Performance is slightly deteriorated.The plot on the right represents an execution with values $[\lambda_1 = 0, \lambda_2 = 1.5]$. Note the almost random algorithm performance

Figure 3 plots population performance measures (best, mean, worse fitness) for a canonical GA on the left and also for our Parisian approach on the right. While these measures are descriptive of the dynamics of our population, the importance is on the *aggregate solution fitness* measure. In this respect, our approach slightly outperforms a canonical methodology in terms of solution quality. However, these results are made more relevant when considering the computational cost involved in fitness evaluation. For our studied object, evaluation based on the bundle adjustment of a complete network of 9 cameras is over 15 times more costly than that of a 3 camera subnetwork. Accordingly, by virtue of our problem decomposition, the total **execution time of the algorithm is reduced 10 times**. Clearly, a significant benefit in performance has been achieved.

The choice of mixing values λ_1, λ_2 is an important aspect in the performance of the algorithm, as they determine the magnitude of the global fitness adjustment given to each individual. In order to exemplify this, we have carried out different experiments varying the ratio and magnitude of these values. Experiments show a fairly robust behavior for similarly scaled values under 1.0. In general, performance deteriorates as the magnitude and the ratio among parameters increases. The right plot of Figure 4 illustrates the scenario where constraint satisfaction is completely predominant over function optimization. As a result, the fitness value of aggregate solutions is decreased by the inclusion of weaker subnetworks that are unreasonably enhanced by the global fitness evaluations.

6 Conclusions and Future Research

The Parisian approach to evolutionary computation offers an efficient way to address the design of photogrammetric networks. Experimental results illustrate its favorable performance against canonical evolutionary algorithms applied to our problem. In fact, solution quality is slightly improved with a 10 times reduction in computational effort. Aspects crucial to our methodology such as problem decomposition and representation, as well as the integration of local and global fitness evaluation have been discussed. However, further characterization of our algorithm behavior still is needed. Particularly, aspects like determining a suitable problem decomposition granularity and population size, the assignment of global fitness mixing values λ_1, λ_2 and the effect of diversity preservation techniques on the evolutionary process need to be addressed.

Acknowledgments

This research was funded by CONACyT and INRIA through the LAFMI project 634-212. First author supported by scholarship 142987 from CONACyT.

References

1. Olague, G.: A Comprehensive Survey of Sensor Planning for Robot Vision. Available as CICESE Research Report No. 25259.
2. Wong C., Kamel M.: Comparing Viewpoint Evaluation Functions for Model-Based Inspectional Coverage. 1st Canadian Conference on Computer and Robot Vision (CRV'04). May 2004. pp. 287-294.
3. Firoozfam P., Negahdaripour S.: Theoretical Accuracy Analysis of N-Ocular Vision Systems for Scene Reconstruction, Motion Estimation, and Positioning. 2nd International Symposium on 3D Data Processing, Visualization, and Transmission, (3DPVT'04). September 2004. pp. 888-895
4. Dunn, E. and Olague, G.: "Evolutionary Computation for Sensor Planning: The Task Distribution Plan". EURASIP Journal on Applied Signal Processing 2003:8,748-756.
5. Olague G. Mohr R.: "Optimal Camera Placement for Accurate Reconstruction". Pattern Recognition, 35(4):927-944 p.
6. Saadatseresht M., Fraser C., Samadzadegan F., Azizi A.: Visibility Analysis In Vision Metrology Network Design. The Photogrammetric Record 19(107):219-236. September 2004.
7. Mason S.: "Heuristic Reasoning Strategy for Automated Sensor Placement. Photogrammetric Engineering & Remote Sensing, 63(9):1093-1102 p.
8. Olague G.: "Automated Photogrammetric Network Design Using Genetic Algorithms". Photogrammetric Engineering & Remote Sensing, 68(5):423-431. Awarded "2003 First Honorable Mention for the Talbert Abrams Award", by ASPRS.
9. McGlone C. (ed.): Manual of Photogrammetry. Published by American Society for Photogrammetry and Remote Sensing. Bethesda, Maryland 20814. 1151pp.
10. Louchet J., Guyon M., Lesot M. and Boumaza A.: Dynamic Flies: a new pattern recognition tool applied to stereo sequence processing. In Pattern Recognition Letters, No. 23 pp. 335-345. (2002)
11. Raynal F., Collet P., Lutton E. and Schoenauer M.: Polar IFS + Parisian Genetic Programming = Efficient IFS Inverse Problem Solving. In Genetic Programming and Evolvable Machines Journal, Volume 1, Issue 4, pp. 339-361. (2000)

Design of Fast Multidimensional Filters Using Genetic Algorithms

Max Langer, Björn Svensson, Anders Brun,
Mats Andersson, and Hans Knutsson

Department of Biomedical Engineering,
Linköpings universitet, SE-581 85 Linköping, Sweden
{maxla, bjosv, andbr, matsa, knutte}@imt.liu.se

Abstract. A method for designing fast multidimensional filters using genetic algorithms is described. The filter is decomposed into component filters where coefficients can be sparsely scattered using filter networks. Placement of coefficients in the filters is done by genetic algorithms and the resulting filters are optimized using an alternating least squares approach. The method is tested on a 2-D quadrature filter and the method yields a higher quality filter in terms of weighted distortion compared to other efficient implementations that require the same ammount of computations to apply. The resulting filter also yields lower weighted distortion than the full implementation.

1 Introduction

A common approach to designing efficient 2-D filters is decomposition into separable 1-D components using singular value decomposition (SVD) [1], [2]. This results in a sum of cascaded 1-D filters, requiring significantly fewer operations than the full 2-D implementation. The resulting implementation also lends itself to parallel implementation, further increasing the computational gain over the full 2-D implementation. Genetic algorithms has been used to optimize filter coefficients in SVD decompositions [3].

In this paper a different approach is investigated. The filters are divided into simpler components using filter networks [4], but instead of restricting component filters to 1-D components, the coefficients can be sparsely scattered in n-D. A generalized convolver [5], whose computational complexity is only dependent on the number of non-zero coefficients and not the spatio-temporal extent of the filter, is used to take advantage of the sparse structure of the component filters. Genetic algorithms [6]-[8] is used to optimize the spatial locations of the coefficients. Finding optimal coefficient values is a multi-linear least squares problem and can be solved by an alternating least squares approach [9]. The method is described for the n-D case but only examples in 2-D are tested.

F. Rothlauf et al. (Eds.): EvoWorkshops 2005, LNCS 3449, pp. 366–375, 2005.

2 Filter Networks

Filter networks provide a structure for optimization and implementation of multiple output filter banks. The layered structure makes it possible to decompose complex filters into simple filter components and intermediary results may contribute to multiple output nodes due to the interconnectivity of the network.

A general filter network is depicted in fig. 1(a). The network consists of N layers with M output nodes. The component filters are located in the arcs connecting two successive layers and the nodes are summation points.

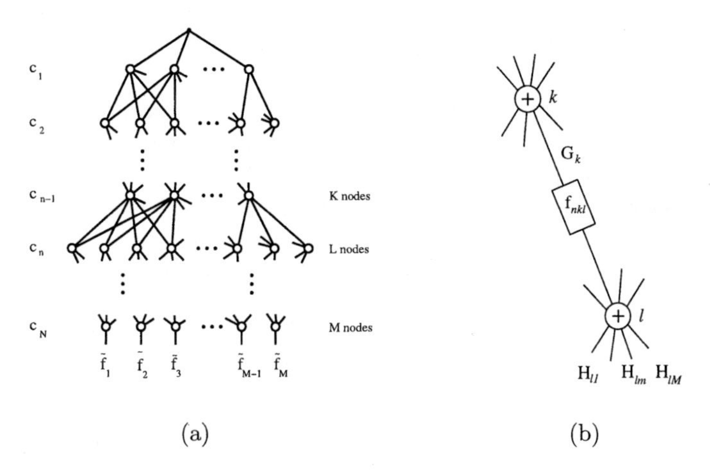

Fig. 1. (a) A general filter network of N layers with M nodes in the final layer. The filters are located in the arcs and the nodes are summation points. (b) Filter (arc) between node k and l at layer n in a general network. Note that there can be any number of connecting arcs to node k and any number of outgoing arcs from node l. G_k is the transfer function from the root node to node k and H_{lm} is the transfer function from node l to output node m

The *structure* of the network is the organization of nodes and arcs in the network and can be read out from figures like fig. 1(a) and the *internal properties* of the network are the number of coefficients for each filter and their spatial positions. There is no restriction on coefficient placement, they can be sparsely scattered or concentrated on a line. A maximum spatial size is usually set for each filter.

At each output node we define an ideal transfer function f_m in the Fourier domain. These are our target filters for the optimization. We also define a Fourier weighting function $W_m(\mathbf{u})$, defining the importance of a close fit for different frequencies for each transfer function. Ideally, coefficient values, coefficient placement and even network structure should be optimized simultaneously. This problem is very complex and a general optimal solution is unlikely to be found.

3 Filter Optimization

If the internal properties of the network are set, i.e. the coordinates for the nonzero filter coefficients are set, it is possible to optimize all filters on the same layer of the network. This is done with respect to the ideal filter functions and the current state of the network. The coefficients of the other filters are kept constant.

The task is to find the filter coefficients that minimize the weighted difference between the resulting filters $\tilde{f}_m$ and the ideal filter functions f_m

$$\min \epsilon^2 = \sum_{m=1}^{M} \left\| W_m(\mathbf{u})\big(f_m(\mathbf{u}) - \tilde{f}_m(\mathbf{u})\big) \right\|^2 \tag{1}$$

To express this minimization problem in the filter coefficients in one layer of the network we first consider one general filter (arc) $f_{nkl}(\mathbf{u})$ at layer n, between node k and l in the network (fig. 1(b)). For each filter in the network we calculate a Fourier transform matrix B_{nkl} where only the Fourier basis functions corresponding to the nonzero coefficient locations are included. The introduction of the Fourier transform matrices implies sampling of the Fourier domain. The Fourier transform of each filter is now computed as

$$f_{nkl} = B_{nkl} c_{nkl} \tag{2}$$

where c_{nkl} is a column vector containing the nonzero coefficients of the current filter. To get an expression on how the current filter affects the resulting transfer functions in the network we need to compute the transfer function $G_k(\mathbf{u})$ from the root node to node k in the previous layer and the transfer functions $H_{lm}(\mathbf{u})$ from node l to each output node. For the initial layer the transfer function G degenerates to an identity matrix. In the same way H will be an identity operator in the final layer. It is now possible to express the transfer functions $\tilde{f}_m(\mathbf{u})$ from the root node through $f_{nkl}(\mathbf{u})$ to the output nodes as a function of the filter coefficients.

$$\min \epsilon^2 = \sum_{m=1}^{M} \left\| W_m\big(f_m - G_k H_{lm} B_{nkl} c_{nkl}\big) \right\|^2 \tag{3}$$

To simplify notation, $G_k(\mathbf{u})$ and $H_{lm}(\mathbf{u})$ are reshaped as matrices with the transfer functions located in the main diagonal. For the same reason, the frequency coordinate $\mathbf{u}$ is dropped from now on.

The final step is to express the problem in all coefficients in one layer. This is accomplished by combining the coefficients of one layer in one vector.

$$\mathbf{c}_n = \big(c_{n11}, c_{n21}, \ldots, c_{nkl}, \ldots, c_{nKL}\big)^T \tag{4}$$

We also define

$$\begin{aligned}\mathbf{B}_m = \big(&G_1 H_{1m} B_{n11},\ G_2 H_{1m} B_{n21},\ \ldots \\ &G_k H_{lm} B_{nkl},\ \ldots,\ G_K H_{LM} B_{nKL}\big)\end{aligned} \tag{5}$$

In both equations 4 and 5 we have

$$k = [1 \dots K] \quad l = [1 \dots L] \quad m = [1 \dots M]$$

where K is the number of nodes in layer $n-1$ (the previous layer), L is the number of nodes in layer n (the current layer) and M is the number of output nodes. We can now express (eq. 1) in terms of all filter coefficients at layer n in the network:

$$\min \epsilon^2 = \sum_{m=1}^{M} \left\| W_m (f_m - \mathbf{B}_m \mathbf{c}_n) \right\|^2 \tag{6}$$

The entire network can be optimized using an alternating least squares approach [9]. The coefficients c_{nkl} of layer n are updated and another layer is selected for optimization. This is repeated through all layers until a convergence criterion is met or a maximum number of iterations is reached. The filter network optimization procedure is summarized below.

1. Initialize coefficient values for given network structure and internal properties
2. Select layer n to optimize
3. Calculate transfer functions G_k from the root node to all nodes in the previous layer
4. Calculate transfer functions H_{lm} from all nodes in the current layer to all output nodes
5. Optimize coefficients in the current layer
6. Repeat from step 2 until convergence criterion is met

4 Genetic Algorithm

The concept of genetic algorithms was first presented in 1975 by John Holland [6]. It is a stochastic search method which is inspired by evolution in biological systems where the search is conducted directly in the solution space. Each solution is encoded in a certain way and is called an *individual*. The search is parallel in the sense that a *population* of individuals is maintained and the quality of the individuals is calculated by a *fitness function*. The population is improved by *crossover*, recombination of genetic material from different individuals. This is based on a hypothesis that a good solution can be built up from shorter partial solutions [6]-[8]. Genetic diversity is maintained by a mutation operation, making random changes in the individuals. To summarize, genetic algorithm consists of five components:

1. A chromosomal representation of solutions
2. a way to create an initial population of solutions
3. a fitness function
4. genetic operators (selection, crossover, mutation)
5. parameter values for the genetic algorithm (population size, probabilities for applying genetic operators etc).

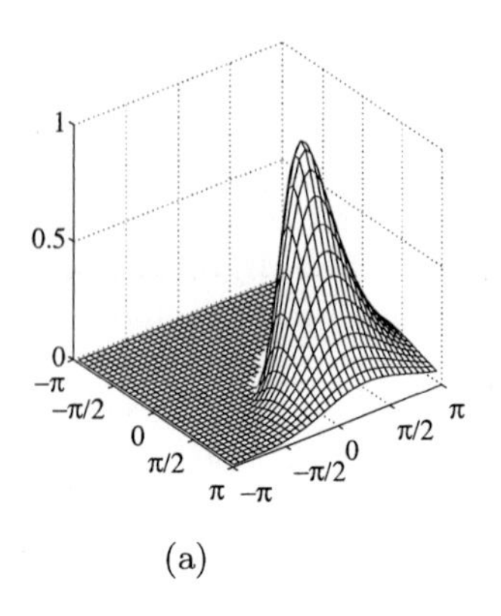

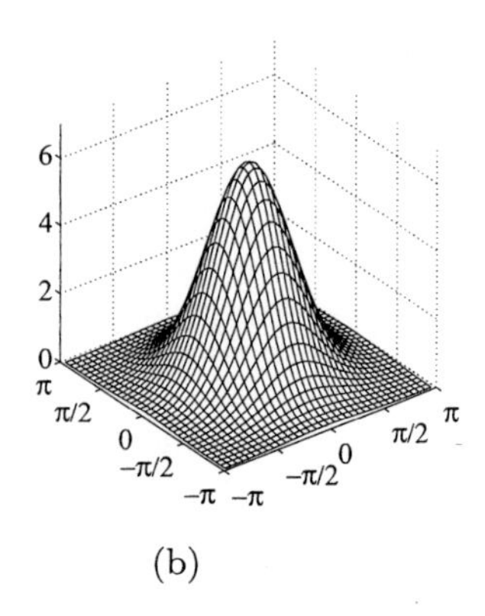

Fig. 2. (a) Ideal transfer function of the first experiment in the Fourier domain, a diagonal quadrature filter. (b) Ideal transfer function of the second experiment in the Fourier domain, a Cartesian separable truncated Gaussian low-pass filter

The implementation of the genetic algorithm is inspired by the algorithm used by Weber et al. [10] in their optimization of sparse ultrasound transceivers. A many-valued encoding strategy is utilized where each solution is represented by four integer strings. Each position contains an index to a coordinate, the index running column-wise from top left to bottom right. If the encoded filter is limited to $M \times M$ in spatial size, the indices can be in the interval $[1 \ldots (M^2-1)/2]$.

Since the ideal functions used here are real-valued in the Fourier domain the resulting filter function must either be real and even or have an even real part and an odd imaginary part [11]. A consequence of this is that filter coefficients must be quadrantly symmetrically distributed. Based on this fact, only symmetric coefficient placement is allowed. This means we only have to encode half the spatial domain.

The population is set up by randomly distributing coefficients in the kernels. Each network is evaluated using the filter network optimization procedure described above. A normalized distortion measure is used as fitness function.

$$D = \sum_{m=1}^{M} \sqrt{\frac{||W_m(f_m - \tilde{f}_m)||^2}{||W_m f_m||^2}} \tag{7}$$

It is calculated as the ratio between the RMS error and the RMS value of the ideal functions f_m. Using only the RMS error is difficult since it is an absolute measure and not directly related to the quality of the resulting filters.

Crossover is implemented as a swapping of complete arcs (kernels) in the same position in the network structure. This is based on the the assumption that the shortest, low order schema relevant to the problem is a complete component filter (an arc in the network) [8].

Mutation is implemented by moving coefficients to random positions in the kernel. A correcting function is then applied to make sure there is only one

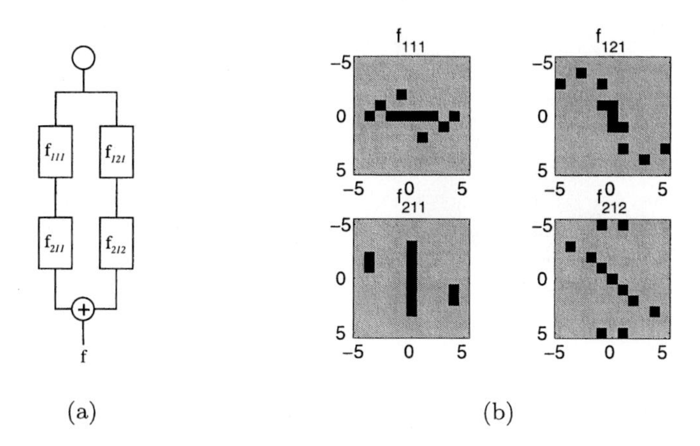

Fig. 3. (a) Network structure used in quadrature filter example. (b) Best coefficient placement in filter network designed by genetic algorithm

coefficient in each spatial position. If the new position is already occupied, a new one is generated randomly and this is repeated until a free position is found. A mutation operator where there is a higher probability to move coefficients to locations closer to the original position was tried but did not perform as well as the pure random mutation.

In the following examples, the population size is set to 40 individuals. The 8 best filters are selected as parents and survive unchanged to the next generation. From these parents, 13 children are generated using the crossover operator described above. These children replace the 13 worst individuals in the previous generation. The remaining 19 individuals are mutated, the mutation rate is set to give an expected number of two mutations per individual.

5 Experiments

Two experiments are set up to test the implemented genetic algorithm. Throughout the examples, component filters will be limited to 11×11 in spatial size. A Fourier domain uniformly sampled in a 35×35 grid is used.

The first example is a quadrature filter along one diagonal with ideal transfer function given in fig. 2(a). The properties of this class of filters can be found in [12]. In this example, we use a filter network with four component filters connected in two parallel branches (fig. 3(a)). Since the ideal function is not symmetric or anti-symmetric, the filter is complex in the spatial domain [11]. Due to this fact, coefficients are allowed to be complex in all component filters.

A frequency weighting function favoring a close fit for low frequencies, based on an estimate of the expected signal spectrum, is used. The same weigthing function is used in all examples below. Note that if the component filters are 1-D in the spatial domain placed along the x and y axis respectively with two orthogonal filters in each branch, this network describes a two term low rank

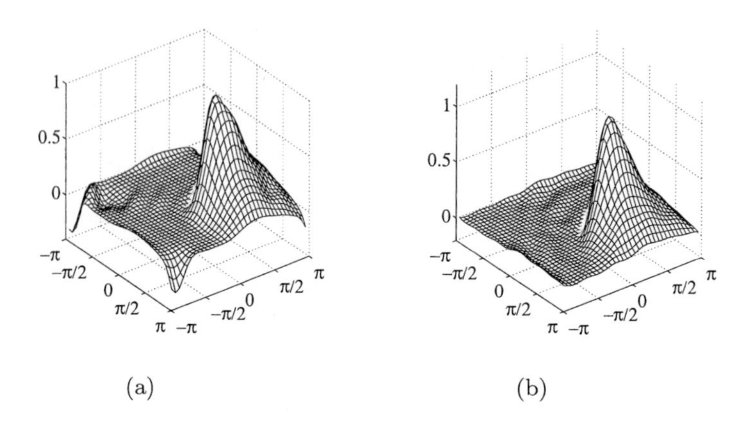

Fig. 4. (a) Transfer functions of two term truncated low rank approximation of the quadrature filter. (b) Transfer function of full 11×11 quadrature filter

approximation decomposition and can be solved by other methods. This filter will be included as reference. Another design example using the same network structure, where coefficients are placed along the diagonals in one of the branches are also included to compare the results of the genetic algorithm to placement of coefficients by hand.

The second example investigated is a truncated Gaussian low-pass filter (fig. 2(b)). The transfer function is constructed by convolving two orthogonal 1-D Gaussian filters in the spatial domain and transforming the result to the Fourier domain. Since the ideal function is symmetric, the filter will be real-valued in the spatial domain and only real-valued coefficients are allowed in the filter network. This filter is truly separable in the spatial domain and hence, as opposed to the first example, there exists a known global optimal solution. The purpose of this example is to investigate convergence properties of the genetic algorithm experimentally.

6 Results

When evaluating the quadrature filter example, the genetic algorithm was allowed to run for 1000 iterations. The diagonal and the low rank filter networks were also evaluated. The results are presented in table 6. The solution found by the genetic algorithm performs better than a full 11×11 filter in terms of distortion.

Since the filter coefficients are complex, the full filter requires 242 multiply and add operations for each element in the signal. Although the other filter implementations contain 44 coefficients, the intermediary results are complex so the second layer has to be applied twice, once for the real part and once for the imaginary part. This gives 66 multiply and add operations per pixel.

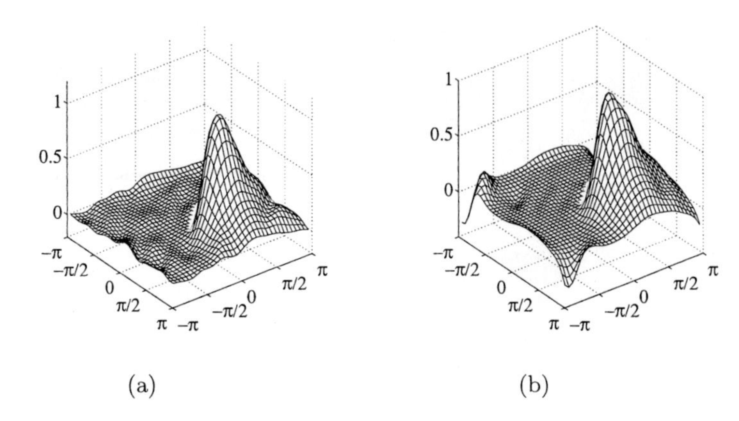

Fig. 5. (a) Transfer function of genetic algorithm optimized filter network. (b) Transfer function of filter network with diagonal coefficient placement. Note that the behaviour close to the corners of the Fourier domain is due to the frequency weighting function

The optimal coefficient placement found by the genetic algorithm is presented in fig. 3(b). This filter network outperforms the networks designed with the other two methods in terms of the weighted distortion measure used above (eq. 7), and also the full 11×11 implementation of the filter. The results for the diagonal and low rank approximation filter networks are presented in fig. 4(a) and 5(b) for comparison. The transfer function of the full 11×11 filter is included in fig. 4(b).

Examining the component filters, the resulting coefficient placement in each branch and the sum of branches (fig. 6(a)) it is hard to draw any general conclusions about the distribution of coefficients. Not surprisingly there is a bias towards the corresponding main direction of the filter in this case, but no guidelines for placement of coefficients by hand are easily created based on the optimization results.

In the truncated Gaussian example, the genetic algorithm was allowed to run for 500 iterations or until the global optimum was found (this yields zero distortion). The algorithm quickly converged towards a low distortion area but convergence rate close to the optimal solution is slow. The optimal decomposition was found in about 200 iterations in all test runs.

Table 1. Quadrature filter optimization results for different coefficient placement strategies. Distortion is measured according to (eq. 7)

	Weighted distortion	Multiply and add per pixel
Genetic algorithm	5.4 %	66
Full 11×11	6.4 %	242
Diagonal	10.9 %	66
Low rank	11.9 %	66

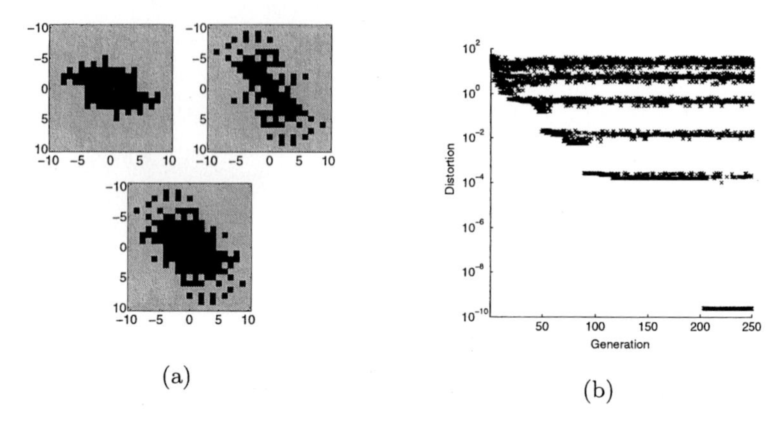

Fig. 6. (a) Coefficient placement in resulting filters clockwise from top left: resulting coefficient placements from convolution of f_{111} and f_{211}, convolution of f_{121} and f_{212} and total resulting coefficient placements from fig. 3. Note that the resulting filter is limited to 21×21 in spatial size due to convolution of two 11×11 filters. (b) Convergence plot for Gaussian example. The points are the fitness of each individual at each generation. Note that the optimal solution is found after about 200 generations

Tests were also made with slightly different implementations of the genetic algorithm. Without the requirement of symmetric component filters, the search converged on local optima. The optimal decomposition was not found in any of the test runs with this implementation. Tests were also made with the alternative mutation operator described above the genetic algorithm but none of these experiments converged to the optimal solution in less than 400 iterations.

7 A Note on Optimality

It is a known fact that genetic algorithms give no guarantee that a global optimum will be found. Safeguards against local optima can be implemented but there are still no guarantees. However, as stated by Goldberg [8], the most important goal of optimization is improvement. In this application with the given limitations in network structure, the genetic algorithm has found a better solution than any previously found.

8 Conclusion

A genetic algorithm for optimizing coefficient placement in filter networks has been presented. The algorithm has been tested on a simple but nontrivial filter network and the optimization shows promising results. For the diagonal quadrature filter the optimized filter outperforms other decomposition methods as well as the full implementation of the filter in terms of weighted distortion. This is partially due to that the corresponding resulting impulse response is in fact limited to 21×21 in spatial size. The genetic algorithm was also tested on a truncated Gaussian low-pass filter with a known global optimum. In this example, the genetic algorithm finds the global optimum in all test runs.

Acknowledgement. The authors would like to thank Johan Wiklund at the Department of Electrical Engineering for the implementation of the filter network optimizer.

References

1. Twogood, R. E., Mitra, S. K.: Computer-Aided Design of Separable Two-Dimensional Digital Filters, IEEE Trans. Accoustics, Speech, and Signal Processing **25**, (1977) 165-169
2. Lu, W.-S., Antoniou, A.: New method for weighted low-rank approximation of complex-valued matrices and its application for the design of 2-D digital filters, Proc. of the 2003 Int. Symp. Circuits and Systems, ISCAS '03, **3** (2003) 694-697
3. Williams, T., Ahmadi, M., Hashemian R., Miller, W. C.: Design Of High Throughput 2-D Fir Filters Using Singular Value Decomposition (SVD) And Genetic Algorithms, IEEE Pacific Rim Conference on Communications, Computers and Signal Processing, **2** (2001) 571-574
4. Andersson, M., Wiklund, J., Knutsson, H.: Filter Networks, Reprint from IASTED International Conference on Signal and Image Processing (1999)
5. Wiklund, J., Knutsson, H.: A Generalized Convolver, Proceedings of the 9th Scandinavian Conference on Image Analysis (1995)
6. Holland, J. H.: Adaptation in Natural and Artificial Systems, MIT Press, USA (1992)
7. Davis, L.: Genetic Algorithms and Simulated Annealing, Pitman Publishing, UK (1987)
8. Goldberg, D. E. (1989): Genetic Algorithms in Search, Optimization, and Machine Learning, Addison-Wesley Publishing Company, USA. (1989)
9. Knutsson, H., Andersson, M., Wiklund, J.: Advanced Filter Design, Proceedings of the Scandinavian Conference on Image analysis (1999)
10. Weber, P. K., Peter, L., Austeng, A., Holm, S., Aakvaak, N.: 1D- and 2D-Sparse-Array-Optimization, 2. Symposium on Quantitative Sonography in Clinic and Research (1998)
11. Bracewell, R. N.: The Fourier Transform and Its Applications, third edition, McGraw-Hill, USA (2000)
12. Granlund, G. H., Knutsson, H.: Signal Processing for Computer Vision, Kluwer Academic Publisher, the Netherlands (1995)

Genetic-Fuzzy Optimization Algorithm for Adaptive Learning of Human Vocalization in Robotics

Enzo Mumolo[1], Massimiliano Nolich[1,2], and Graziano Scalamera[1]

DEEI, University of Trieste, Via Valerio 10, Trieste, Italy
AIBS Lab, Via del Follatoio 12, Trieste, Italy

Abstract. We present a computational model of human vocalization which aims at learning the articulatory mechanisms which produce spoken phonemes. It uses a set of fuzzy rules and genetic optimization. The former represents the relationships between places of articulations and speech acoustic parameters, while the latter computes the degrees of membership of the places of articulation. That is, the places of articulation are considered as fuzzy sets whose degrees of membership are the articulatory features. Subjective listening tests of sentences artificially generated from the articulatory description resulted in an average phonetic accuracy of about 76 %. Through the analysis of a large amount of natural speech, the algorithm can be used to learn the places of articulation of all phonemes.

1 Introduction

Human-robot interactions and dialogue modalities have been widely studied in recent years in robotics and AI communities. As speech is the most natural communication mean for humans, conversational interfaces are one of the most promising methods of human-robot communication from the viewpoint of efficiency of information transfer. Basically, conversational interfaces are built around two technologies: speech synthesis from unrestricted text and speech recognition.

A step further in human-robot interaction is based on humanoid robotics. Its goal is to create a robot designed to work with humans as well as for them. Inevitably, humanoid robots tend to imitate somehow the form and the mechanical functions of the human body in order to emulate some simple aspects of the physical (i.e. movement), cognitive (i.e. understanding) and social (i.e. communication) capabilities of the human beings. Significant contributions have been made within the field of advanced robotics and humanoids like the Cog project at MIT [1].

Human infants learn to speak through interaction with their care-givers. The aim of our study is to build a robot that acquires a vocalization capability in a way similar to human development. More precisely, the speech learning mechanism using our algorithm works as follows: the operator, acting as care-giver,

F. Rothlauf et al. (Eds.): EvoWorkshops 2005, LNCS 3449, pp. 376–385, 2005.

pronounces a word and the robot generates an artificial replica of the word based on the articulatory and acoustic estimation. This process iterates until the artificial word matches the original one according to the operator judgement. At this point the robot has learnt how to pronounce those words in terms of articulatory movements. The operator must repeat this process for a large number of words. After this phases, the speech learning process is completed.

Many research results in vocal tract estimation has been reported so far. For instance, in [2] and [3], the shape of the vocal tract is estimated for driving the shape of a mechanical vocal tract. In this way the shape of the acoustic filter is modified and the resulting signal is modulated by the acoustic filter function of the vocal tract. In [3] the mechanical vocal tract is excited by a mechanical vibrator that oscillates at particular frequencies and acts as artificial larynges. The vocal tract shape is obtained using a neural network.

In our work, we developed a model that relates formant frequencies dynamics of vocal tract and the articulation of the phonation organs of the vocal tract using fuzzy rules that have as input the articulatory parameters and as output the dynamics of acoustic and articulatory parameters together with the corresponding artificial vocalization. The optimal articulatory parameters are estimated using a genetic algorithm.

As compared with other works in acoustic to articulatory mapping, which generally compute the vocal tract area functions from actual speech measurements, our work present a method to estimate the place of articulation of input speech through the development of a novel computational model of human vocalization.

2 Problem Formulation

The goal of this work is to learn automatically the place of articulation of spoken phonemes in such a way that the robot learns how to speak through articulation movements. The block diagram of the system for adaptive learning of human vocalization is depicted in Fig. 1.

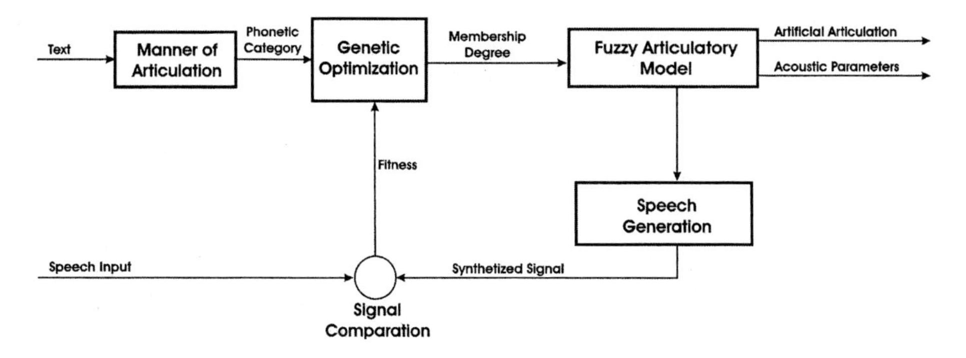

Fig. 1. Block diagram of the genetic-fuzzy optimization algorithm

The algorithm is based on the following assumptions: the degrees of membership of a place of articulation estimated by means of the genetic optimization process is directly related to the physical configuration of the phonatory organs. For example, if a phoneme is characterized by a degree of opening equal to 0.6, it is assumed that the mouth is opened at a 60% degree of the maximum opening width. Even if no direct experimental evidence of that is given in this paper, this assumption can be indirectly verified: the overall model produces good synthetical vocalizations.

3 The Fuzzy Articulatory Module

Usually, phonemes are classified in terms of manner and place of articulation. The manner of articulation is concerned with the degree of constriction imposed by the vocal tract on the airflow, while place of articulation refers to the location of the most narrow constriction in the vocal tract. The following six categories of the manner of articulation have been considered in this work: vowel, in which air flows throw the vocal tract without constrictions; liquid, similar to the vowels but that use the tongue as an obstruction; nasal, which is characterized by a lowering of the velum, allowing airflow out of the nostril; fricative, which employ a narrow constriction in the vocal tract which introduces turbulence in the air flow; plosive, involving a complete closure and subsequent release of a vocal obstruction; affricate, which is a plosive followed by a fricative.

Using manner and place of articulation, any phoneme can be fully characterized in binary form. However, a certain degree of imprecision, due to the lack of knowledge, is involved in this characterization, which thus should be fuzzy rather than strictly binary. For example, it may be that the /b/ phoneme, classically described as plosive, bilabial and voiced, involve also a certain degree of anteriority and rounding, as well as some other features.

The possibility of facing the vagueness involved in the interpretation of phonetic features using methods based on fuzzy logic has been realized in the past, when approaches to speech recognition via phonetic classification were proposed [4, 5].

For simplicity, only a subset of the phonemes in Italian language were considered in this work. This subset is sufficient for achieving a complete intelligibility of a general text. Their classification in terms of the manner of articulation is as follows (using IPA symbols):

vowel : /a/, /e/, /i/, /o/, /u/, /SIL/, /$/
liquid : /l/, //, /r/
nasal : /m/, /n/, /j/
fricative : /f, /v/, /s/, /z/, / R /
plosive : /p/, /b/, /t/, /d/, /k/, /g/
affricate : /dR/, /di/, /dz/, /ts/

Clearly, all the quantities involved, namely phonemes and control parameters, are fuzzified, as described in the following.

3.1 Phoneme and Control Parameters Fuzzification

The phonemes are classified into broad classes by means of the manner of articulation; then, the place of articulation is assigned to them. Therefore, each phoneme is described by an array of nineteen articulatory features, six of them are boolean variables and represent the manner of articulation and the remaining twelve are fuzzy and represent the place of articulation. In this way, the phonetic description appears as an extension of the classical binary definition described for instance by Fant in [6], and a certain vagueness in the definition of the place of articulation of the phonemes is introduced. Representing the array of features as (vowel, plosive, fricative, affricate, liquid, nasal, any, rounded, open, anterior, voiced, bilabial, labiodental, alveolar, prepalatal, palatal, vibrant, dental, velar), the /a/ phoneme, for example, can be represented by the array:

$$[1,0,0,0,0,0|1,0.32,0.9,0.12,1,0,0,0,0,0,0,0,0]$$

indicating that /a/ is a vowel, with a degree of opening of 0.9, of rounding of 0.32, and it is anterior at a 0.12 degree.

The array reported as an example has been partitioned for indicating the boolean and the fuzzy fields respectively. Such arrays, defined for each phoneme, are the membership values of the fuzzy articulatory features of the phonemes.

All the fuzzy sets for the acoustic parameters have trapezoidal membership functions and have been defined as follows:

- Duration D(p). The global range of this fuzzy variable is 0-130 ms.
- Initial Interval I(p). As D(p), this fuzzy variable is divided into a 0-130 ms interval.
- Final Interval F(p). The numeric range is 0-130 ms.
- Locus L(p). The fuzzy values of this variable depend on the actual variable to be controlled.

3.2 Fuzzy Rules and Defuzzification

By using linguistic expressions which combine the above linguistic variables with fuzzy operators, it is possible to formalize the relationship between articulatory and acoustic features. For example, for a transition towards a vowel, the opening and the anteriority of the target phoneme determine the values of the first two formants, and this knowledge can be formalized as follows:

```
IF target phoneme IS Open         THEN  L(F1) IS Medium;
IF target phoneme IS Anterior     THEN  L(F2) IS Medium High;
IF target phoneme IS Not Anterior THEN  L(F2) IS Low;
IF target phoneme IS Round        THEN  L(F3) IS Low;
IF target phoneme IS Not Round    THEN  L(F3) IS Medium.
```

In general, however, the rules involve the actual and the future phonemes; thus, in these last rules, each of the expressions should be read as: IF actual phoneme IS Any AND target phoneme ... where the vocalic phonemes membership to the 'Any' variable is equal to 1.

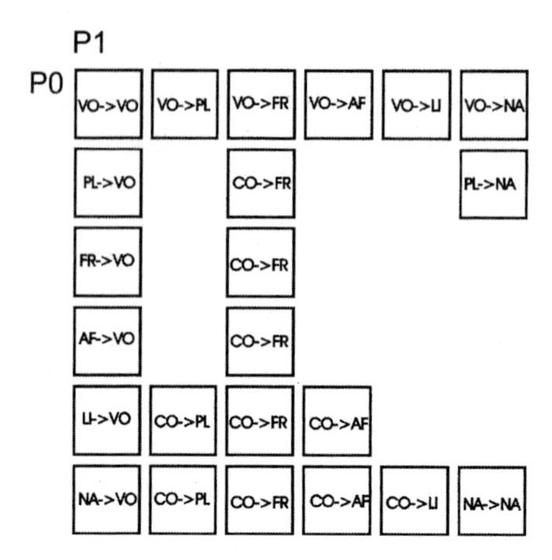

Fig. 2. Outline of the bank of fuzzy rules. P0 and P1 represent the actual and target phonetic categories

Moreover, in general the fuzzy expressions involve the fuzzy operators AND, NOT and OR. Since the manner of articulation well partitions the phonemes in separated regions, the rules have been organized in banks, one for each manner. That is, calling P0 and P1 the actual and the future phonemes respectively, the set of rules is summarized in Fig. 2. The rule decoding process is completed by the defuzzification operation, which is performed with the fuzzy centroid approach.

4 Speech Generation Module

The synthesis of the output speech is performed using a reduced Klatt formant synthesizer [7].

The I(p) control feature determines the starting point of the transition, whose slope and target values are given by the D(p) and L(p) features. The parameter holds the value specified by their locus for an interval equal to F(p) ms; however, if other parameters have not completed their dynamic, the final interval F(p) is prolonged. The I(p), F(p), and D(p) parameters are expressed in milliseconds, while the target depends on what synthesis control parameter is involved; for example, for frequencies and bandwidths the locus is expressed in Hz, while for amplitudes in dB.

5 Genetic Optimization of Articulatory and Acoustic Parameters

Let us take a look at Fig. 1. Genetic optimization aims at computing the optimum values of the degrees of membership for the articulatory features used to generate an artificial replica of the input signal.

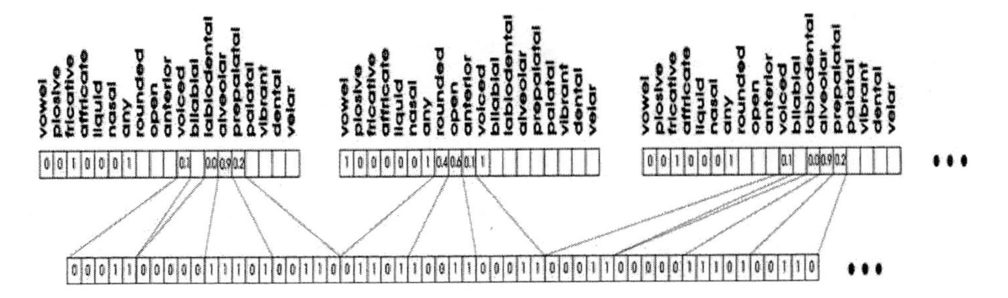

Fig. 3. The binary chromosome obtained by coding

5.1 Genetic Optimization Module

The optimal membership degrees of the articulatory places minimize the distance from the uttered signal; the inputs are the number of phonemes of the signal and their classification in terms of manner of articulation.

One of the main part of the genetic algorithm is coding. The chromosome used for genetic optimization for a sequence of three phonemes is shown in Fig. 3. It represents the binary coding of the degrees of membership. The genetic algorithm uses only mutations of the chromosome. This means that each bit of the chromosome is changed at random and the mutation rate is constant to 2%.

5.2 Fitness Computation and Articulatory Constraints

An important aspect of this algorithm is the fitness computation, which is represented by the big circle symbol in Fig. 1. The fitness, which is the distance measure between original and artificial utterances and is optimized by the genetic algorithm, is an objective measure that reflects the subjective quality of the artificially generated signal. The Modified Bark Spectral Distortion (MBSD) measure has been used [8, 9]. Such measure is based on the computation of the pitch loudness, which is a psycho-acoustical term defined as the magnitude of the auditory sensation. In addition to this, a noise masking threshold estimation is considered. This measure is used to compare the artificial signal generated by the fuzzy module and the speech generation module against the original input signal.

The MBSD measure is frame based. That is, the original and the artificial utterances are first aligned and then divided into frames and the average squared Euclidean distance between spectral vectors obtained via critical band filters is computed. The alignment between the original and artificial utterances is performed by using dynamic programming [10], with slope weighting as described in [11] and shown in Fig. 4.

Therefore, using the mapping curve between the two signals obtained with DTW, the MBSD distance D between original and artificial utterances represented respectively with X and Y is computed as follows:

$$D(X,Y) = \frac{1}{L_{\Phi}} \sum_{k=0}^{T} \left[\sum_{j=0}^{K} M(\Phi_y(k), j) \left| L_x(\Phi_x(k), j) - L_y(\Phi_y(k), j) \right| m(k) \right]$$

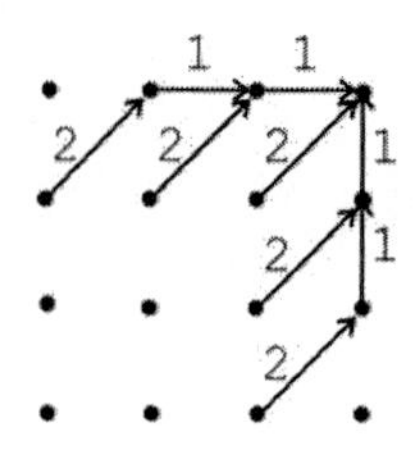

Fig. 4. Slope weighting

where T is the number of frames, K is the number of critical bands, $\Phi = (\Phi_x, \Phi_y)$ is the non-linear mapping, L_Φ is the length of the map, $L_x(i,j)$ is the Bark spectrum of the i-th frame of the original utterance, $L_y(i,j)$ is the Bark spectrum of the i-th frame of the artificial utterance, $M(i,j)$ is the indicator of perceptible distorsion at the i-th frame and j-th critical band, and $m(k)$ are the weights as shown in Fig. 4. The coefficient $M(i,j)$ is a noise masking threshold estimation which determine if the distortion is perceptible by comparing the loudness of the original and artificial utterances. Thus, the fitness function of the Place of Articulation (PA) is:

$$Fitness\,(PA) = \frac{1}{D(X,Y)}.$$

The goal of the genetic optimization is to find the membership values that lead to a maximization of the fitness, i.e. the minimization of the distance $D(X,Y)$, namely $PA = argmax\{Fitness(PA)\}$, $PA = \bigcup PA_i$, $i = 1, \ldots, 12 \cdot N$, where PA_i is the degree of membership of the i-th place of articulation, N is the number of phonemes of the input signal. However, in order to correctly solve the inverse articulatory problem, the following constraints have been added to the fitness:

- it is avoided that a plosive phoneme is completed dental and velar simultaneously;
- it is avoided that a nasal phoneme is completely voiced;
- it is avoided that all the membership degrees are simultaneously less than a given threshold;
- it is avoided that two or more degrees of membership are simultaneously greater than another threshold.

In conclusion, our optimization problem can be formalized as follows:

$$PA = argmax\left\{\frac{1}{D(X,Y)} + \sum_{j=1}^{N_c} P_j\right\}$$

where P_j is the penalty function and N_c is the number of constraints.

6 Experimental Results

In Fig. 5 a typical convergence behaviour is represented, where $D(X,Y)$ against number of generation is shown.

Basing on the results shown in Fig. 5, the experimental results presented in the following are obtained with a population size of 200 elements and a mutation rate equal to 0.02.

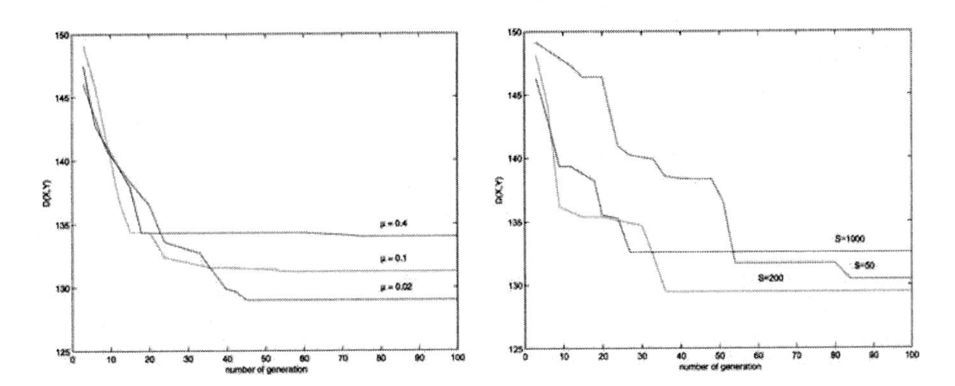

Fig. 5. Convergence diagram, i.e. distance $D(X,Y)$ versus number of generation. In the left panel the mutation rate μ is varied and the population size is maintained constant to 200 elements. In the right panel the population size S is varied and the mutation rate is maintained constant to 0.02

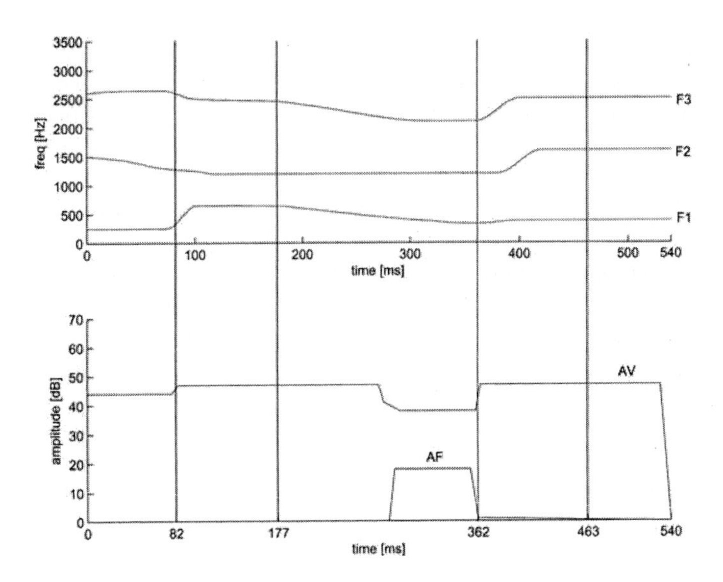

Fig. 6. Acoustic analysis of the Italian word "nove" obtained with fuzzy model and genetic optimization

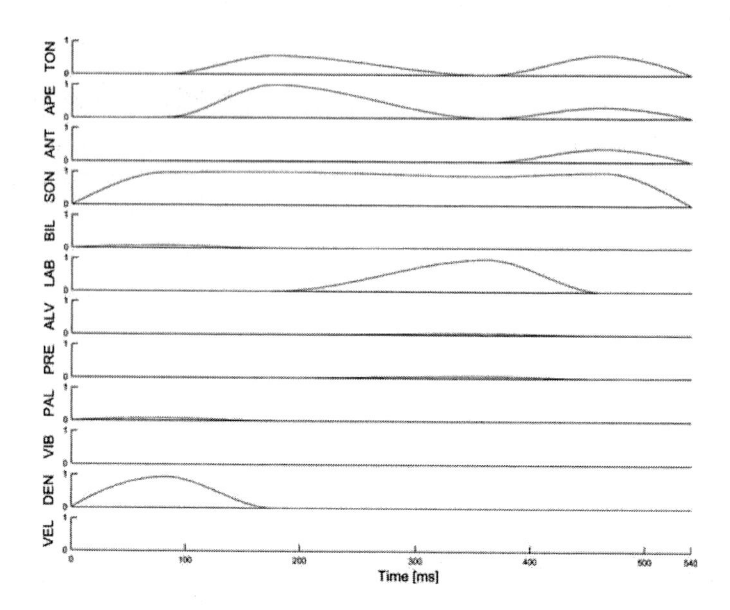

Fig. 7. Articulatory places of articulation of the Italian word 'nove' estimated with genetic optimization

In Fig. 6 and in Fig. 7 some experimental results related to the analysis of the Italian word 'nove' ('nine') are shown. In the upper panel of Fig. 6 the dynamic behaviour of the first three formant frequencies is reported. The vertical lines denote the temporal instants of the stationary part of each phoneme. It is worth noting that this segmentation is done on the synthetic signal but it can be related to the original signal using the non–linear mapping between the original and synthetic word obtained by dynamic programming. In the lower panel of Fig. 6 the behaviour of low and high frequencies amplitudes are shown. In Fig. 7 the dynamic of the membership degrees of the articulatory places of articulation is reported.

In Fig. 8, finally, some subjective evaluation results related to a phonetic listening test are shown: the phonetic categories used in this test are quite critical from a correct comprehension point of view. However, the subjective rate ranges from 70% to 85% and therefore it is quite promising for future developments.

Phonetic Categories	Number of signals	Exact recognitions [%]
plosive	193	70
fricative	105	85
affricate	47	70
liquid	83	78
Total	428	76

Fig. 8. Subjective evaluation results

7 Final Remarks and Conclusions

In this paper a novel approach for the estimation of articulatory features from an input speech signal is described. The approach uses a set of fuzzy rules and a genetic algorithm for the optimization of the degrees of membership of the places of articulation. One interesting property of the fuzzy model is that the fuzzy rules can be quite easily modified and tuned. The membership values of the place of articulation of the spoken phonemes have been computed by means of genetic optimization. Many sentences have been generated on the basis of this articulatory estimation and their subjective evaluations show that the quality of the artificially generated speech is quite good.

Many possible extension and applications of this work are possible, from estimating acoustic parameters and signal segmentation at a low level, to studies of inner mechanism involved in vocalization, to applications in humanoid robotics at higher levels. Current work is directed toward the development of a vocalization system driven by the described algorithm.

References

1. Brooks, R., Breazeal, C., Marjanovic, M., Scassellati, B., Wiliamson, M.: The Cog Project: Building a Humanoid Robot. Lecture Notes in Artificial Intelligence, Springer–Verlag (1998)
2. Higashimoto, T., Sawada, H.: Speech Production by a Mechanical Model Construction of a Vocal Tract and its Control by Neural Network. Proc. IEEE Int. Conf. on Robotics and Automation, pp.3858–3863 (2002)
3. Yoshikawa, Y., Koga, J., Asada, M., Hosoda, K.: Primary Vowel Imitation between Agents with Different Articulation Parameters by Parrot-like Teaching. Proc. IEEE–RSJ Int. Conf. on Intelligent Robots and System, pp.149–154 (2003)
4. De Mori, R.: Computer Models of Speech Using Fuzzy Algorithms. Plenum, New York (1983)
5. De Mori, R., Laface, P.: Use of Fuzzy Algorithms for Phonetic and Phonemic Labeling of Continuous Speech. IEEE Transactions on Pattern Anal. and Machine Intell., v.2, pp.136-148 (1980)
6. Fant, G.: Speech Sounds and Features. MIT Press (1973)
7. Klatt, D.H.: Review of Text-to-Speech Conversion in English. JASA, pp.737-793, (1987)
8. Wang, S., Sekey, A., Gersho, A.: An Objective Measure for Predicting Subjective Quality of Speech Coders. IEEE J. on Select. Areas in Comm., v.10 (1992)
9. Wonho Yang, Dixon, M., Yantorno R.: A modified bark spectral distortion measure which uses noise masking threshold. IEEE Workshop on Speech Coding For Telecommunications Proceeding, pp.55-56 (1997)
10. Sakoe, H., Chiba, S.: Dynamic Programming Algorithm Optimization for Spoken Word Recognition. IEEE Trans. Acoust. Speech Signal Processing, v.26, pp.43-49 (1978)
11. Rabiner, L., Juang, B.-H.: Fundamentals of Speech Recognition. Prentice Hall (1993)

Evolving Parameters of Surveillance Video Systems for Non-overfitted Learning

Óscar Pérez, Jesús García, Antonio Berlanga, and José Manuel Molina

Universidad Carlos III de Madrid. Departamento de Informática,
Avenida de la Universidad Carlos III, 22 Colmenarejo 28270. Madrid. Spain
{oscar.perez.concha, antonio.berlanga}@uc3m.es
jgherrer@inf.uc3m.es, molina@ia.uc3m.es

Abstract. This paper presents an automated method based on Evolution Strategies (ES) for optimizing the parameters regulating video-based tracking systems. It does not make assumptions about the type of tracking system used. The paper proposes an evaluation metric to assess system performance. The illustration of the method is carried out using three very different video sequences in which the evaluation function assesses trajectories of airplanes, cars or baggage-trucks in an airport surveillance application. Firstly, the optimization is carried out by adjusting to individual trajectories. Secondly, the generalization problem (the search for appropriate solutions to general situations avoiding overfitting) is approached considering combinations of trajectories to take into account in the ES optimization. In both cases, the trained system is tested with the rest of trajectories. Our experiments show how, besides an automatic and reliable adjustment of parameters, the optimization strategy of combining trajectories improves the generalization capability of the training system.

1 Introduction

A minimal requirement for automatic video surveillance system is the capacity to track multiple objects or groups of objects in real conditions [1].

One of the main points of this research consists in the evaluation of surveillance results, defining a metric to measure the quality of a proposed configuration [2]. The truth values from real images are extracted and stored in a file [3-4]. To do this, the targets are marked and positioned in each frame with different attributes. Using this metric in an evaluation function, we can apply different techniques to assess suitable parameters and, then, to optimize them. ES are selected for this problem [5-9] because they present high robustness and immunity to local extremes and discontinuities in fitness function. This paper demonstrates that ES are well chosen for this optimization problem as the desired results are reached once an appropriate fitness function has been defined: adjust automatically the tracker performance according to all specifications considered). Furthermore, one of the principal points of this study is the non assumption about the type of tracking system used.

The ES optimized tracker must show a high degree of generalization to be used with sequences that differ greatly. Mitchell [10] defines generalization as a process

F. Rothlauf et al. (Eds.): EvoWorkshops 2005, LNCS 3449, pp. 386–395, 2005.

that uses a great number of specific observations, and extracts and retains the important features that characterize the classes of these given observations. Therefore, generalization can be redefined as a problem of search. The set of data (learning examples) that optimize the search is defined as the "ideal trainer". The ideal trainer is a set of trajectories that represents different movements of cars, trucks or airplanes.

In the next section, the whole image processing system is outlined, indicating the effect of parameters open to the designer. The third section presents the application of ES to the multiple tracking in video sequences and explanation of the proposed evaluation metric. Finally, the fourth section presents the analysis of the fitness function that is used to achieve the desired property of generality. Furthermore, the fourth section shows the results of the optimization process with different set of trajectories and combination alternatives. Finally, some conclusions end the study.

2 Surveillance Video System

This section describes the structure of an image-based tracking system.

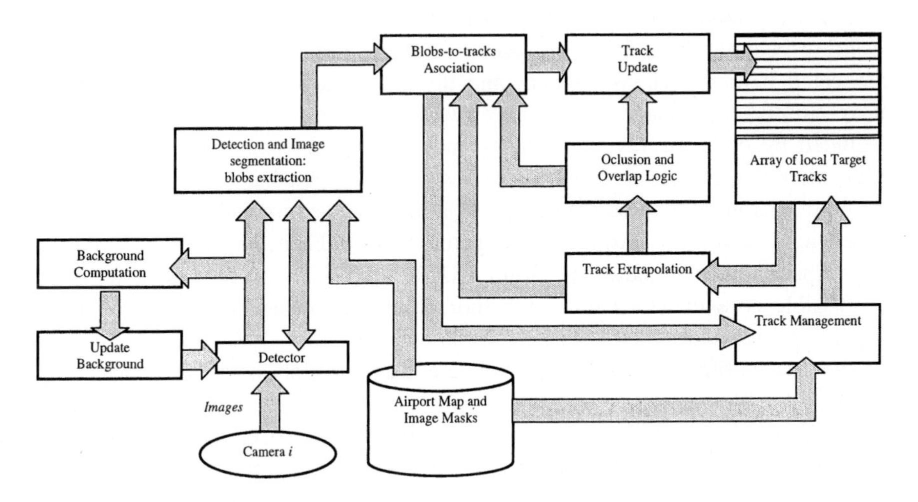

Fig. 1. Structure of video surveillance system

The system architecture is a coupled tracking system where the detected objects are processed to initiate and maintain tracks. These tracks represent the real targets in the scenario and the system estimates their location and cinematic state. The detected pixels are connected to form image regions referred to as blobs.

The association process assigns one or several blobs to each track, while not associated blobs are used to initiate tracks [4].

2.1 Detector and Blobs Extraction

The positioning/tracking algorithm is based on the detection of targets by contrasting with local background, whose statistics are estimated and updated with the video sequence. Then, the pixel level detector is able to extract moving features from background, comparing the difference with a threshold:

$$\text{Detection}(x,y):=[\text{Image}(x,y) - \text{Background}(x,y)]>\text{THRESHOLD}*\sigma \quad (1)$$

being σ the standard deviation of pixel intensity. This parameter determines the first filter on the data amount to be processed in following phases.

Finally, the algorithm for blobs extraction marks with a unique label all detected pixels connected, by means of a clustering and growing regions algorithm [11]. Then, the rectangles which enclose the resulting blobs are built, and their centroids and areas are computed. In order to reduce the number of false detections due to noise, a minimum area, *MIN_AREA,* is required to form blobs. This parameter is a second data filter which avoids noisy detections from the processing chain.

2.2 Blobs-to-Track Association

The association problem lies in deciding the most proper grouping of blobs and assigning it to each track for each frame processed. Due to image irregularities, shadows, occlusions, etc., a first problem of imperfect image segmentation appears, resulting in multiple blobs generated for a single target. So, the blobs must be re-connected before track assignment and updating. However, when multiple targets move closely, their image regions may overlap. As a result, some targets may appear occluded by other targets or obstacles, and some blobs can be shared by different tracks. For the sake of simplicity, first a rectangular box has been used to represent the target. Around the predicted position, a rectangular box with the estimated target dimensions is defined, (xmin, xmax, ymin, ymax). Then, an outer gate, computed with a parameter defined as a margin, MARGIN_GATE, is defined. It represents a permissible area in which to search more blobs, allowing some freedom to adapt target size and shape.

The association algorithm analyses the track-to-blob correspondence. It firsts checks if the blob and the track rectangular gates are compatible (overlap), and marks as conflictive those blobs which are compatible with two or more different tracks. After gating, a grouping algorithm is used to obtain one "pseudoblob" for each track. This pseudoblob will be used to update track state. If there is only one blob associated to the track and the track is not in conflict, the pseudoblob used to update the local track will be this blob. Otherwise, two cases may occur:

1) A conflict situation arises when there are overlapping regions for several targets (conflicting tracks). In this case, the system may discard those blobs gated by several tracks and extrapolate the affected tracks. However, this policy may be too much restrictive and might degrade tracking accuracy. As a result, it has been left open to design by means of a Boolean parameter named CONFLICT which determines the extrapolation or not of the tracks.

2) When a track is not in conflict, and it has several blobs associated to it, these will be merged on a pseudoblob whose bounding limits are the outer limits of all associated blobs. If the group of compatible blobs is too big and not dense enough, some blobs (those which are further away from the centroid) are removed from the

list until density and size constraints are held. The group density is compared with a threshold, MINIMUM_DENSITY, and the pseudo-blob is split back into the original blobs when it is below the threshold.

2.3 Tracks Filtering, Initiation and Deletion

A recursive filter updates centroid position, rectangle bounds and velocity for each track from the sequence of assigned values, by means of a decoupled Kalman filter for each Cartesian coordinate, with a piecewise constant white acceleration model [12]. The acceleration variance that will be evaluated, usually named as "plant-noise", is directly related with tracking accuracy. The predicted rectangular gate, with its search area around, is used for gating. Thus it is important that the filter is "locked" to real trajectory. Otherwise tracks would lose its real blobs and finally drop. So this value must be high enough to allow manoeuvres and projection changes, but not too much, in order to avoid noise. As a result, it is left as an open parameter to be tuned, VARIANCE_ACCEL.

Finally, tracking initialization and management takes blobs which are not associated to any previous track. It requires that non-gated blobs extracted in successive frames accomplish certain properties such as a maximum velocity and similar sizes which must be higher than a minimum value established by the parameter MINIMUM_ TRACK_AREA. In order to avoid multiple splits of targets, established tracks preclude the initialization of potential tracks in the surrounding areas, using a different margin than the one used in the gating search. This value which allows track initialization is named MARGIN_INITIALIZATION.

3 Application of ES to Optimization of Video Tracking System

In this section, an evaluation metric to assess the quality of a specific configuration is proposed. This configuration is defined by the values of the set of parameters described above. Firstly, the evaluation criterion is carried out for a specific trajectory by assessing the similarity of estimated target behaviour to the stored ground truth.

3.1 Definition of ES to Video Tracking Optimization

In this specific problem, one individual will represent a set of parameters of the whole tracking system, described in the previous section. The parameters have an influence on the detector and on the different stages of the tracking system, such as tracks updating and initialization, and on the association procedure.

The eight parameters which are going to be adjusted are those previously indicated:

THRESHOLD, MINIMUM_BLOB_AREA, MARGIN_GATE, MINIMUM _DENSITY, CONFLICT, VARIANCE_ACEL, MINIMUM_TRACK_AREA and MARGIN_INITIALIZATION.

We have implemented ES for this problem with a size of 6+6 individuals and mutation factor $\Delta\sigma$=0.5. Regarding the operators, the type of crossover used in this work is the discrete one and the replacement scheme which is used to select the individuals for the next generation is (μ+λ)-ES.

The main point of this optimization is the search for the ideal combination of tracks to obtain the ideal trainer and the optimal combination of performance evaluation over each trajectory to obtain the best fitness function.

3.2 Evaluation of Video Tracking System Using a Trajectory

One of the most important points of our study is to calculate some figures of merit which allow the evaluation of the performance of the tracking system. To achieve this goal, the measurements given for the tracking system are compared to the ideal output or ground truth. The ground truth is defined as a set of rectangles that form the trajectory of each target.

The evaluation system calculates a number which constitutes the measurement of the quality level for the tracking system using as a reference a certain trajectory. The output track should be as similar as possible to the ground truth trajectory. Thus, the next step is the comparison of the ideal trajectories with the detected ones so that a group of performance indicators can be obtained to analyse the results and determine the quality of our tracking process. The optimization of the evaluation outcome will be the goal for the evolutionary strategy program. We have a deterministic evaluation for individuals in all generations, enhancing system comparison to analyze different parameters within the optimization loop.

The performance evaluation is computed by giving a specific weight to each of the next indicators, divided into 'accuracy metrics' and 'continuity metrics'.

Accuracy Metrics:

1) Error in area (in percentage): The difference between the ideal area and the estimated area is computed. If more than one real track corresponds to an ideal trajectory, the best one is selected (although the multiplicity of tracks is annotated as a continuity fault).
2) X-Error and Y-Error: The difference among the x and y coordinates of the bounding box of an estimated object and the ground truth.
3) Overlap between the real and the detected area of the rectangles (in percentage): The overlap region between the ideal and detected areas is computed and then compared, in percentage, with the original areas. The program takes the lowest value to assess the match between tracking output and ground truth.

Continuity Metrics:

4) Commutation: The first time the track is estimated, the tracking system marks it with an identifier. If this identifier changes in subsequent frames, the track is considered a commuted track.
5) Number of tracks: It is checked if there is not a single detected track matched with the ideal trajectory. Multiple tracks for the same target or lack of tracks for a target indicate continuity faults. There are two counters to store how many times the ground

truth track is matched with more than one tracked object data and how many times the ground truth track is not matched with any track at all.

Finally, in order to normalize the evaluation for different tracks, all values are normalized by the track lifetime, measured with the difference between the last and first frames in which the ideal track appears. To obtain the final result, the addition of each indicator is calcultated.

4 Experiments

This section shows how the analysis of the evaluation method and the subsequent use of ES improve considerably the performance of our tracking system. The eight parameters explained above are going to be studied in order to see the effects of them in the optimization of the tracking system.

The experiments have been tested over a set of three videos which represent common situations in the airport surveillance application domain:

- The first video is a multiple-blob reconnection scenario. There is an aircraft moving with partial occlusions due to stopped vehicles and aircraft in parking positions in front of the moving object. There are multiple blobs representing a single target that must be re-connected, and at the same time there are four vehicles (vans) moving on parallel roads.
- In the second scenario, there are three aircraft moving in parallel taxiways and their images overlap when they cross.
- Finally the third video presents two aircraft moving on inner taxiways between airport parking positions. Both aircraft are turning during the conflict interval, changing their orientations. A third aircraft appears at the end, overlapping with one of the other aircraft.

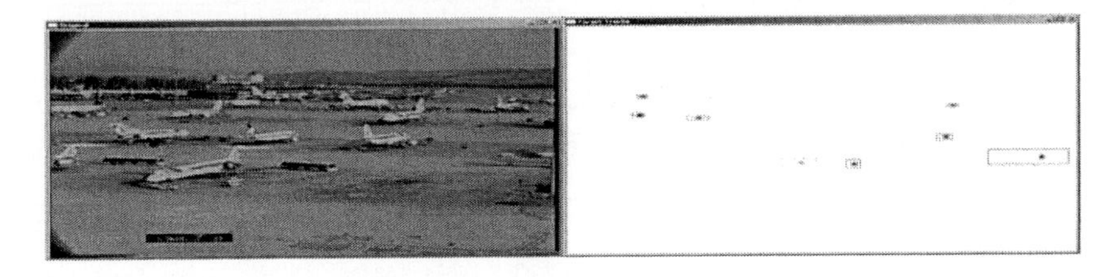

Fig. 2. Shot of the second video and display of the outcome of the surveillance video system

4.1 Applying ES to Optimize Tracking System over a Trajectory

The first experiment gives us a benchmark table with which subsequent experiments can be compared. Here, the optimization has been carried out for each individual trajectory. This means that the best parameters are obtained in order to have the best performance for a specific track.

The results obtained for each target (the configuration parameters) are represented in each column, and so there are 11 sets. Then, these parameters have been tested for the rest of tracks and videos in order to check how they solve other situations. These last results are called cross-fitness values. The next table shows how the cross fitness values do not fit as well as the ones which have been directly calculated for that case. The over fitted results are placed in the diagonal of the table. In some situations, the obtained parameters provide no track for a certain trajectory, and it cannot be evaluated. For instance, the parameters obtained with scenario 1, target 4 (a small van), do not provide any track for target 5 in that video (a big aircraft), since the particular density threshold obtained is adapted to very regular targets with a single blob. This is due to the fact that the aircraft needs systematic blob re-connection to have its track built and regularly updated. This case is marked as a '-'in the table. Partial sums for each scenario and total budgets for all tests (sum of evaluations and faults) indicate the "global" quality for each solution achieved.

Table 1. Cross evaluation for parameters optimized for a single target by using the evaluation metric explained in section 3.2

	designed scenario	Video 1					Video 2			Video 3		
evaluation scenario	Track id	1	2	3	4	5	2	3	4	1	2	3
Video 1	1	**1679,9**	10493	10010		5252,1	17140	8389,9	5851,7	2309,1	3068,1	1786,7
	2	2834,9	**2816,2**	2838,5	2837,9	2823,7	2838,5	2837,8	2836,9	2838,0	2838,2	2837,4
	3	1574,1	7686,8	**4102,2**	978,3	7987,1	7802,9	7569,4	6350,4	7486,8	12602	11494
	4	3753,0	5867,0	4106,5	**257,9**	4947,8	7100,0	4772,6	5155,3	3596,9	3625,1	2080,2
	5	221,4	6581,2	5451,0	-	**47,6**	7250,1	5141,1	7293,3	156,4	4809,6	810,4
	sum	10063	33444	26508	4074,2	21058,2	42131	28710	27487	16387	26943	19009
Video 2	2	1244,2	6879,2	493,5	2501,3	572,8	**248,9**	498,1	470,7	3757,2	529,7	592,4
	3	9091,7	8209,0	1343,0	6238,2	2393,5	1510,0	**731,4**	1521,7	6765,8	6905,9	6962,8
	4	2946,6	7946,6	759,8	768,7	1496,9	752,6	770,4	**743,1**	786,1	761,8	802,3
	sum	13282	23034	2596,2	9508,1	4463,1	2511,6	1999,9	2735,4	11309	8197,5	8357,5
Vdieo 3	1	2720,7	7029,5	9959,7	583,1	57546,1	6508,1	5862,6	6157,4	**172,5**	6756,8	8558,7
	2	7706,9	8104,1	13271	305,5	10189,5	9771,7	5080,6	12383	5079,4	**91,2**	-
	3	11388	9460,7	12319	-	14943,6	14421	11711	16524	9445,9	13398	**459,8**
	sum	21815	24594	35550	888,6	82679,2	30701	22654	35065	14697	20246	9018,5
	total sum	4516	81073	64655	14470	108200	75344	53365	65288	42394	55387	36385
	faults	0	0	0	2	0	0	0	0	0	0	1

4.2 Applying ES to Optimize Tracking System Using Sets of Targets

The next step in our study optimizes all the targets in a video. This means that the best parameters are obtained in order to have the best performance for all the targets at the same time in a specific video. The method proposed to do this is a mini-max strategy, which consists of training with all the tracks of a video and taking the maximum fitness value from the track set for each evaluation. Subsequently, this value is used as the outcome for the ES process.

The results can be observed in the left side of table 2. Now there are only three sets of optimum parameters (three columns) corresponding to the training scenarios. When the parameters for the best performance are computed by this first method, we obtain a fitness number per track which is remarkably better than all the previous results, which were calculated by training over a single track.

Table 2. Cross evaluation for parameters optimized for a whole scenario with a minimax operation (table on the left side) and cross evaluation for parameters optimized for a whole scenario with a sum operation (table on the right side)

evaluat. scenario	designed scenario	Video 1	Video 2	Video 3
Video 1	1	2148,38	6467,07	7118,28
	2	2816,06	2838,03	2829,93
	3	808,49	7571,34	7722,86
	4	611,94	4296,65	2346,35
	5	219,49	6012,59	6898,39
	sum	6604,36	27185,68	26915,81
Video 2	2	7761,07	501,94	498,34
	3	4774,85	735,06	1709,53
	4	5057,37	746,73	1901,62
	sum	17593,2	1983,73	4109,49
Video 3	1	327,80	8511,76	1321,73
	2	352,39	6639,13	4445,21
	3	-	10007,40	2724,84
	sum	680,19	25158,29	8491,78
	total sum	24877	54327,70	39517,08
	faults	1	0	0

evaluat. scenario	designed scenario	Video 1	Video 2	Video 3
Video 1	1	1691,76	2650,92	8046,10
	2	2843,30	2842,96	4232,38
	3	1141,46	1684,53	13112,50
	4	541,82	-	5826,94
	5	177,96	-	2024,31
	sum	6396,30	7178,41	33242,23
Video 2	2	1744,78	284,92	-
	3	9155,52	501,81	-
	4	4403,98	784,92	8095,01
	sum	15304,28	1571,64	8095,01
Video 3	1	7608,18	-	631,79
	2	6906,09	-	2354,02
	3	-	-	6432,97
	sum	14514,2	-	9418,78
	total sum	36214,85	8750,05	50756,02
	faults	1	5	2

The second method proposed consists of training with all the tracks of a video and taking the sum of all trajectory fitness values (right side of table 2). The remarkable difference between both methods is that the cross fitness values get worse in this last experiment, which means that the parameters are over fitted and solutions are clearly less generic, with plenty of bad solutions not providing tracks in some evaluations (a total of 6).

In any way, the improvement in generalization capability is shown in the fact that a lower number of faults appear (one now vs. three in the benchmark table), and the total sum is also lower (notice that two solutions are only comparable with total sum only when their number of faults is the same).

4.3 Applying ES to Optimize Tracking System Using Video Sets

The final step lies in taking for every iteration the maximum or the sum of the fitness values which have been calculated by evaluating the three sets of videos together.

The results are shown in the next two tables. The best aspect of this global optimization lies in the validity of the parameters to all the set of videos, avoiding the over fitted values obtained in the previous experiments. The first table shows the results for the case in which the maximum fitness has been taken in each iteration loop (the minimax strategy). The results applied to a specific scenario are good, but not better than the results obtained in the previous section where each video was trained with itself.

Nevertheless, the total fitness sum for all the scenarios results much better than in all the previous cases. This indicates that the parameters fit well in all videos.

Table 3. Cross evaluation for parameters optimized for all scenarios with minimax (table on the left side) and Cross evaluation for parameters optimized for all scenarios sum (table on the right side)

evaluation scenario	designed scenario	All videos
Video 1	1	2347,60
	2	2820,85
	3	1280,23
	4	3416,05
	5	1146,61
	sum	11011,34
Video 2	2	494,70
	3	2095,89
	4	787,59
	sum	3378,18
Video 3	1	5766,68
	2	5136,36
	3	3168,68
	sum	14071,72
	total sum	**28461,24**

evaluation scenario	designed scenario	All videos
Video 1	1	2243,12
	2	2855,57
	3	7683,49
	4	1676,22
	5	105,63
	sum	14564,03
Video 2	2	7506,24
	3	10970,60
	4	4523,21
	sum	23000,05
Video 3	1	3465,03
	2	6181,07
	3	4363,25
	sum	14009,35
	total sum	**51573,43**

The second table shows the fitness values obtained when the sum of each fitness number is taken every iteration loop as the argument of the ES. All the results can be applied to each video, but are not as good as the ones obtained in the table on the left.

5 Conclusions

In this work, we apply ES to optimize the performance of a video tracking system. A set of parameters of the tracking system should be adjusted to obtain a good performance in very different situations. In this case, the set of examples to apply the ES should produce a general solution of the tracking system. This problem is known as the "ideal trainer". Our ideal trainer is composed by a set of trajectories, and the evaluation of the tracking system should be a combination of the performance of the tracking system over the whole set of trajectories.

We have presented a process to evaluate the performance of a tracking system using a trajectory based on the extraction of information from images filmed by a camera. The ground truth tracks, which have been previously selected and stored by a human operator, are compared to the estimated tracks. The comparison is carried out by means of a set of evaluation metrics which are used to compute a number that represents the quality of the system.

Then, the proposed metric constitutes the argument to be introduced to the evolutionary strategy (ES) whose function is the optimization of the parameters that rule the tracking system. The study tests several videos and shows the improvement of the results for the optimization of three parameters of the tracking system.

References

1. Rosin, P.L., Ioannidis, E.: Evaluation of global image thresholding for change detection, Pattern Recognition Letters, vol. 24, no. 14, (2003) 2345-2356
2. Black, J., Ellis, T., Rosin, P.: A Novel Method for Video Tracking Performance Evaluation, Joint IEEE Int. Workshop on Visual Surveillance and Performance Evaluation of Tracking and Surveillance (VS-PETS) (2003)
3. Piater, J. H., Crowley, J. L.: Multi-Modal Tracking of Interacting Targets Using Gaussian Approximations. IEEE International Workshop on Performance Evaluation of Tracking and Surveillance (PETS) (2001)
4. Pokrajac, D., Latecki, L.J.: Spatiotemporal Blocks-Based Moving Objects Identification and Tracking, IEEE Int. W. Visual Surveillance and Performance Evaluation of Tracking and Surveillance (VS-PETS) (2003)
5. Rechenberg, I.: Evolutionsstrategie'94. frommannholzboog, Stuttgart (1994).
6. Schwefel, H.P.: Evolution and Optimum Seeking: The Sixth Generation. John Wiley & Sons, Inc. New York, NY, USA (1995)
7. Bäck, T.: Evolutionary Algorithms in Theory and Practice, Oxford University Press, New York (1996)
8. Bäck, T., Fogel, D.B., Michalewicz, Z.: Evolutionary Computation: Advanced Algorithms and Operators, Institute of Physics, London (2000)
9. Bäck, T., Fogel, D.B., Michalewicz, Z.: Evolutionary Computation: Basic Algorithms and Operators, Institute of Physics, London (2000)
10. Mitchell, T.M: Generalization as search. Artificial Intelligence, 18 (1982) 203-226
11. Sanka, M., Hlavac, V., Boyle, R.: Image Processing, Analysis and Machine Vision, Brooks/Cole Publishing Company (1999)
12. Blackman, S., Popoli, R.: Design and Analysis of Modern Tracking Systems. Artech House (1999)

A Multistage Approach to Cooperatively Coevolving Feature Construction and Object Detection

Mark E. Roberts and Ela Claridge

School of Computer Science,
University of Birmingham, B15 2TT, UK
{M.E.Roberts, E.Claridge}@cs.bham.ac.uk

Abstract. In previous work, we showed how cooperative coevolution could be used to evolve both the feature construction stage and the classification stage of an object detection algorithm. Evolving both stages simultaneously allows highly accurate solutions to be created while needing only a fraction of the number of features extracting as in generic approaches. Scalability issues in the previous system have motivated the introduction of a multi-stage approach which has been shown in the literature to provide large reductions in computational requirements. In this work we show how using the idea of coevolutionary feature extraction in conjunction with this multi-stage approach can reduce the computational requirements by at least two orders of magnitude, allowing the impressive performance gains of this technique to be readily applied to many real world problems.

1 Introduction

The problem of detecting objects in an image is inherently difficult, and one which has taken humans millions of years of evolution to perfect. The difficulty of this task is largely due to the infinite combinations of scale, orientation, viewpoint, lighting and many other factors that are present in a typical visual scene. In spite of this, the human visual system has evolved to deal with this problem both flexibly and robustly by evolving both the ability to construct and extract useful features from the scene and the ability to infer meaning from those extracted features. The simultaneous coevolution of these stages creates a synergistic system which is far more capable than either would be on its own.

In previous work [1], we have shown an approach which seeks to exploit this type of synergy by using cooperative coevolutionary techniques to simultaneously coevolve both a set of feature constructors, and a classifier that uses those features. This approach was successful on a variety of scale and rotationally invariant problems. In this work we show how a multi-stage approach similar to that proposed by Howard et. al. [2, 3] can be used alongside this framework. The multi-stage approach is actually a remarkably simple idea (see section 3

F. Rothlauf et al. (Eds.): EvoWorkshops 2005, LNCS 3449, pp. 396–406, 2005.

for an overview) which is based around the concept of first creating a quick classifier to deal with a large proportion of the data, and then training secondary classifiers to deal with the errors caused by the first classifiers estimative nature.

This work addresses the problem of *object detection*. This term usually refers to problems of the type "find all of the X in this image". Object detectors often take the form of classifiers applied at every pixel in the image, which have the task of deciding whether or not the pixel belongs to a target object. A popular approach is to extract a number of features at every pixel based on the statistics of the surrounding pixel values. These features and the pixel's class are then used to train a classifier. This classifier could be of any type, from decision tree to neural network, or as in this work, produced by genetic programming.

The idea of coevolving both feature construction and the object detector is motivated by the fact that it is impossible to tell, for a particular dataset, what features would be useful to an object detector. A "feature" is normally a value derived from local pixel statistics (PS) around the current pixel e.g. the mean of the pixels in the 5x5 square surrounding the pixel, or the standard deviation of the 3x3 square 8 pixels to the left of the current pixel etc. In general, generic sets of features are extracted which are often quite large, with obvious consequences for the efficiency of the complete solution. The alternative situation is that highly domain specific features are extracted, which makes it hard to apply the system

Table 1. Pixel statistics used in previous work

Author	Statistic set
Tackett [4]	Means and standard deviations of small and large rectangular windows plus the global mean and standard deviation. A total of 7 zones
Daida [5]	Pixel value, 3x3 area mean, 5x5 area mean, 5x5 Laplacian response, 5x5 Laplacian response of a 3x3 mean. 5 in total.
Howard [2]	Rotationally invariant statistics of 4 concentric rings. For each ring the statistics measured mean and standard deviation, number of edges and a measure of edge distribution. 16 in total
Howard [3]	3x3, 5x5, 7x7 and 9x9 means, 5x5, 7x7, 9x9 perimeter means and variances, and the differences between a 3x3 area mean and 5x5, 7x7 and 9x9 perimeter means. 13 in total
Winkeler [6]	Means and variances of 26 zones. Zone 1 was the entire 20x20 area, zones 2-5 were the four 10x10 quadrants, 6-10 were 5 20x4 horizontal strips, and 11-26 were 16 5x5 pixel squares. 52 in total
Ross [7]	Features related to cross and plane polarized microscopy output. Angle of max. gradient, angle of max. position, max. colour during rotation, min. intensity, and min. colour. 9 in total.
Zhang [8]	Means and variances of 4 large and 4 small squares, plus the means and variances of 2 long and 2 short line profiles, passing horizontally and vertically through the origin. 20 in total
Zhang [9]	Used 3 different terminals sets using means and variances of each zone. Set 1 used two different sized squares, set 2 had four concentric squares, and set 3 had 3 concentric circles. 4, 8, and 6 in total

to other problems. Ideally, we want to find the smallest set of features that enables the classifier to solve the problem. It can be seen from Table 1 that many different sets of features have been used in previous work, illustrating the fact that choosing sets of pixel statistic features is a difficult task which often results in using very large sets of generic features.

Most work in this area has focused on the evolution of the detection algorithms and have not studied the choice of pixel statistics (except [9] which compared the performance of 3 different statistic sets, and [2] which compares simple statistics with DFT based ones). In this work, we show that instead of using these fixed statistic sets, we can use cooperative coevolution [10] to allow the pixel statistic sets to adapt to the problem domain. We show how the pixel statistic sets which produce the inputs to the detectors, can be simultaneously coevolved alongside the detectors themselves, allowing the system to optimise both the feature construction and object detection stages.

2 System Architecture

We can conceptually divide the system implementation into two areas. The first is the basic mechanics of the coevolutionary system used and the representations and interactions between the multiple populations. This is covered in Section 2.1. Within that first area, there is a step - "calculate fitness", which corresponds to the second conceptual area. This area deals with how the multi-stage evolution is performed and this is used to calculate the fitness value for a particular solution. These steps are shown in Section 3.

2.1 Coevolutionary System

The system used for these experiments coevolves normal tree based object detection algorithms, alongside the pixel statistics that they use as terminals. In order to evolve these, the system has two main components. The first is a population of object detection algorithms (ODAs), in which each individual is a "normal" GP tree. The terminals used in the ODAs are the results of the calculation of various pixel statistics. These statistics are supplied by a number of pixel statistic populations (PSPs) in which each individual is a single pixel statistic zone (PSZ). One PSZ is selected from each PSP to form the entire pixel statistic set (PSS) which the ODA is evaluated with. This basic architecture and the interaction between the population is inspired by the work of Ahluwalia [11, 12] on "Coevolutionary ADFs", as in effect, the pixel statistics are "functions" used by the ODAs.

Pixel Statistics. Each individual in the PSP is a single pixel statistic zone (PSZ), which is applied as a moving window over the image and calculates some statistic about the pixels beneath it. It is represented using several values which describe its shape and behaviour. The origin of the PSZ is located at the pixel that the classifier is currently looking at. Its parameters are as follows -

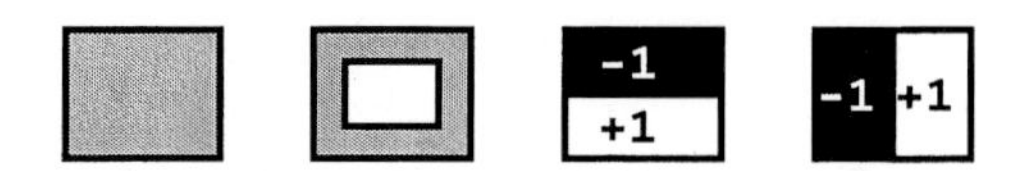

Fig. 1. Zone types used. Two simple statistic zones returing either mean or standard deviation, and two weighted sum zones

- The type of zone, one of the four types shown in Figure 1. The first two just calculate the statistics of the pixels they cover, while the second two perform a weighted sum.
- The distance from the origin in pixels, and the angle in radians. These polar coordinates define the zone's centre.
- The orientation of the shape
- The width and height of the zone.
- The statistic (mean or standard deviation) that the zone returns.

The system uses a set of populations to represent the pixel statistics. One zone from each population will eventually be used to form a complete pixel statistic set (PSS), used for terminal calculations. These populations are evolved in a normal generational model, with tournament selection (of size 3), which allows the PSZs to change their size, position, shape and statistic, allowing them to adapt to the data they are being trained on. Mutation of a PSZ involves a small random change to one of these components. The shape and statistic fields are changed less often, with only 30 percent of selections resulting in a change to one of these fields. Crossover simply interpolates the position and size of two zones, and creates one child. Each population is evolved independently of all other populations i.e. there is no cross-breeding between populations.

ODA Population. This is a population of object detection algorithms which are normal tree based representations of programs using the operators add, subtract, multiply, divide, power, min, max, and a conditional operator (IFLT) which compares two branches and returns one of two other branches depending on the result of the comparison. The terminals used in these trees $(F_1, F_2, \ldots, F_n)$ refer to the outputs of each PSZ in the PSS currently being evaluated. Crossover and mutation are applied to these trees in the normal fashion.

Evaluation. The ODAs and PSZs can only be evaluated in the context of each other, which means we cannot evaluate any of the populations independently. To calculate fitness values the system first creates a pixel statistic set (PSS) i.e. one zone from each PSP, which is paired with an ODA. This pairing is then evaluated on the training images. However, just using one of these collaborations is not usually enough to assess fitness, as one of the components may drag down the fitness of the others. In order to form a good estimate of the fitness of each of the components, we must evaluate each individual several times using different collaborators. Choosing collaborators and the subsequent credit assignment problem is not a well understood task, analysed in detail by Wiegand et. al. [13].

The basic evolutionary process is as follows.

```
begin
  foreach individual in the ODA population
    repeat subset_size times
     randomly select an individual from each PSP to create a PSS
     foreach training image
       extract features described by PSS
       apply ODA to features and calculate fitness (section 3.2)
     end foreach
     record fitness with ODA and each PSZ
    end repeat
   end foreach
  calculate final fitnesses for each ODA
  calculate final fitnesses for each zone in each PSP
  breed each population independently
repeat until termination criteria met
```

The parameter *subset_size* is the collaboration pool size, i.e. the number of pairings to try for each ODA, and should be as high as possible within the bounds of evaluation time. At the end of each generation we use the "optimistic" credit assignment method [13], whereby each ODA and PSZ are assigned the best fitness value of any evaluation it has been involved in.

3 Multi-stage Approach

The multi-stage approach to object detection was introduced by Howard et. al. [2, 14, 15, 3, 16] and is based on the idea that instead of training a classifier using every point in an image, you can train a "quick" classifier on a random sample of the points, apply the resulting classifier to all of the points, and then train second stage classifiers to deal with the errors produced in the first stage. This produces considerable savings in computational requirements.

In the first stage we take all pixels belonging to targets, and a random sample of non-target points (see section 3.1). The size of this random sample is normally around the square root of the number of pixels in the image. An object detector is trained on this subset of points, after which it is applied to every point in the images. This will of course generate a number of false positives, but in general, this number is very small relative to the size of the image i.e. the first stage deals with most of the pixels. In the second stage, classifiers are trained using this set of FPs and all of the target points, and have to discriminate between the two classes. If a target is hit at least once, all of its points are removed so that subsequent stages can concentrate on hitting the other targets. Several second stages successivley refine the solution. Any point that is marked in the first stage, and in any second stage is considered as a "final guess". The set of final guesses is then used to calculate fitness (see Section 3.2). The final output of the training is therefore a complete solution which contains one first stage classifier, which is applied to every pixel, and a set of second stage classifiers which are applied to any pixel that the first stage marked.

3.1 First Stage Point Selection

Ideally, the sample of non-target points chosen for the first stage training will be representative of the pixels in the image. Obviously, as Howard [3] states, it is impossible to pick a truly representative sample of pixels. However, in some cases choosing pixels with uniform probabilities will under-represent important parts of the image. For example consider an image of a grassy field containing cows and sheep, and the task of evolving a "cow detector". If we sample 1000 non-cow points uniformly, we will probably get about 950 grass pixels, and only 50 sheep pixels. This is a problem as distinguishing cows from grass is easy so we don't really need a large sample of them, but distinguishing cows from sheep is much harder so we would prefer more pixels from that class, and sampling uniformly will probably not give us enough sheep pixels to evolve a cow/sheep discriminator. It is also very wasteful, as grass pixels are quite uniform, and we probably only need a small sample of them in order to distinguish them from cows. So, overall, we have more points than we need (which results in longer training time) and we have under-represented an important class.

Although it would be impossible to pick a truly representative sample, a simple heuristic can offer a significant improvement over the random approach. In this work, we use a simple scheme to pick the non-target sample. For each pixel, we calculate 8 attributes, such as the local mean, variance, maximum edge value and direction, line and centre surround responses, etc. These attributes place each pixel into an 8 dimensional space to which we apply k-means clustering to divide the points into a number of classes (10 in this case). We then randomly sample an equal number of points from each of these classes to form our final sample. This scheme is very simple, but is good enough to provide a basic division. This scheme ensures that each class is represented to approximately the same degree. It is not a perfect representative sample, but it minimises the situation described above, and allows us to use less points in total.

3.2 Fitness Function

The set of final guesses produced by a detector is used to calculate the fitness value that will be assigned back to its constituent parts. The fitness function used is defined using the following terms and is shown below.

hits The number of targets which have at least one of their pixels hit.
false negatives (FN) The number of targets that were not hit at all.
false positives (FP) the number of pixels incorrectly marked as targets.

$\boldsymbol{N_t}$ The total number of targets present.
$\boldsymbol{N_b}$ The total number of non-targets pixels in the set.

$$f = \left(\left(1 + \alpha\frac{FN}{N_t}\right)\left(1 + \beta\frac{FP}{N_b}\right)\right) - 1 \tag{1}$$

The function produces a value between 0 and 1, with 0 being a perfect score, by mutipliying a target term and non-target term. This multiplication forms a multi-objective fitness function, where both terms must be minimised in order to achieve a good fitness. The ratio between the parameters α and β, allow us to tune the behaviour of the system. In the first stage α is higher than β so that we maximise hits. In the second stage β is much higher than α in order to minimise false positives. These parameters can be altered to reflect the relative importance of false positives and false negatives as, for example, in some applications (such as medical screening applications) a false negative could be critically important but a few false positives would not be the end of the world.

We also use a *figure-of-merit* to sum up the overall performance of a solution, in a more understandable way than the fitness function. This produces a figure between 0 and 1, with a value of 1 meaning we have no misses or false positives.

$$FOM = \frac{hits}{N_t + FP} \tag{2}$$

4 Experiments

Experiments to test the multi-stage method have been carried out on two datasets. The first of these is an automatically generated dataset. Each image consists of a number of random triangles and circles of different scales and orientations. The image is heavily corrupted by noise. The task in this case is to evolve a "triangle detector". The second dataset consists of images of different pasta shapes on a textured background. The task in this case is to detect only the spiral shaped pasta. Examples of both datasets can be seen in Fig 2.

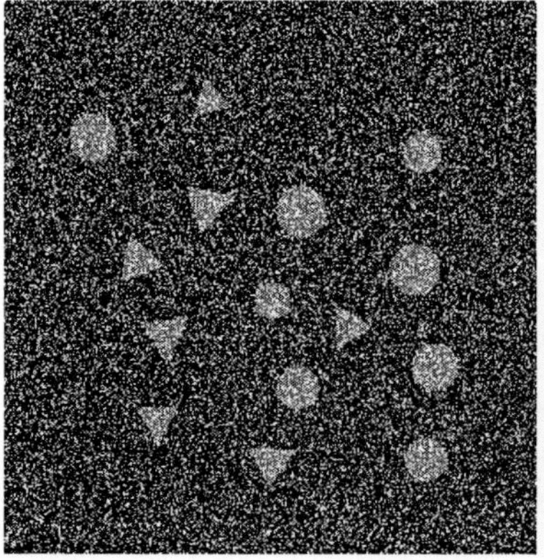
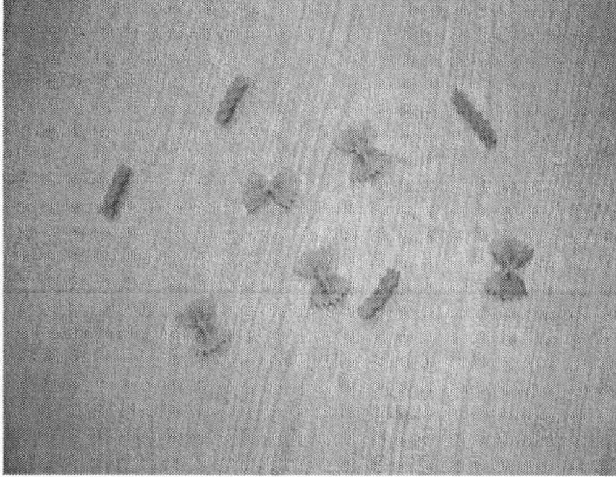

Fig. 2. An example from each dataset

For each dataset we train the multi-stage coevolutionary system on 50 images from the set. The ODA popualtion size is 2000 and the PSZ population sizes are 50. Two PSZ populations are used in each stage i.e. the are 2 features available to all ODAs. The parameters α and β are set to 1.7 and 1.0 in the first stage and 1.0 and 6.0 in the second stages. All stages are run for 50 generations or until a fitness of 0.001 is found. After training, the best combined detector is applied to a set of 50 unseen images, and the mean FOM recorded.

5 Results

5.1 Triangle Detection

On the triangle detection task, a first stage solution was evolved after 50 generations which hit all targets and produced 14117 false positives. The second stages were trained on these false positives and the target points, and after all stages there were only 12 false positives but this was at the expense of missing 18 of the 360 targets. Overall this produced a FOM of 0.92. When applied to the unseen test set, the solution performed similarly producing a FOM of 0.88. This is very good result on such noisy unseen data. An example of the output can be seen in Figure 4.

5.2 Pasta Detection

On the pasta detection task, a first stage solution was evolved after 24 generations which hit all targets and produced 1487 false positives. The second stages were trained on these false positives and the target points, and after all 3 second stages there were only 11 false positives and all of the 195 targets were hit. Overall this produced an average FOM of 0.95. When applied to the unseen test set of 50 images, the solution missed 15 of the 221 targets and generated 18 false positives producing a FOM of 0.86. This means only around 6% of all targets

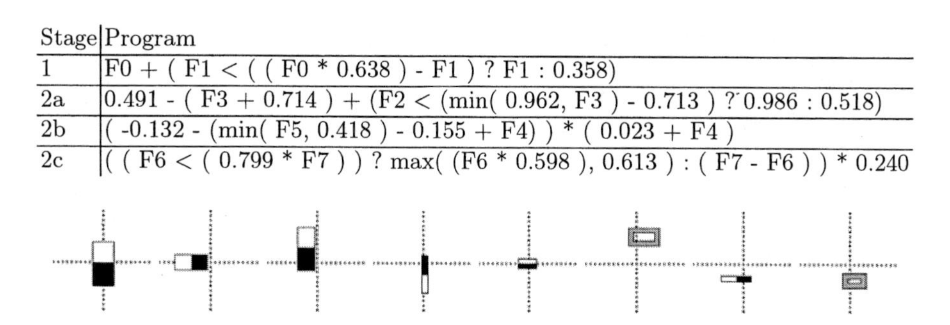

Stage	Program
1	F0 + (F1 < ((F0 * 0.638) - F1) ? F1 : 0.358)
2a	0.491 - (F3 + 0.714) + (F2 < (min(0.962, F3) - 0.713) ? 0.986 : 0.518)
2b	(-0.132 - (min(F5, 0.418) - 0.155 + F4)) * (0.023 + F4)
2c	((F6 < (0.799 * F7)) ? max((F6 * 0.598), 0.613) : (F7 - F6)) * 0.240

Fig. 3. The solution for the pasta detector. The programs for each stage are shown, along with the coevolved features F0-F7 (scale is 32 pixels). Note that the programs use a C style inline conditional operator e.g. *condition ? then-clause : else-clause*

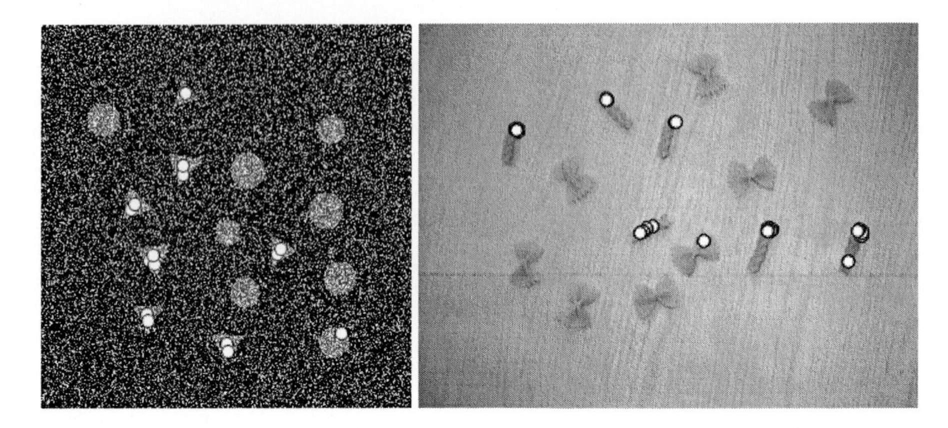

Fig. 4. Detector outputs shown on an unseen image from each dataset. The white circles represent the detector's "guesses". There is one FP in each image. Note also the very difficult triangle detection made at the top of the image. Some targets are hit multiple times by different stages, but this is counted as only one hit in FOM and fitness calculations

were missed, with a relatively small number of false positives. The complete solution is shown in Figure 3 and example output can be seen in Figure 4.

6 Conclusions

The main conclusion of this work is that cooperatively coevolving feature extractors and object detectors is now a feasible method of creating domain independent object detection algorithms. These techniques do not require any specific knowledge about the objects they are trying to detect, and can be used by domain experts (rather than image processing experts) as they simply have to mark the target points for training.

The use of the multi-stage technique shows great potential in this area. Although the "full image" approach shown in [1] actually produced more accurate, compact and efficient solutions, it used an almost unfeasibly large amount of computational power. Due to the way the multi-stage approach works, its overall solutions are larger (as they are the sum of several small solutions), but the computational load is much smaller. As an illustration of this consider the pasta example - the full image solution shown in [1] took around 2000 CPU-hours to evolve a good solution, whereas the solution evolved using the multistage technique took around 5 CPU-hours. This dramatic speed increase comes at the expense of only a small, but still acceptable, decrease in accuracy (the FOM shown in [1] was 0.98), and a small increase in the run time of a complete solution from 0.55 seconds to 0.6 seconds on the pasta images.

References

1. Roberts, M.E., Claridge, E.: Cooperative coevolution of image feature construction and object detection. In et. al., X.Y., eds.: Parallel Problem Solving from Nature - PPSN VIII. Volume 3242 of LNCS., Birmingham, UK, Springer (2004) 899–908
2. Howard, D., Roberts, S.C.: Evolving object detectors for infrared imagery: a comparison of texture analysis against simple statistics. In Miettinen, K., eds.: Evolutionary Algorithms in Engineering and Computer Science, Chichester, UK, John Wiley & Sons (1999) 79–86
3. Howard, D., Roberts, S.C., Brankin, R.: Target detection in imagery by genetic programming. Advances in Engineering Software **30** (1999) 303–311
4. Tackett, W.A.: Genetic programming for feature discovery and image discrimination. In: Proceedings of the 5th International Conference on Genetic Algorithms, ICGA-93, Morgan Kaufmann (1994) 303–309
5. Daida, J.M., Bersano-Begey, T.F., Ross, S.J., Vesecky, J.F.: Computer-assisted design of image classification algorithms: Dynamic and static fitness evaluations in a scaffolded genetic programming environment. In Koza, J.R., Goldberg, D.E., Fogel, D.B., Riolo, R.L., eds.: Genetic Programming 1996: Proceedings of the First Annual Conference, Stanford University, CA, USA, MIT Press (1996) 279–284
6. Winkeler, J.F., Manjunath, B.S.: Genetic programming for object detection. In Koza, J.R., Deb, K., Dorigo, M., Fogel, D.B., Garzon, M., Iba, H., Riolo, R.L., eds.: Genetic Programming 1997: Proceedings of the Second Annual Conference, San Francisco, CA, USA, Morgan Kaufmann (1997) 330–335
7. Ross, B.J., Fueten, F., Yashkir, D.Y.: Edge detection of petrographic images using genetic programming. In Whitley, D., Goldberg, D., Cantu-Paz, E., Spector, L., Parmee, I., Beyer, H.G., eds.: Proceedings of the Genetic and Evolutionary Computation Conference, San Francisco, USA, Morgan Kaufmann (2000) 658–665
8. Zhang, M.: A Domain Independent Approach to 2d Object Detection Based on Neural Networks and Genetic Paradigms. PhD thesis, Department of Computer Science, RMIT University, Melbourne, Victoria, Australia (2000)
9. Zhang, M., Bhowan, U.: Program size and pixel statistics in genetic programming for object detection. In Raidl, G.R., et. al., eds.: EvoIASP 2004. Volume 3005 of LNCS., Coimbra, Portugal, Springer Verlag (2004) 377–386
10. Potter, M.A., De Jong, K.A.: A cooperative coevolutionary approach to function optimization. In: Proceedings of the Third Conference on Parallel Problem Solving from Nature, Springer (1994) 249–257 Lecture Notes in Computer Science.
11. Ahluwalia, M., Bull, L.: Coevolving functions in genetic programming. Journal of Systems Architecture **47** (2001) 573–585
12. Ahluwalia, M., Bell, L., Fogarty, T.C.: Co-evolving functions in genetic programming: A comparison in ADF selection strategies. In Koza, J.R., et. al., eds.: Genetic Programming 1997: Proceedings of the Second Annual Conference, San Francisco, CA, USA, Morgan Kaufmann (1997) 3–8
13. Wiegand, R.P., Liles, W.C., Jong, K.A.D.: An empirical analysis of collaboration methods in cooperative coevolutionary algorithms. In Spector L., et. al. eds.: Proceedings of the Genetic and Evolutionary Computation Conference (GECCO). Morgan Kaufmann (2001) 1235–1245

14. Howard, D., Roberts, S.C.: A staged genetic programming strategy for image analysis. In Banzhaf, W., et. al., eds.: Proceedings of the Genetic and Evolutionary Computation Conference. Vol 2., San Francisco, Morgan Kaufmann (1999) 1047–52
15. Howard, D., Roberts, S.C., Brankin, R.: Evolution of ship detectors for satellite SAR imagery. In Poli, R., Nordin, P., Langdon, W.B., Fogarty, T.C., eds.: Genetic Programming, Proceedings of EuroGP'99. Volume EuroGP'99, part of poli:1999:GP of 3-540-65899-8., Berlin, Springer-Verlag (1999) 135–148
16. Howard, D., Roberts, S.C., Brankin, R.: Target detection in SAR imagery by genetic programming. In Koza, J.R., ed.: Late Breaking Papers at the Genetic Programming 1998 Conference, Stanford, USA, Stanford University Bookstore (1998)

An Implicit Context Representation for Evolving Image Processing Filters

Stephen L. Smith, Stefan Leggett, and Andrew M. Tyrrell

Department of Electronics, The University of York, Heslington, York YO10 5DD, UK
{sls5, amt}@ohm.york.ac.uk
http://www.elec.york.ac.uk/intsys/

Abstract. This paper describes the implementation of a representation for Cartesian Genetic Programming (CGP) in which the specific location of genes within the chromosome has no direct or indirect influence on the phenotype. The mapping between the genotype and phenotype is determined by self-organised binding of the genes, inspired by enzyme biology. This representation has been applied to a version of CGP developed especially for evolution of image processing filters and preliminary results show it outperforms the standard representation in some configurations.

1 Introduction

A form of genetic programming (GP) [1] termed Cartesian Genetic Programming (CGP) [2,3] has been successfully adapted for the evolution of simple image processing filters [4,5] and subsequently implemented in hardware [6,7]. A criticism of CGP (and GP in general) is that the location of genes within the chromosome has a direct or indirect influence on the resulting phenotype [8]. In other words, the order in which specific information regarding the definition of the GP is stored has a direct or indirect effect on the operation, performance and characteristics of the resulting program. Such effects are considered undesirable as they may mask or modify the role of the specific genes in the generation of the phenotype (or resulting program). Consequently, GPs are often referred to as possessing a direct or indirect context representation.

An alternative representation for GPs in which genes do not express positional dependence has been proposed by Lones and Tyrrell [8-12]. Termed *implicit context representation,* the order in which genes are used to describe the phenotype (or resulting program) is determined after their self-organised binding, based on their own characteristics and not their specific location within the genotype. The result is an implicit context representation version of GP termed *Enzyme Genetic Programming.*

This paper describes the first application of implicit context representation to CGP, and more specifically, a version of CGP for implementing image processing filters. CGP has a number of beneficial characteristics that overcome problems often associated with conventional GP, such as restrictive programming paradigm and uncontrolled growth of the program (known as bloat), but still expresses the undesirable characteristics of an indirect context representation.

F. Rothlauf et al. (Eds.): EvoWorkshops 2005, LNCS 3449, pp. 407–416, 2005.

Section 2 of the paper gives a brief introduction to the use of CGP for evolving image processing filers. Section 3 introduces implicit context representation and Section 4 describes the implementation of implicit context representation with CGP; the results obtained for this are presented in Section 5. Conclusions and planned further work are presented in Section 6.

2 Cartesian Genetic Programming for Evolving Image Processing Filters

Cartesian Genetic Programming (CGP) was first proposed by Miller [2,3] as an alternative representation for genetic programming which does not require the use of a parse tree based programming language and does not exhibit uncontrolled expansion commonly termed bloat [13]. As opposed to the rigid tree structure representation of traditional GP, CGP permits the arrangement of functions in a far more flexible, typically rectangular format, referenced by conventional Cartesian co-ordinates.

An extension of CGP for evolving image processing filters was proposed by Sekanina [4,5] and subsequently implemented in hardware by Zang et al. [6,7] as shown in Figure 1.

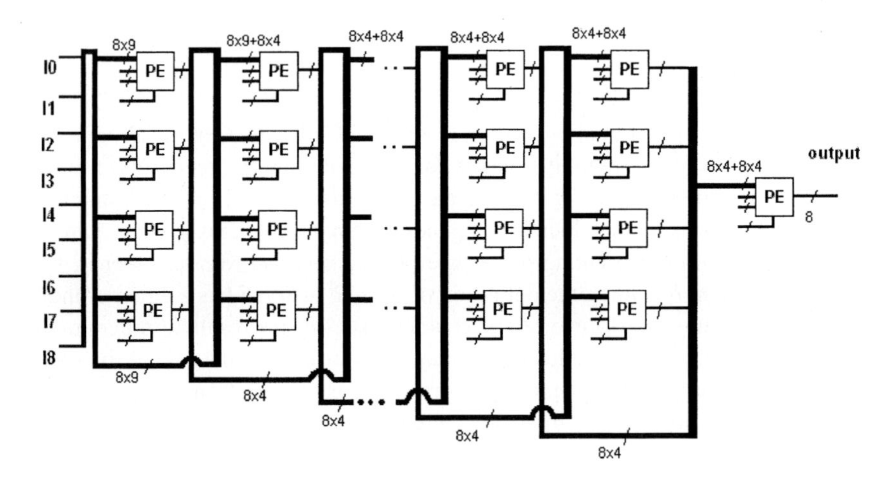

Fig. 1. Extended Cartesian Genetic Programming for evolution of image processing filters

A number of processing elements (PEs) are arranged in a rectangular format, each connected to a data bus. The inputs I0 to I8 are the pixel values obtained from a conventional 3 x 3 neighborhood image filter; these are manipulated by the PEs and the output replaces the pixel of interest in the processed image. The structure of the PE, shown in Figure 2, comprises two multiplexers and a functional block. The multiplexers can be configured, according to the values of cfg1 and cfg2 respectively, to select the output of another PE or image pixel input I0 to I8, as long as it is connected to the same data bus. In the specific hardware representation considered here, this requires that the PE or input be located in the two columns immediately preceding the PE containing the multiplexer in question. The outputs of the two multiplexers are then provided as inputs to the functional block; the function applied to them is determined by cfg3 and selected from the available functions listed in Table 1.

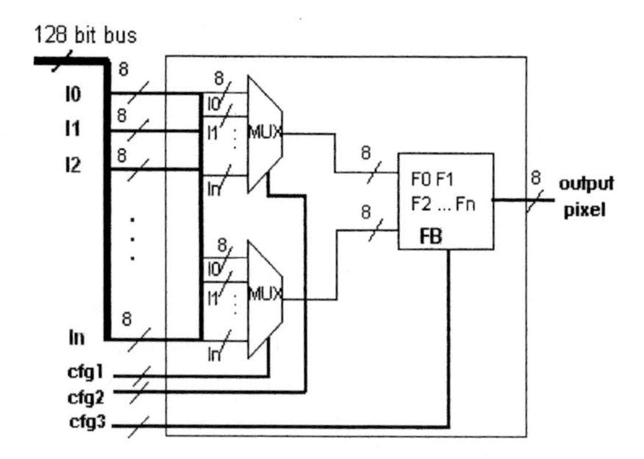

Fig. 2. Architecture of the processing element

Table 1. Functions available for configuration of the processing element's functional block

Code	Function	Code	Function
F0: 0000	X >> 1	F8: 1000 (3)	(X+Y+1) >> 1
F1: 0001	X >> 2	F9: 1001	X & 0x0F
F2: 0010	~ X	F10: 1010	X & 0xF0
F3: 0011	X & Y	F11: 1011	X I 0x0F
F4: 0100	X I Y	F12: 1100	X I 0x F0
F5: 0101 (1)	X ^ Y	F13: 1101 (4)	(X&0x0F) I (Y&0xF0)
F6: 0110	X + Y	F14: 1110	(X&0x0F) ^ (Y&0xF0)
F7: 0111 (2)	(X+Y) >> 1	F15: 1111	(X&0x0F) & (Y&0xF0)

7 4 3 4 0 3 6 1 3 8 5 3 2 11 3 2 1 3 0 4 3 12 3 3 10 9 1 12 16 3 9 13 1
14 11 3 17 18 3 19 20 3 20 20 3 15 20 3 17 19 4 24 19 2 19 24 4 21 22 3
22 26 4 27 22 3 28 21 1 25 23 3 28

Fig. 3. Example chromosome for configuration of the extended CGP

The PEs within the architecture are configure by means of a chromosome, an example of which is given in Figure 3.

The chromosome consists of a string of integer values, arranged logically in groups of three, providing values for cfg1 (multiplex 1 input), cfg2 (multiplex 2 input) and cfg3 (functional block function index), respectively, for each PE in the representation.

A number of these chromosomes form the individuals of a population which are initialised with random values. Each chromosome is then used to configure the hardware representation which, in turn, is used to process a test image. The image resulting from this operation is compared with an ideal (uncorrupted image), and a fitness score derived, which is then associated with the respective individual's chromosome. After all the individuals in the population have been evaluated in this manner, the fittest is retained and used as the parent for a subsequent generation of individuals. These new individuals are generated by simply mutating the parent in a non-deterministic manner.

Yang et al demonstrated that the image filters evolved in this way out performed conventional median and Gaussian image filters [6,7].

3 Implict Context Representation

The extended CGP considered in Section 2 and CGP in general can be described as an *indirect context representation*; the position a particular gene occupies in the chromosome has an influence on the resulting phenotype, or in this case, the configuration of the hardware representation. Ideally, the evolution of a system should be independent of the position of genes within the chromosome, but should still be a result of the values of those genes. This is termed an *implicit context representation* by Lones and Tyrrell [9], who have developed a form of GP that exploits this representation, called *Enzyme Genetic Programming* (EGP). The biological inspiration for Enzyme GP is the metabolic pathway, and the role of enzymes which express computational characteristics. This is not dissimilar to the logic network employed in this work to evolve the image filters described in Section 2 [8].

Lones and Tyrrell have developed an enzyme model comprising a shape, activity and specificities (or binding sites) [11], as shown in Figure 4.

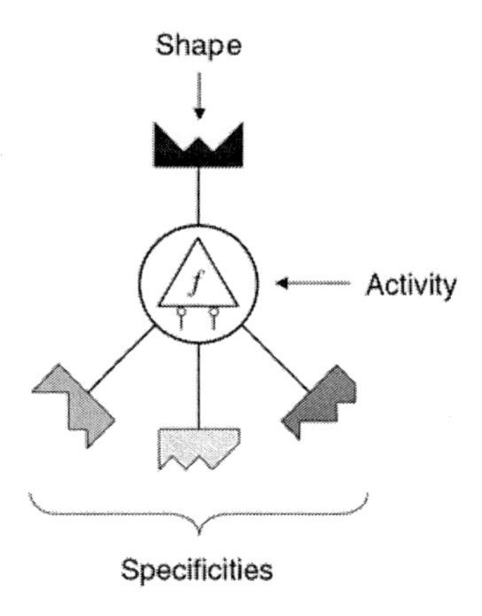

Fig. 4. Enzyme model illustrating shape, activity and specificities (binding sites) [11]

Along with inputs and outputs, the enzyme model can be considered a program component from which a genetic program may be constructed. The shape describes how the enzyme is seen by other program components. Similarly, the binding sites determine the shape (and hence type) of program component the enzyme wishes to bind to. Finally, the activity determines the logical function the enzyme is to perform. A typical EGP will comprise a set number of inputs and outputs and a number of enzyme models or components. Values for each component's binding sites and logical function are initialized non-deterministically; the component's shape, however, is derived from a combination of its binding site's shapes and logical function as shown in Figure 5.

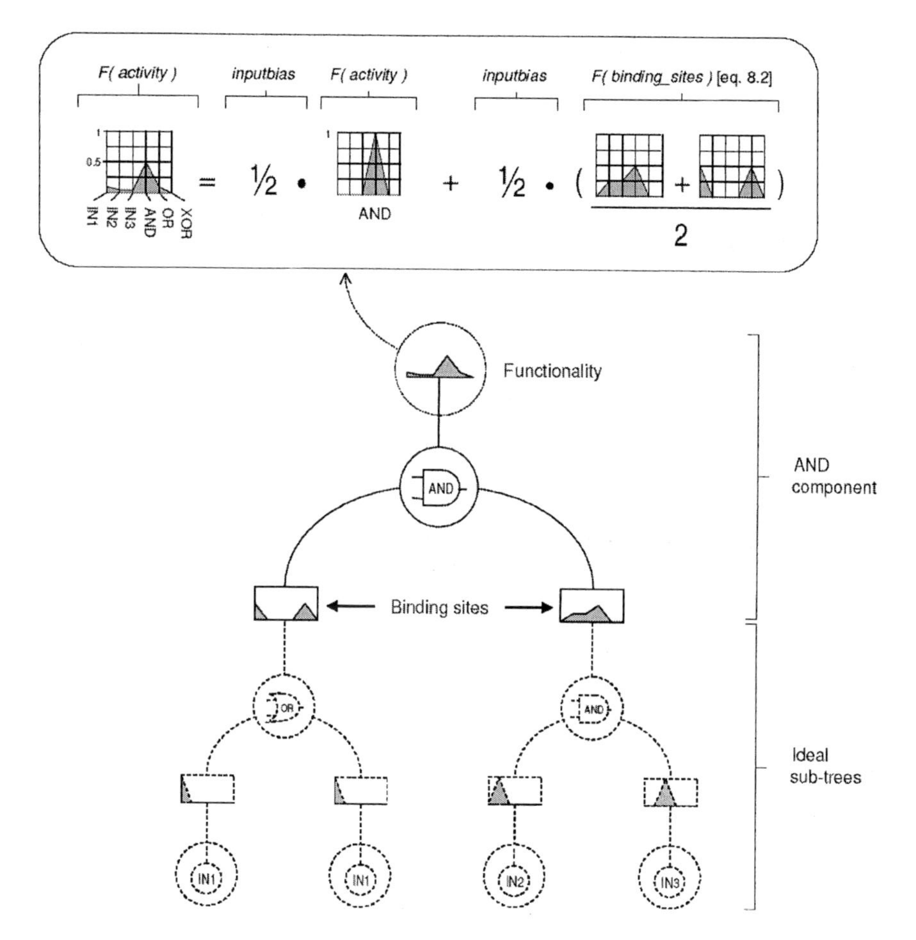

Fig. 5. Calculation of a component's shape from its binding site shapes and logical function [8]

Once initialized, components are bound together to form a network. The order in which components are bound is determined by the closeness of match between one component's binding site and another component's shape. The best matching components are bound first and the process is repeated until a network has formed in which no further binding is possible, as illustrated in Figure 6.

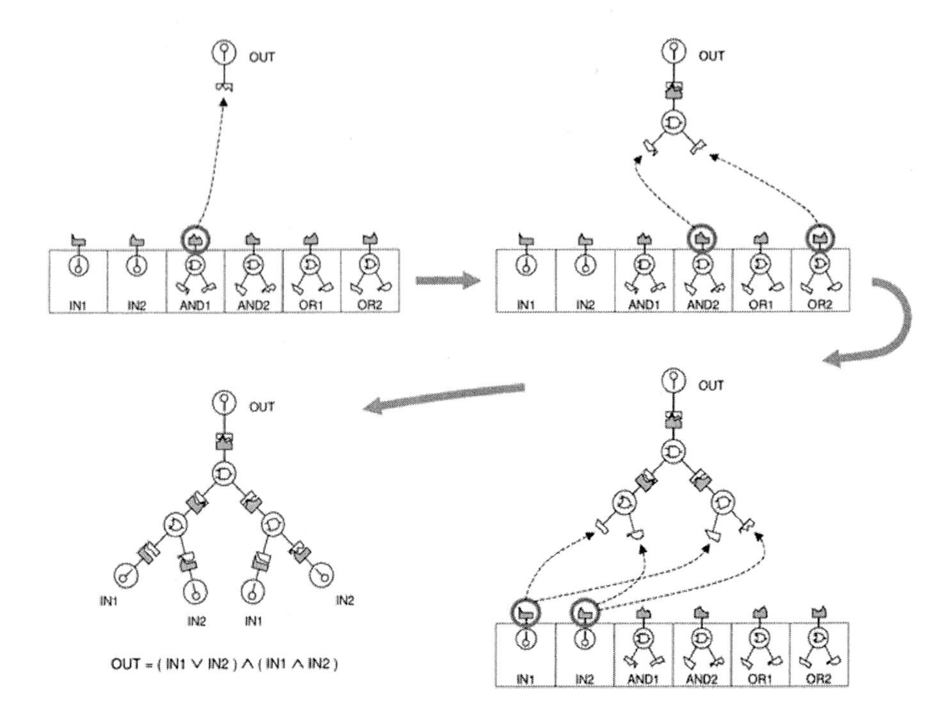

Fig. 6. Binding of components in a network is on the basis of closest match [8]

Over time, components may evolve through mutation. Mutation is applied to the component's binding sites and logical function with a pre-determined probability. When this occurs, a new component shape is derived accordingly and may lead to different binding between components occurring. This in turn may result in a modified network.

Lones and Tyrrell implemented EGP in a parse-tree configuration and a diffusion model genetic algorithm called the *Network GA* [8,11] and applied it to several problems involving symbolic regression achieving some promising results.

4 Implementaion

The aim of the work described here is to combine the benefits of an implicit context representation (described in Section 3) with the extended Cartesian genetic program for evolving image filters described in Section 2.

The processing elements within the extended CGP are particularly suited to representation by the enzyme model used in enzyme genetic programming. However, in stead of employing a parse tree arrangement, the existing CGP Cartesian arrangement is maintained. The significant difference is the manner in which components are selected and interconnected within the representation.

Formation of the CGP network begins with the assignment of an output component; this will ultimately provide a new value for the pixel under consideration in the filtered image. The binding sites of the output component are then made active and will bind to components according the closeness of match between their respective shapes. Once bound, component's binding sites will also become active and will bind to other components in the same way. Binding between components is always undertaken on a "best-fit first" basis until no further binding is possible.

The physical hardware places constraints in the manner with which the formation of the network takes place. Successful binding of a new component may only take place if there is sufficient space for that component in the hardware representation. Typically, this means that any newly bound component must be placed in one of the two columns to the left of the existing component. Similarly, input components I0 to I8 (holding the image pixel values) may only be bound to components one or two columns to their right in the representation.

Once all possible binding has completed, the resulting network is applied to a test image and the resulting filtered image compared with the original, uncorrupted image. A fitness score for the individual that network is described by equation (1) which is identical to that used in previous image filtering evolution by Sekanina [4,5] and Yang [6,7].

$$fitness = 255.(H-2).(W-2) - \sum_{i=1}^{H-2}\sum_{j=1}^{W-2} ideal(i,j) - filt(i,j) \qquad (1)$$

Where:

i,j	are the image co-ordinates
filt(i,j)	is the image resulting from the resulting filter operation
ideal(i,j)	is the ideal (original uncorrupted) image
H, W	is the height and width of the image respectively

A 1+λ evolvable strategy is adopted for formation of the subsequent generation. Once a fitness score has been calculated for all individuals within the population, the best individual is preserved and the next generation is formed mutated versions of this individual. Two separate mutation operations are performed according to predefined probabilities: (i) to the binding sites of the components and, (ii) to the index that selects the component's function from those available (as defined in Table 1). Once these mutations have been performed, new shapes for each component are derived as described in Section 3.

5 Results

The implicit representation Cartesian genetic program (IRCGP) was tested as closely as possible to the conventional extended Cartesian genetic program (ECGP) undertaken by Yang et al [6,7]. A list of the parameters used is given in Table 2 which is the same for both implementations.

Fig. 7. Original image

Fig. 8. Image corrupted with added Gaussian noise $\sigma = 16$

Fig. 9. Corrupted image after conventional median filtering

Fig. 10. Corrupted image after conventional Gaussian filtering

Fig. 11. Corrupted image after applying extended CGP evolved filter

Fig. 12. Corrupted image after applying implicit representation CGP evolved filter

Table 2. Common parameters for both ECGP and IRCGP experiments

Parameter	Value
Population size	15
Number of generations	500
Number of available functions	4
Available functions	(A+B+1)>>1
	(A&0x0F)|(B&0xF0)
	(A+B)>>1
	A^B

Results are provided for 20 runs of the algorithms on a version the 'Lena' image corrupted with Gaussian noise $\sigma = 16$. A segment of the original and corrupted images (Lena's face) are shown in Figures 7 and 8 respectively. The best resulting image from the IRCGP is shown in Figure 12. For the purpose of comparison, the best image resulting from ECGP is also shown in Figure 11, and those resulting from conventional mean and Gaussian smoothing filters are shown in Figures 9 and 10 respectively.

Values for the fitness score of the IRCGP and ECGP over the 20 runs for two hardware configurations are given in Table 3. Fitness scores for traditional median and Gaussian filters are 97.50% and 97.24%, respectively. These results are represented as a percentage of the best possible fitness score (i.e. in comparison to the original image).

Table 3. Fitness score for 20 runs of IRCGP and ECGP experiments

Hardware Configuration			IRCGP (%)	ECGP (%)
Rows	Columns			
4	3	Overall Best	97.52	97.48
		Average Best	97.45	97.38
4	6	Overall Best	97.30	97.53
		Average Best	96.88	97.51

It can be seen that the best run of the IRCGP exceeds the performance of the median and Gaussian filters, and for the smaller (4x3) hardware configuration, outperforms the ECGP.

6 Conclusion

This paper reports the first software implementation of an implicit context representation of extended CGP for the purpose of evolving image processing filters. Initial results for the evolution of noise removal filters have shown that in some configurations this approach outperforms the conventional extended CGP and that of selected traditional image processing filters.

Further work is currently under way to characterize this evolutionary algorithm in terms of its performance and efficiency. Implementation of the algorithm in hardware is also being considered.

References

1. J. Koza: Genetic Programming: On the Programming of Computers by Means of Natural Selection, MIT Press (1992)
2. J. Miller and P. Thomson: Cartesian genetic programming. in Third European Conf. Genetic Programming, R. Poli, W. Banzhaf, W. B. Langdon, J. F. Miller, P. Nordin, and T. C. Fogarty (eds.). Vol. 1802 Lecture Notes in Computer Science, Springer (2000)
3. J. F. Miller, D. Job, and V. K. Vasilev: Principles in the evolutionary design of digital circuits—Part I. Genetic Programming and Evolvable Machines, Vol. 1 (2000) 7–36
4. L. Sekanina, V. Drabek: Automatic Design of Image Operators Using Evolvable Hardware. Fifth IEEE Design and Diagnostic of Electronic Circuits and Systems (2002) 132-139
5. L. Sekanina. Image Filter Design with Evolvable Hardware. Applications of Evolutionary Computing – Proceedings of the 4th Workshop on Evolutionary Computation in Image Analysis and Signal Processing (EvoIASP'02), Lecture Notes in Computer Science. Vol. 2279 Springer-Verlag, Berlin (2002) 255-266
6. Yang, Z., Smith, S.L. and Tyrrell, A.M.: Intrinsic Evolvable Hardware in Digital Filter Design. Lecture Notes in Computer Science. Vol. 3005. Springer-Verlag, Berlin (2004) 389-398
7. Yang, Z., Smith, S.L. and Tyrrell, A.M.: Digital Circuit Design using Intrinsic Evolvable Hardware. Proceedings of 2004 NASA/DoD Conference on Evolvable Hardware, Seattle (2004)
8. M. A. Lones: Enzyme Genetic Programming. PhD Thesis, University of York, UK, (2003)
9. M. A. Lones and A. M. Tyrrell: Enzyme genetic programming. Proc. 2001 Congress on Evolutionary Computation, J.-H. Kim, B.-T. Zhang, G. Fogel, and I. Kuscu (eds.), IEEE Press. Vol. 2 (2001) 1183–1190
10. M. A. Lones and A. M. Tyrrell: Crossover and Bloat in the Functionality Model of Enzyme Genetic Programming. Proc. Congress on Evolutionary Computation 2002 (CEC2002) (2002) 986-992
11. M. A. Lones and A. M. Tyrrell: Biomimetic Representation with Enzyme Genetic Programming. Journal of Genetic Programming and Evolvable Machines. Vol. 3 No. 2 (2002) 193-217
12. M. A. Lones and A. M. Tyrrell: Modelling biological evolvability: implicit context and variation filtering in enzyme generic programming. BioSystems (2004)
13. W. Langdon: Quadratic bloat in genetic programming. In D. Whitley, D. Goldberg, and E. Cantu-Paz, editors, Proceedings of the 2000 Genetic and Evolutionary Computation Conference (2000) 451-458

Learning Weights in Genetic Programs Using Gradient Descent for Object Recognition

Mengjie Zhang and Will Smart

School of Mathematics, Statistics and Computer Science,
Victoria University of Wellington,
P. O. Box 600, Wellington, New Zealand,
{mengjie, smartwill}@mcs.vuw.ac.nz

Abstract. This paper describes an approach to the use of gradient descent search in tree based genetic programming for object recognition problems. A weight parameter is introduced to each link between two nodes in a program tree. The weight is defined as a floating point number and determines the degree of contribution of the sub-program tree under the link with the weight. Changing a weight corresponds to changing the effect of the sub-program tree. The weight changes are learnt by gradient descent search at a particular generation. The programs are evolved and learned by both the genetic beam search and the gradient descent search. This approach is examined and compared with the basic genetic programming approach without gradient descent on three object classification problems of varying difficulty. The results suggest that the new approach works well on these problems.

1 Introduction

Since the early 1990s, there have been a number of reports on applying genetic programming (GP) techniques to object recognition problems [1, 2, 3, 4, 5, 6, 7]. Typically, these GP systems used either high level or low level image features and random numbers to form the terminal set, arithmetic and conditional operators to construct the function set, and classification accuracy, error rate or similar measures as the fitness function. During the evolutionary process, selection, crossover and mutation operators were applied to the genetic beam search to find good solutions. While some of these GP systems achieved reasonable even good results, others did not perform well. In addition, they usually spent a long time for learning good programs for a particular task. One reason for this is that they did not directly use the existing heuristics in the individual programs, for example, the gap between the actual outputs of the programs and the target outputs.

Gradient descent is a long established search technique and is commonly used to train multilayer feed forward neural networks [8]. A main property of this search is that it can use gradients in a neural network or other methods effectively. This algorithm can also guarantee to find a local minima for a particular task. While the local minima might not be the best solution, it often meets

F. Rothlauf et al. (Eds.): EvoWorkshops 2005, LNCS 3449, pp. 417–427, 2005.

the requirement of the task. A main characteristic of gradient descent search is that the solutions can be improved locally but steadily.

Gradient descent search has been applied to numeric terminals [9] or constants [10] of genetic programs in GP. In these approaches, gradient descent search is applied to individual programs locally at a particular generation and the constants in a program are modified accordingly.

The goal of this paper is to develop a new approach to the use of gradient descent search in genetic programming for multiclass object classification problems. Instead of searching for certain good constants or numeric terminals in a genetic program, gradient descent in this approach will be used to learn partial structures in genetic programs. We will investigate how to apply gradient descent to genetic programs so that better sub programs can be learned, how the gradient descent search cooperates with the genetic beam search, whether this new approach can do a good enough job on a sequence of object classification problems of increasing difficulty, and whether this new approach outperforms the basic GP approach.

2 GP Applied to Object Classification

In the basic GP approach, we used the tree-structure to represent genetic programs [11]. The ramped half-and-half method was used for generating programs in the initial population and for the mutation operator [12]. The tournament selection mechanism and the reproduction, crossover and mutation operators [14] were used in the learning and evolutionary process.

We used pixel level, domain independent statistical features (referred to as *pixel statistics*) as terminals and we expect the GP evolutionary process can automatically select features that are relevant to a particular domain to construct good genetic programs. Four pixel statistics are used in this approach: the average intensity of the whole object cutout image (see section 4), the variance of intensity of the whole object cutout image, the average intensity of the central local region, and the variance of intensity of the central local region. In addition, we also used some constants as terminals. These constants are randomly generated using a uniform distribution.

In the function set, we used four arithmetic operators and a conditional operator $\{+, -, \times, pdiv, if\}$. The $+$, $-$, and $\times$ operators have their usual meanings — addition, subtraction and multiplication, while *pdiv* represents "protected" division which is the usual division operator except that a divide by zero gives a result of zero. Each of these functions takes two arguments. The *if* function takes three arguments. The first argument, which can be any expression, constitutes the condition. If the first argument is negative, the *if* function returns its second argument; otherwise, it returns its third argument.

We used classification accuracy on the training set as the fitness function. To translate the single floating point value of the program output to a class label, we used a variant version of the *program classification map* [13] to perform object classification. This variation situates class regions sequentially on the floating point number line. The object image will be classified to the class of the region that the program output with the object image input falls into. Class region boundaries start at some negative number, and end at the same positive number. Boundaries between the starting point and the end point are allocated with an identical interval of 1.0. For example, a five class problem would have the following classification map.

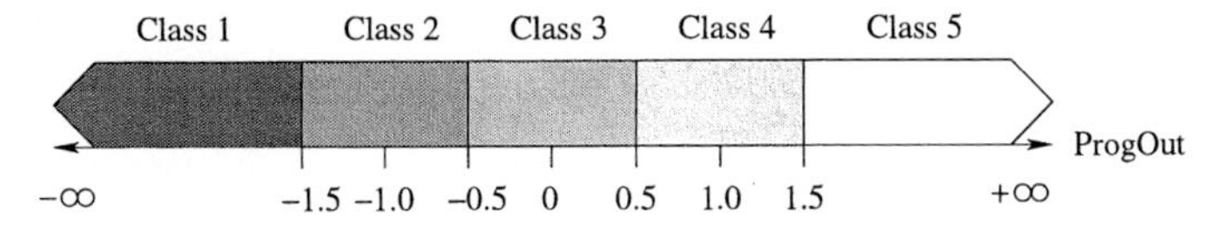

3 Gradient Descent Applied to Weights in Genetic Program Structure

Our new GP approach used similar program representation, terminals, functions, fitness function and genetic operators to the basic GP approach as described in section 2. However, the new approach also introduced a new parameter, *weight*, to each link between two nodes in a genetic program so that the gradient descent search can be applied in order to learn better programs locally. Due to the introduction of the weight parameter, the program trees, functions and the genetic operators need some corresponding changes. This section describes the weight parameter, the corresponding changes, and the method of using gradient descent search to modify the weights.

3.1 Genetic Programs with Weights

Unlike the standard genetic programs, this approach introduces a weight parameter between every two adjacent nodes, as shown in figure 1.

The weights operate in a way similar to the weights in neural networks. Before being passed on to the higher part of the tree, the result of each subtree in the evolved program is multiplied by the value of the weight associated with the link. For example, if w_{ij} is the weight between node i (parent) and node j (child),

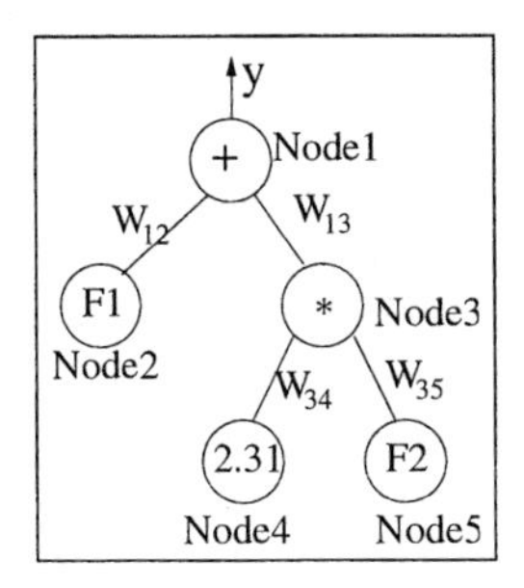

Fig. 1. An example evolved program with weights

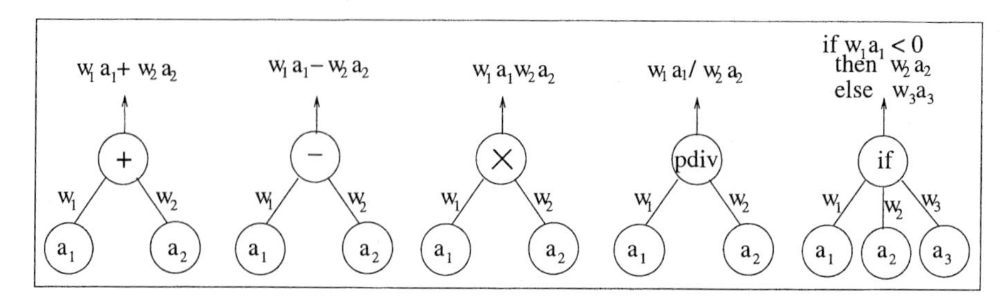

Fig. 2. Functions with the new program structure and new definitions of the functions

then the input of node i from the branch of node j will be the weight w_{ij} times the output of node j. A value 1.0 for the weight corresponds to the standard structure of genetic programs in the basic GP approach.

Due to the introduction of the weights, the program structures are changed. Accordingly, the functions we used in the basic GP approach will also need corresponding changes. For example, if a node of the addition function (+) has two child nodes a_1 and a_2 and corresponding weights w_1 and w_2, the output of the addition function would be $w_1a_1+w_2a_2$. The new definitions of the functions are shown in figure 2.

To cope with this change, the genetic operators in this approach are also redefined in a similar way. In the new crossover operators, the two sub-programs including the nodes and the weights are swapped after two crossover points are selected from the two parent programs. In mutation, after a mutation point is chosen, the sub-program was replaced by a new sub-program with randomly mutated nodes and corresponding weights.

Based on this structure, changing a weight in a program will change the performance of the program. In this approach, all the weights are initialised to 1.0 as the standard GP approach, then are learned and modified through the gradient descent algorithm described in the next sub section.

3.2 Gradient Descent Applied to Program Weights

In this approach, gradient-descent is applied to changing the values of the weights. It is assumed that a continuous cost surface C can be found to describe the performance of a program at a particular classification task for all possible values for the weights. To improve the system performance, the gradient descent search is applied to taking steps "downhill" on the C from the current group of weights.

The gradient of C is found as the vector of partial derivatives with respect to the parameter values. This gradient vector points along the surface, in the direction of maximum-slope at the point used in the derivation. Changing the parameters proportionally to this vector (negatively, as it points to "uphill") will move the system down the surface C. If we use w_{ij} to represent the value of the weight between node i and node j, then the distance that the weight moved (the change of w_{ij}) should therefore be:

$$\Delta w_{ij} = -\alpha \cdot \frac{\partial C}{\partial w_{ij}} \tag{1}$$

where α is a search factor.

Cost Surface C. We used sum-squared error over the N training object examples as the cost surface C, as shown in equation 2.

$$C = \frac{\sum_{k=1}^{N} (y_k - Y_k)^2}{2} \tag{2}$$

where Y_k is the desired program output for training example k, y_k is the actual calculated program output for training object example k.

The desired output Y_k for training object example k is calculated as follows:

$$Y_k = \mathbf{class_k} - \frac{\mathbf{numclass} + 1}{2} \tag{3}$$

where $\mathbf{class}_k$ is the class label of the object k and **numclass** is the total number of classes. For example, for a five class problem as described in section 2, the desired outputs are $-2, -1, 0, 1$ and 2 for object classes 1, 2, 3, 4 and 5, respectively.

Accordingly, the partial derivative of the cost function with respect to the weight between node i and node j in the genetic program would be:

$$\frac{\partial C}{\partial w_{ij}} = \frac{\partial(\frac{\sum_{k=1}^{N} (y_k - Y_k)^2}{2})}{\partial y_k} \cdot \frac{\partial y_k}{\partial w_{ij}} = \sum_{k=1}^{N} ((y_k - Y_k) \cdot \frac{\partial y_k}{\partial w_{ij}}) \tag{4}$$

Partial Derivative $\frac{\partial y_k}{\partial w_{ij}}$. Any genetic program will be an expression consisting of constants and sub-expressions described in section 3.1. Since these expressions are all differentiable with respect to the weights, we can readily construct the partial derivatives of the output of any particular program with respect to the weights. For example, given the program shown in figure 1, if we use O_i to represent the output of node i, then the partial derivative of the genetic program output y with respect to the weight w_{35} between node 3 and node 5 will be:

$$\begin{aligned} \frac{\partial y}{\partial w_{35}} &= \frac{\partial O_1}{\partial w_{35}} = \frac{\partial O_1}{\partial O_3} \cdot \frac{\partial O_3}{\partial w_{35}} \\ &= \frac{\partial(w_{12}O_2 + w_{13}O_3)}{\partial O_3} \cdot \frac{\partial(w_{34}O_4 \times w_{35}O_5)}{\partial w_{35}} = w_{13} \cdot w_{34} \cdot O_4 \cdot O_5 \end{aligned}$$

The values of the partial derivatives can be calculated for each training example given the current values of the weights. In this way the appropriate derivative for any weight, in a program of any depth, can be obtained using the chain rule and the derivatives of the functions, as shown in table 1.

Search Factor α. The search factor α in equation 1 was defined to be proportional to the inverted sum of the square gradients on all weights along the cost surface, as shown in equation 5.

Table 1. Derivatives of the functions used in this approach

Operation	$\frac{\partial o}{\partial a_1}$	$\frac{\partial o}{\partial a_2}$	$\frac{\partial o}{\partial w_1}$	$\frac{\partial o}{\partial w_2}$	$\frac{\partial o}{\partial a_3}$	$\frac{\partial o}{\partial w_3}$
+	w_1	w_2	a_1	a_2		
−	w_1	$-w_2$	a_1	$-a_2$		
×	$w_1 w_2 a_2$	$w_2 w_1 a_1$	$a_1 w_2 a_2$	$a_2 w_1 a_1$		
pdiv	$\frac{w_1}{w_2 a_2}$	$-\frac{w_1 a_1}{a_2^2 w_2}$	$\frac{a_1}{w_2 a_2}$	$-\frac{w_1 a_1}{w_2^2 a_2}$		
if	0	if $(w_1 a_1 < 0)$ then $w2$ else 0	0	if $(w_1 a_1 < 0)$ then a_2 then 0	if $(w_1 a_1 < 0)$ then 0 else $w3$	$if(w_1 a_1 < 0)$ then 0 else a_3

$$\alpha = \frac{\eta}{\sum_{i=1}^{M} (\frac{\partial C}{\partial w_i})^2} \tag{5}$$

where M is the number of weights in the program, and η is a learning rate defined by the user. In this way, the gradients of the weights will move small steps where the cost surface is steep and move large steps where the surface is shallow. We expect that this measure can improve search of good weights.

Summary of the Gradient Descent Algorithm. The gradient descent algorithm in this approach is summarised as follows.

- Evaluate the program, save the outputs of all nodes in the program.
- Calculate the partial derivatives of the program output with respect to each weight $\frac{\partial y}{\partial w_{ij}}$ using the chain rule and table 1.
- Calculate the partial derivative of the cost function with respect to each weight $\frac{\partial C}{\partial w_{ij}}$ using equations 4 and 3.
- Calculate the search factor α using equation 5.
- Calculate the change of each weight using equation 1.
- Update the weights using $(w_{ij})_{new} = w_{ij} + \Delta w_{ij}$.

Note that this algorithm is an offline scheme, that is, the weights are updated after all the patterns in the training set are presented. Also notice that the weight learning and updating process only happens locally for individual programs at a generation, and the genetic beam search still controls the whole evolutionary process globally between different generations. We expect that this hybrid genetic beam-gradient descent approach can improve the system performance.

4 Data Sets

We used three data sets providing object classification problems of varying difficulty in the experiments. Example images are shown in figure 3.

The first set of images (figure 3a) was generated to give well defined objects against a relatively clean background. The pixels of the objects were produced using a Gaussian generator with different means and variances for each class. Three classes of 960 small objects were cut out from those images to form the

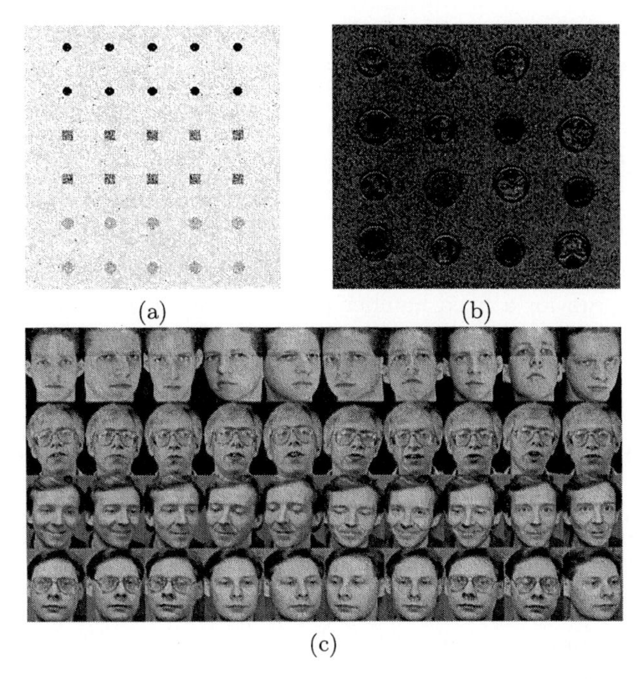

Fig. 3. Sample image data sets. (a) Shape; (b) Coin; (c) Face

classification data set. The three classes are: black circles, grey squares, and light circles. For presentation convenience, this dataset is referred to as *shape*.

The second set of images (figure 3b) contains scanned 5 cent and 10 cent New Zealand coins. The coins were located in different places with different orientations and appeared in different sides (head and tail). In addition, the background was cluttered. We need to distinguish different coins with different sides from the background. Five classes of 801 object cutouts were created: 160 5-cent heads, 160 5-cent tails, 160 10-cent heads, 160 10-cent tails, and the cluttered background (161 cutouts). Compared with the *shape* data set, the classification problem in this data set is much harder. Although these are still regular, man-made objects, the problem is very hard due to the noisy background and the low resolution.

The third data set consists of 40 human faces (figure 3c) taken at different times, varying lighting slightly, with different expressions (open/closed eyes, smiling/non-smiling) and facial details (glasses/no-glasses). These images were collected from the first four directories of the ORL face database [15]. All the images were taken against a dark homogeneous background with limited orientations. The task here is to distinguish those faces into the four different people.

For the shape and the coin data sets, the objects were equally split into three separate data sets: one third for the training set used directly for learning the genetic program classifiers, one third for the validation set for controlling overfitting, and one third for the test set for measuring the performance of the learned program classifiers. For the face data set, due to the small number of images, ten-fold cross validation was applied.

5 Experimental Results and Discussion

5.1 Experiment Configuration

The parameter values used in this approach are shown in table 2. The evolutionary process is run for a fixed number (*max-generations*) of generations, unless it finds a program that solves the classification perfectly (100% accuracy on training set), at which point the evolution is terminated early.

Table 2. Parameters used for GP training for the three datasets

Parameter Names	Shape	coin	face	Parameter Names	Shape	coin	face
population-size	300	500	500	reproduction-rate	10%	10%	10%
initial-max-depth	5	5	5	crossover-rate	60%	60%	60%
max-depth	6	6	6	mutation-rate	30%	30%	30%
max-generations	100	100	100	cross-term	15%	15%	15%

5.2 Results

Each experiment was repeated 50 runs and the average results on the test set at the best results of the validation set were reported. Table 3 shows the results of the new GP approach (GP-gradient) with different learning rates against the basic GP approach. The results at the learning rate "off" are for the basic GP approach, where no gradient descent search is used.

As can be seen from table 3, different learning rates in the GP-gradient approach resulted in different performance. This is consistent with the nature of the gradient descent search — different local minima can be reached from different learning rates. However, for all the three data sets investigated here, it was always possible to find certain learning rates at which the GP-gradient approach outperformed the basic GP approach. It seemed that a learning rate of 3.0 achieved the best classification performance, suggesting that 3.0 could be used as a starting point.

The new GP-gradient method almost always used fewer generations to find a good program classifier than the basic GP approach, and in some cases very much so. This suggests that the gradient descent search locally applied to individual programs did reduce the number of evaluations and made the evolution converge faster.

In terms of the training time, it seems that the GP-gradient method spent more time than the basic GP approach in the easy shape data set and the coin data set. However, this is not the case for the relatively difficult face data set, where the new method used less training time to find a good program classifier.

Table 3. A comparison of the results of the GP-gradient and the basic GP approach

Dataset	Method	Learning rate	Generations	Time (s)	Test Accuracy (%)
Shape	GP-basic	off	15.56	0.85	99.66
	GP-gradient	0.5	10.30	2.31	99.53
		1.0	6.38	1.39	99.69
		2.0	4.58	0.97	99.71
		3.0	5.52	1.22	99.80
Coin	GP-basic	off	43.46	1.68	90.89
	GP-gradient	0.5	43.56	7.12	89.15
		1.0	43.34	7.48	90.22
		2.0	43.22	7.77	90.02
		3.0	43.10	7.53	92.14
Face	GP-basic	off	8.32	0.37	85.00
	GP-gradient	0.5	8.07	0.35	86.50
		1.0	7.30	0.32	86.25
		2.0	6.23	0.27	86.30
		3.0	6.72	0.31	88.15

Notice that the best test accuracies were achieved with a learning rate of 3.0. It is not clear whether better object classification results can be achieved if a larger learning rate is applied — further investigation needs to be carried out.

6 Conclusions

The goal of this paper was to investigate an approach to the use of gradient descent search in tree based genetic programming for multiclass object classification problems. The goal was successfully achieved by introducing a weight parameter to the link between every two adjacent nodes in a genetic program tree and applying gradient descent search to the weight parameter. In this way, changing of a weight corresponds to changing of the effect of the sub-program tree. At a particular generation, learning the changes of a weight parameter was done locally by gradient descent search, but the whole evolution was still carried out across different generations globally by the genetic beam search. Unlike the previous approaches which also combines genetic beam search and gradient descent search but applies gradient descent to the numeric terminals/constants only, this approach can learn better program structures by updating the weights of the sub programs.

This approach was examined and compared with the basic genetic programming approach without gradient descent on three object classification problems of varying difficulty. The results suggest that the new approach outperformed the basic approach on all of these problems under certain learning rates.

The results also showed that different learning rates resulted in different performance. It was always possible to find certain learning rates with which better performance could be achieved. It does not seem to have a reliable way to obtain a good learning rate, which usually needs an empirical search through experiments. However, if this could improve the system performance, such a short search is a small price to pay. According to our experiments, a learning rate of 3.0 seems a good starting point.

This work only used the offline learning scheme when modifying the weights. We will investigate the effectiveness of the online scheme for modifying the weights in this approach in the future.

References

1. Andre, D.: Automatically defined features: The simultaneous evolution of 2-dimensional feature detectors and an algorithm for using them. In Kinnear, K.E., ed.: Advances in Genetic Programming, MIT Press (1994) 477–494
2. Howard, D., Roberts, S.C., Brankin, R.: Target detection in SAR imagery by genetic programming. Advances in Engineering Software **30** (1999) 303–311
3. Loveard, T., Ciesielski, V.: Representing classification problems in genetic programming. In Proceedings of the Congress on Evolutionary Computation. Volume 2., Seoul, Korea, IEEE Press (2001) 1070–1077
4. Song, A., Ciesielski, V., Williams, H.: Texture classifiers generated by genetic programming. In Proceedings of the 2002 Congress on Evolutionary Computation CEC2002, IEEE Press (2002) 243–248
5. Tackett, W.A.: Genetic programming for feature discovery and image discrimination. In Proceedings of the 5th International Conference on Genetic Algorithms, ICGA-93, University of Illinois at Urbana-Champaign, Morgan Kaufmann (1993) 303–309
6. Winkeler, J.F., Manjunath, B.S.: Genetic programming for object detection. In Genetic Programming 1997: Proceedings of the Second Annual Conference, Stanford University, CA, USA, Morgan Kaufmann (1997) 330–335
7. Zhang, M., Ciesielski, V.: Genetic programming for multiple class object detection. In Proceedings of the 12th Australian Joint Conference on Artificial Intelligence (AI'99), Sydney, Australia. Lecture Notes in Artificial Intelligence (LNAI Volume 1747). Springer-Verlag Berlin Heidelberg (1999) 180–192
8. Rumelhart, D.E., Hinton, G.E., Williams, R.J.: Learning internal representations by error propagation. In Parallel distributed Processing, Explorations in the Microstructure of Cognition, Volume 1: Foundations. The MIT Press, Cambridge, Massachusetts, London, England (1986)
9. Zhang, M., Smart, W.: Genetic programming with gradient descent search for multiclass object classification. In Genetic Programming 7th European Conference, EuroGP 2004, Proceedings. Volume 3003 of LNCS., Coimbra, Portugal, Springer-Verlag (2004) 399–408
10. Ryan, C., Keijzer, M.: An analysis of diversity of constants of genetic programming. In Proceedings of the Sixth European Conference on Genetic Programming (EuroGP-2003). Volume 2610 of LNCS., Springer Verlag (2003) 404–413
11. Koza, J.R.: Genetic programming : on the programming of computers by means of natural selection. Cambridge, Mass. : MIT Press, London, England (1992)
12. Banzhaf, W., Nordin, P., Keller, R.E., Francone, F.D.: Genetic Programming: An Introduction on the Automatic Evolution of computer programs and its Applications. San Francisco, Calif. : Morgan Kaufmann Publishers (1998)

13. Zhang, M., Ciesielski, V., Andreae, P.: A domain independent window-approach to multiclass object detection using genetic programming. EURASIP Journal on Signal Processing, Special Issue on Genetic and Evolutionary Computation for Signal Processing and Image Analysis **2003** (2003) 841–859
14. Koza, J.R.: Genetic Programming II: Automatic Discovery of Reusable Programs. Cambridge, Mass. : MIT Press, London, England (1994)
15. Samaria, F., Harter, A.: Parameterisation of a stochastic model for human face identification. In: 2nd IEEE Workshop on Applications of Computer Vision, Sarasota (Florida) (1994)
16. Poli, R.: Discovery of symbolic, neuro-symbolic and neural networks with parallel distributed genetic programming. In: Artificial Neural Nets and Genetic Algorithms: Proceedings of the International Conference, ICANNGA97. Norwich, UK, Springer-Verlag (1997)

Open Problems in Evolutionary Music and Art

Jon McCormack

Centre for Electronic Media Art,
School of Computer Science and Software Engineering,
Monash University, Clayton 3800, Australia
jonmc@csse.monash.edu.au
http://www.csse.monash.edu.au/~jonmc

Abstract. Applying evolutionary methods to the generation of music and art is a relatively new field of enquiry. While there have been some important developments, it might be argued that to date, successful results in this domain have been limited. Much of the present research can be characterized as finding ad-hoc methods that can produce subjectively interesting results. In this paper, it is argued that a stronger overall research plan is needed if the field is to develop in the longer term and attract more researchers. Five 'open problems' are defined and explained as broad principle areas of investigation for evolutionary music and art. Each problem is explained and the impetus and background for it is described in the context of creative evolutionary systems.

1 Introduction

Education is not the filling of a pail, but the lighting of a fire.

— W.B. Yeats

Music, art, and indeed creativity in general are defining traits of the human condition. Moreover, they are one of the primary reasons why we consider the richness of living experience to be more than just one of survival and reproduction, even though, ironically, they may have biological origins and purposes [1, 2, 3, 4]. In recent years, evolutionary computing (EC) methods have been applied to problems in music composition and art.

The premise in creating and researching evolutionary music and art (from here on EMA) is that creative problems are *non-trivial*. Creativity is considered a positive and sought-after trait in all human cultures [5]. Various hypotheses have been put forward for this, for example musical ability has been hypothesized as the result of sexual selection [6]. Moreover, creativity encompasses a broad scope of tasks in terms of psychology, problem solving, judgment and action [7].

Research in EMA has covered a variety of problems in aesthetics, creativity, communication and design. Broadly speaking however, much of the research and results have been ad-hoc; common methodologies have included: 'use technique *X* from complexity research to make images or music'; 'use aesthetic selection to evolve *X*'; 'devise a suitable fitness function to automate the evolution of *X*'; and so on. Certainly there have been many successes using such strategies. However, even in this

F. Rothlauf et al. (Eds.): EvoWorkshops 2005, LNCS 3449, pp. 428–436, 2005.

glib set of scenarios there is a sense that these strategies are simply combinatory products of a set of well-explored ideas from other disciplines (admittedly in a different context).

The aim of this paper is to define a small set of 'open problems' in EMA. The goal is not to be critical of previous work, but to provide a well-defined set of challenges for the EMA research community. Such an approach has been successful in other disciplines [8].[1] After defining an important distinction in modes of research, the remainder of this paper presents five open problems specific to EMA and discusses some current research and background for each of them.

1.1 Evolution and Art

Before introducing the set of open problems, a key difference in research goals needs to be introduced. In classifying approaches, I make a primary distinction between *(i)* research where the resultant music and artwork is intended to be recognized by humans as creative (i.e. art) and *(ii)* research which explores the concept of creativity in general.

The first case is seemingly more straightforward so we will examine it first. In this case the results generated by the system and/or methodology are intended for human appreciation as art. That is, they exhibit properties that humans recognize as displaying some form of creative intention or aesthetic judgment by the creator (we conveniently ignore whether by person or machine). People who use such techniques might call or consider themselves 'artists' in addition to (or as opposed to) 'researchers'.

The second case is different. Here, creativity is considered in a more open context, that is, it is not limited to being recognized by people as creative or aesthetic. In broad terms, creativity is not found exclusively in human behaviour. Bowerbirds, for example, create elaborate aesthetic constructs that serve no direct survival advantage, rather act as displays to attract mates. In the simulation context, a similar parallel exists in artificial life research. In Langton's seminal paper [9], the concept of 'life-as-it-could-be' is introduced. Life-as-it-could-be represents a broader set of living systems, beyond the 'life-as-we-know-it' life observed on Earth. This broader definition of life admits the possibility of other kinds of life, radically different than what we currently know as life. For example, in Fredrick Hoyle's classic novel *The Black Cloud* the Earth is visited by a radically different life-form that is unaware of the destruction it is causing on Earth until it is contacted by astronomers. Of course, this is science fiction, and it has been argued that life-as-it-could-be can never be that different from life-as-we-know-it because we could never recognize it as life [10]. Even more speculative is the idea that life-as-it-could-be might create its own art: *art-as-it-could-be*. That is, artistic products or systems created and analyzed by synthetic autonomous agents. The second case is dealt with in more detail in Section 2.4.

---

[1] At least in terms of getting the authors of the paper numerous citations. By posing difficult problems, most have not been solved, ensuring the longevity of the paper. This strategy is employed in this paper as well.

2 The Open Problems

This section introduces the open problems. A discussion surrounding each is also presented.

2.1 Searching for Interesting Phenotypes

A common practice in EMA is one of *search for interesting phenotype*. In this scenario, the artist or programmer designs some form of parameterized system. The system generates output, typically in the form of sound or image. In most cases, the number of parameters is very large, making an incremental or ordered search of the entire parameter space intractable. Hence the use of other search techniques such as genetic algorithms or aesthetic selection.

In this mode of EMA there are two primary considerations:

1. the design of the generative system and its parameterization;
2. the evaluation of the fitness of phenotypes produced by the system.

In the case of aesthetic selection, the fitness evaluation is implicit, being performed by the user of the system. I will return to the second consideration in a later section, for now let us examine the first point in more detail.

The well-known system of Karl Sims generated images using Lisp expressions evolved by aesthetic selection [11]. In essence these expressions were a combination of basic arithmetic operations and standard mathematical functions such as trigonometric and fractal functions. Even with a limited number of such expressions, the range or gamut of possible images is extremely large. However, it turns out that all of the images produced by such a system are of a certain 'class' — that is they all look like images made using mathematical expressions. While there might exist a Lisp expression for generating the *Mona Lisa* for example, no such expression has been found by aesthetic selection..[2]

Steven Rooke extended the aesthetic selection system of Karl Sims [12]. He did not change the basic methodology (evolving images created from expressions by aesthetic selection), rather he added a range of additional functions to further increase the gamut of possibilities. Certainly his images looked different and more complex than those of Karl Sims, but they were still of a certain class (images made using an expanded set of mathematical functions).

Indeed, in all uses of aesthetic selection the results produced are 'of a certain class', that is they exhibit strong traits of the underlying formalized system that created them (the parameterized system). A natural, but unsuccessful strategy has been to increase the scope and complexity of the parameterized system, giving an even larger gamut of possibilities in the phenotype. Systems of more than trivial complexity cannot be exhaustively searched. In all systems to date, this process is limited by the

[2] In one version of Sims' system, scanned images could form part of an expression tree, allowing 'real' images to be manipulated and processed by the system. This does not change the problem discussed here, however.

creativity of the artist or programmer in that they must use their creativity to come up with representations and parameterizations they think will lead to interesting results. The search process has shifted up a level (from parameters to mechanisms), but it is still a search problem that needs to be undertaken by humans: it cannot (yet) be formalized, and hence, automated.

What is needed then is a system capable of introducing novelty *within itself.* The physical entities of the Earth were capable of such a task, in that they were able to create an emergent physical replication system. This was achieved from the bottom up, in a non-teleological process of self-assembly and self-organization. It was possible because atoms, molecules, genes, cells and organisms are all physical entities and part of the same system. Generative systems for EMA could use such a mechanism. This brings us to state the first open problem:

Open problem #1: To devise a system where the both the genotype, phenotype and the mechanism that produces phenotype from genotype are capable of automated and robust modification, selection, and hence evolution.

That is, a system that does not produce images of mathematical functions or biomorphs or any particular class of phenotype, due to a fixed parameterized representation. Rather, the genotype, its interpretation mechanism and phenotype exist conceptually as part of a singular system, capable of automated modification. Any such system must be 'robust' in the sense that it is tolerant of modification without complete breakdown or failure. A similar challenge has been posed in artificial life research for the evolution of novel behaviors [13].

It might be argued that the phenotypes produced by DNA are 'of a certain class' (i.e. biological organisms), however DNA is able to build organisms, which in the appropriate environment are capable of open-ended creative behaviour. These systems exploit dynamical hierarchies to achieve their complexity.

2.2 The Problem of Aesthetic Selection

Aesthetic selection of images carried the promise of being able to search for the most beautiful or interesting phenotypes in any parameterized system. In practical terms however, it can only perform a limited search within a certain class of phenotypes, not all possible phenotypes that can be generated by the system. Therefore, the methodology itself tells us little about creativity in general, and does not really offer *the* most beautiful or interesting images from any system.

This limitation of aesthetic selection leads us to ask why it is does not achieve its goals and what other methods might be better. Aesthetic selection has several problems:

1. Population size is limited by the ability of people to perform subjective comparisons on large numbers of objects (simultaneously comparing 16 different phenotypes is relatively easy, comparing 10,000 would be significantly more difficult). In the case of visual phenotypes, the available display size may also limit the number and complexity of phenotypes that can be simultaneously shown in order to perform subjective comparison.

2. The subjective comparison process, even for a small number of phenotypes, is slow and forms a bottleneck in the evolutionary process. Human users may take hours to evaluate many successive generations that in an automated system could be performed in a matter of seconds.
3. Genotype-phenotype mappings are often not uniform. That is, a minor change in genotype may produce a radical change in phenotype. Such non-uniformities are particularly common in tree or graph based genotype representations such as in evolutionary programming, where changes to nodes can have a radical effect on the resultant phenotype. This problem is not limited to EMA applications and has been widely studied in the EC community.
4. The size and complexity of genotypes is limited. In general, simple expressions generate simple images. Complex images require more resources to compute and in a real-time system genotypes that consume too much time or space are usually removed before they can complete. In general, it is difficult to distinguish a genotype that takes a long time to do nothing (such as a recursive null-op) and one that takes a long time to do something interesting (this is analogous to the halting problem). Fractal and IFS functions are often found in aesthetic image systems, as they are an easy way of generating complexity in an image with minimal time and space complexity. The problem is that this is not a general complexity, but a fractal one, with characteristic shapes and patterns.

These limitations are indicative of why we can't find the Lisp expression that generates the *Mona Lisa* by aesthetic selection – the human doing the selecting is limiting population size and diversity to such an extent that the genetic algorithm has little change of finding anything more than local sub-optima. Moreover, the generation scheme, its mapping and complexity, is limited by representation and resources.

Such difficulties have lead researchers to try to devise schemes that remove some or all of these limitations while still providing the ability to find interesting phenotypes within the parameterized system's gamut of possibilities. One approach has been to change the interface and selection relationship between user(s) and phenotype [14] rather than removing human subjectivity from the process completely. However, this technique while successful for the situation in which it was devised is not generally applicable to all aesthetic selection problems.

Genotype-Phenotype mapping has also been researched. One interesting approach has been to evolve genotypes that represent some computational process, which is itself generative. That is, the genotype specifies the process of construction and then the construction process builds the phenotype. As the construction process itself is evolvable rather than fixed, more complex outcomes are possible [15].

To address the problems of subjective fitness evaluation by humans, a different approach has been to try to formalize the fitness function, so it can be performed by computer rather than human. This introduces the second open problem:

Open problem #2: To devise formalized fitness functions that are capable of measuring human aesthetic properties of phenotypes. These functions must be machine representable and practically computable.

Aesthetics, while well studied in art theory and philosophy, has yet to be fully understood by science. While there have been some noble attempts to measure aesthetic properties, many consider the proposition itself doomed to failure. The mathematician G. D. Birkhoff famously proposed an 'aesthetic measure', equal to order divided by complexity. Birkhoff defined ways of measuring order and complexity for several different categories of object, including two-dimensional polygons and vases. While somewhat successful for simple examples, it failed to capture aesthetic qualities with any generality, being described more as a measure of 'orderliness' [16].

Neuroscientists have also studied human aesthetic response in order to gain understanding about what makes us consider things beautiful. Ramachandran proposes 'ten laws of art which cut across cultural boundaries'. These include 'peak shift' where exaggerated features exemplify learned classifications, grouping, contrast, isolation, symmetry, repetition, rhythm, balance and metaphor [17].

Birkhoff's measure and subsequent aesthetic measures of its lineage focus on *measurable features* of aesthetic objects. These are commonly geometric properties, dimension, proportion, fixed feature categories, organizational structure, etc. The basis being that any such feature or property can be objectively measured directly. However, there are many things considered important to aesthetic theory that cannot be measured directly. Such features or properties are generally interpreted rather than measured, often in a context-sensitive way. For example, much has been made of harmonious proportions (such as the golden ratio) in nature, art and music [18]. While such measures are interesting and revealing properties of many different types of structure, they say nothing about the semantics of the structure itself. It not only matters that ancient Greek temples exhibit similar geometric golden ratios, but the context of their form in relation to Greek and human culture, the meaning and significance to the observer, and the perceptual physicality (the interpreted physical relation between observer and observed). It seems that such easily measurable properties are used at the expense of details that are more specific. That is, they are at a too high level of abstraction, where other important features and specific details are ignored. Scientific theories deliberately choose levels of abstraction applicable for physical laws to be 'universal'. This has been a reasonably successful strategy for the physical universe. For aesthetic laws, however, there are not necessarily such direct abstractions or physical measures.

2.3 What Is Art?

In answer to the question 'what is art?' Frieder Nake proposed that anything exhibited in art galleries is art [19]. That is, in general terms, people (usually experts) feel that a work has qualities that deem it appropriate to be exhibited in a place recognized for the exhibition and appreciation of art. While there have been many exhibitions of 'computer generated' art and even more specifically EMA art, many of these works are primarily selected because they are created by computer, rather than because they are art.

If EMA art is to mature, it needs to become recognized as art for what it is, in addition to how it was made.

Open problem #3: To create EMA systems that produce art recognized by humans for its *artistic* contribution (as opposed to any purely technical fetish or fascination).

One might consider this a new version of the Turing test, where artistic outcomes of EMA systems might be compared alongside those done by humans. If the audience cannot tell the difference, or at least considers both worthy of the title 'art' then the test has been passed.

The idea of this test does not discount the possibility that EMA might have its own new aesthetic qualities or be part of a wider 'movement' in machine–based or generative art.[3] Indeed, Western art is characterized by continuing change and innovation, with movements and styles fluctuating in acceptance and popularity. However, human systems of art theory and appreciation do consider these factors (along with many others) in deciding what is art – so if EMA is really art these systems should be able to accommodate it.

2.4 Artificial Creativity

I now turn to the creative activity of artificial systems. As discussed earlier, this differs fundamentally from those systems designed to produce art that is recognized and appreciated by humans. Artificial creativity extends Langton's idea of artificial life being 'life-as-it-could-be'. Artificial agent and creature simulations are a common tool in artificial life research. More recently, some researchers have begun to look at creative behavior in artificial systems.

A number of definitions exist for creativity and creative behavior. In developing computational models of creativity, Partridge and Rowe require that creativity involve production of something *novel* and *appropriate* [21]. In addition, novelty may exist relative to the individual (Boden's P-creativity), and for society or the whole of human culture (H-creativity in Boden's terminology) [22]. For their computation model of creativity, Partridge and Rowe see novelty involving the creation of new representations through *emergent memory*.

Rob Saunders evolved artificial agents capable of 'creative' behavior using a co-evolutionary strategy of creative agents and critics [23]. Agents responded in terms of a psychological theory of interest to novel behavior. Most systems involved in the generation of novelty do so by appropriate recombination of basic primitives. This is Cariani's *combinatoric emergence* where wholes that are more complex are constructed by combinations of irreducible primitives; the important point is that the total set of primitives and their function are fixed. In the case of *creative emergence* fundamentally new primitives enter the system [24]. Cariani and others have made a case that creative emergence is what we observe in nature. Clearly, this distinction relates to open problem #1, where we want the emergence of new primitives in our system, not just the combination of a fixed set.

Open problem #4: To create artificial ecosystems where agents create and recognize their own creativity. The goal of this open problem is to help understand creativity and emergence, to investigate the possibilities of 'art-as-it-could-be'.

[3] For an interesting survey of historical precedents to generative art see [20].

Computational creativity has largely relied on psychological theories of creativity. As neuroscience advances our understanding of creative behavior, this may lead to new models. The challenge for researchers in EMA is to convincingly demonstrate the autonomous emergence of agents capable of generating and recognizing novelty in their interactions.

2.5 Theories of Evolutionary Music and Art

Finally, any research involving music or art must be mindful of theories related to such practices from the disciplines themselves. Even studying these theories from an anthropological perspective is likely to shed light on the nature of creativity and aesthetics. Human culture and art is constantly changing and evolving – practices accepted today as art may not have received such acceptance in the past. Evolutionary and generative art is no exception. If this art is to progress, there must be critical theories to contextualize and evaluate it and its practitioners.

Open problem #5: To develop *art* theories of evolutionary and generative art.

It is important to distinguish between art theory and art criticism. Art criticism is based on how to evaluate art within some critical framework. Art theory is not like scientific theory in that it's use for prediction or general explanation is minimal. There does not seem to be any laws of art that will predict artists' behaviors, or that explain the 'evolution' of art history by detailing what 'succeeds' in making a work beautiful or significant. For the products of EMA to be accepted as art, there must be some artistic theory that is associated with them. Some developments have begun in this area [25].

3 Conclusion

This paper has presented five grand challenges for evolutionary music and art research. There is little doubt that these are hard problems and will probably not be solved in the immediate future. However, there is no theoretical reason that prohibits their solution eventually. What is even more profound than the solution of the problems themselves, is the impact their solution will have on society and our understanding of ourselves and our creativity. This is certainly a worthy research agenda.

References

1. Dissanayake, E.: Homo Aestheticus: Where Art Comes from and Why. University of Washington Press, Seattle (1995)
2. Dissanayake, E.: What Is Art For? University of Washington Press, Seattle (1988)
3. Pinker, S.: How the Mind Works. Penguin Press, Middlesex, England (1997)
4. Miller, G.F.: The Mating Mind: How Sexual Choice Shaped the Evolution of Human Nature. William Heinemann, London (2000)
5. Brown, D.E.: Human Universals. McGraw-Hill, New York (1991)

6. Miller, G.F.: Evolution of Human Music through Sexual Selection. In Wallin, N.L., Merker, B., Brown, S. (eds.) The Origins of Music, MIT Press, Cambridge, MA (2000) 329-360
7. Dartnall, T. (ed.) Creativity, Cognition, and Knowledge: An Interaction. Praeger, Westport, Connecticut; London (2002)
8. Bedau, M.A., McCaskill, J.S., Packard, N.H., Rasmussen, S., Adami, C., Green, D., Ikegami, T., Kaneko, K., Ray, T.S.: Open Problems in Artificial Life. Artificial Life **6**:4 (2000) 363-376
9. Langton, C.G.: Artificial Life. In Langton, C.G. (ed.) Artificial Life, SFI Studies in the Sciences of Complexity, Addison-Wesley (1989) 1-47
10. Bonabeau, E.W., Theraulaz, G.: Why Do We Need Artificial Life? Artificial Life **1**:3 (1994) 303-325
11. Sims, K.: Artificial Evolution for Computer Graphics. Proceedings of SIGGRAPH '91 (Las Vegas, Nevada, July 28 - August 2, 1991) In Computer Graphics **25**:4 ACM SIGGRAPH, New York, NY (1991) 319-328
12. Rooke, S.: Eons of Genetically Evolved Algorithmic Images. In Bentley, P.J., Corne, D.W. (eds.) Creative Evolutionary Systems, Academic Press, London (2002) 339-365
13. Taylor, T.: Creativity in Evolution: Individuals, Interactions, and Environments. In Bentley, P.J., Corne, D.W. (eds.) Creative Evolutionary Systems, Academic Press, London (2002) 79-108
14. McCormack, J.: Evolving Sonic Ecosystems. Kybernetes **32**:1/2 (2003) 184-202
15. Kitano, H.: Designing Neural Networks Using Genetic Algorithms with Graph Generation System. Complex Systems **4**:4 (1990) 461-476
16. Birkhoff, G.D.: Aesthetic Measure. Harvard University Press, Cambridge, MA (1933)
17. Ramachandran, V.S.: The Emerging Mind. Reith Lectures; 2003. BBC in association with Profile Books, London (2003)
18. Doczi, G.: The Power of Limits: Proportional Harmonies in Nature, Art and Architecture. Shambhala (Distributed by Routledge & Kegan Paul), London (1981)
19. Nake, F.: How Far Away Are We from the First Masterpiece of Computer Art? In Brunnstein, K., Raubold, E. (eds.) 13th World Congress 94, Elsevier Science, B.V., North-Holland (1994) 406-413
20. Zelevansky, L.: Beyond Geometry: Experiments in Form, 1940s-70s. MIT Press, Cambridge, MA (2004)
21. Partridge, D., Rowe, J.: Computers and Creativity. Intellect Books, Oxford, England (1994)
22. Boden, M.A.: What Is Creativity? In Boden, M.A. (ed.) Dimensions of Creativity, MIT Press, Cambridge, MA (1994) 75-117
23. Saunders, R., Gero, J.S.: Artificial Creativity: A Synthetic Approach to the Study of Creative Behaviour. In Gero, J.S. (ed.) Proceedings of the Fifth Conference on Computational and Cognitive Models of Creative Design, Key Centre of Design Computing and Cognition, Sydney (2001)
24. Cariani, P.: Emergence and Artificial Life. In Langton, C.G., Taylor, C., Farmer, D., Rasmussen, S. (eds.) Artificial Life II, SFI Studies in the Sciences of Complexity, Addison-Wesley, Redwood City, CA (1991) 775-797
25. Whitelaw, M.: Metacreation: Art and Artificial Life. MIT Press, Cambridge, MA (2004)

Genetic Paint: A Search for Salient Paintings

J.P. Collomosse and P.M. Hall

Department of Computer Science, University of Bath, Bath, U.K.
{jpc, pmh}@cs.bath.ac.uk

Abstract. The contribution of this paper is a novel non-photorealistic rendering (NPR) algorithm for rendering real images in an impasto painterly style. We argue that figurative artworks are salience maps, and develop a novel painting algorithm that uses a genetic algorithm (GA) to search the space of possible paintings for a given image, so approaching an "optimal" artwork in which salient detail is conserved and non-salient detail is attenuated. We demonstrate the results of our technique on a wide range of images, illustrating both the improved control over level of detail due to our salience adaptive painting approach, and the benefits gained by subsequent relaxation of the painting using the GA.

1 Introduction

Paintings are abstractions of photorealistic scenes in which salient elements are emphasised. In the words of art historian E.H. Gombrich, "works of art are not mirrors" [1] — artists commonly paint to capture the structure and elements of the scene that they consider to be important; remaining detail is abstracted away in some differential style. This *differential level of emphasis* is evident in all artwork, from the sketches of young children to works of historical importance.

Processing images into artwork remains an active area of research within the field of non-photorealistic rendering (NPR). This paper presents a novel automatic NPR technique for rendering images in an impasto painterly style. Our approach contrasts with those before us in that we seek to emulate the aforementioned *differential emphasis* practised by artists — automatically identifying salient regions in the image and concentrating painting detail there.

Our algorithm makes use of a new image salience measure [2], that can be trained to select features interesting to an individual user, and which performs global analysis to simultaneously filter and classify low-level features to detect artifacts such as edges and corners. This enables us both to adaptively vary level of detail in painted regions according to their salience, and to vary brush stroke style according to the classification of salient artifacts. Further, we use a genetic algorithm (GA) to search the space of possible paintings for the given image, and so approach an optimal painting. A painting is deemed "better" if its level of detail coincides more closely with the salience magnitude of the original image, resulting in conservation of salient detail and abstraction of non-salient detail. Although we are not the first to propose relaxation approaches to painting [4, 5],

F. Rothlauf et al. (Eds.): EvoWorkshops 2005, LNCS 3449, pp. 437–447, 2005.

our approach is novel in that we converge toward a *globally* defined minimum distance between salience and corresponding detail in the painting.

1.1 Related Work and Context

The development of automated painterly renderers arguably began to gain momentum with Haeberli's semi-automatic painting environments [6]. Fully automatic data dependent approaches were later presented, driven by heuristics based on local image processing techniques that estimated stroke attributes such as scale or orientation. Litwinowicz [7] employed short, linear paint strokes, which were clipped to thresholded edges. Treavett and Chen [8] proposed the use of local statistical measures, aligning strokes to axes of minimum intensity variance. A similar approach using chromatic variance was proposed in [9]. Hertzmann proposed a coarse-to-fine approach to painting [10] and was the first to automatically place curved (β-spline) strokes rather than dabs of paint. Our stroke placement algorithm is based firmly upon this technique. Gooch *et al.* [11] also use curved strokes fitted to skeletons extracted from locally connected regions.

A commonality exists between all of these algorithms; the attributes of each brush stroke are determined independently, by heuristics that analyse small pixel neighbourhoods local to that stroke's position. Rendering is, in this sense, a *spa-*

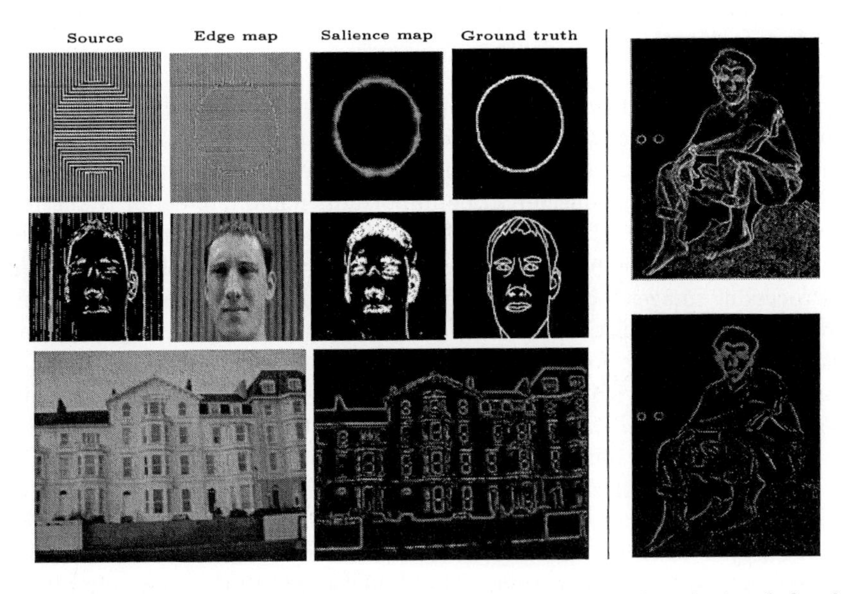

Fig. 1. Left: Examples of images edge detected, salience mapped, and a hand-sketched ground truth. We observe that the global, rarity based salience maps [2] are qualitatively closer to sketches, and can "pick out" the circle and face where local methods such as edge detection fail. The salience measure estimates salience magnitude and also classifies artifacts into trained categories (bottom row). Edges are red, ridges green, and corners blue. Right: Sobel edges (top) and salience map (bottom), corresponding to Fig. 6b. Salient edges are discriminated from non-salient high frequency texture

tially local process. The heuristics typically seek to convey the impression of an artistic style whilst preserving content such as edges, and other artifacts contributing to the upper frequencies of the Fourier spectrum. Indeed, existing relaxation-based painting algorithms [4, 5] actively seek to maximise conservation of high-frequency content from the original image. Measures of variance [8, 9], or more commonly, simple edge detectors (such as Sobel) [7, 10] drive these heuristics. This results in a painting in which all fine detail is emphasised, rather than only the *salient* detail. Arguably this disparity contributes to the undesirable impression that such paintings are of machine rather than natural origin.

In Fig. 1 (left) we demonstrate that not all fine scale artifacts are salient; indeed in these images, salient and non-salient artifacts are of similar scale. Such examples make the case for some other measure of salience incontrovertible. When one speaks of the salience of image regions, one implicitly speaks of the importance of those regions relative to the image as a whole. It follows that *global* image analysis is a prerequisite to salience determination, rather than restricting attention to *spatially local* image properties.

Our desire to control level of detail in NPR is most strongly aligned with recent techniques, which appeal to user interaction to control emphasis. De Carlo uses a gaze-tracker [12] to guide level of detail in painting. Masks, specified manually or *a priori*, have also been used to interactively reduce level of detail [10, 13]. Yet the problem of automatically controlling painting emphasis remains; this paper presents a solution.

2 A Salience Measure for Painting

Salience is subjective; faces photographed in a crowd will hold different levels of salience to friends or strangers. User training is one way in which subjectivity can be conveyed to an automated salience measure, although current Computer Vision restricts general analysis to a lower level of abstraction than this example.

We make use of a trainable salience measure, described more fully elsewhere [2], that combines three operators to estimate the *salience map* of an image — a scalar field in which the value of any point is directly proportional to the perceived salience of the corresponding image point. The first of the three operators performs unsupervised global statistical analysis to evaluate the relative rarity of image artifacts (after Walker *et al.*[14] who observe that salient features are uncommon in an image). Salient artifacts must also be visible, and a second operator filters detected artifacts to enforce this constraint. Finally, certain classes of artifact, for example edges or corners, may be more salient than others. This observation is accommodated by a third operator that is trained by the user highlighting salient artifacts in photographs. Signals corresponding to these artifacts are clustered to produce a classifier which may be applied to to estimate salience in novel images. This definition holds further advantage in that *classes* of salient features may be trained and classified independently.

This trainable salience measure is well suited to our NPR painting application for two reasons. First, the salience maps produced have been shown to be measurably closer to human figurative sketches of scenes than edge maps and a number of other prescriptive salience measures [3]. Second, the ability to estimate both the salience and the classification of image artifacts simultaneously allows us to vary stroke style according to the class of artifact encountered (Fig. 2). We begin by applying the salience measure to the source image; obtaining both a salience map and a classification probability for each pixel. An intensity gradient image is also computed using Gaussian derivatives, from which a gradient direction field is obtained. The source image, direction field, salience map and classification map are used in subsequent stages of our painting algorithm.

3 Painting as a Search

Our observations of artists lead us to assert that the level of detail in a painting should closely correlate with the salience map of its source image. In this sense, the *optimality criterion* for our paintings is a measure of the strength of this correlation (defined in subsection 3.2, step I). We treat the painting process as a search for the "optimal" painting under this definition. Our search strategy is genetic algorithm (GA) based. When one considers the abstraction of a painting as an ordered list of strokes [6] (comprising control points, thickness, etc. with colour as a data dependent function of these), the space of possible paintings for a given source image is very high dimensional, and our optimality criterion makes this space extremely turbulent. Stochastic searches that model evolutionary processes, such as GAs [15], are often cited among the best search strategies in such situations; large regions of problem space can be covered quickly, and local minima more likely to be avoided [16, 17].

Our algorithm accepts as input a source image I; paintings derived from I are points in our search space. We begin by initialising a fixed size population of individuals. Each individual is single point in our search space, represented by an ordered list of strokes that, when rendered, produces a painting from I. Having initialised the population, the iterative search process begins. We now describe the initialisation and iteration stages of the search in detail.

3.1 Initialising the Painting Population

We initialise the search by creating an initial population of fifty paintings, each derived from the source image via a stochastic process. We now describe this derivation process for a single painting.

A painting is formed by compositing curved spline brush strokes on a 2D canvas of identical size to the source image. We choose piecewise Catmull-Rom splines for ease of control since, unlike β-splines (used in [10, 11]), control points are interpolated. A collection of "seed points" are scattered over the canvas stochastically, with a bias toward placement in more salient regions. Brush strokes are then grown to extend bi-directionally from each seed; each end grows independently until halted by one or more preset criteria. Growth proceeds in a manner similar to [10] in that we hop between pixels in the direction tangential

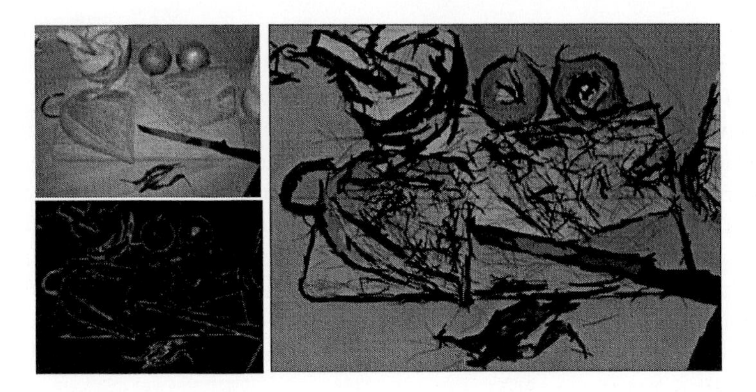

Fig. 2. Left: a still-life composition and corresponding salience map. Right: a loose and sketchy painting, exhibiting differential stroke style determined by local feature classification. Edges are drawn with hard, precise thick strokes; ridges with a multitude of light, inaccurate strokes. Rendered prior to the relaxation step of subsection 3.2

to intensity gradient. The list of visited pixels forms the control points for the stroke. However, noise forms a component of any real image, and hop directions are better regarded as samples from a stochastic distribution. We have observed this noise to obey the central limit theorem [18], and so model this distribution as a zero centred Gaussian, $G(0, \sigma)$; we determine σ empirically (see subsection 3.1.1). Given a locally optimal direction estimate θ we select a hop direction by adding Gaussian noise $G(0, \sigma)$. The magnitude of the hop is also Gaussian distributed; $G(\mu', \sigma')$, both parameters being inversely proportional to the local value of the salience map. The growth of a stroke end is halted when either the curvature between adjacent pixels, or the difference between the colour of the pixel to be appended and the mean colour of visited pixels exceeds a threshold. This method initially yields a sub-optimal trajectory for the stroke with respect to our optimality criterion. However, for a "loose and sketchy" painting this is often desirable (see Fig. 2).

The degrees of freedom available from each of the many hops combine to create a range of stroke loci, at least one of which will result in the maximum conservation of salient detail. The combination of these optimally positioned strokes comprises the optimal painting, and it is by means of breeding the fittest paintings to create successively superior renderings, that we later search for such a painting in our iterative process. This process can out-perform stroke placements based purely on local estimates of direction.

3.1.1 Calibration for Image Noise

The choice of σ significantly influences stroke growth, and later the relaxation process. A value of zero forces degeneration to a loose and sketchy painting system; a high value will lengthen the relaxation process unnecessarily and also may introduce unnecessary local minima. We propose a one time user calibration process to select σ as follows.

The user is asked to draw a window around an image region where direction of image gradient is perceived to be equal; i.e. along which they would paint strokes of similar orientation. Gradient direction within this window is sampled, and σ computed as twice the unbiased standard deviation of the sampled angles. We typically obtain similar σ values for similar imaging devices, which allows us to perform this calibration very infrequently. A typical σ ranges from around 2 to 5 degrees. This variation allows between 12 and 30 degrees of variation per hop which, given the number of hops per stroke, yields a wide range of stroke loci. This adds credence to our argument for a relaxation process taking into account image noise; local variations due to uncompensated image noise will likely cause inaccurate stroke placements in single iteration painterly systems [7, 10, 9, 11].

3.1.2 Rendering a Painting

At this stage we may render one of the paintings in our initial population to produce a "loose and sketchy" painting (Fig. 2). Alternatively we may proceed to the iterative search stage of subsection 3.2 to locate a more optimal painting — each iteration also requires paintings to be rendered to evaluate fitness. We now describe how paintings are formed from individuals in the population.

A painting is formed by scan-converting and compositing its list of curved spline brush strokes. Stroke thickness is set inversely proportional to *stroke salience*; taken as the mean salience over each control point. Stroke colour is uniform and set according to the mean of all pixels encompassed in the footprint of the thick paint stroke. During rendering, strokes of least salience are laid down first, with more salient strokes being painted later. This prevents strokes from non-salient regions encroaching upon salient areas of the painting. The ability of our salience measure to differentiate between classes of salient feature (e.g. edge, ridge) also enables us to paint in context dependent styles. In Fig. 2 the classification probability of a feature is used as a parameter to interpolate between three stroke rendering styles *flat*, *edge* and *ridge*.

3.2 Iterative Relaxation by GA

Genetic algorithms (GAs) simulate the process of natural selection by breeding successive generations of individuals through cross-over, fitness-proportionate reproduction and mutation [17]. In our implementation such individuals are paintings; their genomes being ordered lists of strokes. We now describe a single iteration of the GA search, which is repeated until the improvements gained over the previous few generations are marginal (the change in both average and maximum population fitness over a sliding time window fall below a threshold).

I. Fitness and Selection. The entire population is rendered, and edge maps of each painting are produced using by convolution with Gaussian derivatives, which serve as a quantitative measure of local fine detail. The generated maps are then compared to a precomputed salience map of the source image. The mean squared error (MSE) between maps is used as the basis for determining the fitness of a particular painting; the lower the MSE, the better the painting. In this manner, individuals in the population are ranked according to fitness. The bottom 10% are culled, and the top 10% percent pass to the next generation; this promotes convergence. The top 90% percent are used to produce the remainder of the next generation. Two individuals are selected stochastically with a bias to fitness, and bred via *cross-over* and *mutation* to produce a novel offspring for the successive generation. This process repeats until the population count of the new generation equals that of the current generation.

II. Cross-over. Two difference images, A and B, are produced by subtracting the edge maps of the parents from the salience map of the source image. By computing the binary mask $A > B$, and likewise $B > A$, we are able to determine which pixels in one parent contribute toward the fitness criterion to a greater degree than those in the other. Since the primitives of our paintings are thick strokes rather than pixels, we apply dilation to both masks. Strokes seeded within

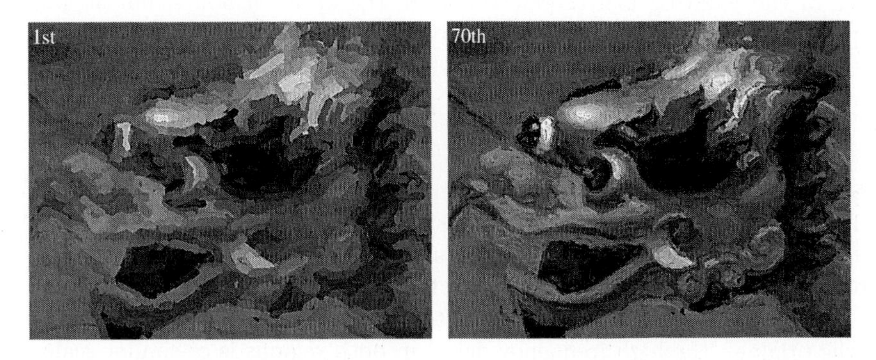

Fig. 3. Relaxation by GA search. Detail in the salient region of the 'dragon' painting sampled from the fittest individual in the 1st, and 70th generation of the relaxation process. Strokes converge to tightly match contours in salient regions of the image thus conserving salient detail

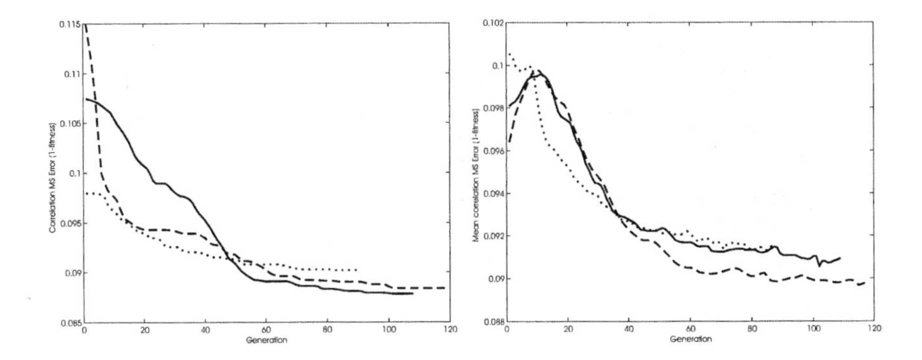

Fig. 4. Left: Three runs of the relaxation process; dotted line corresponds to the model (Fig. 6a), dashed line the dragon (Fig. 3) and solid line the truck (Fig. 6g). MSE of the fittest individual is plotted against time. Right: MSE averaged over each generation

the set regions in each mask are copied from the respective parent to the new offspring. A binary AND operation between masks yields mutually preferred regions, for which the contributing parent is decided arbitrarily.

III. Random Mutation. A "temporary" painting individual is synthesised as described in subsection 3.1. A binary mask is produced containing several small discs of stochastic number, location and radius. Strokes seeded within set regions of the mask are substituted for those in the temporary painting. On average, mutation occurs over approximately 4% of the canvas area.

Implementation Notes. The evaluation step is the most lengthly part of the GA process, and rendering is farmed out to several machines concurrently. In our implementation we distribute rendering over a small heterogeneous (Pentium III/UltraSPARC) cluster. The typical time to render a 50 painting generation at 1024×768 resolution is on average 15 minutes over 6 workstations. Relaxation of the painting can therefore take in the order of hours, but significant improvements in stroke placement can be achieved, as can been seen in Fig. 3.

4 Results and Discussion

We present a gallery of rendered paintings in Fig. 6. The painting of the model in Fig. 6b converged after 92 generations. Thin precise strokes have been painted along salient edges, while ridges and flats have been painted with coarser strokes. Observe that non-salient high-frequency texture on the rock has been attenuated, yet tight precise strokes have been used to emphasise salient contours of the face. In the original, the high frequency detail in both regions is of similar scale and magnitude; existing painterly techniques would, by contrast, assign both regions equal emphasis. With current techniques, one might globally increase the kernel scale of a low-pass filter [10] or raise thresholds on Sobel edge magnitude [7] to reduce emphasis on the rock. However this would cause a similar drop in the level of detail on the face (Fig. 6a). Conversely, by admitting detail on the face

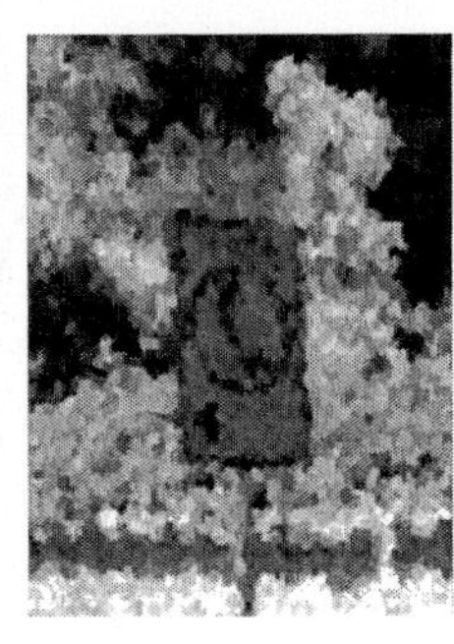
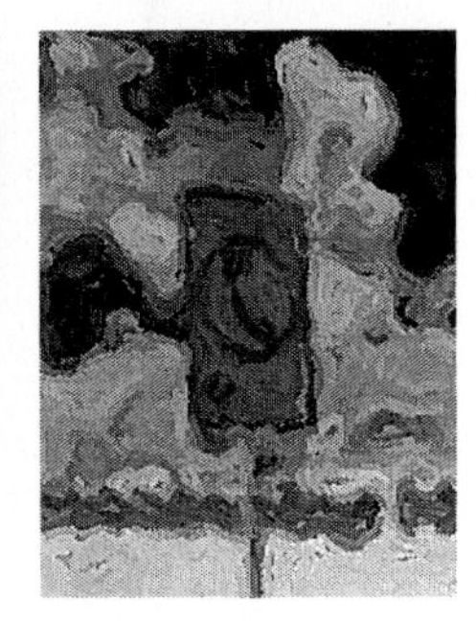

Fig. 5. Detail from Fig. 6g, region A. Using our adaptive approach, salient detail (sign-post) is conserved, and non-salient texture (shrubbery) is abstracted away. Left: original. Middle: existing approach [7]. Right: our proposed GA approach

Fig. 6. A gallery of paintings illustrating application of our algorithm. Higher resolution electronic versions of all our paintings are included in the material accompanying this paper

one would unduly emphasise the rock (Fig. 6c). We automatically differentiate between such regions using a perceptual salience map (Fig. 1) – contrast this with the Sobel edge field where no such distinction can be made.

We present a still-life in Fig. 6e which achieved convergence after 110 generations. In Fig. 6f we present a similar painting prior to relaxation, demonstrating differential rendering style; strokes with a high probability of being edges are darkened to give the effect of a holding line. Further examples of level of detail adaptation to salience are given in Fig. 6g. In region A, the salient sign is emphasised whilst non-salient texture of the background shrubbery is not (see Fig. 5). For the purposes of demonstration we have manually altered a portion of salience map in region B, causing all detail to be regarded as non-salient.

All of our experiments have used populations of 50 paintings per generation. We initially speculated that population level should be set in order of hundreds to create the diversity needed to relax the painting. Whilst convergence still occurs with such population limits, it requires, on average, 2 to 3 times as many iterations to achieve. Such interactions are often observed in complex optimisation problems employing GAs [17]. We conclude that the diversity introduced by our mutation operator is sufficient to warrant the lower population limit.

As regards rendering, we might choose to texture strokes to produce more realistic brush patterns — however, we have concentrated on stroke placement rather than media emulation, and leave such implementation issues open. We believe the most productive avenues for future research will explore both new fitness functions and alternative uses for salience measures in image-based NPR.

References

1. Gombrich, E.H.: Art and Illusion. Phaidon Press Ltd., Oxford (1960)
2. Hall, P.M., Owen, M., Collomosse, J.P.: A trainable low-level feature detector. In: Proc. Intl. Conf. on Pattern Recognition (ICPR). (2004) 1:708–711
3. Hall, P., Owen, M., Collomosse, J.P.: Learning to detect low-level features. In: Proc. 15th British Machine Vision Conf. (BMVC). Volume 1. (2004) 337–346
4. Sziranyi, T., Tath, Z.: Random paintbrush transformation. In: Proc. 15th Intl. Conf. on Pattern Recognition (ICPR). Volume 3., Barcelona (2000) 155–158
5. Hertzmann, A.: Paint by relaxation. In: Proc. Comp. Graph. Intl. (2001) 47–54
6. Haeberli, P.: Paint by numbers: abstract image representations. In: Proc. ACM SIGGRAPH. Volume 4. (1990) 207–214
7. Litwinowicz, P.: Processing images and video for an impressionist effect. In: Proc. ACM SIGGRAPH, Los Angeles, USA (1997) 407–414
8. Treavett, S., Chen, M.: Statistical techniques for the automated synthesis of non-photorealistic images. In: Proc. 15th Eurographics UK Conf.. (1997) 201–210
9. Shiraishi, M., Yamaguchi, Y.: An algorithm for automatic painterly rendering based on local source image approximation. In: Proc. ACM NPAR (2000) 53–58
10. Hertzmann, A.: Painterly rendering with curved brush strokes of multiple sizes. In: Proc. ACM SIGGRAPH. (1998) 453–460
11. Gooch, B., Coombe, G., Shirley, P.: Artistic vision: Painterly rendering using computer vision techniques. In: Proc. ACM NPAR (2002) 83–90
12. DeCarlo, D., Santella, A.: Abstracted painterly renderings using eye-tracking data. In: Proc. ACM SIGGRAPH. (2002) 769–776

13. Hertzmann, A., Jacobs, C.E., Oliver, N., Curless, B., Salesin, D.H.: Image analogies. In: Proc. ACM SIGGRAPH. (2001) 327–340
14. Walker, K.N., Cootes, T.F., Taylor, C.J.: Locating salient object features. In: Proc. 9th British Machine Vision Conf. (BMVC). Volume 2. (1998) 557–567
15. Holland, J.: Adaptation in Natural and Artificial Systems. U. Michigan (1975)
16. de Jong, K.: Learning with genetic algorithms. Machine Learning **3** (1988) 121–138
17. Goldberg, D.: GAs in Search, Optimization, and Machine Learning. Add.W. (1989)
18. Collomosse, J.P.: Higher Level Techniques for the Artistic Rendering of Images and Video. PhD thesis, University of Bath, U.K. (2004)

Artificial Life, Death and Epidemics in Evolutionary, Generative Electronic Art

Alan Dorin

Centre for Electronic Media Art, Computer Science & Software Engineering,
Monash University, Clayton, Australia 3800
aland@csse.monash.edu.au

Abstract. This paper explores strategies for slowing the onset of convergence in an evolving population of agents. The strategies include the emergent maintenance of separate agent sub-populations and migration between them, and the introduction of virtual diseases that co-evolve parasitically within their hosts. The method looks to Artificial Life and epidemiology for its inspiration but its ultimate concerns are in studying epidemics as a process suitable for application to generative electronic art. The simulation is used to construct a prototype art-work for a fully interactive stereoscopic virtual-reality environment to be exhibited in a science museum.

1 Motivation and Past Work

Evolutionary, ecological simulations often converge with the population predominantly of similar genetic composition. For generative electronic art this may have undesirable aesthetic consequences, especially if the desired result is an exploration of diverse audio or visual solutions within the constraints of the work. For instance, it may be desirable that diverse visual and sonic forms be present in a simulated environment simultaneously and so the tendency of the population towards genetic impoverishment must be averted.

This paper focuses primarily on a notable omission from many Artificial Life models and publications, *disease*[1]. Typical ecological simulations model creatures competing for food, mating, fighting, and dieing. Yaeger's *PolyWorld* is a seminal example in which agents interact utilizing colour vision [1]. Todd has noted strategies for removing creatures from a population subject to a genetic algorithm but stops short of exploring different reasons for death in the population [2] (for example disease or suicide). Mascaro et al. have dealt specifically with suicide in a population of simple agents [3]. Ray's *Tierra* simulation eliminates elderly or ineffective population members with a "reaper". Also of interest is the emergence of "parasitic" code in his system [4].

The Artificial Life literature has much to say on co-evolution as a means of improving a genetic algorithm's performance through increased population diversity

[1] Little has been written in the Artificial Life literature on disease's counterpart *decay* either for that matter — a subject for future investigation.

F. Rothlauf et al. (Eds.): EvoWorkshops 2005, LNCS 3449, pp. 448–457, 2005.

[5, 6]. This work is similarly inspired, only the simulations model agents in virtual worlds and do not optimize explicit fitness functions.

This paper borrows generally from ideas presented in all of the above literature, but its concerns are initially aesthetic and it specifically examines the *process* of epidemics — the transmission of disease through a population of susceptible individuals. Hence, most of the ideas for the research did not emerge from the publications above, but from ideas extracted from the literature of epidemiology and from art history. Both of these influences are surveyed below.

1.1 A Brief Note on the Epidemic in Art

It may seem a little crass or at best gothic to study the aesthetics of epidemics, yet for centuries plagues and disease have inspired artists to depict their horrors [7] and the fear of the ghastly death that often ensues [8]. Literary works such as *The Masque of the Red Death* (Poe 1842), the novel *The Andromeda Strain* (Crichton 1969), the film *The Crazies* (Romero 1973), and countless derivatives, maintain the cultural visibility and gory fascination of the epidemic. Contemporary art is no less concerned with epidemics. In particular, HIV/AIDS has driven many to artistic expression [9].

In the context of generative electronic art, the spread of infection is a biological process that may initially be treated without consideration of its emotional connotations. Such studies may lend themselves to a more thorough investigation of the potential of disease for application to software-based art. After the mechanisms of epidemics are understood this knowledge may be re-coupled with the emotional impact of the epidemic. The value of investigating biological processes and artificial life in the context of art are considered elsewhere [10].

1.2 An Introduction to Models from Epidemiology

A fully-cited history of the mathematical theory of epidemics is beyond the scope of this paper. The history leading to the classic model discussed below is provided in [11, 12].

At least since the 1920's, stochastic models of epidemics have been utilized. The standard model is based on a population of individuals who are either susceptible to a specific disease (*susceptibles* denoted **S**) or infected with the disease and capable of transmitting it to others (*infectives* denoted **I**). Population members who overcome a disease may become immune to further infection[2] or may become susceptible once again depending on the particular disease. Population members who are immune to a disease or remain infected but through isolation cannot transmit it, are considered *removed* (denoted **R**). The model as described is known as an *SIR* model. It may be modified slightly to provide fresh susceptibles through birth or immigration.

Some pertinent parameters of epidemic models are as follows. The period of time during which a disease exists entirely within an organism is known as the disease's *latent period*. The organism is not infective during this period. An *incubation period*

[2] Following a bout of a disease a victim may be deceased, alternatively their immune system may prevent repeat infiltration by the same virus.

often follows latency. During incubation the organism may not show outward sign of infection but is nevertheless infective. Usually once the incubation period is over, the victim of the disease is clearly marked by symptoms and can therefore be avoided by susceptibles.

Probabilistic epidemiological models that operate in discrete time steps are particularly suited to implementation in software.[3] At any time step, the probability of a new case of the disease appearing is proportional to the number of susceptibles multiplied by the number of infectives. This basic model assumes random mixing of individuals in the population and does not allow for the complex interactions between physically separated sub-populations, nor for variable incubation or latent periods of a disease. Various extensions to allow for these phenomena have been added over the last fifty years. Some mathematical models and computer simulations deal with the spatial distribution of susceptibles along a line, across a lattice or over a network to overcome the inaccuracies due to the assumption of random mixing of the population. Cellular-automata and other discretized versions of the SIR method have been utilized also [13, 14]. Some of these models have also incorporated disease *carriers* (e.g. some viruses are transferred by mosquito), and non-homogeneous populations.

The current threats of biological warfare and terrorism have raised the stakes in Western society for epidemiology. The U.S. National Institute of General Medical Sciences has devoted $1.6 billion to a fledgling agent-based study of epidemics [15].

Like the U.S. project, this paper adopts agent modelling to represent the principles of epidemiology in an intuitive but realistic fashion. The motivation is, in this case, purely aesthetic. As shall be shown, the process of epidemic spread offers a means of varying the genetic and phenotypic diversity of a population and of capping its growth and density.

1.3 Relevant Consequences of Basic Epidemic Theory

There are two theories of epidemiology that are particularly relevant here. The first of these is known as the *Threshold Theorem* [16]: a disease cannot take hold in a population of susceptibles unless the population density is above a particular threshold. This value relates to the infectivity of a disease and the death and recovery rates it induces. If population density passes beyond the threshold, the disease will reduce the population to a level as far below the threshold as it was above it prior to the epidemic.

The *Threshold Theorem* has many consequences, one of which has come to be known as *Herd Immunity* [11, pp. 27-31]. This theory indicates that a calculable number less than the full population needs to be immunized to prevent an epidemic. Unfortunately the theory has been shown to provide inaccurate figures in practice, due to its assumption of random mixing in a population. Nevertheless, it highlights an important aspect of epidemics, namely that the spread of a disease is not dependent on the percentage of a population who are immune, but on the contact between

[3] It is interesting to note that in the 1920's two American epidemiologists Reed and Frost demonstrated a discrete *mechanical* model in which coloured balls represented susceptibles and infectives.

susceptibles and infectives. When a population does not mix uniformly, the supply of susceptibles may be similarly irregular.[4]

2 An Agent-Based Simulation of Infectious Disease Epidemics

The present simulation runs in discrete time steps during which a population of agents (represented visually as coloured boxes of different sizes) moves freely about a continuous-space, toroidal world. The essential features of the model are described below, followed by a description of a prototype artwork that employs the new model.

2.1 Agent Composition, Behaviour and Evolution

Agents (coloured boxes) are able to detect the presence, colour and size of other boxes within their individual visual range. Boxes may move at their own speed towards or away from a box they see, or they may wander randomly. Their strategy for interacting with the world is based on a set of parameters that describe a path to travel based on the presence of a neighbour they "like", "dislike", or the absence of any visible neighbours. The like/dislike functions relate to each box's preferences for others of particular sizes and colours. The natural speed of movement of a box is inversely proportional to its volume. The box may speed up or slow down with respect to this value depending on how attractive or repulsive its goal appears.

Boxes meeting in space may find one another mutually attractive. If so, they may mate with one another and produce a single child per time step. The characteristics of the child are specified in terms of the genotype of each parent: an array of floating-point values coding an agent's dimensions and colour. The floating-point genotype also includes two colour templates, one that the agent seeks out in its companions (and potential mates) and the other from which it flees. A gene specifying a target partner size for the agent's likes and dislikes has also been implemented and operates in a similar manner.

The system behaves equally well with various schemes for crossover and mutation of the agent's genotype. This aspect of the system is not critical to its success but the method used employed a single crossover point and mutation of a gene in every child by a random amount between +/- 5%.

New births are subject to an overflow test of the available computer memory. If a birth would cause an overflow the request is refused. Following an unsuccessful request, a random member of the population may (or may not) be eliminated from the simulation to make room for future requests.

During each time step of the simulation, agents expend an amount of energy proportional to their volume to move and metabolise. Agent reproduction requires energy from each parent. This acts as a contribution to the initial energy of the

---

[4] For example, if a socio-economic group is immunized against a disease, and these people do not mix randomly with people from other groups, an epidemic may still occur within the latter groups whilst the former is immunized. I.e. sub-group mixing is *important* in considering the spread of a disease.

offspring. Energy is gained by an agent each time step from the environment in an amount proportional to its upper surface area as if each box was equipped with a rooftop solar cell charging a battery. Boxes exhausting their energy supply are removed from the simulation (they die). Hence reproduction and movement both shorten the lifespan of a box if it is not able to glean sufficient energy to support itself.

Boxes age throughout a simulation. Those younger than a preset maturity age are unable to reproduce. Boxes that reach the end of their lifespan are removed from the simulation. These parameters are subject to evolution in the same way as the parameters discussed earlier.

2.2 Disease Behaviour and Evolution

The agents in the model are provided with features implementing infection and the transmission of diseases to others. Agents may visually detect the symptoms of a disease in others although this feature was not utilized in the current experiments.

A simulation disease may only occur within a host agent i.e. disease does not persist in the simulation environment. An agent is infected by a disease when it meets an infective and it is determined to be susceptible. A box of a particular colour is susceptible to a disease according to its match with the *colour-signature* of the disease. The closer the match between the two colours, the higher the probability the disease will be successfully transferred from the host to a susceptible in a time step during which the agents are in contact. The presence of an active disease in an agent blocks secondary infection.

Simulation diseases possess a *devastation* value. Highly devastating diseases remove large amounts of energy from their hosts every time step. Devastation is scaled by the match between a disease's colour-signature and the colour of the host. The closer the match, the more the devastation of the disease is scaled upwards. In addition, the higher the devastation and the closer the colour match to a susceptible, the more likely that agent will be infected.

A parameter determines the lifespan of a simulation disease in a particular host. The longer this is, the more energy a host needs to invest in total to overcome a disease. If a disease is overcome without the death of the host, the agent acquires immunity to the strain of the disease by adding it to an *immunity list*. Any further contact with this disease will result in an *immune response* that prevents the disease from infecting the agent a second time.

A given instance of a disease is killed when its host is killed, irrespective of the lifespan of the disease. In addition to an individual lifespan, each disease has parameters indicating its dormant and incubation periods.

Real diseases such as viruses replicate and mutate within a host much more rapidly than the hosts themselves reproduce, circumventing the response of a host's auto-immune system. Consequently, it is possible for humans to repeatedly catch viruses such as the common cold and flu. In order to model this aspect of biology, within a specific host the simulation diseases undergo reproduction during every time step of their lifespan. This is modelled asexually and may mutate any aspect of the disease including its colour-signature, devastation, lifespan, incubation and latent periods.

Mutation may also alter the parameter of each disease that determines how frequently it mutates during reproduction. Together all of these parameters allow the diseases to co-evolve with the more slowly evolving agent population.

The parameters for the disease and agents fully specify the features of the epidemic models discussed above. In conjunction with the agents' evolutionary model, a complex and flexible simulation has been devised that allows for studies of epidemics in non-homogeneous populations with non-random mixing. Fig. 1 illustrates the visualization scheme employed for the simulation.

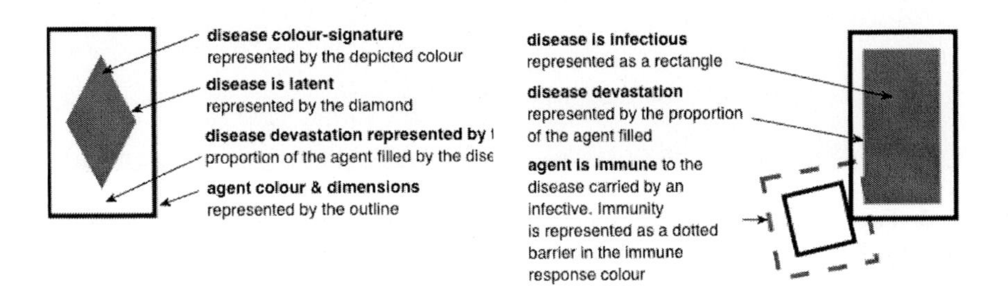

Fig. 1. The visualization scheme for agents and infection

3 Results

It is beyond the scope of this paper to delve into the quantitative results of the simulation, this would be more fitting in the epidemiology literature. Instead, in keeping with the present aesthetic exploration, the results shall be described qualitatively.

As indicated in the motivation for this work, it had been noted that ecological simulations in which agents competed for resources (including mates and energy) often resulted in a genetically impoverished, homogeneous population. In the context of electronic media art, the resulting genetic drift through the phenotypic space was perceived by the artist to be quite beautiful, but nevertheless, perhaps a little boring. It was hoped that by introducing the element of disease to the population, diversity might be encouraged and uniformity exploited and eliminated by infection.

Happily, the disease did indeed exploit the population's uniformity when it arose. Disease also exploited populations of agents that clustered tightly together. In the absence of disease, boxes of particular colours and sizes often dominated a simulation, forming large colonies of potential mates. A typical screen shot after 14,000 time steps of the simulation without disease is reproduced in fig. 2(a). Fig. 2(b) illustrates a run after 14000 time steps with identical initial conditions, but in which the disease model was introduced. The diversity in dimensions, colour and spread of the population is far greater in fig. 2(b) than in fig. 2(a). In fact, after as few as 2500 time steps, the non-diseased model converges to homogeneity and does not break from this condition but drifts gently through genetic space. The population model incorporating disease seems to maintain its diversity indefinitely.

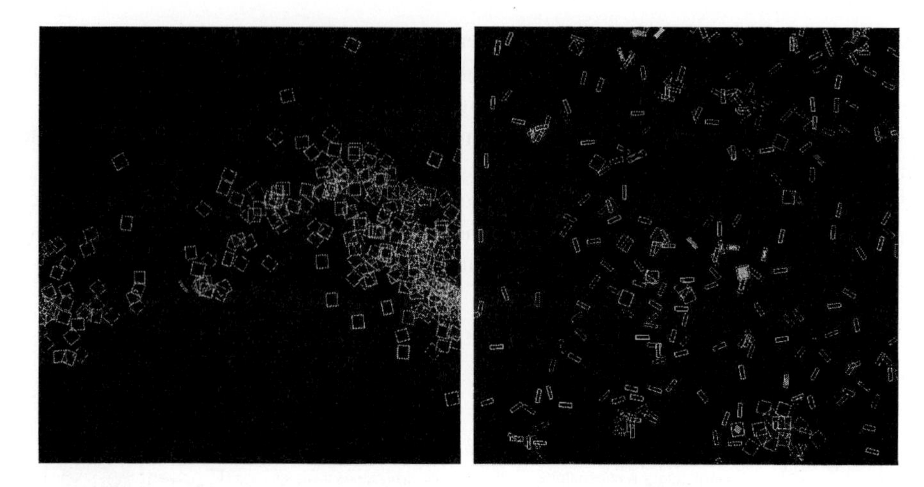

Fig. 2. Two simulation screenshots after 14,000 time steps: (a) without the epidemiological model; (b) with the epidemiological model

A disease simulation run involves the spontaneous appearance of a disease on average once every one-hundred-thousand agent updates. This new disease is generated with a colour-signature that matches the colour of the randomly selected agent. It is then infected with the disease and left to continue its travels. Apart from the colour-signature, all other disease parameters for the infection are randomly generated.

Depending on the parameters of the new disease, the traits of the infected agent and the population as a whole, the new disease may or may not cause an epidemic. The likelihood of an epidemic is specified by the Threshold and Herd Immunity theories described above. Some observed outcomes are described below along with the conditions giving rise to them in the present simulation environment.

Disease elimination (immediate). If the disease is insufficiently long-lived, or the population is insufficiently dense, or the host does not co-habit with others of a similar colour to itself, then the disease may fail to contact any susceptibles before it dies within the host. The disease will be eliminated from the population immediately.

Disease spread (immediate). A disease may mutate sufficiently within a host to infect susceptibles of a colour significantly different to the original host. If the host mixes amongst others of its kind they may become infected with the disease also. Occasionally the stochastic mechanism allows for a disease to infect a host coloured differently to its own signature. In this case, the devastation of the disease will be low in the infected host but the host nevertheless is able to infect other susceptibles. Such a host may be considered a "carrier" of the disease.

Disease elimination (eventual). If the disease manages to take a hold in the population it may nevertheless die out eventually if the number of susceptibles is reduced. This may happen when a sizeable proportion of the agents encountered by infectives is immune to the disease (even though the population as a whole may not have a significant number of immune members – see footnote 4 above). Circumstances like this arise when agents overcome the disease and acquire immunity, or when the disease is so devastating that it rapidly wipes out the supply of susceptibles before the agents are able to produce many offspring.

Disease spread (continual). A disease well-suited to its environment has sufficient lifespan to ensure it is passed from one susceptible agent to another. Such a disease also needs to be sufficiently devastating that it can be transferred successfully, but not so devastating that it kills off its supply of susceptibles. Diseases that fit these criteria also have to be sufficiently stable to avoid unwanted mutations that would render them ineffective, but sufficiently mutable so that they can keep infecting an evolving population of hosts. The simulation has given rise to diseases that meet all of these criteria and persist in the population for long periods of time.

Of particular interest are diseases that sustain themselves indefinitely when they are able to utilize susceptibles that are prolific breeders. Such diseases are able to spread through contact between mates who seek one another out (sexually transmitted diseases?) and also by contact between a parent and its newly born. Newly born agents may have traits slightly different to their parents so that occasionally one tends to wander off to seek its own preferred companions, taking the disease to infect others. As long as the disease remains latent for a sufficiently long interval, it will not kill or weaken the agent prior to its immigration to a further enclave.

4 Artwork Prototype

The eventual goal for this research is to establish an interactive, stereoscopic, virtual environment. The viewer's position will be monitored by a camera aimed in front of the viewing screen and mapped into a virtual location overlaid with the physical space. The overlay region is a section of the virtual world extending visually beyond the stereoscopic screen and therefore conceptually and apparently forming the interface between the virtual and real environments of the visitor's experience.[5] See fig. 3 for a model and schematic of the system. The final work will have a more complex geometric model for the agents than the present wire-frame box.

By moving within the overlay region, visitors may attract the attention of the agents in the neighbourhood. Depending on the visual characteristics of the visitors, and the likes and dislikes of the agents, the virtual creatures may flee, approach, or remain disinterested. It is anticipated that the traits of the audience that will be perceived by the agents include coarse approximations of an audience member's size, position and the presence of any colour as determined by processing a live video image of the overlay region for each screen of the visualization system. The eight-screen stereoscopic system for viewing the environment has been constructed and is currently situated in the Museum of Victoria in Melbourne, Australia. Software to

[5] Due to the nature of stereoscopic projection, the viewer will be unable to see agents behind or to the side of their own position but these will be simulated all the same. Agents in front of the viewer and apparently on either side of the screen will be visible.

monitor the video image has been partially written but has not yet been configured to operate with the simulation.

The prototype of the system above has been implemented employing virtual-human agents in place of the visitors. The apparent size of the virtual humans is normalized to match the dimensions of the box agents. Any colouration of the virtual agents (corresponding to any coloured clothing of the real humans on the video image) is treated as the presence of an infective disease in the human agent. Human agents do not die although they may leave the overlay region and thus the simulation. Box agents may mate with human agents if the box agents find the humans attractive. Offspring traits are determined by combining what is known about the human agent (based on their appearance on the video image) with the genes of the box agent.
If a "diseased" human agent encounters a susceptible box agent, the box may catch the disease and carry it into the virtual environment. Hence, the presence of human viewers wearing coloured clothes spreads disease that directly alters the evolution of the virtual population, culling populations of colour closest to the colours worn by the visitors.

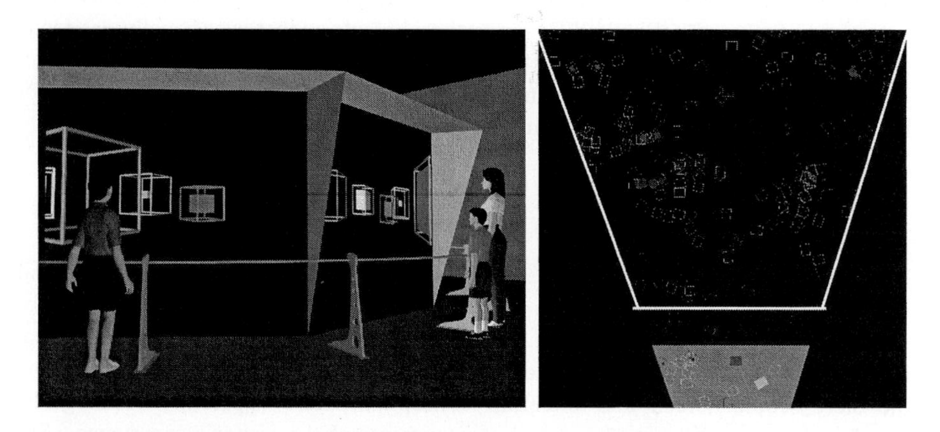

Fig. 3. Artwork prototype: (a) installation visualization (agents will not be rendered as boxes in the final version); (b) schematic, overhead representation of a single screen. Humans (drawn as filled squares) and agents interact in the grey region "outside" the screen

5 Conclusions and Future Work

A model of epidemics has been introduced to an evolutionary, agent-based simulation in order to increase the aesthetic interest of the population at any one time. The model improved the overall diversity of the population as desired and also encouraged its spread across the available virtual space. The behaviour of the epidemics modelled was in and of itself interesting to watch. A wide variety of disease outcomes emerged from the simulation, each an apparently plausible model of real-world outbreaks.

Much remains to be done. Besides the countless experiments that might be carried out with the epidemiological model for its own sake, the interactive artwork is still in its early stages. The video-processing software needs to be altered to work with the simulation (a time-consuming task) and more interesting visual forms need to be designed for the simulation agents. Nevertheless, even as it stands the simulation provides a fascinating experience for the informed viewer who understands the visualization system employed, and can interpret the coloured patterns appearing on the screen.

References

1. Yaeger, L.: Computational Genetics, Physiology, Metabolism, Neural Systems, Learning, Vision and Behavior or Polyworld: Life in a New Context. Artificial Life III, Addison-Wesley (1992) 263-298
2. Todd, P.M.: Artificial Death. Jahresing 41 (German modern art yearbook) (1994) 90-107
3. Mascaro, S., Korb, K.B., Nicholson, A.E.: Suicide as an Evolutionary Stable Strategy. Advances In Artificial Life, 6th European Conference. Prague, Springer (2001) 120-133
4. Ray, T.S.: An Approach to the Synthesis of Life. Artificial Life II. Santa Fe, New Mexico, Addison Wesley (1990) 371-408
5. Cartlidge, J., Bullock, S.: Caring Versus Sharing: How to Maintain Engagement and Diversity in Coevolving Populations. 7th European Conference on Artificial Life. Dortmund, Germany, Springer-Verlag (2003) 299–308
6. Hillis, W.D.: Co-Evolving Parasites Improve Simulated Evolution as an Optimization Procedure. Artificial Life II. Santa Fe, New Mexico, Addison-Wesley (1990) 313-324
7. Boeckl, C.M.: Images of Plague and Pestilence : Iconography and Iconology. Truman State University Press Kirksville (2000)
8. Wolbert, K.: Memento Mori : Der Tod Als Thema Der Kunst Vom Mittelalter Bis Zur Gegenwart. Das Landesmuseum Darmstadt (1984)
9. Atkins, R., Sokolowski, T.W.: From Media to Metaphor: Art About Aids. Independent Curators Incorporated New York (1991)
10. Dorin, A.: The Virtual Ecosystem as Generative Electronic Art. 2nd European Workshop on Evolutionary Music and Art, Applications of Evolutionary Computing: EvoWorkshops 2004. Coimbra, Portugal, Springer-Verlag (2004) 467-476
11. Ackerman, E., Elvebeck, L.R., Fox, J.P.: Simulation of Infectious Disease Epidemics. Charles C Thomas Springfield, Illinois (1984)
12. Bailey, N.T.J.: The Mathematical Theory of Infectious Diseases and Its Applications. 2. Charles Griffin & Company London (1975)
13. Willox, R., Grammaticos, B., Carstea, A.S., Ramani, A.: Epidemic Dynamics : Discrete-Time and Cellular Automaton Models. Physica A 328 (2003) 13-22
14. Martins, M.L., Ceotto, G., Alves, S.G., Bufon, C.C.B., Silva, J.M., Laranjeira, F.F.: A Cellular Automata Model for Citrus Variagated Chlorosis. eprint arXiv : cond-mat/0008203 (2000) 10
15. Pilot Projects for Models of Infectious Disease Agent Study (Midas). Webpage. National Institute of General Medical Sciences (NIGMS) http://grants.nih.gov/grants/guide/rfa-files/RFA-GM-03-008.html (Accessed: October 2004)
16. Kermack, W.O., McKendrick, A.G.: Contributions to the Mathematical Theory of Epidemics. Proceedings of the Royal Society, London 115 (1927) 700-721

The Electric Sheep Screen-Saver: A Case Study in Aesthetic Evolution

Scott Draves

Spotworks, San Francisco CA, USA

Abstract. Electric Sheep is a distributed screen-saver that harnesses idle computers into a render farm with the purpose of animating and evolving artificial life-forms known as *sheep*. The votes of the users form the basis for the fitness function for a genetic algorithm on a space of fractal animations. Users also may design sheep by hand for inclusion in the gene pool. This paper describes the system and its algorithms, and reports statistics from 11 weeks of operation. The data indicate that Electric Sheep functions more as an amplifier of its human collaborators' creativity rather than as a traditional genetic algorithm that optimizes a fitness function.

1 Introduction

Electric Sheep [5] [6] was inspired by SETI@Home [1] and has a similar design. Both are distributed systems with client/server architecture where the client is a screen-saver installable on an ordinary PC. Electric Sheep distributes the rendering of fractal animations. Each animation is 128 frames long and is known as a *sheep*.

Besides rendering frames, the client also downloads completed sheep and displays them to the user. The user may vote for the currently displayed sheep by pressing the up arrow key.

Each sheep is specified by a genetic code comprised of about 160 floating-point numbers. The codes are generated by the server according to a genetic algorithm where the fitness is determined by the collective votes of the users. This is a form of aesthetic evolution, a concept first realized by Karl Sims [9] and analyzed by Alan Dorin [3].

This is how Electric Sheep worked until March 2004, when a new source of genomes appeared: Apophysis [10]. Apophysis is a traditional, stand-alone Windows GUI to the sheep genetic code primarily intended for still-image design, but useful for key-frame animation. Besides a traditional direct manipulation interface where the user drags sliders and types numbers into labeled fields, it includes a Sims-style mutation explorer.

In March, Townsend and Draves connected this application to the Electric Sheep server. A simple menu command causes the current genome to be posted to the server, rendered, and distributed to all active clients. If the resulting animation receives votes it may reproduce and interbreed with the artificially evolved population.

Not surprisingly, these posted genomes proved much more popular than the purely random ones. And as they are subject to mutation and crossover, the genetic algorithm creates variants of them. One can compare the total amount of quality animation to the

F. Rothlauf et al. (Eds.): EvoWorkshops 2005, LNCS 3449, pp. 458–467, 2005.

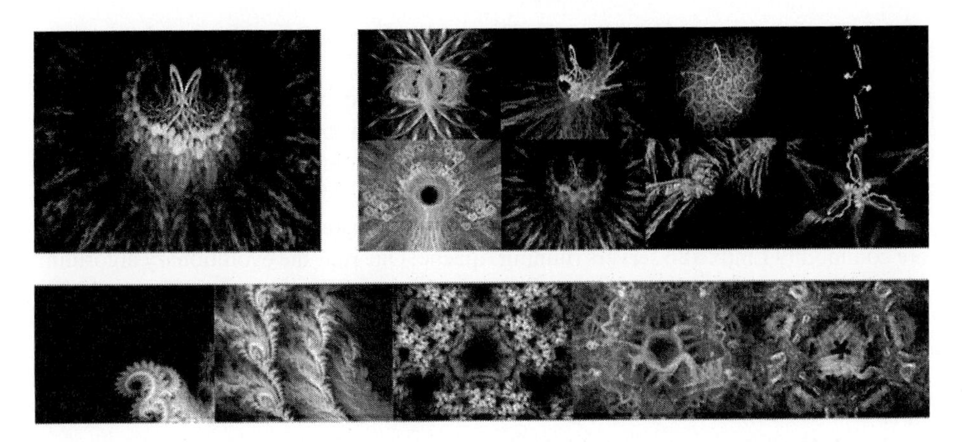

Fig. 1. Sheep 15875, on the top-left, was born on August 16 and died 24 hours later after receiving one vote. It was one of 42 siblings. It was reincarnated on October 28 as sheep 29140, received a peak rating of 29, lived 7 days, and had 26 children, 8 of which appear to its right. Below are five generations of sheep in order starting on the left. Their numbers are 1751, 1903, 2313, 2772, and 2975. The last is a result of mutation, the previous three of crossover, the first was posted by etomchek

amount that was human designed. This ratio is the *creative amplification factor* of the system, as discussed in Section 6.1.

The rest of this paper is structured as follows: Section 2 describes the architecture and implementation of Electric Sheep. Section 3 briefly explains the concept and artistic goals of the project. Section 4 surveys the genetic code on which all sheep rendering and evolution is based, and Section 5 explains the genetic operators and the specifics of the evolutionary algorithm. Section 6 reports empirical results of running this system for over 11 weeks during which time more than 6000 sheep were born. Section 7 puts this work in context of past research, and Section 8 concludes.

2 Architecture and Implementation

Electric Sheep has a client/server architecture. The client initiates all communication between them, and if no client were running the server would not run at all.

The client has three main threads. One thread downloads sheep animations from the server to a local disk cache. It downloads those with highest rating first. The default size of the cache is 300Mbytes (enough for 65 animations) but the user may change it. Another thread reads the sheep from the cache and displays them in a continuous sequence on the screen. The third thread contacts the server, receives a genome specifying a frame to render, renders the frame, then uploads the resulting JPEG file.

The server maintains several collections of sheep. Sheep are numbered as they are created and are identified by this sequence number. Freshly conceived genomes start out in the render queue. When all the frames of a sheep have been uploaded, they are compressed into MPEG and deleted, and the sheep is made available for download and voting. Sheep average 4.6Mbytes each. Eventually the sheep dies (Section 5.2 explains when) and the MPEG file is deleted.

All these sheep are referred to collectively as a generation. Each time the server is reset the database is wiped, all sheep are deleted from the server and from all client caches, the generation number is incremented, and evolution starts fresh. The sheep that are the subject of this paper are members of generation 165.

The server is implemented with two machines in separate colocation facilities. Both are commodity Linux x86 servers running Apache. One runs the evolutionary algorithm, collects frames and votes, compresses frames, and sends genomes to clients for rendering. The other only serves the completed MPEGs. The first server receives 221Kbit/s from the clients and transmits 263Kbit/s to them (measured average of July to October 2004). The MPEG server's bandwidth allocation has varied from 15 to 20Mbit/s, and it uses all of it.

The MPEG server is currently the bottleneck in the system. A future version of Electric Sheep will use a P2P network to distribute this bandwidth load much as the computation load already is.

The client runs on Linux, OSX, and Windows. It uses only the HTTP protocol on port 80 and it supports proxies. However, it does require a broadband, always-on connection to the internet. When the server is not reachable the client's sheep display still works but no new sheep appear.

All the code is open source and is licensed under the GPL (General Public License). The fractal flame utilities are written in C and the server is written in Perl. The clients are written in C, C++, and Objective-C.

3 Concept and Motivation

Electric Sheep is an attention vortex. It illustrates the process by which the longer and closer one studies something, the more detail and structure appears. Electric sheep investigates the role of experiencers in creating the experience. If nobody ran the client, there would be nothing to see. The sheep system exhibits increasing returns on each of its levels:

- As more clients join, more computational muscle becomes available, and the resolution of the graphics may be increased, either by making the sheep longer, larger, or sharper. The more people who participate, the better the graphics look. These adjustments are made manually on the server or with new client releases.
- Likewise, as developers focus more of their attention on the source code, the client and server themselves become more efficient, grow new features, and are ported into new habitats. The project gains momentum, and attracts more developers.
- And as more users vote for their favorite sheep, the evolutionary algorithm more quickly distills randomness into eye candy.

The votes tell the server which sheep are receiving the most attention. Those sheep are elaborated, expanding the variety and detail of those parts of the fractal space that are most interesting.

There is a deeper motiviation however: I believe the free flow of code is an increasingly important social and artistic force. The proliferation of powerful computers with high-bandwidth network connections forms the substrate of an expanding universe. The electric sheep and we their shepherds are colonizing this new frontier.

4 The Genetic Code

Each image produced by Electric Sheep is a fractal flame [4], a generalization and refinement of the Iterated Function System (IFS) category of fractals [2]. The genetic code used by Electric Sheep is just the parameter set for these fractals. It consists of about 160 floating-point numbers.

A classic IFS consists of a recursive set-equation on the plane:

$$S = \bigcup_{i=0}^{n-1} F_i(S)$$

The solution S is a subset of the plane (and hence a two-tone image). The F_i are a small collection of n affine transforms of the plane.

A fractal flame is based on the same recursive equation, but the transforms may be non-linear and the solution algorithm produces a full-color image. The transforms are linear blends of a set of 18 basis functions known as *variations*. The variations are composed with an affine matrix, like in classic IFS. So each transform F_i is:

$$F_i(x,y) = \sum_j v_{ij} V_j(a_i x + b_i y + c_i, d_i x + e_i y + f_i)$$

where v_{ij} are the 18 blending coefficients for F_i, and a_i through f_i are 6 affine matrix coefficients. The V_j are the variations, here is a partial list:

$$V_0(x,y) = (x,y) \qquad V_3(x,y) = (r\cos(\theta+r), r\sin(\theta+r))$$
$$V_1(x,y) = (\sin x, \sin y) \qquad V_4(x,y) = (r\cos(2\theta), r\sin(2\theta))$$
$$V_2(x,y) = (x/r^2, y/r^2) \qquad V_5(x,y) = (\theta/\pi, r-1)$$

where r and θ are the polar coordinates for the point (x,y) in rectangular coordinates. V_0 is the identity function so this space of non-linear functions is a superset of the space of linear functions. See [4] for the complete list.

There are 3 additional parameters for density, color, and symmetry, not covered here. Together these 27 (18 for v_{ij} plus 6 for a_i to f_i plus 3 is 27 total) parameters make up one transform, and are roughly equivalent to a gene in biological genetics. The order of the transforms in the genome does not effect the solution image. Many transforms have visually identifiable effects on the solution, for example particular shapes, structures, angles, or locations.

Normally there are up to 6 transforms in the function system, making for 162 (6×27) floating-point numbers in the genome. Note however that most sheep have most variational coeffients set to zero, which reduces the effective dimensionality of the space.

4.1 Animation and Transitions

The previous section described how a single image rather than an animation is defined by the genome. To create animations, Electric Sheep rotates over time the 2×2 matrix part (a_i, b_i, d_i, and e_i) of each of the transforms. After a full circle, the solution image returns to the first frame, so sheep animations loop smoothly. Sheep are 128 frames long, and by default are played back at 23 frames per second making them 5.5 seconds long.

The client does not just cut from one looping animation to another. It displays a continuously morphing sequence. To do this the system renders transitions between sheep in addition to the sheep themselves. The transitions are genetic crossfades based on pair-wise linear interpolation, but using a spline to maintain C^1 continuity with the endpoints.

Transitions are also 128 frames long. For each sheep created, three transitions are also created: one from another random flock member to the new sheep, one from the new sheep to a random flock member, and another one between two other random members. Most of the rendering effort is spent on transitions.

5 The Genetic Algorithm

There are three parts of the genetic algorithm: the rating system that collects the votes and computes the fitness of individual sheep, the genetic operators used to create new genomes, and the main loop that controls which live and die.

As already mentioned, users can vote for a sheep they like by pressing the up arrow key. If the sheep is alive its rating is incremented. Pressing the down arrow key decrements the rating. Votes for dead sheep are discarded. Users may also vote for or against a sheep by pressing buttons on its web page.

The ratings decay over time. Each day the ratings are divided by four with integer arithmetic rounding down.

5.1 Genetic Operators

There are four sources of genomes for new sheep: random, mutation, crossover and posts from Apophysis. The parents for mutation and crossover operators are randomly picked from the current population weighted by rating. The probability of being selected is proportional to the rating. Sheep that have received no votes have rating zero and so cannot be selected.

Random. The affine matrix coefficients are chosen with uniform distribution from [-1, 1]. The variational coefficients are set to zero except for one variation chosen at random that is set to one.

Crossover. The crossover operation has two methods chosen equally often. One method creates a genome by alternating transforms (genes) from the parents. The other method does pair-wise linear interpolation between the two parent genomes where the blend factor is chosen uniformly from [0, 1].

Mutation. The mutation operator has several different methods: randomizing just the variational coefficients, randomizing just the matrix coefficients of one transform, adding noise (-10 decibels, or numbers from [-0.1, 0.1]) to all the matrix coefficients, changing just the colors, and adding symmetry.

When applying these three automatic operators, the server renders a low-resolution frame and tests if the image is too dark or too bright. The operator is iterated until the resulting genome passes. For random genomes, 43% are rejected (in a test run 177 tries were required to get 100 passing genomes).

Post. Human designers may post genomes to the server with Apophysis. The designer is required to submit a password with the genome, but its value is shared on a public email list and is common knowledge there. The genome is checked for syntactic correctness, but the image it creates is not tested.

The server has a queue of sheep and transitions that are currently being rendered. Posted genomes go into this queue. When the queue is left with fewer than 12 sheep, it is filled with genomes derived with one of the three automatic operators. 1/4 of these genomes are random and have no ancestor. The remaining 3/4 are divided equally between mutation and crossover.

5.2 The Main Loop

The server maintains a single flock of sheep and continuously updates their ratings, creates new sheep, and kills off old ones. The server has 510MB of disk space for storing sheep animations, enough for 28 sheep and 83 transitions. Each time a sheep is born, the sheep with the lowest rating is killed to make room. If several sheep are tied for worst, then the oldest is taken (usually several sheep have received no votes and are tied with a rating of zero).

Killing a sheep removes the animation file from the server, but not from clients who may have allocated more disk space to their caches. The other records, including the peak rating, parentage, genome, 16 thumbnails, and the first frame are kept. This archive may be browsed on the server either sorted by peak rating, or as extended family trees.

This on-line or steady-state approach contrasts with the more traditional genetic algorithm's off-line main loop that divides the population into generations and alternates between rating all the individuals in a generation and then deriving the next generation from the ratings. Note that Electric Sheep does have 'generations', but it means something else, see Section 2.

6 Empirical Results

Three main datasets are analyzed below. The voting and posting data are from the web server log files from August 7th to November 4th 2004, 90 days later. Another dataset has daily aggregate usage reports from the server June 18th to October 31st (139 days).

The primary dataset was collected from the server's database starting May 13th until October 13th, 153 days later. May 13th is when version 2.5 became operational. Previous versions of the server did not keep a record of the sheep: when they died they were completely deleted from the server. And though there are large collections of sheep collected by clients from before May 13th, they are not complete and they lack fields for ratings and parentage.

During this time the server was subject to performance optimization but the basic algorithm remained fixed with one exception: until July 13th there was no limit on how many votes a user could make. An increasing incidence of users voting many times in rapid succession instigated a limit of 10 votes per user per day. The numbers below are from after that change.

The data we have collected are a starting point to understanding the system and its behavior. However, they are somewhat confounded:

- IP addresses are equated with users, but because of the prevalence of Network Address Translation (NAT), several or many users may appear as the same IP address. Worse, some computers are assigned an IP address dynamically, so one user may appear under many addresses.
- The client uses the ratings to prioritize downloading. Now that the server is busy enough that some clients cannot download all the sheep, this causes a snowball effect where a high rating itself causes more votes.
- The audience is fickle: sheep with identical genomes regularly receive completely different ratings (See Figure 1). Presumably the audience becomes fatigued by repeated exposure to variations of a successful genome, and stops voting for them. Even once popular sheep reintroduced much later do not necessarily fare well.
- Designers may vote for their own sheep, post many similar sheep, or post alterations of the results of automatic evolution. The administrator occasionally kills sheep, explicitly directs mating, mutation, reincarnation, and votes without limit.

There are many ways of measuring the number of users. Figure 2 shows four of them, and a graph of the rate of new users trying the system. The daily downloaders linear growth rate is 9 users per day, far fewer than the hundreds of new addresses per day. When Electric Sheep is first installed and run it takes quite some time (often about

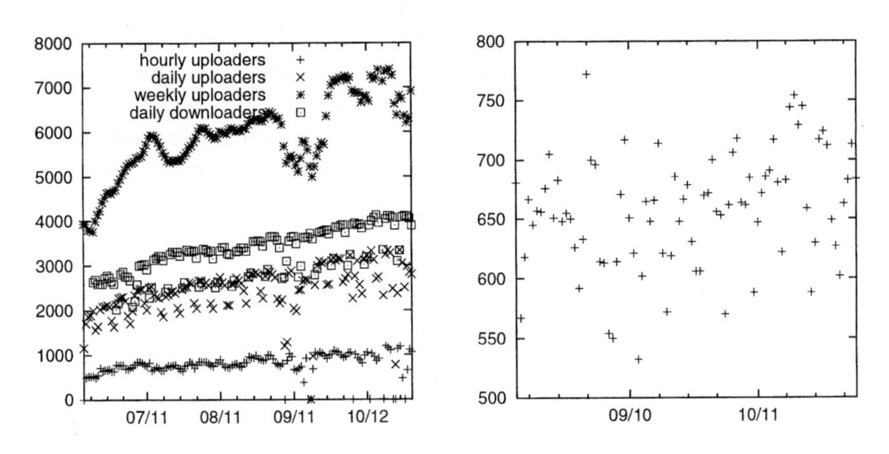

Fig. 2. On the left is a graph of number of users over time. Downloaders refers to clients downloading sheep, uploaders refers to clients uploading rendered frames. The dip from 9/2 to 9/19 coincides with server outages. On the right is a graph of the number of new client addresses over time. The total number of unique downloaders is 62000

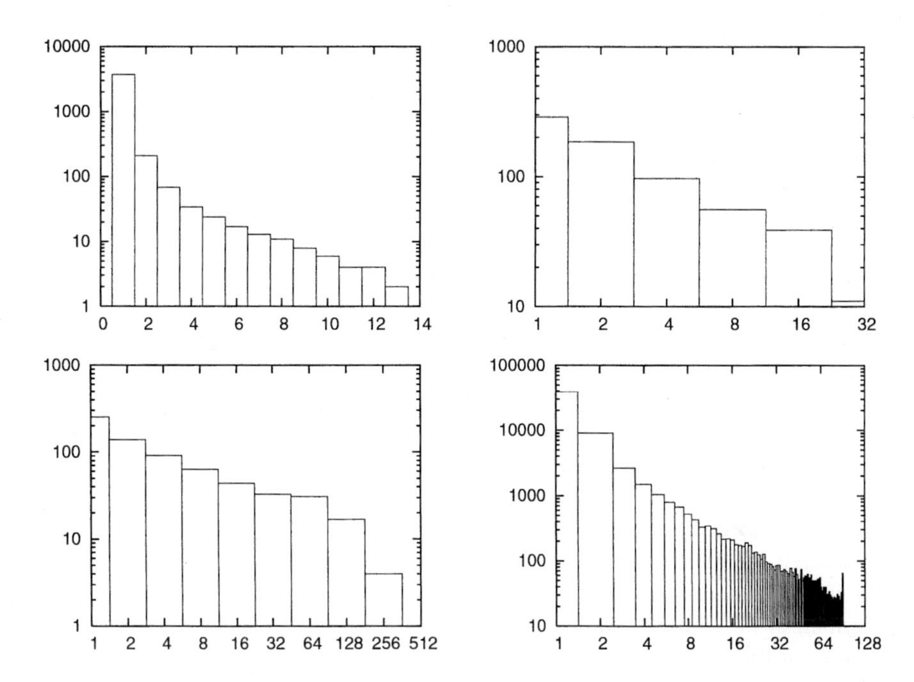

Fig. 3. On the top-left is a histogram of lengths of lineages, and on the top-right is a histogram of the ratings of the sheep. On bottom-left is a histogram of the sum of ratings of all descendents. Children of two parents contribute half of their sum to each parent. On the bottom-right is a histogram of number of days of activity out of 90 total possible by each IP address. 1839 clients were active half the days or more, and 65 were active every day (the upturn on the far right)

ten minutes but according to some reports hours or even days) to download and display the first sheep. Perhaps many people decide the software is broken and remove it. Or it could result from miscounting dynamic IP addresses.

There were on average 166 downloads of the client installers per day from the home-page during the first 8 days of December. But the installers are distributed on CD-ROM and mirrored on high-volume sites such as nonags.com, debian.org, and freebsd.org from which no statistics are available. These secondary sites also lack preview graphics and explanation so users from them may be likely to remove it.

Figure 3 shows the distribution of how many clients had a given number of days of activity (at least one attempted download). 1839 clients were active half the days or more, and 65 were active every day. Using data collected from November 2 to November 7 2004, there were 626 votes made, 76% with the arrow keys on client, and 24% with the web site.

The average number of valid votes per day is 111. During the 90 day period votes were recieved from a total of 1682 different client IP addresses. The average number of posts per day is 10.8 from a total of 64 different client IP addresses.

6.1 Amplification of Creativity

In a system with human-computer collaboration, the creative amplification is the ratio of total content divided by the human-created content. If we compare the posted genomes with their evolved descendents we can measure how much creative amplification Electric Sheep provides.

In the primary dataset there were 21% hand-designed, posted sheep and 79% evolved sheep. If the sum is weighted by rating, then we get 48% to 52%, for an amplification factor of 2.08 (1+52/48). One could say the genetic algorithm is doubling the output of the human posters.

Of the 79% evolved sheep, 42% of them result from the totally random genetic operator. Their fraction of total ratings is only 3.8%.

There are some caveats to this metric. For example, if the genetic algorithm just copied the posted genomes, it might receive some votes for its 'creativity'. Or if it ignored the posted genomes and evolved on its own, it would receive some votes but they would not represent 'amplification'.

Figure 3 shows the distribution of lengths of lineages of the sheep. The lineage length of a sheep is the maximum number of generations of children that issue from it. Sheep with no children are assigned one, and sheep with children are assigned one plus the maximum of the lineage lengths of those children. Instead of fitness increasing along lineage, we find it dying out: the rating of the average parent is 6.7 but the average maximum rating of direct siblings is only 3.8.

The decay in ratings may result from the audience losing interest in a lineage because it fails to change fast enough, rather than a decay of absolute quality of those sheep. The viewpoint of watching the screensaver and seeing sheep sequentially is different from the viewpoint of browsing the archive and comparing all the sheep. Neither can be called definitive.

Genetic algorithms normally run for many tens to hundreds or thousands of generations. In contrast, the lineages (number of generations) of the sheep are very short: the longest is 13.

7 Related Research

There are now many distributed screen-savers. Most are scientific (SETI@Home, climateprediction.net), cryptographic (distributed.net) or mathematical (zetagrid.net), rather than graphical or artistic. The Golem@Home project [7] has a evolutionary algorithm for evolving locomotion in electro-mechanical assemblages.

The aesthetic selection used by Electric Sheep is inspired by Karl Sims [9]. His supercomputer is replaced by internet-connected commodity PCs. The sheep voting community is much larger but much less focused than his user-base. Sims used Lisp expressions on pixel coordinates, but also included a primitive for IFS.

The International Genetic Art IV project [8] uses a web-based Java client to evolve images following the technique of Sims. Since the images are rendered on the clients, the computation is distributed, but users do not share votes or the gene pool. Its previous incarnation, International Genetic Art II, ran on a single server with a web interface so the computation was not distributed but the voting and gene pool were shared by all users.

8 Summary and Conclusion

Electric Sheep is a distributed screen-saver for animating and evolving fractal flames, a kind of iterated function system. The animations are shared among the clients and displayed in parallel with rendering. The evolution is guided by the will of the audience. The genetic code is made of geometric transforms, each one containing 6 coefficients for an affine transform of the plane and 18 coefficients for blending the variational functions.

The genetic algorithm employed does not converge on an optimal solution, but follows the attention of the users. And while the genetic algorithm is not competitive with the human designs, it does serve to effectively elaborate and amplify human designs.

Acknowledgements

Many thanks to: The anonymous reviewers, Tamara Munzner, and Matthew Stone for providing feedback on drafts of this paper, Dean Gaudet for help with performance optimization and server hosting, Tristan Horn for hosting the high-bandwidth server, Matt Reda and Nick Long for porting the client to OSX and Windows, respectively. And to all those who sent in patches, reported bugs, voted for their favorite sheep, or just ran the software once.

References

1. David Anderson et al: SETI@home: An Experiment in Public-Resource Computing. Communications of the ACM **45** (2002) 56–61.
2. David Barnsley: Fractals Everywhere. Academic Press, San Diego, 1988.
3. Alan Dorin: Aesthetic Fitness and Artificial Evolution for the Selection of Imagery from The Mythical Infinite Library. Advances in Artificial Life, Proc. 6th European Conference on Artificial Life, (2001) 659–668.
4. Scott Draves: The Fractal Flame Algorithm. Forthcoming, available from `http://flam3.com/flame.pdf`.
5. Scott Draves: The Interpretation of Dreams. YLEM Journal **23;6** (2003) 10–14.
6. Scott Draves: The Electric Sheep web site at `http://electricsheep.org`.
7. Hod Lipson and Jordan B. Pollack: Automatic Design and Manufacture of Robotic Lifeforms, Nature **406** (6799), 2000 August 31, 945–947.
8. John Mount, Scott Reilly, Michael Witbrock: International Genetic Art IV. Published on the web at `http://mzlabs.com/gart`.
9. Karl Sims: Artificial Evolution for Computer Graphics. Proceedings of SIGGRAPH 1991, ACM Press, New York.
10. Mark Townsend: Apophysis v2 software distributed from `http://apophysis.org`. 2004.

Swarm Tech-Tiles

Tim Blackwell[1] and Janis Jefferies[2]

[1] Department of Computing
[2] Department of Visual Arts, Goldsmiths College,
New Cross, London SE14 6NW, UK
{t.blackwell, j.jefferies}@gold.ac.uk,
http://www.timblackwell.com

Abstract. This paper describes an exploration of visual and sonic texture. These textures are linked by a swarm of "tech-tiles", where each tech-tile is a rectangular element of an image or a sequence of audio samples. An entire image can be converted to a single tech-tile, which can be performed as a composition, or a swarm of small tiles can fly over the image, generating a sonic improvisation. In each case, spatial (visual) structure is mapped into temporal (sonic) structure. The construction of a tech-tile from an image file or a sound clip and the swarm/attractor dynamics is explained in some detail. A number of experiments report on the sonic textures derived from various images.

1 Introduction

The word texture (15th Century, from the Latin texere, to weave) most commonly refers to the tactile appearance of a surface, especially that of woven fabric. The texture of a textile arises from colored material, woven from yarn that is laid out lengthways (the warp) and width ways (the weft). The resulting fabric has large-scale structure – the design itself – as well as local structure at the scale of the yarn. Upon closer inspection, there is also a micro-texture evident in the twisted fibers that constitute a strand of yarn.

A concept of texture exists in numerous specialized domains. In each case, the concept reflects the common meaning of the term but adds (albeit loosely) domain-specific connotations. For example, in painting, *visual texture* would refer to the representation of the nature of a surface. Visually, large scale structure can emerge from dense, intricate patterning of smaller elements; the works of Jackson Pollock present a good example.

The warp/weft idea exists informally in the idea of *musical texture*. A musical work can be described at a music-theoretic level as the interleaving of "horizontal" individual parts or voices, and "vertical" strands of harmony. An introductory book on composition even illustrates this idea with pictures of textured fabric (silk, polyester yarns and cheesecloth) [7].

In contrast, *sonic texture* refers to descriptions at the sound level [11]. The breakdown of tonality in 20th Century Western art music has led to a heightened awareness of texture within musical composition. This awareness reaches its zenith in the genres

F. Rothlauf et al. (Eds.): EvoWorkshops 2005, LNCS 3449, pp. 468–477, 2005.

of Sonic Art and Improvised Music [14, 2], where music theoretic descriptions become almost meaningless. Ascension by John Coltrane, a remarkable abstraction within jazz, is a good illustration of sonic and musical texture [6]. In this work, the improvising ensemble (comprising five saxophonists, two trumpeters, two bassists, a pianist and a drummer) create a highly detailed and emotional sound-world. This seems to be a direct analogy of the abstraction pursued by Pollock.

Sound texture in the domain of signal-processing refers to sounds with no perceptible long-term structure [1]. A short sample of a sound texture gives no indication of when it actually occurred within the whole sound, in contrast to musical instruments which produced time-dependent envelopes. Examples of sound textures are running water and traffic noise.

Another meaning of texture is to be found within image processing. The segmentation of a digital image into regions of different *image texture* (for example the segmentation of an aerial image into fields and forests) is achieved by a quantifiable texture measure, although there are many to choose from [8].

These domain specific applications of surface texture share a more general idea: texture is concerned with the structure and relationship of the component parts of something.

In order to understand this meaning, it is suggested in this paper that large scale sonic structure might emerge from the interactions of sonically dense micro-textures. To be precise, each sonic micro-texture is a sound "grain", where, arguing by analogy, the grain is constructed from a small element or micro-texture of the image. Small rectangular tiles from an image are mapped to sound grains of short, but perceivable, duration. The grains are emitted asynchronously from a synthesizer in dense clouds in a technique known as granulation [13]. The constituency of the micro-textures and their interactions is determined by a virtual swarm that flies over the textured image. Dynamic sonic macro-texture emerges from the self-organization of the swarm around attractors of high micro-texture.

Three uses of Swarm Tech-tiles are demonstrated here. Firstly, a "tech-tile" of a whole image can be rendered into sound. This produces a short piece of sonic art. Secondly, a user can explore interesting regions of micro-texture by selecting tile locations with a mouse click. More interesting, though, is to allow a swarm/attractor system search for interesting micro-texture, producing an improvisation of indeterminate length. Swarm techniques are pertinent in this context for a number of reasons, not least because different starting configurations lead to different outputs, so that many pieces can be generated from a single image (i.e. the output is improvisatory). Also, the self-organizing properties of swarms will endow the improvisation with temporal order, so that the piece exhibits musical structure.

The following section describes the tech-tile. This is the image-sound map which is used to construct grains (for an improvisation) or a single complete texture (for a composition). Section 3 explains the novel particle swarm used in Swarm Tech-tile. A discussion of how the swarm interprets the local micro-texture of an image and of how particle positions are interpreted as dynamic parameters of a sound stream follows in Sect. 4. The following section continues with a description of some experiments and the final section summarizes the findings of this paper.

2 Texture Tiles

There is no necessary connection between images and sounds since each stimulates a different sense (modality). Moreover there are infinitely many possible maps from the two spatial dimensions of an image into the single time dimension of a sound. Some guiding principles, however, are at hand. At the least, a spectator should be able to derive correlations between visual and sonic texture. This principle, in a different context, is known as transparency [5].

A computer generates sound by rendering an audio stream at the sampling rate. An audio stream is, in turn, a succession of samples or digitized elements of a pressure wave. Alternatively, a digital image is an arrangement of adjacent pixels, where each pixel is a digitization of color, for example the three RGB pixel-values, one for each primary color. The simplest map is therefore an association between a single pixel value (R, G or B) and an audio sample. (The map has to stipulate a scaling between positive pixel values and audio samples which can be positive or negative, and with a much larger range of values.)

Locality would imply that closeness in image space relates to closeness in time so that image micro-textures are mapped to sonic micro-textures. Smoothly varying patches of similar color might correspond in a local map to continuous, harmonic waves, perceivable as tones. Edges between colors, on the other hand, which occur at thread boundaries on a textile, might map to buzzy, non-harmonic waves (square, saw-tooth and other discontinuous wave-forms). A very speckled texture might map to a noise stream.

The audio rate determines how quickly samples are rendered, and the simplest choice is to move between pixels at a time interval equal to the reciprocal of the audio rate. But there is a problem of exactly how to move (by horizontal, vertical, or diagonal steps in any combination), and what action to take at a tile edge. In order to solve this, we were guided by the manufacture of an actual textile. Textiles are made be weaving threads length-ways (warp) and width-ways (weft). An obvious way to preserve this feature is to scan the image vertically (along the warp) and horizontally (along the weft). However, in order to preserve locality at the tile edges, each warp/weft scan is continued in the opposite direction, Fig. 1.

The scanning time can be calculated by dividing the number of pixels in the image by the sampling rate. A 64 x 64 tile contains 4096 pixels/samples, and would therefore take 93 ms to scan at a sampling rate of 44100Hz, certainly long enough to hear as a micro-texture and not just as a click. A digital image of a large textile is shown in Figure 2, [9]. The original image, of size 2646 x 1760 = 4.7 megapixels, would produce 106s of sound if rendered as a single tile.

In summary, a texture tile – or tech-tile – is, visually, a rectangular portion (a tile) of an image, or aurally, it is a local parameterization of a sonic stream. The image to sound map is made by simultaneously scanning at the audio rate along the warp and the weft. R, G and B pixel values are picked up and scaled into audio samples. In order to preserve locality, and to prevent anomalous discontinuities (these would be heard as clicks in an otherwise smooth sound), the scan doubles back at the tile edges. Six audio streams are produced; for convenience the warp and weft streams are sent to separate stereo channels.

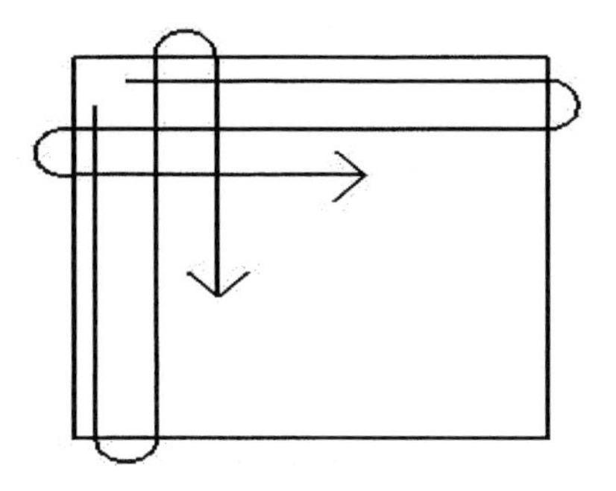

Fig. 1. Warp and weft scan lines

Fig. 2. Uniform and Laundry by Janis Jefferies. *Photo: David Ramkalawon*

3 Swarms and Stigmergy

Swarms of interacting particles, moving in a *d*-dimensional real space, are a more abstract realization of the A-Life swarm, herd and flock animations initially studied by Reynolds [12]. Reynolds' demonstrated that the collective behavior of these animal groups can be explained through local (rather than globally scripted) interactions. These swarms typically exhibit self-organization (SO). SO emerges from direct particle interactions and also through indirect environment-mediated interactions known as stigmery [3]. Particle Swarm Optimization (PSO) and Ant Colony Optimization are practical applications of these ideas [10, 3].

The creative use of particle swarms has also been investigated [4, 5]. In the swarm/attractor systems of these papers, particles are drawn towards special points in space known as attractors. These systems are interactive: attractors are positioned as a result of input with an external system (human, or another swarm) and the output from the swarm is an interpretation of particle positions as grains of sound (Swarm Granulator) or musical events (Swarm Music). Coherence and structuring of the respective outputs, and correlations to the inputs are manifestations of SO, induced by the stigmergetic interaction of particles with attractors.

However, in PSO, attractors derive not from interaction but from the evaluation of a fitness function at each particle location [10]. The swarm described below incorporates function evaluation with biologically plausibility particle dynamics.

3.1 Particle Dynamics

The particle dynamics for this swarm implementation have been modified from the system described in [5]. Here, particles' perceive attractors and each other within a local hyper-spherical neighborhood, and not over the entire space[1]. This more plausible feature favors the development of subswarms – breakaway groups of particles

[1] This modification was suggested to one of the authors by a participant of the EvoMUSART 2004 workshop.

searching for new attractors. Spring-like interactions have also been replaced by fixed magnitude accelerations which, together with velocity clamping, lead to constant magnitude velocity vectors. This moderates the (somewhat biologically unrealistic) tendency of particles to oscillate about an attractor.

The particle positions $x_i \in [0, x_{max}]^2$ and velocity v_i are updated by determining the local neighborhood of other particles and attractors, where the neighborhoods are determined with respect to the perception radius $r = 0.25x_{max}$. The acceleration towards a particle at x_j and an attractor at p_k is given by

$$a_i(x_j) = Q^2 \frac{(x_i - x_j)}{\left|x_j - x_i\right|^3} + \frac{C}{N}\frac{(x_j - x_i)}{\left|x_j - x_i\right|} + \frac{C}{M}\frac{(p_k - x_i)}{\left|p_k - x_i\right|} \tag{1}$$

where the first term is a collision avoiding repulsion and the second and third terms are attractions towards x_j and p_k respectively. Constants Q and C determine the strength of these accelerations and are set to $x_{max}/32$ and $x_{max}/128$ respectively. N and M are the number of particles/attractors in the neighborhood. The accelerations are added to the current velocity, and the velocity is clamped to $v_{max} = x_{max}/32$. The position is finally updated by adding on the new velocity, and reflecting the particle from the sides of $[0, x_{\max}]^2$ if necessary. $x_{\max}$ is an arbitrary scale, fixed at 128.

3.2 Attractor Stigmergetics

There are important differences between this swarm and PSO in the treatment of attractors. PSO uses a cognitive-social model for attractor placement. In PSO, each particle has a memory of the best position (as measured by an objective function) it has attained (cognitive model) and has knowledge of the best position obtained by other particles (social model) in a topological neighborhood. Although good for optimization, the social model of PSO is not plausible from a swarm perspective (but does make more sense for the social networks found in human culture).

Stigmergy is a biological term for environment-mediated interaction [3]. In Swarm Tech-tiles, particles make decisions based on the attractors that they can *actually see* i.e. in their immediate spatial neighborhood – this implementation of stigmergy is more faithful to biological swarms. Decisions to create or move attractors depend on the value of a micro-texture measure T at the particle position. The particle only has access to values of T at visible attractor positions, but retains a memory of the best texture that it has encountered. However, the longevity of attractors and particle memory is limited in a way that will now be described.

Attractor death. The initial perception radius $r_{init} = 0.25\, x_{\max}$ of an attractor shrinks by a decay constant $\lambda \in [0, 0.1]$ at each particle visit, eventually leading to attractor annihilation when $r < r_{crit} = 0.5r_{init}$. The biological parallel is a food source that is progressively consumed on each visit. Dynamically this means that a swarm (or sub-swarm) cannot stagnate around an old attractor.

Attractor movement. Suppose that a particle can see one or more attractors $\{p\}$, and it is currently at a better position x than all these attractors. In this case the best attractor, $p_{best,}$ will be moved to x. The particle always stores $T(p_{best})$, irrespective of movement.

Attractor creation: Suppose that a particle at x cannot currently see any attractors, and $T(x)$ improves upon its memory, $T(p_{best})$. The particle then deposits a new attractor at x and remembers $T(x)$. This rule is necessary to compensate for attractor death. This rule produces fascinating cooperative behavior. A breakaway particle or small sub-swarm can generate new attractors along a fluctuation, leading to particle "streamers". This is an example of positive feedback which is an important ingredients of SO [3].

Particle forgetfulness. A particle will forget $T(p_{best})$ if no attractor is visible for a fixed number of interpretation iterations, even though its current $T(x)$ is no improvement. Any location will then become tempting and new attractors can be created. This rule counteracts this scenario: suppose the swarm finds a very good location, and each particle sets an attractor in the neighborhood of this location. After the attractors have been consumed, particles may then wander aimlessly, never depositing a new attractor since $T(p_{best})$ is never exceeded.

The observational effect of these attractor stigmergetics is an alternation between phases of exploration (breakaway subswarms, attractor creation and particle forgetfulness) and exploitation (attractor movement and consumption).

4 Interpretation: Swarm/Environment Interface

A general architecture for a swarm/attractor system has been given in [5]. Each software module corresponds to a mathematical function. The input interpretation function P explains how the system represents the environment as a pattern of dynamic attractors' p and with an objective function T. The output interpretative function Q explains how the system tries to modify the environment by generating external events $q = Q(x)$ from particle positions x. The internal workings of the system are given by the particle dynamics, $x(t) = f(p, x, v)$ where the t is real time and all the arguments of f are evaluated in the interval $[t\text{-}\delta t, t]$ (refer to [5] for a full explanation).

The patterning module has been specified in the previous section. This section continues with a description input/output modules P and Q.

Q maps a particle position x onto a tech-tile $q(\xi, w, h)$ of width w and height h. q is centered at image coordinates ξ where ξ is obtained by a simple re-scaling of $[0, x_{max}]^2$ into the image space $[0, w_{max}] \times [0, h_{max}]$. If q extends beyond the image in either dimension, it is translated so that all scan lines fall inside the image. The warp/weft scans are then performed according to the description of Sect. 2. Q runs in a separate thread to f, pausing by Δt after each interpretation. The particle update time interval, δt, is chosen to give smooth animations and is typically set to 10 ms. Each 64 x 64 tile takes 93 ms to render, giving up to 10 overlapping grains at δt = 10 ms. Tiles up to dimension 20 x 20 can be rendered without overlap, but tiles this small have no discernible timbre. Overlapping is desirable in granulation since this gives a continuous, sonically dense stream. Too many overlaps can strain the processor and lead to unwanted audio fragmentation. In practice, a grain rate Δt of 10-50 ms is used for a 1.7 GHz processor with 250 MB memory.

P concerns attractor placement. In Swarm Tech-tiles, an objective function determines the desirability of any sampled location of the textured environment by calculating a measure of micro-texture T for the tile at ξ. T should be chosen to lead the swarm towards "interesting" texture. This is clearly an arbitrary and aesthetically-driven choice. However, certain measures of image texture are commonly used for image segmentation [8] and a couple of these have been tested in experiments.

The first measure is a generalization of the grayscale entropy of the co-occurrence matrix P_{ij} of the image I,

$$S = -\sum_{i \in I} \sum_{j \in I} P_{ij} \log(P_{ij}) , \tag{2}$$

where P_{ij} is the probability that two pixels in I, separated by a displacement vector d, will have grayscale pixel values i and j. The generalization is $T_{entropy} = max\{S_R, S_G, S_B\}$ where $S_{R,G,B}$ is the entropy of the R, G or B co-occurrence matrix over the tile at ξ. In effect $T_{max\ entropy}$ measures the randomness of the RGB distribution. In order not to discriminate between the two scan directions, a displacement vector that compares pixels at different warps and wefts, $d = (1, 1)$, is used.

The second measure is a simple statistical measure, $T_{stat} = max\{\sigma^2_{R}, \sigma^2_{G}, \sigma^2_{B}\}$ where $\sigma^2_{R,G,B}$ is the variance of the R, G or B values over the tile at ξ. This measure also favors inhomogeneity and has the advantage that it is efficiently computed. However, T_{stat} does not discriminate between different scales of texture.

Figs 3-6 show two images and their *max entropy* maps. The images of these texture maps of have been equalized in order to highlight differences. Fig. 3 shows the tech-tile for a 5 second multiphonic saxophone tone. This was produced by mapping the complete tone, recorded in 8 bit mono, onto a single tile so that $R = G = B = sample + 128$, and preserving the warp/weft scan of Fig 1. A multiphonic is a technique whereby a single fingering produces multiple pitches. Since the pitches are harmonic, the samples vary smoothly in time, evident in the shallow texture of Fig. 3, and from the largely uniform texture map of Fig. 4. (The non-uniformities are due to edge effects.)

Figs 5 and 6 show part of the textile of Fig. 2 and the corresponding map. The texture of the textile image is far coarser than the multiphonic tone, showing greater inhomogeneity with a noticeable change in texture between the top right and bottom left of the image. The *max variance* maps of Figs 3 and 5 are broadly similar to the *max entropy* maps. Texture maps can be used as a look-up table for $T(x)$, hence saving a costly computation, but precision is lost due to the conversion of the measure into pixel values.

5 Experiments

Experiments were performed on images of a sunset, a calm seascape, a Eucalyptus tree, the Jefferies textile, recorded saxophone and voice, and synthetic images of pure tones, white noise, color rainbows and an image with an island of noise centrally placed on a constant color background. Some of these images and sonic tech-tiles are available for download at www.timblackwell.com.

Fig. 3. Techtile of a multiphonic saxophone tone

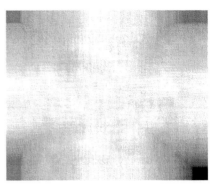

Fig. 4. *Max entropy* texture map of Fig. 3

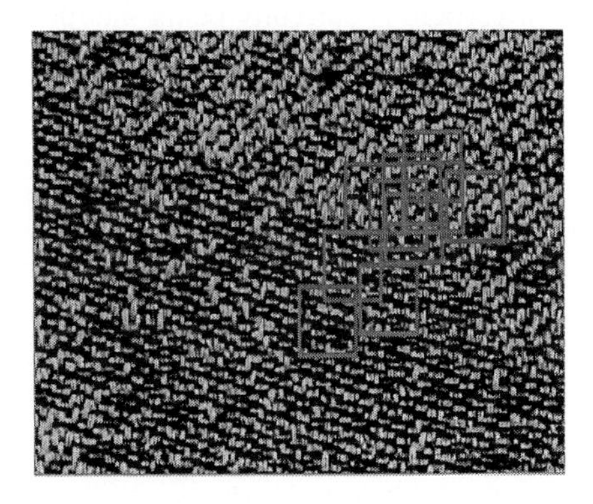

Fig. 5. 595 x 413 part (top left) of the textile of Fig. 2. The red squares show the last ten techtiles rendered by the granulator

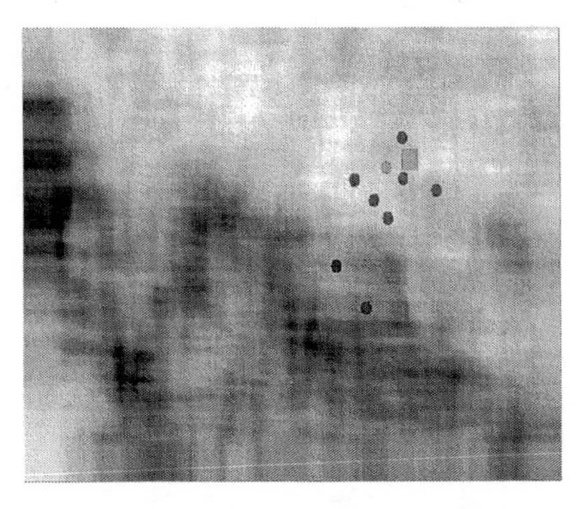

Fig. 6. What the swarm sees when flying above the textile of Fig. 5. Particles (red and blue discs) position attractors (boxes) on regions of high entropy

To begin, a single tech-tile, chosen from any of the images by a mouse click, was rendered into sound. The size of this tile varied from 1 x 1 (which is too brief to be audible) up to the size of the entire image. Tiles below about 40 x 40 are heard as clicks with no discernable timbre. Tiles larger than 64 x 64 were used in the following trials since these are capable of probing both sonic and image micro-texture.

Images with areas of blended color, such as the reddening sky around a setting sun produce quiet, pulsating sounds. This is because the warp/weft scan picks up neighboring pixel values that are only gradually changing along a color gradient. As the scan doubles back and retraces the color gradient in the opposite sense, the sample values match this oscillation and harmonic tones at frequencies around 44100/(64 x 2) ≈ 345 Hz are audible.

When the tile contains edges, the color discontinuity corresponds to a sample jump, so that the scan produces discontinuous waves. If the tile is placed over rough visual texture, for example when placed over the branches of the Eucalyptus image, a loud buzzing and a lesser noisy component is audible. When placed over the leaves on the same image, corresponding to finer texture, the sound contains more noise. The Jefferies textile produces a gentler and less buzzy sound, with more rattle and noise.

A sonic composition of some 106 seconds results from representing the Jefferies textile as a single tile. Although slight, changes in noise spectrum are audible against a background rattle (which has a rather pleasing loom-like hum). Regions of different visual texture produce intervals of varying sonic texture, which confirms that a single large tech-tile preserves large-scale structure.

Textural improvisations were also produced by allowing a ten particle swarm to fly over the image. Attractors were deposited and observed to move to regions of high micro-texture (statistical or entropic). Figs 5 and 6 show a typical configuration during one such improvisation. The swarm flies over the texture map of Fig. 6, where white pixels correspond to tiles of high micro-texture. The red particles of this figure can see an attractor (green box) and the green particle is currently being interpreted. The blue particles are searching for an attractor. There is an old attractor (red box) which has not yet been consumed. Fig. 5 shows the ten tech-tiles sampled from the swarm.

The influence of each attractor shrinks with each particle visit until the attractor evaporates completely. The perception decay rate λ was varied in a number of trials. Larger rates produce more diversity in the population and a more varied sonic output but the surrounding texture is not explored in any detail and the swarm wanders aimlessly above the image. Finally, it was confirmed that a swarm is able to discover the textured region in the island-of-noise image, developing pockets of sonic micro-texture between short intervals of silence.

6 Conclusions

Domain-specific definitions of texture are hard to pin down, but they all refer to a literal tactile meaning of the term. Analogous meanings can be investigated with mappings from *actual* texture onto the domain-specific texture. Although such mappings are arbitrary since they cross three modalities – touch, vision and hearing - they should preserve large-scale structure and possess a degree of "transparency". Transparency means, in this context, that an intuitive link can be established between, for example, roughness of actual texture and roughness of sound in a sonic texture.

The tech-tile map, inspired by the construction of textiles, establishes just such a map. Experiments on a number of synthetic images, images of natural texture and a textile image confirm that this map preserves structure and relates qualities such as smoothness (glowing sky during a sunset ↔ quiet, pulsating sounds), roughness (branches, twigs ↔ harsh buzzing sounds) and fine detail (Jefferies textile, leaves ↔ rattles and noise).

It is suggested that a swarm/attractor system in conjunction with a sound granulator can be used to develop a textural improvisation of indeterminate length. The improvisation itself arises from the exploration of the image for regions of high micro-texture.

Mechanisms for attractor creation, movement and death, and particle forgetfulness, ensure that the improvisation is sonically diverse.

Swarm Tech-tiles is still in development and future research topics includes the investigation of more sophisticated texture measures (for example using Fourier analysis to quantify harmonicity and noise) and the use of cameras to extract 3D data from a textile. This would close the gap between our tactile experience of texture and the mapping onto sound. Finally, an exciting prospect is to use eye tracking equipment to extract information about how a user views a textured surface. Attractors can then be positioned accordingly and the viewers will be able to hear what they are seeing.

References

1. Athineos, M. and Ellis, D.: Sound Texture Modeling with Linear Prediction in both Time and Frequency Domains. Proceedings of ICASSP-03, Hong Kong (2003)
2. Bailey, D.: Improvisation: its Nature and Practice in Music. The British Library Sound Archive, London (1992)
3. Bonabeau, E., Dorigo M. and Theraulaz T: From Natural to Artificial Swarm Intelligence. Oxford University Press, New York (1999)
4. Blackwell T: Swarm Music: Improvised Music with Multi-Swarms. Proc. AISB '03 Symposium on Artificial Intelligence and Creativity in Arts and Science (2003), 41-49
5. Blackwell, T. and Young M.: Self-Organised Music. Organised Sound **9**(2), Cambridge University Press (2004) 123–136
6. Coltrane, J. Ascension. Impulse! Records (2000)
7. Howard, J.: Learning to Compose. Cambridge University Press, Cambridge (1990)
8. Jain, R., Kasturi, R., Schunck, B.: Machine Vision. McGraw-Hill, New York 1995)
9. Jefferies, J: Uniform and Laundry (Restaged I). Produced at the Centre for Contemporary Textiles, Montreal (2003)
10. Kennedy, J. and Eberhart, R: Swarm Intelligence. Morgan Kauffman (2001)
11. Macy, L. (ed): 'Texture', Grove Music Online (Accessed 3 November 2004), http://www.grovemusic.com
12. Reynolds, C. 1987. Flocks, Herds, and Schools: A Distributed Behavioural Model. SIGGRAPH '87 **21**(4) (1987) 25-34
13. Roads, C: Microsound. MIT Press, Cambridge, MA (2001)
14. Wishart, T. On Sonic Art. (Revised Edition.) Harwood Academic, Amsterdam (1996)

Evolutionary Methods for Ant Colony Paintings

Gary Greenfield

Mathematics & Computer Science,
University of Richmond, Richmond VA 23173, USA
ggreenfi@richmond.edu
http://www.mathcs.richmond.edu/~ggreenfi/

Abstract. We investigate evolutionary methods for using an ant colony optimization model to evolve "ant paintings." Our model is inspired by the recent work of Monmarché et al. The two critical differences between our model and that of Monmarché's are: (1) we do not use an interactive genetic algorithm, and (2) we allow the pheromone trail to serve as both a repelling and attracting force. Our results show how different fitness measures induce different artistic "styles" in the evolved paintings. Moreover, we explore the sensitivity of these styles to perturbations of the parameters required by the genetic algorithm. We also discuss the evolution and interaction of various castes within our artificial ant colonies.

1 Introduction

Monmarché et al [1] recently described an interactive genetic algorithm involving ant colony optimization (ACO) methods for the purpose of evolving aesthetic imagery. Although it was never well documented in the literature, it should also be noted that the digital image special effects developed by Michael Tolson using populations of neural nets in order to breed intelligent brushes may be an historical precedent for using a non-interactive approach involving ACO methods for evolving aesthetic imagery [10]. Thanks to the early favorable publicity garnered by Dawkin's *Biomorphs* [3], Sim's *Evolving Expressions* [11], and Latham's *Mutator* [13], *interactive* genetic algorithms have long played a central role in the evolution of aesthetic imagery. The use of more traditional non-interactive genetic algorithms to produce aesthetic imagery — the computational aesthetics approach — has received much less attention, no doubt due in large measure to the inherent difficulty in formulating fitness criteria to assess images on the basis of their aesthetic merits. Research previously published along these lines includes approaches involving neural nets [2], evolving expressions [6] [7] [8], and dynamical systems [12].

It has been suggested that the evolution of images by means of organisms that are evolved for their *aesthetic* potential is less about evolution and more about the search for novelty [4]. In fact, psychologists are only just beginning to understand the neurological underpinnings of visual aesthetics [9] [14]. Be that as it may, the goal of this paper is to consider techniques for evolving ant colony paintings using a non-interactive genetic algorithm. Our objective is to show how

F. Rothlauf et al. (Eds.): EvoWorkshops 2005, LNCS 3449, pp. 478–487, 2005.

different fitness criteria used to evaluate the aesthetic contributions of the ants determine different painting styles as well as influence ant colony formation.

2 The Basic Model

The basic framework for our ACO ant painting model follows that of Monmarché et al [1]. In their model individual ants possess *attributes*: an RGB color (C_R, C_G, C_B) for the ant to deposit; an RGB color (F_R, F_G, F_B) for the ant to follow; a vector of probabilities (P_l, P_r, P_a) satisfying $P_l + P_r + P_a = 1$ used to determine the probability that the ant veers left, veers right, or remains on course by moving directly ahead; a movement type D which takes the value o or d according to whether an ant veering off course veers at a 45° angle or a 90° angle; and a probability P_f for changing direction when "scent" is present. Thus an individual ant genome is simply a vector of the form $(C_R, C_G, C_B, F_R, F_G, F_B, P_l, P_r, P_a, D, P_f)$. Since ant behavior is determined using a simple move-deposit-sense-orient sequence, scent trails form from the colors the ants deposit as they explore a toroidal grid. Deposited colors are allowed to seep throughout the 3×3 neighborhood the ant is currently occupying by invoking a convolution filter defined using a 4:2:1 ratio such that immediately adjacent cells to the one occupied receive half as much color as the occupied one receives, while diagonally adjacent cells receive one-quarter as much color as the occupied one receives. The unusual and surprising feature of the Monmarché model is that the metric used for detecting scent depends on a *luminance* calculation. Specifically, if L_W is the luminance of the neighboring cell W that the ant is sensing, and L_F is the luminance of the color (F_R, F_G, F_B) that the ant is attempting to follow, then the scent value detected by the ant is $\Delta_W = |L_W - L_F|$. Of course the smaller Δ_W is the stronger the scent is.

Using the same sensing constraint as Monmarché, namely that scent following behavior should not be invoked unless Δ_W falls below the threshold value `MAXS` of 40, thanks to the table of genomes accompanying the four hand-crafted examples appearing in Figure 1 of [1], when using a 200×200 toroidal grid where each

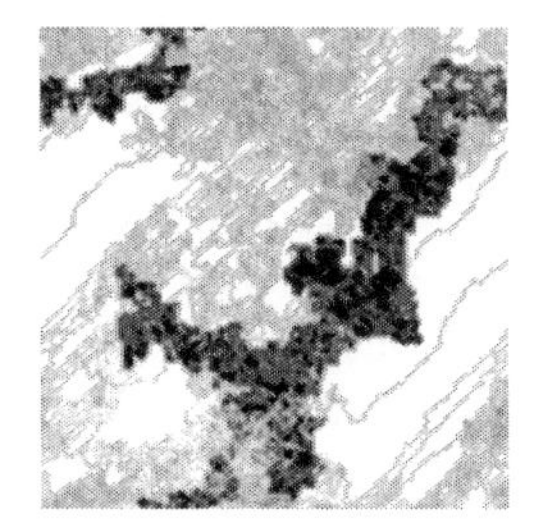
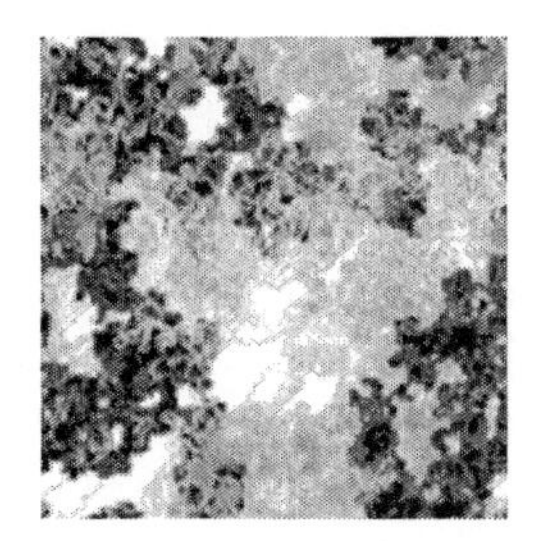

Fig. 1. Images reprising the three-ant, black-and-white Monmarché example that were obtained using randomly generated initial positions and directions. The exploration times were (left to right) 12000, 24000, and 96000 times steps respectively

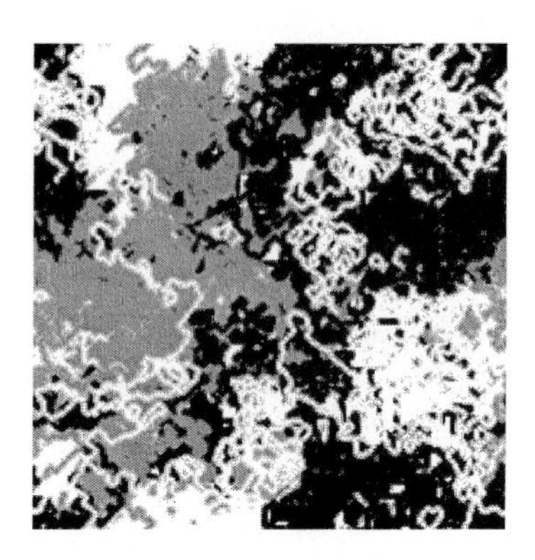

Fig. 2. The effect of replacing, as opposed to blending, using the deposited color. The image on the left uses the five-ant, multi-color Monmarché genomes, and the one on the right the three-ant, black-and-white genomes

ant was allowed to explore for 24000 time steps we were essentially able to duplicate the paintings of that figure. Figure 1 shows the results from a reverse engineering experiment using the three-ant, black-and-white example of [1] in order to estimate an appropriate value for the exploration time parameter.

Based on the description given in [1], it was not initially clear to us whether an ant should *replace* the color of the square it currently occupied with the color it was depositing, or blend (using a 1:4 ratio) the color it was depositing. Figure 2 shows what happened when we tried using the replacement strategy for both the three-ant, black-and-white example and the five-ant, multi-color example given in [1]. Such tests convinced us that replacement was not the method intended and that the following color printed as $(255, 0, 0)$ in Table 1 of [1] for the five-ant, multi-color example was probably meant to be $(255, 153, 0)$. This correction is consistent with the assertion that those examples were hand-crafted so that each ant was seeking a color that another ant was depositing.

3 Scent Modification in the Basic Model

Recall that the luminance L of an RGB color (X_R, X_G, X_B) is defined to be $L = 0.2426X_R + 0.7152X_G + 0.0722X_B$. Determining scent on the basis of luminance has two implications. First, ants become abnormally sensitive to the green component of the color they are following. Second, colors perceived as different RGB colors by humans may be perceived as virtually identical colors by ants. While it is true that in nature a swarming species such as, say, bees may perceive colors differently than humans, for the most part there is still a comparable basis for color *differentiation*. Consider, for example, ultraviolet photographs of flowers compared to photographs of those same flowers made using the "visible" spectrum. Since ant paintings are not *imaged* solely on the basis of the luminance channel, the ant paintings we view are not the same as the ones the ants view. For this reason, henceforth we will measure scent using a supremum norm by defining

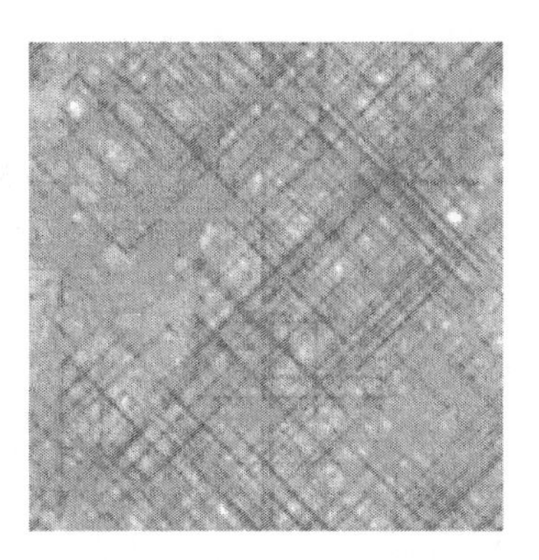

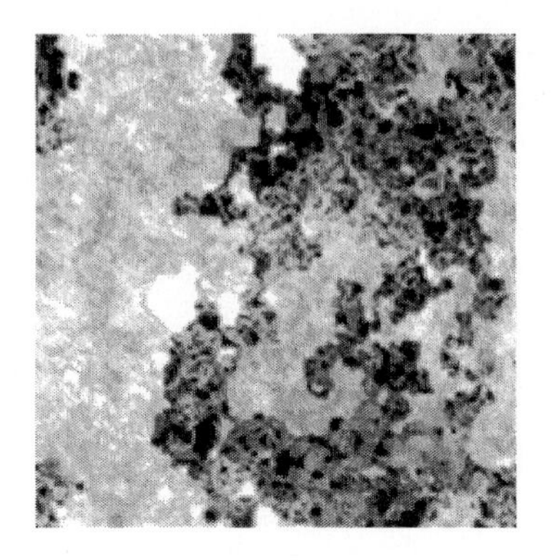

Fig. 3. The effect of redefining the scent metric. On the left the four-ant, red-component Monmarché example and on the right the three-ant, black-and-white example

$$\delta_W = \max(|W_R - F_R|, |W_G - F_G|, W_B - F_B|).$$

Under this metric, with the scent threshold MAXS still set to 40, Figure 3 reprises two of the four examples of [1]. Indeed, when we ran all the examples of [1] using this metric we discovered that most of the following behavior that ants exhibited was due to following their own scent, and nearly half of the ants exhibited no following behavior at all.

To further reduce the opportunity for ants to follow themselves we introduced a repelling force, under the control of the threshold constant MINR, to inhibit ants from following scent in a neighboring cell W whenever

$$\psi_W = \max(|W_R - C_R|, |W_G - C_G|, W_B - C_B|)$$

falls below this threshold. To test the parameters MAXS and MINR using our new metric, we generated several five-ant, multi-color examples for which depositing and following colors were randomly and independently chosen. Figure 4 shows representative results from these tests.

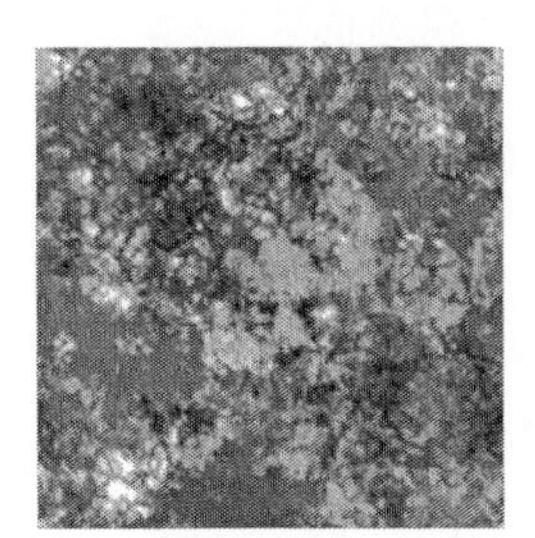

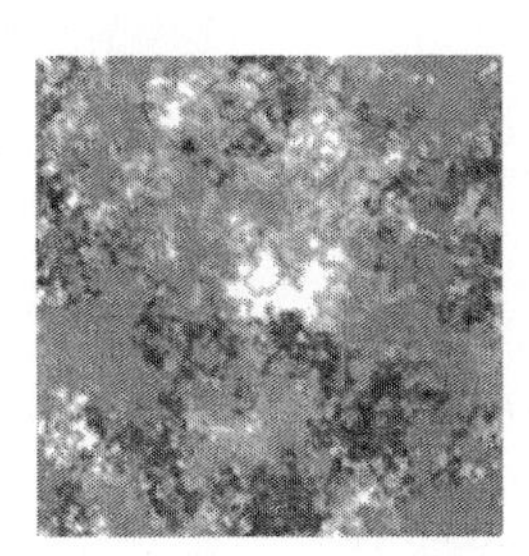

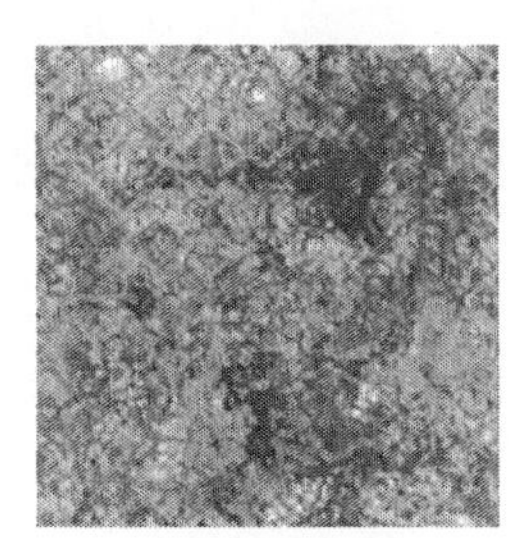

Fig. 4. Sample images from five-ant, multi-color examples testing the attracting and repelling thresholds. Left MAXS = 40, MINR = 60. Center MAXS = 60, MINR = 120. Right MAXS = 60, MINR = 40

4 Adding Non-interactive Fitness to the Model

For consistency, while exploring fitness criteria to use for ants, we fixed MAXS at 80 and MINR at 40. On a 200×200 grid, with ants in motion for 24000 time steps, each ant in the population can hope to visit at most sixty percent of the grid. Ant fitness criteria were based on two measurements recorded during this exploration period, the number of distinct cells visited by an ant, denoted N_v, and the number of times cells were visited by following scent, denoted N_f. Our initial population consisted of twelve randomly generated and randomly positioned ants. We quickly discovered we could not breed replacements for too many ants after each generation because ants became overly sensitive to the background color i.e. evolution quickly evolved monochrome paintings that were dependent wholly on the settings of the scent thresholds MAXS and MINR. This explains why after each generation we chose to replace only the four least fit ants. The replacement scheme invoked uniform crossover coupled with a point mutation scheme. We replaced least fit ants two at a time by randomly mating a pair randomly chosen from the eight ant breeding pool. During some runs we introduced "mortality" by replacing the three least fit ants plus one ant randomly chosen from the remainder of the population. Evolution proceeded for twenty generations. Ant paintings were preserved every other generation.

We discovered that failure to reset the grid to the background color after each generation introduced monochromatic degeneracies. Letting ant fitness depend solely on the number of squares visited, N_v, also caused monochromatic degeneracies. This occurred because ants were rewarded for being able to simulate a random walk by *ignoring* an overpowering scent arising from the *average* blended color — yet another instance of organisms exploiting a flaw in the "physics" used to model the optimization task.

Figure 5 shows two ant paintings from separate runs using ant fitness function $A_1(a) = N_f$. They reveal how rewarding ants solely on the basis of their

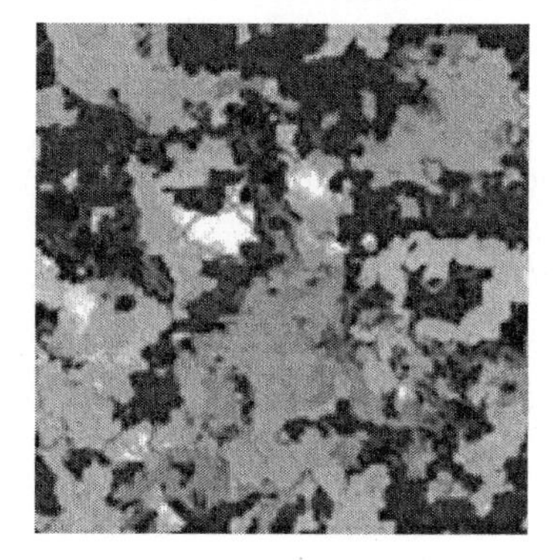

Fig. 5. Examples of the blotchy style using fitness function A_1 where fitness was determined solely by the ability to detect and follow scent. An image from generation #8 is on the left and generation #10 is on the right

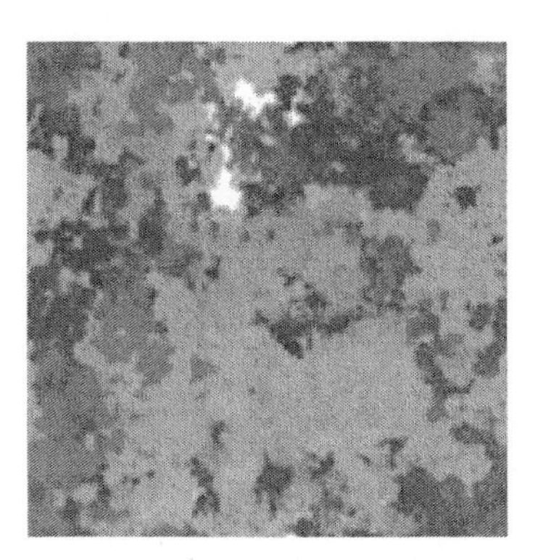
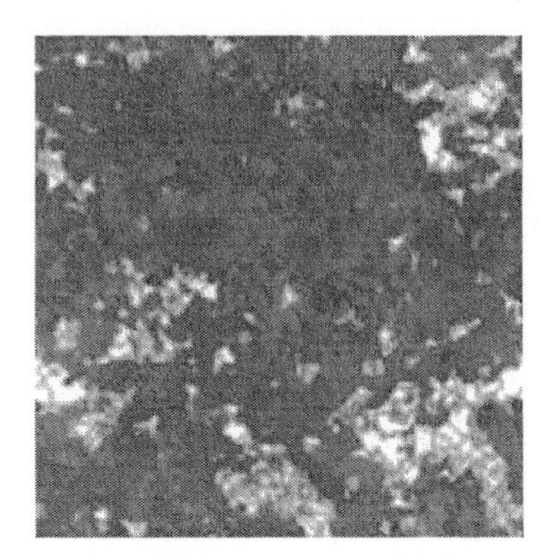

Fig. 6. Examples of a bi-level style using fitness function A_2 where fitness was determined solely by the ratio of the ability to follow scent and the ability to explore. Both images are from generation #16 of their respective runs

ability to follow scent leads to a blotchy style where paintings seem to be dominated by trails of polka-dots. "Convergence" usually occurred quickly, within ten generations.

Figure 6 shows two ant paintings from separate runs using the fitness function $A_2(a) = 100N_f/N_v$. The style that results is a bi-level style. Often, two castes of ants evolved each depositing a different color, but with both seeking essentially the same scent trail. It appears that the `MINR` threshold operating in tandem with the exploration penalty in the fitness function caused ants to evolve a plowing forward behavior so that ants in the two castes could mutually support each other. In some runs more complex behavior emerged due to deposit color "shades" evolving within the two different castes. Unfortunately, no examples of paintings in this style with good color aesthetics were ever evolved.

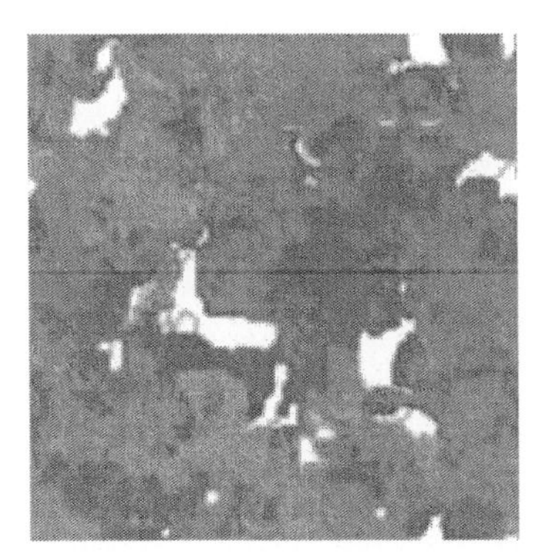

Fig. 7. Examples of a dramatic, organic style obtained using fitness function A_3, a linear combination of terms measuring the ability to follow scent and the ability to explore. Both images are from generation #20

By letting the fitness function be $A_3(a) = N_f + N_v$, we achieved our most dramatic style. The evolved paintings had an organic form and the stark color contrasts gave shading highlights. The key feature of the ant populations that made these paintings is the formation of one dominant caste that made up over half the population and provided the base color for the paintings. Interestingly enough, even though the evolved paintings reveal that significant portions of the background are never visited, the underlying probability vectors reveal that the ants possess more or less random exploration tendencies. This indicates that scent following is tightly coupled with the evolved behavior. Figure 7 gives examples of the evolved paintings we obtained.

Our most impressive ant paintings, from both a composition and color standpoint, were obtained by letting the fitness function be $A_4(a) = N_f \cdot N_v/1000$. As Figure 8 demonstrates, shading and detail received equal emphasis. Figure 8

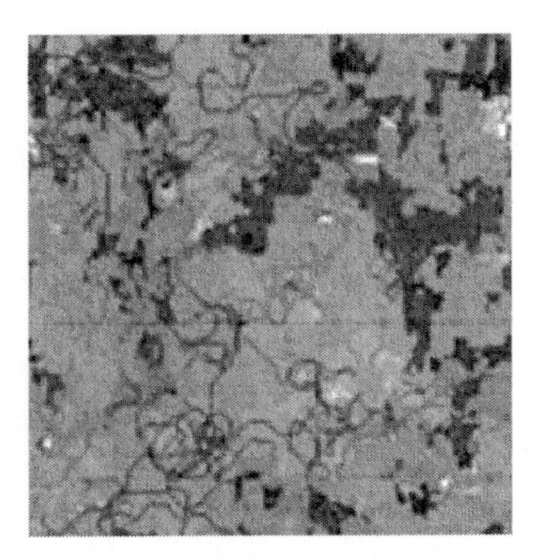
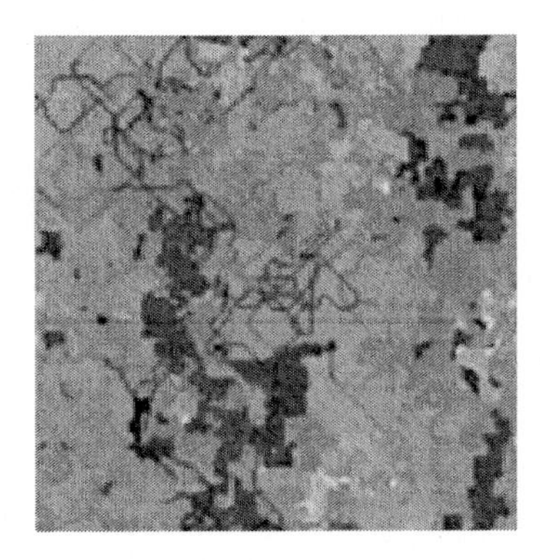

Fig. 8. Examples of the balanced style using fitness function A_4, a product of terms measuring the ability to follow scent and the ability to explore. Left image from generation #12 and right image from generation #14 of the same run

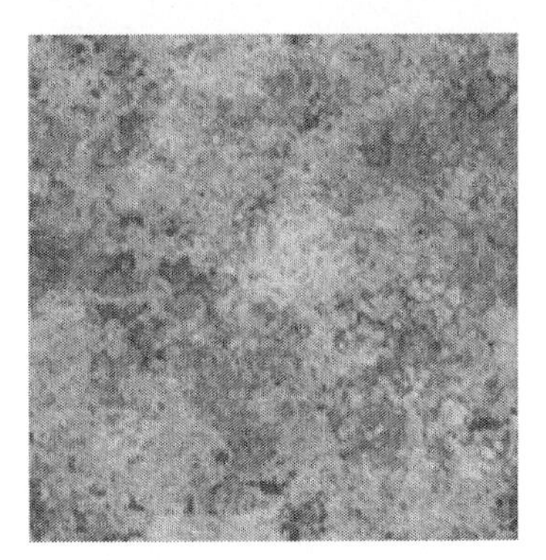

Fig. 9. Using a neutral background color the "degenerate" fitness function $A_5(a) = N_v$ evolved non-degenerate paintings. Both images are from generation #20

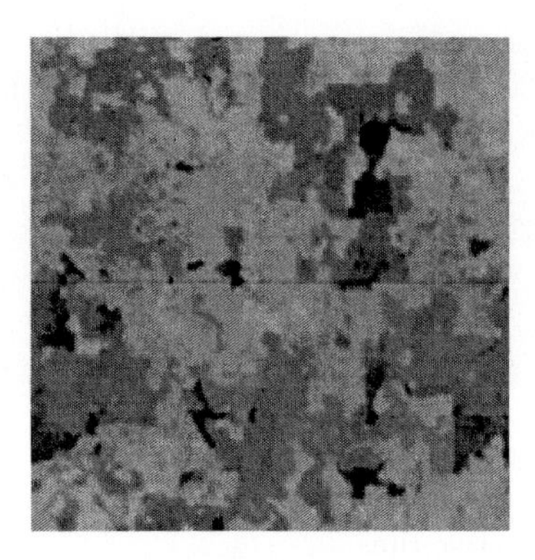
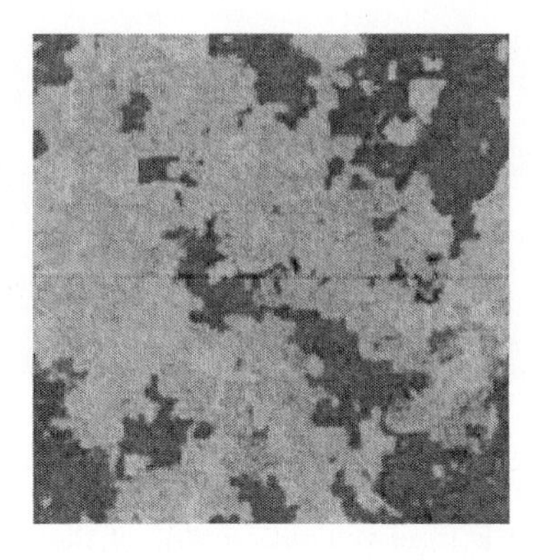

Fig. 10. The fitness function A_4 was the most robust with respect to shifts in background color. Image on the left using a black background and on the right using a neutral gray background. Both images are from generation #20

shows an example where one of three extant castes went extinct while passing from generation #12 to generation #14.

Attempting to change ant behavior in such a way that evolved paintings more closely mimic the style of the images found in Monmarché by raising the value of the repelling threshold `MINR` to 60, 80, 100, or even 120 did indeed reproduce their style for the first few generations, but as evolution progressed the paintings always degenerated to dark, monochromatic paintings.

The effect of using background colors other than white was difficult for us to assess. We remarked earlier that the fitness function $A_5(a) = N_v$ evolved monochrome paintings with either a black or white background. However, Figure 9 shows two examples that we obtained after 20 generations using the neutral gray RGB background color (128, 128, 128). Figure 10 shows why we believe our most consistent fitness function $A_4(a) = N_f \cdot N_v/1000$ was also the most robust in this regard by showing examples evolved using both a black and neutral gray background.

5 Adding Initial Conditions to the Model

All of the ant paintings described above used random positioning of ants at the start of each generation. This meant that it was equiprobable any cell on the grid would be visited and, given the exploration time, highly probable almost all cells would be visited. It seemed plausible that more organized paintings would result if ants explored the grid by always starting from fixed central locations. Therefore we experimented with placing ants at the same fixed locations within a central 20×20 "cluster" and pointing them in the same fixed directions from those locations at the start of each generation. We also allowed longer evolution

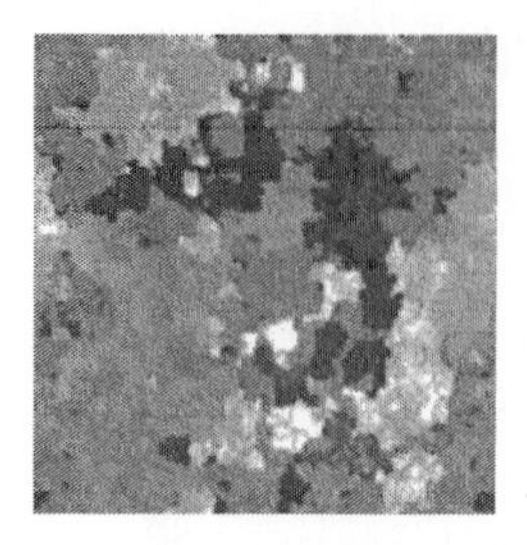  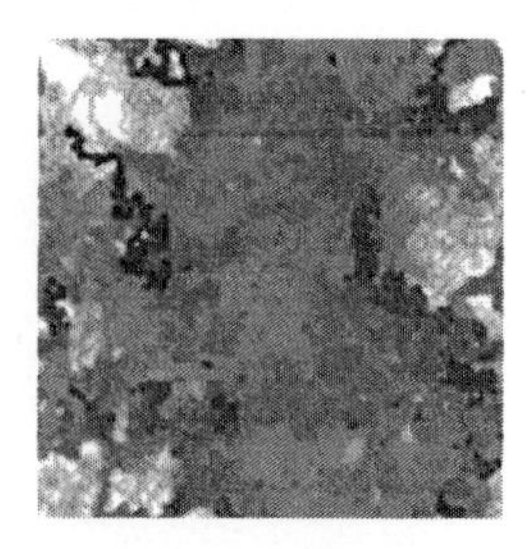

Fig. 11. Ant paintings evolved using fitness function A_4, initial clustering of ants, and longer evolution times. The two on the left are from generation #30 and the one on the right is from generation #35 of their respective runs

times and preserved paintings after every five generations. Figure 11 shows paintings representative of the results we obtained when using our preferred fitness function $A_4(a) = N_f \cdot N_v/1000$.

6 Summary and Conclusion

We have introduced a more carefully reasoned and more sophisticated model for evolving ant paintings while exploring the problem of automating their evolution. We have shown how different styles of ant paintings can be achieved by using different fitness criteria, and we have investigated the effects of varying the simulation parameters controlling evolution. While the resulting paintings are still aesthetically primitive, there is reason to hope that ants possessing additional sensory capabilities and given better guidance for color aesthetics could produce more complex and interesting paintings.

References

1. Monmarché, N., Aupetit, S., Bordeau, V., Slimane, M., Venturini, G.: Interactive evolution of ant paintings. 2003 Congress on Evolutionary Computation Proceedings (eds. B. McKay et al), IEEE Press, **2** (2003) 1376–1383.
2. Baluja, S., Pomerleau, D., Jochem, T.: Towards automated artificial evolution for computer-generated images. Connection Science **6** (1994) 325–354.
3. Dawkins, R.: The evolution of evolvability. Artificial Life (ed. C. Langton), Addison Wesley, Reading MA, (1989) 201–220.
4. Dorin, A.: Aesthetic fitness and artificial evolution for the selection of imagery from the mythical infinite library. Advances in Artificial Life — ECAL 2001 Proceedings (eds. J. Keleman and P. Sosik), Springer-Verlag, Berlin, LNAI 2159, (2001) 659–658.
5. Graf, J., Banzhaf, W.: Interactive evolution of images. Genetic Programming IV: Proceedings of the Fourth Annual Conference on Evolutionary Programming (eds. J. McDonnell et al), MIT Press, (1995) 53–65.

6. Greenfield, G.: Art and artificial life — a coevolutionary approach. Artificial Life VII Conference Proceedings (eds. M. Bedau et al), MIT Press, Cambridge MA, (2000) 529–536.
7. Greenfield, G.: Color dependent computational aesthetics for evolving expressions. Bridges 2002 Conference Proceedings (ed. R. Sarhangi), Central Plains Book Manufacturing, Winfield KS, (2002) 9–16.
8. Greenfield, G.: Evolving aesthetic images using multiobjective optimization. 2003 Congress on Evolutionary Computation Proceedings (eds. B. McKay et al), IEEE Press, **3** (2003) 1902–1909.
9. Ramachandran, V., Hirstein, W.: The science of art: a neurological theory of aesthetic experience. Journal of Consciousness Studies **6** (1999) 15–52.
10. Robertson, B.: Computer artist Michael Tolson. Computer Artist **2** (1993) 20–23.
11. Sims, K.: Artificial evolution for computer graphics. Computer Graphics **25** (1991) 319–328.
12. Sprott, J.: The computer artist and art critic. Fractal Horizons (ed. C. Pickover), St. Martin's Press (1996) 77–115.
13. Todd, S., Latham, W.: Evolutionary Art and Computers, Academic Press, San Diego CA (1992).
14. Zeki, S.: Inner Vision, An Exploration of Art and the Brain. Oxford University Press, New York NY (1999).

Evolutionary Search for Musical Parallelism

Søren Tjagvad Madsen[1,2] and Gerhard Widmer[1,2]

[1] Austrian Research Institute for Artificial Intelligence,
Freyung 6/6, A-1010 Vienna, Austria
[2] Department of Computational Perception, University of Linz, Austria
soren@oefai.at, gerhard.widmer@jku.at

Abstract. This paper presents an Evolutionary Algorithm used to search for similarities in a music score represented as a graph. We show how the graph can be searched for similarities of different kinds using interchangeable similarity measures based on *viewpoints*. A segmentation algorithm using the EA for automatically finding structures in a score based on a specific-to-general ordering of the viewpoints is proposed. As an example a fugue by J. S. Bach is analysed, revealing its extensive use of inner resemblance.

1 Musical Similarity

Repetition or parallelism is a fundamental feature of western tonal music. Ockelford cites musicologists arguing that music is a self-contained art form [1]. Music can not refer to the phenomenal world as well as other art forms are able to, so the mind's longing for reference can only be satisfied through repetition.

This paper addresses the search for parallelism in a symbolic representation of music. Looking at the music as symbols or events, parallelism can appear as repetition of phrases, but also as systematic changes in the quantitative information of the notes/events (pitch and duration), or through elaborations and simplifications of phrases (inserting/deleting notes). The artistic compositional effects can be varied infinitely.

Our approach is an attempt to see how far one can go by analysing western tonal music based on musical similarities solely calculated from information present in a score. We will demonstrate this with a fugue by J. S. Bach.

West et al. introduce the term *transformation* when a musical object (composed of a number of events/notes) can be inferred from another musical object in a perceptible way, and when it is possible to specify the transforming function [2]. This function determines the type of their relation and describes in what way they can be said to be different inflections of the same musical idea. Our strategy will be to carefully define a set of transformation functions, and then to use them in searches.

What types of transformation functions could we expect? The problem is often divided in two. The easiest task is describing similarities involving a one-to-one correspondence between the notes in the respective phrases – transformations that do not change the number of notes, but only their parameters. We

F. Rothlauf et al. (Eds.): EvoWorkshops 2005, LNCS 3449, pp. 488–497, 2005.

will call these simple transformations. This could for example be the transposition of a phrase. The real challenge is to find phrases that sound similar while allowing elaboration/simplification (inserting/deleting notes). To detect such a non-simple parallelism there is not a single strategy, since there are numerous ways of embellishing the music and consequently many transformation functions. Although the non-simple transformations are quite important in music, we will in this paper concentrate on the simpler case of defining (and finding) similarities between equal-numbered groups of notes. Simple transformation functions are quite common means of creating parallelisms, and can be thought of as systematic changes of the notes like transpositions (of various types) and inversions. A small set of these transformations are able to account for a great deal of similarities in western tonal music.

Describing musical similarity and proposing similarity measures is a field which has already been given much attention, due to its central position in automatic music analysis (see, e.g., [3, 4, 5], to name but a few). Segmentation based on similarities has also been studied by [6]. We have built our similarity measures upon the ideas by Conklin et al. [7, 8]. Their method of computing similarities are defined on sequences of notes solely. We have adopted this sequential approach to the problem, which turns out to work fine for the fugue analysed here.

2 Data Acquisition and Representation

To be able to analyse music in the best way, we decided to depend mainly on a reasonably detailed music source: the MuseData format (`www.musedata.org`). This source contains information about enharmonic pitch spelling. We can use MIDI as input too, but MIDI lacks the diatonic information (relating every note to a 7 step scale) that is crucial when searching for harmonically related phrases.

Musical events are put into a graph structure – the *music graph*. Each note/rest or 'event' in the score is represented as a vertex containing the pitch (spelling, octave, alterations), duration, start time in the score, key, and time signatures etc. The temporal relations of the notes/rests are present as directed edges of the graph. Vertices representing events following each other in time (a note/rest that starts immediately where another ends) are connected by an edge of type *follow*. Notes with the same start time are connected by *simultaneous* edges. The graph representation does not bias the representation of the music to be mainly 'vertical' (homophonic) or 'horizontal' (polyphonic), but any connected subset of the notes – a subgraph – can be considered as an entity of the music. The features of this graph representation is explained in detail in [9]. For the experiments explained here, *follow* edges only exist between vertices belonging to the same voice in the score. The MuseData representation and scores in general are by nature divided into such voices or parts.

For practical reasons, subgraphs can either be sequential (graphs of vertices connected by edges of type *follow* only) or non-sequential. The similarity measures presented in this paper are defined for sequences of notes and the music analysed is mainly polyphonic, so in this paper we are searching for similar sequential subgraphs. [9] presents a way to extend the similarity measures to the more general case of comparing any (equal sized) subgraphs. We will from now on refer to sequential subgraphs as simply subgraphs.

3 The Search Algorithm

The EA maintains a population of *Similarity Statements*. A similarity statement (SS) is a guess that two subgraphs of the same size (same number of vertices) are similar. The population is initialised with SSs each pointing at two random subgraphs of the 'mother' graph. By doing crossover, mutation, and selection in terms of altering the guesses it is possible to change the SSs to point at increasingly more similar subgraphs, dynamically adjusting their size and position.

A new generation is composed of three parts. A given percentage is chosen through tournament selection (tournament size of 2), another percentage is created by crossover of two selected individuals, and the remaining percentage is created through mutation of the selected or crossbred individuals. Finally, after a new population is created, the mutation rate is also used to determine a number of random mutations that are applied to the new generation. The most fit similarity statement in each generation always survives to the next (elitism). (The numbers used in the presented experiment were: selection: 0.45, crossover: 0.05, and mutation 0.5).

Mutation on a similarity statement with sequential subgraphs s_1 and s_2 can take several forms. The different mutation operations add and remove edges and vertices to/from the subgraphs. It should be noted that subgraphs are implemented as pointers to a subset of the vertices and edges in the 'mother' graph. Mutations on subgraphs do not change the graph structure itself. Common to all operators is that they must preserve the constraints that make s_1 and s_2 sequential (connected and only including edges of type 'follow').

- **Extension** Extend both s_1 and s_2 once, either at the 'beginning' of the subgraph (against the *follow* edge direction) or at the 'end' (along the follow edge direction), chosen at random and independently for each graph. The size of both subgraphs is increased by 1. (Applied with probability 0.3)
- **Shortening** Shorten both s_1 and s_2. As with extension, a subgraph can be shortened at either of the ends chosen at random and independently for each graph. The size is decreased by 1. (Probability 0.3)
- **Slide** Slide both subgraphs once along or against the follow edge direction (s_1 and s_2 do not need to slide the same way). The size of the subgraphs is not altered by a slide. (Probability 0.3)
- **Substitution** Substitute either s_1 or s_2 with a new and randomly generated subgraph of the same size. (Probability 0.1)

The crossover operation takes two SSs and combines them by picking one subgraph from each (chosen at random). The two parent statements most often have different sizes, so the smallest of them is extended at random until the sizes match. However, the chances that this will improve the fitness are low, because often the two subgraphs that are chosen are very different. As a result, the crossover parameter is often set low (in this experiment to 0.1).

Viewpoint	View
Absolute MIDI Pitch	[48,52,50,53,52,53,55,47,48]
MIDI Pitch Interval	[4,-2,3,-1,1,2,-8,1]
Pitch Class	[0,4,2,5,4,5,7,11,0]
Diatonic Absolute Pitch (A)	[C3,E3,D3,F3,E3,F3,G3,B2,C3]
Diatonic Pitch Interval (T)	[2,-1,2,-1,1,1,-5,1]
Diatonic Interval mod 7 (M)	[2,6,2,6,1,1,2,1]
Diatonic Inversion Interval	[-2,1,-2,1,-1,-1,5,-1]
Absolute Duration	$[\frac{1}{8},\frac{1}{8},\frac{1}{8},\frac{1}{8},\frac{1}{8},\frac{1}{8},\frac{1}{8},\frac{1}{8},1]$

Fig. 1. Example of the viewpoints

3.1 Evaluation

The fitness function evaluates the degree of similarity of two equal sized subgraphs according to an interchangeable similarity measure. The fitness function has to balance four conflicting goals:

- Optimise musical similarity
- Prefer larger matches to smaller ones
- Prefer phrases that conform to grouping structure rules
- Forbid overlap between subgraphs in any SS

The size of the graphs in a SS contribute in the fitness calculation to make the algorithm explore larger matches even if a perfect one is already found. A decreasing function of the size of the subgraphs is added to the fitness to make the EA prefer longer matches.

An evaluation of both subgraphs' agreement with *grouping structure* (a quick implementation of some simple rules suggested by [10]) is also included in the evaluation. The rules essentially propose a computable way of telling how well the two phrases individually correspond to phrase boundaries in the music. Even though only a few rules are implemented, the effect is noticeable. There is unfortunately no space here to go into details about this.

No overlap between two subgraphs in a similarity statement is permitted. We want phrases to be non-overlapping so we later can substitute the subgraphs in an unambiguous way. Overlap result in a bad evaluation.

We have used the notion of *viewpoints* presented in [7] when calculating the musical similarities between subgraphs. A viewpoint is a function taken on a sequence of musical objects, yielding a *view* of the sequence. The sequence is in our case a sequential subgraph, and the view is a list of values. Figure 1 shows a note sequence and examples of viewpoints and corresponding views. As shown, values from both the pitch and time domains can be included (other information available could also be used). Viewpoints can either be absolute – using values from each note – or relative – calculated from the relation between notes. The list is by no means complete.

To compare if two subgraphs of equal length are equal under a given viewpoint, we compare their views. This is simply done by counting the number of pairwise disagreeing values in the view vectors. When this *view difference* evaluates to zero, the sequences are equal under this view.

A motif and its exact repetition have the same view under the absolute pitch viewpoint A (the identity transformation, 'Absolute') and the pitch interval viewpoint T ('Transposition'). A motif and its transposed version will however have different views under A, but equal under T. A is said to be more specific than T. In the segmentation of the fugue presented in Sec. 5, we chose to use three viewpoints from the pitch domain – the ones denoted as A, T and M in Fig. 1. They can be ordered by pitch in this way:

$$DiatonicAbsPitch(A) \leq DiatonicInt(T) \leq DiatonicIntMod7(M) \quad (1)$$

If two patterns are found similar (regarding pitch) under a viewpoint, they will also be equal under a less specific viewpoint.

The fitness function combines evaluations of the above mentioned parameters and selected viewpoints into a single value. The global optimal solution is thus based on a mixture of melodic similarities and the size and boundaries of the subgraphs. In the measures used for the fugue segmentation we have adjusted the weighting of these parameters in such a way that pitch is the dominating factor.

4 The Segmentation Algorithm

We now perform multiple runs of the EA to discover and categorise different kinds of similarities inside a piece. Segmenting a performance consists in iteratively using the EA to find similar passages in the piece. When an EA terminates, the most fit SS is evaluated against a threshold, determining if the fitness is 'good enough' for the music to be claimed similar.

The EA is also used to search for more occurrences of a pattern – the occurrence search. It works by keeping one subgraph fixed in each SS (the original pattern) while doing crossover and mutations on the other, never changing the size of the graphs. The multiple EA searches are done with different similarity measures in a special order, controlled by the segmentation algorithm. The goal of the segmentation algorithm is to find patterns and derivations and categorise motifs with as specific a measure as possible.

When the segmentation algorithm finds a subgraph sufficiently similar to another, it removes it from the original graph and replaces it with a *compound vertex* (CV) now representing the pattern. This CV is labelled with the similarity measure it was evaluated with (its relation), as well as an identifier representing its 'type'. The edges to and from the subgraph that was substituted now connect the CV. A subsequent EA run on the updated graph now needs to take account of CVs in the graph and compare these on equal terms with other vertices. CVs can then be included in other subgraphs making further nesting possible.

A CV can only be similar to another CV if it has the same type (representing the same motif), and CVs should only be included when they were not found found with a more general measure than the current. This is necessary to pre-

```
SimilaritySegmenter(array of similarityMeasures, MusicGraph) {
  i=-1
  for(SimMeasure sm_i = 1..k){
      //search for new repeated patterns
      while(it is possible to find similar patterns with sm_i){
          let s_1 and s_2 be the most similar subgraphs of a EA run with sm_i
          if( SimMeasure(s_1,s_2) is within the similarity threshold){
              i++
              substitute s1 and s2 in the graph with CVs of type i and measure sm_i
              for(SimMeasure sm_j = i..k){
                  while(there are more occurrences of one of s_1 and s_2 to find with sm_j){
                      let s_n be the graph found in an Occurrence search for s_1 with sm_j
                      if( SimMeasure(s_1,s_n) within threshold)
                          substitute s_i in the graph with a CV of type i and measure sm_j
                      let s_n be the graph found in an Occurrence search for s_2 with sm_j
                      if( SimMeasure(s_2,s_n) within threshold)
                          substitute s_i in the graph with a CV of type i and measure sm_j
                  }
              }//advance to next SimMeasure in the occurrence search
          }//no more patterns to be found with this SimMeasure
      }//advance to next SimMeasure
} }
```

Fig. 2. The segmentation algorithm

serve the relation of the measures that the CVs were labelled with. Otherwise it would, e.g., be possible to allow transposed phrases (labelled T) to be part of a compound labelled with an absolute measure A, which is wrong since the phrases are not copies of each other.

The segmentation algorithm resembles the one proposed by [11] in the way all derivations of a recurring pattern are found before a new pattern is addressed. The idea is to iteratively find new patterns with as general a measure as possible (in successively weaker order) and next to find all derivations of it in successively weaker order. A new pattern is not considered before all derivations are believed to be found (when the search for more occurrences fails). The algorithm (shown in Fig. 2) takes as input the graph, and a list of (ordered) similarity measures. The algorithm iterates through a double for-loop. Both of the loops start EA searches with the similarity measures in the order given. The outer loop searches for new patterns and the inner searches for occurrences of these. The similarity measure sm_i advances one step (becomes more general) every time the while-condition is not fulfilled: when it is not possible to find a pattern with sm_i. In the inner loop sm_j advances when it is not possible to find more occurrences with sm_j. The segmentation terminates when the EA fails to find a 'good' repeated pattern with the most general similarity measure.

An alternative to the strict segmentation order would be to calculate all pitch related viewpoints in every evaluation and grade the fitness according to similarity in highest pitch specificity. That evaluation approach resembles the one presented in [12]. A more drastically but less controllable change would be to always evaluate all implemented viewpoints and see which one that 'clicks' and based on this information calculate a rating or a description of the similarities of the subgraphs (e.g. "s_1 is a rhythmically augmented diatonic inversion of s_2"). We plan to do experiments along these lines.

5 Experiments: Segmenting a Fugue

We will present a single segmentation by the algorithm: an analysis of J.S. Bach's Fugue in C minor, BWV 847 from WTK Book 1. This will illustrate how well the search algorithm does the job of selecting phrases and finding derivations.

The threshold was set so that only patterns having identical views regarding pitch (under the given viewpoint) were considered similar. Duration and grouping were given less importance. The EA was generously given 800 generations for finding new patterns and 500 generations for finding occurrences. The population size was 120 similarity statements. The segmentation was set to terminate when the best pattern found was of size 2 or it's fitness above the threshold. The segmentation algorithm found 19 patterns (some of them extending others) of sizes 3-26 notes/rests and made a total of 72 substitutions.

A graphical representation of the segmentation is shown in Fig. 3, giving an overview of the musical material found in the piece. The figure shows the entire three part fugue. Each part is represented as a row of boxes. Every note in the score is represented by a box. The width of the box shows the relative duration of the note. Rests are not printed but present as 'missing boxes'. Every compound found in the segmentation is labelled with a number and a letter indicating the measure with which it was found similar to another pattern. The beginning of a compound is also indicated by a black triangle. Every compound type is given a unique shade of grey (unfortunately not too clear in the figure). White boxes are thus notes which have not been found belonging to any pattern. Subpatterns in a nested compound are shown in the background. Bar numbers are printed above the vertical bar lines.

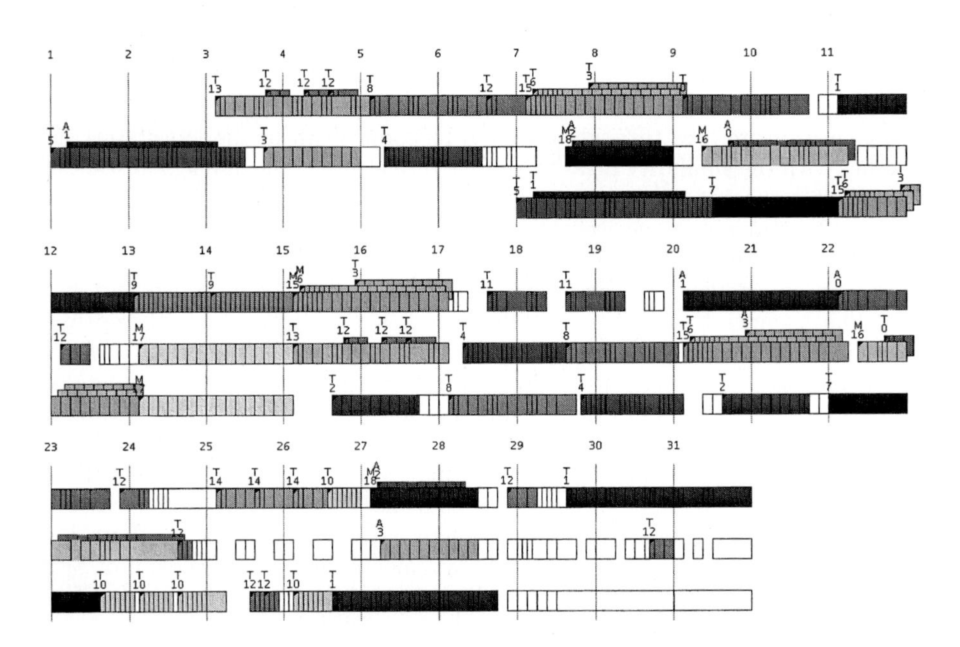

Fig. 3. A segmentation of J. S. Bach, Fugue in C minor

Fig. 4. Dux, iteration 1

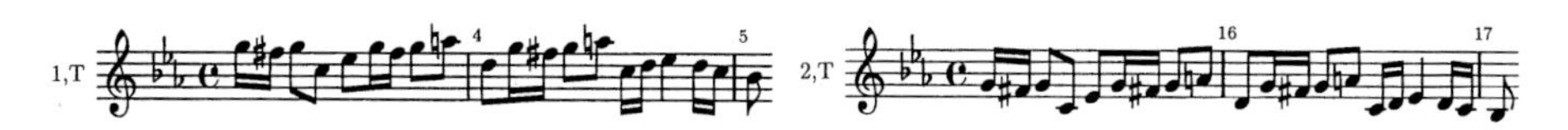

Fig. 5. Comes, iteration 13

Fig. 6. Counterpoint, iteration 15

The pattern found in the second iteration (iteration 1) is exactly the main theme, traditionally called *dux*. It has six occurrences in the fugue, and all were found (see Fig. 4 – the numbers and letters indexing the patterns refer to voice (1, 2 or 3) and similarity measure (A, T or M)). The only two occurrences that are exactly the same in pitch (A) were found first (beginning in bars 1 and 20) – the rest were found as transpositions (T). The ending dux (bar 29) differs only on the ending note, which has been transposed half a tone. The occurrences starting in bar 7 and 26 are octave transposition of the theme. The dux in bar 11 is a 'major' version of the theme, starting on *e♭* in stead of *c*.

A harmonic variant of the theme was found in iteration 13. It differs in one note (or two jumps) from the dux, and is traditionally called *comes*. It was found as an original pattern, since we did not allow a single note difference in the search. It has two occurrences which were correctly found as transpositions of each other – they occur in different octaves (Fig. 5).

One more important pattern in a fugue is the *counterpoint*. It is an accompanying figure to both dux and comes. It has six occurrences, but not all were found in their entirety. The pattern found in iteration 3 is a subpart of this figure and was found in all 6 places. In iteration 15, four counterpoints were found – extending the 3 pattern (Fig. 6). It could be argued that the note preceding this pattern musically belongs to the counterpoint figure. All occurrences are transpositions of each other, starting on *e♭*, *g*, *b* and *e♭* again. One was found with the measure M. This is because of the difference in jump between the 7th and the 8th note. In three of the cases this is an octave and a third, but in bar 15 it is only a third.

The simultaneous occurrences of dux/comes and counterpoint can be seen from Fig. 3. Also worth noticing is bars 17-19 where the patterns 8 and 4 occur simultaneously twice in succession in the two lowest voices – switching place. Furthermore the pattern labelled '0' occurs four times – each time accompanying itself. This happens in bars 9-11 and 22-24 (voices 1 and 2).

The presented segmentation is quite representative for the behavior of the algorithm. Running 10 segmentations with the given parameters, 7 of them found all occurrences of dux in the same iteration. Every time the comes was found. The counterpoint was in each segmentation recognised in 3-6 occurrences of varying extension.

6 Conclusion

The best we can do so far to evaluate a segmentation is to make the similarities visible/audible and compare to a manual analysis of the piece. The previous section showed that the algorithm did a fairly good job in finding the common motifs and derivations as well as categorising them correctly. It was not perfect, but taking into account that it depends on a nondeterministic algorithm which can been tuned, we can hope that even better results might be possible.

The overall segmentation certainly shows some structural dependencies in the fugue. A mechanical way of discovering when patterns occur simultaneously would be possible when allowing motifs to be non-sequential and thus spanning over notes from more parts. The data structure and search mechanism support this, but we will need some more effective and efficient similarity measures to be able to do this in practice.

The strength of the EA is it's ability to select the similar motifs in the music of any length. The search for occurrences might however have been more efficient with a deterministic algorithm in this simple sequential graph.

A crucial factor in the segmentation process seems to be the bounding of patterns. It is hard to evaluate the effect of the grouping structure rules. We did not give it much importance in this experiment. More focused experiments will be needed to study the relationship between musical grouping structure and meaningful musical patterns.

Also important is the choice of similarity measures. A larger set of transformation functions would be required to analyse other types of music. For example the MIDI pitch views might be more relevant when segmenting non-diatonic music. The rhythmical aspect should be explored. Fortunately it is easy to integrate new similarity measures into the evaluation function.

The idea of using the MuseData scores is to take advantage of the diatonic information. The different nature of the diatonic pitch viewpoints allows for some variation within the same the view. In our segmentation we only allowed phrases that had exactly equal views to be similar. A natural extension would be to allow for example one or more notes to differ while searching for 'new' themes. One could expect motifs to sound similar even when some notes are different (as for example the dux and comes). Allowing this introduces some uncertainties. We cannot be sure that what we find in every case also will be perceived as similar – especially when the note sequences are short. This approach would produce less 'correct' segmentations, but is an unavoidable next step that we have to examine.

Acknowledgements. This research was supported by the Austrian FWF (START Project Y99) and the Viennese Science and Technology Fund (WWTF, project CI010). The Austrian Resarch Institute for Artificial Intelligence acknowledges basic financial support from the Austrian Federal Ministries of Education, Science and Culture and of Transport, Innovation and Technology.

References

1. Ockelford, A.: 4: The Role of Repetitions in Percieved Musical Structures. In: Representing Musical Structure, Eds.: Peter Howell, Robert West and Ian Cross. Academic Press (1991) 129–160
2. West, R., Howell, P., Cross, I.: 1: Musical Structure and Knowledge Representation. In: Representing Musical Structure, Eds.: Peter Howell, Robert West and Ian Cross. Academic Press (1991) 1–30
3. Cambouropoulos, E., Widmer, G.: Automatic motivic analysis via melodic clustering. Journal of New Music Research **29(4)** (2000) 303–317
4. Grachten, M., Arcos, J.L., de Mantaras, R.L.: Melodic similarity: Looking for a good abstraction level. In: Proceedings of 5th International Conference on Music Information Retrieval (ISMIR'04), Barcelona, Spain (2004)
5. Rolland, P.Y.: Discovering patterns in musical sequences. Journal of New Music Research **28(4)** (1999) 334–350
6. Grilo, C., Cardoso, A.: Musical pattern extraction using genetic algorithms. In: Proceedings of the First International Symposium on Music Modelling and Retrieval (CMMR'03). (2003)
7. Conklin, D., Witten, I.: Multiple viewpoint systems for music prediction. Journal of New Music Research **24** (1995) 51–73
8. Conklin, D.: Representation and discovery of vertical patterns in music. In Anagnostopoulou, Ferrand, S., ed.: Proceedings of Second International Conference on Music and Artificial Intelligence (ICMAI'02), Edinburgh, Scotland, Springer (2002)
9. Madsen, S.T., Jørgensen, M.E.: Automatic discovery of parallelism and hierarchy in music. Master's thesis, University of Aarhus, Århus, Denmark (2003)
10. Lerdahl, F., Jackendoff, R.: A Generative Theory of Tonal Music. MIT Press (1983)
11. Smaill, A., Wiggins, G., Harris, M.: Hierarchical music representation for analysis and composition. Computers and the Humanities **27** (1993) 7–17
12. Lubiw, A., Tanur, L.: Pattern matching in polyphonic music as a weighted geometric translation problem. In: Proceedings of 5th International Conference on Music Information Retrieval (ISMIR'04), Barcelona, Spain (2004)

Developing Fitness Functions for Pleasant Music: Zipf's Law and Interactive Evolution Systems

Bill Manaris[1], Penousal Machado[2], Clayton McCauley[3], Juan Romero[4], and Dwight Krehbiel[5]

[1,3] Computer Science Department, College of Charleston, 66 George Street, Charleston, SC 29424, USA
{manaris, mccauley}@cs.cofc.edu
[2] Instituto Superior de Engenharia de Coimbra, Qta. da Nora, 3030 Coimbra, Portugal
machado@dei.uc.pt
[4] Creative Computer Group - RNASA Lab - Faculty of Computer Science, University of A Coruña, Coruña, Spain
jj@udc.es
[5] Psychology Department, Bethel College, North Newton KS, 67117, USA
krehbiel@bethelks.edu

Abstract. In domains such as music and visual art, where the quality of an individual often depends on subjective or hard to express concepts, the automating fitness assignment becomes a difficult problem. This paper discusses the application of Zipf's Law in evaluation of music pleasantness. Preliminary results indicate that a set of Zipf-based metrics can be effectively used to classify music according to pleasantness as reported by human subjects. These studies suggest that metrics based on Zipf's law may capture essential aspects of proportion in music as it relates to music aesthetics. We discuss the significance of these results for the automation of fitness assignment in evolutionary music systems.

1 Introduction

Interactive Evolution (IE) is one of the most popular approaches in current evolutionary music generation systems. In this paradigm the user assigns fitness to the generated pieces, guiding evolution according to his/hers aesthetic preferences. In the field of music, IE has been used for the evolution of rhythmic patterns, melodies, Jazz improvisations, composition systems, and many other applications (a comprehensive survey can be found in [1]).

In spite of its popularity, IE has several shortcomings that become particularly severe in time-based domains like music. Listening to all generated pieces is a tedious and demanding task; it leads to user fatigue and inconsistency in evaluation, and imposes severe limits on population size and number of generations. To overcome this shortcoming, some researchers (e.g. [2, 3, 4]) resort to Artificial Neural Networks (ANNs). The ANNs can be trained using a set of user-evaluated pieces created by an IE system [3]; scores of well-known musicians [2]; rhythmic boxes [4]; etc.

F. Rothlauf et al. (Eds.): EvoWorkshops 2005, LNCS 3449, pp. 498–507, 2005.

Although appealing, this approach has several shortcomings (see e.g. [2, 3]), most notably the difficulty of identifying a representative training set and, consequentially, of avoiding shortcuts – ways of creating false maximums.

Our research explores the connection between Zipf's law and music in the context of developing fitness functions for evolutionary music systems. We begin by performing an analysis of the music by extracting several Zipf-based measurements. These measurements serve as input for ANNs. We have successfully performed several validation experiments for author and style identification. In this paper, we describe a similar experiment, in the context of predicting music pleasantness.

The next sections discuss Zipf's law and its connection to music, and present results demonstrating how Zipf's law may be used to quantify music pleasantness. These results suggest that Zipf's law is a useful tool for developing fitness functions for evolutionary music.

1.1 Zipf's Law

Zipf's law reflects the scaling properties of many phenomena in human ecology, including natural language and music [5, 6]. Informally, it describes phenomena where certain types of events are quite frequent, whereas other types of events are rare. In English, for instance, short words (e.g., "a", "the") are quite frequent, whereas long words (e.g., "anthropomorphologically") are quite rare. In music, consonant harmonic intervals are more frequent, whereas dissonant harmonic intervals are quite rare, among other examples. In its most succinct form, Zipf's law is expressed in terms of the frequency of occurrence (quantity) of events, as follows:

$$F \sim r^{-a} \tag{1}$$

where F is the frequency of occurrence of an event within a phenomenon, r is its statistical rank (position in an ordered list), and a is close to 1.

Another formulation of Zipf's law is

$$P(f) \sim 1/f^{n} \tag{2}$$

where $P(f)$ denotes the probability of an event of rank f and n is close to 1. In physics, Zipf's law is a special case of a *power law*. When n is 1 (Zipf's ideal), the phenomenon is called $1/f$ or *pink noise*. Interestingly, when rendered as audio, $1/f$ (pink) noise is perceived by humans as balanced, whereas $1/f^{0}$ or *white noise* is perceived as too random, and $1/f^{2}$ or *brown noise* as too correlated [6].

In the case of music, we may study the frequency of occurrence of pitch events, duration events, melodic interval events, and so on. For instance, consider Chopin's "Revolutionary Etude." To determine if its melodic intervals follow Zipf's law, we count the different melodic intervals in the piece, e.g., 89 half steps up, 88 half steps down, 80 unisons, 61 whole steps up, and so on. Then we plot these counts against their statistical rank on log-log scale. This plot is known as *rank-frequency* distribution (see Fig. 1).

In general, the slope of the distribution may range from 0 to $-\infty$, with -1 denoting Zipf's ideal. This slope corresponds to the exponent n in (2). The R^2 value may range from 0 to 1, with 1 denoting a straight line. The straighter the line, the more reliable

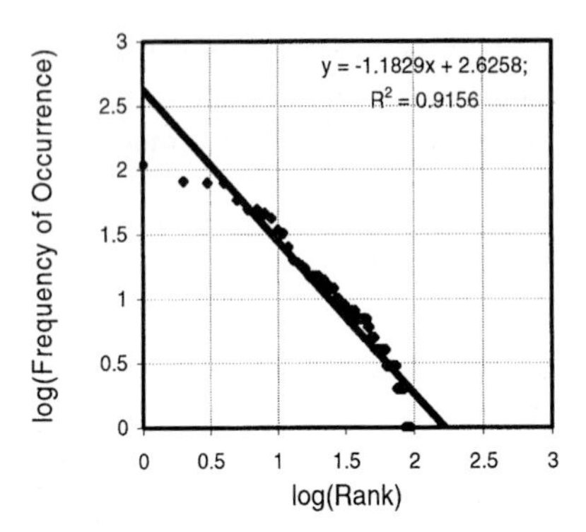

Fig. 1. The rank-frequency distribution of melodic intervals for Chopin's "Revolutionary Etude," Op. 10 No. 12 in C minor

the measurement. For example, melodic intervals in Chopin's "Revolutionary Etude" approximate a Zipfian distribution with slope of –1.1829 and R2 of 0.9156.

2 Experimental Studies

Earlier studies indicate that Zipfian distributions abound in socially-sanctioned music [7]. By *socially-sanctioned* we mean music that is sanctioned by a large enough musical subculture to be published/recorded, and thus survive over time; this is consistent with Zipf's use of the term (see [5], p. 329)

Currently, we have a set of 40 metrics based on Zipf's law [8]. We have used these metrics to extract features from MIDI-encoded music pieces. Specifically, these metrics count occurrences of various types of events and calculate the slope and R^2 value of the corresponding Zipf distribution. Table 1 shows a subset of these metrics.

The features extracted from these metrics (i.e., slope and R^2 values) have been used to train ANNs to classify these pieces in terms of composer, style, and pleasantness. To perform these classification studies, we compiled several corpora, whose size ranged across experiments from 12 to 758 music pieces [8]. These pieces are MIDI-encoded performances, the majority of which come from the Classical Music Archives [9]. We applied Zipf metrics to extract various features per music piece. The number of features per piece varied across experiments, ranging from 30 to 81.

These feature vectors were separated into two data sets. The first set was used for training the ANN. The second set was used to test the ANN's ability to classify new data. We experimented with various architectures and training procedures using the Stuttgart Neural Network Simulator [10].

In terms of author attribution, we conducted five experiments: *Bach vs. Beethoven*, *Chopin vs. Debussy*, *Bach vs. four other composers*, and *Scarlatti vs. Purcell vs. Bach vs. Chopin vs. Debussy* [11, 12]. The average success rate across the five author attribution experiments ranged from 95% to 100%.

Table 1. A sample of metrics based on Zipf's law [8]

Metric	Description
Pitch	Rank-frequency distribution of the 128 MIDI pitches
Chromatic tone	Rank-frequency distribution of the 12 chromatic tones
Duration	Rank-frequency distribution of note durations
Pitch duration	Rank-frequency distribution of pitch durations
Pitch distance	Rank-frequency distribution of length of time intervals between note (pitch) repetitions
Harmonic interval	Rank-frequency distribution of harmonic intervals within chord
Harmonic consonance	Rank-frequency distribution of harmonic intervals within chord based on music-theoretic consonance
Melodic interval	Rank-frequency distribution of melodic intervals within voice
Harmonic bigrams	Rank-frequency distribution of adjacent harmonic interval pairs
Melodic bigrams	Rank-frequency distribution of adjacent melodic interval pairs

We conducted several experiments for style identification tasks, using different ANN architectures and parameters. A detailed description and analysis of these results is awaiting publication. The average success rate across experiments, which required discerning between seven different styles, ranged from 91% to 95%.

These studies suggest that Zipf-based metrics may be used effectively for ANN classification, in terms of authorship attribution and style identification. These two tasks are relevant to evolutionary music composition, as it may contribute to fitness functions for composing music that is similar to a certain composer or music style. The next session presents ANN results related to music pleasantness.

3 Pleasantness Prediction

Much psychological evidence indicates that *pleasantness* and *activation* are the fundamental dimensions needed to describe human emotional responses [13]. Following established standards, we conducted an experiment in which we asked 21 subjects to classify music in terms of pleasantness and activation. The subjects were college students with varied musical backgrounds. The experiment was double blind, in that neither the subjects nor the people administering the experiment knew which of the pieces presented to the subjects were presumed as pleasant or unpleasant.

3.1 Data Collection Methodology

The subjects were presented with 12 MIDI-encoded musical performances. Our goal was to provide six pieces that an average person might find pleasant, and six pieces that an average person might find unpleasant. A member of our team with extensive music theory background helped identify 12 such pieces (see Table 2). From these pieces, we extracted excerpts up to two minutes long, in order to lessen fatigue for the human subjects and thus increase the consistency of the collected data.

Table 2. Twelve pieces used for music pleasantness classification study. Subjects rated the first six pieces as "pleasant", and the last six pieces as "unpleasant"

Composer	Piece	Duration
Beethoven	Sonata No. 20 in G. Opus 49. No. 2	(1:00)
Debussy	Arabesque No.1 in E (Deux Arabesques)	(1:34)
Mozart	Clarinet Concerto in A. K.622 (1. Allegro)	(1:30)
Schubert	Fantasia in C minor. Op.s 15	(1:58)
Tchaikovsky	Symphony 6 in B minor. Opus 36. Movement 2	(1:23)
Vivaldi	Double Violin Concerto in A minor. F.1. No. 177	(1:46)
Bartok	Suite. Op. 14	(1:09)
Berg	Wozzeck (trans. for piano)	(1:38)
Messiaen	Apparation de l'Eglise Eternelle	(1:19)
Schönberg	Pierrot Lunaire (5. Valse de Chopin)	(1:13)
Stravinksy	Rite of Spring. Movement 2 (tran. for piano)	(1:09)
Webern	Five Songs (1. Dies ist ein Lied)	(1:26)

While listening to the music, the subjects continuously repositioned the mouse in a 2D selection space to indicate their reaction to the music. The horizontal dimension represented *pleasantness* while the vertical dimension represented *activation* or arousal. The system recorded the subject's cursor coordinates once per second. Positions were recorded on 0 to 100 scales with the point (50,50) representing emotional indifference or neutral reaction.

Similar methods for continuous recording of emotional response to music have been used elsewhere [14].

3.2 ANN Training Methodology

For the ANN experiment, we divided each music excerpt into segments. All segments started at 0:00 and extended in increments of four seconds. That is, the first segment extended from 0:00 to 0:04 seconds, the second segment from 0:00 to 0:08 seconds, the third segment from 0:00 to 0:012 seconds, and so on. We applied Zipf metrics to extract 81 features per music increment. Each feature vector was associated with a target output vector (x, y), where x and y ranged between 0.0 and 1.0. Target vectors were constructed from the exact ratings (averaged over subjects) at each point in time in the piece. Target vector (1.0, 0.0) corresponded to most pleasant, (0.0, 1.0) corresponded to most unpleasant, and (0.5, 0.5) corresponded to neutral. This generated a total of 210 training vectors.

We conducted a 12-fold, "leave-one-out," cross-validation study. This allowed for 12 possible combinations of 11 pieces to be "learned" and 1 piece to be tested. The ANN had a feed-forward architecture with 81 elements in the input layer, 18 in the hidden layer, and 2 in the output layer. Internally, the ANN was divided into two 81x9x1 "Siamese-twin" pyramids both sharing the same input layer. One pyramid was trained to recognize pleasant music, the other unpleasant. Classification was based on the average of the two outputs.

Table 3. ANN results from all 12 experiments for the human-training, human-testing condition

Composer	Cycles	Test Rate	Test MSE	Train Rate	Train MSE
Beethoven	11200	100.00%	0.002622	100.00%	0.011721
Debussy	151000	100.00%	0.086451	100.00%	0.001807
Mozart	104000	100.00%	0.003358	100.00%	0.005799
Schubert	194000	100.00%	0.012216	100.00%	0.002552
Tchaikovsky	4600	100.00%	0.002888	100.00%	0.019551
Vivaldi	2600	100.00%	0.002026	94.05%	0.046553
Bartók	20200	100.00%	0.006760	100.00%	0.008813
Berg	4600	80.95%	0.100619	100.00%	0.015412
Messiaen	35200	100.00%	0.001315	100.00%	0.008392
Schönberg	4400	100.00%	0.013170	99.49%	0.024644
Stravinksy	10800	100.00%	0.000610	100.00%	0.015685
Webern	6400	100.00%	0.006402	100.00%	0.015366
Average	45750	98.41%	0.019870	99.46%	0.014691
Std	66150	0.0549	0.034775	0.0171	0.012118

During each training cycle the ANN was presented with every training vector once, in random order. Using back-propagation, the ANN weights were adjusted to reduce output mean standard error (*train MSE*). Every 200 cycles, the ANN was tested against the test data keeping track of the output mean standard error (*test MSE*). If the test MSE did not improve after a number of cycles, the ANN was considered stuck at a local minimum. Using a simulated annealing schedule, the ANN weights were "jogged" (adjusted by adding small amounts of random noise to the original weights). This forced the ANN to explore neighboring areas in the search space. The ANN weights were jogged with decreasing frequency as training progressed. The back-propagation part of the training focused on minimizing the train MSE, whereas the simulated-annealing part focused on minimizing the test MSE. By combining back-propagation with simulated annealing, we aimed at finding the best possible fit of the training data given the test data.

3.3 Experimental Results

The ANN performed extremely well with an average success rate of 98.41%. All pieces were classified with 100% accuracy, with one exception: Berg's piece was classified with 80.95% accuracy (see Table 3). The ANN was considered successful if it rated a music excerpt within one standard deviation of the average human rating; in other words it came within 68% of the range of human responses (i.e., 32% of the humans were outside of this range). There are two possibilities for the decrease in accuracy of the ANN with regard to Berg: Either our metrics fail to capture some essential aspects of Berg's piece, or the other 11 pieces do not contain sufficient information to enable the interpretation of Berg's piece.

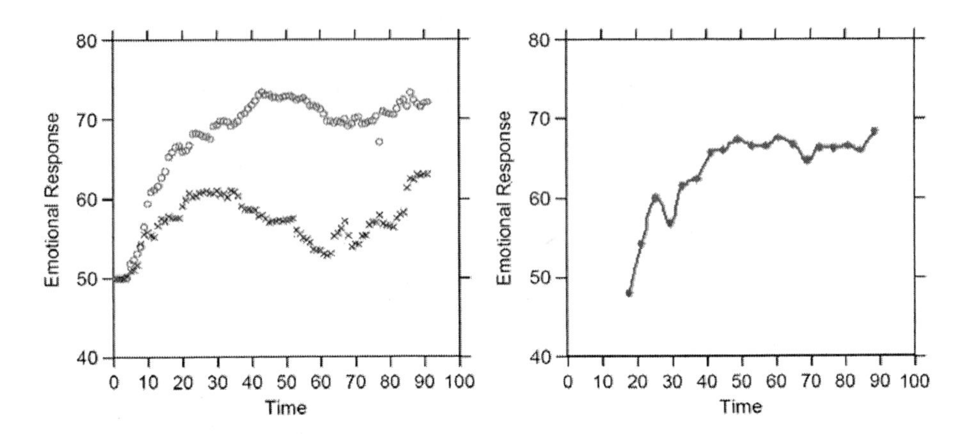

Fig. 2. The average pleasantness (o) and activation (x) ratings from 21 human subjects for the first 1:30 seconds of Mozart's "Clarinet Concerto in A" (K.622). A rating of 50 denotes neutral response

Fig. 3. Pleasantness classification by ANN of the same piece having been trained on the other 11 pieces

Fig. 2 displays the average human ratings for the excerpt from Mozart's "Clarinet Concerto in A" K.622. Fig. 3 shows the pleasantness ratings predicted by the ANN for the same piece. The ANN prediction approximates the average human response.

Additionally, we performed three control experiments to validate the results produced by the ANN. In specific, all values in the Human-training, Human-testing (HH) data were replaced by values generated using a uniform-distribution random number generator. These and the original values were then combined into three data sets for the control experiments: Random-training and Random-testing (RR), Random-training and Human-testing (RH), and Human-training and Random-testing (HR). Each of the control experiments was a complete 12-fold cross-validation study, just like the human data experiment.

Fig. 4 shows the Test MSE per piece across all four conditions (HH, RR, RH, and HR). The reader should recall that the first six pieces were pleasant and the last six unpleasant. Fig. 5 shows the average Test MSE across the four experiments.

3.4 Discussion

The ANN was able to discover strong correlations between the human pleasantness data and Zipf-based metrics (HH condition). Also, as expected, the ANN did not discover any correlations between random data and Zipf-based metrics (HR and RR conditions).

However, the ANN performed relatively well when trained against random data and tested against human data (RH condition). This may be surprising at first, however, it simply demonstrates the effect of *peeking at the test data* while training (see [15], p. 661) – as mentioned above, we used simulated annealing to "jog" the weights when the ANN appeared stuck in local minima relative to the test MSE. In other

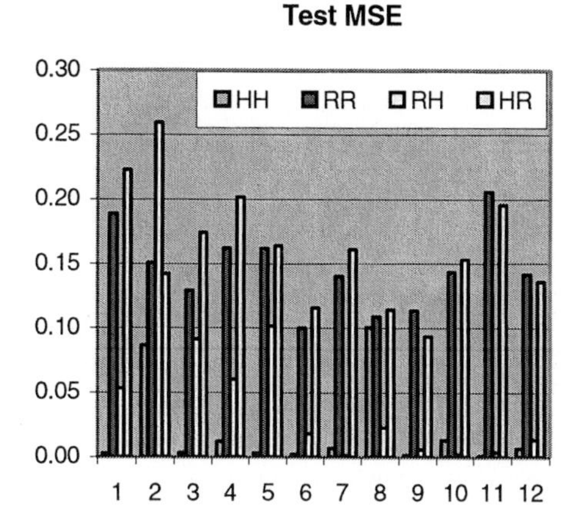

Fig. 4. ANN Test MSE for each piece across all conditions

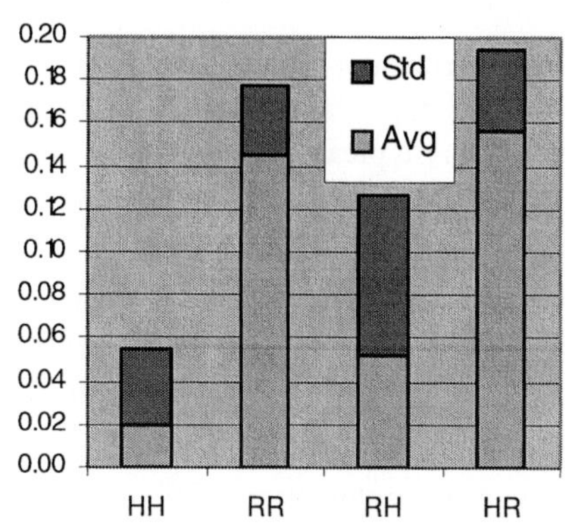

Fig. 5. Average Test MSE and standard deviation across all conditions

words, the ANN was trained to minimize both the test and the train MSEs. This indicates that the ANN is actually able to learn something about the human data, even though it was trained on random "noise.". While the ANN does succeed in classifying the data, its error rate is more than double than when it was trained with actual human data.

Reassuringly, this peeking effect produced no convergence in all 12 experiments of the RR condition (random training, random testing). This strongly suggests that there is a correlation between Zipf metrics and human pleasantness data, and no correlations with random data.

Analysis of the ANN weights associated with each metric suggests that *harmonic consonance* and *chromatic tone* were consistently relevant for "pleasantness" prediction, across all 12 experiments. Other relevant metrics include chromatic-tone distance, pitch duration, harmonic interval, harmonic and melodic interval, harmonic bigrams, and melodic bigrams.

The HH (and RH) results indicate that the ANN is able to identify patterns that are relevant to human reporting of pleasantness. The feature extractor and ANN evaluator used in this experiment can easily be incorporated into an evolutionary music system as part of fitness evaluation. Our results suggest that such a fitness function has strong potential to guide the evolutionary process towards music that sounds pleasant to humans. However, given the statistical nature of the metrics, we expect that additional structural, music-theoretic metrics may be required to discourage evolution from finding shortcuts – ways of creating false maxima. In other words, we suspect that ANN-based fitness functions, such as the one reported in the pleasantness study, at best, define a necessary but not sufficient (pre)condition for pleasant music. To evaluate this hypothesis, we are in the process of developing an evolutionary music system, called NevMusE, that will be used to generate music guided by such ANN-based "pleasantness" fitness functions.

4 Conclusions

The experimental results attained show that the considered set of metrics captures important music attributes, facilitating not only accurate prediction of author and style, but also pleasantness of musical pieces.

We propose that this approach may be applied successfully in the scope of a fully- or partially-automated system to assign fitness according to:

- compliance to a given musical style or styles;
- similarity to the works of some composer(s); and
- predicted pleasantness of the piece

There are several differences, and potential advantages over previous works dealing with the automation of fitness assignment. For instance, by using a set of well-known pieces instead of ones generated through IE, we ensure that the training set is unbiased towards the scores typically generated by the system. Also, the tasks of author and style identification do not involve subjective criteria. The output vector of the ANN can be seen as a set of distances to particular styles and authors, which opens new possibilities in terms of fitness assignment. Finally, the ANNs trained for predicting the pleasantness of pieces appear to capture fundamental principles of aesthetics. This contrasts with other approaches where the ANNs, when successful, capture only some of the preferences of an individual user.

Similarly to other approaches there is always the possibility of errors in classification and prediction. As such, using a totally automated system may result in convergence to false optimums. Taking into account the current state of development, we believe that it is probably wiser and more interesting to use a partially interactive system. The system would run on its own using the ANNs to assign fitness. However, the user can interfere at any point of the evolutionary run assigning fitness to the individuals, thus overriding the automatic evaluations.

We have already used this scheme in a partially interactive visual art evolutionary system [16]. The experimental results show that user intervention was enough to overcome the deficiencies of the fitness assignment scheme, which, in that case, where quite severe. Nevertheless, due to the generic properties of the extracted features, it is expected that, in the case of music, our approach results in more generic and robust fitness assignment.

Acknowledgements

This project has been partially supported by an internal grant from the College of Charleston and a donation from the Classical Music Archives. We thank Timothy Hirzel, Robert Davis and Walter Pharr for various comments and contributions. William Daugherty and Marisa Santos helped conduct the ANN experiments. Giovanni Garofalo helped collect human emotional response data for the ANN pleasantness experiment.

References

1. Burton, A. R., Vladimirova, T.: Applications of Genetic Techniques to Musical Composition. Computer Music Journal, Vol. 23, 4 (1999) 59-73
2. Spector, L., Alpern, A.: Induction an Recapitulation of Deep Musical Structure. IJCAI-95 Workshop on Artificial Intelligence and Music (1995) 41-48
3. Biles, J. A., Anderson, P. G., Loggi, L.W.: Neural Network Fitness Function for a Musical GA. International ICSC Symposium on Intelligent Industrial Automation (IIA'96) and Soft Computing (SOCO'96) (1996) B39-B44
4. Burton, A. R., Vladimirova T.: Genetic Algorithm Utilising Neural Network Fitness Evaluation for Musical Composition. 1997 International Conference on Artificial Neural Networks and Genetic Algorithms (1997) 220-224
5. Zipf, G.K.: Human Behavior and the Principle of Least Effort. Addison-Wesley Press, New York (1949)
6. Voss, R.F., and Clarke, J.: 1/f Noise in Music and Speech. Nature, Vol. 258 (1975) 317-318
7. Manaris, B., Vaughan, D., Wagner, C., Romero, J. and Davis, R.B.: Evolutionary Music and the Zipf–Mandelbrot Law – Progress towards Developing Fitness Functions for Pleasant Music. EvoMUSART2003 – 1st European Workshop on Evolutionary Music and Art, Essex, UK, LNCS 2611, Springer-Verlag (2003) 522-534
8. Manaris, B., Romero, J., Machado, P., Krehbiel, D., Hirzel, T., Pharr, W., and Davis, R.B.: Zipf's Law, Music Classification and Aesthetics. Computer Music Journal, Vol. 29, 1, MIT Press, Cambridge, MA (2005)
9. Classical Music Archives [online]: http://www.classicalarchives.com (2004).
10. Stuttgart Neural Network Simulator [online]: http://www-ra.informatik.uni-tuebingen.de/SNNS/ (2004)
11. Machado, P., Romero, J., Manaris, B., Santos, A., and Cardoso, A.: Power to the Critics - A Framework for the Development of Artificial Critics. Proceedings of 3rd Workshop on Creative Systems, 18th International Joint Conference on Artificial Intelligence (IJCAI 2003), Acapulco, Mexico (2003) 55-64
12. Machado, P., Romero, J., Santos, M.L., Cardoso, A., and Manaris, B.: Adaptive Critics for Evolutionary Artists. EvoMUSART2004 – 2nd European Workshop on Evolutionary Music and Art, Coimbra, Portugal, *Lecture Notes in Computer Science*, Applications of Evolutionary Computing, LNCS 3005, Springer-Verlag (2004) 437-446
13. Barrett, L.F., and J. A. Russell, J.A.: The Structure of Current Affect: Controversies and Emerging Consensus. Current Directions in Psychological Science, Vol. 8, 1 (1999) 10-14
14. Schubert, E.: Continuous Measurement of Self-report Emotional Response to Music. In Music and Emotion – Theory and Research, Juslin, P.N. and J.A. Sloboda (eds). Oxford University Press, Oxford, UK (2001) 393-414
15. Russell, S., and Norvig, P.: Artificial Intelligence – A Modern Approach, 2nd ed. Prentice Hall, Upper Saddle River, NJ (2003)
16. Machado, P., Cardoso, A.: All the Truth about NEvAr. Applied Intelligence, Vol. 16, 2 (2002) 101–119

Understanding Expressive Music Performance Using Genetic Algorithms

Rafael Ramirez and Amaury Hazan

Music Technology Group, Pompeu Fabra University,
Ocata 1, 08003 Barcelona, Spain
Tel:+34 935422165, Fax:+34 935422202
rafael@iua.upf.es, ahazan@iua.upf.es

Abstract. In this paper, we describe an approach to learning expressive performance rules from monophonic Jazz standards recordings by a skilled saxophonist. We use a melodic transcription system which extracts a set of acoustic features from the recordings producing a melodic representation of the expressive performance played by the musician. We apply genetic algorithms to this representation in order to induce rules of expressive music performance. The rules collected during different runs of our system are of musical interest and have a good prediction accuracy.

1 Introduction

Expressive performance is an important issue in music which has been studied from different perspectives (e.g. [8]). The main approaches to empirically study expressive performance have been based on statistical analysis (e.g. [26]), mathematical modelling (e.g. [27]), and analysis-by-synthesis (e.g. [7]). In all these approaches, it is a person who is responsible for devising a theory or mathematical model which captures different aspects of musical expressive performance. The theory or model is later tested on real performance data in order to determine its accuracy.

In this paper we describe an approach to investigate musical expressive performance based on evolutionary computation. Instead of manually modelling expressive performance and testing the model on real musical data, we let a computer use a genetic algorithm [14] to automatically discover regularities and performance principles from real performance data (i.e. Jazz standards example performances).

The rest of the paper is organized as follows: Section 2 describes how the acoustic features are extracted from the monophonic recordings. In Section 3 our approach for learning rules of expressive music performance is described. Section 4 reports on related work, and finally Section 5 presents some conclusions and indicates some areas of future research.

2 Melodic Description

In this section, we summarize how the melodic description is extracted from the monophonic recordings. This melodic description has already been used to

F. Rothlauf et al. (Eds.): EvoWorkshops 2005, LNCS 3449, pp. 508–516, 2005.

characterize monophonic recordings for expressive tempo transformations using CBR [12]. We refer to this paper for a more detailed explanation.

We compute descriptors related to two different temporal scopes: some of them related to an analysis frame, and some other features related to a note segment. All the descriptors are stored into a XML document. A detailed explanation about the description scheme can be found in [11].

The procedure for description computation is the following one. First, the audio signal is divided into analysis frames, and a set of low-level descriptors are computed for each analysis frame. Then, we perform a note segmentation using low-level descriptor values. Once the note boundaries are known, the note descriptors are computed from the low-level and the fundamental frequency values. We refer to [10, 12] for details about the algorithms.

2.1 Low-Level Descriptors Computation

The main low-level descriptors used to characterize expressive performance are instantaneous energy and fundamental frequency. Energy is computed on the spectral domain, using the values of the amplitude spectrum. For the estimation of the instantaneous fundamental frequency we use a harmonic matching model, the Two-Way Mismatch procedure (TWM) [18]. First of all, we perform a spectral analysis of a portion of sound, called analysis frame. Secondly, the prominent spectral peaks of the spectrum are detected from the spectrum magnitude. These spectral peaks of the spectrum are defined as the local maxima of the spectrum which magnitude is greater than a threshold. These spectral peaks are compared to a harmonic series and an TWM error is computed for each fundamental frequency candidates. The candidate with the minimum error is chosen to be the fundamental frequency estimate. After a first test of this implementation, some improvements to the original algorithm where implemented and reported in [10].

2.2 Note Segmentation

Note segmentation is performed using a set of frame descriptors, which are energy computation in different frequency bands and fundamental frequency. Energy onsets are first detected following a band-wise algorithm that uses some psycho-acoustical knowledge [17]. In a second step, fundamental frequency transitions are also detected. Finally, both results are merged to find the note boundaries.

2.3 Note Descriptor Computation

We compute note descriptors using the note boundaries and the low-level descriptors values. The low-level descriptors associated to a note segment are computed by averaging the frame values within this note segment. Pitch histograms have been used to compute the pitch note and the fundamental frequency that represents each note segment, as found in [19]. This is done to avoid taking into account mistaken frames in the fundamental frequency mean computation.

2.4 Implementation

All the algorithms for melodic description have been implemented within the CLAM framework [1]. They have been integrated within a tool for melodic description, *Melodia*. This tool is available under GPL license. of the melodic description tool.

3 Learning Expressive Performance Rules in Jazz

In this section, we describe our inductive approach for learning expressive performance rules from Jazz standard performances by a skilled saxophone player. Our aim is to find note-level rules which predict, for a significant number of cases, how a particular note in a particular context should be played (e.g. longer than its nominal duration). We are aware of the fact that not all the expressive transformations regarding tempo (or any other aspect) performed by a musician can be predicted at a local note level. Musicians perform music considering a number of abstract structures (e.g. musical phrases) which makes of expressive performance a multi-level phenomenon. In this context, our ultimate aim is to obtain an integrated model of expressive performance which combines note-level rules with structure-level rules. Thus, the work presented in this paper may be seen as a starting point towards this ultimate aim.

The training data used in our experimental investigations are monophonic recordings of four Jazz standards (*Body and Soul, Once I loved, Like Someone in Love* and *Up Jumped Spring*) performed by a professional musician at 5 different tempos around the nominal one (i.e. the nominal, 2 slightly faster and 2 slightly slower).

In this paper, we are concerned with note-level expressive transformations, in particular transformations of note duration, onset and energy. The note-level performance classes which interest us are *lengthen, shorten, advance, delay, louder* and *softer.* A note is considered to belong to class *lengthen* if its performed duration is 20% or more longer that its nominal duration, e.g. its duration according to the score. Class *shorten* is defined analogously. A note is considered to be in class *advance* if its performed onset is 5% of a bar earlier (or more) than its nominal onset. Class *delay* is defined analogously. A note is considered to be in class *louder* if it is played louder than its predecesor and louder then the average level of the piece. Class *softer* is defined analogously.

Each note in the training data is annotated with its corresponding class and a number of attributes representing both properties of the note itself and some aspects of the local context in which the note appears. Information about intrinsic properties of the note includes the note duration and the note's metrical position, while information about its context includes the duration of previous and following notes, and extension and direction of the intervals between the note and both the previous and the subsequent note.

[1] http://www.iua.upf.es/mtg/clam

Using this data, we applied a genetic algorithm to automatically discover regularities and music performance principles. A genetic algorithm can be seen as a general optimization method that searches a large space of candidate hypothesis seeking one that performs best according to a fitness function. The genetic algorithm we used for this investigation is the standard algorithm (reported in [6]) with parameters r, m and p respectively determining the fraction of the parent population replaced by crossover, the mutation rate, and population size. We set these parameters as follws: $r = 0.8$, $m = 0.05$ and $p = 200$. During the evolution of the population, we collected the rules with best the fitness for the classes of interest (i.e. shorten, same and lengthen). It is worth mentioning that although the test was running over 40 generations, the fittest rules were obtained around the 20th generation.

Hypothesis Representation. The hypothesis space of rule preconditions consists of a conjunction of a fixed set of attributes. Each rule is represented as a bit-string as follows: the previous and next note duration are represented each by five bits (i.e. much shorter, shorter, same, longer and much longer), previous and next note pitch are represented each by five bits (i.e. much lower, lower, same, higher and much higher), metrical strength by five beats (i.e. very weak, weak, medium, strong and very strong), and tempo by three bits (i.e. slow, nominal and fast). For example in our representation the rule

"if the previous note duration is much longer and its pitch is the same and it is in a very strong metrical position then lengthen the duration of the current note"

is coded as the binary string

00001 11111 00100 11111 00001 111 001

The exact meaning of the adjectives which the particular bits represent are as follows: previous and next note durations are considered much shorter if the duration is less than half of the current note, shorter if it is shorter than the current note but longer than its half, and same if the duration is the same as the current note. Much longer and longer are defined analogously. Previous and next note pitches are considered much lower if the pitch is lower by a minor third or more, lower if the pitch is within a minor third, and same if it has same pitch. Higher and much higher are defined analogously. The note's metrical position is very strong, strong, medium, weak, and very weak if it is on the first beat of the bar, on the third beat of the bar, on the second or fourth beat, offbeat, and in none of the previous, respectively. The piece was played at slow, nominal, and fast tempos if it was performed at a speed slower of more than 15% of the nominal tempo (i.e. the tempo identified as the most natural by the performer), within 15% of the nominal tempo, and faster than 15% of the nominal tempo, respectively.

Genetic operators. We use the standard single-point crossover and mutation operators with two restrictions. In order to perform a crossover operation of two parents the crossover points are chosen at random as long as they are on the attributes substring boundaries. Similarly the mutation points are chosen randomly as long as they do not generate inconsistent rule strings, e.g. only one class can be predicted so exactly one 1 can appear in the last three bit substring.

Fitness function. The fitness of each hypothesized rule is based on its classification accurracy over the training data. In particular, the function used to measure fitness is

$$tp^{1.15}/(tp + fp)$$

where tp is the number of true positives and fp is the number of false positives.

Despite the relatively small amount of training data some of the rules generated by the learning algorithms have proved to be of musical interest and correspond to intuitive musical knowledge. In order to illustrate the types of rules found let us consider some examples of duration rules:

RULE1: 01000 11100 01111 01110 00111 111 010

"If the previous note is slightly shorter and not much lower in pitch, and the next note is not longer and has a similar pitch (within a minor third), and the current note is not on a weak metrical position, then the duration of the current note remains the same (i.e. no lengthening or shortening)."

RULE2: 11111 01110 11110 00110 00011 010 001

"In nominal tempo, if the duration of the next note is similar and the note is in a strong metrical position then lengthen the current note."

RULE3: 00111 00111 00011 01101 10101 111 100

"If the previous and next notes durations are longer (or equal) than the duration of the current note and the pitch of the previous note is higher then shorten the current note."

These simple rules turn out to be very accurate: the first rule predicts 90%, the second rule predicts 92% and the third rule predicts 100% of the relevant cases. The rules were collected during 10 independent runs of the genetic algorithm. The mean accuracy of the 10 best rules collected (one for each run of the algorithm) for "shorten", "same" and "lengthen" was 81%, 99% and 64%, respectively. We implemented our system using the evolutionary computation framework GAlib [9].

4 Related Work

4.1 Evolutionary Computation

Evolutionary computation has been considered with growing interest in musical applications. Since [15], it has often been used in a compositional perspective, either to generate melodies ([4]) or rhythms ([28]). In [22] the harmonization subtask of composition is addressed, and a comparison between a rule-based system and a genetic algorithm is presented.

Evolutionary computation has also been considered for improvisation applications such as [3], where a genetic algorithm-based model of a novice Jazz musician learning to improvise was developed. The system evolves a set of melodic ideas that are mapped into notes considering the chord progression being played. The fitness function can be altered by the feedback of the human playing with the system.

Nevertheless, few works focusing on the use of evolutionary computation for expressive performance analysis have been done. The issue of annotating correctly a human Jazz performance regarding the score is addressed in [13], where the weights of the edit distance operations are optimized with genetic algorithm techniques.

4.2 Other Machine Learning Techniques

Previous research in learning sets of rules in a musical context has included a broad spectrum of music domains. The most related work to the research presented in this paper is the work by Widmer [29, 30]. Widmer has focused on the task of discovering general rules of expressive classical piano performance from real performance data via inductive machine learning. The performance data used for the study are MIDI recordings of 13 piano sonatas by W.A. Mozart performed by a skilled pianist. In addition to these data, the music score was also coded. The resulting substantial data consists of information about the nominal note onsets, duration, metrical information and annotations. When trained on the data, the inductive rule learning algorithm named PLCG [31] discovered a small set of 17 quite simple classification rules [29] that predict a large number of the note-level choices of the pianist.In the recordings the tempo of a performed piece is not constant (as it is in our case). In fact, of special interest to them are the tempo transformations throughout a musical piece.

Other inductive machine learning approaches to rule learning in music and musical analysis include [5], [2], [21] and [16]. In [5], Dovey analyzes piano performances of Rachmaniloff pieces using inductive logic programming and extracts rules underlying them. In [2], Van Baelen extended Dovey's work and attempted to discover regularities that could be used to generate MIDI information derived from the musical analysis of the piece. In [21], Morales reports research on learning counterpoint rules. The goal of the reported system is to obtain standard counterpoint rules from examples of counterpoint music pieces and basic musical knowledge from traditional music. In [16], Igarashi et al. describe the analysis of respiration during musical performance by inductive logic programming. Using a respiration sensor, respiration during cello performance was measured and rules were extracted from the data together with musical/performance knowledge such as harmonic progression and bowing direction.

5 Conclusion

This paper describes an evolutionary computation approach for learning expressive performance rules from Jazz standards recordings by a skilled saxophone player. Our objective has been to find note-level rules which predict, for a significant number of cases, how a particular note in a particular context should be played (e.g. longer or shorter than its nominal duration). In order to induce expressive performance rules, we have extracted a set of acoustic features from the recordings resulting in a symbolic representation of the performed pieces and then applied a genetic algorithm to the symbolic data and information about the context in which the data appear.

Future Work: This paper presents work in progress so there is future work in different directions. We plan to increase the amount of training data as well as experiment with different information encoded in it. Increasing the training data, extending the information in it and combining it with background musical knowledge will certainly generate a more complete set of rules. Another short-term research objective is to compare expressive performance rules induced from recordings at substantially different tempos. This would give us an indication of how the musician note-level choices vary according to the tempo. We also intend to incorporate structure-level information to obtain an integrated model of expressive performance which combines note-level rules with structure-level rules. A more ambitious goal of this research is to be able not only to obtain interpretable rules about expressive transformations in musical performances, but also to generate expressive performances. With this aim we intend to use genetic programming to evolve an initial population of rule trees and interpret these trees as regression trees.

Acknowledgments: This work is supported by the Spanish TIC project ProMusic (TIC 2003-07776-C02-01). We would like to thank Emilia Gomez, Esteban Maestre and Maarten Grachten for processing the data, as well as the anonymous reviewers for their insightful comments and pointers to related work.

References

1. Agrawal, R.T. (1993). Mining association rules between sets of items in large databases. International Conference on Management of Data, ACM, 207,216.
2. Van Baelen, E. and De Raedt, L. (1996). Analysis and Prediction of Piano Performances Using Inductive Logic Programming. International Conference in Inductive Logic Programming, 55-71.
3. Biles, J. A. (1994). GenJam: A genetic algorithm for generating Jazz solos. In ICMC Proceedings 1994.
4. Dahlstedt, P., and Nordahl, M. Living Melodies: Coevolution of Sonic Communication First Iteration Conference on Generative Processes in the Electronic Arts, Melbourne, Australia, December 1-3 1999.

5. Dovey, M.J. (1995). Analysis of Rachmaninoff's Piano Performances Using Inductive Logic Programming. European Conference on Machine Learning, Springer-Verlag.
6. De Jong, K.A. et al. (1993). Using Genetic Algorithms for Concept Learning. Machine Learning, 13, 161-188.
7. Friberg, A. (1995). A Quantitative Rule System for Musical Performance. PhD Thesis, KTH, Sweden.
8. Gabrielsson, A. (1999). The performance of Music. In D.Deutsch (Ed.), The Psychology of Music (2nd ed.) Academic Press.
9. The GAlib system. lancet.mit.edu/ga
10. Gómez, E. (2002). Melodic Description of Audio Signals for Music Content Processing. Doctoral Pre-Thesis Work, UPF, Barcelona.
11. Gómez, E., Gouyon, F., Herrera, P. and Amatriain, X. (2003). Using and enhancing the current MPEG-7 standard for a music content processing tool, Proceedings of the 114th Audio Engineering Society Convention.
12. Gómez, E. Grachten, M. Amatriain, X. Arcos, J. (2003). Melodic characterization of monophonic recordings for expressive tempo transformations. Stockholm Music Acoustics Conference.
13. Grachten, M., Luis Arcos, J., and Lopez de Mantaras, R. (2004). Evolutionary Optimization of Music Performance Annotation.
14. Holland, J.H. (1975). Adaptation in Natural and Srtificial Systems. University of Michigan Press.
15. Horner, A., and Goldberg, 1991. Genetic Algorithms and Computer-Assisted Music Composition, in proceedings of the 1991 International Computer Music Conference, pp. 479-482.
16. Igarashi, S., Ozaki, T. and Furukawa, K. (2002). Respiration Reflecting Musical Expression: Analysis of Respiration during Musical Performance by Inductive Logic Programming. Proceedings of Second International Conference on Music and Artificial Intelligence, Springer-Verlag.
17. Klapuri, A. (1999). Sound Onset Detection by Applying Psychoacoustic Knowledge, Proceedings of the IEEE International Conference on Acoustics, Speech and Signal Processing, ICASSP.
18. Maher, R.C. and Beauchamp, J.W. (1994). Fundamental frequency estimation of musical signals using a two-way mismatch procedure, Journal of the Acoustic Society of America, vol. 95 pp. 2254-2263.
19. McNab, R.J., Smith Ll. A. and Witten I.H., (1996). Signal Processing for Melody Transcription,SIG working paper, vol. 95-22.
20. Mitchell, T.M. (1997). Machine Learning. McGraw-Hill.
21. Morales, E. (1997). PAL: A Pattern-Based First-Order Inductive System. Machine Learning, 26, 227-252.
22. Phon-Amnuaisuk, S., and A. Wiggins, G. (1999?) The Four-Part Harmonisation Problem: A comparison between Genetic Algorithms and a Rule-Based System.
23. Quinlan, J.R. (1993). C4.5: Programs for Machine Learning, San Francisco, Morgan Kaufmann.
24. Ramirez, R. Hazan, A. Gómez, E. Maestre, E. (2004). Understanding Expressive Transformations in Saxophone Jazz Performances Using Inductive Machine Learning. Sound and Music Computing '04, IRCAM, Paris.
25. Ramirez, R. Hazan, A. Gómez, E. Maestre, E. (2004). A Machine Learning Approach to Expressive Performance in Jazz Standards MDM/KDD'04, Seattle, WA, USA.

26. Repp, B.H. (1992). Diversity and Commonality in Music Performance: an Analysis of Timing Microstructure in Schumann's 'Traumerei'. Journal of the Acoustical Society of America 104.
27. Todd, N. (1992). The Dynamics of Dynamics: a Model of Musical Expression. Journal of the Acoustical Society of America 91.
28. Tokui, N., and Iba, H. (2000). Music Composition with Interactive Evolutionary Computation.
29. Widmer, G. (2002). Machine Discoveries: A Few Simple, Robust Local Expression Principles. Journal of New Music Research 31(1), 37-50.
30. Widmer, G. (2002). In Search of the Horowitz Factor: Interim Report on a Musical Discovery Project. Invited paper. In Proceedings of the 5th International Conference on Discovery Science (DS'02), Lbeck, Germany. Berlin: Springer-Verlag.
31. Widmer, G. (2001). Discovering Strong Principles of Expressive Music Performance with the PLCG Rule Learning Strategy. Proceedings of the 12th European Conference on Machine Learning (ECML'01), Freiburg, Germany. Berlin: Springer Verlag.

Toward User-Directed Evolution of Sound Synthesis Parameters

James McDermott, Niall J.L. Griffith, and Michael O'Neill

University of Limerick, Ireland
jamesmichaelmcdermott@gmail.com, Niall.Griffith@ul.ie,
Michael.O'Neill@ul.ie

Abstract. Experiments are described which use genetic algorithms operating on the parameter settings of an FM synthesizer, with the aim of mimicking known synthesized sounds. The work is considered as a precursor to the development of synthesis plug-ins using evolution directed by a user. Attention is focussed on the fitness functions used to drive the evolution: the main result is that a composite fitness function – based on a combination of perceptual measures, spectral analysis, and low-level sample-by-sample comparison – drives more successful evolution than fitness functions which use only one of these types of criterion.

1 Introduction

1.1 Motivation

A first-time user of synthesis software is typically overwhelmed with options. Synthesizer plug-ins with 30, 40, or more parameters are not uncommon, and the problem is often compounded by their non-linear interactions. Even an experienced user, while composing with a complex synthesizer, might sometimes prefer to pursue a desired sound through an intuitive process with immediate feedback rather than switching into analytical, "parameter-setting" mode. This is partly because the acoustic and psychoacoustic effects of moving within a synthesizer's parameter space are not well-understood.

EAs are often thought of as good methods for searching poorly-understood or oddly-shaped search spaces. Recalling that an EA can be driven by a user (as in, for example, [13]), rather than by a computer-calculated fitness function, we can say that EAs have the potential to control synthesis parameters "on behalf of" a user.

1.2 Previous Work

Several authors (see [11] for an overview) have applied the techniques of Evolutionary Algorithms (EAs) to musical problems, including composition, musicology, and of most relevance here, synthesis.

Johnson [9] created a stand-alone graphical interface to the CSound FOF synthesis algorithm, which allows the user to direct the evolution of a population

F. Rothlauf et al. (Eds.): EvoWorkshops 2005, LNCS 3449, pp. 517–526, 2005.

of 9 sounds. He reports good results, including that the system allows easy and intuitive exploration of the FOF algorithm's possibilities.

Horner, Beauchamp and Haken [5] used Genetic Algorithms (GAs) in attempting to emulate the spectra of real instruments using FM synthesis. The GA was used to determine the best carrier-to-modulator frequency ratios and (time-invariant) modulation indices. They achieved good results, especially when using several carriers.

Horner, Cheung and Beauchamp [6] used GAs to find additive synthesis envelope breakpoints, and found that a GA performed better than a greedy algorithm, a "sensible" equal-spacing algorithm, and a simple random search algorithm in all cases.

Takala et al. [13] used LISP-like tree expressions ("Timbre Trees") to generate sounds. Evolving the sounds was then an application of Genetic Programming (GP), where evolution is directed not by an explicit fitness function but by user choices, made with reference to accompanying animations.

Blackwell and Young [1] used a particle swarm algorithm to control granular synthesis: their system is also capable of "interpretation", so that live musicians can improvise with the system.

Miranda [12] used cellular automata to control granular synthesis parameters in the Chaosynth program.

1.3 Method

In this paper we use unconstrained synthesizer representations which are closer to those seen by the typical end-user: filters, envelopes, amplitudes, etc, are controlled by continuously-variable parameters, in contrast to the free-form tree representation used by Takala et al. Thus, our approach is more general than [5] or [6], but more applicable to real synthesizer design than [13]. [1] and [12] are not directly comparable, since they do not use GA's; and finally, [9] is a user-directed system, which therefore does not use automatic fitness functions as developed here.

We see the development of synthesis software controlled by a user-directed EA as a three-step process. The first step, to be reported here, is simply to mimic known sounds using EA techniques. Several fitness functions, which measure the success of candidate sounds with respect to the target, are implemented and compared. The functions can be considered as moving from low-level, detail-oriented ones towards higher-level and perceptually-oriented ones, so leading towards our second step: attempting to generate sounds with a purely human fitness function (i.e. under user control). Here, the user will be attempting to generate sounds "to order" - whether the target is a sound with a particular metaphorical or verbally-described quality, or just "what sounds good". Success in this will be have to be measured in terms of both human satisfaction with the results, and psychoacoustically-based measures of timbre.

Finally, we hope to use the software in the design of experiments investigating timbral invariance: the phenomenon that psychoacoustic attributes such as *centroid* and *roughness* constitute a many-to-one mapping from sounds to real values.

2 Experimental Setup

2.1 Synthesizers

We choose to work with three different synthesizers (*simple additive*, *granular*, and *FM*), partly to ensure that any idiosyncracies of a single one don't have too large an effect on the results; however in this paper, for clarity, we report results for the FM synthesizer only, partly because FM is a more familiar and intuitive method of synthesis than granular. We use a single-carrier, single-modulator FM synthesizer with a peaking EQ filter: including envelopes and LFOs, it has 24 continuously-variable parameters.

2.2 Target Sounds

The target sound for preliminary experiments was a 3-second sine wave at 440Hz, with an amplitude of 0.305 (where digital full scale is 1.0). For the main body of experiments, we used targets with 1, 4, 8, 16, and finally 50 partials (the upper limit for a 440Hz fundamental and a sampling rate of 44.1kHz), where the amplitude of partial n is given by $1/n$. The 50-partial target is therefore a bandlimited sawtooth wave.

We shorten the targets to 0.5 seconds for efficiency: this is justified since we treat length as a user-specified input parameter (see section 2.4).

As a preliminary test, we confirmed by hand that the synthesizer was capable of closely matching the simplest, 1-partial target sound.

2.3 Genotypic Representation

Individual genomes were represented in GALib [14] as arrays of floating-point numbers, where the length and ordering of the array is fixed for the synthesizer. The genotype-to-phenotype mapping is performed by the synthesizer as it parses the genome into its own parameter format, and generates the corresponding sound.

2.4 Fitness Functions

A fitness function, in this context, is a measure of similarity between the individual sound and a target. Several fitness functions were implemented and tested, each returning results of the form $1/(1+d)$ ($\in [1/2, 1]$), where d ($\in [0, 1]$) is the *distance* between the two sounds and is calculated in a different way for each function.

Uniform Metric. Digital audio signals can be thought of as discrete versions of real-valued functions of time. The *uniform metric* on L_1, the space of such functions, is defined as

$$d_U(x, y) = \frac{\sup_{t=0}^{T} |x(t) - y(t)|}{2} \tag{1}$$

for two functions or sequences x and y; the same expression can be used whether t varies discretely or continuously. We divide by a factor of 2 since the audio signal varies in $[-1, 1]$. The Uniform metric is commonly used in analytical mathematics but is too "severe" for this application, and is not used in this study.

Pointwise Metric. The most obvious definition for d, and the simplest generalisation of the Uniform metric, might be

$$d_P(x, y) = \frac{\sum_{t=0}^{T} |x(t) - y(t)|}{2T} \tag{2}$$

which is of course the discrete equivalent to integrating the difference between the two functions. We divide by a factor of 2 for the same reason as in the Uniform metric, and by T in order to keep $d_P(x, y) \in [0, 1]$. We'll call this the *pointwise metric.*

Discrete Fourier Transform Metrics. We define DFT metrics as follows:

$$d_{\mathrm{D}_L}(x, y) = \frac{\sum_{j=0}^{N} \left(\sum_{i=0}^{L/2} |X_j(i) - Y_j(i)| \right)}{C_L} \tag{3}$$

where L is the transform length, X_j and Y_j are the normalised outputs from the jth transforms of the input signals x and y, and N, the number of transforms for each sound, is determined on the basis of 2×-overlapping Hann windows. C_L, a normalisation factor, is determined by experiment.

Initial experiments showed that the transform of length $L = 256$ gave the best results.

Perceptual Metric. Research including [3], [8], and [10] shows that timbral attributes such as *centroid*, *harmonicity*, *attack time* and so on can be defined, measured, and used to measure the degree of human-perceived similarity between pairs of sounds. We define a simple *perceptual metric* (named d_{H_1} to indicate that we intend to define further human-perception oriented measures later) as follows:

$$d_{H_1}(x, y) = (1/3)(d_A(x, y) + d_C(x, y) + d_M(x, y)) \tag{4}$$

with the *attack time*, *centroid* and *mean amplitude* metrics defined as follows:

$$d_A(x, y) = \frac{|\mathrm{attack}(x) - \mathrm{attack}(y)|}{C_A} \tag{5}$$

$$d_C(x, y) = \frac{\sum_{j=0}^{N} |\mathrm{centroid}(X_j) - \mathrm{centroid}(Y_j)|}{C_C} \tag{6}$$

$$d_M(x, y) = |\overline{\mathrm{amp}}(x) - \overline{\mathrm{amp}}(y)| \tag{7}$$

where the normalisation constants C_A and C_C are determined by experiment. Finally, we define

$$\mathrm{attack}(x) = \min\{t \;:\; x(t) = \sup_{s=0}^{T} |x(s)|\} \tag{8}$$

$$\mathrm{centroid}(X) = \frac{\sum_{i=0}^{L/2} f(i)X(i)}{\sum_{i=0}^{L/2} X(i)} \tag{9}$$

$$\overline{\mathrm{amp}}(x) = \frac{\sum_{t=0}^{T} |x(t)|}{N} \tag{10}$$

where $f(i)$ is the centre frequency of the ith frequency bin in the DFT X.

Composite Metric. We define a composite metric by summing the weighted results of several simpler measures:

$$d_{\mathrm{Comp}}(x,y) = (1/9)d_{D_{256}}(x,y) + (1/9)d_{D_{1024}}(x,y) + (1/9)d_{D_{4096}}(x,y) + (1/3)d_{H_1}(x,y) + (1/3)d_P(x,y) \tag{11}$$

2.5 Pitch and Length Parameters

Early experiments showed that a fitness function based on the pointwise metric led fairly consistently to the evolution of silence. This happens because unless the candidate individual and the target are very close in frequency, they will go in and out of phase over the length of the sounds. Using a target sine wave 3 seconds long with a frequency of 440Hz, we found that candidate sine waves have fitness values as shown in fig. 1.

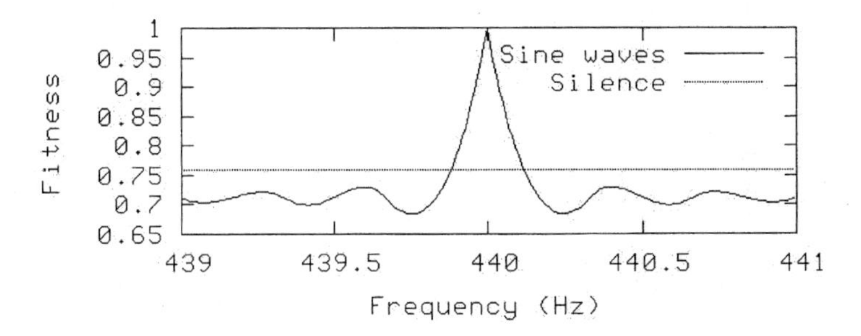

Fig. 1. Fitness values for silence and for sine waves, using a pointwise fitness function and a 440Hz sine wave target

Clearly, there is only a very small area of the entire frequency axis on which the GA can follow an upward gradient towards the target frequency. Evolution becomes a random search, and since silence scores higher than any wave outside this area, populations converge on silence from various directions.

We solve this problem by deeming pitch to be a fixed parameter of the synthesis. That is, we fix it at a certain value (here, 440Hz) and do not place it under the control of the GA.

A similar problem occurs in the case of the length of the sounds. It is not clear how to compare two sounds of different lengths - the obvious approaches (zero-padding the shorter is one; truncating the longer is another) both constitute "loopholes" that a GA can exploit in producing individuals which score highly but have obvious defects, such as trivial length. So, we deem length to be a fixed parameter also.

These decisions are justified since both pitch and note-length are typically controlled by the player of an instrument (eg by choosing a key to strike, and by holding it for a certain length of time), rather than by the sound designer. Even when these two jobs are performed by the same person, they're usually performed at different times and in different contexts.

2.6 GA Parameters

We use a steady-state GA with 100 generations, population size 300, and replacement probability 0.5. The crossover probability is 0.5, while the Gaussian mutation operator is applied with a per-gene probability of 0.1. Selection is by a roulette wheel scheme.

2.7 Running the GA

For each target waveform, and for each of the 4 fitness functions *pointwise*, *perceptual*, *DFT 256*, and *composite*, we run the evolution 30 times. Each evolution is driven by a single fitness function: at the end of each run we have a single best individual, and its fitness according to the fitness function driving that run. For the purpose of comparison, we also evaluate the fitness of the best individual under each of the other 3 fitness functions.

3 Results

3.1 Results Using 4 Fitness Functions

Figs 2–5 show the averaged best results over 30 runs using the FM synthesizer and targets consisting of 1, 4, 8, 16, and 50 partials. Each graph shows evolution *driven* by each of the four fitness functions *evaluated* by the function indicated in the caption. Error bars indicate the standard deviation for each data set.

Since our purpose is to compare the fitness functions (and find "the best"), we must be careful not to compare the fitness values they report against each other directly. Each fitness function performs better than all others when it is used as the evaluator.

The results are open to interpretation: there does not seem to be a single right way to objectively interpret them to decide which fitness function is better than the others. However, the main result is that the composite and perceptual fitness functions score quite well overall. They also exhibit smaller standard deviations, as indicated by the error bars, while the DFT-based fitness function gives very high standard deviations.

In general all fitness functions are successful for the simplest targets, and performance drops off for the complex targets composed of many partials. Also, there is a strong anti-correlation between the perceptual and pointwise functions:

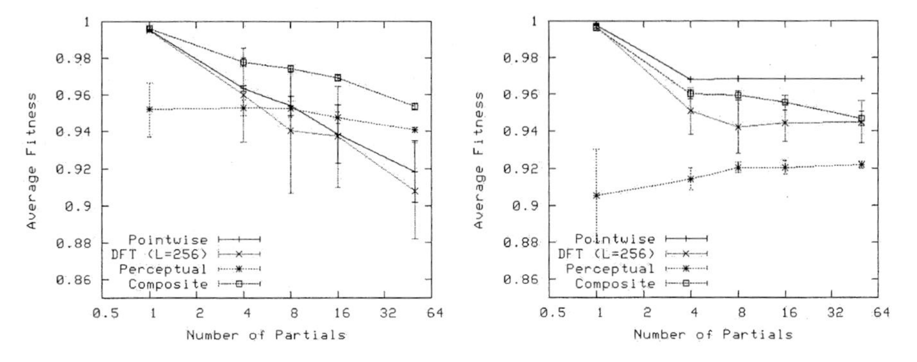

Fig. 2. Composite evaluation

Fig. 3. Pointwise evaluation

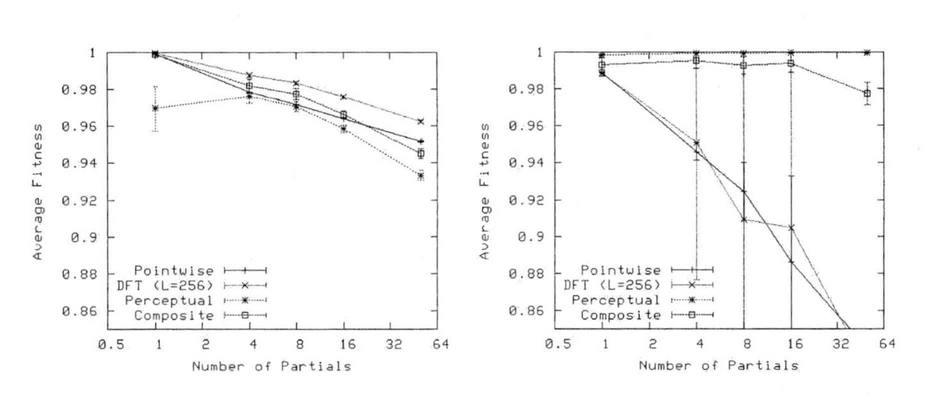

Fig. 4. DFT evaluation

Fig. 5. Perceptual evaluation

the perceptual function scores well when evaluated by itself, and badly when evaluated by the pointwise function – and vice versa. This reflects the different kinds of information exploited by the two functions.

3.2 Scatter Plots Showing Contrasting Fitness Scores for Individuals Under Different Measures

Each of figs 6–8 shows 600 random individuals and 600 "best" individuals (30 for each of the 5 targets and 4 methods of driving evolution), evaluated under the *two* fitness measures indicated in the captions.

The fact that these plots do not exhibit a strong concentration on the bottom-left–top-right diagonal is evidence that the fitness functions are not strongly correlated: i.e. an individual scoring highly in one measure doesn't necessarily score highly in another. This is expected: if all fitness functions were highly correlated there would be no need to use more than one.

Where random individuals clump together, we can infer that large areas of the search space contain individuals whose scores coincide under one measure and under another: these features warrant further investigation. Where "best"

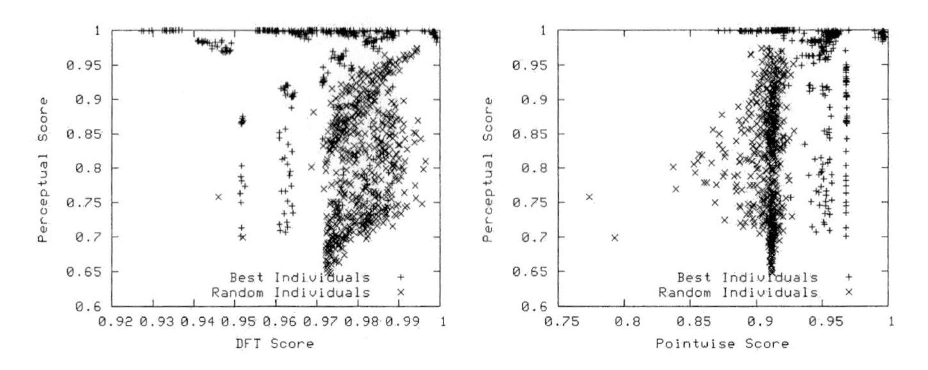

Fig. 6. Perceptual v. DFT

Fig. 7. Perceptual v. Pointwise

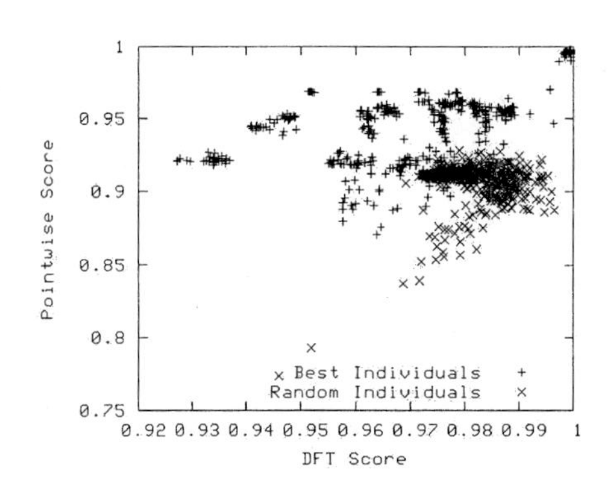

Fig. 8. Pointwise v. DFT

individuals clump together, we can infer either of two things: local maxima in the search space, or patterning according to the target being pursued.

The vertical and horizontal bands, especially in the first two plots, indicate the non-uniqueness of the perceptual measure. (see section 1.2).

The small concentrations of best individuals in the top-right corner of each graph are probably the collections of individuals which were very successful in imitating the 1-partial target. For this target (only), all fitness functions do seem to be correlated.

4 Conclusions and Further Work

We consider that evolution has been successful enough to justify further work. Investigation is required to understand the paths which the GA follows towards its targets, and to explain some patterns in the search space suggested by the scatter plots.

Informal, subjective listening tests do suggest that all measures can drive successful evolution; and that the perceptual and combined measures might be the best by a slight margin. However, more rigorous listening experiments will be required to test this.

The perceptual measure performs adequately and of the three fitness measures it is the most transferable to evolving novel sounds rather than mimicking existing sounds. The version used in our experiments is a very simple one, being based on only 3 of the more important perceptual measures: attack time, centroid, and mean amplitude. It could be extended by adding components based on *harmonicity*, *roughness*, *irregularity*, the *odd/even harmonic ratio*, and perhaps others. Also, the components are currently weighted equally, but listening experiments performed with human subjects are expected to show that some components should be given more weight than others. After making these improvements, we hope to show that a perceptually-motivated measure can drive more successful evolution than any other measure.

The GA parameters mentioned in section 2.5 have fairly generic values. Tweaking one or more of them may help evolution to proceed more efficiently. Of particular interest will be the changes required to make evolution practical when under user control.

Other planned work includes applying the same methods to the parameters of a non-linear filter, instead of a synthesizer; implementing synthesizer and filter plug-ins with GUIs; and running experiments in the area of timbral invariance (as explained in section 1.2).

Acknowledgements

Co-author James McDermott is supported by IRCSET grant no. 9262003.

References

1. Blackwell, Tim, Young, Michael: (2004) Swarm Granulator. *EvoWorkshops 2004 Proceedings*, 399–408. Berlin: Springer-Verlag.
2. Goldberg, D.: (1989) Genetic Algorithms in Search, Optimization and Machine Learning. Reading, MA: Addison-Wesley.
3. Grey, John M.: (1976) Multidimensional perceptual scaling of musical timbres. *J. Acoust. Soc. Am.*, **61** (5)
4. Grey, John M., Moorer, James A.: (1977) Perceptual evaluations of synthesized musical instrument tones. *J. Acoust. Soc. Am.*, **62** (2)
5. Horner, A., Beauchamp, J., and Haken, L.: (1993) Machine Tongues XVI: Genetic Algorithms and their Application to FM Matching Synthesis. *Computer Music Journal* **17** (4)
6. Horner, A., Cheung, N.-M., and Beauchamp, J.: (1995) Genetic Algorithm Optimization of Additive Synthesis Envelope Breakpoints and Group Synthesis Parameters. *Proceedings of the 1995 International Computer Music Conference.* San Francisco: International Computer Music Association.

7. Horner, A., and Goldberg, D.: (1991) Genetic Algorithms and Computer-Assisted Music Composition. *Proceedings of the 1991 International Computer Music Conference*, 479–482. San Francisco: International Computer Music Association.
8. Jensen, K.: (1999) Timbre Models of Musical Sounds. Ph.D. Thesis, Dept. of Computer Science, University of Copenhagen.
9. Johnson, C. G.: (1999) Exploring the sound-space of synthesis algorithms using interactive genetic algorithms. *Proceedings of the AISB'99 Symposium on Musical Creativity*, 20–27. Brighton, UK: Society for the Study of Artificial Intelligence and Simulation of Behaviour.
10. McAdams, S. and Cunibile, J.-C.: (1992) Perception of timbral analogies. *Philosophical Transactions of the Royal Society* **336**, London Series B
11. Miranda, E. R.: (2004) At the Crossroads of Evolutionary Computation and Music: Self-Programming Synthesizers, Swarm Orchestras and the Origins of Melody. *Evolutionary Computation* **12** (2), 137–158
12. Miranda, E. R.: (2000) The Art of Rendering Sounds from Emergent Behaviour: Cellular Automata Granular Synthesis. *Proceedings of the 26th EUROMICRO Conference*, 350–355. Maastricht, The Netherlands: IEEE.
13. Takala, T., Hahn, J., Gritz, L., Geigel, J., and Lee, J.W.: (1993) Using physically-based models and genetic algorithms for functional composition of sound signals, synchronized to animated motion. *Proceedings of the 1993 International Computer Music Conference*, 180–185. San Francisco: International Computer Music Association.
14. Wall, M.:GALib, a C++ Genetic Algorithm Library. `http://lancet.mit.edu/ga/`, viewed 26 October 2004.

Playing in the Pheromone Playground: Experiences in Swarm Painting

Paulo Urbano

Faculdade de Ciências da Universidade de Lisboa,
Campo Grande 1749-016 Lisboa, Portugal
pub@di.fc.ul.pt

Abstract. This paper is about collective artistic work inspired by natural phenomena, namely the use of pheromone substances for mass recruitment in ants. We will describe two different uncoordinated groups of very simple virtual micro-painters: the Colombines and the Anti-Colombines. These painters have very limited perception abilities and cannot communicate directly with other individuals. The virtual canvas, besides being a computational space for depositing paint, is also a pheromone medium (that mirrors the painting patterns) influencing the painters' behaviour. Patterns are the emergent result of interaction dynamics involving the micro-painters and their pheromone medium.

1 Introduction

The study of biological self-organization [1] has revealed that numerous sophisticated pattern formation, decision-making, and collective behaviour, are the emergent result of the interaction of very simply behaviours performed by masses of individuals relying only on local information. In particular, successful problem solving by social insects made models of their collective mechanisms particularly attractive [2,3]. The emphasis of this paper is on the design of micro-painters swarms, which are able to create interesting patterns in artistic terms. There are already examples of collective paintings inspired by social insects: L. Moura [4] has used a small group of robot-painters inspired by ants' behaviour, that move randomly in a limited space. Stimulated by the local perception of the painting they may leave a trace with one of their coloured pens. The painters rely on stigmergic interaction [6] in order to create chaotic patterns with some spots of the same colour. Colour has the pheromone role: a spot dominated by a certain colour has the capacity to stimulate the painter-robot to add some paint of the same colour. Monmarché et al. [5] have also designed groups of painters inspired by ants' pheromone behaviour. It is based on a competition between ants: the virtual artists try to superimpose their colours on traces made by others, creating a dynamic painting which is never finished. Ants have the capability to "sniff" the painted colour and react appropriately. The societies are composed by a small number of individuals (less than 10). We will show and analyze the types of patterns resulting from the interaction dynamics of a pheromone environment (reflecting the painted and non-painted spots) and the mass of simple micro-painters, which

F. Rothlauf et al. (Eds.): EvoWorkshops 2005, LNCS 3449, pp. 527–532, 2005.

cannot communicate directly with each other. The artists have only local perception (they never see the "tableau" as a whole), there isn't any type of social coordination, interactions are stigmergic: the painters modify the painting area which influences their movement and painting behaviour. We have designed two societies of micro-painters: the Colombines and the Anti-Colombines. The first are attracted to non-painted spots and, in contrast, the Anti-Colombines are attracted to painted ones. In both, there is a chemical medium which reflects the bi-dimensional canvas state and it is the chemical that attracts the artists (the painters try to prefer to go to patches with more chemical). They were implemented in Starlogo [7].

One of the differences from the other ant-paintings is that the ants are not charged for pheromone production, (the environment is responsible for that task). More, the diffusion process does not occur on any of the ant paintings we have referred. We introduced also populations of numerous agents: we have experimented with groups composed of up to 2000 individuals working in the same artistic piece.

The remainder of the paper is organized as follows: In section 2 we describe in detail the Colombines and making some variations on the basic painter behaviours. In section 3 we focus on the Anti-Colombines. Finally we conclude, discussing the results and pointing future directions.

2 The Colombines

The Colombines are a swarm of small artificial micro-painters, individually very simple, which are able to paint a bi-dimensional virtual canvas, composed of small cells.

The canvas is bi-dimensional space with a toroidal format, divided in small squared sections, called patches or cells, it is a kind of grided paper, with no borders, folded in every direction, in which two types of virtual materials coexist: paint and a chemical signal. Each patch can have a certain colour and can have a certain quantity of chemical. There is a fixed colour (usually grey) for the background. Any other colour corresponds to paint. As we said before, our goal is that the non-painted cells have more attraction power (more chemical). Therefore, every cell has the potential ability to release chemical, but only the non-painted cells (the background ones) are chemical producers. The squared canvas is a kind of chemical medium where every cell is permanently diffusing chemical to their immediate neighbours, independently of being painted or not. The chemical evaporates at a constant rate. Without evaporation, the attraction power decay of recently painted spots will last more time, disorientating the painters, attracting them to painted spots. Foremost, the evaporation phenomenon increases the painters' efficiency: the painting will be completed sooner.

The cells behaviour is the following: 1) if it is not painted increase its own chemical quantity by a certain amount, otherwise the chemical level is maintained intact; 2) diffuses a percentage of its chemical to their 8 immediate neighbours; 3) delete a percentage of its chemical (evaporation. The chemical constant produced by non-painted cells, the evaporation and diffusion taxes are parameters modifiable by the user.

Initially, we launch these painters in a non-painted background, each one occupying a particular cell, and they will move along, depositing a trace of ink, until the canvas is completely fulfilled. Note that each painter is constrained to paint only non-painted cells and when there isn't any non-painted cell left, the artistic work cannot change and is considered finished. Our micro-painters have a very limited perception field—they have an orientation and have access just to the three cells in front of them. Each painter is created with a particular colour and they never change to another colour. It's the empty spots that guide the painters. They prefer to move towards empty spots.

If each Lilliputian painter just acted on its own, without any interactions, either with the world or with the others, interesting phenomena would never arise. They do no more than moving on the virtual canvas, visiting preferentially cells with more amount of chemical, (preferring to move towards non-painted spots) and painting cells still unpainted, leaving traces of colour behind them. In case of identical chemical values in their neighbouring cells they have a tendency to preserve its current direction. Each Colombine has a position (real Cartesian coordinates), an orientation (0..360), and can only inhabit one cell, the one that corresponds to their coordinates. They see just their own cell and also the three cells immediately in front of them. On the other hand, the painters are created with a particular colour that is never changed. The behaviour of each Colombine is the following: 1) he senses the three immediate cells in front of him and chooses the one with more chemical, changing his orientation towards that winning cell and moving to it; 2) if that cell is not yet painted, stamps his colour on it, otherwise, does not paint it. In detail, the painter senses his three forward neighbouring cells and if there is no better patch than the one in front he remains with the same orientation and go forward one step (rounding his coordinates). If the left path is the most attractive he rotates 45 degrees to the left and moves forward one unity, rounding both position coordinates; the same happens when he prefers the right cell: he rotates to the right 45 degrees, moving forward one unity, rounding his coordinates. The round operation influences the patterns generated, as we will see later.

The evolution of the collective artistic work happens the following way. Initially, the virtual canvas is grey and each patch has an identical quantity of chemical (normally 0). We create a colony of Colombines, each one with its own colour and orientation, distributing them in the environment. The painting process will begin in a sequence of iterations until every patch is painted completing the plastic work. Each iteration is divided in two steps: in the first, every cell executes its behaviour (chemical production, diffusion and evaporation); in the second step, the Colombines move, attracted by chemical, depositing paint.

2.1 Dynamics Responsible for Pattern Emergence

The canvas can be seen as a dynamical chemical landscape, in permanent mutation—there is a constant interaction between chemical distribution and the painters' behaviour. The chemical world is information floating both in the painted and background patches. There is a strong circularity: On one hand, the chemical information guides the movement of the Colombines, attracting them toward non-painted spots, On the

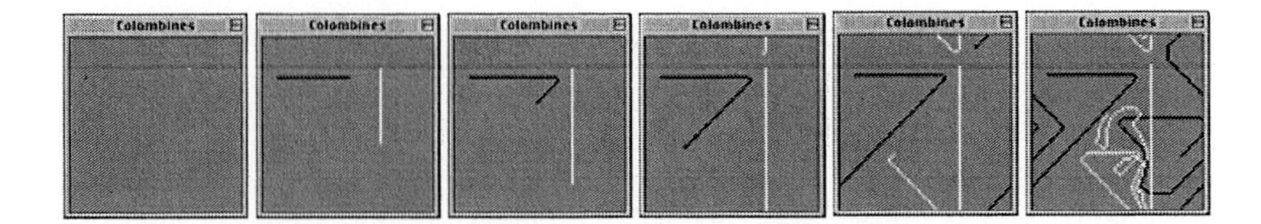

Fig. 1. The interaction between two painters. Illustration of the tendency to conserve direction and to avoid painted patches. The coordinates rounding effect is also visible: traces suffer only rotations of 45 or 90 degrees

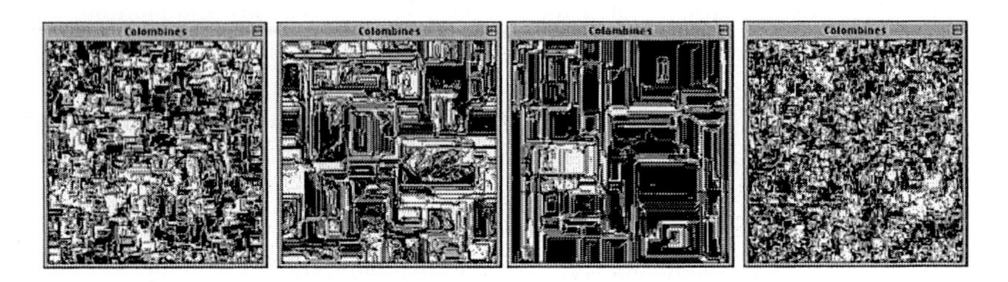

Fig. 2. Colombine black and white paintings

other hand, their painting activity change the information landscape, in an permanent auto-catalytic interaction. The patterns, the coloured forms, are the by-product of the collaboration between the Colombines and their chemical environment. Figure1 illustrates the pattern emergence.

We have two painters, one white and one black. They have an initial orientation (black moves east and white goes south). They both tend to preserve their directions. The black suddenly changes direction, avoiding the trace left by the white painter. After a while the white painter reaches his own trace and avoids it, changing direction and having to avoid later the black trace and the painting history goes on. Sometimes, the painters have to cross already painted spots, due to the fact that the three immediate neighbours are painted. Notice that we can find spots with the same colour due to the fact that a painter can be on a non-painted area which is surrounded by traces, constraining him to be inside, painting that enclosed spot.

In figure 2 we show four examples of finished paintings made by groups of Colombines of different sizes (from left to right, 1000, 100, 50 and 2000 painters) in a world of 125*125 pixels. There are only black and white painters equitably distributed by each of the colours and which are randomly scattered on the "tableau". The painters were created with random orientations. If we increase the number of micro-painters, the possibility of encountering traces also increases. The resulting effect is that the spots with the same colour have a smaller area and we find less rectilinear traces.

In reality, these paintings are only declared finished when there are no grey patches, but, alternatively, we could finish the collective work after a random or fixed number of iterations.

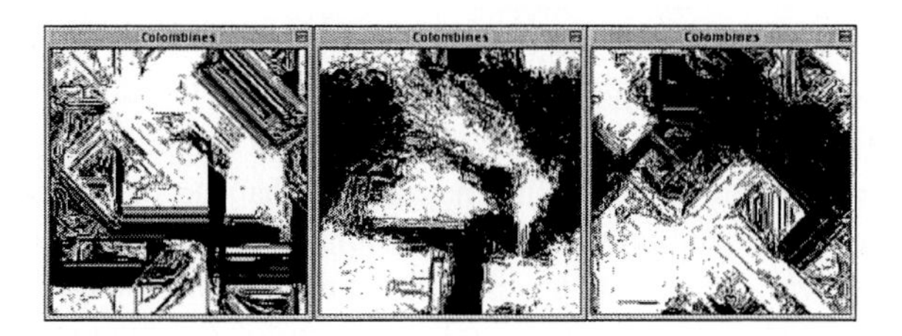

Fig. 3. Three paintings, from left to right: 2000 Colombines divided in groups of 200, 300, and 100 individuals

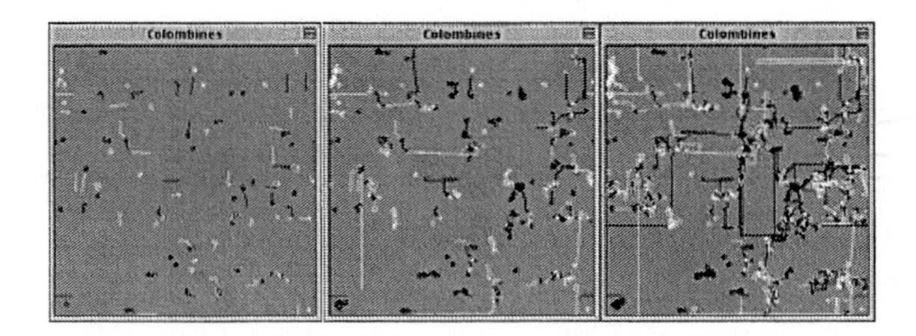

Fig. 4. Evolution of a painting made by 100 b&w Anti-Colombines, not clustered, with initial random positions and orientations

2.2 Variation: Clustering Painters

Now, we will make a slight variation on the Colombines initial settings. We will divide them in groups (the groups number is a global parameter) and put each group in the same patch. The painters in the same group have the same colour but have different directions. We introduce also a variation on the painters' behaviour. We do not want that the group will remain in the same patch forever. Each painter will make a small random jump (for example, up to three cells) to a neighbouring patch every time he senses another mate in his cell. This behaviour allows a kind of dispersion inside a group before a group invades painted zones. Figure 3 shows the final paintings of other clusters of Colombines. As we can see, these initial clusters will make the painting with more spots of the same colour.

3 The Anti-colombines

We want to introduce an inversion on the patches behaviour. Our goal now is that, instead of being attracted by the empty spots, the painters shall be attracted by the painted ones. The new painters, the Anti-Colombines, will maintain their behaviour (they continue to go up-hill in the gradient field) but now the non-painted patches are not chemical producers—the producers are the painted ones. Since the painters are attracted to the painted spots it is not wise to expect that the canvas is going to be fulfilled with paint. Thus we are obliged to choose a random or fixed number of iterations, and so after finishing them the painting is considered finished.

In figure 4 we show three snapshots of a plastic work made by a colony of 300 Anti-Colombines, scattered randomly on the grey "tableau" (they have initial random orientations and positions). The painters do not apply a round operator to their coordinates after making their moves. What happens is that the painters will be trapped inside painted spots, circling around, and occasionally they can get out of these traps slowly enlarging them. When they have the three neighbouring patches (their perception field) non-painted they will choose one of them for occupation. The probability of getting out of the painted traps decreases with the increase of their areas.

4 Future Work

We have designed two types of swarm micro-painters, the Colombines and Anti-Colombines, relying only on local perception and with no coordination, being able to produce interesting patterns, which can be seen as a kind of artificial art. Due to the regularities of the resulting patterns, we dare to say that there is what we can name, a Colombine or Anti-Colombine style. We have only showed black and white paintings for obvious reasons, but we could have made experimentations with different colours.

We are already working with what we may call consensual painters. The painters are able to interact with each other and achieve a decentralized consensus about certain attributes, like the colour they are painting, the position they are occupying, the velocity they possess or the direction they are facing, creating different type of structure and patterns. These consensual painters are able to randomly alter their attributes, shifting from a unanimous situation forcing the others to converge to a new consensus, in cycles consensus, cycles of order and chaos reflected on the patterns formed.

References

1. Camazine, S., Deneubourg, J.-L., Franks, N., Sneyd J., Theraulaz, Z., Bonabeau,.E.: *Self-Organization in Biological Systems*. Princeton University Press. (2001)
2. Bonabeau,.E, Dorigo, M., Theraulaz, Z.: Swarm Intelligence: From Natural to Artificial Systems. Santa Fe Institute, Studies in the Sciences of Complexity. (1999)
3. Dorigo, M., Maniezzo, V., Colorni, A.: "The Ant System: Optimization by a Colony of Cooperating Agents." *IEEE Trans. Syst. Man. Cybern. B* **26** (1996):29-41.
4. Moura, L.: Swarm Paintings. Architopia: Art, Architecture, Science. (ed. Maubant) Institut d'Art Contemporaine (2002)
5. Monmarché, N., Slimane, Mohamed, Venturini, G.: "Interactive Evolution of Ant Paintings." IEEE Congress on Evolutionary Computation. (2003)
6. Grassé, P.-P.: "Termitologia, Tome II." *Fondation des Sociétés*. Construction. Paris: Masson, (1984)
7. Resnick, M.: *Turtles, Termites and Traffic Jams: exploration sin massively parallel microworlds*. MIT Press (1994)

Convergence Synthesis of Dynamic Frequency Modulation Tones Using an Evolution Strategy

Thomas J. Mitchell and Anthony G. Pipe

Faculty of Computing, Engineering & Mathematical Sciences,
The University of the West of England, Frenchay, Bristol, BS16 1QY, UK
{thomas.mitchell, anthony.pipe}@uwe.ac.uk

Abstract. This paper reports on steps that have been taken to enhance previously presented evolutionary sound matching work. In doing so, the convergence characteristics are shown to provide a synthesis method that produces interesting sounds. The method implements an Evolution Strategy to optimise a set of real-valued Frequency Modulation parameters. The development of the evolution is synthesised as optimisation takes place, and the corresponding dynamic sound can be observed developing from initial disorder, into a stable, static tone.

1 Background

Horner et al. have presented a collection of evolutionary dynamic-tone matching systems applied to a variety of synthesis techniques: Frequency Modulation Synthesis (FM) [1], [2], Wavetable synthesis [3], and Group Synthesis [4]. Throughout Horner's work, a Genetic Algorithm (GA) is employed to optimise a set of static wavetable *basis-spectra*[1], which are, as in wavetable synthesis, combined in time-varying quantities to match dynamic target tones.

The Evolution of modular synthesis arrangements and interconnections has formed the subject of two independent studies. Originally, Wehn applied a GA to 'grow' synthesis 'circuits' [5]. Later, Garcia used Genetic Programming to evolve tree structures that represent the synthesis topology [6].

A further matching technique has been put forward by Manzolli [7]. The Evolutionary Sound Synthesis Method (ESSynth) evolves waveforms directly applying the principles of Evolutionary Computation to recombine and mutate waveform segments. At each generation, ESSynth passes the 'best' waveform to an output buffer for playback allowing users to experience the evolution as it takes place. Wehn also observed this phenomenon, noting that sounds produced throughout the search phase can be 'quite entertaining', and occasionally 'rich and strange' [5].

The convergence synthesis technique, presented here, is built upon the work introduced above, and is concerned with the dynamic sounds that are produced as an Evolutionary Algorithm converges upon an optimal target match. The

[1] Where *basis-spectra* refers to the harmonic content of a wavetable.

F. Rothlauf et al. (Eds.): EvoWorkshops 2005, LNCS 3449, pp. 533–538, 2005.

synthesis model at the heart of this process is based upon Horner's early FM matching experiments and provides initial steps towards a real-valued extension to his work.

2 Tone Matching with Frequency Modulation Synthesis

FM is an effective musical synthesis tool [8]. Despite the efficiency with which complex tones can be synthesised, FM is often regarded as unintuitive and cumbersome. Consequently, parameter estimation, specifically for the reproduction of acoustic musical instrument tones, has formed the subject of numerous academic studies. Recent advances in sound matching with FM have been presented by Horner [2]. Horner's early work utilised a unique FM arrangement referred to as *formant* FM. The *formant* FM model is ideal for matching harmonic instrument tones, as the carrier frequency can only be set to integer multiples of the modulating frequency (which is tied to f_0). Restriction of the synthesis parameters in this way significantly reduces the search complexity, as non-harmonic solutions are excluded from the model space. However, with the omission of non-integer variables, the algorithm is only able to match a discrete number of harmonic sounds, and could not be applied directly to regular FM models, as the majority of the sound space would be inaccessible.

Throughout Horner's FM matching experiments, a GA was applied to optimise a set of FM synthesis variables. GAs' perform their genetic operations on bit-strings, which are naturally suited to integer based combinatorial search problems. With the intention of expanding the search into the full parameter space, a reduced static-tone version of Horner's *formant* FM matching model was developed without limiting the synthesis parameters to integer numbers. Whilst GAs' can be modified to represent real-valued numbers, with specialised operators that permit arithmetic crossover and ordinal mutation, an Evolution Strategy was chosen, for the work presented here, as it provides a powerful optimisation paradigm that is naturally real-valued [9].

Objective Function. In previous sound matching work, objective calculations are often performed in the frequency domain using the *Squared Spectral Error* (SSE) (1), or a variant thereof [1] - [7]. The target spectrum for the evolutionary algorithm is obtained via the spectral analysis of the target waveform, prior to the execution of the search. A complete run of the objective function can be summarised as follows:

1. Insert candidate solution into the FM model,
2. Subject the corresponding synthesised waveform to spectrum analysis,
3. Calculate error between target and synthesised candidate spectra.

The SSE is given by the equation:

$$SSE = \sqrt{\sum_{b=0}^{N_{bin}} (T_b - S_b)^2} \ . \tag{1}$$

T = The target spectrum amplitude coefficients
S = The synthesised candidate spectrum amplitude coefficients
N_{bin} = The number of frequency bins produced by spectrum analysis

As will be demonstrated in section 3, a modified SSE metric is sufficient for this work.

3 Real Valued Static Formant FM Matching

An initial step towards the creation of a real-valued extension to Horner's FM matching algorithm has been developed for matching exclusively static tones. The matching synthesiser employs a single carrier/modulator FM arrangement, similar to Horner's *formant* model. The synthesis parameters, *Modulation Index* and *Carrier Frequency Multiple*, can be set to any value within the range [0, 15]. The object parameters are permitted to search regions of the sound space that would be unavailable with an integer restricted model. An ES, with strategy parameters[2] (5/5,25) is employed to optimise the synthesis parameters. To ensure that globally-optimal matches are consistently achieved, target tones are generated by the matching synthesiser. *Contrived* target tones are useful for testing purposes as they are known to exist within the matching sound space. Successful matches, therefore, yield parameters identical to those with which the target tone was produced.

Landscapes. The ES was found to become trapped at locally-optimal points of the SSE object landscape, see figure 1(a). Very rugged, Multi-modal landscapes, such as this, are problematic for any optimisation engine (including EC). To overcome this problem, the spectrum of the target and synthesised tones (T and S), are modified according to (2), to produce *windowed* spectra TW and SW which then replace T and S in (1) to provide the *Windowed Squared Spectral Error* (WSSE) metric.

$$XW_b = \sum_{n=0}^{N_{bin}} \sum_{b=0}^{w} \left(\frac{w-b}{w} \right)^2 (X_{n+b} + X_{n-b}) \ . \tag{2}$$

X = The spectrum amplitude coefficients
XW = The 'windowed' spectrum amplitude coefficients
w = The bandwidth of the window

Windowing, allows spectrum error to be measured across a band, which has a smoothing effect on the object landscape. The landscapes of SSE and WSSE are plotted in Fig. 1(a) and (b) respectively. The location of the optimum is immediately obvious in the WSSE plot as the surrounding landscape slopes downwards isotropically. With a (5/5, 25) evolution strategy using the WSSE objective

[2] In the standard ES format (μ/ρ,λ) where μ is the parent size, ρ is the mixing number and λ is the offspring size.

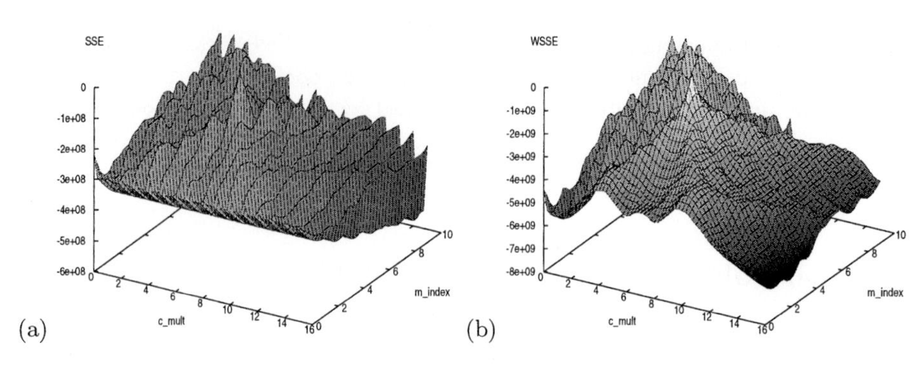

Fig. 1. Exhaustive SSE & WSSE landscape plots *Carrier Frequency Multiple*/c_mult = 7.00 and *Modulation Index*/m_index = 4.00

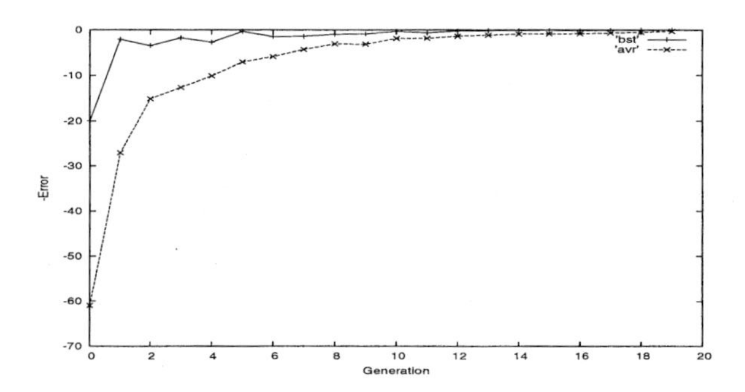

Fig. 2. Typical convergence plot for the evolutionary matching process

function, a globally optimal solution is located, for any target parameter setting, within 20 generations. A typical convergence plot is provided in Fig. 2.

4 Convergence Synthesis

Running an FM synthesiser program in parallel with the ES allows convergence to be monitored aurally in real time. At the turn of each generation, the strongest offspring (or, in fact, any other) is passed to the synthesiser and a corresponding tone can be heard. Prior to the evolutionary process, the user is required to select a target tone by adjusting the synthesis parameters, *Modulation Index* and *Carrier Frequency Multiple*, via their respective sliders on the program interface. The target provides the final static tone upon which the ES will converge. The application responds to midi messages, enabling the user to set the modulator frequency (f_0) and inspect the target tone before the evolutionary synthesis process begins. To ensure that there are no audible clicks as the synthesiser is played, a short fade-in (*attack*) function is applied to the carrier amplitude on receipt of a MIDI *note-on* message, and equivalent fade-out (*release*) function on corresponding *note-off* messages. Following the selection of a satisfactory target tone, the evolutionary synthesis process begins. The *note-on* message, in addition to initiating the carrier fade-in function, triggers the initialisation of the ES. Individuals are randomly seeded and optimisation begins. The synthesis parameters are controlled by the ES as they progress towards their target values. Random reseeding ensures that the evolutionary path to the optimum is never the same twice.

The frequency plots, of the sound produced as a (20/20,25) evolution strategy converges upon it's FM target tone, are illustrated in Fig. 3. An initial stochastic period can be observed giving rise to the smooth harmonic partials of the target tone. After approximately 3.5 seconds the strategy has converged and the tone is stable.

Evolutionary Synthesis Parameters. The temporal characteristics of the sounds produced by convergence synthesis are closely coupled with the exogenous strategy parameters μ, ρ and λ. The variables that control the ES are, themselves, synthesis parameters that now control the sound directly. As μ and λ are varied, their values and respective ratio affect the dynamic characteristics of the synthesised tone. The former adjusts the selection pressure and, thus, the generational rate of convergence. The latter controls the period of each generation, as the objective function is called λ times. As the ES progresses, the offspring become increasingly similar until each λ is identical. This homogenising of the population is apparent in the convergence plot of Fig. 2. By passing the nth *fittest* λ to the synthesiser, the rate at which the tone stabilises can again be prolonged, increasing the duration of the initial transient.

5 Results and Future Work

Matching can be carried out on any tone within the synthesis parameter space. The matching process itself provides time-variant parameter control to produce dynamic sounds. From one run of the ES to the next, the path to the target is different, providing a tone evolution that is varied yet predictable. This effect can be observed in Fig. 3, where two differing convergence sonograms can be seen converging upon the same target. There is considerably more high frequency content visible between 0.0s and 0.1s on the Fig. 3(a), which takes slightly longer to converge than Fig. 3(b). It would be desirable to expand the FM synthesis model

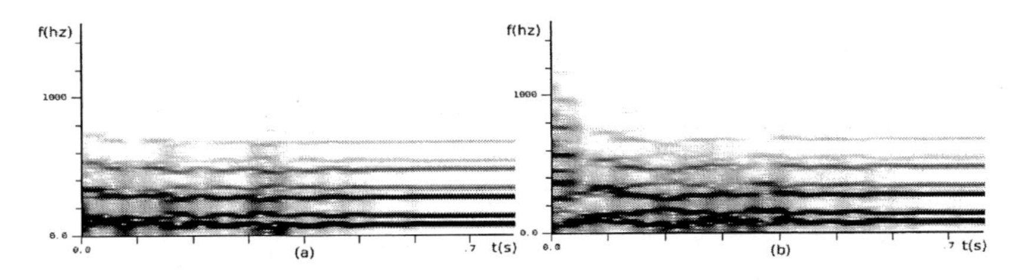

Fig. 3. Sonograms providing differing convergence paths for the same target tone

with the use of multiple carrier/modulator, nested modulator or even feedback arrangements. The ES would then be required to optimise a significantly more complicated multi-dimensional landscape. Optimisation of parameter envelopes may also enable the matching of dynamic target sounds, both harmonic and non-harmonic.

6 Conclusions

An evolutionary synthesis method has been presented that produces interesting dynamic sounds whilst approaching the match of a static target tone. The technique has emerged from the early developmental stages of a real-valued FM synthesis parametric optimisation process. The algorithm itself uses a basic Evolution Strategy, to optimise a set of *formant* FM synthesis parameters that most closely match a given static target tone. The sounds produced can be observed evolving from a stochastic initial transient, as a result of the random seeding of the initial population, into the static target tone that is chosen by the user at the beginning of the process. This convergence process also provides an alternative means by which an object landscape and convergence can be observed aurally.

References

1. Horner, A.: Spectral Matching of Musical Instrument Tones, PhD dissertation, University of Illinois Computer Science Department, Report No. UIUCDCS-R-93-1805 UILU-ENG-93-1720. (1993)
2. Horner, A.: Nested Modulator and Feedback FM Matching of Instrument Tones, IEEE Transactions on Speech and Audio Processing, 6(6), 398-409. (1998)
3. Wavetable Matching Synthesis of Dynamic Instruments, Audio Engineering Society Preprint No. 4389, pp. 1-23. (presented at the 101st Audio Engineering Society Convention). (1996)
4. Cheung, Ngai-Man, and Horner, A.: Group Synthesis with Genetic Algorithms, Journal of the Audio Engineering Society, 44(3), 130-147. Abstract on JAES Homepage. (1996)
5. Wehn, K.: Using Ideas from Natural Selection to Evolve Synthesized Sounds. Proc.Digital Audio Effects (DAFX98), Barcelona. (1998)
6. Garcia, R.: Growing Sound Synthesizers using Evolutionary Methods. European Conference in Artificial Life ECAL2001. Artificial Life Models for Musical Applications. University of Economics, Prague, Czech Republic. (2001).
7. Manzolli, J., Maia, A., Fornari, J., Damiani, F.: The evolutionary sound synthesis method. Proceedings of the ninth ACM international conference on Multimedia, Ottawa, Canada (2001)
8. Chowning, J.: The synthesis of complex audio spectra by means of frequency modulation. Journal of the Audio Engineering Society, vol. 21, pp. 526-534. (1973)
9. Beyer, H., Schwefel, H.: Evolution Strategies A Comprehensive Introduction. Natural Computing: an international journal, v.1 n.1, p.3-52, (2002)

Granular Sampling Using a Pulse-Coupled Network of Spiking Neurons

Eduardo Reck Miranda and John Matthias

Computer Music Research, Faculty of Technology and Faculty of Arts,
University of Plymouth, Smeaton 206, Drake Circus, Plymouth PL4 8AA, UK
{john.matthias, eduardo.miranda}@plymouth.ac.uk
http://cmr.soc.plymouth.ac.uk/

Abstract. We present a new technique for granular sampling using a pulse-coupled network of spiking artificial neurons to generate grain events. The system plays randomly selected sound grains from a given sound sample when any one of a weakly coupled network of up to 1000 neurons fires. The network can exhibit loosely correlated temporal solutions and also collective synchronised behaviour. This leads to very interesting sonic results, particularly with regard to rhythmic textures which can be controlled with various parameters within the model.

1 Brief Introduction to Granular Synthesis

Granular synthesis works by generating a rapid succession of very short sound bursts called grains that together form larger sound events. The notion behind it is largely inspired by a sound representation method published in a paper by Dennis Gabor back in the 1940s [1]. Gabor's point of departure was to acknowledge the fact that the ear has a time threshold for discerning sound properties. Below this threshold, different sounds are heard as clicks, no matter how different their spectra might be. The length and shape of a wavecycle define frequency and spectrum properties, but the ear needs several cycles to discern these properties. Gabor referred to this minimum sound length as an acoustic quantum and estimated that it usually falls between 10 and 30 milliseconds, according to the nature of both the sound and the subject.

2 Approaches to Granular Synthesis

As far as the idea of sound grains is concerned, any synthesiser capable of producing rapid sequences of short sounds may be considered as a granular synthesiser. Three general approaches to granular synthesis can be identified as follows [2]: sequential, scattering and granular sampling approaches. The sequential approach works by synthesising sequential grain streams. The length of the grains and the intervals between them are controllable, but the grains must not overlap. The scattering approach uses more than one generator simultaneously to scatter a fair amount of grains, not necessarily in synchrony, as if they were the 'dots' of a 'sonic spray'. The expression 'sound clouds' is usually employed by musicians to describe the outcome of the scattering approach.

F. Rothlauf et al. (Eds.): EvoWorkshops 2005, LNCS 3449, pp. 539–544, 2005.

Granular sampling employs a granulator mechanism that extracts small portions of a sampled sound and applies an envelope to them. The granulator may produce the grains in a number of ways. The simplest method is to extract only a single grain and replicate it many times. More complex methods involve the extraction of grains from various portions of the sample. In this case, the position of the extraction can be either randomly defined or controlled by an algorithm.

Thus far, most granular synthesis systems have used stochastic methods to control the production of the grains; for example, a probability table holding waveform parameters can be called to provide synthesis values for each grain during the synthesis process. As an alternative method, we have devised Chaosynth, a granular synthesiser that uses cellular automata to manage the spectrum of the sound grains [3]. Chaosynth explored the emergent behaviour of cellular automata to produce coherent grain sequences with highly dynamic spectra.

The challenge with granular sampling to find interesting and new controllable ways of playing back the grains, which are taken from the input sound sample. Grain events which are triggered independently will produce randomised signals which can have very interesting flow textural properties [4]. However in order to go beyond this one needs to look at introducing some kind of correlation in grain parameters whilst maintaining the inherent stochastic element which has been so effective in granular synthesis algorithms thus far. Chaosynth utilised the emergent behaviour of a cellular automata model in order to do this. Other attempts have looked at the collective properties of a large number of interacting particles, or swarms, to generate grain events [5]. In this work we have used the correlated firing properties of a large collection of pulse-coupled artificial neurons.

Spiking neural networks have a very rich dynamics and the relevant timescales are of the same order as those relevant to granular synthesis. (i.e., on the level of milliseconds and tens of milliseconds). This makes them very suitable to use as a triggering mechanism for a granular sampler. There is great variety in the dynamics at the level of the single neuron and this becomes even more interesting when we look at networked systems. The single neurons can show regular spiking, bursting (very fast spiking) so-called chattering and resonant behaviour. When connected they exhibit collective excitations on timescales larger than the inherent responses of single neurons (Fig. 1). Such collective excitations include synchronization of the firing times of large numbers of neurons in groups and repetition of signals over very large time scales (of the order of seconds) which have become known as Cortical Songs [6].

3 Spiking Neural Networks

Essentially one can visualise a neuron as an object that fires a spike signal when its input voltage exceeds a certain threshold [7]. The amplitude of the spikes of real neurons is of the order of 100mV (millivolts) and the duration of the spikes are of the order of 1-2ms. The spikes then travel to all the other (post-synaptic) neurons to which this (pre-synaptic) neuron is connected. The time taken for these spike signals to reach the post-synaptic neurons is also of the order of milliseconds. When one of the post-synaptic neurons receives a spike it will fire a spike in turn if its current voltage state plus the signal of the spike are above its threshold and so on. Most of this processing takes place in the cortex. Each neuron in a mammalian cortex is typically connected to up to 10 000 other neurons so one can see quickly how complicated the resulting dynamics would be from simply sending out a single spike to just one neuron.

The firing patterns of individual cortical neurons are known to be very varied. In Fig. 1 we can see twelve of the most common forms of mammalian cortical neuron firing. As far as the interaction of the neurons with other neurons is concerned, the class of spiking neuron models in which we are interested are called Pulse Coupled Neural Networks (PCCN). Essentially, this means that when a neuron receives a spike it updates its connection with all the neurons with which it is connected. In such models, the connections between neurons are modelled by a matrix of synaptic connections $S = (s_{ij})$, and these synaptic connections are used inherently in the dynamics. In this paper, we have used the simple condition that when the jth neuron fires, the membrane potential, v_i of the all the connected neurons immediately increases by s_{ij} [8].

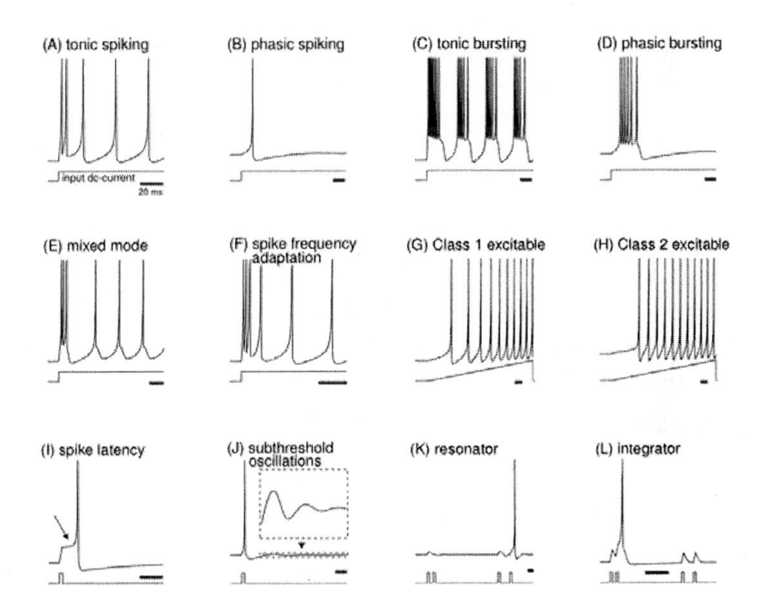

Fig. 1. Twelve of the different types of firing patterns exhibited by single neurons in the mammalian cortex. (This figure is reproduced with permission from Eugene Izhikevich.)

4 Izhikevich's Pulse-Coupled Neural Model

It has been recently discovered that surprisingly simple mathematical models of spiking neurons with random connections can produce realistic organised collective behaviour. The model of Eugene Izhikevich [8] [10] contains enough detail to produce the rich firing patterns found in cortical neurons (Fig 1), yet is also computationally very efficient.

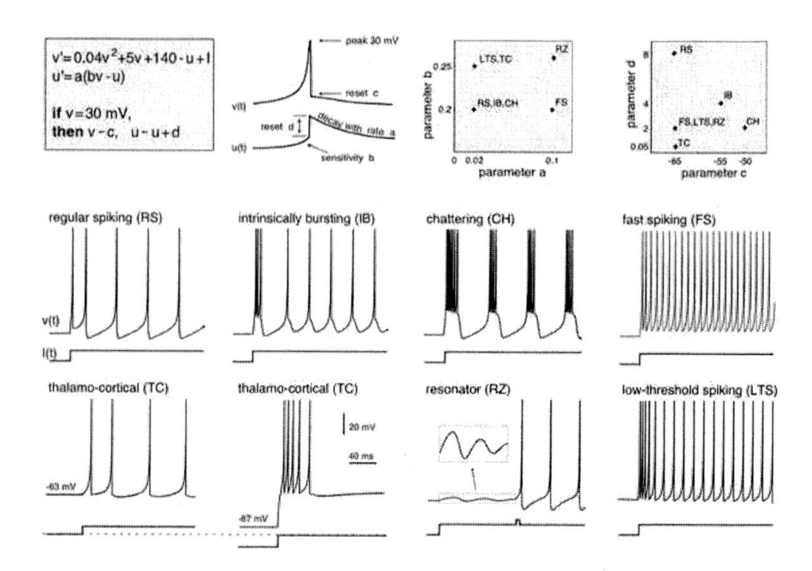

Fig. 2. Known types of neurons correspond to different values of the parameters *a*, *b*, *c* and *d* in Izhikevich's model. Each inset shows a voltage response of the mode neuron to a step of dc-current *I*=10 (bottom). (This figure is reproduced with permission from from Eugene Izhikevich.)

The temporal firing patterns of the network show both stochastic and synchronised behaviour depending on the values of various parameters, the number of neurons, the matrix of synaptic connections and the history of the behaviour of the network. The frequencies of collective modes of the system are between 1 and 40Hz and present a very interesting case for controlling a granular sampler, particularly in terms of rhythmic structure. We therefore use Izhikevich's model along with a granular sampler such that grains of sounds (taken from a recording) are triggered when any of the neurons fire.

The model contains N neurons, each of which are described by two dimensionless variables v_i and u_i where v_i represents the membrane potential of the ith neuron and u_i represents a membrane recovery variable, which provides negative feedback to v_i. The system is then described by the following coupled ordinary (nonlinear) differential equations:

$$\frac{dv_i}{dt} = 0.04v_i^2 + 5v_i + (140 - u_i) + I_i \tag{1}$$

$$\frac{du_i}{dt} = a(bv_i - u_i) \tag{2}$$

with the following auxilliary after spike resetting; if $v_i \geq 30$ millivolts then $v_i \rightarrow c$ and $u_i \rightarrow (u_i + d)$. Essentially, the first of these conditions means that when a neuron receives a spike input then its membrane potential is immediately reset.

The neurons are coupled to one another through a matrix of synaptic connection weights. These synaptic connection weights are given by the matrix $S = (s_{ij})$, such that the firing of the jth neuron instantaneously changes variable v_i by s_{ij}. We have used a version of Izhikevich's model where the matrix S is a random matrix. However, in other versions of the model, S can updated itself according to various learning algorithms such as 'Spike Timing Dependent Plasticity' in which connections between neurons are reinforced according to temporal correlations. Synaptic currents or injected dc-currents (currents which come from either other neurons or from sensory information) are encompassed within variable I (which in our version is also a random variable) and, a, b, c and d are parameters whose effects are summarised in Fig 2. Essentially, different values of these parameters produce different individual intrinsic neuron firing patterns such that complex spiking, bursting or chattering of cortical and thalamic neurons can be simulated.

5 Controlling a Granular Sampler

The algorithm by which the granular sampler works is straightforward: When a neuron in the network fires at time t, a sound grain of random length (between 10-50ms) and random amplitude is taken from a random place in a recorded sample of sound and played back. The sound grain is convoluted within a Hanning envelope [11]. Effectively, the neural network plays a granular sampler. Synchronized firing of neurons sound like a pulse, whilst networks containing only a few neurons have a very interesting sparse rhythmic quality (between completely random and correlated. The system therefore has a very wide variety of temporal patterns and behaviours, which can be controlled according to the parameters in the mathematical model. One can control the parameters a, b, c and d, which determine the intrinsic properties of the neurons and one can control the number and type of neurons. In the current version, the connections are completely noisy in the sense that the matrix S is a random matrix and all current inputs are noisy. However it would be straightforward to extend the model by varying the connections and the input ('thalamic') current. Generally speaking, increasing the number of neurons in the model means more firing and therefore more sonic texture, although when the solutions exhibit synchronous behaviour increasing the number of neurons tends to lower the frequency of the collective response. It is interesting in itself that such random (noisy) inputs can produce synchronous pulses of sound.

Generally speaking, in this version of the model (without any temporal correlation such as Spike Timing Dependent Plasticity [7]) one gets interesting sounds if we have either rather few (up to 10) or very many (over 500) neurons. The result with up to 10 neurons sounds very sparse but one can hear rhythms, which appear and then transiently die away. They do not repeat exactly; the network is effectively isolated from any sensory input (unlike real neurons in a mammalian cortex) and therefore not stimulated by correlated information. The synchronous solution appears in the dynamics if all the neurons selected are the same and if there are more than 500 of them. This sounds like a very gritty pulse, especially if the selected grain size is short.

6 Concluding Remarks

The technique we have introduced successfully fulfills the object of our enquiry in that its domain lies right in between the completely random and completely correlated in its temporal behaviour. Given a set of initial parameters, outputs are not predictable fully due to the large number of noisy elements in the model, but do follow discernable dynamical patterns especially when the system is in a dynamically synchronised state. The output is also controllable to a large extent. There would seem to be much profitable study from looking at this method further.

Acknowledgements

We would like to thank Stephen Coombes, University of Nottingham's Institute of Neuroscience, for stimulating discussions.

References

1. Gabor, D. (1947). "Acoustical quanta and the theory of hearing", *Nature*, 159(4044):591-594.
2. Miranda, E. R. (2002). *Computer Sound Design: Synthesis Techniques and Programming*. Oxford (UK): Elsevier/Focal Press.
3. Miranda, E. R. (1995). "Granular Synthesis of Sounds by Means of a Cellular Automaton", *Leonardo*, 8(4):297-300.
4. Truax, B. (1988). "Real Time Granular Synthesis with a Digital Signal Processor", *Computer Music Journal*, 12(2):14-26.
5. Blackwell T. M. and Young M. (2004). "Swarm Granulator", *Proceedings of Applications of Evolutionary Computing EuroWorkshops 2004*, LNCS 3005, Springer-Verlag, pp. 399-408.
6. Ikegaya Y., Aaron G., Cossart R., Aronov D., Lampl L., Ferster D. and Yuste R. (2004), "Synfire chains and cortical songs: Temporal modules of cortical activity", *Science*, 304:559-564.
7. Gerstner, W and Kistler, W. M. (2002). *Spiking Neuron Models*. Cambridge (UK): Cambridge University Press.
8. Izhikevich, E. M. (2003). "A simple model of a spiking neurons", *IEEE Transactions on Neural Networks*, 14:1469
9. Hoppenstadt, F. C. and Izhikevich, E. M. (1997). *Weakly connected neural networks*. New York (NY): Springer-Verlag.
10. Izhikevich E. M. (in print). "A Simple model of a spiking network' *IEEE Transactions on Neural Networks*".
11. Abramovitz M. and Stegun I. (1972) *Handbook of Mathematical Functions*. New York (USA): Dover.

Growing Music: Musical Interpretations of L-Systems

Peter Worth and Susan Stepney

Department of Computer Science, University of York, York YO10 5DD, UK

Abstract. L-systems are parallel generative grammars, used to model plant development, with the results usually interpreted graphically. Music can also be represented by grammars, and it is possible to interpret L-systems musically. We search for simultaneous 'pleasing' graphical and musical renderings of L-systems.

1 Introduction

L-systems are parallel generative grammars [12], originally defined to model plant development. Starting from an axiom string, or 'seed', the grammar rules are applied in parallel to each element of the string, for several iterations or generations. For example, consider the following L-system [12]:

ω: X $\qquad$ p_1: X → F[+X][-X]FX $\qquad$ p_2: F → FF

Starting from the axiom ω, successive generation strings are:

0: X
1: F[+X][-X]FX
2: FF[+F[+X][-X]FX][-F[+X][-X]FX]FFF[+X][-X]FX

and so on. The resulting string is typically rendered graphically, by interpreting the elements as turtle graphics commands [10]. For example, interpreting F as 'forward distance d, drawing a line', ± as 'turn through ± δ degrees', [ ] as 'start/end branch', and X as null, then after 5 generations the example L-system renders as a 'leaf':

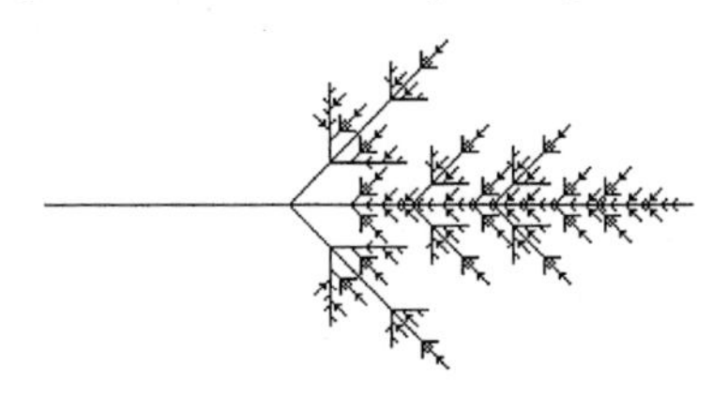

Non-graphical renderings can be considered. Here we consider musical renderings, and ask: "is it possible to have simultaneous 'pleasing' graphical and musical renderings of L-systems?"

2 Musical Grammars

The idea of generating music algorithmically is not new. The earliest recorded work was by the Italian monk Guido D'arezzo in 1026. Demand for his Gregorian chants

F. Rothlauf et al. (Eds.): EvoWorkshops 2005, LNCS 3449, pp. 545–550, 2005.

was so high that he devised a system to systematically create them from liturgical texts. Mozart, Haydn, and C.P.E. Bach had an interest in generative music; Mozart invented *Musikalisches Würfelspiel* (musical dice game), which involved using dice to decide which of a set of pre-defined musical phrases came next in the piece [9].

Heinrich Schenker's (1868–1935) work on the analysis of tonal master works provides an insight into the formal organisation of music. He broke pieces down into their background, middleground, and foreground [4]. These are structural levels, each of which intuitively fits the idea of description by a formal grammar. The *fundamental line* (*urlinie*) gives the tonal progression of the piece which is generally part or all of a scale. This low level structure can be embellished by expanding the components into more complicated sections until the foreground is reached.

Music as grammar has been widely investigated, eg [3][6][13]. [1] describes a context sensitive grammar for generating European melodies; these are structured around a *kernel*, the sequence of all the notes in a scale between arbitrarily chosen first and last notes, and the melody is the way the notes move around the kernel.

[11] maps the turtle drawing into musical score, by using a lookup table to map y co-ordinates to notes, and line lengths to note durations. [15] maps branching angles to changes in pitch, and distance between branches to duration. [14] maps the turtle's 3D movement, orientation, line length, thickness, colour, programmably into pitch, duration, volume, and timbre. [7] uses L-system grammars directly to represent pitch, duration and timbre, without going via a graphical rendering. This gives a better separation of concerns than deriving the music from the graphical rendering, and we follow that approach here.

3 Plants to Music : Finding a Rendering

First, we experiment with existing L-systems that produce pleasing-looking 'plants', and try to discover pleasing musical renderings of these.

(All the examples below are taken from [12], unless otherwise stated, and all the musical examples discussed here can be listened to at the website http://www-users.cs.york.ac.uk/~susan/bib/ss/nonstd/eurogp05.htm).

Music is essentially sequential in time: we do not want a temporal branching interpretation. We define a *sequential rendering*: interpret [/] as 'push/pop current state *except* the time'; F as 'play a note of duration 1'; a sequence of n Fs as 'play a single note of duration n'. So a note is broken by a change in pitch, by a new branch, or by the current branch ending. The *sequential rendering* of the 4-generation 'leaf' L-system is rhythmically interesting, and makes sense melodically:

Although the sequential rendering produces pleasant results, it can be improved to capture a Schenkerian background/middleground/foreground hierarchy. Jonas [4]

uses the term "the flowerings of diminution" in describing the sonata form. This suggests an interpretation in which only the ends of the plant (leaves, flowers) are "heard". The middleground and background (the stem and branches) are not actually heard in Schenkerian analysis: they just give the structure from which the foreground appears.

In our *Schenkerian rendering*, interpret F as 'increase note duration by a quarter note', ± as 'move up/down one note in the chosen scale', [as 'push current state and set note duration to 0',] as 'play note according to current state, and pop', and X as null.

Under this rendering, the 'leaf' now plays as:

Despite not appearing to fit into a 4/4 framework, this melody sounds very musical, with a quite distinctive tune, even with a metronome beating 4/4 time behind it.

4 Stochastic L-Systems

Plants are all different: stochastic L-systems are used to generate plants from the same "family" but with different details. A musical rendering should similarly generate a variety of pieces in the same "style". Consider the following simple stochastic L-system (where the subscript on the arrow gives the probability that rule is chosen).

$$\omega: F \quad p_1: F \rightarrow_{1/3} F[+F]F[-F]F \qquad p_2: F \rightarrow_{1/3} F[+F]F \qquad p_3: F \rightarrow_{1/3} F[-F]F$$

The Schenkerian renderings of three different productions of this stochastic L-system (3 iterations deep) are:

These sound 'random' but well structured, and not overly complex (as one would expect from the fairly simple nature of the rules). They do sound 'similar' to each other, but different enough to be used perhaps at different points in the same piece of music, or when combined.

5 Context-Sensitive L-Systems

Context sensitivity in L-systems gives more power as parts of the string or plant can grow differently depending on what is around them. This could be useful in music since a generated piece could build to a climax or break down at certain points. In a context-sensitive Lsystem the production rule is applied to symbol only if it appears in a specific context (between other symbols). The notation A<B>C means the string B with A to the left, and C to the right.

Consider the following context-sensitive L-system, from [2].

ω: F1F1F1

p_1: $0<0>0 \rightarrow 0$

p_2: $0<0>1 \rightarrow 1[+F1F1]$

p_3: $0<1>0 \rightarrow 1$

p_4: $0<1>1 \rightarrow 1$

p_5: $1<0>0 \rightarrow 0$

p_6: $1<0>1 \rightarrow 1F1$

p_7: $1<1>0 \rightarrow 0$

p_8: $1<1>1 \rightarrow 0$

p_9: $+ \rightarrow -$

p_{10}: $- \rightarrow +$

This melody, and others derived similarly, sound fairly 'random' (despite being deterministic); they are reminiscent of jazz solos. They do not fit well into 4/4 score notation because many of the notes are offbeat, but this just adds to their "freeform" sound. Yet the tunes always return to a main motif or phrase, that is sometimes transposed or played at a different point in the bar. For example, in the score above, the series of notes in the 1st bar is repeated in the 9th bar, but very offbeat (moved forward a quarter of a beat) and raised by 2 semitones. This kind of repetition mirrors how music is normally composed or improvised.

6 Music to Plants

Previously we started from existing L-system 'plants', and tried interpreting them as music. Here we take the opposite approach, of starting from musical grammar notations, and trying to produce L-system versions.

We combine the ideas of Jones [4] and Baroni *et al* [1] to write a formal grammar that generates music by recursively splitting up an event space (initially one long note) into 2 or 3 shorter, different notes. After a number of recursions we have a melody that is the length of the initial event space. Insertion rules [1] provide tonal information (we add an 'identity' insertion that does nothing), and halving note duration rules provide the rhythm. These insertion rules were initially written for analysis; adding probabilities as in stochastic musical grammars [5] allows them to be used for production. Variations of the rules of insertion and the rhythm grammar are given below.

We interpret d as ‘halve the duration’. We get the following grammar:

Identity:	F $\rightarrow_{1/2}$ F		
Repetition:	F $\rightarrow_{1/26}$ [dFF]		
Appogiatura1:	F $\rightarrow_{1/26}$ [d-F+F]	Appogiatura2:	F $\rightarrow_{1/26}$ [+F-F]
Neighbour note1:	FF $\rightarrow_{1/26}$ [Fd+F-F]	Neighbour note2:	FF $\rightarrow_{1/26}$ [Fd-F+F]
Skip1:	F+F $\rightarrow_{1/26}$ [Fd++F-F]	Skip2:	F+F $\rightarrow_{1/26}$ [Fd+++F--F]
Skip3:	F+F $\rightarrow_{1/26}$ [Fd-F++F]	Skip4:	F+F $\rightarrow_{1/26}$ [Fd--F+++F]
Skip5:	F-F $\rightarrow_{1/26}$ [-Fd++F-F]	Skip6:	F-F $\rightarrow_{1/26}$ [-Fd+++F--F]
Skip7:	F-F $\rightarrow_{1/26}$ [-Fd-F++F]	Skip8:	F-F $\rightarrow_{1/26}$ [-Fd--F+++F]

Starting from the axiom F++F++F+++F---F--F--F, using the *sequential rendering* and the classical turtle graphical rendering, after 4 iterations we get

The tune is pleasant. The graphical rendering (to its left) looks *somewhat* plantlike, but is not very aesthetically appealing. Starting from the musical grammars, it is unclear how to add the necessary branching instructions to get pleasing-looking plants.

7 Conclusions and Further Work

We present two musical renderings that produce pleasant sounds from classic ‘plant’ L-systems. The sequential rendering is relatively naïve, yet works well. The Schenkerian rendering is inspired by an analogy between the musical theory concepts of fore/middle/background and the components of a plant, and produces very pleasant pieces.

These examples have been evaluated to a depth of 3 or 4 iterations only. There seems to be enough information in a typical L-system to create only a short melody and still be interesting. At longer derivations, the melodies begin to get dull: the same bit of music is repeating continually, albeit normally transposed in some way. Stochastic L-systems may help, by enforcing some kind of structure on the score but giving varied melodies. The context-sensitive L-systems seem to offer the best potential for creating longer pieces of music, since identical parts of the string in different places can grow differently, so the piece can actually “go somewhere” rather than repeat the same pattern.

Starting from musical grammars and producing L-systems from them works well musically. However, the attempt to get simultaneously pleasing graphics starting from a musical grammar has been less successful: the branching necessary for graphics is not an intrinsic part of existing musical theory, and it is not clear how to add it in. More work on the Schenkerian rendering from a music theory point of view may be valuable here.

More powerful L-systems, such as parametric L-systems, could be used to generate more complex and realistic music. One exciting possibility is the use of L-systems with environmental inputs [8]. These have been developed to model environmental effects on plant growth (sun, shade, etc), but might be applicable to music generation, to allow two L-systems growing their music together as different "instruments" to react to each other.

References

[1] M. Baroni, R. Dalmonte, C. Jacobini. Theory and Analysis of European Melody. In A. Marsdon, A. Pople, eds, *Computer Representations and Models in Music*, 187-206. Academic Press, 1992

[2] P Hogeweg, B. Hesper. A Model Study On Biomorphological Description. *Pattern Recognition* **6** 165-179, 1974

[3] S. R. Holtzman. Using Generative Grammars for Music Composition. *Computer Music Journal* **5**(1):51-64, 1981

[4] O. Jonas. Introduction to the Theory of Heinrich Schenker. Longman 1972

[5] K. Jones. Compositional Applications of Stochastic Processes. *Computer Music Journal* **5**(2):45-61, 1981

[6] F. Lerdahl, R. Jackendoff. *A Generative Theory of Tonal Music*. MIT Press, 1983

[7] J. McCormack. Grammar-Based Music Composition. In Stocker *et al*, eds. *Complex Systems 96: from local interactions to global phenomena*, 321-336. IOS Press, 1996

[8] R. Měch, P. Prusinkiewicz: Visual models of plants interacting with their environment. *Proc. SIGGRAPH 1996*, 397-410. ACM SIGGRAPH 1996

[9] A. Moroni, J. Manzolli, F. Von Zuben, R. Gudwin. Vox Populi: Evolutionary Computation for Music Evolution. In P. J. Bentley, D. W. Corne, eds, *Creative Evolutionary Systems*. Academic Press 2002

[10] S. A. Papert. *Mindstorms*. Harvester Press 1980

[11] P. Prusinkiewicz. Score Generation with L-Systems. *Proc. Intl. Computer Music Conf '86*, 455-457,1986

[12] P. Prusinkiewicz, A. Lindenmayer. *The Algorithmic Beauty of Plants*. Springer 1990

[13] C. Roads. Grammars as Representations for Music. *Computer Music Journal* **3**(1):48-55, 1979

[14] D. Sharp. LMUSe, 2001. http://www.geocities.com/Athens/Academy/8764/ lmuse/

[15] F. Soddell, J. Soddell. Microbes and Music. *PRICAI 2000*, 767-777. LNAI 1886, Springer 2000

Exploring Rhythmic Automata

Andrew R. Brown

Queensland University of Technology, Brisbane, 4059, Australia

Abstract. The use of Cellular Automata (CA) for musical purposes has a rich history. In general the mapping of CA states to note-level music representations has focused on pitch mapping and downplayed rhythm. This paper reports experiments in the application of one-dimensional cellular automata to the generation and evolution of rhythmic patterns. A selection of CA tendencies are identified that can be used as compositional tools to control the rhythmic coherence of monophonic passages and the polyphonic texture of musical works in broad-brush, rather than precisely deterministic, ways. This will provide the composer and researcher with a clearer understanding of the useful application of CAs for generative music.

1 Introduction

Algorithms utilising discrete states and rule-based transitions are particularly suited to digital computers, and thus the developments of Cellular Automata (CA) have paralleled the history of the computer. CA are spatially specific finite state systems where cells represent discrete values. The cells are arranged in grids of any dimension, with 1 or 2 dimensions being the most common. A transition function determines the subsequent state of each cell depending upon its previous state and that of its neighbours. Each cell in has a state that is most commonly either 1 or 0. Transition functions are typically made up of rules such as "If the sum of my and my neighbours state is less than two then set my state to 0." Normally all cell transitions are computed in parallel and updated synchronously. Over the years, musicians have utilised CA because of parallels with sound patterns and structure, and because CA provide a great deal of complexity and interest from quite simple initial setup. Composers and sound designers have mapped CA states to musical form and macrostructure, note-level patterns, and parameters of sonic microstructure, see for instance [1-9]. For a detailed overview of the musical application of CA see the review of CA music research Burraston and his colleagues [10]. Many previous studies work systematically through known CA rules and are most often concerned with pitch mapping. This study differs in that it is concerned with rhythmic mappings of the generated data and the exploration is driven more by musical or aesthetic necessities than by systematic or exhaustive searches. The result is that formal models are explored in an informal way and new system of rule description that embeds musical knowledge is described.

F. Rothlauf et al. (Eds.): EvoWorkshops 2005, LNCS 3449, pp. 551–556, 2005.

The software used for this exploration of rhythmic automata was specially written in Java using the jMusic library [11]. The study was limited to one-dimensional CA where two states, 1 and 0, were 1 equalled a note and 0 a rest. Notes and rests were all considered to be of equal duration, usually a sixteenth note. A single row of cells represented a step sequence of notes and rests. Each generation of the row was treated as a musical phrase and performed in order using unpitched percussion samples triggered via MIDI playback. The length of a row was kept to sixteen cells. The correspondence between the binary and common practice notation representations of the grid is shown in Figure 1, while the visual patterning of a sequence of row evolutions is quite apparent in Figure 2 which has a grid of rows ordered temporally from top to bottom. The representation of the grid evolution used in Figure 2 is well established; each row in the grid is a single generation with earlier generations at the top.

1 0 0 1 1 1 0 0 0 1 1 0 0 0 1 1

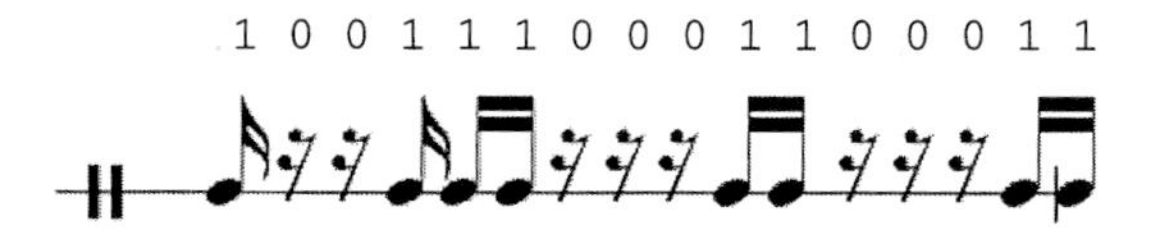

Fig. 1. The binary and notational representations of the same CA rhythm

```
1 0 1 1 0 0 1 1 0 0 0 0 1 1 1 1
0 1 0 0 1 1 0 0 1 0 0 1 0 0 0 0
1 1 1 1 0 0 1 1 1 1 1 1 1 0 0 0
1 0 0 0 1 1 0 0 0 0 0 0 0 1 0 1
0 1 0 1 0 0 1 0 0 0 0 0 1 1 1 0
1 1 1 1 1 1 1 1 0 0 0 1 0 0 0 1
0 0 0 0 0 0 0 0 1 0 1 1 1 0 1 0
```

Fig. 2. An extended sequence where the generational patterning is visually evident

The initial explorations used Totalistic rules; where a transformation depends only on the sum of 1s in its neighbourhood, regardless of their position. The choice of Totalistic rules (a subset of all possible CA rules) was based on a compositional desire to work with rhythmic density where decisions were based on the more or less 'crowded' or 'busy' a section. The range of cell totals evidently depends on how many neighbours are taken into account. The neighbourhood scope used in these explorations was either one cell to either side. A variety of transformations were allowed for,. Set to a *rest* (0) = R, Set to a *note* (1) = N, *Flip* current status = F, Leave *unchanged* = U.

The mapping between the summed value of 1s in the cell neighbourhood and a transformational outcome defines a transition. The totals and outcome symbols can be combined to express the transition such as 3 -> U, which can be read as "if the sum of this and neighbouring cells equals 3 then leave the current cell state as it was." A set of transition expressions is a CA rule, for example 0 -> R, 1 -> N, 2 -> R, 3->R. For those more familiar with established CA rule numbering, here is a simple comparison.

The CA rule number 22 would be described as 0 -> R, 1 -> N, 2 -> R, 3->R. The use of only four rules indicates that the scope of neighbours is one either side of the cell. It should be noted that the *flip* and *unchanged* transitions are not used in the established CA rule numbering system, however, rules in the notation established in this paper are a subset of established system and do have an equivalent CA rule number.

The CA evolution is also influenced how boundaries are handled. In the explorations reported here boundaries wrap around, linking the first and last cell to effectively create a circular row. The starting condition of a CA also has a significant influence on its development. In the explorations reported here, understanding the tendencies of rule across a range of starting conditions was of interest, so randomised starting conditions were generally used while occasionally simple start conditions with a single note were used.

2 Rules and Techniques

Stephen Wolfram [12], classified the transition rules into four classes of behaviour according to their stability and complexity. Class 1 CA eventually evolve to a stable state, class 2 CA result in patterned output dependent upon initial conditions, class 3 CA are aperiodic, and class 4 CA produce complex but stable evolution. These classifications informed the explorations of CA for musical rhythm described in this paper, however as with all such classifications of CA's, there is no mathematical proof or law that defines rules into these categories.

Given that much of the potential output for generative processes is quite chaotic, it is useful to observe that some CA rules-sets can assist in providing order, as with Wolfram's class-one transition rules. Because of their repetition, stabilising rules can be utilised in many musical situations. For example, the following rule will produce a stable rhythm pattern after a few generations.

0 -> N, 1 -> U, 2 -> R, 3 -> N

Delayed Stability. Some rules result in a slower evolution towards stability which can be very useful for providing both variation and cohesion. One example is the following, a rule.

0 -> N, 1 -> F, 2 -> F, 3 -> U

Under these conditions CA tend toward interesting variations with global patterns emerging for some time until a stable point is reached. From randomised starting points, they take between 10 to 100 generations to stabilise (in general). The stable point can be a rhythm (notes and rests) or silence (all rests) depending upon the outcome specified for the zero-neighbour state. Musically this could be useful for sections of change and stability could be detected and responded to.

Balancing the need for stability in musical rhythms is the desire for novelty and variation. This section lists some rules that provide particularly useful variability.

Rhythmic Inversion. In order to change all notes to rests and vice versa, all states are set to *flip*. This is the equivalent to using CA rule number 51. Over time the continual inversion of the pattern results in a perception of a double-length stable pattern This simple technique may be surprisingly musical provided that it is used in moderation. An inversion rule is:

0 -> F, 1 -> F, 2 -> F, 3 -> F

Density Thinning. A rule where most transitions result in rests will provide a thinning of dense or complex rhythms. An assumption, here, is that long sections with only notes or only rests are undesirable and that an even distribution is more interesting. An example is the rule:

0 -> U, 1 -> R, 2 -> N, 3 -> R

Most CA evolution results in changing patterns, even though stable states are not uncommon. This continual variety provides rich opportunities for musical rhythms. Rules that result in large scale repetitions are particularly interesting, even though they may not be considered strictly evolutionary in the sense of continually varying development and change. When an evolving pattern has a zero-neighbour rule resulting in a *note* or *flip* outcome, a perpetual cycle is almost guaranteed.

Evolving Inversion. In a CA where most outcomes are set to *flip* except the zero-neighbour rule with a *note* or *rest* outcome, the pattern largely inverts between generations but is irregularly mutated. This type of rule can produce some slowly changing double-length patterns. An example rules is:

0 -> N, 1 -> F, 2 -> F

Emergent Cycles. With some rules large-scale patterns emerge. For example, the rule 0-> N, 1 -> R, 2 -> R, 3 -> U produces patterns that often fall into cycles that repeat over several generations. However, the establishment of cyclic patterns is quite sensitive to initial conditions and cyclic patterns are less likely with small row sizes. As a result this behaviour is insecure for real-time applications but adequate for the composer who can generate and select.

While the CA process can provide tendencies toward particular density and pattern distribution, aperiodic tendencies and use of circular boundary conditions often result in a lack of pulse or metre. While this might be intellectually fascinating it is only occasionally successful from the perspective of a common aesthetic. The resulting rhythmic structures have a weak sense of pulse on several counts, 1) patterns can be too long for human perception to appreciate, 2) pulse is undermined by the circularity of the cell neighbour checking, and 3) any sense of regularity is smeared by the constant variation imposed by the pattern change at each generation. One solution is to modify the CA rules to account for cell position within the row. In this way musically 'stronger' and 'weaker' positions within the row are specifiable. This is a strong form of what is sometimes referred to as *elementary* rule, where a cell ignores its neighbourhood context. As well, a convention of applying the same rule to all cells is also broken in the beat prioritisation.

Probabilistic Transitions. To achieve beat prioritisation an elemental probabilistic rule is applied to regularly spaced cells, for example in positions 0, 4, 8, and 12 only, such that they would transition to a note with a specified probability. This can provide a sense of pulse or metre. The beat prioritising with a low probability can also help consolidate an otherwise unstable CA rhythm.

A side effect of prioritising beats is that the increase in note activity extends evenly forwards and backwards through the row when it would be more musical to start or finish the rhythm activity *on* the beat, rather than *around* it. To counter this Totalistic rules were modified to take cell position into account.

Position Sensitive Rules. A simple implementation of this position orientation is to only count neighbours to one side of the current cell. Only cells to either the left or right can be considered where calculating the neighbourhood sum. For example, three cells – 1 1 0 – would total two when 'looking left' or total one when 'looking right'. The effect of adding this feature to rules with beat priority is to distort the angular patterns as desired, however the clustering tendency of notes and rests can persist. These rules are represented as in this example:

0 -> U, 1 -> N, 2 -> N, 3 -> R: Pos. 1, 5, 9, 13 -> Prob. 0.7: Look left

Rhythmic automata become particularly exciting when several patterns are played together. Subtle changes in the rules are amplified over time and so some careful hand tweaking of rules can pay significant dividends for the patient composer. Some obvious possibilities exist for polyrhythmic automata works, including where an evolving CA can be played against a static one, or when two different evolving automata can be used, or when the same rule can be used with different starting conditions in each part. Traditional cannonic effects can be achieved by starting CA patterns at different times.

Rhythmic Phasing. Phasing effects, such as those employed by Steve Reich, use the same material at slightly different speeds. A variation on this that does not require tempi adjustment is to counterpoint rhythmic automata that have rules which vary only slightly. Carefully chosen, these will start together and drift into quite independent rhythms over several generations. An example is the rules 0 -> R, 1 -> N, 2 -> *, 3 -> R; with the two-neighbour outcome (*) set to *rest* in the first example and to *flip* in the second.

Alternate Starting Points. A cannonic effect can be mimicked without the need for delayed starting times, by using starting states that are equivalent but bit-shifted along (around) the row. The resultant synchronisation of phase offset provides an effective tension between musical coherence and contrast.

3 Conclusion

This paper has reported on an exploration of CA for the generation of rhythmic patterns. The approach to CA has been somewhat unorthodox in a number of ways, particularly in the way rules have been described. The rule formation is more descriptive than the conventional numbering system and usefully limits the CA rule space by at times ignoring (totalistic) or generalising (look left and right) neighbour position when specifying rules. In addition, probabilistic elemental rules that emphasise beat or accent locations, have been trialed. This approach incorporates musical knowledge into the rule representation system and has the advantage over the conventional numbered CA rule system that it provides a more musically meaningful way of specifying and changing the rules to providing indirect control over density and metre, compositional outcomes. A series of rhythmic automata tendencies useful for musical applications have been identified and example rules described. This research enables a directed search of the space of CA rule possibilities for rhythmic material and, it is hoped, will stimulate further utilisation of rhythmic automata.

Acknowledgements

This research has been partially supported by the Australasian CRC for Interaction Design.

References

1. Beyls, P.: The Musical Universe of Cellular Automata. International Computer Music Conference. ICMA, Columbus Ohio (1989) 34-41.
2. Burraston, D. and Edmonds, E.: Global Dynamics Approach to Generative Music Experiments with One Dimensional Cellular Automata. Australasian Computer Music Conference. ACMA, Wellington: (2004)
3. Hoffman, P.: Towards an "Automated Art": Algorithmic Processes in Xenakis' Compositions. Contemporary Music Review, Vol. 21 No. 2/3 (2002) 121-131
4. Millen, D.: Cellular Automata Music. http://comp.uark.edu/~dmillen/cam.html (1988 - 2004)
5. Millen, D.: Cellular Automata Music. International Computer Music Conference. ICMA, Glasgow (1990) 314-316
6. Miranda, E. R.: Composing Music with Computers. Focal Press, Oxford (2001)
7. Miranda, E. R.: On the Music of Emergent Behavior: What Can Evolutionary Computation Bring to the Musician? Leonardo, Vol. 36 No. 1 (2003) 55-59
8. Reiners, P.: Autonomous Monk http://www.automatous-monk.com/ (2004)
9. Reiners, P.: Cellular Automata and Music: Using the Java language for Algorithmic music composition. IBM, New York http://www-106.ibm.com/developerworks/java/library/j-camusic/ (2004)
10. Burraston, D., Edmonds, E., Livingstone, D. and Miranda, E. R.: Cellular Automata in MIDI based Computer Music. International Computer Music Conference. ICMC, Miami (2004)
11. Sorensen, A. and Brown, A. R.: Introducing jMusic. Australasian Computer Music Conference. ACMA, Brisbane (2000) 68-76
12. Wolfram, S.: Universality and Complexity. Physica D, Vol. 10 (1984) 1-35.

Extra-Music(ologic)al Models for Algorithmic Composition

Alice C. Eldridge

Creative Systems Lab. University of Sussex, East Sussex, BN1 9QH, UK
alicee@sussex.ac.uk

Abstract. This paper addresses design approaches to algorithmic composition and suggests that music-theoretic tenets alone are unsuitable as prescriptive principles and could be profitably complemented by attempts to represent and recreate dynamical structures of music. Examples of ongoing work using adaptive dynamical processes for generating dynamic structures are presented.

1 Aims, Assessment and Approaches

In [16], Pearce et al bring attention to the diversity of motivations and techniques within the field of automated composition and suggest that research aims needs to be more explicit in order to implement suitable assessment procedures. Conversely, clarity in aim and assessment focuses consideration of design approaches. This paper is concerned with designing Algorithmic Composition systems, with the aim of producing music to "expand the compositional repertoire for human listeners", [16]. As suggested, the output can be usefully assessed in the same way that all other composition is appraised: through listener's response to publicly disseminated material (concerts, recordings etc). This is a seemingly obvious, but important point as it implicitly states that our primary concern is the development of a system capable of producing music which serves the same *function* as music created in more traditional ways: something that people can have some "some degree of meaningful or gratifying perceptual engagement with" [6].

How then should we best approach the design of these systems? A common approach is to embed music-theoretic principles, either explicitly in knowledge based systems [9] or implicitly in evolutionary or learning algorithms [17]. This is attractive as it ensures our system follows the 'rules' of music theory, but results are often described as lacking 'life' or 'musical logic' [7]. At the other extreme mathematical models, selected on the basis that they represent some aspect of a musical phenomena, are used to generate data which is then mapped onto musical parameters. These are often described as an 'extra-musical' and seen to be inherently less musical because "their 'knowledge' about music is not derived from human works" [15] p 2.

In this paper we briefly consider the nature of musicological thinking and suggest that the principle focus differs from that of the average listener's immediate aural experience. An alternative understanding of music from the listener's

F. Rothlauf et al. (Eds.): EvoWorkshops 2005, LNCS 3449, pp. 557–562, 2005.

perspective is presented, highlighting the fundamental importance of motion in music. Examples of 'extra-musical' algorithms designed to create dynamic structures are presented, to suggest that 'extra-musical' algorithms can at times capture important musical characteristics that evade purely theoretic approaches, and are perhaps misnamed.

2 Musicological Versus Musical Perspectives

2.1 The Analyst's Music

Music theories, represent attempts to *understand* music in a Musicological sense; the analyses aim to achieve possible coherent sets of principles and ideas with which to rationalise, investigate and analyse the structurally functional aspects of music. This act is neither exhaustive, nor aimed primarily at describing music in terms of the listener's perception. "A formal analysis is a kind of mechanism whose input is the score, and whose output is a determination of coherence...In other words, it purports to establish or explain what is significant in music while circumventing the human experience through which such significance is constituted; ... it aims at 'deleting the subject'" [6] p.241

In [6] Cook argues that there is an important and inevitable discrepancy between the experience of music aurally, and the ways in which it is thought about. He draws a useful distinction between 'musical listening' which is concerned with aesthetic gratification in a non-dualistic sense, and 'musicological listening' for the purpose of establishing facts or formulating theories about music.

The discrepancy between analytic and experiential musical realities is illustrated by two experiments. In one, two versions of short piano pieces were played to music students: their original form, which began and ended in the same key, and an altered form which had been modified so as to modulate to and end in a different and unrelated key, sometimes as distant as the minor second [4]. Theoretically, tonal closure is the very core of standard classical music forms. However in these trials, Cook reports no significant differences in preference for the original over the altered forms. In another set of tests [5], music students were played the first movement of a Beethoven sonata. The performance was broken off just before the final two chords, yet many predicted that the music would carry on for another minute or more. Theoretically, the recapitulation and coda are key functional structures, signifying the close of a piece: these results suggest that they can be insignificant in an aural presentation.

More conclusive studies are needed to make strong claims, but it is common for musicologists to differentiate between the aural and analytic aspects of a piece. Kathryn Bailey describes Webern's symphony as "two quite different pieces - a visual, intellectual piece and an aural, immediate piece, one for the analyst and another for the listener." [2] p.195. Thomas Clifton expresses this more incisively: "For the listener, musical grammar and syntax amount to no more than wax in his ears." [3] p.71.

These discrepancies stem in part from the contrasting nature of time in aural and written music. Schutz suggests that attempts to describe the musical

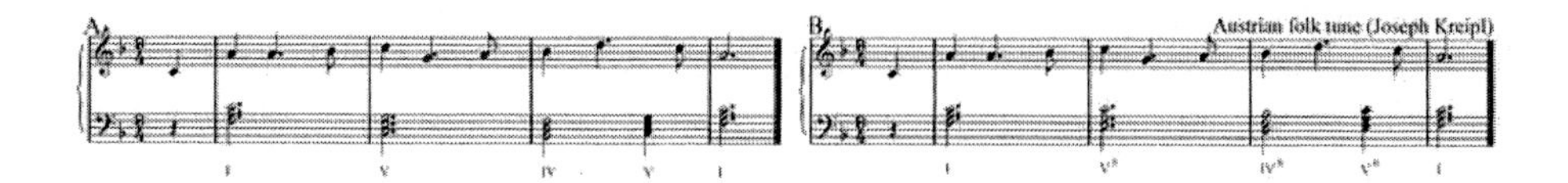

Fig. 1. A Natural harmonisation of a simple phrase using I (tonic), V (dominant) and IV (sub-dominant)(left) and appropriate chordal inversions (IV^betc) (right)

experience in the 'outer' time of reflection and notation, poses a variant of the Eleatic paradox - ie that the flight of Zeno's arrow cannot be described because the ongoing quality of its motion cannot be represented [18]. And indeed many composer-theorists who are primarily concerned with the listener's experience focus on the fundamentally dynamic and continuous nature of music eg [20], [19].

2.2 The Listener's Music

In [19] Toch gives an account of how all musical writing must respond to the listener's psychological needs. If harmonic structure is the cornerstone of traditional music theory, Toch sees the movement of melodic 'impulses' as the central force of music from the listener's perspective.

In an example that is not unlike some algorithmic composition tasks set eg for Genetic Algorithms (GA), Toch presents a phrase from a folk tune, that invites a simple I, IV V harmonisation (Fig.1A). Even the inversions required to produce a smoother chordal structure (Fig.1B) could potentially be found by implementing theoretic axioms such as minimising the number of steps that each note must take into membership of its adjacent harmonies. The apparent simplicity and efficacy of this kind of 'rule' is precisely what is attractive to the algorithmic composer, but as Toch warns: "While this axiom seems a simple expedient for the beginner, it implants in him a dangerous misconception, namely the view point of rigidly preconceived harmony as a fixed unit, within the frame of which each voice seeks to take up its appropriate place." [19] p5.

The point is illustrated by comparing a typical Chorale harmonisation, which concedes to all the traditional rules of harmony, with the alternative harmonisations, which Toch arrives at by a more general principle that he calls 'linear voice leading'. In contrast to the 'appropriate' harmonisations, some of these voice-led harmonies go against every rule in the book: consecutive fifths, cross relations, arbitrary dissonances etc. And yet, he argues, "they convey a certain organic logic and life." He continues: "The truth is that the melodic impulse is primary, and always preponderates over the harmonic; that the melodic, or linear impulse is the force out of which germinates not only harmony but also counterpoint and form. For the linear impulse is activated by *motion* and motion means life, creation, propagation and formation." ibid p.10.

Toch's approach stands in stark contrast to the way in which the 'harmonisation problem' is sometimes addressed in algorithmic composition: "We apply the following criteria: we avoid parallel fifths, we avoid hidden unison, we forbid progression from diminished 5th to perfect 5th; we forbid crossing voices...", an approach that does not seem to capture the organic logic of successful music: "...from an aesthetic perspective, the results are far from ideal: the harmonisation produced by the GA has neither clear plan or intention." [17] p.5.

This is an extreme, although not untypical application of music theory to the design of algorithmic systems. Even if Cook's suggestion that theoretical structures are aurally irrelevant is not upheld, it seems likely that a theory derived largely from a static symbolic score may not tell us everything we need to know to design systems that can emulate the dynamic immediate nature of music that stimulates the listener's subjective experience. Over-reliance on such theories may be one reason why we frequently see comments such as: "while conforming to classical triadic harmony, the music seems lifeless." [7] p.21.

The musical subject is making more frequent appearances in musicological research [8], and practitioners are beginning to employ psychological measures such as harmonic tension in developing algorithmic composition systems [14]. We suggest that as well as consideration of emotive qualities, algorithmic composition design could benefit from considering ways of representing and generating the dynamic structures of music.

3 Adaptive Systems Music

In an ongoing project, various dynamical system models are being explored as possible mechanisms for creating a sense of musical motion and progression. Models based on differential equations enable the representation of temporal flow as the system is defined in terms of change in state over time. In a sense, the Eleatic paradox is resolved. Both systems described below are small networks of interconnected nodes. In each case the state of each node is a function (directly or indirectly) of the activity in the rest of the network. An attractive property of this class of dynamical system then, is that as well as enabling the representation of dynamic structures, a certain logical structure is present. One possibility being examined, is whether these structures can create musically coherent relationships.

3.1 Neural Oscillator Networks

A small network of neural oscillators as described by [13], was built. The oscillator system consists of two simulated neurons arranged in mutual inhibition. If an oscillatory input is applied, the pair will entrain the input frequency, although the phase may differ. The nodes are arranged in a network such that the input signal to one pair is the output from one or more other pairs. This creates a collection of continuous periodic output signals, with differing phase and form which are synchronised. The outputs are mapped onto frequency changes.

Although not restricted to any particular key or time signature, the periodic form - similar to the wave-like form of many melodies [19]- produces a sense of movement. Because all outputs are synchronised, but exhibit different forms and phases, a sense of ensemble is achieved with voices moving in unison or opposition according to phase [1]. In this current form, the outputs - being periodic - are very repetitive. This is not presented as music in itself, but as a possible mechanism for conveying movement, and dynamic relations between parts.

3.2 Homeostatic Harmonies

In previous work [10], the potential for generating harmonies using principles of homeostasis has been explored. The model, based on Ashby's homeostat [1], can be conceived as a number of interconnected nodes, where the output of each node is a function of the weighted sum of all connected node outputs. If any one output exceeds a critical value, all the weights in the network are re-randomised. This means that the system converges to a stable state, and reacts to environmental changes, creating local directed dynamics and global contrasts between stability and exploration. Such dynamics are comparable in abstract form to equilibrium-disturbance-reaction schemes of narrative form, or the basic repetitions and variations ubiquitous in musical forms.

Various live performances, compositions and installations made with versions of this basic system have been well received. Although not pertaining to any particular musical style or genre, participants in a listening test agreed that it was 'musical', elaborating their choice with comments such as: 'sense of melody', 'there were definite harmonies if unusual at times' ... 'sense of harmonic structure and melodic progression'. This reference to structure was made by several listeners: 'structure and development on different time scales/resolutions'. Listeners' comments also suggest that the output had emotive qualities: 'tension building and resolution of tension.' [10].

4 Summary and Discussion

Contrasting Cook's views on the nature of musicological thinking with Toch's ideas on compositional techniques with the listener's experience in mind, we propose that music-theoretic principles may be being misused as exclusive prescriptive design principles for algorithmic composition systems. It seems that as well as focusing on perspectives such as emulation of emotive qualities or 'colours' in music [14], research may benefit from an exploration of methods of representing and recreating the dynamic qualities or flux in music.

Examples of attempts to capture a sense ensemble or harmonic progression as an 'emergent' property of melodic movement using simple dynamical models were given. Although the relations between the nodes do not follow any musical model, it is an interesting possibility that the internal logical coherence creates musical structures that listeners describe as harmonic or melodic progressions. Whilst it is too early to make any claims in this area, investigations into the use of sound to analyse complex dynamic systems, suggest that the dynamics (eg

[1] Example output from can be found on-line at [12]

chaotic, complex, ordered) of some systems can be appreciated when presented aurally [11]. This suggests that attempts to create dynamical models of musical structures may be a fruitful avenue of research which could potentially contribute to our broader understanding of existing music as well as affording possibilities for creating new musical styles.

References

1. W.R. Ashby. *Design for a Brain; The Origin of Adaptive Behaviour.* Chapman and Hall Ltd and Science Paperbacks, 1965.
2. K. Bailey. Webern's Opus 21: Creativity in tradition. *Journal of Musicology*, 2:184–195, 1983.
3. T. Clifton. An application of Goethe's concept of *Steigerung* to the morphology of diminution. *Journal of music Theory*, 14:165–189, 1970.
4. N. Cook. Musical form and the listener. *Journal of Aesthetics and Art Criticism*, 46:23–29, 1987.
5. N. Cook. Perception of large scale tonal structure. *Music Peception*, 5:197–205, 1987.
6. N. Cook. *Music, Imagination and Culture.* Clarendon, Oxford, 1990.
7. D. Cope. Computer modelling of musical intelligence in EMI. *Computer Music*, 16(2):69–83, 1992.
8. M. Costa, P. Fine, and P.E.R Bitti. Interval distributions, mode and tonal strength of melodies as predictors of perceived emotion. *Music Perception*, 22(1):1–14, 2004.
9. K. Ebcioglu. An expert system for harmonizing four-part chorales. *Computer Music Journal*, 12(1):43–51, 1988.
10. A.C. Eldridge. Adaptive systems music: Algorithmic process as musical form. In *Proceedings of the 2002 Generative Art Conference*, Milan, December 2002.
11. A.C. Eldridge. Issues in auditory display: How to listen to your simulation. *Artificial Life (forthcoming)*, 2005.
12. http://www.informatics.sussex.ac.uk/users/alicee/AdSyM.
13. K. Matsuoko. Sustained oscillations generated by mutually inhibiting neurons with adaptation. *Biological Cybernetics*, 52:367:376, 1985.
14. A.F. Melo and G. Wiggins. A connectionist approach to driving chord progressions using tension. In *Proceedings of the AISB'03 Symposium on Creativity in Arts and Science*, Aberystwyth, Wales, 2003.
15. G. Papadopolous and G.A. Wiggins. AI methods for algorithmic composition: a survey, a critical review and future prospects. In *Proceedings of the AISB '99 Symposium on Musical Creativity*, pages 110–117, Brighton, UK, 1999. SSAISB.
16. M. Pearce, D. Meredith, and G. Wiggins. Motivations and methodologies for the automation of the compositional process. *Musicae Scientiae*, 2002.
17. A. Phon-Amnuaisuk, S. Tuson and G. Wiggins. Evolving musical harmonisation. In *Proceedings of ICANNGA*, 1999.
18. J.R. Schutz. Fragments of the phenomenology of music. In J. Smith, editor, *In Search of Musical Method*, pages 5–71. London, 1982.
19. E. Toch. *The Shaping Forces of Music.* Dover Publications, New York, 1948.
20. T. Wishart. *On Sonic Art.* Imagineering Press, York, 1985.

The Memory Indexing Evolutionary Algorithm for Dynamic Environments

Aydın Karaman, Şima Uyar, and Gülşen Eryiğit

Istanbul Technical University,
Computer Engineering Department, Maslak Istanbul TR34469 Turkey
karamanayd@itu.edu.tr
{etaner, gulsen}@cs.itu.edu.tr

Abstract. There is a growing interest in applying evolutionary algorithms to dynamic environments. Different types of changes in the environment benefit from different types of mechanisms to handle the change. In this study, the mechanisms used in literature are categorized into four groups. A new EA approach (MIA) which benefits from the EDA-like approach it employs for re-initializing populations after a change as well as using different change handling mechanisms together is proposed. Experiments are conducted using the 0/1 single knapsack problem to compare MIA with other algorithms and to explore its performance. Promising results are obtained which promote further study. Current research is being done to extend MIA to other problem domains.

1 Introduction

Evolutionary algorithms (EA) are heuristic search algorithms applied in both stationary and non-stationary (dynamic) problems. For the rest of the paper, EAs designed for dynamic problems will be called dynamic evolutionary algorithms (DynEA). Very successful implementations of EAs in stationary environments exist in literature but there are additional challenges in dynamic environments, such as having to adapt to the new environmental conditions and tracking the optima. In order to accomplish these, new algorithms have been introduced. Detailed surveys and discussions of EAs and dynamic environments can be found in [4, 13, 18]. DynEAs are categorized under three headings in [4] as *Type-1*: DynEAs reacting on change, *Type-2*: DynEAs maintaining diversity throughout the run and *Type-3*: DynEAs based on memory. Each type of DynEA employs a single approach or a set of approaches (models), while adapting to the new environments. These models can be classified as *operator-based*, *memory-based*, *population-based* and *initialization-based*. A similar but more detailed classification can be found in [18]. *Operator-based models* use adaptation or modification of some EA operators, especially those that are responsible for diversity such as mutation and selection. The main aim is to diversify the current genetic material by using genetic operators, when change occurs. Hyper-mutation [5], Variable Local Search [17], Random Immigrants [7] and Thermo-Dynamical Genetic Algorithms [11] are examples of such DynEAs. The first three of these approaches

F. Rothlauf et al. (Eds.): EvoWorkshops 2005, LNCS 3449, pp. 563–573, 2005.

work with the mutation operator, while the last one uses a different selection scheme based on a free energy function. *Memory-based models* use extra memory in order to preserve extra genetic material that may be useful in later stages of the run. Extra memory usage can be implemented either implicitly or explicitly. Algorithms that use redundant representations are among the well known implicit memory implementations. The most common of these representations are ones that use diploid chromosomes with a dominance scheme [10, 14, 16]. Explicit memory is implemented by allocating extra space for preserving currently available genetic material to be used in later runs. A good example of this approach is discussed in [2]. It can be seen that, memory based techniques perform best when the environment oscillates between several states. *Population-based models* use more than one population, each of which may be assigned different responsibilities. The aim is to use the available number of individuals in a more effective way. For example, the memory-enhanced algorithm introduced in [2] uses two populations. One of these populations is responsible for remembering good old solutions and the other is responsible for preserving diversity. A more recently proposed approach [3] uses multiple sub-populations distributed throughout the search space to watch over previously found optima (local or global), thus increasing the diversity in the overall population. *Initialization-based models* use problem specific knowledge in order to initialize the first population of the current environment so that individuals are in the locality of the new optima. The Case-Based Initialization of Genetic Algorithms introduced in [15] uses a similar approach. In that study, a model of the environment is maintained and is updated after environment changes. The good solutions found for previous environments are inserted into the GA population when similar environments are encountered. All of these models have their strengths and weaknesses. Studies show that different types of dynamic problems need different types of models. It can be seen that it is better to have a combined model for better performance under different types of change. In [18], the author develops a formal framework for classifying different types of dynamic environments and tries to perform a mapping between problem classes, DynEAs and performance criteria.

For the purposes of this paper, a DynEA will be defined as an evolutionary algorithm designed for working in dynamic environments, consisting of a set of the models mentioned above. This definition implies that, it might be advantageous to partition an evolutionary algorithm into sub-parts, each of which is designed for different purposes. Most algorithms found in literature have been developed to only include one or two of the models. For instance, the hyper-mutation approach [5] uses only the operator-based model while the memory-enhanced algorithm [2] uses the population-based model together with the memory-based model. When thinking of designing an efficient evolutionary algorithm in terms of these models, it can be seen that the evolutionary algorithm should include an operator-based model in order to supply the needed diversity, a memory-based model to benefit from the previously discovered genetic material, a population-based model to use the limited number of individuals efficiently and an initialization-based model that uses problem specific knowledge to guide the individuals towards the new optima.

In this paper, a new DynEA approach called Memory Indexing Algorithm (MIA) is proposed. MIA has been designed with the above considerations in mind. This study is a preliminary look into the design and performance of such an algorithm and the results obtained give a preliminary overview as to the applicability and the performance of the approach. Currently, for ease of analysis, a very simple test problem, namely the 0/1 single knapsack problem, has been selected as the benchmark. Experiments are performed to understand the operations of the basic mechanisms of MIA as well as to compare its performance with similar state of the art DynEA approaches in literature. Research continues to extend MIA to other problem domains. This paper is organized as follows: Section 2 introduces MIA and explores its mechanisms and outline. Section 3 gives details of the experimental setup, presents and discusses the test results. Section 4 concludes the paper and proposes possibilities for future work.

2 The Memory Indexing Algorithm

The Memory Indexing Algorithm (MIA) uses problem specific information for working with the appropriate incorporated mechanisms. This information is based on a measure that identifies environments and is used by MIA to index encountered environments. For the 0/1 single knapsack problem, the indexing mechanism used to differentiate environments is termed as the environment quality (EQ) and is explored in greater detail in [8]. As can be seen in the next subsections, MIA uses concepts originating from explicit memory based DynEAs, the hyper-mutation mechanism and also estimation of distribution approaches (EDAs) [9]. MIA can be put in *Type-1* of algorithms because it acts on change and initializes the next generation according to a distribution array (DA) if a similar environment has been encountered before. If there is no previous information regarding the new environment, the algorithm uses a standard technique. In this study, as an example, the hyper-mutation method is applied. MIA can also be said to be of *Type-3*, since it uses a DA that can be seen as a form of an external memory. Due to this, MIA is compared with the hyper-mutation approach of *Type-1*, and Memory Enhanced Algorithm (MEA) of *Type-3*. Experiments and results are given in section 3.

The distribution array (DA) is a list of ratios representing the frequency of each allele at each gene position in the population. Each value in the DA is used by MIA as the probability of initializing the corresponding gene location to the corresponding allele when a similar environment to the one in which the DA was calculated is re-encountered. This idea is similar to the population initialization technique used in the PBIL [1] approach. There is a DA for each different environment group which has been encountered during the run of MIA. To the authors' best knowledge, such a memory and population re-initialization scheme has not been applied to dynamic environments before. The DA values are calculated as given in Eq. 1, using the individuals in the generation when the change occurs (assuming no changes within a generation).

$$DA_i(j) = \frac{N_{ij}}{PopSize}, \; i = 1, 2, ..., L_C \; and \; j = 1, 2, ..., L_A \tag{1}$$

where $DA_i(j)$ is the ratio of allele-j at ith gene location, and N_{ij} is the number of genes having allele-j at ith gene location in the population, L_C is the chromosome length and L_A is the number of different alleles. Grouping of environments requires problem specific criteria to differentiate between different types of environments. For example in [15] the example task is a two-agent game of cat-and-mouse. They use four aspects of the environment (namely distribution of speed values and turn values of the target agent, the radius within which the target may detect the tracker and the size of the target) to categorize the environments. Using fitness distributions may also be a simple but effective measure for this purpose. However, since this paper mainly focuses on the benefits of applying a DA-based re-initialization for similar environments, for simplicity, when applying MIA to the single constraint 0/1 knapsack problem, environments are grouped according to the EQ. Better and more practical groupings are the focus of another study. The grouping used in this study is also controlled by an additional parameter called the *tuner parameter* (TP). TP can be thought of as the number of cuts in the environment space and determines the number of environment groups. In this study, for the chosen example problem, the proportion of feasible points in the search space to the whole search space is used to differentiate between different environments and is assigned as the EQ for each environment. Theoretically, this EQ value and the group id which is based on this, is calculated as in Eq. 2.

$$EQ_i = \frac{m_i}{T} \quad \Rightarrow \quad G_i = \lfloor TP * EQ_i \rfloor \tag{2}$$

where EQ_i is the environment quality of the ith environment, m_i is the number of feasible individuals in the ith environment, T is the total number of all possible individuals in the search space, G_i is the group id of the ith environment and TP is the tuner parameter. Normally the search space of the problems encountered are large and thus a complete enumeration of all points is not possible. An approximation measure is needed. So instead of using all the points in the search space to calculate the EQ value, k points are sampled randomly and the estimated EQ value denoted as EQ' is determined using the chosen samples. As part of MIA, a standard evolutionary algorithm runs until a change is detected. When change occurs, the DA of the population is calculated and is recorded for the corresponding group of the previous environment. The EQ'value for the new environment and its group id is determined. If the DA for the new environment was previously calculated, a percentage of the population is re-initialized using the corresponding DA, otherwise a standard hyper-mutation method is applied. For this current implementation of MIA, for the chosen test problem, a standard EA is run during the stationary periods and a standard hyper-mutation technique is applied if there is no current information to apply the indexing and re-initialization mechanism of MIA. However, other EA approaches and other change handling techniques can be experimented with.

3 Experiments and Results

The results obtained in this section give a preliminary overview as to the applicability and the performance of the MIA approach. It is shown in [12] that good change detection techniques are effective in the performance of EAs in dynamic environments. In this study, to rule out any effects that may be the result of faulty change detection mechanisms, changes are explicitly made known to all tested algorithms. Since MIA uses the hyper-mutation method (HMT) when the new environment group has not been encountered before, it is compared with HMT [5] and since MIA also employs an external memory, it is also compared with the Memory Enhanced Algorithm (MEA) [2]. For all tested algorithms: there are 50 individuals in each population; the chromosome length is 100; uniform crossover rate is 0.8; parent selection is via binary tournaments; generational replacement is used with elitism where the best individual of the previous population replaces the worst individual of the new population (except for right after a change); binary mutation rate is $0.6/ChromosomeLength$; there are 50 environments for each run and results are averaged over 50 runs. For HMT, hyper-mutation rate is 0.1 and this higher mutation rate is applied for 2 EA generations after a change occurs. For MEA, memory size is 10 individuals and the storing period is 10 EA generations. For MIA, TP is chosen as 10 and for practical purposes, the estimated EQ'values are used to classify environments rather than using the actual EQ values; k (number of sampled points) is chosen as 100 individuals. These randomly sampled 100 points is called the test population. The individuals of the test population are determined randomly at the start of the EA run. The same test population is applied once to each environment that is encountered right after a change to determine the corresponding EQ'value and its group id. Tests were performed in [8] to show that a simple random sampling technique is sufficient to provide the heuristic information needed for MIA. Offline performance [4] measure is used to evaluate the performance of selected algorithms. All results are reported as averaged over 50 runs against the same problem instance. T-tests with a significance level of 0.05 were performed on the offline performances. The results of the tests (omitted here for lack of space) showed that all observed differences, except for the comparison of MIA1 and MIA2 in Test 1 of Experiment 2 and the comparison between MIA2 and MIA3 in Test 2 of Experiment 3, are statistically significant. For each encountered environment, MIA applies the environment to the test population once to determine the group id. For performance comparisons, this means 5000 more fitness evaluations corresponding to 50 generations. Each run consists of 50 randomly generated environments and the change period is determined separately for each test, so the maximum number of generations for each test is $(50 * changePeriod)$. The same single constraint 0/1 knapsack instance which is randomly generated prior to the experimental runs, is used in all experiments.

There are 100 items with profits between 1000 and 5000, and weights between 1 and 100. For simplicity of the analysis, changes are allowed to occur only in the capacity constraint of the knapsack. All capacity constraints are produced randomly if not otherwise stated. A penalty-based method is used for infeasible individuals and the penalty function proposed for multiple knapsack problems in [6] is adapted for a single knapsack problem as given in Eq. 3.

$$penalty = \frac{p_{max} + 1}{w_{min}} * CV \quad , \quad w_{min} > 0 \tag{3}$$

where p_{max} is the maximum profit, w_{min} is the minimum resource consumption (weight), and CV is the constraint violation of the individual. For testing different properties of MIA, in some of the experiments in the next sections, a *training stage* and a *test stage* is defined. In the training stage, controlled environments are introduced to MIA. This stage is defined in such a way that MIA gets a chance to calculate a DA for each of the possible environment groups. The training stage consists of a series of stationary periods. The length of the stationary periods determines the number of generations the EA will run for each environment. The length of these stationary periods in the training stage will be referred to as the *training period* and will be given in EA-generations units. At the end of the training stage, a DA for all possible environment groups is initialized. The number of these environments depends on the selected TP parameter as explained above. In the testing stage, environments are generated and introduced to the EA randomly. A *testing period* is also determined similarly to the one explained above for the training stage.

3.1 Experiment 1

Experiment 1, composed of 3 tests, is for studying the algorithms under different settings of the training period and test period. In the first 2 tests, first 11 environments (TP=10) are selected such that DAs for all groups are initialized once during the training stage. The performances of the algorithms are calculated only for the test stage which begins right after the training stage is completed. The last test in this experiment does not have a training stage. Environments included in the testing stages are generated randomly in all test instances. In the offline performance figures of Test 1 and Test 2, generations are shown to start from zero, however this value corresponds to the beginning of the testing stage. The aim of Tests 1 and 2 in this experiment is to observe the effect of the length of the training period on the performance of MIA. In Test 1, training period and test period are chosen as 100 EA generations and 20 EA generations respectively. The training period is considered to be long, while the test period is not. Under this setting, MIA is expected to perform its best compared to the other settings in the following tests. This is because of the fact that, a long training period means that the EA has more time to converge to a local/global optimum and thus may be able to calculate more accurate distribution arrays. This will allow the EA to start off from a better point in the search space in the similar environments encountered during the test phase. HMT and MEA are expected

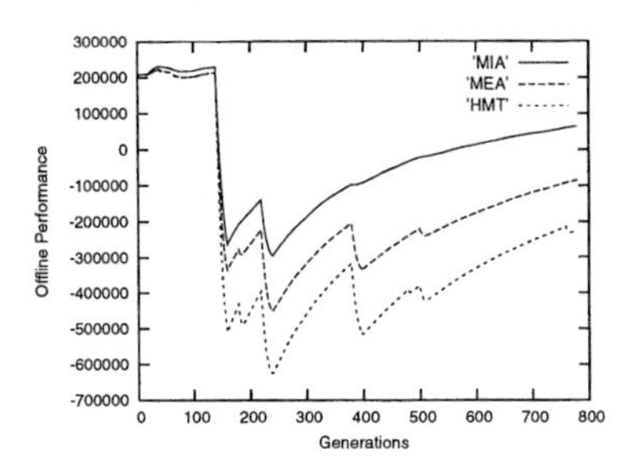

Fig. 1. Offline Performances: Exp.1 Test 1

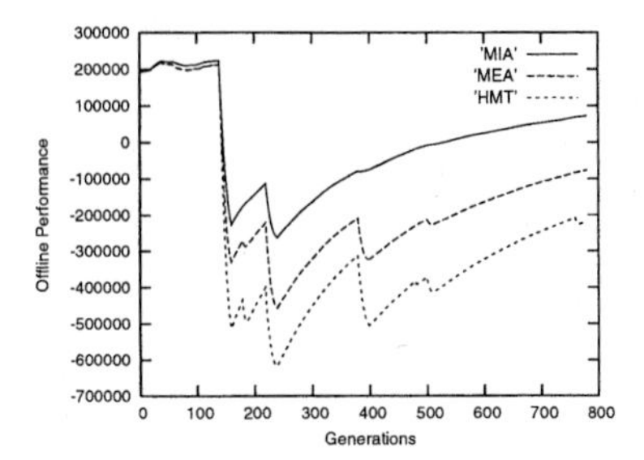

Fig. 2. Offline Performances: Exp.1 Test 2

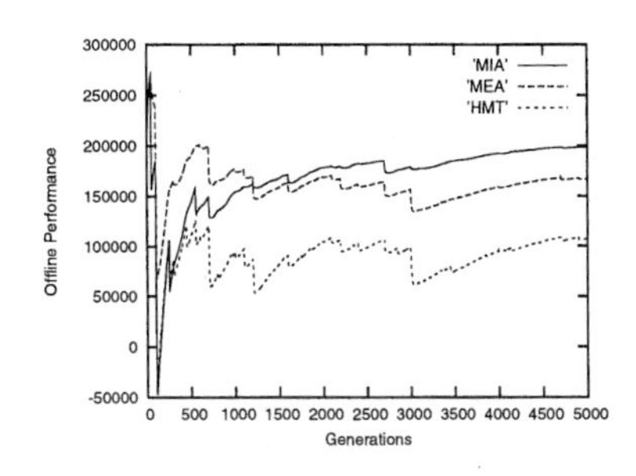

Fig. 3. Offline Performances: Exp.1 Test 3

not to recover as well as MIA, due to the fact that MIA is expected to start from better points in the search space than the others due to the initialization using a DA. In Test 2, both the training period and the test period is chosen as 20 EA generations. The purpose of Test 3 is to evaluate the performance of MIA against the performances of HMT and MEA when there is no training stage. There are a total of 100 environments produced at random where the period of change is every 50 EA generations. Offline performances of algorithms for Test 1, Test 2 and Test 3 are given in Fig. 1, Fig. 2, and Fig. 3 respectively. Fig. 1, Fig. 2, and Fig. 3 show that performance of the MIA is the best whereas the performance of the HMT is the worst. Comparing Fig. 1 and Fig. 2, it can be said that, although the training period is decreased from 100 to 20, this does not much effect the relative performance of MIA. Moreover from Fig. 3, it can be said that although there is no training stage, MIA outperforms the others after some time. The reason for the delay of MIA to be the best performer is that MIA is more likely to run hyper-mutation at the beginning of the evolution. As a result, MIA is expected to perform as well as hyper-mutation in the beginning but better in the later generations after several environments have been encountered and DAs for these environments have been recorded. All figures show that MIA gets better as time passes.

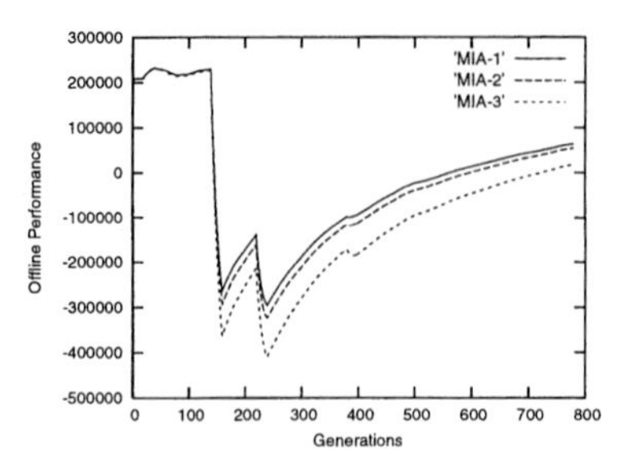

Fig. 4. Offline Performances: Exp.2 Test 1

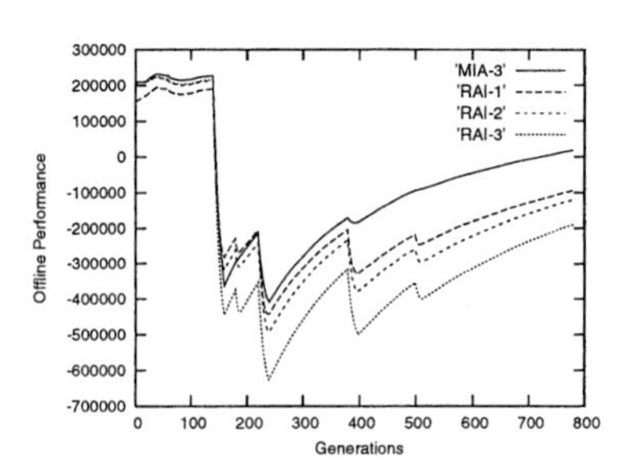

Fig. 5. Offline Performances: Exp.2 Test 2

3.2 Experiment 2

Experiment 2 includes 2 tests with exactly the same parameter settings and the problem instance, including the environments with Test 1 of Experiment 1. The aim of Test 1 is to observe the effect of the initialization ratio on the performance of MIA and the aim of Test 2 is to prove that initialization of the population with a DA is a more powerful technique than a pure random initialization of the population when change occurs. As stated previously, if the new environment is in a group whose DA is initialized previously, the first population in this environment is re-initialized partly using the DA. Initialization ratio is the parameter which determines the ratio of the new individuals to insert into the current population. Parameter settings and the problem instances of the following tests, including the environments, are same as Test 1 of Experiment 1. Test 1 is conducted to test MIA with different initialization ratios. In Fig. 4, MIA1, MIA2, and MIA3 are the MIAs with initialization ratios of 1, 0.5 and 0.2 respectively. The individuals to be replaced by new individuals are determined randomly. In Test 2, performance of MIA is compared with a pure random initialization (RAI). In Fig. 5, RAI1, RAI2, and RAI3 are the random initialization with ratios 1.0, 0.5, and 0.2 respectively. Fig. 4 depicts the MIAs with different initialization ratios. It is seen that performance of MIA3 is the worst, while MIA1 is the best. Moreover, performance of MIA2 is much nearer to MIA1 than MIA3. Therefore, initialization ratio should be greater than 0.5. Fig. 5 shows the performances of RAIs and MIA3. Although MIA3 is the worst of the MIAs compared in Test 1, it outperforms all of the RAIs, which means that MIA's performance is largely due to the use of a DA to initialize the population.

3.3 Experiment 3

This experiment is conducted to observe the effect of the tuner parameter TP on the performance of MIA. In the figures of this experiment, MIA1, MIA2, MIA3 and MIA4 are the MIAs with TP equal to 5, 10, 20 and 60 respectively. Parameter settings and the problem instance of these tests, including the environments are same with the Test 3 in Experiment 1. Results of Test 1 are shown in Fig. 6, which depicts the performances of MIAs with different TPs when there are 100

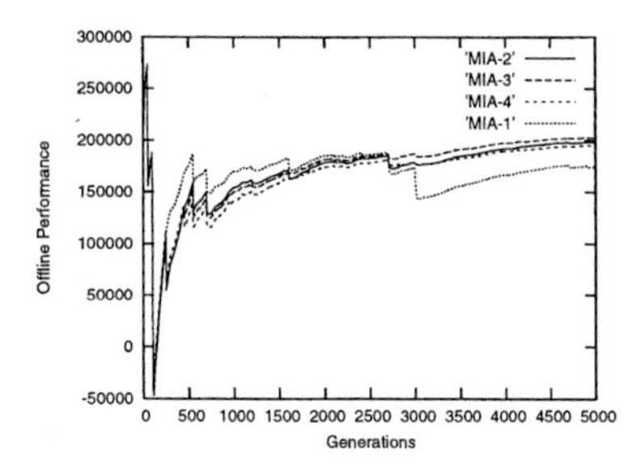

Fig. 6. Offline Performances: Exp.2 Test 1

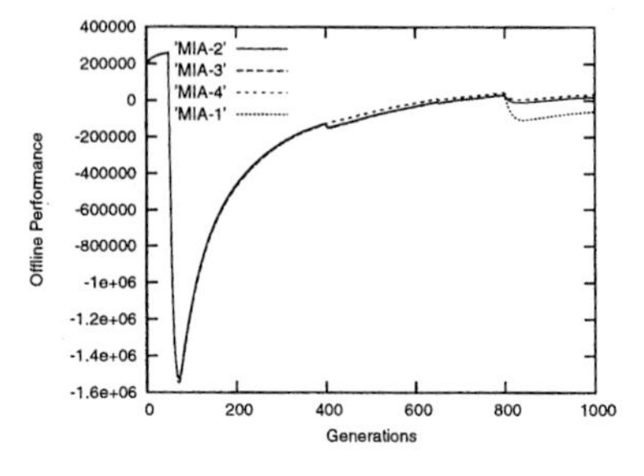

Fig. 7. Offline Performances: Exp.2 Test 2

environment changes. In Test 2, number of environment changes is decreased from 100 to 20. Results are shown in Fig. 7. In both Fig. 6 and Fig. 7, it is apparent that when TP is decreased from 10 to a lower number, performance of MIA decreases drastically. On the other hand, if TP is increased from 10 to a greater value, MIA's performance may decrease or increase. This indefiniteness is due to the fact that, if TP increases, there is a smaller number of environments in each group and thus the groups get more homogeneous. Therefore, DAs of each group will lead to more accurate initializations leading to a better performance. However, since the number of groups is increased by TP, the new environment detected after a change is less likely to fall into a group whose DA was previously initialized. This means that the EA may end up having to use hyper-mutation most of the time. In conclusion, these two contradictory factors determine the performance of MIA when TP increases.

As it is mentioned before, MIA becomes more successful as generations pass due to the initialization-based approach it uses. Each time MIA encounters an environment similar to a previous one, the population is re-initialized using the recorded DA. During the current stationary environment MIA gets a chance to work on the same (or similar) environment for more generations, possibly to find a better solution. At the end of the stationary period, the newly calculated DA is recorded for this environment group. Thus each time MIA works on a representative of a group, it has a chance of finding and recording a better DA for the group.

4 Conclusions and Future Work

Studies in literature introduce many different approaches to dynamic problems. This paper defines dynamic evolutionary algorithms as being a set of approaches working with each other. A DynEA should benefit from operator-based approaches to obtain needed diversity, from memory-based approaches to remember useful genetic material, from population-based approaches to use limited number of individuals effectively, and from initialization-based approaches, if possible, to start from the locality of the optimum of the current environment.

The Memory Indexing Algorithm (MIA) is introduced and compared with hypermutation and Memory Enhanced Algorithms through a set of experiments. All experiments show that MIA outperforms others for the tested cases on the selected test problem and is more robust to environment changes. MIA can be further improved by modification in the calculation of the distribution array. Additionally, in this study a very simple problem was selected as the benchmark, however extending the applicability of MIA to other problem domains is needed. Since, MIA uses problem specific information in order to create an index for each possible environment, it is an interesting study to explore the ways of defining and differentiating environments of other dynamic problems. This study is currently being conducted. Furthermore, MIA only uses three of the identified approaches for dealing with change. In its current implementation, it does not use population-based techniques explicitly, though each indexed environment and its DA may be interpreted similar to the sub-populations in the SOS [3] approach. This aspect of the approach will be explored as a future study.

References

1. Baluja S., Caruana R., "Removing the Genetics from the Standard Genetic Algorithm", in *Proc. of the Intl. Conf. on Machine Learning*, Morgan Kaufmann, pp. 38-46, 1995.
2. Branke J., "Memory Enhanced Evolutionary Algorithms for Changing Optimization Problems", in *Proc. of Cong. On Evolutionary Computation*, pp 1875-1882, IEEE, 1999.
3. Branke J., Kaussler T., Schmidt C., and Schmeck H., "A Multi-Population Approach to Dynamic Optimization Problems", Adaptive Computing in Design and Manufacturing, pp. 299-308, Springer, 2000.
4. Branke J., *Evolutionary Optimization in Dynamic Environments*, Kluwer, 2002.
5. Cobb H. G., "An Investigation into the Use of Hypermutation as an Adaptive Operator in Genetic Algorithms having Continuous, Time-Dependent Non-stationary Environments", NCARAI Report: AIC-90-001, 1990.
6. Gottlieb J., "Evolutionary Algorithms for Constrained Optimization Problems", PhD Thesis, Tech. Univ. of Clausthal, Germany, 1999.
7. Grefenstette J. J., "Genetic Algorithms for changing environments", in *Proc. of Parallel Problem Solving from Nature*, pp. 137-144, Elsevier, 1992.
8. Karaman, Uyar S., "A Novel Change Severity Detection Mechanism for the Dynamic 0/1 Knapsack Problem", in *Proceedings of Mendel 2004: 10th Intl. Conf. on Soft Computing*, 2004.
9. Larranaga P., Lozano J. A., *Estimation of Distribution Algorithms: A New Tool for Evolutionary Computation*, Kluwer, 2001.
10. Lewis J., Hart E., Graeme R., "A Comparison of Dominance Mechanisms and Simple Mutation on Non-Stationary Problems", in *Proc. of Parallel Problem Solving from Nature*, Springer, 1998.
11. Mori N., Kita H., and Nishikawa Y., "Adaptation to a Changing Environment by Means of the Feedback Thermodynamical Genetic Algorithm", in *Proc. of Parallel Problem Solving from Nature*, pp. 149-158, 1998.

12. Morrison R. W., De Jong K. A., "Triggered Hypermutation Revisited", in *Proceedings of Cong. On Evolutionary Computation*, pp 1025-1032, IEEE, 2000.
13. Morrison R. W., *Designing Evolutionary Algorithms for Dynamic Environments*, Springer, 2004.
14. Ng K. P., Wong K. C., "A New Diploid Scheme and Dominance Change Mechanism for Non-Stationary Function Optimization", in *Proc. of the Sixth Intl. Conf. on Genetic Algorithms*, 1995.
15. Ramsey C. L., Grefenstette J. J., "Case-Based Initialization of Genetic Algorithms", in *Genetic Algorithms: Proceedings of Fifth International Conference*, pp. 84-91, San Mateo: Morgan Kaufmann, 1993.
16. Ryan C., "The Degree of Oneness", in *Proc. of the 1994 ECAI Workshop on Genetic Algorithms*, Springer, 1994.
17. Vavak F., Jukes K. A., Fogarty T. C., "Performance of a Genetic Algorithm with Variable Local Search Range Relative to Frequency of the Environment Change", in *Genetic-Programming 1998: Proc. of the Third Annual Conf.*, pp. 602-608, Morgan Kaufmann, 1998.
18. Weicker K., *Evolutionary Algorithms and Dynamic Optimization Problems*, Der Andere Verlag, 2003.

Dynamic Decentralized Packet Clustering in Networks

Daniel Merkle, Martin Middendorf, and Alexander Scheidler

Department of Computer Science,
University of Leipzig, Augustusplatz 10-11, D-04109 Leipzig, Germany
{merkle, middendorf, scheidler}@informatik.uni-leipzig.de

Abstract. In this paper we study a dynamic version of the Decentralized Packet Clustering (DPC) Problem. For a network consisting of routers and application servers the DPC problem is to find a good clustering for packets that are sent between the servers through the network. The clustering is done according to a data vector in the packets. In the dynamic version of DPC the packets data vector can change. The proposed algorithms to solve the dynamic DPC are inspired by the odor recognition system of ants. We analyze the new algorithms for situations with different strengths of dynamic change and for different number of routers in the network.

1 Introduction

The Decentralized Packet Clustering (DPC) problem is to find a clustering for information packets that are sent between application processes running on servers in a network ([9]). The clustering is done based on a data vector that each information packet has. The cluster number is stored within every packet. The DPC problem occurs in situations where the application processes handle the information packets more appropriately when they know their type (i.e. the cluster number) of an arriving packet. Note, that DPC is a general problem with only few assumptions and does not depend on a specific application problem. The aim of [9] was to find a decentralized algorithm for DPC where the clustering is done by the routers such that each router neither needs much computational power nor much memory for running the clustering algorithm.

In this paper we investigate a dynamic version of DPC (d-DPC) where the data vectors of the information packets in the network can be changed (e.g. by the application processes) over time. We aim to find decentralized algorithms for d-DPC that are able to provide a good clustering at any time. The proposed algorithms have some similarity with a nature inspired clustering algorithm named ANTCLUST [4, 5]. ANTCLUST is inspired by the chemical recognition system of ants. Due to the limited space we can neither give an overview over dynamic clustering techniques nor over nature inspired and decentralized clustering techniques. But it should be noted that dynamic clustering problems have been studied for example for mobile ad-hoc networks (see, e.g., [7]) and stream data

F. Rothlauf et al. (Eds.): EvoWorkshops 2005, LNCS 3449, pp. 574–583, 2005.

(see, e.g., [8]). An overview over relevant decentralized clustering techniques and nature inspired clustering algorithms is given in [9].

In Section 2 we define the Dynamic Decentralized Packet Clustering (d-DPC) problem. The proposed algorithms for solving d-DPC and the algorithms that are used as comparison are presented in Section 3. The experimental setup and the used cluster validity measures are described in Section 4. Results are presented in Section 5 and a conclusion is given in Section 6.

2 Dynamic Decentralized Packet Clustering

The DPC problem is defined as follows. Given is a network where the nodes are routers. The DPC is to find for a set of packets $\mathcal{P} = \{P_1, P_2, \ldots, P_n\}$ that are send through the network a good (with respect to some criterion) clustering, i.e., a partitioning $\mathcal{C} = \{C_1, \ldots, C_{|\mathcal{C}|}\}$ of $\mathcal{P}$. Each packet $P_i \in \mathcal{P}$ contains a data vector v_i that is used for clustering. Instances of DPC differ with respect to network topology, number of routers, and how packets are send through the network. In the dynamic version called d-DPC that is studies in this paper the data vector of a packet can change at every time step, where in one time step (iteration) $|\mathcal{P}|$ randomly chosen packets chose a random router to pass.

3 Algorithms

In this section we describe the four cluster algorithms that are considered in this paper. The k-means algorithm is a classical centralized cluster algorithm, which is used for comparison. The DPClust algorithm was proposed in [9] for the static DPC. The new algorithms that are proposed in this paper for d-DPC are called d-DPClust$_{cz}$ and d-DPClust$_{zc}$.

3.1 k-Means

A standard algorithm for clustering data vectors is the k-means algorithm. This is an iterative algorithm that starts with a set of k initial vectors, called center points. Each data vector is assigned to its nearest (e.g., with respect to Euclidean distance) center point. All data vectors that are assigned to the same center point form a cluster. For each cluster its geometric centroid is computed and these centroids form the new center points for the next iteration. The algorithm stops when some convergence criterion has been met, e.g., the center points have not changed or a maximal number of iterations has been done. The selection of the initial center points has a great influence on the results of the algorithm and different methods have been proposed for choosing these initial points (one is to simply select the center points randomly from the given data vectors).

For d-DPC we initialize the centroid estimation of k-means with the centroids that were found one time step before and then perform k-means until it converges. Note, that k-means is a centralized algorithm with global knowledge of the data vectors and that for dynamic problems we run k-means until it converges before the test instance might change again in the next time step.

3.2 DPClust

Each packet $P_i = (v_i, c_i, z_i)$, $i = 1, \ldots, n$ has in addition to its data vector v_i, a cluster number c_i and a vector z_i that is an estimation of the centroid of its cluster. When a packet passes a router, a modified (how this is done is described in the following) copy of the packet is stored in the router. This copy of the packet $P = (v, c, z)$ and the next packet $P_i = (v_i, c_i, z_i)$ passing the router are compared. If both packets are in the same cluster (i.e. $c_i = c$), the estimation of the centroid for the corresponding cluster is updated according to $z_i := z_i \cdot (1 - \beta) + \beta v$, where β is a parameter that determines how strong the data vector v of the packet copy stored in the router and the estimated centroid z_i influences the new value of z_i. If both packets are not in the same cluster, then the passing packet decides, if it should change its cluster. Taking the distances of the data vector v_i to the centroid estimation in the packet (z_i) and the centroid estimation stored in the router (z) into account, this is done iff $v_i - z_i > v_i - z$. For further details see [9]. Note, that some inspiration comes from real ants that exchange chemical clues (data vector) when they meet nestmates (packets from same cluster) resulting in a typical colony odor (centroid estimation) (cmp. [4]). The odor is also used to detect intruders (packets from other clusters).

3.3 d-DPClust$_{cz}$

For algorithm d-DPClust$_{cz}$ each packet $P_i = (v_i, c_i)$ consists of a data vector v_i and a cluster number c_i only. Each router r stores a vector of estimated centroids $\mathcal{Z}_r = (z_r^1, \ldots, z_r^{|\mathcal{C}|})$. A packet P_i that arrives at a router r determines its cluster number c_i by using the distances of its data vector v_i to the estimated centroids z_r^j, $j = 1, \ldots, |\mathcal{C}|$ that are stored in the router. A packet belongs to the cluster for which this distance is minimal, i.e. $c_i = \text{argmin}_j ||v_i - z_r^j||$. Then, in the router the centroid estimation for cluster c_i is modified according to $z_r^{c_i} = (1-\beta) \cdot z_r^{c_i} + \beta \cdot v_i$.

Note, that in algorithm d-DPClust$_{cz}$ the centroid estimations of two routers r_1 and r_2 may have a different order (in the sense that $z_{r_1}^i$ corresponds to the centroid estimation $z_{r_2}^j$ with $i \neq j$). Therefore, the routers $r = 2, \ldots, |R|$ iteratively reorder their centroids after every $e \geq 1$ time steps, so that reordering is done according to a permutation π for which $\sum_{i=1}^{|\mathcal{C}|} ||z_r^{\pi(i)} - z_{r-1}^i||$ gets minimal, where e is a parameter of the algorithm.

3.4 d-DPClust$_{zc}$

In algorithm d-DPClust$_{zc}$ the packets $P_i = (v_i, c_i)$ consist of a data vector v_i and a cluster number c_i only. The algorithm works similar to d-DPClust$_{cz}$. The difference is that in d-DPClust$_{zc}$ the router centroid estimation is modified *before* a packet determines its cluster number. Hence, the modification of the centroid estimation is applied to z_r^j where $j = c_i$. The new cluster number for

P_i is determined similarly to d-DPClust$_{cz}$ (i.e., it is the cluster for which $c_i = \text{argmin}_j ||v_i - z_r^j||$). In contrast to d-DPClust$_{cz}$ its new cluster is computed when the packet leaves a router (after the modification of the estimated centroids). A router exchange step as in d-DPClust$_{cz}$ is not applied.

4 Experiments

4.1 Test Instances

As test instances for the d-DPC we used the same type of clustering instances for 2-dimensional data vectors that have been used in other papers on ant-based clustering before (e.g. [1, 2]). A set of data vectors consists of vectors from several classes. The data vectors of a class are generated by a two-dimensional normal distribution with standard deviation 2 in both dimensions. Test instance of type T_1 has four classes with four different center points $(0, 0)$, $(0, 10)$, $(10, 0)$, and $(10, 10)$ and 250 vectors from each class. For test instance T_2 consists of two vector classes with center points $(0, 0)$, $(10, 0)$ and 500 vectors in each class. After each time step all data vectors v_i of a class are moved according to $v_i = v_i + \Delta v_i \cdot v$ where $\Delta v_i = (\Delta v_i^1, \Delta v_i^2) \in [-1, 1]^2$ is a direction value that has been chosen for each class and parameter v is used to adjust the strength of the dynamics (or the speed of the moving classes). If the center point of a class leaves a predefined cluster area A in a dimension k, then the sign of Δv_i^k is changed. For test instance T_1 the initial moving directions of the four classes were defined randomly using a uniform distribution ($\Delta v_i \in [-1, 1]^2$) and the cluster area was defined as $A = [-10, 20]^2$. In Figure 1 an example for the d-DPC test instance of type T_1 is given for different time steps. For test instance T_2 the class with center point $(0, 10)$ does not move, and the second class initially moves to the right along the horizontal axis ($\Delta v_i = (1, 0)$). The cluster area for instance T_2 was $A = [0, 20]^2$.

We have done experiments that are similar to experiments presented in [1, 9] on static test instances, where the number of elements of each class is not the same. The test instances $Size_k$ are based on test instance T_1 with 1000 data vectors and 4 classes, but the ratio of the number of elements in one large class and the size of three small classes (which are of equal size) is k.

4.2 Cluster Validity

To test our algorithms several well known clustering validity measures have been used as explained in the follwoing. Let v_i, $i = 1, \dots, n$ be a data vector and $\mathcal{C} = \{C_1, \dots, C_{|\mathcal{C}|}\}$ be a clustering.

Silhouette coefficient: In [3] the often used silhouette coefficient is defined as follows. Let $\gamma_1(v_i, \mathcal{C})$ (resp. $\gamma_2(v_i, \mathcal{C})$) be the cluster which is the nearest (resp. second nearest) cluster to data vector v_i (with respect to the distance $d(v_i, C_{\gamma_1(v_i, \mathcal{C})})$ of the data vector to the geometric centroid of a cluster). The silhouette coefficient s_i for data vector v_i is defined as the normalized difference:

$$s_i := \frac{d(v_i, C_{\gamma_2(v_i, \mathcal{C})}) - d(v_i, C_{\gamma_1(v_i, \mathcal{C})})}{\max\{d(v_i, C_{\gamma_1(v_i, \mathcal{C})}), d(v_i, C_{\gamma_2(v_i, \mathcal{C})})\}}$$

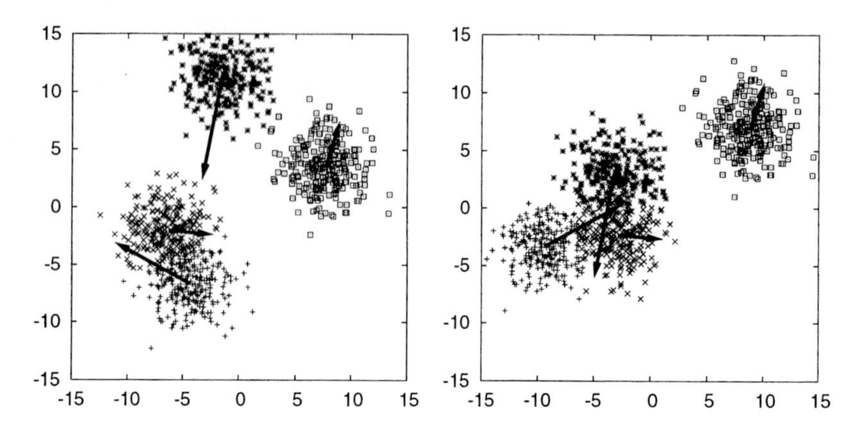

Fig. 1. Test instance T_1 dynamic case ($v = 0.1$); 100 time steps between both figures; used algorithm for clustering was d-DPClust$_{cz}$; arrows indicate direction and speed of the classes

The silhouette coefficient SC is defined as the average value over all s_i, $i = 1, \ldots, n$. The measure is influenced by the cohesion of the clusters and the separation between clusters. Empirical studies as in [3] show that $s_i > 0.7$ indicates an excellent separation between the clusters, a value between 0.5 and 0.7 indicates a clear assignment of data points to cluster centers, values between 0.25 and 0.5 indicate that there are many data points that cannot be clearly assigned, and $s_i < 0.25$ indicates that it is practically impossible to find significant cluster centers. For dynamic test instances the average silhouette coefficient over all t time steps is used, i.e. $SC_\varnothing := \sum_{i=1}^{t} SC_i/t$, where SC_i is the silhouette coefficient of a clustering in time step i.

Dunn Index: The Dunn index measures the minimal ratio between cluster diameter and inter-cluster distance for a given partitioning. If C_i and C_j are the closest clusters according to the average distance, and C_h is the cluster with the largest diameter, then the Dunn index DI can be computed as $DI = d(C_i, C_j)/diam(C_h)$, where $d(C_i, C_j)$ is the average distance of all pairs of elements in C_1 and C_2. A low Dunn index value indicates a fuzzy clustering, whereas a value close to 1 indicates a near-crisp clustering. The Dunn index tries to identify well separated and compact clusters. $DI_\varnothing$ denotes the average DI value over all time steps.

Sum of Squares: Let $\hat{v}_l$ be the geometric centroid of cluster C_l. The sum of squares criterion is defined as $SS = \frac{1}{c}\sum_{l=1\ldots|C|}(\sum_{v_i \in C_l} ||v_i - \hat{v}_l||^2/|C_l|)$ and measures the compactness of a clustering. The smaller value SS is, the more compact is the clustering. For dynamic test instances, with $SS_\varnothing$ we denote the average Sum of Squares value over all time steps. In contrast to $SC_\varnothing$ and $DI_\varnothing$, which have to be maximized, the sum of squares value $SS_\varnothing$ has to be minimized.

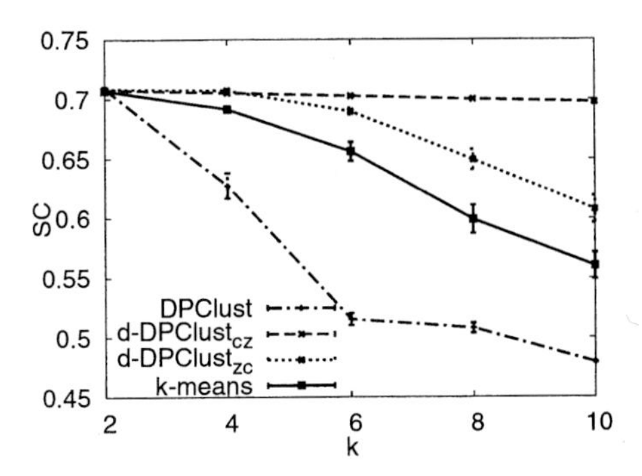

Fig. 2. Silhouette coefficient for k-means, DPClust, d-DPClust$_{cz}$, and d-DPClust$_{zc}$ on $Size_k$, $k \in \{2, 4, 6, 8, 10\}$ with different ratios between cluster sizes; one router; vertical bars show the standard error (50 test runs)

5 Results

In this section the experimental results for the four described clustering algorithms on instances of DPC and d-DPC are presented. Algorithm k-means is used as a reference. If not stated otherwise parameter $\beta = 0.1$ is used for all algorithms and $e = 10$ for d-DPClust$_{cz}$. The number of iterations is 20000 per test run. All dynamic test runs were started with the same initial random seed for the different algorithms, so that the situation after a fixed number iterations is identical for all algorithms.

5.1 Static Problem Instances

To evaluate the new algorithms we first present results for static test instances which are also the basis for the dynamic test instances. In Figure 2 the silhouette coefficient SC of the four algorithms on test instances $Size_k$, $k \in \{2, 4, 6, 8, 10\}$ is shown for a network with a single router (results are averaged over 50 test runs per k). Note, that d-DPClust$_{cz}$ outperforms k-means. For large values of k the initial center points lay with high probability in the large class of data vectors. Thus, k-means with high probability starts with a partition of the large class into several smaller clusters but might combine several smaller classes into one cluster. In contrast to k-means, algorithm d-DPClust$_{cz}$ has the ability to escape from this situation.

Although d-DPClust$_{cz}$ leads to very good results on the given static test instances, its convergence speed gets worse if the ratio between number of routers and packets gets too large. This can be seen in Figure 3. Depicted is SC for algorithms DPClust, d-DPClust$_{cz}$, and d-DPClust$_{zc}$ for 1, 10, 100, and 1000 routers (results are averaged over 50 test runs). Hence, d-DPClust$_{cz}$ applicability in dynamic situations may be bad for a large number of routers. It should be

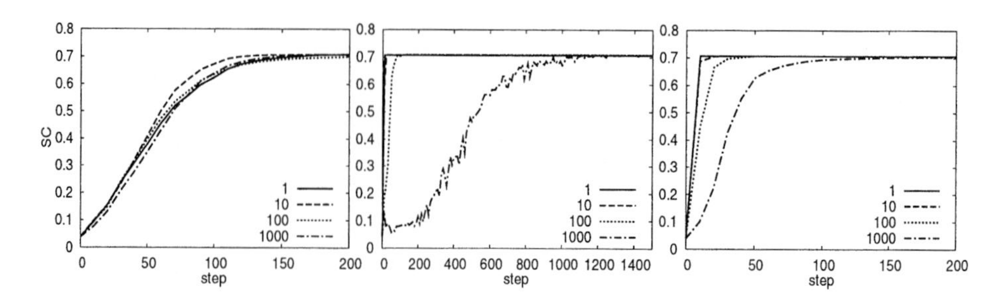

Fig. 3. Convergence Speed of DPClust (left), d-DPClust$_{cz}$ (middle), and d-DPClust$_{zc}$ (right) on problem instance $Size_1$; different lines correspond to different number of routers; shown is the silhouette coefficient SC in different time steps; each line is averaged over 50 test runs

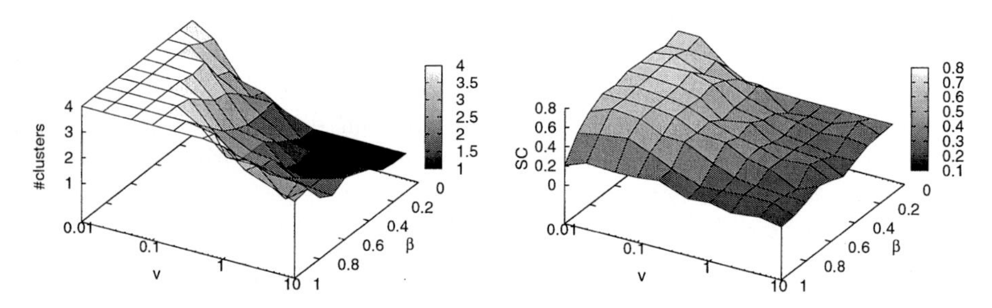

Fig. 4. Average number of clusters (left) and SC (right) after 20000 steps for DPClust on test instance T_1; results are averaged over 50 test runs

noted that the reordering of the routers may be time consuming if the number of clusters gets too large.

5.2 Dynamic Problem Instances

In Figure 4 it can be seen that algorithm DPClust performs bad in dynamic situations. Even for small dynamic changes ($v \gtrapprox 0.02$) it may happen that during a test run a cluster is lost, i.e., there are no more packets which belong to a certain cluster. In algorithm DPClust (in contrast to the other algorithms) this is definitely irreversible. To a small extent this effect can be reduced by adapting β. The bad behaviour of DPClust is also shown by the value of SC that becomes smaller for larger values of v. We exclude DPClust from further investigations on dynamic instances.

To analyze how good the algorithms can find a good clustering in a dynamic situation, we investigated them on instances T_1 and T_2 for a situation with one router. In Figure 5 the values of the average silhouette coefficient

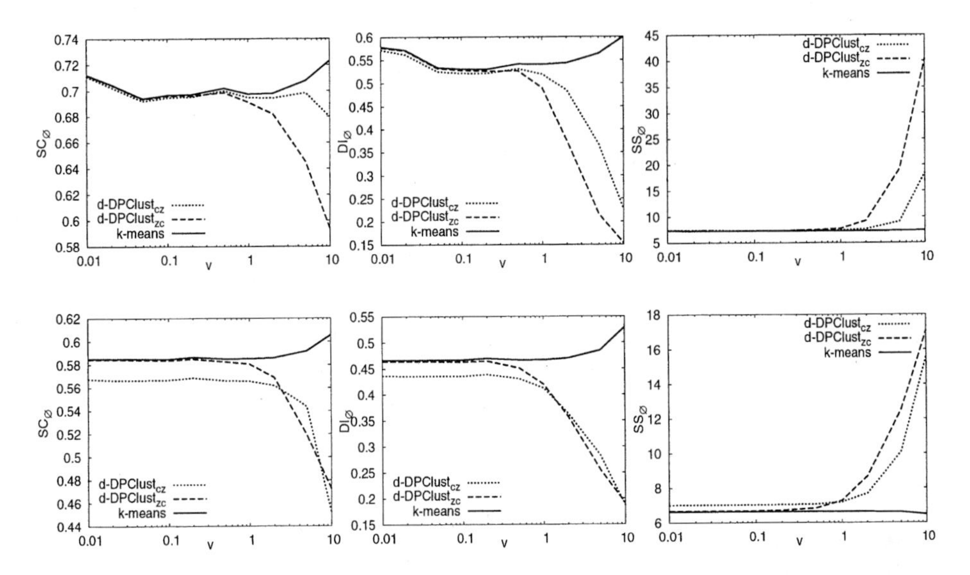

Fig. 5. Performance of DPClust, d-DPClust$_{zc}$, and d-DPClust$_{cz}$: Given are average silhouette coefficient $SC_{\varnothing}$ (left), average Dunn index $DI\varnothing$ (middle) and average sum of squares $SS_{\varnothing}$ (right) for test instance T_1 (top) and T_2 (bottom); dynamic change $v \in \{0.01, 0.02, 0.05, 0.1, \ldots, 10\}$

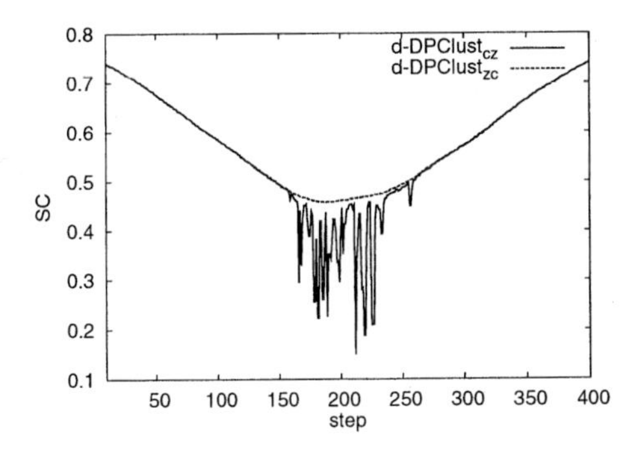

Fig. 6. Silhouette coefficient for d-DPClust$_{cz}$ and d-DPClust$_{zc}$ given over time steps $t = 10, \ldots, 400$ when the one cluster in test instance T_1 crosses the second cluster; v=0.05

$SC_{\varnothing}$, the average Dunn index $DI_{\varnothing}$, and the average sum of squares $SS_{\varnothing}$ are depicted for algorithms d-DPClust$_{cz}$, d-DPClust$_{zc}$, and (as reference) for k-means (as described in Section 3). The extent of dynamic changes varied with $v \in \{0.01, 0.02, 0.05, 0.1, \ldots, 10\}$. Note, that $v \approx 5$ or larger is a very strong dynamic change in every iteration. With respect to all three validity measures both variants of d-DPClust show a very good clustering behavior for $v \lesssim 1$ on both

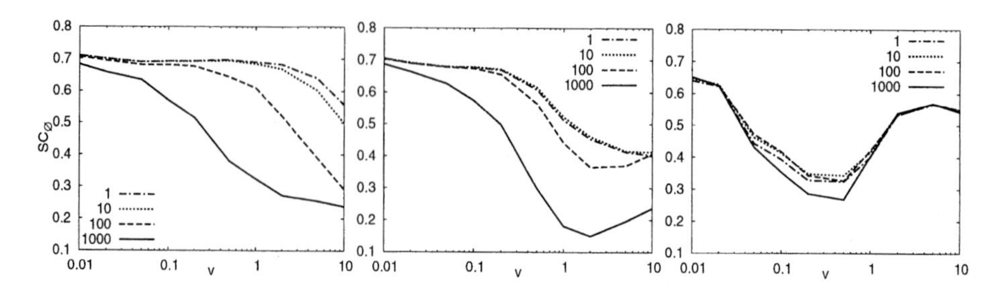

Fig. 7. Performance with respect to $SC_{\varnothing}$ of d-DPClust$_{cz}$ with different number of routers — 10 routers (left), 100 routers (middle) and 1000 routers (right) — each for different values of parameter $e \in \{1, 10, 100, 1000\}$; test instance T_1

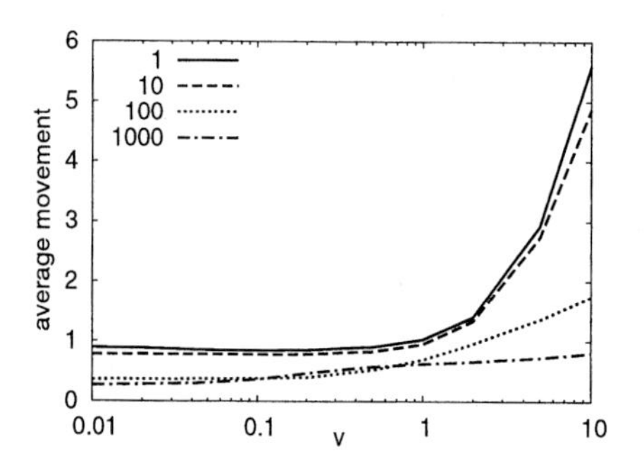

(a) d-DPClust$_{cz}$: average movement of the estimated cluster centroids of the routers for different values of v and different number of routers (1, 10, 100, 1000); $e = 10$

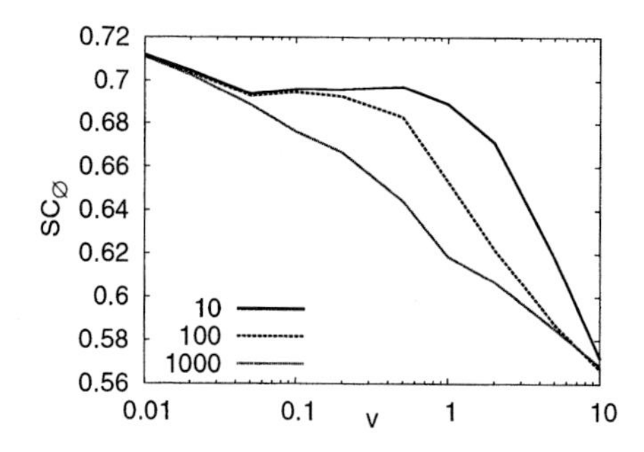

(b) Performance of d-DPClust$_{zc}$: Given is the silhouette coefficient $SC_{\varnothing}$ on test instance T_1 for 10, 100, and 1000 routers

Fig. 8. Results for d-DPClust$_{cz}$ and d-DPClust$_{zc}$

test instances, they perform only slightly worse than the centralized k-means algorithm. Compared to d-DPClust$_{zc}$ the performance of d-DPClust$_{cz}$ is better for very dynamic situations ($v \gtrapprox 1$). On test instance T_2 algorithm d-DPClust$_{cz}$ performs slightly worse than d-DPClust$_{zc}$ for $v < 1$. The reason for this is illustrated in Figure 6. When one class crosses the other class, their centers are nearly identical (the chance for this event in instance T_1 with $v > 0$ is much smaller than in instance T_2). In contrast to d-DPClust$_{zc}$ this very small distance between two class centers leads to a small SC value.

Now we investigate the d-DPClust variants in dynamic situations with more than one router. In Figure 7 the measure $SC_{\varnothing}$ is given for d-DPClust$_{cz}$ with different number of routers. As suggested before the performance of d-DPClust$_{cz}$ gets bad for a very large number of routers ($r = 1000$). For 10 and 100 routers better performance can be achieved with a higher frequency of router exchange steps. The increasing values of $SC_\varnothing$ for 1000 routers with $v \gtrapprox 2$ can be explained with Figure 8(a). Shown is the average movement of the estimated cluster centroids of the routers, defined as $1/(t \cdot |\mathcal{C}| \cdot |\mathcal{R}|)) \cdot \sum_{k=1}^{t-1} \sum_{R_i \in \mathcal{R}, C_j \in \mathcal{C}} ||\hat{v}_{ij}^{k+1} - \hat{v}_{ij}^{k}||$, where $\hat{v}_{ij}^{k}$ is the estimated cluster centroid in router R_i of cluster C_j at time step k. The nearly constant values for the centroid movement show, that the cluster centroids remain nearly stable. Hence the algorithm does not follow the moving classes. Nevertheless, data vectors close to a non-moving geometric cluster center are assigned to that cluster. Although there is no tracking behavior, this leads to an increasing value for $SC_\varnothing$. The results for d-DPClust$_{zc}$ are given in Figure 8(b). The results are very promising since in all cases, even with strong dynamics like $v = 10$, it holds $SC_\varnothing > 0.55$.

6 Conclusion

We have studied a dynamic version of the Decentralized Packet Clustering (DPC) Problem. Two algorithms have been proposed for this problem that were inspired by the chemical recognition system of ants. We analyzed the new algorithms called d-DPClust$_{cz}$ and d-DPClust$_{zc}$ for different strengths of dynamic change and showed that both algorithms cope in general well with dynamic situations. While d-DPClust$_{cz}$ mostly has a better average performance, d-DPClust$_{zc}$ can better handle situations where the number of routers is large.

References

1. J. Handl, J. Knowles, and M. Dorigo: Strategies for the increased robustness of ant-based clustering. Postproc. of the First International Workshop on Engineering Self-Organising Applications (ESOA 2003), LNCS 2977, 90–104 (2003).
2. J. Handl, J. Knowles, and M. Dorigo: On the performance of ant-based clustering. In: Proc. 3rd Int. Conf. on Hybrid Intell. Systems (HIS 2003), IOS Press (2003).
3. L. Kaufman, P.J. Rousseuw. Finding Groups in Data: An Introduction to Cluster-Analysis. Wiley, New York (1990).
4. N. Labroche, N. Monmarch, G. Venturini. A new clustering algorithm based on the chemical recognition system of ants. Proc. Eur. Conf. on AI, 345–349 (2002).
5. N. Labroche, N. Monmarch, G. Venturini: AntClust: Ant Clustering and Web Usage Mining. Proc. of GECCO-2003, Springer, LNCS 2723, 25–36 (2003).
6. N. Labroche, N. Monmarch, G. Venturini. Visual clustering with artificial ants colonies. Proc. 7th International Conference on Knowledge-Based Intelligent Information & Engineering Systems (KES 2003), Springer, LNCS 2773, 332–338 (2003).
7. G. Lambertsen, N. Nishio. Dynamic Clustering Techniques in Sensor Networks. Proc. 7th JSSST SIGSYS Workshop on Systems for Programming and Applications (SPA2004), (2004).
8. J. Yang. Dynamic Clustering of Evolving Streams with a Single Pass. Proc. 19th International Conference on Data Engineering, (2003).
9. Daniel Merkle, Martin Middendorf, and Alexander Scheidler: Decentralized Packet Clustering in Router-based Networks. International Journal of Foundations of Computer Science. To appear (2005).

MOEA-Based Approach to Delayed Decisions for Robust Conceptual Design

Gideon Avigad, Amiram Moshaiov, and Neima Brauner

School of Mechanical Engineering,
The Iby and Aladar Fleischman Faculty of Engineering,
Tel-Aviv University, Israel

Abstract. This paper presents a modification of our multi-objective concept-based EC method to enhance the simultaneous development of a robust conceptual front. It involves a hierarchy of abstractive descriptions of the design, and a structured EC approach. The robustness, treated here by a novel MOEA approach, is for uncertainties, which result from delaying decisions during the conceptual design stage. The suggested procedure involves a robust non-dominancy sorting algorithm.

1 Introduction

Recently we have suggested a concept-based Interactive Evolutionary Computation (IEC) methodology, which supports multi-objective human-computer assessments of conceptual designs [1], [2]. The current paper extends our work to include human uncertainties in the conceptual design stage. Our treatment is for robustness to uncertainties concerning the concept abstraction level, which occur due to delayed decisions. The uncertainties treated here are different from those defined in [3]. Andersson, [4], handles concepts' front and its robustness using a MOEA approach. In addition to his use of an EC approach, which is different from the one presented here, it is also concerned with a dissimilar robustness. The suggested EC approach, described below, performs a simultaneous evolution of a robust conceptual front.

2 Methodology

2.1 Design Space and Robustness

We choose to consider the space of some possible conceptual designs as represented by an 'and/or' hierarchical tree, which reflects decision making during the conceptual design stage. The nodes involve Sub-Concepts (S-Cs). S-Cs are abstract descriptions of generic parts of the conceptual solutions. The 'or' operation designates the alternative S-Cs, which allows the extraction of different 'and' trees (concepts), out of the 'and/or' tree. It is noted that the number of hierarchies, and the number of alternative nodes within a hierarchy, may change from one branch to the other. The hierarchical tree permits the representation of concepts at different level of abstraction, by pruning branches of the tree at different hierarchies. The more pruning, the less detailed is a

F. Rothlauf et al. (Eds.): EvoWorkshops 2005, LNCS 3449, pp. 584–589, 2005.

abstractive description of a concept. A concept, represented by extracting an 'and' tree from the pruned 'and/or' tree is hereby termed Higher-Level Concept (HL-C).

We explore the design space by examining concepts using their associated clusters of preliminary designs (e.g., [1]). Our procedure is based on the assumption that each concept, described by a complete extracted 'and' tree (un-pruned), has an associated model and thus termed Modeled-Concept (M-C). The model allows spanning the M-C into a set of preliminary designs, and the calculation of their design performances. A change of a S-C results in a change of the concept and its model. It should be noted that a S-C could be shared by more than one M-C. Therefore, replacing a S-C in the design space might change more than one M-C and the related models. In contrast to M-C, HL-C does not have a model due to its insufficient description. HL-Cs might be related to several M-Cs. This happens, whenever an 'or' operation exists in the 'and/or' tree under the pruning location. In this case the concept (HL-C) is characterized by more than one detailed representation and therefore has more then one model associated with it. Such HL-Cs are termed Multi-Model Concepts (MM-Cs).

The problem discussed here, involves the evaluation of concepts with an uncertainty associated with undecided (delayed) S-Cs. During conceptual design decisions are occasionally postponed (e.g., [5]). These delays, lead to a more abstractive concept and causes the decided concept to be an MM-C. Each MM-C is associated with several M-Cs, hence with several sets of preliminary designs, and an associated front in the objective space. This front is composed of the non-dominated set of the associated preliminary solutions. Finding the robust front of an MM-C, should be based on the best performances of its worst M-C. The robust conceptual front of several MM-Cs is a combination of the best parts of the robust MM-Cs fronts. Figure 1 depicts an example of two MM-Cs (indexed 1 & 2), their related M-Cs, and the associated sets of preliminary designs (designated by different legends).

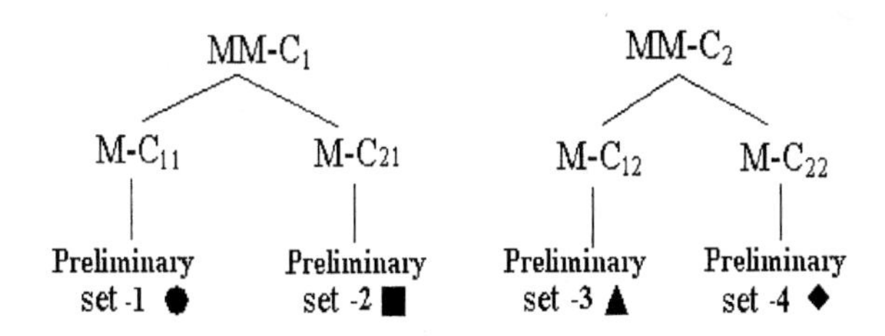

Fig. 1. MM-Cs, their related M-Cs & preliminary sets

The four M-Cs' fronts, in a bi-objective space, are depicted in figure 2a, which also includes the complete sets of the preliminary designs of the example. The two fronts of M-C_{11} and M-C_{21}, which are related to MM-C_1, are designated by continuous lines, while those related to MM-C_2, are designated by broken lines. The worst fronts of MM-C_1 and MM-C_2 are depicted in figure 2b, designated by their different lines. The combined robust conceptual front is depicted in figure 2c.

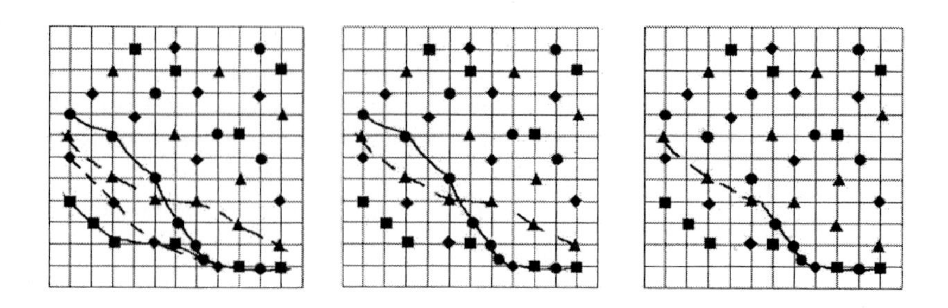

Fig. 2a. M-Cs fronts **Fig. 2b.** Robust MM-Cs' fronts **Fig. 2c.** Robust front

2.2 EC Implementation

The idea of a Compound-Individual, (C-I), has been introduced in [1]. The C-I code corresponds to the 'and/or' tree of the design space. The S-Cs located at the 'or' nodes are competing with each other to win. The winning S-Cs, point to decoded design parameters of a preliminary design. Using these parameters results in the performances of the chosen preliminary design, which are represented as a single point in the objective space. In the case of an MM-C, the competition is restricted to S-Cs located at nodes that are left after the pruning, whereas all possible paths (combinations), below the pruning, are considered. Thus, an MM-C is associated with more then one decoded preliminary design and its C-I may involve several points in the objective space.

Commonly in design application, the performances of a preliminary design are used as a base for the fitness assignment. Here, the problem is that a C-I of an MM-C may have more then one representative in the objective space. The C-I's fitness is determined, according to its relative success, by robustness considerations, using its worst representative (or representatives). A Robust Non-dominancy Sorting (RNS) procedure, which assigns ranks to the C-Is and ensures that high-fitted robust C-Is will have more offspring in the next generation, is introduced in the following pseudo-algorithm.

Pseudo Algorithm-1: The RNS procedure

```
1. Calculate performances of all decoded C-Is
2. Reverse problem's optimum demands, set k=0, and for all C-Is set
   R(C-I) = 0
  While not all C-Is' solutions sorted
   3. Find non-dominancy front of the reversed problem
   4. Set the rank of C-Is belonging to the front to be R(C-I) = k
   5. Remove all C-Is belonging to the front from unranked list of C-
        Is
   Set k = k+1
  End
qr = k
6. Reverse the C-Is' ranks r(C-I) = qr-R(C-I)
```

In step #1 of the suggested procedure, the performances related to all C-Is are calculated. For example, figure 3a shows the performances of 20 C-Is, with a legend as in figure 1. The representatives of each of the C-Is are encircled. In this example, the uncertainty is related to two M-Cs within each MM-C of each decoded C-I. Therefore, this step produces 40 representatives within the objective space. Step # 2 reverses the optimization. The loop of steps 3 to 5 assigns temporary ranks to each of the C-Is. In step #3 the non-dominancy front (in the reversed problem) is found. Due to the inversing of the problem, each C-I with multi representations will have a rank according to its worst performance. Once the C-I is ranked, it is no longer in the ranking loop, hence its better representatives are not accounted. In step #4, C-Is belonging to the resulting front are assigned with a temporary rank R(C-I), increasing by one, for each subsequent front. Step #5 removes the C-Is belonging to the already assigned rank. The procedure continues with the remaining population until all of the C-Is are ranked.

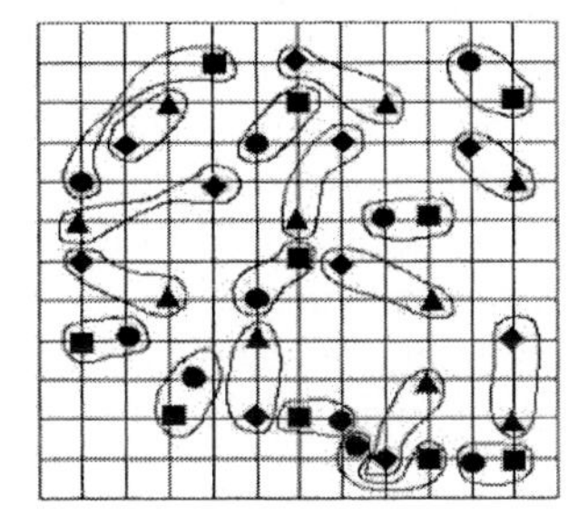

Fig. 3a. C-I's representatives

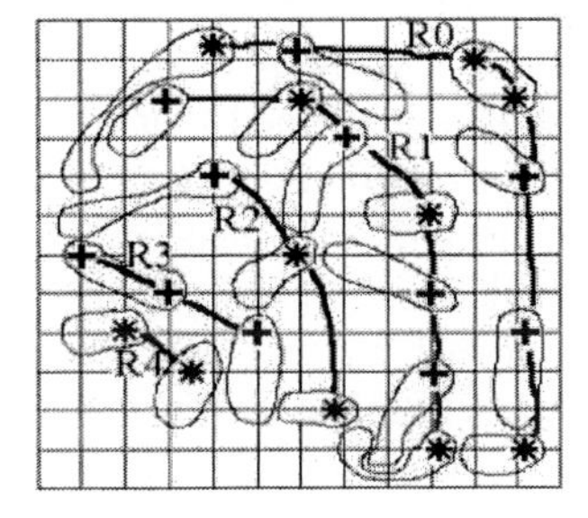

Fig. 3b. Non-dominancy sorting

Step #6 reverses the ranking to return to the original problem. In figure 3b, intermediate results, before step #6, are depicted, with the temporary ranks from the most non-dominated front (R0) to the most dominated front (R4). The representatives of the C-Is related to MM-C_1 are distinguished by stars while those related to MMC_2 by pluses. In figure 3b only the worst representative/representatives is/are indicated within each of the C-I's clusters. In summary, the RNS procedure ranks the C-Is according to their success in the objective space using their worst performances, to account for robustness.

Pseudo Algorithm-2: Evolutionary Process

```
Initialize population of C-Is
  While stopping criterion not reached
    Use RNS procedure (pseudo algorithm 1)
    Compute fitness, f (see ref. # 2) for all C-Is
    Reproduce C-I, Recombine C-I, and Mutate C-I
   End
Perform non-dominancy sorting for every C-I with r(C-I) = 1
Present most dominated sets of all C-Is (just sorted) as
                                     the robust conceptual front
```

The ranks, r, which are obtained using the procedure described above, are used to assign fitness to each C-I, in a procedure that is outlined in [2]. The fitness is degraded according to "Front-based concept sharing" and "In-concept front niching." The degraded fitness, f, is used for the evolutionary process as summarized in the pseudo algorithm #2.

3 Case Study

A bi-objective academic example is used. The objectives are:

$$f_1 = x^2 + 5b + 10 \quad ; \quad f_2 = (x-4)^2 + c(b-1)x \qquad (1)$$

The problem domain, x, is defined within the interval [-10, +10]. The parameters b and c serve to define the S-Cs of the problem. The design space consists of four concepts, which are composed of some predefined combinations of two out of four predefined S-Cs. The predefined four S-Cs are: b=2, b=3, c=-3, c=-4. The four concepts are defined by the following four combinations of the S-Cs: c=-3 and b=2; c=-4 and b=2; c=-4 and b=3; c=-3 and b=3. The different values of these parameters define different models for the objectives of each of the concepts. Figure 4a shows the performances of each concept, as well as the resulting front, which holds three out of the four concepts (1, 2, and 3). Concept #4, marked by a triangle, did not survive due to its inferior performances. Two of these concepts (1, 2), share the same section of the front (upper part). The mutual survival of the concepts on the front is due to front-based concept sharing.

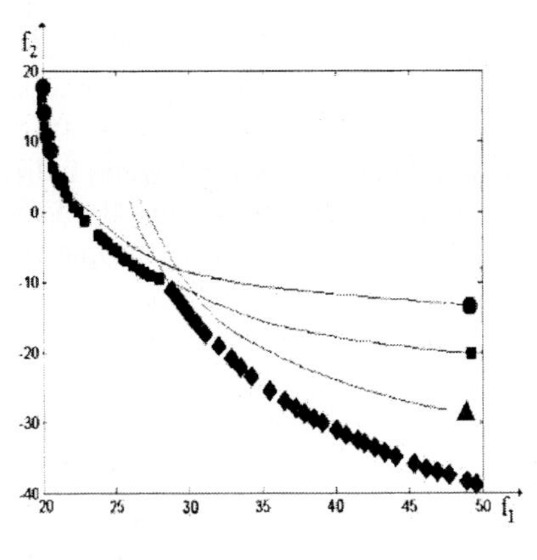

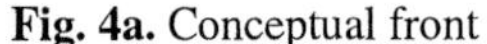
Fig. 4a. Conceptual front

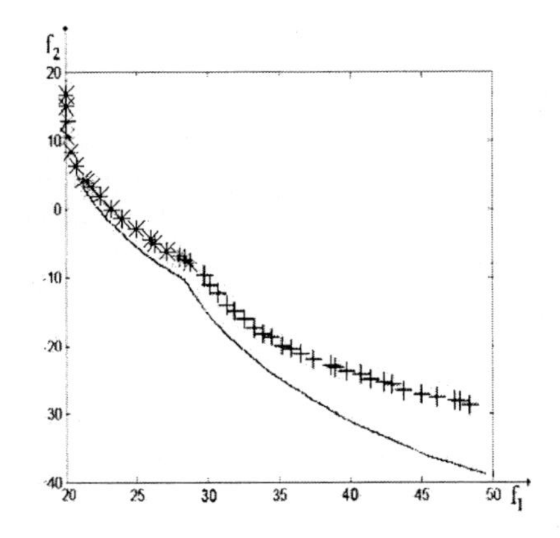

Fig. 4b. Robust conceptual front

Now, suppose that the designers have uncertainty about the use of the S-Cs associated with 'c'. Then two HL-Cs are involved in the design. The design space is represented by the pruned tree depicted in figure 3b, containing the two HL-Cs. A decision on a robust concept, having either b=2 or b=3, has to be made, while the decision on 'c' is postponed. The procedure, introduced in the methodology, is followed to evolve the front depicted in figure 4b. In this figure, the HL-

C related to b=2 is designated with a star while that of b=3 with a plus. The conceptual front holds the combination of the worst fronts belonging to b=2 (circles and squares) and to b=3 (rhombus and triangles). The resulting front of stars and pluses (shown in figure 4b) is not as good as the original one shown by a line, but it is a robust one. Once the robust front is achieved, a selection of an HL-C can be made. The selection ensures the robustness of the selected concept to the uncertainty about the 'c' S-Cs. The GA parameters were maintained as detailed in [2].

4 Summary

In this paper we have introduced a new use of MOEA for robust design. The robustness of a conceptual design with respect to early design stage uncertainties has been encountered using EC. The result of the suggested evolutionary process is a concept-based robust front that provides clusters of robust preliminary designs (related to robust concepts), to the designers, to choose from. The front is the best set of solutions that can be achieved taking into account the worst performances of solutions that might rise from the delaying of decisions in the conceptual design stage.

References

1. Avigad, G., Moshaiov, A., and Brauner N.: *Concept-based Interactive Brainstorming in Engineering Design,* J. of ACIII, vol. 8 no.5, (2004) 454-459.
2. Avigad, G., Moshaiov, A., and Brauner N. (2004). : Pareto-directed Interactive Concept-based Evolution, submitted to the J. of GPEM. http://www.eng.tau.ac.il/~moshaiov/GPEM.pdf.
3. Deb, K. and Gupta, H. Introducing Robustness in Multi-Objective Optimization. *KanGAL Report No. 2004016*, (2004).
4. Andersson J.: Sensitivity analysis in Pareto optimal design. Proc. of the 4th Asia- Pacific Conf. on Simulated Evolution and Learning, SEAL 02, Singapore, (2002) 18-22.
5. Cristiano J.J., White C.C., and Liker J.K. :Application of multiattribute decision analysis to quality function deployment for target setting. IEEE Trans. on SMC – Part C - Applications and Reviews. vol. 31, no. 3, (2001) 366-382.

Unified Particle Swarm Optimization in Dynamic Environments

K.E. Parsopoulos and M.N. Vrahatis

Computational Intelligence Laboratory (CI Lab), Department of Mathematics,
University of Patras, GR–26110 Patras, Greece
{kostasp, vrahatis}@math.upatras.gr
University of Patras Artificial Intelligence Research Center (UPAIRC),
University of Patras, GR–26110 Patras, Greece

Abstract. A first investigation of the recently proposed Unified Particle Swarm Optimization algorithm on dynamic environments is provided and discussed on widely used test problems. Results are very promising compared to the corresponding results of the standard Particle Swarm Optimization algorithm, indicating the superiority of the new scheme.

1 Introduction

Particle Swarm Optimization (PSO) is a stochastic optimization algorithm that belongs to the category of *swarm intelligence* methods [1,2]. PSO has attained increasing popularity due to its ability to solve efficiently and effectively a plethora of problems in diverse scientific fields [3, 4]. Most of these problems involve the minimization of a static objective function, i.e., the main goal is the computation of a global minimizer that does not change.

Dynamic optimization problems, where the global minimizer moves in the search space, arise very often in engineering applications. In contrast to the static optimization case, the main goal in dynamic problems is to track the orbit of the minimizer [5, 6, 7]. Many algorithms that address efficiently static problems, fail when applied to dynamic problems due to their inability to adapt and respond to changes in the environment. Therefore, studies on static environments are usually insufficient to reveal an algorithm's performance when the problem is dynamic. Carlisle and Dozier [8, 9, 10] conducted a thorough investigation of PSO on a large number of dynamic test problems. Modifications of PSO that can tackle dynamic problems efficiently have been recently proposed [3, 11, 12].

A Unified PSO (UPSO) scheme has been recently introduced [13]. This scheme harnesses the local and global variant of PSO, combining their exploration and exploitation abilities without imposing additional requirements in terms of function evaluations. Convergence in probability was proved for the new scheme, and experimental results on widely used static benchmark functions justified its superiority against the standard PSO [13].

In this paper, the performance of UPSO in dynamic environments is investigated and compared with both the local and the global variant of the stan-

F. Rothlauf et al. (Eds.): EvoWorkshops 2005, LNCS 3449, pp. 590–599, 2005.

dard PSO algorithm. A test suite of five widely used test problems is employed. The movement of the global minimizer is simulated by adding to its position a random vector. Numerous experiments are performed and analyzed to justify UPSO's superiority and provide empirical rules regarding the most promising parameter configuration. The paper is organized as follows: PSO and UPSO are described in Section 2. Experimental results are reported in Section 3 and the paper concludes in Section 4.

2 Unified Particle Swarm Optimization

PSO was introduced by Eberhart and Kennedy [1, 14]. Similarly to the evolutionary algorithms, PSO exploits a population of potential solutions to probe the search space simultaneously. However, its dynamic is based on laws that govern socially organized colonies rather than natural selection. PSO adheres to the five basic principles of swarm intelligence [15, 16], therefore it is categorized as a swarm intelligence algorithm.

In PSO's context, the population is called a *swarm* and its individuals (search points) are called *particles*. Each particle moves in the search space with an adaptable velocity. Moreover, each particle has a memory where it retains the best position it has ever visited in the search space, i.e., the position with the lowest function value. Also, the particles share information among them. More specifically, each particle has an index number, and, according to this index, it is assigned a neighborhood of particles with (usually) neighboring index numbers. In the *global* variant of PSO, the neighborhood of each particle is the whole swarm. In the *local* variant, the neighborhoods are strictly smaller and they usually consist of a few particles.

Assume an n–dimensional function, $f : S \subset \mathbb{R}^n \rightarrow \mathbb{R}$, and a swarm, $\mathbb{S} = \{X_1, X_2, \ldots, X_N\}$, of N particles. The i–th particle, $X_i \in S$, its velocity, V_i, as well as its best position, $P_i \in S$, are n–dimensional vectors. A neighborhood of radius m of X_i consists of the particles $X_{i-m}, \ldots, X_i, \ldots, X_{i+m}$. The particles are usually assumed to be organized in a cyclic topology with respect to their indices. Thus, X_N and X_2 are the immediate neighbors of the particle X_1.

Assume g_i to be the index of the particle that attained the best previous position among all the particles in the neighborhood of X_i, and t to be the iteration counter. Then, according to the *constriction factor* version of PSO, the swarm is updated using the equations [17],

$$V_i(t+1) = \chi\Big[V_i(t) + c_1 r_1 \big(P_i(t) - X_i(t)\big) + c_2 r_2 \big(P_{g_i}(t) - X_i(t)\big)\Big], \qquad (1)$$

$$X_i(t+1) = X_i(t) + V_i(t+1), \qquad (2)$$

where $i = 1, \ldots, N$; χ is the constriction factor; c_1 and c_2 are positive constants, referred to as *cognitive* and *social* parameters, respectively; and r_1, r_2 are random vectors with components uniformly distributed in $[0, 1]$. All vector operations in Eqs. (1) and (2) are performed componentwise.

The constriction factor was introduced as means of controlling the magnitude of the velocities, in order to avoid the "swarm explosion" effect that was detrimental for the convergence of early PSO versions, and it is determined through the formula [17, 18],

$$\chi = \frac{2\kappa}{|2 - \phi - \sqrt{\phi^2 - 4\phi}|}, \tag{3}$$

for $\phi > 4$, where $\phi = c_1 + c_2$, and $\kappa = 1$. This selection is based on the stability analysis provided in [17].

There are two main characteristics of a population–based algorithm that affect its performance, namely *exploration* and *exploitation*. The first is the ability to probe effectively the search space, while the latter is the ability to converge to the most promising solutions with the smallest possible computational cost. A proper trade–off between exploration and exploitation is necessary for the efficient and effective operation of the algorithm. The global variant of PSO promotes exploitation since all particles are attracted by the same best position, thereby, converging faster towards the same point. On the other hand, local variant has better exploration properties, since the information regarding the best position of each neighborhood is communicated slower to the rest of the swarm through neighboring particles. Therefore, the attraction to specific points is weaker, thus, preventing the swarm from getting trapped in local minima. Obviously, the proper selection of neighborhood size affects the trade–off between exploration and exploitation. However, there are no general rules regarding the selection of neighborhood size, and it is usually based on the experience of the user.

The *Unified Particle Swarm Optimization* (UPSO) scheme was recently proposed as an alternative that combines the exploration and exploitation properties of both the local and global PSO variants [13]. For completeness purposes, a brief description of UPSO is provided in the following paragraphs. The presented scheme is based on the constriction factor version of PSO, although it can be straightforwardly defined also for the inertia weight version. Let $\mathcal{G}_i(t+1)$ and $\mathcal{L}_i(t+1)$ denote the velocity update of the i–th particle, X_i, for the global and local PSO variant, respectively [13],

$$\mathcal{G}_i(t+1) = \chi \left[V_i(t) + c_1 r_1 \big(P_i(t) - X_i(t)\big) + c_2 r_2 \big(P_g(t) - X_i(t)\big) \right], \tag{4}$$

$$\mathcal{L}_i(t+1) = \chi \left[V_i(t) + c_1 r_1' \big(P_i(t) - X_i(t)\big) + c_2 r_2' \big(P_{g_i}(t) - X_i(t)\big) \right], \tag{5}$$

where t denotes the iteration number; g is the index of the best particle of the whole swarm (global variant); and g_i is the index of the best particle in the neighborhood of X_i (local variant). The search directions defined by Eqs. (4) and (5) are aggregated in a single equation, resulting in the main UPSO scheme [13],

$$\mathcal{U}_i(t+1) = (1-u)\,\mathcal{L}_i(t+1) + u\,\mathcal{G}_i(t+1), \quad u \in [0,1], \tag{6}$$

$$X_i(t+1) = X_i(t) + \mathcal{U}_i(t+1). \tag{7}$$

We named the parameter u, *unification factor*. This factor balances the influence of the global and local search directions in the final scheme. The standard global

PSO variant is obtained by setting $u = 1$ in Eq. (6), while $u = 0$ corresponds the standard local PSO variant. All values of $u \in (0, 1)$, correspond to composite variants of PSO that combine the exploration and exploitation characteristics of its global and local variant.

Besides the aforementioned scheme, a stochastic parameter that imitates the mutation of evolutionary algorithms can also be incorporated in Eq. (6) to further enhance the exploration capabilities of UPSO [13]. Thus, depending on which variant UPSO is mostly based, Eq. (6) can be written either as [13],

$$\mathcal{U}_i(t+1) = (1-u)\,\mathcal{L}_i(t+1) + r_3\, u\, \mathcal{G}_i(t+1), \tag{8}$$

which is mostly based on the local variant, or

$$\mathcal{U}_i(t+1) = r_3\,(1-u)\,\mathcal{L}_i(t+1) + u\, \mathcal{G}_i(t+1), \tag{9}$$

which is mostly based on the global variant, where $r_3 \sim \mathcal{N}(\mu, \sigma^2 I)$ is a normally distributed parameter, and I is the identity matrix. Although r_3 imitates mutation, its direction is consistent with the PSO dynamics. For these UPSO schemes, convergence in probability was proved in static environments [13]. Experimental results on widely used test problems justified the superiority of UPSO against the standard PSO, for various configurations of the PSO parameters proposed in the relative literature [13, 18].

3 Results and Discussion

UPSO's performance was investigated on the most common DeJong test suite, which consists of the Sphere, Rosenbrock, Rastrigin, Griewank and Schaffer's F_6 function [9,17,18]. We will refer to these problems as Test Problem (TP) 1–5, respectively. Test Problems 1–4 were considered in 30 dimensions, while Test Problem 5 was considered 2–dimensional. The initialization ranges were $[-100, 100]^{30}$, $[-30, 30]^{30}$, $[-5.12, 5.12]^{30}$, $[-600, 600]^{30}$, and $[-100, 100]^{2}$, respectively.

The aforementioned static optimization problems were transformed to dynamic problems by moving their global minimizer. In order to make the simulation more realistic, we considered the global minimizer moving randomly and unbounded in the search space. This was performed by adding to the global minimizer a normally distributed random vector with mean value equal to zero and three different values of the standard deviation (denoted as MStD), 0.01, 0.10 and 0.50. Moreover, the movement was considered to be asynchronous, i.e., at each iteration, the global minimizer moved with a probability equal to 0.5.

Regarding PSO, we used the constriction factor version with the standard default parameter values, namely, $\chi = 0.729$, $c_1 = c_2 = 2.05$. Since the global minimizer was moving without constraints, no bounds were posed on velocities and particles. The best positions of the particles were re–evaluated after each movement of the global minimizer (a technique for the detection of changes in the environment is proposed in [9]). In order to fully exploit the exploration abilities of local PSO, a neighborhood of radius 1 was used for the computation of the search direction $\mathcal{L}$ of the local PSO variant.

The values $u = 0.2$ and $u = 0.5$ of the unification factor have been proved to be very efficient in static optimization problems [13]. Preliminary experiments on dynamic problems were in accordance with the results for static problems. Thus, initially, we considered the UPSO approach defined by Eqs. (6) and (7) for $u = 0.0$ (standard local PSO), $u = 0.2$, $u = 0.5$, and $u = 1.0$ (standard global PSO). For each test problem, 100 experiments were conducted and the algorithm was allowed to perform 1000 iterations per experiment. Since the main goal in dynamic environments is to track the orbit of the global minimizer rather than to acquire it [7], the quality assessments of static problems, such as the position with the smallest function value, are not valid in our case [6]. Instead, at each iteration (out of 1000) of an experiment, the mean function value of the particles' best positions was recorded. This value provides a more robust measure of the true quality of the particles [6]. Thus, 1000 such values

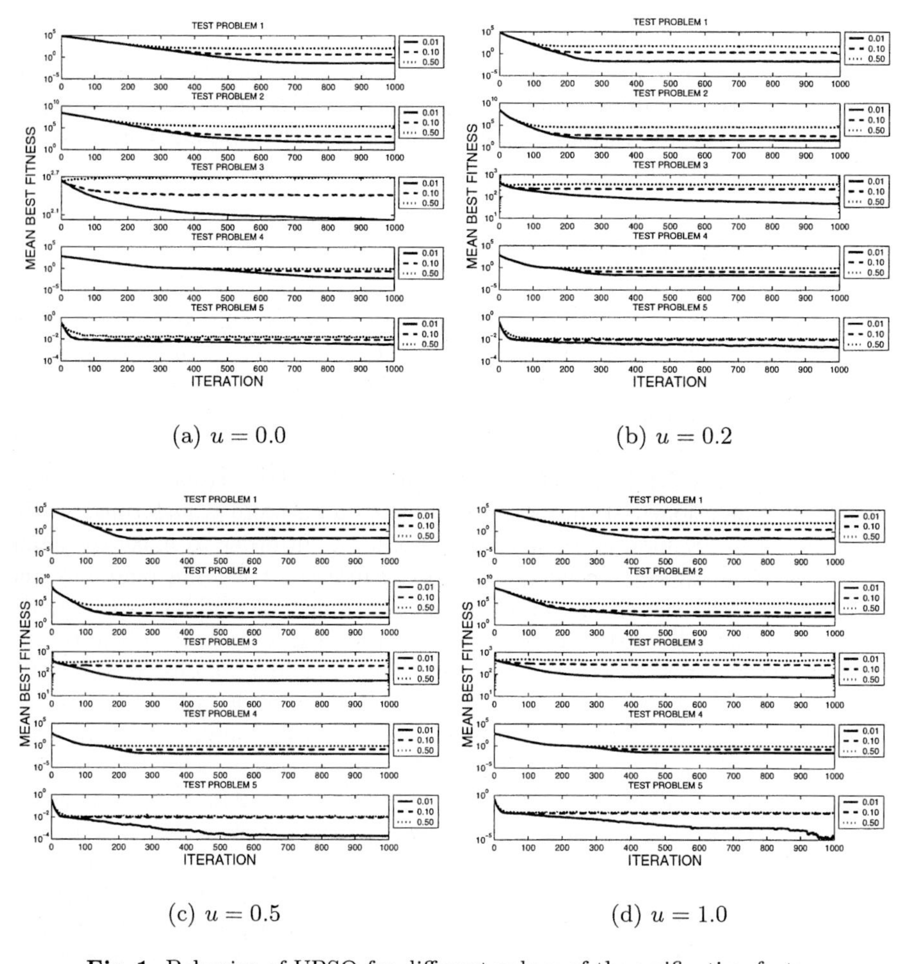

(a) $u = 0.0$

(b) $u = 0.2$

(c) $u = 0.5$

(d) $u = 1.0$

Fig. 1. Behavior of UPSO for different values of the unification factor

Table 1. Results for Test Problems 1–3

TP	MStD	u	Mean	StD	Min	Max
1	0.01	0.0	3.1655×10^3	4.8187×10^2	1.9664×10^3	4.3169×10^3
		0.2	9.1959×10^2	1.2136×10^2	6.3073×10^2	1.3489×10^3
		0.5	7.2230×10^2	9.7704×10^1	5.0892×10^2	1.0408×10^3
		1.0	1.6995×10^3	3.0792×10^2	1.0917×10^3	2.4738×10^3
	0.10	0.0	3.1021×10^3	5.3588×10^2	1.8875×10^3	4.4504×10^3
		0.2	9.1283×10^2	1.2416×10^2	5.8145×10^2	1.2892×10^3
		0.5	7.2875×10^2	1.0179×10^2	5.1276×10^2	9.3233×10^2
		1.0	1.6461×10^3	3.1279×10^2	9.9135×10^2	2.7464×10^3
	0.50	0.0	3.2536×10^3	4.7804×10^2	2.2689×10^3	4.4672×10^3
		0.2	9.5902×10^2	1.2177×10^2	6.6160×10^2	1.3458×10^3
		0.5	7.8051×10^2	9.9330×10^1	5.4604×10^2	1.0057×10^3
		1.0	1.8109×10^3	3.2161×10^2	1.1427×10^3	2.7648×10^3
2	0.01	0.0	8.6011×10^6	2.2275×10^6	3.0879×10^6	1.3499×10^7
		0.2	1.9842×10^6	3.8800×10^5	1.1650×10^6	2.9605×10^6
		0.5	1.4565×10^6	2.9872×10^5	8.9335×10^5	2.4064×10^6
		1.0	4.7003×10^6	1.5037×10^6	1.4236×10^6	9.8277×10^6
	0.10	0.0	8.5898×10^6	2.1055×10^6	3.8289×10^6	1.3550×10^7
		0.2	1.9888×10^6	3.1969×10^5	1.4178×10^6	3.0690×10^6
		0.5	1.4439×10^6	3.1333×10^5	7.1354×10^5	2.5139×10^6
		1.0	4.6714×10^6	1.4462×10^6	2.0045×10^6	8.4536×10^6
	0.50	0.0	8.7243×10^6	2.1533×10^6	3.6296×10^6	1.4842×10^7
		0.2	2.0783×10^6	4.5228×10^5	1.3253×10^6	4.2118×10^6
		0.5	1.4959×10^6	3.2928×10^5	7.0034×10^5	2.5733×10^6
		1.0	5.1983×10^6	2.0042×10^6	2.3351×10^6	1.1613×10^7
3	0.01	0.0	1.5523×10^2	1.9187×10^1	1.0409×10^2	2.1289×10^2
		0.2	9.7620×10^1	1.4826×10^1	6.9749×10^1	1.2776×10^2
		0.5	7.5825×10^1	1.2203×10^1	4.9258×10^1	1.0963×10^2
		1.0	1.0740×10^2	1.7714×10^1	6.4876×10^1	1.5081×10^2
	0.10	0.0	2.7790×10^2	6.1293×10^0	2.6168×10^2	2.9438×10^2
		0.2	2.3185×10^2	4.1795×10^0	2.2192×10^2	2.4213×10^2
		0.5	2.3210×10^2	6.5381×10^0	2.1548×10^2	2.4953×10^2
		1.0	2.8625×10^2	9.3753×10^0	2.6475×10^2	3.0655×10^2
	0.50	0.0	4.9098×10^2	1.8086×10^1	4.4882×10^2	5.4354×10^2
		0.2	3.6957×10^2	8.9636×10^0	3.4994×10^2	3.9674×10^2
		0.5	3.9097×10^2	1.5980×10^1	3.5973×10^2	4.2600×10^2
		1.0	4.5141×10^2	1.8168×10^1	4.1512×10^2	5.1713×10^2

were obtained per experiment. The behavior of UPSO for each test problem and unification factor, are illustrated in Fig. 1, for the three levels of MStD. Each line style corresponds to a different value of MStD and it stands for the mean value of the particles' best position per iteration, averaged over 100 experiments.

Table 2. Results for Test Problems 4 and 5

TP	MStD	u	Mean	StD	Min	Max
4	0.01	0.0	2.8443×10^{1}	4.7740×10^{0}	1.6007×10^{1}	3.9408×10^{1}
		0.2	8.5797×10^{0}	1.0504×10^{0}	6.2604×10^{0}	1.1255×10^{1}
		0.5	6.6704×10^{0}	9.2881×10^{-1}	5.0150×10^{0}	9.5947×10^{0}
		1.0	1.5233×10^{1}	3.1764×10^{0}	6.4421×10^{0}	2.3072×10^{1}
	0.10	0.0	2.8802×10^{1}	4.3512×10^{0}	1.8535×10^{1}	3.9056×10^{1}
		0.2	8.6834×10^{0}	1.1195×10^{0}	6.5475×10^{0}	1.2439×10^{1}
		0.5	6.8352×10^{0}	8.0967×10^{-1}	4.9374×10^{0}	8.6670×10^{0}
		1.0	1.5398×10^{1}	2.7152×10^{0}	8.5176×10^{0}	2.2396×10^{1}
	0.50	0.0	2.9844×10^{1}	4.6330×10^{0}	1.6961×10^{1}	3.9251×10^{1}
		0.2	9.2982×10^{0}	1.0643×10^{0}	7.4027×10^{0}	1.1693×10^{1}
		0.5	7.6360×10^{0}	1.0219×10^{0}	5.4295×10^{0}	1.0125×10^{1}
		1.0	1.6169×10^{1}	3.2054×10^{0}	9.3358×10^{0}	2.6588×10^{1}
5	0.01	0.0	8.0300×10^{-3}	3.3491×10^{-3}	1.6961×10^{-3}	1.3762×10^{-2}
		0.2	6.2539×10^{-3}	3.2677×10^{-3}	6.4010×10^{-4}	1.2861×10^{-2}
		0.5	3.3174×10^{-3}	1.6358×10^{-3}	9.0027×10^{-4}	1.2290×10^{-2}
		1.0	3.4830×10^{-3}	1.6805×10^{-3}	1.0990×10^{-3}	1.0637×10^{-2}
	0.10	0.0	1.2160×10^{-2}	1.0920×10^{-3}	1.0468×10^{-2}	1.6584×10^{-2}
		0.2	1.1681×10^{-2}	7.3334×10^{-4}	1.0157×10^{-2}	1.3808×10^{-2}
		0.5	1.1334×10^{-2}	6.7662×10^{-4}	9.9373×10^{-3}	1.3156×10^{-2}
		1.0	1.1156×10^{-2}	5.1947×10^{-4}	1.0125×10^{-2}	1.2706×10^{-2}
	0.50	0.0	2.1850×10^{-2}	2.0692×10^{-3}	1.5954×10^{-2}	2.8011×10^{-2}
		0.2	1.4647×10^{-2}	9.3145×10^{-4}	1.2654×10^{-2}	1.6941×10^{-2}
		0.5	1.3380×10^{-2}	9.3176×10^{-4}	1.1682×10^{-2}	1.8421×10^{-2}
		1.0	1.3399×10^{-2}	7.7779×10^{-4}	1.1639×10^{-2}	1.6081×10^{-2}

For statistical comparison purposes, the mean function values obtained per experiment, were averaged over the 1000 iterations. Thus, a single averaged mean function value was obtained for each experiment. Therefore, for each test problem, we obtained a total of 100 such averaged means. The mean, standard deviation (StD), minimum (Min) and maximum (Max) of the sample of these 100 averaged means are reported for all test problems and unification factor values in Tables 1 and 2. We observe that UPSO always outperformed both the local and global variant of PSO, which correspond to the values $u = 0.0$ and $u = 1.0$, respectively. The unification factor with the most promising behavior, with respect to the reported mean, is $u = 0.5$, which exhibits the smallest means in most cases, and the best overall behavior for both small and large values of MStD, which is an indication of its robustness.

The good performance of $u = 0.5$ triggered our interest on its behavior using the UPSO schemes of Eqs. (8) and (9). These schemes enhanced significantly UPSO's performance in static optimization problems [13]. For each test problem, 100 experiments were conducted using the UPSO schemes that incorporate $r_3 \sim \mathcal{N}(\mu, \sigma^2 I)$ either on the term of $\mathcal{G}$ (cf. Eq. (8)) or on the term of $\mathcal{L}$ (cf. Eq. (9)), in the update of UPSO's search direction, $\mathcal{U}$. Two different vectors, $\mu = (0, \ldots, 0)^\top$ and $\mu = (1, \ldots, 1)^\top$, were investigated (for simplicity purposes we use the notion $\mu = 0$ and $\mu = 1$, respectively), while a small standard deviation, $\sigma = 0.01$, was selected to alleviate deterioration of UPSO's dynamics [13]. The results are reported in Tables 3 and 4. The addition of r_3 improved further the performance

Table 3. Results for $u = 0.5$ using $r_3 \sim \mathcal{N}(\mu, \sigma^2 I)$ for Test Problems 1–3

TP	MStD	μ	Position	Mean	StD	Min	Max
1	0.01	0.0	on $\mathcal{G}$	3.1805×10^{2}	1.2188×10^{2}	1.8903×10^{2}	7.1865×10^{2}
			on $\mathcal{L}$	5.0574×10^{3}	1.2547×10^{3}	2.3535×10^{3}	8.7807×10^{3}
		1.0	on $\mathcal{G}$	7.2196×10^{2}	9.3121×10^{1}	4.1754×10^{2}	9.3062×10^{2}
			on $\mathcal{L}$	7.3134×10^{2}	9.8959×10^{1}	4.9692×10^{2}	9.5854×10^{2}
	0.10	0.0	on $\mathcal{G}$	3.3886×10^{2}	1.1305×10^{2}	2.2138×10^{2}	9.6920×10^{2}
			on $\mathcal{L}$	5.1672×10^{3}	1.2852×10^{3}	2.5525×10^{3}	7.8607×10^{3}
		1.0	on $\mathcal{G}$	7.3028×10^{2}	1.0014×10^{2}	5.2279×10^{2}	9.9433×10^{2}
			on $\mathcal{L}$	7.2599×10^{2}	9.0081×10^{1}	5.3669×10^{2}	1.0628×10^{3}
	0.50	0.0	on $\mathcal{G}$	1.4828×10^{3}	2.7719×10^{2}	9.0263×10^{2}	2.2291×10^{3}
			on $\mathcal{L}$	6.8694×10^{3}	1.5433×10^{3}	3.7164×10^{3}	1.0982×10^{4}
		1.0	on $\mathcal{G}$	7.7892×10^{2}	1.0623×10^{2}	5.5643×10^{2}	1.0218×10^{3}
			on $\mathcal{L}$	7.8379×10^{2}	9.5146×10^{1}	5.7097×10^{2}	1.1437×10^{3}
2	0.01	0.0	on $\mathcal{G}$	7.2742×10^{5}	2.0365×10^{5}	4.2395×10^{5}	1.3934×10^{6}
			on $\mathcal{L}$	3.0387×10^{6}	1.2173×10^{6}	7.5792×10^{5}	6.5612×10^{6}
		1.0	on $\mathcal{G}$	1.4257×10^{6}	3.3186×10^{5}	8.3978×10^{5}	2.5889×10^{6}
			on $\mathcal{L}$	1.4241×10^{6}	2.9498×10^{5}	8.4386×10^{5}	2.5711×10^{6}
	0.10	0.0	on $\mathcal{G}$	7.2379×10^{5}	1.7563×10^{5}	3.6936×10^{5}	1.2948×10^{6}
			on $\mathcal{L}$	3.3831×10^{6}	1.3056×10^{6}	1.0576×10^{6}	7.5686×10^{6}
		1.0	on $\mathcal{G}$	1.4670×10^{6}	3.0998×10^{5}	8.2882×10^{5}	2.5353×10^{6}
			on $\mathcal{L}$	1.4631×10^{6}	2.9592×10^{5}	8.4093×10^{5}	2.2752×10^{6}
	0.50	0.0	on $\mathcal{G}$	2.0289×10^{7}	1.1652×10^{7}	6.5949×10^{6}	6.8726×10^{7}
			on $\mathcal{L}$	5.7990×10^{7}	2.7477×10^{7}	1.6168×10^{7}	1.4867×10^{8}
		1.0	on $\mathcal{G}$	1.5569×10^{6}	3.4020×10^{5}	9.4104×10^{5}	2.5788×10^{6}
			on $\mathcal{L}$	1.4809×10^{6}	3.0238×10^{5}	8.8182×10^{5}	2.6824×10^{6}
3	0.01	0.0	on $\mathcal{G}$	8.7090×10^{1}	2.0353×10^{1}	5.2522×10^{1}	1.6865×10^{2}
			on $\mathcal{L}$	1.7782×10^{2}	2.5795×10^{1}	1.3637×10^{2}	2.6563×10^{2}
		1.0	on $\mathcal{G}$	7.4911×10^{1}	1.3349×10^{1}	4.8584×10^{1}	1.1738×10^{2}
			on $\mathcal{L}$	7.3238×10^{1}	1.2202×10^{1}	5.0768×10^{1}	1.0567×10^{2}
	0.10	0.0	on $\mathcal{G}$	3.2810×10^{2}	1.7360×10^{1}	2.9064×10^{2}	3.7078×10^{2}
			on $\mathcal{L}$	4.2494×10^{2}	2.8040×10^{1}	3.6882×10^{2}	5.0882×10^{2}
		1.0	on $\mathcal{G}$	2.3175×10^{2}	7.3102×10^{0}	2.1569×10^{2}	2.5508×10^{2}
			on $\mathcal{L}$	2.3111×10^{2}	7.1345×10^{0}	2.1828×10^{2}	2.5152×10^{2}
	0.50	0.0	on $\mathcal{G}$	1.9161×10^{3}	3.8433×10^{2}	1.3110×10^{3}	3.0697×10^{3}
			on $\mathcal{L}$	2.1162×10^{3}	3.9866×10^{2}	1.3790×10^{3}	3.5007×10^{3}
		1.0	on $\mathcal{G}$	3.9092×10^{2}	2.4602×10^{1}	3.4996×10^{2}	5.1264×10^{2}
			on $\mathcal{L}$	3.8603×10^{2}	1.8018×10^{1}	3.4986×10^{2}	4.2635×10^{2}

Table 4. Results for $u = 0.5$ using $r_3 \sim \mathcal{N}(\mu, \sigma^2 I)$ for Test Problems 4 and 5

TP	MStD	μ	Position	Mean	StD	Min	Max
4	0.01	0.0	on $\mathcal{G}$	2.8650×10^{0}	1.1803×10^{0}	1.7891×10^{0}	9.3589×10^{0}
			on $\mathcal{L}$	4.5350×10^{1}	1.2000×10^{1}	1.8176×10^{1}	8.4527×10^{1}
		1.0	on $\mathcal{G}$	6.7278×10^{0}	1.0156×10^{0}	4.7195×10^{0}	1.0510×10^{1}
			on $\mathcal{L}$	6.6302×10^{0}	8.5156×10^{-1}	4.8633×10^{0}	9.1128×10^{0}
	0.10	0.0	on $\mathcal{G}$	3.5698×10^{0}	1.0864×10^{0}	2.4488×10^{0}	7.1177×10^{0}
			on $\mathcal{L}$	4.5699×10^{1}	1.1305×10^{1}	2.3487×10^{1}	7.9473×10^{1}
		1.0	on $\mathcal{G}$	6.9523×10^{0}	8.7108×10^{-1}	5.5159×10^{0}	9.5926×10^{0}
			on $\mathcal{L}$	6.7420×10^{0}	8.7237×10^{-1}	4.5845×10^{0}	9.8724×10^{0}
	0.50	0.0	on $\mathcal{G}$	4.0269×10^{0}	9.8038×10^{-1}	3.0680×10^{0}	7.9478×10^{0}
			on $\mathcal{L}$	4.7370×10^{1}	1.0754×10^{1}	2.7856×10^{1}	7.0848×10^{1}
		1.0	on $\mathcal{G}$	7.5279×10^{0}	8.9793×10^{-1}	5.4762×10^{0}	1.0165×10^{1}
			on $\mathcal{L}$	7.6153×10^{0}	9.1264×10^{-1}	5.7358×10^{0}	9.8793×10^{0}
5	0.01	0.0	on $\mathcal{G}$	8.6723×10^{-3}	3.1477×10^{-3}	3.1126×10^{-3}	1.9211×10^{-2}
			on $\mathcal{L}$	7.6079×10^{-3}	5.4494×10^{-3}	1.4328×10^{-3}	3.8073×10^{-2}
		1.0	on $\mathcal{G}$	3.2053×10^{-3}	1.1405×10^{-3}	8.4245×10^{-4}	6.2444×10^{-3}
			on $\mathcal{L}$	2.9865×10^{-3}	1.3656×10^{-3}	9.8465×10^{-4}	1.0661×10^{-2}
	0.10	0.0	on $\mathcal{G}$	1.7457×10^{-2}	8.4832×10^{-3}	1.1007×10^{-2}	5.7459×10^{-2}
			on $\mathcal{L}$	2.0828×10^{-1}	1.1152×10^{-1}	5.0655×10^{-2}	4.8215×10^{-1}
		1.0	on $\mathcal{G}$	1.1240×10^{-2}	6.5926×10^{-4}	1.0022×10^{-2}	1.3759×10^{-2}
			on $\mathcal{L}$	1.1246×10^{-2}	6.7370×10^{-4}	9.4405×10^{-3}	1.3298×10^{-2}
	0.50	0.0	on $\mathcal{G}$	1.9668×10^{-1}	6.3972×10^{-2}	6.6878×10^{-2}	3.3205×10^{-1}
			on $\mathcal{L}$	4.6819×10^{-1}	2.8537×10^{-2}	4.0353×10^{-1}	5.2920×10^{-1}
		1.0	on $\mathcal{G}$	1.3523×10^{-2}	1.1829×10^{-3}	1.1475×10^{-2}	2.1030×10^{-2}
			on $\mathcal{L}$	1.3627×10^{-2}	1.3526×10^{-3}	1.1788×10^{-2}	2.1753×10^{-2}

of UPSO. For a given level of MStD, the best mean and minimum value both correspond to the same value of μ in all cases, with $\mu = 1$ being marginally better than $\mu = 0$, overall. The best behavior was obtained when r_3 was incorporated in the term of $\mathcal{G}$. Finally, we must notice that for large values of MStD (i.e., 0.5) $\mu = 1$ proved to be the best choice in all test problems with the exception of Test Problem 4.

Summarizing the results, the UPSO scheme of Eq. (8) can be considered a good default choice in unknown dynamic environments when no additional information is available.

4 Conclusions

We investigated the performance of the new, Unified Particle Swarm Optimization (UPSO) scheme in dynamic environments. Experiments on widely used benchmark problems were conducted with very promising results. UPSO outperformed both the local and global PSO variant. Guidelines regarding the most promising UPSO scheme are derived by analyzing the results, and support the claim that, besides static optimization problems, UPSO is a promising scheme also in dynamic environments.

References

1. Eberhart, R.C., Kennedy, J.: A new optimizer using particle swarm theory. In: Proc. 6th Symp. Micro Mach. Hum. Sci., IEEE Service Center (1995) 39–43
2. Kennedy, J., Eberhart, R.C.: Swarm Intelligence. Morgan Kaufmann Publ. (2001)
3. Parsopoulos, K.E., Vrahatis, M.N.: Recent approaches to global optimization problems through particle swarm optimization. Natural Computing **1** (2002) 235–306
4. Parsopoulos, K.E., Vrahatis, M.N.: On the computation of all global minimizers through particle swarm optimization. IEEE Trans. Evol. Comp. **8** (2004) 211–224
5. Branke, J.: Evolutionary Optimization in Dynamic Environments. Kluwer Academic Publishers (2001)
6. Bäck, T., U., H.: Evolution strategies applied to perturbed objective functions. In: Proc. IEEE Congr. Evol. Comput. (1994) 40–45
7. Angeline, P.J.: Tracking extrema in dynamic environments. In: Proc. Evolutionary Programming VI. (1997) 335–345
8. Carlisle, A., Dozier, G.: Adapting particle swarm optimization to dynamic environments. In: Proc. Int. Conf. Artif. Intell., Las Vegas (NV), USA (2000) 429–434
9. Carlisle, A.: Applying the Particle Swarm Optimizer to Non–Stationary Environments. PhD thesis, Auburn University, Auburn, Alabama, USA (2002)
10. Carlisle, A., Dozier, G.: Tracking changing extrema with adaptive particle swarm optimizer. In: Proc. 2002 World Autom. Congr., Orlando (FL), USA (2002)
11. Blackwell, T., Branke, J.: Multi–swarm optimization in dynamic environments. In: Lecture Notes in Computer Science. Volume 3005. Springer-Verlag (2004) 489–500
12. Janson, S., Middendorf, M.: A hierarchical particle swarm optimizer for dynamic optimization problems. In: Lecture Notes in Computer Science. Volume 3005. Springer-Verlag (2004) 513–524
13. Parsopoulos, K.E., Vrahatis, M.N.: UPSO: A unified particle swarm optimization scheme. In: Lecture Series on Computer and Computational Sciences, Vol. 1, Proc. Int. Conf. Comput. Meth. Sci. Eng. (ICCMSE 2004), VSP International Science Publishers, Zeist, The Netherlands (2004) 868–873
14. Kennedy, J., Eberhart, R.C.: Particle swarm optimization. In: Proc. IEEE Int. Conf. Neural Networks. Volume IV., IEEE Service Center (1995) 1942–1948
15. Millonas, M.M.: Swarms, phase transitions, and collective intelligence. In Palaniswami, M., Attikiouzel, Y., Marks, R., Fogel, D., Fukuda, T., eds.: Computational Intelligence: A Dynamic System Perspective. IEEE Press (1994) 137–151
16. Eberhart, R.C., Simpson, P., Dobbins, R.: Computational Intelligence PC Tools. Academic Press (1996)
17. Clerc, M., Kennedy, J.: The particle swarm–explosion, stability, and convergence in a multidimensional complex space. IEEE Trans. Evol. Comput. **6** (2002) 58–73
18. Trelea, I.C.: The particle swarm optimization algorithm: Convergence analysis and parameter selection. Information Processing Letters **85** (2003) 317–325
19. Matyas, J.: Random optimization. Automatization and Remote Control **26** (1965) 244–251

Shaky Ladders, Hyperplane-Defined Functions and Genetic Algorithms: Systematic Controlled Observation in Dynamic Environments

William Rand and Rick Riolo

University of Michigan,
Center for the Study of Complex Systems,
Ann Arbor, MI 48109-1120, USA
wrand@umich.edu

Abstract. Though recently there has been interest in examining genetic algorithms (GAs) in dynamic environments, work still needs to be done in investigating the fundamental behavior of these algorithms in changing environments. When researching the GA in static environments, it has been useful to use test suites of functions that are designed for the GA so that the performance can be observed under systematic controlled conditions. One example of these suites is the hyperplane-defined functions (hdfs) designed by Holland [1]. We have created an extension of these functions, specifically designed for dynamic environments, which we call the shaky ladder functions. In this paper, we examine the qualities of this suite that facilitate its use in examining the GA in dynamic environments, describe the construction of these functions and present some preliminary results of a GA operating on these functions.

1 Introduction

Previous work has shown the GA to be useful in dynamic environments but more needs to be done to fully understand the underlying process [2] [3] [4]. It is often not obvious whether applications based results are particular to that application or if they reflect a more fundamental quality of the GA. There also has been work in examining theoretical questions regarding dynamic optimization [5] [6] [7], but the GA has so many components that to understand it on a theoretical level requires many assumptions, which makes it difficult to apply the results to "real" GAs. However, the middle ground of systematic, controlled observation allows the researcher the ability to contribute to both theory and practice. Methodically accumulating observations that support conjectures provides guidance for theoretical exploration. Also, these same observations can be used to make recommendations to GA applications practitioners. Thus, our goal is to use systematic observations to better understand the dynamics of populations being modified by a GA in a non-static environment.

In order to examine the efficacy of the GA this paper develops a test suite of functions to use as a benchmark. Other test suites for EAs in dynamic environments exist, such as the dynamic knapsack problem, the moving peaks problem

F. Rothlauf et al. (Eds.): EvoWorkshops 2005, LNCS 3449, pp. 600–609, 2005.

and more [3]. The test suite presented here is similiar to the dynamic bit matching functions utilized by Stanhope and Daida [8] among others. However the difference is that the underlying representation of this test suite is schemata, which are also the basis of the GA. By utilizing a test suite that is specific to the GA and that reflects the way the GA searches, the performance of the GA can be easily observed. Hence, we propose a suite that is a subset of the test functions created by John Holland, the hyperplane-defined functions (hdfs) [1], which we have extended to allow for dynamic fitness functions.

2 Shaky Ladder Functions

The test functions that we will be utilizing to explore the GA in dynamic environments is a subset of the hdfs as described by Holland [1]. Holland created these functions in part to meet criteria developed by Whitley [9]. The hdfs are designed to represent the way the GA searches by combining building blocks, hence they are appropriate for understanding the operation of the GA. We begin by describing these functions, then we describe a subset called the building block hdfs. We then examine the strengths and weaknesses of the hdfs and go on to describe the shaky ladder hdfs.

2.1 The Hyperplane-Defined Functions

Holland's hdfs are defined over the set of all strings (usually binary) of a given length n. The fitness of each string is determined by the *schemata* present in the string. A *schema* is a string defined over the alphabet of the original string plus one more character, $*$, and is of length n. The $*$ character represents a *wildcard* that will match against either a 1 or a 0. Any position in the string which is not a wildcard is a *defining locus* or *defining bit.* The *length* of a schema $l(b)$ is the distance between the first and last defining loci in the schema, and the *order* $o(b)$ is the number of defining loci in the schema. An argument string x *matches* (or *contains*) a schema if for every position i from 0 to n, the character at that position in the schema is the wildcard or matches the character at that same position in string x. Each schema s is assigned a fitness contribution, or utility, $u(s)$ that can be any real number. Thus we can now define the hdf fitness, $f(x)$, as the sum of the fitness contributions of all of the schemata x matches [1].

2.2 Building Block hdfs

Holland states that the most interesting hdfs are those built from ground level schemata. We select a group of *elementary* (also called *base* or *first-level*) schemata which have a short length relative to n and a low order. We then combine pairs of these schemata to create *second-level* schemata. We then combine these second-level schemata and so on, to create the *intermediate* schemata, repeating this process until we generate a schema with length close to n. We place all of these schemata within a set B and associate a positive fitness contribution with each of them. These are sometimes called the *building block* schemata, since they are combined to make progress toward the maximal fitness for the function. We also generate a set P of schemata with negative fitness contributions called *potholes*, since they are local depressions in the landscape. The potholes are created by using one elementary schema as a basis and adding defining bits from a supplementary schema. We will refer to the set of hdfs that are constructed from the sets B and P as the *building block hdfs* or *bb-hdfs.*

2.3 Qualities of the hdfs

Holland originally introduced the hdfs as a set of test functions that met Whitley's guidelines [9] as well as a few of his own. In Holland's words [1] Whitley says that the test functions need to be: "(i) generated from elementary building blocks, (ii) nonlinear, nonseperable, and nonsymmetric (and, so, resistant to hillclimbing), (iii) scalable in difficulty, and (iv) in a canonical form."

Holland [1] goes on to state three additional criteria: "(v) can be generated at random and are difficult to reverse engineer... (vi) exhibit an array of landscape-like features... (vii) include all finite functions in the limit..." All of these guidelines seem to be met by Holland's bb-hdfs.

However there are two difficulties with Holland's bb-hdfs that must be addressed when trying to use them to analyze the performance of GAs in dynamic environments. First, though we can easily generate a random hdf, we have no canonical way of generating another one that is "similar" to it. As a result it is difficult to construct a controllably dynamic (but not chaotic) landscape in which the GA can be tested. The second difficulty with Holland's bb-hdfs is that they are intractably difficult to analyze in the general case, i.e. there is no way to know *a priori* the maximum fitness of a given hdf. This is a problem because it is then hard to know exactly how well the GA is performing.

2.4 Shaky Ladder hdfs

In this section we describe three conditions that restrict the set of all bb-hdfs to a set that does not have the two difficulties described in the previous section.

The first condition we call the *Unique Position Condition.* It requires that all elementary schemata contain no conflicting bits. For instance if both schema s_1 and s_2 have a defining bit at position n, they must specify the same value.

The second condition we call the *Unified Solution Condition.* This condition guarantees that all of the specified bits in the elementary level schemata must be present in the highest level schema. This condition also guarantees that all intermediate schemata are a *composition* of lower level schemata. A *composition* of two schemata is simply a new schema which contains all of the defining loci of both schemata[1]. The Unified Solution Condition means (a) if bit n is specified by s_1 then it also must be specified in the highest level schema s_h, and (b) there can be only one unique highest level schema.

[1] Given the Unique Position Condition we do not have to worry about conflicting loci.

The third condition is the *Limited Pothole Cost Condition*, which states that the fitness contribution of any pothole plus the sum of the fitness contributions of all the building blocks in conflict with that pothole must be greater than zero. A pothole, p, is in *conflict* with a building block, b, if both p and b specify a defining locus with the same value. The idea is that potholes are supposed to be temporary barriers to the search, but that a string should be rewarded if it matches all of the building blocks in conflict with that pothole.

These three conditions guarantee that any string which matches the highest level schema must be a string with optimal fitness[2]. By knowing the optimal set of strings we solve one of the problems with Holland's original hdfs.

2.5 Shaky Ladder hdf Generation

We now describe how to create a set of hdfs with similar structure and resulting fitness landscapes. First generate a set of elementary level schemata in a way similar to that used for the bb-hdfs, while satisfying the Unique Position Condition. Then combine these first level schemata in pairs to create second level schemata, combine those to create third level, and so on, until we create a highest-level schema which contains all of the elementary level schemata and meets the Unified Solution Condition. Then generate potholes in a way similar to the bb-hdfs. However when assigning their cost make sure that they satisfy the Limited Pothole Cost Condition.

This technique makes it possible to generate hdfs that are similar to each other. Once a set of elementary schemata have been established we already know the highest level schema and thus we have created a space of hdfs that have the same elementary level schemata and the same highest level schema. We then can generate new hdfs by generating new intermediate level schemata. In this form the sl-hdf's are meant to represent a class of problems in which problem instances may change, but there exists a universal way to solve the problem, similiar to Hillis' sorting problems [2]. Of course it would be easy to allow the optimal value to change as well, and the new value could still be easily calculated, but that would be a different class of problems.

We call the set of hdfs generated in this way the *shaky ladder hdfs* or *sl-hdfs*. For clarity we call a particular hdf of this set an *sl-hdf*, and we call a set of hdfs with the same elementary schemata and highest-level schema an *sl-hdf equivalence set*. The inspiration for the name comes from an old carnival game where the contestant must climb a rope ladder that is almost parallel to but suspended above the ground, which can twist and turn underneath the contestant. If the contestant makes it to the top without falling off, he or she rings a bell and gets a prize. The starting point is always the bottom of the ladder, in the same way the starting point for an sl-hdf equivalence set is always the same elementary schemata. The ending point of the game is always the top of the ladder. Similarly, the highest point in the search space of these hdf functions is always the same highest level schema. However along the way the ladder can

[2] The proof, which is fairly simple, will appear in a future publication.

shake such that the best place for the contestants to position their feet and hands can change. In the same way it is possible to switch an sl-hdf during GA runs to another sl-hdf from the same equivalence set, changing the intermediate building blocks. Strings will still be rewarded for any of the basic building blocks they contain, but some of the intermediate schemata may no longer be rewarded, while other ones that were not rewarded before will be rewarded now.

The sl-hdfs fulfill the conditions that Holland [1] and Whitley [9] set out, but also have the benefit of an easily identifiable optimal fitness. Moreover, since all of the intermediate schemata are composed of unchanging elementary schemata, it is plausible that the difficulty level should be relatively similar between hdfs of the same equivalence set. The only thing that is changing is the particular combination of elementary schemata that are being rewarded, and since each elementary schemata has a pothole associated with it, no elementary schemata are any better than any other. Of course if the GA is already committed to exploiting one particular combination of schemata it will be difficult to switch to another set, but that is the problem we are interested in. We will further address the similarity of different individuals drawn from the same equivalence set in a future work. In the rest of this paper, we will describe one algorithm for constructing sl-hdfs and then briefly explore how a simple GA performs on a series of sl-hdfs which "shake" at different rates.

3 The Algorithm

This section describes the algorithm we used to create the sl-hdfs for the experiments described below. While there are many algorithms which could be used to construct sl-hdfs, we focus here on one approach which is as simple as possible, while closely following the general guidelines outlined by Holland [1]. For example, the fitness contributions are derived from Holland [1].

The algorithm has four major parts: elementary schemata, highest level schema, potholes, and intermediate schemata. Of course we also need to describe how we "shake the ladder" by changing the intermediate schemata.

The Elementary Schemata: The elementary schemata or building blocks must be created in such a way as to fulfill the Unique Position Condition specified above. Holland also recommends that the elementary schemata have short lengths and low orders. Thus we need to be able to create random schemata with length l and order o that meet the Unique Position Condition. Holland recommends a length equivalent to 1/10 the length of the string and an order of roughly 8. Due to the Unique Position Condition it is fairly difficult to meet both of these requirements for an arbitrary set of schemata. Thus to simplify matters in all of the experiments in this paper we set the order of elementary schemata to 8 and did not worry about the schema length. In future work we will explore the effect of different schemata lengths.

To begin with we create a random schema with order o (8 in this paper). We do this by choosing o random indices in the schema and with equal probability setting them to either a 1 or 0. We then create another list of indices which acts as a cumulative record of which loci can be assigned 1's, by examining the previous schema and adding to the list any places that are still wildcards or that are defined as ones. In a similar way we create a list for zeroes. We then shuffle these lists. To create subsequent schemata we then flip a coin o times and if it is a 1 (0) we pull an index off the list for ones (zeroes), making sure we do not reuse any index we have already set. After we are done with this we add this schema to the list of schemata, update the cumulative list for 1's and 0's, and repeat the process for the next schema. When we are done we have a list of what we call *non-conflicting* schemata since they all meet the Unique Position Condition. We assign a fitness contribution of 2 to each of them.

Highest Level Schemata: Given the elementary schemata, it is a simple matter to create the highest level schema since according to the Unified Solution Condition it contains all of elementary schemata. We examine each index in the string and see if any of the elementary schemata have it defined; if so we set the same location in the highest level schema equal to the same defining bit. If none of the elementary schemata have the bit set we leave the location as a wildcard in the highest level schema. We assign a fitness contribution of 3 to this schema[3].

Potholes: The potholes are created in a way inspired by the bb-hdfs. We simply go through the list of elementary schemata (ordered randomly) and use each one as a base schema and its neighbor in the list as a supplementary schema, which creates one pothole for every elementary schema. We build a pothole by including all of the defined bits from the base schema in the pothole, and then for each defining bit in the supplementary schema, we copy that bit value into the pothole with 0.5 probability. We assign a fitness contribution of -1 to all potholes. Note that it is possible to create a pothole that is simply the union of two elementary schemata, or a pothole that is exactly the same as another schema. That means that when examining the Limited Pothole Cost Condition, we include the fitness contribution of one elementary schema and the highest level schema when making sure the Limited Pothole Cost Condition is met.

Intermediate Schemata: To create the intermediate schemata, we first decide how many schemata, $nNextLevel$, should be at the next level by taking the number of schemata at the previous level and dividing by two. We draw two schemata (without replacement) from a shuffled list of all schemata at the current level, and create a composition of these schemata which we add to the next level of schemata. We repeat this process $nNextLevel$ times. We continue to create new levels until $nNextLevel \leq 1$

Shaking The Ladder: To "shake the ladder" we first delete all of the previous intermediate schemata, and then we create new ones by repeating the process

[3] Holland assigns a fitness contribution of 2 to everything after the second-level, but since we do not differentiate between second and higher-levels we assign 3 to everything after the elementary schemata.

described in the previous paragraph. Since at each level we randomly select partners to create the combinants at the next level, we should get different intermediate schemata most of the time.

4 The Experiment

In order to gain experience with the sl-hdf test suite, we carried out experiments in which we simply varied the length of time between "shaking" the ladder, i.e., we varied the frequency of change. Besides allowing us to explore the dynamics of a GA on a sl-hdf, this experiment also allows us to test the common claim that if the time between changes in a dynamic environment is too short, then the best the GA can do is have a population which reflects an average of the environments it faces. However, if the time between changes is long enough, then the GA can track the changes in the environment, and the best individuals are able to attain high performance after changes [7].

Thus we define a control variable, t_δ which specifies the number of generations which elapse before we change the environment. We run a simple GA using the sl-hdf as its fitness function. This GA uses one-point crossover, per bit mutation, full population replacement, and is similar to the one defined in Chapter 3 of Goldberg [10]. Since the point of this paper is to introduce this test suite and examine some preliminary results we decided to start with the simple GA and some standard parameter settings. In the future we will investigate the effect of the parameters on the GA's performance on the sl-hdfs. For the GA and sl-hdf we use the parameters in Table 1. Every t_δ generations we shake the ladder and switch to another sl-hdf in the same equivalence set. We set t_δ from the set $(1, 5, 10, 25, 50, 100, 900, 1801)$. In the last case the time between changes exceeds the run of the GA and thus it provides a benchmark of the performance of the GA on a static environment.

We have done preliminary experiments varying two parameters of the sl-hdf, the length of the string, and the number of base schemata. In short, increasing the string length makes the problem harder for the GA, however this difficulty eventually levels off at length 500, given the other experimental parameters used. The effect of the number of base schemata on the difficulty of solving the problem is more complicated and is compounded because it also increases the number of

Table 1. Parameters of the Experiment

Population Size	1000
Mutation Rate	0.001
Crossover Rate	0.7
Generations	1800
String Length	500
Selection Type	Tournament, size 3
Number of Elementary Schemata	50
Number of Runs	30

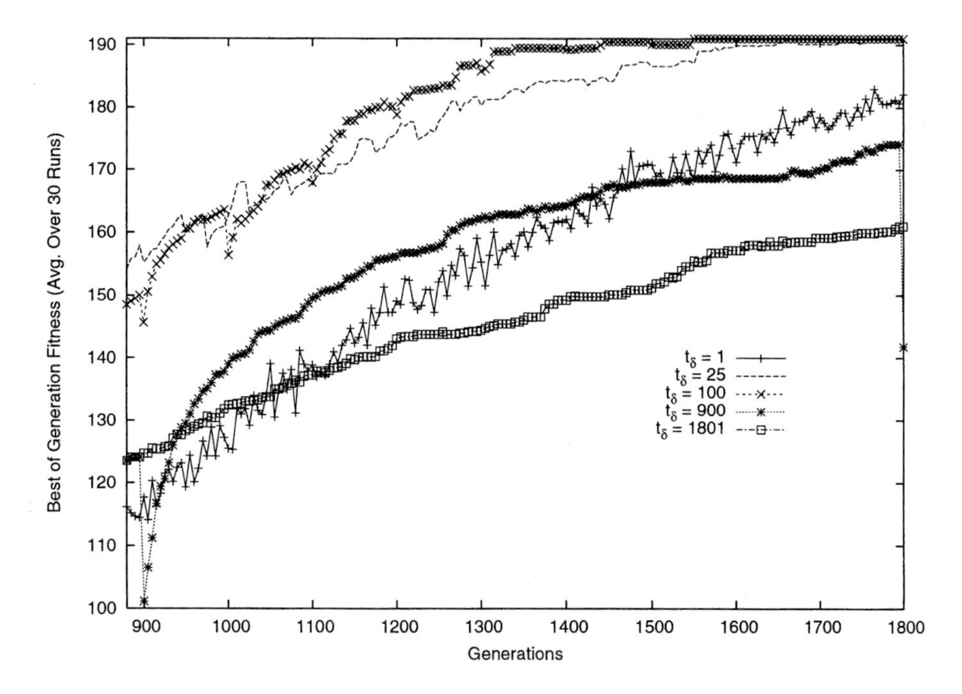

Fig. 1. Results of Experiment after 880 Generations

potholes as explained above. However, our initial investigations indicate that the temporal progress of the GA is roughly the same regardless of the number of base schemata, so we chose 50 to investigate here.

To examine the results, we look at the best fitness found by the GA in each generation, and average those results over 30 runs. The best fitness of the generation is representative of how well the system can do in the current environment, regardless of what has happened in the past.

The results of this experiment for every fifth generation from 880 to 1800 are presented in Figure 1, for $t_\delta = (1, 25, 100, 900, 1801)$ since these capture many of the salient points. The dynamics in the early part of the runs also are interesting, as the populations are identifying the elementary schemata, and we will explore this phase of the search process in future work. The highest possible fitness for these sl-hdfs is 191. Only the runs with $t_\delta = (25, 100)$ were able to attain this result in every run.

5 Results

Figure 1 shows that the GA is able to track the dynamic behavior of the sl-hdf when t_δ is large $(900, 1801)$. When $t_\delta = 1801$ the environment never changes and the system is able to constantly improve its performance. When $t_\delta = 900$ the performance tracks that of $t_\delta = 1801$ until generation 900 at which point it suffers a major set-back, but then performance quickly increases and eventually even surpasses that of the $t_\delta = 1801$ system. However, though performance for both $t_\delta = (900, 1801)$ continues to slowly increase, the GA finds an optimum string in only 1 run of $t_\delta = 900$ and never when $t_\delta = 1800$.

At the other extreme, when $t_\delta = 1$ the system increases performance at a steady pace throughout the run. And despite the noisiness introduced by shaking the ladder every generation, under these conditions the GA's performance eventually surpasses what it achieves with $t_\delta = 1801$ and 901 around 1000 and 1500 generations, respectively. Moreover, at the end of 19 of the runs the GA population contains an optimal string.

The GA attains its best performance at intermediate rates of change in the environment. When $t_\delta = 100$ the system tracks and adapts to changes rapidly, almost constantly improving performance with only small dips for the changes. For $t_\delta = 25$ the results are much better than with $t_\delta = 1$, approaching the performance of $t_\delta = 100$ at the end of the runs. Under both of these conditions the GA finds the optimal strings in all runs within 1800 generations.

In summary, under these conditions the simple GA does worst when the environment never changes, it does a bit better (after initially doing worse) with very fast changes, and it does best for intermediate rates of change. What might explain these results? One explanation is that shaking the ladder helps the GA to avoid "premature convergence" as a result of hitchhiking [1]. With infrequent shaking, the GA makes early gains by focusing on intermediate schemata that are combinations of the base schemata it happens to find first. But it also loses diversity due to hitchhiking of incorrect bit values in the base schemata it has not found. Thus the greatest hits in performance suffered by any system are with the $t_\delta = 900$ GA, in which it has a large investment in intermediate building blocks and when they change it suffers tremendously. On the other hand, when $t_\delta = 1$ the intermediate schemata provide little guidance toward finding the optimal string, since they change each time step. As a result the GA performance improves slowly at the start of the runs with $t_\delta = 1$, though it does continue to improve over the entire run, eventually doing better than the slow/no change cases. However, with intermediate rates of shaking, the GA is able to use the intermediate level schemata to help guide the search for the optimal string. But because the intermediate levels change periodically, the GA does not focus too much on any particular intermediate schemata, and thus it is able to find a wide range of base schemata supporting the different intermediates that are rewarded over time. The diversity of the strings in the population acts as a kind of memory of the intermediate schemata the GA has seen.

Thus it seems that a GA in a faster changing environment is trying to capture as many unchanging elementary building blocks as it can while a GA in a slower changing environment gets misguided down a path that relies on "shaky" intermediate schemata. The GA does track changes in the slower changing environments, and at intermediate rates of change, the GA does in some sense "average" over the equivalence set of the sl-hdf it is searching, in that it is guided by the intermediate schemata it has seen. For the fastest rates of change, the GA simply focuses on the only constant part of the environment, the base schemata, and by discovering those it eventually gains enough components that regardless of what intermediate schemata are being rewarded it is still being partially rewarded. It eventually finds an optimal string in some runs by identifying all of the base schemata.

6 Conclusion and Future Work

Though various researchers have investigated the GA in dynamic environments before, our contribution is to undertake these experiments in a systematic way on a newly defined test suite that allows for the comparison of results both within and between environment types. We are able to do this because of the sl-hdfs which we created to examine the ability of the GA to work in dynamic environments. The experiments presented here will also serve as a basis for the investigation of more complex evolutionary processes beyond the simple GA, in which we explore other system characteristics like robustness, satisficability, and population diversity. Further investigations into the parameter space of the GA and sl-hdf, as well as an examination of the similarity of sl-hdf's is also warranted. Finally, we plan to further test the explanations regarding hitchhiking that we explicate in the discussion.

References

1. Holland, J.H.: Building blocks, cohort genetic algorithms, and hyperplane-defined functions. Evolutionary Computation **8** (2000) 373–391
2. Hillis, W.D.: Co-evolving parasites improve simulated evolution as an optimization procedure. In Langton, C.G., Taylor, C., Farmer, J.D., Rasmussen, S., eds.: Artificial Life II. Volume X of SFI Studies in the Sciences of Complexity., Addison-Wesley (1991) 313–324
3. Branke, J.: Evolutionary algorithms for dynamic optimization problems: A survey. Technical Report 387, Institute AIFB, University of Karlsruhe (1999)
4. Branke, J.: Evolutionary Optimization in Dynamic Environments. Kluwer Academic Publishers (2001)
5. Branke, J.: Evolutionary approaches to dynamic optimization problems - introduction and recent trends. In Branke, J., ed.: GECCO Workshop on Evolutionary Algorithms for Dynamic Optimization Problems. (2003) 2–4
6. Bäck, T.: On the behavior of evolutionary algorithms in dynamic environments. In: 1998 IEEE International Conference on Evolutionary Computation proceedings, New York, IEEE (1998) 446–51
7. Ronnewinkel, C., Wilke, C.O., Martinetz, T.: Genetic Algorithms in Time-Dependent Environments. In: Theoretical Aspects of Evolutionary Computing. Springer (2000)
8. Stanhope, S.A., Daida, J.M.: Optimal mutation and crossover rates for a genetic algorithm operating in a dynamic environment. In: Evolutionary Programming VII. Number 1447 in LNCS, Springer (1998) 693–702
9. Whitley, D., Rana, S.B., Dzubera, J., Mathias, K.E.: Evaluating evolutionary algorithms. Artificial Intelligence **85** (1996) 245–276
10. Goldberg, D.E.: 3. In: Genetic Algorithms in Search, Optimization, and Machine Learning. Addison-Wesley Publishing Company, Inc., Reading, Massachusetts (1989)

A Hierarchical Evolutionary Algorithm with Noisy Fitness in Structural Optimization Problems

Ferrante Neri[1], Anna V. Kononova[2], Giuseppe Delvecchio[3], Marcello Sylos Labini[1], and Alexey V. Uglanov[2]

[1] Department of Electrotechnics and Electronics, Polytechnic of Bari, Via E. Orabona 4, 70125, Bari Italy (Neri, Sylos Labini)
neri@deemail.poliba.it, sylos@deemail.poliba.it

[2] Laboratory of Theoretical-Probabilistic Methods, Yaroslavl' State University, 150000, Yaroslavl', Sovetskaya st., 14, Russia (Kononova, Uglanov)
annk2000@mail.ru, uglanov@univ.uniyar.ac.ru

[3] Technical Area, University of Bari, Piazza Umberto I, 1 70124, Bari, Italy
g.delvecchio@area-tecnica.uniba.it

Abstract. The authors propose a hierarchical evolutionary algorithm (HEA) to solve structural optimization problems. The HEA is composed by a lower level evolutionary algorithm (LLEA) and a higher level evolutionary algorithm (HLEA). The HEA has been applied to the design of grounding grids for electrical safety. A compact representation to describe the topology of the grounding grid is proposed. An analysis of the decision space is carried out and its restriction is obtained according to some considerations on the physical meaning of the individuals. Due to the algorithmic structure and the specific class of problems under study, the fitness function of the HLEA is noisy. A statistical approach to analyze the behavior and the reliability of the fitness function is done by applying the limit theorems of the probability theory. The comparison with the other method of grounding grid design shows the validity and the efficiency of the HEA.

1 Introduction

The grounding grids are important countermeasures to assure the safety and the reliability of the power systems and apparatus. In order to guarantee the safety level conditions, the touch voltages generated by the grounding grid in each point of the soil surface must be lower than the prearranged values fixed by Standards [1]. This requirement causes a significant increase in the cost of both conductor material and ditching. It is therefore fundamental, when a grounding grid has to be designed, to choose a criterion which guarantees both the low cost and the respect of the safety conditions.

The study of grounding grids and their design has been intensively discussed over the years and has been carried out according to different approaches. Empirical approaches have been suggested and some criteria resorting to the "compression ratio" have been given in [2]. Besides, some methods based on the genetic algorithms (GAs) have also been implemented [3], [4], [5].

F. Rothlauf et al. (Eds.): EvoWorkshops 2005, LNCS 3449, pp. 610–616, 2005.

2 Description of the Hierarchical Evolutionary Algorithm

The problem of the design of a grounding grid can be formulated as follows: once leaking current, depth, volume of conductors (i.e. diameter of the cylinder conductors and number of the horizontal and vertical conductors) are prearranged, the topology of the grounding grid such that touch voltages are minimized has to be obtained.

2.1 The Lower Level Evolutionary Algorithm to Calculate U_{Tmax}

Let us consider a fixed topology of grounding grid with a section of conductors S_c which is buried at a depth d in a soil whose resistivity is ρ and leaks a fault current I_F. In order to determine the touch voltage U_T in a point P of the soil surface it is necessary to solve a system of 2nd order PDEs which in many cases cannot be solved theoretically. The Maxwell's subareas method [6] has been therefore implemented.

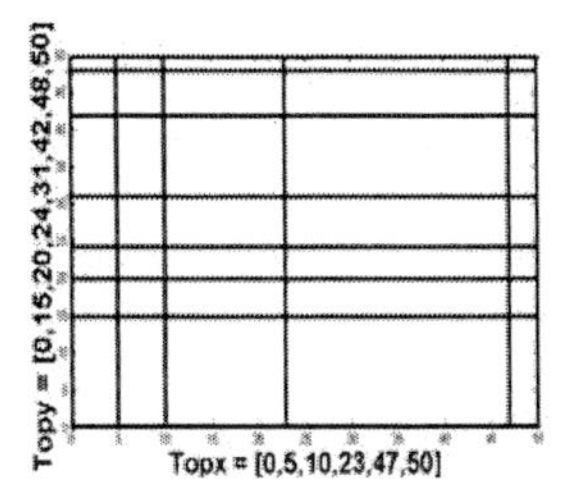

Fig. 1. Example of grounding grid (See paragraph 2.2)

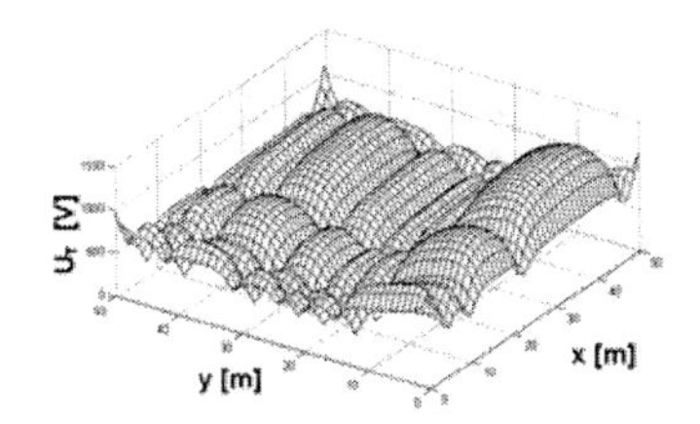

Fig. 2. Example of trend of touch voltage generated by a grounding grid

The calculation of the touch voltage by the Maxwell's method is computationally expensive and, in order to determine the maximum touch voltage U_{Tmax}, an optimization method is required. Since the trend of the touch voltage is, in general, multimodal (See Fig. 2), an approach which makes use of an exact method or a gradient-based method is not acceptable.

The LLEA here proposed is a steady-state GA [7] which works on a population of N_p points P of the soil surface initially sampled pseudo-randomly. Each point $P(x,y)$ is an individual having x and y as chromosomes. Due to the continuous trend of the touch voltage a real encoding has been chosen. In consideration of the rectangular external shape of the grounding grid and therefore of the rectangular shape of the decision space the arithmetic crossover technique [8] has been chosen. The probability of random mutation [8] has been set on 0.1. The algorithm is stopped when at least one of the two following conditions occurs: the difference $\Delta U_{T\ k}$ between the maximum touch voltage and the average value among all the touch voltages $U_{T\ k}(P)$ obtained in the kth iteration is smaller than a pre-arranged value of accuracy ϵ; the number of iterations N_{iter} reaches a pre-arranged value $N_{iter\text{-}max}$.

2.2 The Higher Level Evolutionary Algorithm to Obtain $Grid^{opt}$

The authors propose the HLEA, a steady-state [7] GA, which uses the value of maximum touch voltage U_{Tmax} obtained by the LLEA as a fitness function to find the topology of the grounding grid $Grid^{opt}$ which generates a minimum value of maximum touch voltage. Each individual of the population is a grounding grid *Grid(Topx, Topy)*. This grid (individual) has a genotype composed by two vectors of integer non-negative numbers; each of these numbers identifies a conductor of the grounding grid under study and represents the distance of this conductor from the reference axis. The origin of the reference axis is the lower left corner of the grounding grid (See Fig. 1). The two-point crossover technique [8] has been chosen. Mutation occurs, with a probability equal to 0.04, in the following way. A position of the gene to undergo the mutation is set pseudo-randomly. Another positive or negative random integer value is added to this gene. This operation is equivalent to moving a conductor of the grounding grid under examination of a small predetermined quantity. The stop criterion follows the same logic as that occurred in the LLEA.

2.3 Computational Analysis and Restriction of the Decision Space

The design problem consists in minimizing the function $U_{Tmax}(Topx, Topy)$ in the set $S = \mathbf{N}^{Nx \cdot + Ny}$ where **N** is the set of natural numbers. The number of vertical and horizontal conductors Nx, and Ny as well as the vertical and horizontal size of the grounding grid Ly, Lx are prefixed a priori. The decision space $D \subset S$ has cardinality:

$$card(D) = (Lx+1)^{Nx} (Ly+1)^{Ny}. \tag{1}$$

The position of the gene within the chromosome does not give any further information to the topology of the grounding grid. Consequently, the authors have chosen to add, at each iteration, a sorting cycle for the genes of each vector *Topx* and *Topy*. This choice decreases the cardinality of the decision space and, therefore, the computational complexity of the algorithm.

Proposition. The cardinality of the new decision space D^* is given by:

$$card(D^*) = \frac{(Nx - Lx)!}{Nx! \cdot Lx!} \cdot \frac{(Ny - Ly)!}{Ny! \cdot Ly!} \tag{2}$$

Proof: Let us consider the set of vertical conductors. Let c_l be the number of occurrences of l in the chromosome. Each individual belonging to D^* can be also expressed as a vector of occurrences $[c_0,..,c_{Lx}]$ where $c_0...+ c_{Lx} = Nx$. Applying the one-to-one correspondence of this kind to all the elements of the decision space D^* a new set M^* is generated and $card(D^*) = card(M^*)$. Let us consider a polynomial $\hat{p}(x) = (1 + x + ... + x^m)^n = \sum_{k=0}^{mn} b_k x^k$. In this case b_k is the cardinality of:

$A_k^{m,n+1} = \{[d_0 ,.. d_n] : d_i \in \{0,...,m\}, i = 0,...,n \text{ and } \sum_{i=0}^{n} d_i = k\}$. According to the above definition, $A_{Nx}^{Nx,Lx} = M^*$ and $b_{Nx} = card(A_{Nx}^{Nx,Lx}) = card(D^*)$. Performing the division of the following polynomials we obtain that $\frac{1 - x^{Nx+1}}{1 - x} = 1 + x + x^2 + ... + x^{Nx}$.

Let us consider the Lx^{th} order derivative of the geometric progression taken termwise, the common ratio $|x|$ being less than 1, $\frac{Lx!}{(1-x)^{Lx+1}} = \sum_{i=Lx}^{\infty} \frac{i!}{(i-Lx)!} x^{i-Lx}$; therefore $\frac{1}{(1-x)^{Lx+1}} = \sum_{n=0}^{\infty} \frac{(n+Lx)!}{(Lx)!n!} x^n = \sum_{n=0}^{\infty} \binom{n+Lx}{Lx} x^n$. The $card(A_{Nx}^{Nx,Lx})$ is the Nx^{th} coefficient of the classical expansion of $p(x) = \left(\frac{1-x^{Nx+1}}{1-x}\right)^{Lx+1} = \sum_{n=0}^{\infty} \binom{n+Lx}{Lx} \cdot (\sum_{j=0}^{Lx+1} (-1)^j \binom{Lx+1}{j} x^{n+j(Nx+1)})$ that is $b_{Nx} = \binom{Nx+Lx}{Lx} = \frac{(Nx+Lx)!}{Nx!Lx!}$. The proof performed above is identical in the case of horizontal conductors. The sets of vertical and horizontal conductors are independent of each other, therefore the cardinality of the new decision space is given by (2)[1].

3 Statistical Analysis of the Noisy Fitness and Numerical Results

As previously described the HLEA makes use of the fitness given by the LLEA. Since the latter is a GA, it gives a value of touch voltage that is not deterministic but it takes its own value according to a stochastic process. Therefore the HLEA works with a noisy fitness.

Results shown in Table 1 and Table 2 refer to a grounding grid whose genotype is *Grid* ([0 , 5, 20, 25, 35, 40, 50], [0, 10, 15, 30, 36, 40, 50]) laying in a domain 50m x 50m. The values for the parameters characterizing the problem are the following:

d=0.5m S_c=50 mm^2 ρ=100 Ωm I_F=400 A $N_{iter\text{-}max}$=20 N_{exp} =500

For each population size N_{pop}, a set of values of U_{Tmax} has been obtained. Among these values, the maximum $U_{Tmax}{}^{max}$, the minimum $U_{Tmax}{}^{min}$ and the width of the interval $W = U_{Tmax}{}^{max} - U_{Tmax}{}^{min}$ have been determined. Hence, fixing a percent accuracy $acc\%$ and therefore the interval V=[(100-$acc\%$)/100 $U_{Tmax}{}^{max}$, $U_{Tmax}{}^{max}$], the probability q^* that a value given by the LLEA falls within the interval V has been approximated, according to the law of the big numbers, by $q = N_V/N_{exp}$ where N_V is the number of simulations so that the value falls within V and N_{exp} is the total number of simulations (experiments)-see Table 1.

The accuracy of the approximation of q^* has been verified by the central limit theorem (see Table 2). The probability P that the approximated probability q does not deviate from the real value q^* more than a value δ is expressed by:

$$P\{|q-q^*| \le \delta\} \approx 2\Phi\left(\delta\sqrt{N_{\exp}/q(1-q)}\right)-1 \quad \text{where} \quad \Phi(x) = 1/\sqrt{2\pi} \int_{-\infty}^{x} e^{-\frac{y^2}{2}} dy$$

[1] The problem studied and then the results got are similar to a classic problem of quantum mechanics (Bose-Einstein Statistics).

Table 1. Estimation of the noise and calculation time in the LLEA

N_{pop}	$U_{Tmax}{}^{max}$ [V]	$U_{Tmax}{}^{min}$ [V]	W [V]	Approximated probabilities q that U_{Tmax} determined by the LLEA falls within the interval V (for several values of $acc\%$)						Mean time[2] t [s]
				1%	2%	3%	5%	7%	9%	
40	97.241	83.562	13.679	0.72	0.74	0.75	0.76	0.77	0.99	4.82
70	97.241	88.874	8.387	0.83	0.84	0.85	0.86	0.88	1	11.03
100	97.241	88.874	8.387	0.94	0.97	0.97	0.97	0.97	1	32.64

Table 2. Confidence intervals δ for approximation performed with confidence level P=0.995

N_{pop}	Confidence intervals δ related to the probabilities q for several population sizes N_{pop} and values of accuracy $acc\%$					
	1%	2%	3%	5%	7%	9%
40	0.083	0.080	0.079	0.079	0.077	0.018
70	0.069	0.067	0.066	0.064	0.060	0
100	0.044	0.031	0.031	0.031	0.031	0

Table 3. Comparison between the HEA and other designing methods

Method	Objective	Assumptions	Decision Space
[2]	None	Exponential regularity	Continuous
[4]	Volume	Symmetry, particular proportions	Continuous
[3]	Cost	Symmetry, conductors on the perimeter	$2^{\frac{Lx}{2}+\frac{Ly}{2}+2}$
[5]	Touch voltage	None	$(Lx+1)^{Nx}(Ly+1)^{Ny}$
HEA	Touch voltage	None	$\frac{(Nx-Lx)!}{Nx!\cdot Lx!}\cdot\frac{(Ny-Ly)!}{Ny!\cdot Ly!}$

As the results in Table 1 show, if the size of the population of the LLEA is small, the fitness of the HLEA is quickly calculable but it is noisier; on the contrary, if the size is large the fitness is less noisy but each single evaluation requires a longer time. On the other hand, we should consider that the functions whose values take a large variability demand a large population size in order to find the global optimum [10]. In other words, if the population size of the LLEA is small, the HLEA, due to the noisy fitness, requires a large population size; if population size of the HLEA is small, a low-noisy fitness is required and therefore a large population of the LLEA. According to the obtained results, the authors propose for 50m$<Lx,Ly<$1000m the population sizes $N_{pop\,L}=70$ and $N_{pop\,H}=50$ for LLEA and HLEA, respectively.

[2] The calculation times refer to a PC with a frequency of 3 GHz and 512 Mb RAM.

Table 3 shows the comparison between the HEA and the other designing method found in literature.

Table 4. Numerical results obtained by different designing methods

Method	*Topx Topy*	U_{Tmax} [V]	Decision Space
[2]	[0,9.88,22.89,40,57.11,70.11,80] [0,13.05,30,46.95,60]	905.3762	Continuous
[3]	[0,6,22,40,58,74,80][0,10,30,50,60]	818.2894	$4.7224\text{x}10^{21}$
[5]	[0,7,22,40,58,73,80][0,10,30,50,60]	801.5457	$1.9322\text{x}10^{22}$
HEA	[0,7,22,40,58,73,80][0,10,30,50,60]	801.5457	$4.8265\text{x}10^{16}$

Table 4 shows the results obtained for the following set of parameters [3]:

D=0.5m S_c=69 mm^2 ρ=100 Ωm I_F=5 kA Lx=80m Ly=60m

4 Conclusions

The HEA performed better results in terms of efficiency and computational complexity compared to the other methods found in literature. Besides, the HEA is a completely general and automatic method which does not require any assumption on the topology of the grid. The HEA works in a decision space much smaller than the other designing methods thanks to the compact representation and the sorting cycle implemented. Though the decision space is small it contains not only all the representations of grounding grids considered in [3] but also non-symmetrical solutions that can be optimal in the case of non-constant resistivity. The statistical analysis carried out proves the reliability of the algorithm notwithstanding the noisy fitness. Moreover, the HEA is an extremely flexible method that can be applied in several other structural optimization problems.

References

1. IEEE Standard 80 – 2000: IEEE Guide for Safety in AC Substation Grounding. (2000)
2. Sun, W., He, J., Gao, Y., Zeng, R., Wu, W., Su, Q.: Optimal Design Analysis of Grounding Grids for Substations built in non-uniform soil. Proc. of Powercon. Intern. Conference on Power System Technology, Vol. 3. (2000) 1455-1460
3. Otero, A. F., Cidras, J., Garrido, C.: Genetic Algorithm Based Method for Grounding Grids Design. Proc. of the IEEE Inter. Conference on Evolutionary Computation. World Congress of Computational Intelligence. (1998) 120-123
4. Costa, M. C., Filho, M. L. P., Marechal, Y., Coulomb, J.-L., Cardoso, J. R.: Optimization of Grounding Grids by Response Surface and Genetic Algorithms. IEEE Transactions on Magnetics, Nol. 39, No 3. (2003) 1301-1304
5. Neri, F.: A New Evolutionary Method for Designing Grounding Grids by Touch Voltage Control. Proc. of IEEE Intern. Symposium on Industrial Electronics, (2004) 1501-1505

6. Sylos Labini, M., Covitti, A., Delvecchio, G., Marzano, C.: A Study for Optimizing the Number of Subareas in the Maxwell's Method. IEEE Trans. on Magnetics, Vol. 39, No. 3. (2003) 1159-1162
7. Eshelman, L. J.: The CHC adaptative search algorithm: How to have safe search when engaging in nontraditional genetic recombination. In FOGA-1 (1991) 265-283
8. Michalewicz, Z.: Genetic Algorithms + Data Structures = Evolution Programs. 3rd edn. Springer-Verlag, Berlin Heidelberg New York (1996)
9. Goldberg, D. E., Deb, K., Clark, J. H.: Genetic algorithms, noise, and sizing of populations. Complex Systems 6(4). (1992) 333-362

Assortative Mating in Genetic Algorithms for Dynamic Problems

Gabriela Ochoa, Christian Mädler-Kron, Ricardo Rodriguez, and Klaus Jaffe

Universidad Simon Bolivar, Caracas 89000, Venezuela
gabro@ldc.usb.ve, christian@sauron.sytes.net

Abstract. Non-random mating seems to be the norm in nature among sexual organisms. A common mating criteria among animals is assortative mating, where individuals mate according to their phenotype similarities (or dissimilarities). This paper explores the effect of including assortative mating in genetic algorithms for dynamic problems. A wide range of mutation rates was explored, since comparative results were found to change drastically for different mutation rates. The strategy for selecting mates was found to interact with the mutation rate value: low mutation rates were the best choice for dissortative mating, medium mutation values for the standard GA, and higher mutation rates for assortative mating. Thus, GA efficiency is related to mate selection strategies in connection with mutation values. For low mutation rates typically used in GA, dissortative mating was shown to be a robust and promising strategy for dynamic problems.

1 Introduction

A source of inspiration toward understanding and improving the application of GAs is still natural evolution. Biological evolution is not possible without reproduction, and among sexual organisms, reproduction is not possible without mating or the fusion of two gametes. Mating is very unlikely to be random in nature, and mate selection may be as important in guiding evolution than natural selection. Theoretical studies of mate selection using agent-based simulations [8, 9, 6], suggest that some mating strategies confer higher fitness to individuals, and produce higher population diversity than random mating.

Most previous work on non-random mating in GAs, refers to incest-prevention techniques, where the idea is to prevent recombination between related individuals [2, 1, 3]. Other authors [5, 4] explore the inclusion of assortative mating within GAs. Assortative mating is a form of *non-random mating* common in nature, where individuals of similar phenotype mate more or less often than expected by chance. It is positive if similar organism mate more often, and negative (or dissortative) if dissimilar organisms mate with higher frequencies. All previous research discussed above deal with stationary problems, and in general report that both incest prevention techniques and negative assortative mating maintain a higher population diversity, which in turn allows a broader exploration of the

F. Rothlauf et al. (Eds.): EvoWorkshops 2005, LNCS 3449, pp. 617–622, 2005.

search space, and may speed up the searching process for the global optimum. Since it is believed that population diversity is an important issue for non-stationary problems, it seems worth exploring the effect of non-random mating in this context. The mutation rate has also implications on population diversity, and results can turn upside down with different mutation values. Therefore, in this work, several mutation values were explored. Moreover, both haploid and diploid representations were considered.

2 Methods

All experiments were run using a generational GA with tournament selection (tournament size of 2) and a population of 100 individuals. The genetic operations were 2-point crossover with a rate of 0.8, and the standard bit mutation. Mutation rates were expressed as mutations per genotype; several mutation values were tested (ranging from 0.0 to up to 6.0 mutations per genotype, with a step of 0.5). Both haploid and diploid representation were considered and the GA was run in three modes: (i) using both mutation and recombination in the standard way (**GA**), (ii) implementing dissortative mating (**GA-Dsrt**), and (iii)implementing assortative mating (**GA-Asrt**). Assortative mating was implemented as follows: when selecting two individuals for a crossover, the first parent was selected as usual. To chose the second parent, a set of p (*pool size*) individuals were selected using the GA fitness-based selection method. Thereafter, the similarity between each of these p genotypes and the first parent was computed. For dissortative mating, the genotype with less similarity was chosen. For assortative mating, the genotype closer to the first parent was selected as the second parent. For the experiments reported here, Hamming distance was used as the similarity measure, and the pool size p was set to 3.

2.1 Haploid/Diploid Encoding

We implemented both haploid and diploid genotypes. For the haploid encoding, a single chromosome of length n directly represents the individual's phenotype. The diploid scheme consists of a genome made up by two chromosomes of length n, and a phenotype which is obtained by applying a dominance map between each genotypic allele. The dominance map used, called *additive dominance* was originally proposed by Ryan [10]. Instead of binary digits, each locus in the genotype can be assigned a numeric value in the set $\{2, 3, 7, 9\}$. If the sum of the two alleles in a given locus is more than 10, then the phenotypic expression of this gene is 1, otherwise it is 0. In order to obtain better results when using the diploid encoding, we decided to extend Ryan's additive map with the dominance change mechanism suggested by Lewis et al. [7].

2.2 Test Problem

We used a an oscillating version of the traditional single knapsack problem. The objective is to fill a knapsack with the greatest number of objects from a set of size n, such that the total weight of the included objects is as close as possible to a target t. This implementation is based on the test problem suggested by Lewis et al. [7]. A solution is represented by a binary string of length n. This string represents the phenotype of an individual, were each gene x_i can be 1 or 0, indicating whether the object is included in the knapsack or not. The fitness f of a solution $\overline{x}$ is given by

$$f(\overline{x}) = \frac{1}{1 + |target - \sum_{i=1}^{n} w_i x_i|} \quad (1)$$

For the experiments in this article, 28 objects were used. Each object had a weight $w_i = 2^i$ where i ranged from 0 to 27. This guarantees that each target is reachable by a unique combination of objects in the knapsack. This is a normalized function, values for $f(\overline{x})$ lie between 0 and 1.

Target Changes. We considered both oscillating fixed targets and randomly generated targets. When using fixed targets, every s generations the algorithm switched between two fixed targets. These targets, t_1 and t_2, are initially randomly selected with a Hamming distance of 5. When considering random targets, the algorithm switches every s generations between newly created random targets with no specific Hamming distance. We tested two switching periods s: of 50 and 100 generations. For each experiment 10 cycles were considered, producing runs of 500 and 1000 generations respectively.

2.3 Performance Measures and Plots

For estimating optimal mutation rates in GAs we need to define what an optimal or near-optimal mutation rate is. The working definition used here is: an optimal mutation rate is the one that produces optimal performance. But then, we need a good way of measuring GA performance. We selected offline performance, that is, the mean of the current best fitness values trough the whole run. Averages of 50 runs were considered. These values were plotted for several mutation rates and each reproductive strategy. We will call these plots the *offline-performance* plots.

In order to have a dynamic view of the different strategies' performance, the best fitness (averaged over 50 runs) was plotted for each generation. In these plots, the optimal (or near optimal) mutation rate, as observed from the offline performance plots, was selected for each reproductive strategy. From now onwards we will refer to these plots as: *best-fitness-trace* plots.

3 Results

Figure 1 shows results for the oscillating knapsack with two fixed targets, and a switching period of 100 generations. Ten oscillating cycles, that is, 1000 generations were considered.

From the offline-performance plots (Figure 1 (a) and (c)). We can distinguish 3 different ranges of the mutation rate; for each of these ranges, one of the mating

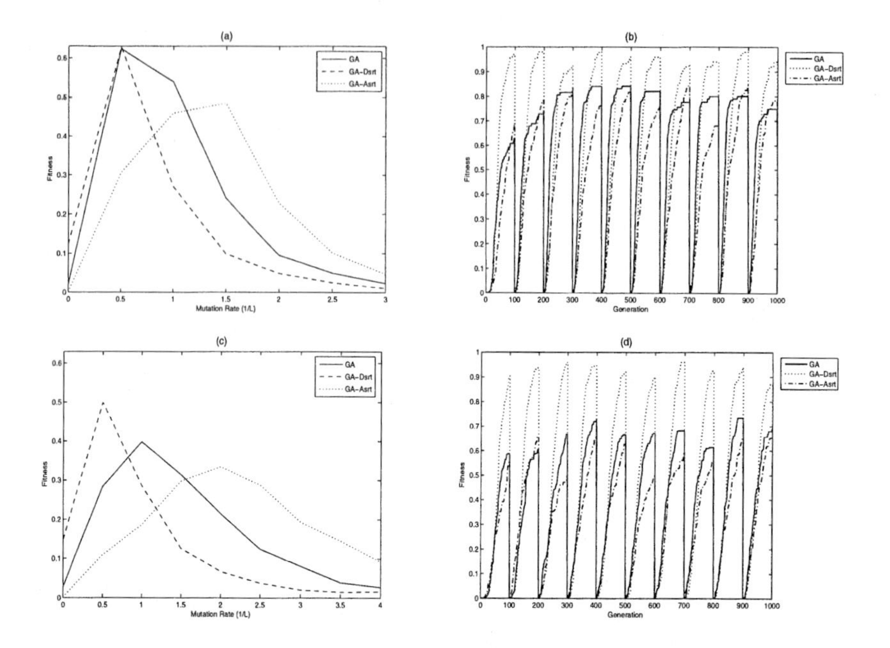

Fig. 1. Performance for fixed oscillating targets with a switching period of 100 generations. (a) Offline-performance for the haploid encoding, (b) Best-performance-trace for the haploid encoding, (c) Offline-performance for the diploid encoding. (d) Best-performance-trace for the diploid encoding. For the trace plots, the optimal mutation rate for each strategy was selected

strategies achieves the higher performance: Low mutation rates represent an advantage for GA-Dsrt, classical GA outperforms other methods at medium mutation rates, while high mutation rates create better conditions for GA-Asrt. Similar results were also found for a shorter switching period of 50 generations.

Notice from the best-fitness-trace plots (Figure 1 (b) and (d)), that the best performance (highest peaks) are delivered by GA-Dsrt in combination with a mutation rate of 0.5. Fitness values obtained with haploid individuals are slightly better than those obtained with diploid individuals. Again, similar results were found for a switching period of 50 generations.

Although the offline-performance plots show sometimes similar performance for the GA strategy as compared to GA-Dsrt (Figure 1 (a)), this was not the case when looking in detail at the corresponding generational best-fitness-trace plot: the GA strategy reaches good results, but its peaks are nonetheless almost always occluded by GA-Dsrt, which in each experiment reaches the highest peaks. This contrast was caused by the different convergence velocities of the rapidly evolving GA strategy vs. the somewhat slower dissortative strategy. Both strategies have their pros and cons: the more traditional GA may have faster convergence, but GA-Dsrt reaches higher peaks due to its inherent capacity to maintain a higher population diversity.

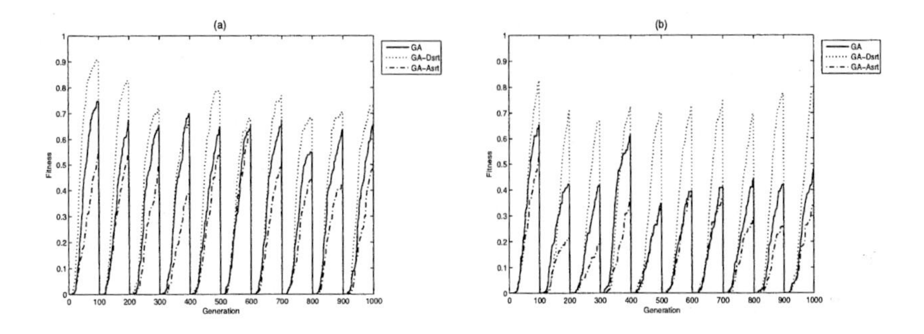

Fig. 2. Best-performance-trace for random oscillating targets with a switching period of 100 generations. (a)haploid encoding, (b) diploid encoding. The optimal mutation rate for each strategy was selected

Figures 2 show the best-performance-trace plots with random changing targets, for a switching period of 100 generations. Again, dissortative mating for both haploid and diploid individuals, is the strategy that achieves the highest fitness values (highest peaks) on each cycle. Similar results where found for the shorter switching period of 50 generations, with the exception of diploid organisms, where performance was very poor since the population had not enough time to adapt.

4 Discussion

Results suggest that the mutation rate parameter interacts with the mating strategy. Optimal mutation rates are different for each strategy; thus, fair comparisons can not be performed without selecting the optimal mutation rate for each case. When considering offline performance, on all explored scenarios dissortative mating consistently required a lower mutation rate (close to $0.5/L$) to perform better, whilst the standard GA needed a somewhat higher mutation rate (between 0.5 to 1.0) mutations per genotype) to produce similar performance. On the other hand, assortative mating produced the worst performance, and its optimal mutation rate was in the range of 1.5 to 2.5 mutations per genotype. If the optimal mutation rate was selected for each strategy, dissortative mating was found to be a good strategy for finding the highest peaks.

Surprisingly, in our experiments the haploid scheme produced better results than the diploid encoding. These results confirm the observations by Lewis et al. [7] that diploid schemes do not seem to be a robust mechanism for non-stationary problems. According to them, adding some form of dominance change considerably improves matters, but the form of the change mechanism can have a significant effect.

Mate selection is a force guiding natural evolution that has not been widely explored within artificial evolution. It is clear that selecting mates have an impact on both the search process and the population diversity. This work is a preliminary assessment of the effect of including mate selection in evolutionary algorithms for dynamic problems. Dissortative mating was shown to be a promising scheme that may improve the algorithm performance. Further work with other non-stationary problems will make it possible to assess if these results can be generalised.

Acknowledgments: The first author thanks the Santa Fe Complex Systems Summer School organizers for providing such a nice and stimulating school. The preliminary experiments and ideas of this paper were developed in the context of SFI CSSS 2002.

References

[1] Agostinho C. da Rosa Carlos Fernandes, Rui Tavares, *niGAVaPS - outbreeding in genetic algorithms*, Proceedings of the 2000 ACM Symposium on Applied Computing (Villa Olmo, Como, Italy), 2000.

[2] Rob Craighurst and Worthy Martin, *Enhancing GA performance through crossover prohibitions based on ancestry*, Proceedings of the 6th International Conference on Genetic Algorithms (San Francisco) (Larry J. Eshelman, ed.), Morgan kaufmann Publishers, July 15–19 1995, pp. 130–135.

[3] Larry J. Eshelman and J. David Schaffer, *Preventing premature convergence in genetic algorithms by preventing incest*, Proceedings of the 4th International Conference on Genetic Algorithms (San Diego, CA) (Lashon B. Belew, Richard K.; Booker, ed.), Morgan Kaufmann, 1991, pp. 115–122.

[4] Carlos Fernandes, Rui Taveres, Cristian Munteanu, and Agostinho C. Rosa, *Using assortative mating in genetic algorithms for vector quantization problems*, Selected Areas in Cryptography, 2001, pp. 361–365.

[5] C.-F Huang, *An analysis of mate selection in genetic algorithms*, Tech. report, Center for the Study of Complex Systems, University of Michgan, 2001.

[6] Klaus Jaffe, *On sex, mate selection and evolution: an exploration*, Comments on Theoretical Biology **7** (2002), no. 2, 91–107.

[7] Jonathan Lewis, Emma Hart, and Graeme Ritchie, *A comparison of dominance mechanisms and simple mutation on non-stationary problems*, Parallel Problem Solving from Nature – PPSN V (Berlin) (Agoston E. Eiben, Thomas Bäck, Marc Schoenauer, and Hans-Paul Schwefel, eds.), Springer, 1998, Lecture Notes in Computer Science 1498, pp. 139–148.

[8] G. F. Miller and P. M. Todd, *The role of mate choice in biocomputation: Sexual selection as a process of search, optimization and diversification*, Lecture Notes in Computer Science **899** (1995), 169–198.

[9] Gabriela Ochoa and Klaus Jaffe, *On sex, mate selection and the red queen*, Journal of Theoretical Biology **199** (1999), 1–9.

[10] Conor Ryan, *The degree of oneness*, 1st Online Workshop on Soft Computing (Nagoya, Japan), Nagoya University, 1996, pp. 100–105.

A Hybrid Approach Based on Evolutionary Strategies and Interval Arithmetic to Perform Robust Designs

C.M. Rocco S.

Universidad Central, Facultad de Ingeniería,
Apartado 47937, Caracas 1040A, Venezuela
crocco@reacciun.ve

Abstract. This paper proposes an approach based on the use of Cellular Evolutionary Strategies (CES) and Interval Arithmetic (IA) as an alternative technique to obtain robust system design. CES are an approach that combines the Evolution Strategy techniques with concepts from Cellular Automata in order to optimise a given function, while IA is used as a checking technique that guarantees the feasibility of the design. IA is able to consider simultaneously the effects of uncertainty of all of the parameters on a performance function and to provide strict bounds (minimum and maximum values) with only one evaluation. CES and IA are used to obtain, by an iterative process, a robust design, that is, the maximum size of each variable deviation that allow to comply with a set of specifications. The proposed approach is an indirect method based on optimisation instead of a direct method based on mapping from the output into the input space. A numerical example, related to an electronic circuit system design, illustrates the application of the approach.

1 Introduction

Sensitivity analysis, uncertainty propagation and uncertainty analysis are techniques that have been used for examining the effects of uncertain inputs within a model [1]. These techniques are usually carried out by determining which parameter or parameters have significant effects on the results of a study. An attempt is then made to increase the precision of these parameters in order to reduce the danger of serious error.

In system design, mathematical models are used to describe the properties of the system to be designed. As an example, consider the temperature controller circuit [2] shown in figure 1. The performance function is:

$$R_{T-on} = \frac{R_1 R_2 (E_2 R_4 + E_1 R_3)}{R_3 (E_2 R_4 + E_2 R_2 - E_1 R_2)}\ \Omega \tag{1}$$

We can evaluate, for example, what are the effects on R_{T-on} if the value of one (local analysis) or more (global analysis) components is changed.

The robustness of a design is defined as the maximum size of component deviation from this design that can be tolerated whereby the product still meets all requirements [3]. For example, what are the maximum possible deviations for each component in the figure 1 consistent with $2.90\ \text{k}\Omega \le R_{T-on} \le 3.10\ \text{k}\Omega$?

F. Rothlauf et al. (Eds.): EvoWorkshops 2005, LNCS 3449, pp. 623–628, 2005.

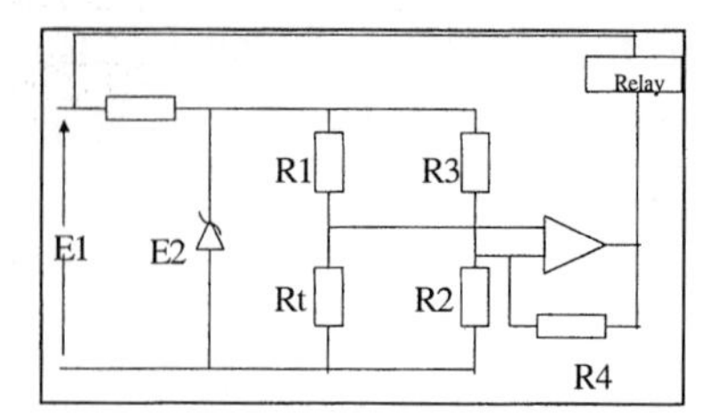

Fig. 1. Temperature Controller Circuit [2]

The problem we address is different from variability or sensitivity analysis [4,5]. In this paper we assess input parameter uncertainty for one or all components, which maintains the output performance function within specified bounds. The approach is based on the use of Cellular Evolutionary Strategies (CES) and Interval Arithmetic (IA). Section 2 contains an overview of Interval Arithmetic. The Evolutionary approach is presented in Section 3. Section 4 presents the general approach used to solve the robust design problem. An example is considered in Section 5, and finally Section 6 presents the conclusions.

2 Interval Arithmetic

Interval "numbers" are an ordered pair of real numbers representing the lower and upper bound of the parameter range. If we have two interval numbers T=[a,b] and W=[c,d] with $a \leq b$ and $c \leq d$ then T+W=[a+c,b+d]. Similar expressions can be defined for the other basic operations and for trascendental functions [6-8].

Only some of the algebraic laws valid for real numbers remain valid for intervals. An important property referred to as sub-distributivity does hold. It is given mathematically by the set inclusion relationship: $T(W+Z) \subseteq TW+TZ$. The failure of the distributive law often causes overestimation. In general, when a given variable occurs more than once in an interval computation, it is treated as a different variable in each occurrence. This effect is called the "dependency problem".

Consider a real valued function f of real variables $t_1,t_2,..,t_n$ and an interval function F of interval variables $T_1,T_2,..,T_n$. The interval function F is said to be an interval extension of f, if $F(t_1,t_2,..,t_n)=f(t_1,t_2,..., t_n)$. The range of a function f of real variables over an interval can be calculated from the interval extension F, changing t_i by T_i. Note that: $f(t_1,t_2,.., t_n) \subseteq F(T_1,T_2,..., T_n)$, for all $t_i \in T_i$ (i=1,.., n) [6].

3 Evolutionary Approach

Evolution Strategies (ES) have been applied to a wide range of problems especially in those cases where traditional optimisation techniques have shown poor performances or simply have failed [9]. Schwefel and Bäck [10] generalised these strategies to the multimember evolution strategy now denoted by ES($\mu+\lambda$) and ES(μ,λ). In a ($\mu+\lambda$)

strategy, the μ best of all $(\mu+\lambda)$ individuals survive to become parents of the next generation. Using the (μ,λ) strategy, selection takes place only among λ offspring.

Cellular Evolutionary Strategies (CES) [11] are an approach that combines the $ES(\mu,\lambda)$ techniques with concepts from Cellular Automata [12] for the parent's selection step, using the concepts of neighbourhood. Each individual is located randomly in a cell of a two-dimension array. To update a specific individual, the parents' selection is performed looking only at determined cells (its neighbourhood) in contrast with the general ES, which search parents in the whole population.

4 Robust Design Methodology

4.1 Problem Description [3]

The robustness of a system design is defined as the maximum size of component deviations from their design values that can be tolerated such that the system still meets all defined specifications. The designer formulates target values on the quality of the product by setting lower and upper bounds on the property $y_i(\mathbf{x})$. The problem is to find a product $\mathbf{x}$ in the experimental region X which fulfils the requirements on the properties.

We define the following slack function $g_i(\mathbf{x})$:

$g_i(\mathbf{x})=UB_i - y_i(\mathbf{x})$ when there is an upper bound requirement or
$g_i(\mathbf{x})= y_i(\mathbf{x})-LB_i$ when there is an lower bound requirement

This results in a product design problem mathematically formulated as: find an element of $F\cap X$, with: $F:=\{\mathbf{x} \in \Re^n \mid g_i(\mathbf{x})\geq 0, i=1,....,r\}$.

4.2 Problem Definition [13]

Suppose the area shown in figure 3a is the feasible zone for a generic design with variables R and Ra. Within the feasible zone any pair (R,Ra) satisfies the specifications. An exact description of the Feasible Solution Set (FSS) (figure 3a) is in general not simple, since it may be a very complex set. Moreover, the FSS could be limited by non-linear functions. For this reason, approximate descriptions are often looked for, using simply shaped sets like boxes or ellipsoids containing (outer bounding, figure 3b) or contained in (inner bounding, figure 3c and 3d) the set of interest. In particular Minimum Volume Outer box (MOB) (figure 3b) and Maximum volume Inner Box (MIB) (figure 3c and 3d) are of interest. In this paper only the MIB determination is presented.

The maximum ranges of possible variations of the feasible values are the sizes (along co-ordinate axis) of the axis-aligned box of minimum volume containing FSS. To obtain the MIB it is required that all the points inside the generated box satisfy the constraints. Then, the mathematical formulation is:

Let B the box defined by: $B:= \{\mathbf{x}, \mathbf{C} \in \Re^n \mid x_i \in [x_{i,lower}, x_{i,upper}], C_i =(x_{i,upper}+x_{i,lower})/2\}$

1. Centre Specified: $$\max_{\mathbf{x}} \prod_{i=1}^{n} abs((x_i - C_i))$$
$$\text{s.t. } \mathbf{x}\in F\cap B$$

2. Centre Unspecified: $\max_{\mathbf{x},\mathbf{C}} \prod_{i=1}^{n} \mathrm{abs}((x_i - C_i))$

$$\text{s.t. } \mathbf{x}, \mathbf{C} \in F \cap B$$

The objective functions represent a quantity that is proportional to the MIB hyper-volume. Note that **x** represents a vertex of the optimal MIB. From here, the range of each variable is easily determined, as described in the next section.

4.3 General Approach

Given r constraint functions $g_i(x_1,x_2,...,x_n)$, lower and upper bounds (LB_i, UB_i), we generate an initial random point $\mathbf{x}=\{x_{1o},x_{2o},..,x_{n0}\}$ (initial vertex). We check if this point is a feasible point, evaluating all the constraints. If the point is infeasible, then we generate a new point and check it again for feasibility. If the generated point is feasible then we have two possible actions depending on the goal established:

1) Inner box, centre specified (C)

Given a feasible vertex, we generate a symmetrical "box" (hyper-rectangle) using the point **C** as symmetry centre and check for the feasibility of the generated box. If the box is feasible, we calculate the associated volume. If the box is not feasible, then we discard the box, and repeat the process with a new feasible initial vertex.

2) Inner box, centre unspecified

In this case, the centre co-ordinates are considered as additional variables. So we generate along with the initial random vertex, the initial random centre co-ordinate $\mathbf{C}=\{C_{1o},C_{2o},..,C_{n0}\}$. As in the previous case, we generate a symmetrical "box" using **C** as symmetry centre, then check the feasibility of the generated box, and, if feasible, we calculate the associated volume.

In both cases, the goal is to maximise the inner volume using CES as the optimisation technique.

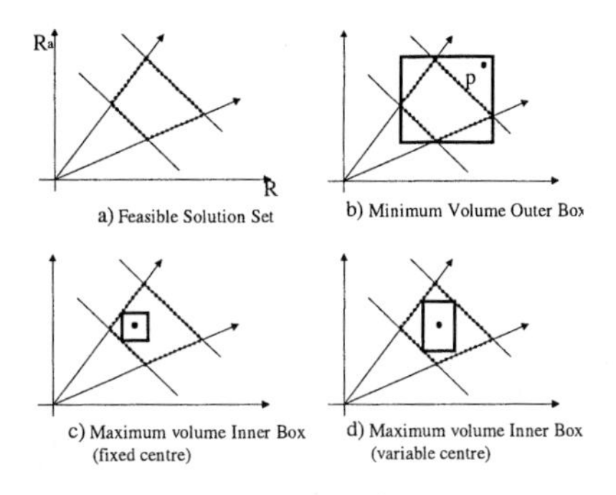

Fig. 2. Feasible Solution Set and approximate descriptions

To check the box feasibility, we may have two problems: 1) To evaluate the function in 2^n vertices and 2) Extreme values are not necessarily at vertices of the generated box. To overcome these two drawbacks, constraint functions are evaluated as interval functions. This means that only one "interval" evaluation is required for each constraint and the exact range of the constraint functions inside the generated box is obtained. Note that if the FSS is non-convex, feasibility check using IA will consider this.

5 Example: A Temperature Controller [14]

We will use the proposed robust design approach in order to define a robust design for the temperature controller circuit [2] shown in figure 1. The design must guarantee that $R_{T\text{-}on}$ belongs to the interval [2.9,3.1] kΩ and that $R_{T\text{-}on}$ evaluated at the mid-point should be 3.00 kΩ. We will define the centre of each variable (midpoint of the starting interval) and obtain the MIB. Because of the stochastic nature of CES, 20 trials were performed and the best solution from among the 20 trials was used as the final solution. All CES runs were performed using 30 generations, 49 individuals in a 7x7 grid, Von Neumann neighbourhood with radius =1 and asynchronous substitution.

Table 2 shows the result obtained using the hybrid approach CES and IA, with a MIB volume of 930296.40. These ranges produce an output $R_{T\text{-}on}$ belonging to: [2.90573,3.09999] kΩ. The MIB volume obtained using a non-linear optimisation program, was 931811.45, that is only 0.163 % greater than the obtained with the CES-IA approach. The average relative error obtained in 20 runs was only 0.272 %.

6 Conclusions

This paper proposes a promising approach based on the use of Cellular Evolutionary Strategies and Interval Arithmetic as an alternative technique to obtain robust system design. CES are used as the optimisation technique while IA is used as a checking technique that guarantees the feasibility of the design.

Table 1. Robust design using CES and IA approach

VARIABLE	Starting Interval	Final Interval
R_1 (kΩ)	[0.5,1.5]	[0.994,1.005]
R_2 (kΩ)	[6,12]	[8.956,9.044]
R_3 (kΩ)	[2,6]	[3.973,4.027]
R_4 (kΩ)	[16,48]	[31.428,32.572]
E_1 (V)	[7.5,9.5]	[8.427,8.573]
E_2 (V)	[4.5,7.5]	[5.946,6.0535]

The excellent results obtained suggest that the CES-IA approach has great potential in dealing with difficult system design problems. It is interesting to note that even if

Interval Arithmetic can overestimate the size of the hyper-box, due to the dependency problem, we are sure that the box obtained is a valid robust solution, which satisfies all the defined constraints. The overestimation can be treated using special techniques. The added burden to the procedure CES-IA for determining the feasibility verification of the generated box is far outweighed by the flexibility provided by such technique (only one "interval" evaluation and guaranteed ranges) in contrast to multiple vertices evaluation.

References

1) M. Granger, M. Henrion: Uncertainty: A Guide to Dealing with Uncertainty in Quantitative Risk and Policy Analysis. Cambridge University Press. 1993.
2) Hadjihassan S., Walter E., Pronzato L., Quality Improvement via Optimisation of Tolerance Intervals During the Design Stage, *in Applications of Interval Computations*, Ed. Kearfott R.B., Kreinovich V., Kluwer Publishers, Dordrecht, 1996
3) E.M. Hendrix, C.J. Mecking and Th.H.B. Hendriks: "Finding Robust Solutions for Product Design Problems", European Journal of Operational Research", 92, 1996, pp28-36
4) Saltelli, M. Scott: Guest editorial: The role of sensitivity analysis in the corroboration of models and its link to model structural and parametric uncertainty, Reliability Engineering And System Safety, 57, 1997, 1-4
5) Constantinides: "Basic Reliability", 1994 Annual Reliability and Maintainability Symposium, Anaheim, California, USA
6) R. Moore: "Methods and Applications of Interval Analysis", SIAM Studies in Applied Mathematics, Philadelphia, 1979
7) A.Neumaier: "Interval Methods for Systems of Equations", Cambridge Univ. Press, 1990
8) E. Hansen: "Global Optimization Using Interval Analysis", Marcel Dekker Inc., New York, 1992
9) F. Kursawe: "Towards Self-Adapting Evolution Strategies", Proc. Of the Tenth International Conference on Multiple Criteria Decision Making, G Tzeng and P. Yu (Eds.), Taipei 1992
10) H.P Schwefel, Th. Back: "Evolution Strategies I: Variants and their computational implementation", in J. Periaux and G. Winter (Eds.), Genetic Algorithm in Engineering and Computer Science, John Wiley & Sons, 1995.
11) M.Medina, N.Carrasquero, J.Moreno: "Estratégias Evolutivas Celulares para la Optimización de Funciones", IBERAMIA'98, 6° Congresso Iberoamericano de Inteligencia Artificial, Lisboa, Portugal, 1998
12) Wolfram: "Cellular automata as models of complexity", Nature 3H, 1984
13) M. Milanese, J. Norton, H. Piet-Lahanier (Eds.): "Bounding Approaches to System Identification", Plenum Press, New York, 1998
14) C. M. Rocco S. "Robust Design using Evolutionary Strategies and Interval Arithmetic", Report 2003-23, Universidad Central de Venezuela, Facultad de Ingenieria (in Spanish)

Author Index

Lecture Notes in Computer Science

For information about Vols. 1–3344

please contact your bookseller or Springer

Vol. 3452: F. Baader, A. Voronkov (Eds.), Logic for Programming, Artificial Intelligence, and Reasoning. XI, 562 pages. 2005. (Subseries LNAI).

Vol. 3449: F. Rothlauf et al. (Eds.), Applications of Evolutionary Computing. XX, 630 pages. 2005.

Vol. 3448: G.R. Raidl, J. Gottlieb (Eds.), Evolutionary Computation in Combinatorial Optimization. XI, 271 pages. 2005.

Vol. 3447: M. Keijzer, A. Tettamanzi, P. Collet, J.v. Hemert, M. Tomassini (Eds.), Genetic Programming. XIII, 382 pages. 2005.

Vol. 3444: M. Sagiv (Ed.), Programming Languages and Systems. XIII, 439 pages. 2005.

Vol. 3443: R. Bodik (Ed.), Compiler Construction. XI, 305 pages. 2005.

Vol. 3442: M. Cerioli (Ed.), Fundamental Approaches to Software Engineering. XIII, 373 pages. 2005.

Vol. 3441: V. Sassone (Ed.), Foundations of Software Science and Computational Structures. XVIII, 521 pages. 2005.

Vol. 3440: N. Halbwachs, L.D. Zuck (Eds.), Tools and Algorithms for the Construction and Analysis of Systems. XVII, 588 pages. 2005.

Vol. 3436: B. Bouyssounouse, J. Sifakis (Eds.), Embedded Systems Design. XV, 492 pages. 2005.

Vol. 3434: L. Brun, M. Vento (Eds.), Graph-Based Representations in Pattern Recognition. XII, 384 pages. 2005.

Vol. 3433: S. Bhalla (Ed.), Databases in Networked Information Systems. VII, 319 pages. 2005.

Vol. 3432: M. Beigl, P. Lukowicz (Eds.), Systems Aspects in Organic and Pervasive Computing - ARCS 2005. X, 265 pages. 2005.

Vol. 3427: G. Kotsis, O. Spaniol, Wireless Systems and Mobility in Next Generation Internet. VIII, 249 pages. 2005.

Vol. 3423: J.L. Fiadeiro, P.D. Mosses, F. Orejas (Eds.), Recent Trends in Algebraic Development Techniques. VIII, 271 pages. 2005.

Vol. 3422: R.T. Mittermeir (Ed.), From Computer Literacy to Informatics Fundamentals. X, 203 pages. 2005.

Vol. 3421: P. Lorenz, P. Dini (Eds.), Networking - ICN 2005, Part II. XXXV, 1153 pages. 2005.

Vol. 3420: P. Lorenz, P. Dini (Eds.), Networking - ICN 2005, Part I. XXXV, 933 pages. 2005.

Vol. 3419: B. Faltings, A. Petcu, F. Fages, F. Rossi (Eds.), Constraint Satisfaction and Constraint Logic Programming. X, 217 pages. 2005. (Subseries LNAI).

Vol. 3418: U. Brandes, T. Erlebach (Eds.), Network Analysis. XII, 471 pages. 2005.

Vol. 3416: M. Böhlen, J. Gamper, W. Polasek, M.A. Wimmer (Eds.), E-Government: Towards Electronic Democracy. XIII, 311 pages. 2005. (Subseries LNAI).

Vol. 3415: P. Davidsson, B. Logan, K. Takadama (Eds.), Multi-Agent and Multi-Agent-Based Simulation. X, 265 pages. 2005. (Subseries LNAI).

Vol. 3414: M. Morari, L. Thiele (Eds.), Hybrid Systems: Computation and Control. XII, 684 pages. 2005.

Vol. 3412: X. Franch, D. Port (Eds.), COTS-Based Software Systems. XVI, 312 pages. 2005.

Vol. 3411: S.H. Myaeng, M. Zhou, K.-F. Wong, H.-J. Zhang (Eds.), Information Retrieval Technology. XIII, 337 pages. 2005.

Vol. 3410: C.A. Coello Coello, A. Hernández Aguirre, E. Zitzler (Eds.), Evolutionary Multi-Criterion Optimization. XVI, 912 pages. 2005.

Vol. 3409: N. Guelfi, G. Reggio, A. Romanovsky (Eds.), Scientific Engineering of Distributed Java Applications. X, 127 pages. 2005.

Vol. 3408: D.E. Losada, J.M. Fernández-Luna (Eds.), Advances in Information Retrieval. XVII, 572 pages. 2005.

Vol. 3407: Z. Liu, K. Araki (Eds.), Theoretical Aspects of Computing - ICTAC 2004. XIV, 562 pages. 2005.

Vol. 3406: A. Gelbukh (Ed.), Computational Linguistics and Intelligent Text Processing. XVII, 829 pages. 2005.

Vol. 3404: V. Diekert, B. Durand (Eds.), STACS 2005. XVI, 706 pages. 2005.

Vol. 3403: B. Ganter, R. Godin (Eds.), Formal Concept Analysis. XI, 419 pages. 2005. (Subseries LNAI).

Vol. 3401: Z. Li, L.G. Vulkov, J. Waśniewski (Eds.), Numerical Analysis and Its Applications. XIII, 630 pages. 2005.

Vol. 3399: Y. Zhang, K. Tanaka, J.X. Yu, S. Wang, M. Li (Eds.), Web Technologies Research and Development - APWeb 2005. XXII, 1082 pages. 2005.

Vol. 3398: D.-K. Baik (Ed.), Systems Modeling and Simulation: Theory and Applications. XIV, 733 pages. 2005. (Subseries LNAI).

Vol. 3397: T.G. Kim (Ed.), Artificial Intelligence and Simulation. XV, 711 pages. 2005. (Subseries LNAI).

Vol. 3396: R.M. van Eijk, M.-P. Huget, F. Dignum (Eds.), Agent Communication. X, 261 pages. 2005. (Subseries LNAI).

Vol. 3395: J. Grabowski, B. Nielsen (Eds.), Formal Approaches to Software Testing. X, 225 pages. 2005.

D. Kazakov, E. Alonso (Eds.), lti-Agent Systems III. VIII, 313 NAI).

ci, U. Montanari, F. Orejas, G. Eds.), Formal Methods in Software ng. XXVII, 413 pages. 2005.

nformation Networking. XVII,

Vol. 3390: R. Choren, A. Garcia, C. Lucena, A. Romanovsky (Eds.), Software Engineering for Multi-Agent Systems III. XII, 291 pages. 2005.

Vol. 3389: P. Van Roy (Ed.), Multiparadigm Programming in Mozart/OZ. XV, 329 pages. 2005.

Vol. 3388: J. Lagergren (Ed.), Comparative Genomics. VII, 133 pages. 2005. (Subseries LNBI).

Vol. 3387: J. Cardoso, A. Sheth (Eds.), Semantic Web Services and Web Process Composition. VIII, 147 pages. 2005.

Vol. 3386: S. Vaudenay (Ed.), Public Key Cryptography - PKC 2005. IX, 436 pages. 2005.

Vol. 3385: R. Cousot (Ed.), Verification, Model Checking, and Abstract Interpretation. XII, 483 pages. 2005.

Vol. 3383: J. Pach (Ed.), Graph Drawing. XII, 536 pages. 2005.

Vol. 3382: J. Odell, P. Giorgini, J.P. Müller (Eds.), Agent-Oriented Software Engineering V. X, 239 pages. 2005.

Vol. 3381: P. Vojtáš, M. Bieliková, B. Charron-Bost, O. Sýkora (Eds.), SOFSEM 2005: Theory and Practice of Computer Science. XV, 448 pages. 2005.

Vol. 3380: C. Priami, Transactions on Computational Systems Biology I. IX, 111 pages. 2005. (Subseries LNBI).

Vol. 3379: M. Hemmje, C. Niederee, T. Risse (Eds.), From Integrated Publication and Information Systems to Information and Knowledge Environments. XXIV, 321 pages. 2005.

Vol. 3378: J. Kilian (Ed.), Theory of Cryptography. XII, 621 pages. 2005.

Vol. 3377: B. Goethals, A. Siebes (Eds.), Knowledge Discovery in Inductive Databases. VII, 190 pages. 2005.

Vol. 3376: A. Menezes (Ed.), Topics in Cryptology – CT-RSA 2005. X, 385 pages. 2005.

Vol. 3375: M.A. Marsan, G. Bianchi, M. Listanti, M. Meo (Eds.), Quality of Service in Multiservice IP Networks. XIII, 656 pages. 2005.

Vol. 3374: D. Weyns, H.V.D. Parunak, F. Michel (Eds.), Environments for Multi-Agent Systems. X, 279 pages. 2005. (Subseries LNAI).

Vol. 3372: C. Bussler, V. Tannen, I. Fundulaki (Eds.), Semantic Web and Databases. X, 227 pages. 2005.

Vol. 3371: M.W. Barley, N. Kasabov (Eds.), Intelligent Agents and Multi-Agent Systems. X, 329 pages. 2005. (Subseries LNAI).

Vol. 3370: A. Konagaya, K. Satou (Eds.), Grid Computing in Life Science. X, 188 pages. 2005. (Subseries LNBI).

'ol. 3369: V.R. Benjamins, P. Casanovas, J. Breuker, A. ngemi (Eds.), Law and the Semantic Web. XII, 249 es. 2005. (Subseries LNAI).

Vol. 3368: L. Paletta, J.K. Tsotsos, E. Rome, G.W. Humphreys (Eds.), Attention and Performance in Computational Vision. VIII, 231 pages. 2005.

Vol. 3367: W.S. Ng, B.C. Ooi, A. Ouksel, C. Sartori (Eds.), Databases, Information Systems, and Peer-to-Peer Computing. X, 231 pages. 2005.

Vol. 3366: I. Rahwan, P. Moraitis, C. Reed (Eds.), Argumentation in Multi-Agent Systems. XII, 263 pages. 2005. (Subseries LNAI).

Vol. 3365: G. Mauri, G. Păun, M.J. Pérez-Jiménez, G. Rozenberg, A. Salomaa (Eds.), Membrane Computing. IX, 415 pages. 2005.

Vol. 3363: T. Eiter, L. Libkin (Eds.), Database Theory - ICDT 2005. XI, 413 pages. 2004.

Vol. 3362: G. Barthe, L. Burdy, M. Huisman, J.-L. Lanet, T. Muntean (Eds.), Construction and Analysis of Safe, Secure, and Interoperable Smart Devices. IX, 257 pages. 2005.

Vol. 3361: S. Bengio, H. Bourlard (Eds.), Machine Learning for Multimodal Interaction. XII, 362 pages. 2005.

Vol. 3360: S. Spaccapietra, E. Bertino, S. Jajodia, R. King, D. McLeod, M.E. Orlowska, L. Strous (Eds.), Journal on Data Semantics II. XI, 223 pages. 2005.

Vol. 3359: G. Grieser, Y. Tanaka (Eds.), Intuitive Human Interfaces for Organizing and Accessing Intellectual Assets. XIV, 257 pages. 2005. (Subseries LNAI).

Vol. 3358: J. Cao, L.T. Yang, M. Guo, F. Lau (Eds.), Parallel and Distributed Processing and Applications. XXIV, 1058 pages. 2004.

Vol. 3357: H. Handschuh, M.A. Hasan (Eds.), Selected Areas in Cryptography. XI, 354 pages. 2004.

Vol. 3356: G. Das, V.P. Gulati (Eds.), Intelligent Information Technology. XII, 428 pages. 2004.

Vol. 3355: R. Murray-Smith, R. Shorten (Eds.), Switching and Learning in Feedback Systems. X, 343 pages. 2005.

Vol. 3354: M. Margenstern (Ed.), Machines, Computations, and Universality. VIII, 329 pages. 2005.

Vol. 3353: J. Hromkovič, M. Nagl, B. Westfechtel (Eds.), Graph-Theoretic Concepts in Computer Science. XI, 404 pages. 2004.

Vol. 3352: C. Blundo, S. Cimato (Eds.), Security in Communication Networks. XI, 381 pages. 2005.

Vol. 3351: G. Persiano, R. Solis-Oba (Eds.), Approximation and Online Algorithms. VIII, 295 pages. 2005.

Vol. 3350: M. Hermenegildo, D. Cabeza (Eds.), Practical Aspects of Declarative Languages. VIII, 269 pages. 2005.

Vol. 3349: B.M. Chapman (Ed.), Shared Memory Parallel Programming with Open MP. X, 149 pages. 2005.

Vol. 3348: A. Canteaut, K. Viswanathan (Eds.), Progress in Cryptology - INDOCRYPT 2004. XIV, 431 pages. 2004.

Vol. 3347: R.K. Ghosh, H. Mohanty (Eds.), Distributed Computing and Internet Technology. XX, 472 pages. 2004.

Vol. 3346: R.H. Bordini, M. Dastani, J. Dix, A.E.F. Seghrouchni (Eds.), Programming Multi-Agent Systems. XIV, 249 pages. 2005. (Subseries LNAI).

Vol. 3345: Y. Cai (Ed.), Ambient Intelligence for Scientific Discovery. XII, 311 pages. 2005. (Subseries LNAI).